Hiranmoy Das
Editor

# Tissue Repair and Regeneration

## Elucidating Cellular and Molecular Mechanisms with Therapeutic Implications

Springer

*Editor*
Hiranmoy Das
Vasc. Bio. and Stem Cell Research Lab.
Texas Tech University
AMARILLO, TX, USA

ISBN 978-3-031-93676-0 ISBN 978-3-031-93677-7 (eBook)
https://doi.org/10.1007/978-3-031-93677-7



This Springer imprint is published by the registered company Springer Nature Switzerland AG
The registered company address is: Gewerbestrasse 11, 6330 Cham, Switzerland

# Preface

The quest to understand life's inherent capacity for repair, renewal, and regeneration has fascinated scientists and philosophers for centuries. From the mythical Prometheus, whose liver is continuously regenerated, to the humble planarian, who can regenerate an entire organism from a fragment, regeneration stands as a testament to the remarkable plasticity of biological systems. Regeneration is not a singular, monolithic process; rather, it encompasses a diverse array of strategies, each tailored to the specific needs and constraints of the organism in question. From the epimorphic regeneration of salamander limbs to the compensatory hypertrophy of mammalian livers, the biological responses to tissue loss are as varied as life itself.

This book delves into the intricate world of regeneration, exploring the diverse mechanisms that underpin this extraordinary process across a spectrum of organisms for their organ and tissues. We traverse the cellular and molecular landscapes, examining the secreted factors from cells and signaling pathways that orchestrate the rebuilding of lost tissues and organs.

This exploration is not merely an academic exercise. Understanding the fundamental principles of regeneration holds immense potential for translational applications. By deciphering the language of tissue repair, we aspire to unlock new avenues for regenerative medicine, offering the possibility of treating debilitating injuries, combating degenerative diseases, and even reversing the effects of aging.

The authors aim to provide a comprehensive overview of regeneration research, highlighting key discoveries, controversial findings, and emerging scientific frontiers. We have striven to integrate insights from diverse fields, including developmental biology, cell biology, immunology, molecular biology, biomaterials, small pharmacological compounds, and material delivery systems, to paint a holistic picture of this complex and fascinating phenomenon.

This book aims to engage a wide audience, including researchers, students, and anyone with a genuine interest in the wonders of biological repair. We hope it will serve as a valuable resource and inspire further exploration into the limitless potential of regeneration. Although the field remains filled with unanswered questions, the progress made so far offers a glimpse into a future where the body's inherent healing capacity can be harnessed to enhance human health and well-being. The journey into the mechanisms of regeneration delves into the very essence of life's resilience. The authors invite you to join us as we investigate the remarkable ability of living organisms to rebuild, renew, and restore.

AMARILLO, TX, USA Hiranmoy Das

# Contents

# Wound Regeneration

# From Injury to Recovery: The Role of miRNAs in Wound Healing

**Madhur Sachan and Mark W. Feinberg**

**Abstract** Restoring tissue integrity after injury involves a coordinated action of four overlapping phases: hemostasis, inflammation, proliferation, and remodeling. Disruptions in any of these phases can progress into chronic wounds with delayed healing. Emerging studies highlight the role of microRNAs (miRNAs) in fine-tuning cellular responses to various physiological processes, including wound healing. This chapter elucidates each phase of the wound healing process and associated miRNAs implicated in the cellular and molecular signaling events important for wound regeneration. Challenges and opportunities for therapeutic intervention using emerging miRNA delivery platforms are highlighted. Further research is needed to explore the stage-specific roles of miRNA therapeutics to promote regenerative responses, especially for the treatment of chronic wounds.

**Keywords** Wound healing · MicroRNAs · Chronic wounds · Signaling · Tissue remodeling

## 1 Introduction

The skin is the body's first line of defense against external threats that disrupt the affected area's integrity and homeostasis (Jiang et al., 2022; Peña & Martin, 2024). It consists of three layers: the epidermis, composed mainly of keratinocytes; the dermis, rich in collagen, fibroblasts, macrophages, lymphocytes, sensory neurons, and blood vessels; and the hypodermis, primarily made of adipose tissue (Mulholland et al., 2017). The coordinated action of these three skin layers is essential to restore the damaged barrier after any injuries such as abrasion, burns, and ulcers. Although the skin has a strong regenerative capacity, millions of individuals suffer from skin

M. Sachan · M. W. Feinberg (✉)
Cardiovascular Division, Department of Medicine, Brigham and Women's Hospital, Harvard Medical School, Boston, MA, USA
e-mail: mfeinberg@bwh.harvard.edu



H. Das (ed.), *Tissue Repair and Regeneration*,
https://doi.org/10.1007/978-3-031-93677-7_1

injuries every year that are often accompanied by infections that are difficult to heal (Sen, 2021). Therefore, finding effective clinical therapeutic approaches against such injuries is critical.

Wound healing is a highly complex and coordinated process in which skin responds to tissue damage within a relatively short timeframe (Peña & Martin, 2024). It comprises four overlapping phases: hemostasis, inflammation, proliferation, and remodeling. Disruption in any of these phases can lead to chronic wounds, such as ulcers, which are common in patients with diabetes. These wounds either fail to progress consistently toward healing or do not heal in a timely, programmed manner, often taking longer than 12 weeks to close (Falanga et al., 2022).

Emerging studies have recognized the involvement of epigenetic regulators such as microRNAs (miRNAs) in wound healing (Mulholland et al., 2017). miRNAs are a class of small noncoding RNAs that orchestrate posttranscriptional gene regulation, primarily by inhibiting multiple target gene expressions and associated proteins and pathways (Bartel, 2004). Several miRNAs are known to be dysregulated during wound progression (Siu et al., 2023). Together, this chapter examines the contribution of miRNAs in different phases of wound healing and their potential as therapeutic agents in the wound healing process.

## 2 Phases of Wound Healing and miRNA Involvement

Wound healing progresses through a coordinated and overlapping series of hemostasis, inflammation, proliferation, and remodeling phases and involves several cell types (Fig. 1) (Peña & Martin, 2024). Shortly after injury, constrictions of blood vessels limit bleeding, inflammatory cells clear the debris, new vessels and tissue are formed, and collagen remodels to restore skin integrity. In contrast to the normal and linear progression of wound regeneration, chronic diabetic wounds suffer from stalled and asynchronous wound healing phases, and therefore, lack a defined timeline (Falanga et al., 2022). This impaired wound healing is primarily due to vascular dysfunction and a weakened immune system, results in prolonged inflammation, reduced angiogenesis, and delayed tissue formation, and involves differential expression of several miRNAs (Ozdemir & Feinberg, 2018). A growing list of miRNAs has been implicated in the stage-specific response to both acute and chronic wound healing that can be classified as follows.

### *2.1 Hemostasis Phase*

Within minutes after the injury, hemostasis begins to stop bleeding and stabilize the wound (Seré & Hackeng, 2003). The process initiates with vasoconstriction to reduce blood flow. Here, plasma VWF (von Willebrand factor), secreted by

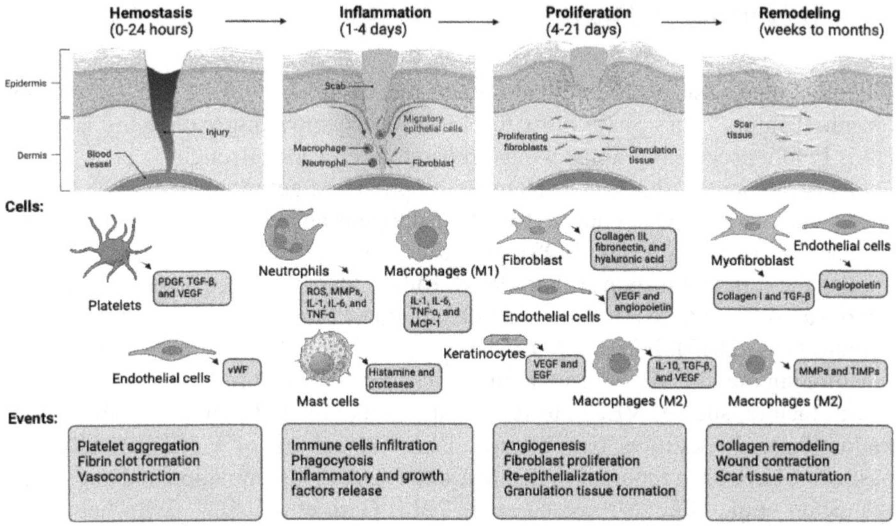


**Fig. 1** Schematic representation of the wound healing process. In hemostasis, platelet aggregation forms a fibrin clot to prevent bleeding, followed by the inflammation phase in which neutrophils, pro-inflammatory M1 macrophages, and mast cells infiltrate to clear the cell debris and secrete inflammatory and other signaling molecules. Around 4 days after the initial injury, fibroblasts, keratinocytes, anti-inflammatory M2 macrophage, and endothelial cells drive granulation tissue formation, angiogenesis, and re-epithelialization. In the final remodeling phase, collagen type III is replaced by the more durable collagen type I to facilitate extracellular matrix deposition and scar tissue maturation that restores tissue integrity. Disruptions in any of these phases can result in chronic wounds characterized by delayed or incomplete wound healing. *PDGF* platelet-derived growth factor, *TGF-β* transforming growth factor beta, *VEGF* vascular endothelial growth factor, *vWF* von Willebrand factor, *ROS* reactive oxygen species, *MMPs* matrix metalloproteinases, *IL-1* interleukin-1, *IL-6* interleukin-6, *TNF-α* tumor necrosis factor-alpha, *MCP-1* monocyte chemoattractant protein-1, *EGF* epidermal growth factor, *IL-10* interleukin-10, *TIMPs* tissue inhibitors of metalloproteinases. (The figure is generated in BioRender with license for publication.)

endothelial cells, promotes the tethering of platelets to exposed collagen at sites of vascular injury (Ruggeri, 2007). Following this, the platelets attach to the exposed collagen and damaged endothelium to release the chemical signals and clotting factors that activate the coagulation cascade (de Groot et al., 2012). This cascade converts prothrombin to thrombin and promotes the formation of fibrin clots consisting of fibrin mesh and aggregated platelets to prevent blood loss. In addition, platelets release growth factors, e.g., platelet-derived growth factor (PDGF), transforming growth factor- β (TGF-β), and vascular endothelial growth factor (VEGF), which recruit inflammatory cells for the subsequent phase of healing (Lynch et al., 1987). Although the miRNAs regulating the hemostasis have not been specifically identified, many miRNAs implicated in thrombus formation may contribute to this coagulation process (Ozdemir & Feinberg, 2018; Teruel-Montoya et al., 2015).

## 2.2 *Inflammatory Phase*

The inflammatory phase follows hemostasis that occurs between 1 and 4 days after the initial injury. This phase aims to eliminate any invading pathogens and cell debris before proceeding to the wound reparatory cascade (Broughton et al., 2006; Goldberg & Diegelmann, 2017). Here, neutrophils first get mobilized to the wound site through diapedesis to phagocytose the pathogens and clear debris, while releasing reactive oxygen species (ROS), proteases, and pro-inflammatory cytokines to amplify the immune response (Peiseler & Kubes, 2019). Subsequently, circulating monocytes differentiate into M1 (pro-inflammatory) macrophages at the wound site (Minutti et al., 2017). Initially, M1 macrophages clear debris and pathogens before transitioning into the M2 (anti-inflammatory) phenotype that releases cytokines and growth factors, such as VEGF, to resolve inflammation and prepare for the proliferation phase. In addition, lymphocytes, including subsets of T cells, B cells, and mast cells, have been reported in this phase to regulate inflammation and angiogenesis (Short et al., 2022; Sîrbulescu et al., 2017; Guth et al., 2024; Tellechea et al., 2016; Moura et al., 2016; Tanno et al., 2017).

Several miRNAs are implicated in the inflammatory phase of wound healing. Emerging studies have shown that miRNAs such as miR-146a/b and miR-155 fine-tune the balance of pro-inflammatory and anti-inflammatory signals (Fig. 2) (Testa et al., 2017). Generally, cytokines, e.g., interleukin-1β (IL-1β) and tumor necrosis factor-α (TNF-α), upregulate the expression of miR-146a/b and miR-155 (Roy & Sen, 2011). miR-146a/b inhibits multiple pro-inflammatory genes such as interleukin-1 receptor-associated kinase 1 (IRAK1), tumor necrosis factor receptor-associated factor 6 (TRAF6), signal transducer and activator of transcription 1 (STAT1), and cyclooxygenase-2 (COX2) and inhibits the activation of toll-like receptor 4 (TLR4) signaling pathway to promote wound repair (Park et al., 2015; Wang et al., 2019a). Interestingly, *Staphylococcus aureus*, a common colonizer of diabetic wounds, increases the expression of miR-155 and impairs the DNA repair process (Jneid et al., 2017; Ramirez et al., 2018). miR-155, a pro-inflammatory miRNA, silences src homology 2 domain-containing inositol polyphosphate 5-phosphatase 1 (SHIP1), suppressor of cytokine signaling 1 (SOCS1), interleukin-12 (IL-12), and musculoaponeurotic fibrosarcoma oncogene homolog (MAF) as well as enhances the differentiation of T helper cells to facilitate immune response to the wound site (Wang et al., 2019b). The miRNAs associated with this phase influence key inflammatory signaling pathways such as toll-like receptor (TLR), nuclear factor kappa B (NF-κB), and janus kinase-signal transducer and activator of transcription (JAK-STAT) (Table 1). These tightly controlled signaling pathways orchestrate the inflammatory response to facilitate or hinder the healing process, as seen in diabetic wounds.

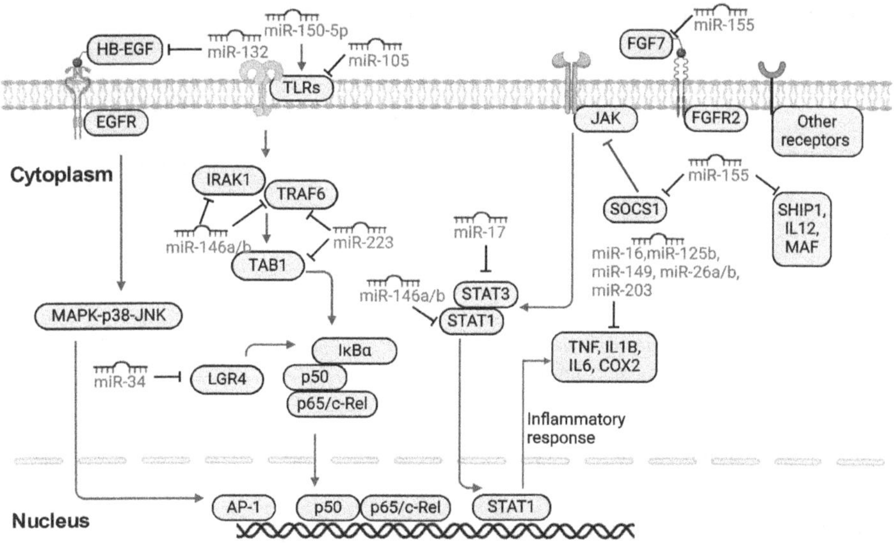


**Fig. 2** miRNAs involved in the inflammation phase of wound healing. In the inflammation phase, miRNAs such as miR-146a/b and miR-155 regulate immune cell recruitment, cytokine production, and the balance between pro-inflammatory and anti-inflammatory responses. By targeting specific genes, these miRNAs influence the activity of neutrophils and macrophages via signaling pathways such as toll-like receptors (TLRs), nuclear factor kappa B (NF-κB), and janus kinase-signal transducer and activator of transcription (JAK-STAT) that are essential to eliminate invading pathogens and cell debris before transitioning to subsequent healing phases. *HB-EGF* heparin-binding EGF-like growth factor, *EGFR* epidermal growth factor receptor, *MAPK* mitogen-activated protein kinase, *p38* mitogen-activated protein kinase 14, *JNK* c-Jun N-terminal kinases, *AP-1* activator protein 1, *IRAK1* interleukin-1 receptor associated kinase 1, *TRAF6* TNF receptor–associated factor 6, *TAB1* TGF-beta activated kinase 1 binding protein 1, *LGR4* leucine-rich repeat-containing G-protein coupled receptor 4, *IκBα* nuclear factor of kappa light polypeptide gene enhancer in B cells inhibitor alpha, *p50* nuclear factor kappa B subunit 1, *p65* nuclear factor NF-kappa-B p65 subunit, *c-Rel* REL proto-oncogene, *TNF* tumor necrosis factor, *IL-B* interleukin-1 beta, *IL-6* interleukin-6, *COX2* cyclooxygenase-2, *SHIP1* SH2 domain containing inositol 5-phosphatase 1, *IL-12* interleukin-12, *MAF* musculoaponeurotic fibrosarcoma oncogene homolog, *FGF* fibroblast growth factors, *FGFR2* fibroblast growth factor receptor 2. (The figure is generated in BioRender with license for publication)

## 2.3 *Proliferative Phase*

The proliferative phase of wound healing begins as the inflammatory phase subsides and is characterized by the generation of new tissue to restore skin integrity (Falanga et al., 2022). At this point, fibroblasts enrich the wound site and synthesize extracellular matrix (ECM) proteins such as collagen type III to form a provisional matrix, termed granulation tissue. Concurrently, endothelial cells stimulate the blood vessel network to supply oxygen and nutrients to the regenerating tissue while M2 macrophages further support tissue formation by secretion of growth factors and promoting fibroblast activity (Krzyszczyk et al., 2018). Further, keratinocytes migrate and

**Table 1** miRNAs in the inflammation phase of wound healing

| miRNA | Function | Target(s) | References |
|---|---|---|---|
| miR-16 | Relieves inflammatory response | COX2 | Yan et al. (2020), Bo et al. (2021), and Shanmugam et al. (2008) |
| miR-17-5p | Inhibits JAK/STAT signaling pathway activation | STAT3 | Yang et al. (2017) |
| miR-19a/b | Inhibits p65 nuclear translocation | SHCBP1 | Li et al. (2020) |
| miR-20a | Inhibits p65 nuclear translocation | SEMA7A | Li et al. (2020) |
| miR-21 | Enhances M1 macrophage polarization; promotes DCs differentiation; restricts TLR4 activity | PDCD4, PTEN | Sheedy et al. (2010), Xie et al. (2022), and Liechty et al. (2020) |
| miR-23b | Decreases pro-inflammatory response while increasing the anti-inflammatory responses | ASK1 | Li et al. (2018b) |
| miR-26a/b | Reduces the inflammatory response | COX2 | Kwon et al. (2015) |
| miR-27b | Reduces ROS production | SHC1 | Wang et al. (2014b) |
| miR-34 | Enhances the activity of the NF-κB signaling pathway | LGR4 | Wu et al. (2019) |
| miR-105 | Decreases the inflammatory response | TLR2 | Benakanakere et al. (2009) |
| miR-125b | Promotes fibroblast migration | TNF, TP53INP1, SIRT7 | Xu et al. (2011); Zhang et al. (2021c), Wang et al. (2023), Zhou et al. (2014), and Xia et al. (2022) |
| miR-132 | Suppresses the activation of NF-κB signaling | HBEGF | Li et al. (2015b) |
| miR-139-5p | Inhibits neutrophils differentiation | EIF4G2 | Zhang et al. (2021b) |
| miR-140 | Modulates PDGF signaling during palatogenesis | PDGF | Eberhart et al. (2008) |
| miR-142 | Promotes neutrophils migration | Small GTPase | Tanaka et al. (2016) |
| miR-146a/b | Inhibits TLR4 signaling pathway activation; dampens the inflammatory response | TRAF6, IRAK1, STAT1, TNF, COX2 | Park et al. (2015) and Wang et al. (2019a) |
| miR-149 | Inhibits secretion of pro-inflammatory cytokines | IL1A, IL1B, IL6 | Lang et al. (2017) |
| miR-150-5p | Activates TLRs' signaling pathway | #NA | Yang et al. (2015) |
| miR-155 | Enhances inflammatory cell recruitment and activation; promotes T helper cells differentiation | FGF7, SOCS1, SHIP1, IL12, MAF | Wang et al. (2019b), Xu et al. 2022a, b), and Moura et al. (2019) |

(continued)

**Table 1** (continued)

| miRNA | Function | Target(s) | References |
|---|---|---|---|
| miR-203 | Alleviates skin inflammation | TNF, IL24 | Primo et al. (2012) |
| miR-223 | Suppresses neutrophil activation; induces M2 macrophage polarization; inhibits NF-κB signaling | MEF2C, CUL1, TRAF6, TAB1 | Johnnidis et al. (2008) and Zhou et al. (2018) |
| miR-375 | Reduces ROS production | TIMP1 | Sheng et al. (2016) |
| miR-497 | Inhibits secretion of pro-inflammatory cytokines | #NA | Ban et al. (2020) |
| miR-5591-5p | Reduces ROS production | AGEs | Li et al. (2018c) |

proliferate to facilitate the covering of the wound site in the re-epithelialization process.

In the proliferation phase, miRNAs are involved in regulating re-epithelialization, angiogenesis, and granulation tissue steps (Fig. 3). (Soliman et al., 2018). In keratinocytes, miR-205 has been shown to suppress SH2-domain containing phosphatidylinositol-3,4,5-trisphosphate 5-phosphatase 2 (SHIP2), a negative regulator of the AKT (protein kinase B) signaling. Together with miR-184, miR-205 has been shown to enhance keratinocyte migration by modulating the RhoA-ROCK pathway and actin cytoskeleton dynamics (Yu et al., 2008, 2010). Another miRNA, miR-21, silences tissue inhibitor of metalloproteinase-3 (TIMP3) and T cell lymphoma invasion and metastasis 1 (TIAM1), an activator of Rho GTPases, and augments keratinocyte migration and re-epithelialization during wound healing by promoting TGF-β1-induced cell motility (Yang et al., 2011). miR-26a expression increased in a diabetic dermal wound healing model in mice, and local delivery of inhibitors to miR-26a rapidly induced angiogenesis, increased granulation tissue, and accelerated wound closure (Icli et al., 2016). The miRNAs associated with the proliferation phase commonly influence key angiogenic and proliferation processes including TGF-β, VEGF, integrin, and hypoxia signaling pathways (Table 2).

## 2.4 *Remodeling Phase*

In the remodeling phase, the immature fibrin matrix and granulation tissue formed during the proliferative phase are reorganized to form scar tissue and restore the wound site (Peña & Martin, 2024). Here, the fibroblasts differentiate into myofibroblasts, and collagen type III is replaced by the more sustainable collagen type I. This collagen remodeling is regulated by the proteolytic matrix metalloproteinases (MMPs). Remodeling continues for months to ensure optimal tensile strength and function. As the remodeling progresses, angiogenesis subsides, and excess blood vessels are pruned by the vascular regression process, which ultimately leads to a mature scar formation.

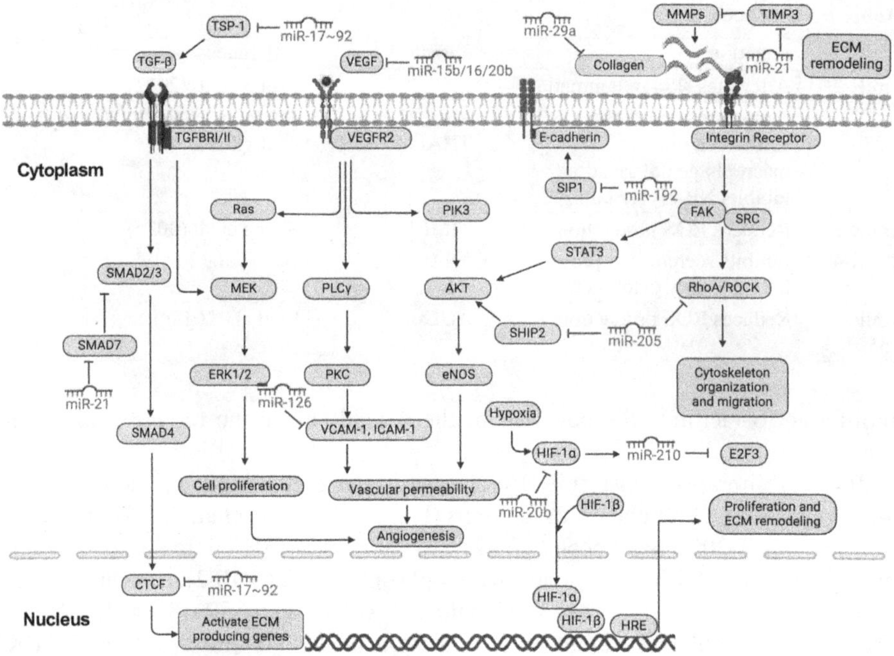


**Fig. 3** miRNAs involved in the proliferation and remodeling phases of wound healing. In the proliferation phase, miRNAs such as miR-21 and miR-205 promote keratinocyte migration, re-epithelialization, angiogenesis, fibroblast proliferation, and extracellular matrix (ECM) deposition. During the remodeling phase, miRNAs including miR-29 and miR-192 regulate collagen remodeling and ECM maturation, facilitating proper scar formation and the restoration of tissue integrity. The miRNAs associated with the proliferation and remodeling phases commonly influence transforming growth factor-beta (TGF-β), vascular endothelial growth factor (VEGF), and integrin-, hypoxia-, and ECM remodeling signaling pathways. *TSP-1* thrombospondin 1, *TGFBRI/II* transforming growth factor beta receptor I/II, *SMAD* (mothers against decapentaplegic homolog), *CTCF* CCCTC-binding factor, *Ras* rat sarcoma virus, *MEK* mitogen-activated protein kinase kinase, *ERK1/2* mitogen-activated protein kinase 1/2, *VEGFR2* vascular endothelial growth factor receptor 2, *PLCγ* phosphoinositide phospholipase C, *PKC* protein kinase C, *VCAM-1* vascular cell adhesion molecule 1, *ICAM-1* intercellular adhesion molecule 1, *PIK3* phosphatidylinositol-4,5-bisphosphate 3-kinase, *AKT* protein kinase B, *eNOS* nitric oxide synthase 3, *SIP1* smad interacting protein 1, *STAT3* signal transducer and activator of transcription 3, *HIF-1α* hypoxia-inducible factor 1-alpha, *HIF-1β* hypoxia-inducible factor 1-beta, *HRE* hypoxia-response element, *FAK* focal adhesion kinase, *SRC* src proto-oncogene, *RhoA/ROCK* ras homolog family member A/rho-associated coiled-coil-containing protein kinase, *E2F3* E2F transcription factor 3, *MMPs* matrix metalloproteinases, *TIMP3* tissue inhibitor of metalloproteinases 3. (The figure is generated in BioRender with license for publication)

Deposition of collagen is a key feature of the remodeling phase of wound healing. miR-29a directly regulates collagen expression to impacting ECM composition (van Rooij et al., 2008). In mammals, fetal skin heals without scarring, but during gestation this shifts toward a scarring phenotype where miR-192, miR- 29b, and miR-29c are strongly induced (Beanes et al., 2002; Beanes et al., 2003; Maurer

**Table 2** miRNAs in the proliferation and remodeling phases of wound healing

| miRNA | Function | Target(s) | References |
|---|---|---|---|
| Proliferation phase | | | |
| miR-15b | Impairs angiogenesis in endothelial cells | VEGF | Hua et al. (2006) |
| miR-16 | Impairs angiogenesis in endothelial cells | VEGF | Hua et al. (2006) |
| miR-17 ~ 92 cluster | Enhances cell survival functions and angiogenesis | TSP1, CTGF, TIMP1 | Landskroner-Eiger et al. (2015), Lucas et al. (2017), Gallant-Behm et al. (2018), Nie et al., (2023), and Banerjee and Sen (2013) |
| miR-17 | Impairs angiogenesis in endothelial cells | JAK1 | Doebele et al. (2010) |
| miR-20a | Impairs angiogenesis in endothelial cells | MAP2K3, VEGF | Hua et al. (2006) and Pin et al. (2012) |
| miR-20b | Impairs angiogenesis and migration in endothelial cells | HIF1A, VEGF | Hua et al. (2006) and Lei et al. (2009) |
| miR-21 | Promotes keratinocyte migration and re-epithelialization | TIMP3, TIAM1 | Yang et al. (2011) |
| miR-26a | Impairs angiogenesis in endothelial cells | SMAD1 | Icli et al. (2016) |
| miR-31 | Promotes keratinocyte proliferation, migration, and re-epithelialization | EMP1, HIF1AN, SPRED1 | Li et al. (2015a), Hu et al. (2015), and Edmonds et al. (2016) |
| miR-92a | Impairs angiogenesis in endothelial cells | ITGA5 | Bonauer et al. (2009) |
| miR-99 | Regulates AKT/mTOR signaling in keratinocytes | IGF1R, MTOR, AKT1 | Jin et al. (2013) |
| miR-126 | Enhances migration and angiogenesis of endothelial cells | VCAM1, SPRED1, PIK3R2 | Harris et al. (2008), Li et al. (2016), Wang et al. (2008), and Kuhnert et al. (2008) |
| miR-130a | Promotes angiogenesis | MEOX2, HOXA5 | Chen and Gorski (2008) |
| miR-155 | Promotes inflammatory cell accumulation and differentiation of T helper cells | FGF7 | Pottier et al. (2009) |
| miR-184 | Impairs miR-205, AKT signaling activity, and promotes keratinocyte apoptosis | #NA | Yu et al. (2008) |
| miR-198 | Impairs keratinocyte migration | DIAPH1, PLAU, LAMC2 | Sundaram et al. (2013) |
| miR-200 | Hypoxia-sensitive miRNA, promotes endothelial angiogenesis | ETS1 | Chan et al. (2011) and Gregory et al. (2008) |
| miR-203 | Promotes keratinocyte proliferation, migration, and re-epithelialization | RAN, RAPH1 | Viticchiè et al. (2012) |

(continued)

**Table 2** (continued)

| miRNA | Function | Target(s) | References |
|---|---|---|---|
| miR-205 | Promotes keratinocyte migration and inhibits apoptosis | SHIP2, RHOA, ROCK1 | Yu et al. (2010) and Yu et al. (2008) |
| miR-210 | Hypoxia-inducible miRNA, impairs keratinocyte proliferation and endothelial angiogenesis | E2F3, ISCU, EFNA3 | Biswas et al. (2010), Ma et al., (2019), and Fasanaro et al. (2008) |
| miR-221 | Inhibits cells proliferation | KIT | Felli et al. (2005) |
| miR-222 | Inhibits cells proliferation | KIT | Felli et al. (2005) |
| miR-296 | Promotes endothelial angiogenesis | HGS | Würdinger et al. (2008) |
| miR-320 | Impairs angiogenesis in diabetic endothelial cells | IGF1 | Wang et al. (2009) |
| miR-378 | Promotes cell survival and angiogenesis | FUS, SUFU | Lee et al. (2007) |
| miR-424 | Hypoxia-inducible miRNA, promotes endothelial angiogenesis | CUL2 | Ghosh et al. (2010) |
| miR-483-3p | Regulates keratinocyte migration and proliferation | MAPKAPK2, MK167, YAP1 | Bertero et al. (2011) |
| miR-503 | Impairs endothelial cell (EC) proliferation, migration, and angiogenesis | CCNE1, CDC25A | Caporali et al. (2011) |
| Remodeling phase | | | |
| miR-21 | Impairs endothelial angiogenesis | PTEN, SMAD7 | Yang et al. (2011) and Wang et al. (2012) |
| miR-29a | Regulates collagen expression during ECM remodeling | TAB1, collagen I/II | Maurer et al. (2010) and Ciechomska et al. (2014) |
| miR-29b | Regulates collagen expression during ECM remodeling | SMADs, β-catenin | Li et al. (2009) |
| miR-29c | Regulates collagen expression during ECM remodeling | SMADs, β-catenin | van Rooij et al. (2008) |
| miR-143/145 | Promotes fibroblasts' differentiation and regulates ECM remodeling | IRS1, PDGF, α-SMA | Davis-Dusenbery et al. (2011), Chivukula et al. (2014), and Wang et al. (2014a) |
| miR-192/ miR-215 | Promotes collagen and E-cadherin expression | CDH1, SIP1 | Beanes et al. 2003 and Wang et al. (2010) |
| miR-198 | Promotes cell proliferation | DIAPH1, PLAU, LAMC2 | Sundaram et al., (2013) and Tan et al. (2011) |

et al., 2010; Banerjee et al., 2011). miR-29b and miR-29c directly influence TGF-β signaling, while miR-192 enhances collagen type I expression by targeting smad interacting protein 1 (SIP1) (Li et al., 2009; Kato et al., 2007). Consequently, by regulating collagen synthesis, miR-29 and miR-192 optimize collagen accumulation in the ECM during tissue remodeling (Table 2).

## 3 miRNA Dysregulation in Chronic Wounds

Chronic conditions like diabetes, vascular disease, or neuropathy exacerbate the skin regeneration capacity, resulting in chronic wounds, e.g., diabetic foot ulcers or DFUs, which afflicts 20% of the diabetic population and may lead to amputation (Akkus & Sert, 2022). Prolonged exposure to hyperinsulinemia and hyperglycemia leads to skin dysfunction and contributes to chronic, nonhealing DFUs that is further exacerbated by peripheral artery disease (PAD) and neuropathy (Chen et al., 2022; Burgess et al., 2021). These chronic diabetic wounds show low-grade persistent inflammation, infection, senescent fibroblasts, nonmigratory keratinocytes, poor angiogenesis, excessive protease activity, and oxidative stress (Frykberg & Banks, 2015). Recent studies suggest that in the inflammatory phase of wound healing, diabetic wounds fail to adequately suppress the pro-inflammatory signals and impair the subsequent transition to a proliferative phase and ECM remodeling (Baltzis et al., 2014; Mirza et al., 2014).

Dysregulation of miRNAs, which are typically expressed during acute wound healing, has often been observed in chronic wound healing. These miRNAs may hold significant therapeutic value in the healing of chronic wounds such as DFUs. For instance, diabetic wound healing shows upregulation of miR-132 and miR-155, and downregulation of miR-146a compared to healthy controls (Fumimoto et al., 2024; Li et al., 2015b; Xu et al., 2022b). miR-132 promotes the transition from inflammation to proliferation during wound healing by targeting heparin-binding EGF-like growth factor (HB-EGF) (Li et al., 2015b). Consequently, exosomes from miR-132-overexpressing adipose stem cells improved diabetic wound healing by reducing inflammation, promoting angiogenesis, and stimulating M2 macrophage polarization by inhibiting NF-kB signaling (Ge et al., 2023). In mice, restoring miR-146a expression via MSC treatment reduced inflammation and improved diabetic wound healing (Xu et al., 2012). In addition, inhibition of miR-155 was shown to restore healing by increasing fibroblast growth factor (FGF7) in a diabetic mouse model (Moura et al., 2019).

## 4 Circulating miRNAs in Wound Healing

Extracellular vesicles (EVs), such as exosomes, microvesicles, and apoptotic bodies containing different miRNAs, originate from various cell types relevant to wound healing including epithelial cells, macrophages, mesenchymal stem cells (MSCs), or fibrocytes (Petkovic et al., 2020). The differential expression of these circulating miRNAs holds significant promise for diagnosing and treating both acute and chronic wounds. For example, in patients with diabetic foot ulcers elevated serum miR-217 expression was found to target hypoxia-inducible factor 1-alpha (HIF-1$\alpha$) and regulate an angiogenic signaling pathway (Lin et al., 2019). In a rodent model, miR-217 inhibition improved angiogenesis by enhancing HIF-1$\alpha$ expression and

VEGF regulation. Similarity in diabetic foot ulcer patients, low serum level of miR-145, an antiangiogenic miRNA, inversely correlated with high TGF-β1 levels, while miR-145 expression was expressed higher in the muscles of amputated limbs (Petkovic et al., 2020; Davis-Dusenbery et al., 2011). In patients with diabetes and critical limb ischemia, a severe form of PAD, the plasma level of miR-4739 was elevated and was suggested as a biomarker of diabetic limb ischemia (Li et al., 2018a). EVs are often enriched with miRNAs, such as miR-21, miR-181, and miR-146a, and could be harnessed to resolve inflammation and promote angiogenesis (Tian et al., 2023). Together, circulating and EV-derived miRNAs offer potential biomarkers and therapeutic opportunities in wound regeneration.

## 5 Therapeutic Potential and Challenges of miRNAs in Wound Repair

Emerging studies show that miRNAs may serve as promising noninvasive biomarkers for acute and chronic diseases as demonstrated by miRNA profiling from plasma or peripheral blood cells (Wu et al., 2023). It is hypothesized that they hold the potential as prognostic biomarkers for chronic ischemic wounds to track disease progression and healing outcomes (Fasanaro et al., 2010). Along with its prognostic value, the dysregulated expression of miRNAs provides a unique advantage compared to the conventional gene therapy in wound healing by impacting a set of genes using specific miRNAs (Rupaimoole & Slack, 2017). This approach involves the delivery (either local or systemic) of protective, pro-healing miRNA mimics to overexpress miRNAs, and the inhibition of detrimental ones in relevant models of disease for wound healing. miRNA mimics are typically synthetic double-stranded RNA molecules that replicate the target miRNA sequence and integrate into the RNA-induced silencing complex (RISC) complex and mediate silencing of target gene(s). In contrast, the inhibition of harmful miRNAs can be facilitated by delivery of inhibitors called anti-miRs, which are complementary oligonucleotides that inhibit miRNA activity most often by preventing the binding of the miRNA seed sequence to its cognate binding site in the 3'untranslated region on target mRNAs.

Despite over 60 miRNA or small interfering RNA (siRNA)-based therapeutics being in various stages of clinical trials, with some already approved, their application in wound healing is still in preliminary stages. This may in part be due to challenges of miRNA therapeutic delivery strategies navigating complex gene-regulatory networks, identifying suitable target cells, and optimizing delivery systems (Seyhan, 2024; Brillante et al., 2024; Zhang et al., 2021a). The success of miRNA mimics or anti-miR-based delivery systems depends on their resistance to tissue degradation and their binding affinity to target miRNAs. Naked miRNAs face nuclease degradation in the physiological milieu and show limited passive cellular uptake due to their hydrophilic nature and negative charge. Therefore, efficient delivery systems are essential for enhancing stability and delivery efficacy. At present, these delivery systems can be categorized into viral and nonviral vectors (Taghdiri & Mussolino, 2024). The viral vectors include adenovirus, adeno-associated virus, retrovirus, and lentivirus; however, the efficiency of these viral vectors is associated with potential mutagenesis and toxicity risks. In contrast, nonviral vectors offer a safer and more efficient alternative. For instance, anti-miR-210 encapsulated lipid nanoparticles have been shown to benefit ischemic wound healing (Ghatak et al., 2016). Similarly, nanoparticle-laden hydrogels containing miR-223 mimic were demonstrated to suppress inflammation and promote wound healing (Saleh et al., 2019). The nonviral vectors can be further classified into inorganic delivery systems (Jiang et al., 2022), e.g., gold, silica, and iron oxide particles (Peña & Martin, 2024), lipid-based nanocarriers, e.g., liposomes (Mulholland et al., 2017); polymers or dendrimers, e.g., poly(lactic-co-glycolic acid) (Sen, 2021); cell-derived membrane vesicles, e.g., exosomes and microvesicles; and three-dimensional (3D) (Falanga et al., 2022) scaffold-based delivery systems, e.g., hydrogels (Fu et al., 2019). In addition, chemical modifications to oligonucleotides, e.g., locked nucleic acid (LNA) modifications, can be combined with these vectors to further facilitate stability, reduce degradation, and enhance target specificity of miRNA therapeutics (Gebert et al., 2014).

## 6 Conclusion and Perspectives

A growing body of literature has demonstrated the therapeutic potential of miRNAs in both acute and chronic wound healing. These miRNAs regulate different phases of wound healing by targeting associated genes and signaling networks. In addition, miRNAs can serve as a significant biomarker of wound progression as they can be detected in blood, which in the future may potentially inform successful therapeutic treatment. However, a major challenge in implementing miRNA therapy is developing an effective delivery system to overcome the extracellular and intracellular barriers without eliciting the immune response. Integrating miRNA therapeutics into wound dressings offers a practical solution to maintain a moist, protective healing environment while efficiently delivering the miRNAs to the affected wound site (Ghosh et al., 2024; Berger et al., 2023). With an optimized delivery system and successful clinical trials, miRNAs could emerge as a cutting-edge third-generation nucleic acid therapy, especially for the treatment of chronic wounds (Kulkarni et al., 2021).

**Acknowledgments** This work was supported in part by funding from the National Institutes of Health and the American Heart Association.

**Conflicts of Interest** None.

## References

Akkus, G., & Sert, M. (2022). Diabetic foot ulcers: A devastating complication of diabetes mellitus continues non-stop in spite of new medical treatment modalities. *World Journal of Diabetes, 13*(12), 1106–1121.

Baltzis, D., Eleftheriadou, I., & Veves, A. (2014). Pathogenesis and treatment of impaired wound healing in diabetes mellitus: New insights. *Advances in Therapy, 31*(8), 817–836.

Ban, E., Jeong, S., Park, M., Kwon, H., Park, J., Song, E. J., et al. (2020). Accelerated wound healing in diabetic mice by miRNA-497 and its anti-inflammatory activity. *Biomedicine & Pharmacotherapy, 121*, 109613.

Banerjee, J., & Sen, C. K. (2013). MicroRNAs in skin and wound healing. *Methods in Molecular Biology, 936*, 343–356.

Banerjee, J., Chan, Y. C., & Sen, C. K. (2011). MicroRNAs in skin and wound healing. *Physiological Genomics, 43*(10), 543–556.

Bartel, D. P. (2004). MicroRNAs: Genomics, biogenesis, mechanism, and function. *Cell, 116*(2), 281–297.

Beanes, S. R., Hu, F.-Y., Soo, C., Dang, C. M. H., Urata, M., Ting, K., et al. (2002). Confocal microscopic analysis of scarless repair in the fetal rat: Defining the transition. *Plastic and Reconstructive Surgery, 109*(1), 160–170.

Beanes, S. R., Dang, C., Soo, C., & Ting, K. (2003). Skin repair and scar formation: The central role of TGF-[beta]. *Expert Reviews in Molecular Medicine, 5*(08), 1.

Benakanakere, M. R., Li, Q., Eskan, M. A., Singh, A. V., Zhao, J., Galicia, J. C., et al. (2009). Modulation of TLR2 protein expression by miR-105 in human oral keratinocytes. *The Journal of Biological Chemistry, 284*(34), 23107–23115.

Berger, A. G., Deiss-Yehiely, E., Vo, C., McCoy, M. G., Almofty, S., Feinberg, M. W., et al. (2023). Electrostatically assembled wound dressings deliver pro-angiogenic anti-miRs preferentially to endothelial cells. *Biomaterials, 300*, 122188.

Bertero, T., Gastaldi, C., Bourget-Ponzio, I., Imbert, V., Loubat, A., Selva, E., et al. (2011). miR-483-3p controls proliferation in wounded epithelial cells. *The FASEB Journal, 25*(9), 3092–3105.

Biswas, S., Roy, S., Banerjee, J., Hussain, S.-R. A., Khanna, S., Meenakshisundaram, G., et al. (2010). Hypoxia inducible microRNA 210 attenuates keratinocyte proliferation and impairs closure in a murine model of ischemic wounds. *Proceedings of the National Academy of Sciences of the United States of America, 107*(15), 6976–6981.

Bo, Y., Liu, B., Yang, L., Zhang, L., & Yan, Y. (2021). Exosomes derived from miR-16-5p-overexpressing keratinocytes attenuates bleomycin-induced skin fibrosis. *Biochemical and Biophysical Research Communications, 19*(561), 113–119.

Bonauer, A., Carmona, G., Iwasaki, M., Mione, M., Koyanagi, M., Fischer, A., et al. (2009). MicroRNA-92a controls angiogenesis and functional recovery of ischemic tissues in mice. *Science, 324*(5935), 1710–1713.

Brillante, S., Volpe, M., & Indrieri, A. (2024). Advances in microrna therapeutics: From preclinical to clinical studies. *Human Gene Therapy, 35*(17–18), 628–648.

Broughton, G., Janis, J. E., & Attinger, C. E. (2006). The basic science of wound healing. *Plastic and Reconstructive Surgery, 117*(7 Suppl), 12S–34S.

Burgess JL, Wyant WA, Abdo Abujamra B, Kirsner RS, Jozic I. Diabetic wound-healing science. . 2021;57(10).

Caporali, A., Meloni, M., Völlenkle, C., Bonci, D., Sala-Newby, G. B., Addis, R., et al. (2011). Deregulation of microRNA-503 contributes to diabetes mellitus-induced impairment of endothelial function and reparative angiogenesis after limb ischemia. *Circulation, 123*(3), 282–291.

Chan, Y. C., Khanna, S., Roy, S., & Sen, C. K. (2011). miR-200b targets Ets-1 and is down-regulated by hypoxia to induce angiogenic response of endothelial cells. *The Journal of Biological Chemistry, 286*(3), 2047–2056.

Chen, Y., & Gorski, D. H. (2008). Regulation of angiogenesis through a microRNA (miR-130a) that down-regulates antiangiogenic homeobox genes GAX and HOXA5. *Blood, 111*(3), 1217–1226.

Chen, L., Sun, S., Gao, Y., & Ran, X. (2022). Global mortality of diabetic foot ulcer: A systematic review and meta-analysis of observational studies. *Diabetes, Obesity & Metabolism, 25*, 36.

Chivukula, R. R., Shi, G., Acharya, A., Mills, E. W., Zeitels, L. R., Anandam, J. L., et al. (2014). An essential mesenchymal function for miR-143/145 in intestinal epithelial regeneration. *Cell, 157*(5), 1104–1116.

Ciechomska, M., O'Reilly, S., Suwara, M., Bogunia-Kubik, K., & van Laar, J. M. (2014). MiR-29a reduces TIMP-1 production by dermal fibroblasts via targeting TGF-β activated kinase 1 binding protein 1, implications for systemic sclerosis. *PLoS One, 9*(12), e115596.

Davis-Dusenbery, B. N., Chan, M. C., Reno, K. E., Weisman, A. S., Layne, M. D., Lagna, G., et al. (2011). Down-regulation of Kruppel-like factor-4 (KLF4) by microRNA-143/145 is critical for modulation of vascular smooth muscle cell phenotype by transforming growth factor-beta and bone morphogenetic protein 4. *The Journal of Biological Chemistry, 286*(32), 28097–28110.

de Groot, P. G., Urbanus, R. T., & Roest, M. (2012). Platelet interaction with the vessel wall. *Handbook of Experimental Pharmacology, 210*, 87–110.

Doebele, C., Bonauer, A., Fischer, A., Scholz, A., Reiss, Y., Urbich, C., et al. (2010). Members of the microRNA-17-92 cluster exhibit a cell-intrinsic antiangiogenic function in endothelial cells. *Blood, 115*(23), 4944–4950.

Eberhart, J. K., He, X., Swartz, M. E., Yan, Y.-L., Song, H., Boling, T. C., et al. (2008). MicroRNA Mirn140 modulates Pdgf signaling during palatogenesis. *Nature Genetics, 40*(3), 290–298.

Edmonds, M. D., Boyd, K. L., Moyo, T., Mitra, R., Duszynski, R., Arrate, M. P., et al. (2016). MicroRNA-31 initiates lung tumorigenesis and promotes mutant KRAS-driven lung cancer. *The Journal of Clinical Investigation, 126*(1), 349–364.

Falanga, V., Isseroff, R. R., Soulika, A. M., Romanelli, M., Margolis, D., Kapp, S., et al. (2022). *Chronic wounds Nature Reviews Disease Primers, 8*(1), 50.

Fasanaro, P., D'Alessandra, Y., Di Stefano, V., Melchionna, R., Romani, S., Pompilio, G., et al. (2008). MicroRNA-210 modulates endothelial cell response to hypoxia and inhibits the receptor tyrosine kinase ligand Ephrin-A3. *The Journal of Biological Chemistry, 283*(23), 15878–15883.

Fasanaro, P., Greco, S., Ivan, M., Capogrossi, M. C., & Martelli, F. (2010). microRNA: Emerging therapeutic targets in acute ischemic diseases. *Pharmacology & Therapeutics, 125*(1), 92–104.

Felli, N., Fontana, L., Pelosi, E., Botta, R., Bonci, D., Facchiano, F., et al. (2005). MicroRNAs 221 and 222 inhibit normal erythropoiesis and erythroleukemic cell growth via kit receptor downmodulation. *Proceedings of the National Academy of Sciences of the United States of America, 102*(50), 18081–18086.

Frykberg, R. G., & Banks, J. (2015). Challenges in the treatment of chronic wounds. *Advances in Wound Care (New Rochelle), 4*(9), 560–582.

Fu, Y., Chen, J., & Huang, Z. (2019). Recent progress in microRNA-based delivery systems for the treatment of human disease. *ExRNA, 1*(1), 24.

Fumimoto, C., Yamauchi, N., Minagawa, E., & Umeda, M. (2024). MiR-146a is mutually regulated by high glucose-induced oxidative stress in human periodontal ligament cells. *International Journal of Molecular Sciences, 25*(19).

Gallant-Behm, C. L., Piper, J., Dickinson, B. A., Dalby, C. M., Pestano, L. A., & Jackson, A. L. (2018). A synthetic microRNA-92a inhibitor (MRG-110) accelerates angiogenesis and wound healing in diabetic and nondiabetic wounds. *Wound Repair and Regeneration, 26*(4), 311–323.

Ge, L., Wang, K., Lin, H., Tao, E., Xia, W., Wang, F., et al. (2023). Engineered exosomes derived from miR-132-overexpresssing adipose stem cells promoted diabetic wound healing and skin reconstruction. *Frontiers in Bioengineering and Biotechnology, 1*(11), 1129538.

Gebert, L. F. R., Rebhan, M. A. E., Crivelli, S. E. M., Denzler, R., Stoffel, M., & Hall, J. (2014). Miravirsen (SPC3649) can inhibit the biogenesis of miR-122. *Nucleic Acids Research, 42*(1), 609–621.

Ghatak, S., Li, J., Chan, Y. C., Gnyawali, S. C., Steen, E., Yung, B. C., et al. (2016). AntihypoxamiR functionalized gramicidin lipid nanoparticles rescue against ischemic memory improving cutaneous wound healing. *Nanomedicine, 12*(7), 1827–1831.

Ghosh, G., Subramanian, I. V., Adhikari, N., Zhang, X., Joshi, H. P., Basi, D., et al. (2010). Hypoxia-induced microRNA-424 expression in human endothelial cells regulates HIF-α isoforms and promotes angiogenesis. *The Journal of Clinical Investigation, 120*(11), 4141–4154.

Ghosh, R., Singh, P., Pandit, A. H., Tariq, U., Bhunia, B. K., & Kumar, A. (2024). Emerging technological advancement for chronic wound treatment and their role in accelerating wound healing. *ACS Applied Bio Materials, 7*(11), 7101–7132.

Goldberg, S. R., & Diegelmann, R. F. (2017). Basic science of wound healing. In R. S. Dieter, R. A. Dieter Jr., R. A. Dieter III, & A. Nanjundappa (Eds.), *Critical limb ischemia* (pp. 131–136). Springer.

Gregory, P. A., Bert, A. G., Paterson, E. L., Barry, S. C., Tsykin, A., Farshid, G., et al. (2008). The miR-200 family and miR-205 regulate epithelial to mesenchymal transition by targeting ZEB1 and SIP1. *Nature Cell Biology, 10*(5), 593–601.

Guth, C., Limjunyawong, N., & Pundir, P. (2024). The evolving role of mast cells in wound healing: Insights from recent research and diverse models. *Immunology and Cell Biology, 102*(10), 878–890.

Harris, T. A., Yamakuchi, M., Ferlito, M., Mendell, J. T., & Lowenstein, C. J. (2008). MicroRNA-126 regulates endothelial expression of vascular cell adhesion molecule 1. *Proceedings of the National Academy of Sciences of the United States of America, 105*(5), 1516–1521.

Hu, J., Chen, C., Liu, Q., Liu, B., Song, C., Zhu, S., et al. (2015). The role of miR-31/FIH1 pathway in TGFβ-induced liver fibrosis. *Clinical Science, 129*(4), 305–317.

Hua, Z., Lv, Q., Ye, W., Wong, C.-K. A., Cai, G., Gu, D., et al. (2006). MiRNA-directed regulation of VEGF and other angiogenic factors under hypoxia. *PLoS One, 1*(1), e116.

Icli, B., Nabzdyk, C. S., Lujan-Hernandez, J., Cahill, M., Auster, M. E., Wara, A. K. M., et al. (2016). Regulation of impaired angiogenesis in diabetic dermal wound healing by microRNA-26a. *Journal of Molecular and Cellular Cardiology, 91*, 151–159.

Jiang, Y., Xu, X., Xiao, L., Wang, L., & Qiang, S. (2022). The role of microRNA in the inflammatory response of wound healing. *Frontiers in Immunology, 21*(13), 852419.

Jin, Y., Tymen, S. D., Chen, D., Fang, Z. J., Zhao, Y., Dragas, D., et al. (2013). MicroRNA-99 family targets AKT/mTOR signaling pathway in dermal wound healing. *PLoS One, 8*(5), e64434.

Jneid, J., Lavigne, J. P., La Scola, B., & Cassir, N. (2017). The diabetic foot microbiota: A review. *Human Microbiome Journal, 5–6*, 1–6.

Johnnidis, J. B., Harris, M. H., Wheeler, R. T., Stehling-Sun, S., Lam, M. H., Kirak, O., et al. (2008). Regulation of progenitor cell proliferation and granulocyte function by microRNA-223. *Nature, 451*(7182), 1125–1129.

Kato, M., Zhang, J., Wang, M., Lanting, L., Yuan, H., Rossi, J. J., et al. (2007). MicroRNA-192 in diabetic kidney glomeruli and its function in TGF-beta-induced collagen expression via inhibition of E-box repressors. *Proceedings of the National Academy of Sciences of the United States of America, 104*(9), 3432–3437.

Krzyszczyk, P., Schloss, R., Palmer, A., & Berthiaume, F. (2018). The role of macrophages in acute and chronic wound healing and interventions to promote pro-wound healing phenotypes. *Frontiers in Physiology, 1*(9), 419.

Kuhnert, F., Mancuso, M. R., Hampton, J., Stankunas, K., Asano, T., Chen, C.-Z., et al. (2008). Attribution of vascular phenotypes of the murine Egfl7 locus to the microRNA miR-126. *Development, 135*(24), 3989–3993.

Kulkarni, J. A., Witzigmann, D., Thomson, S. B., Chen, S., Leavitt, B. R., Cullis, P. R., et al. (2021). The current landscape of nucleic acid therapeutics. *Nature Nanotechnology, 16*(6), 630–643.

Kwon, Y., Kim, Y., Eom, S., Kim, M., Park, D., Kim, H., et al. (2015). miR-26a/–26b-COX-2-MIP-2 loop regulates allergic inflammation and allergic inflammation-promoted enhanced tumorigenic and metastatic potential of cancer cells. *The Journal of Biological Chemistry, 290*(22), 14245–14266.

Landskroner-Eiger, S., Qiu, C., Perrotta, P., Siragusa, M., Lee, M. Y., Ulrich, V., et al. (2015). Endothelial miR-17∼92 cluster negatively regulates arteriogenesis via miRNA-19 repression of WNT signaling. *Proceedings of the National Academy of Sciences of the United States of America, 112*(41), 12812–12817.

Lang, H., Zhao, F., Zhang, T., Liu, X., Wang, Z., Wang, R., et al. (2017). MicroRNA-149 contributes to scarless wound healing by attenuating inflammatory response. *Molecular Medicine Reports, 16*(2), 2156–2162.

Lee, D. Y., Deng, Z., Wang, C.-H., & Yang, B. B. (2007). MicroRNA-378 promotes cell survival, tumor growth, and angiogenesis by targeting SuFu and Fus-1 expression. *Proceedings of the National Academy of Sciences of the United States of America, 104*(51), 20350–20355.

Lei, Z., Li, B., Yang, Z., Fang, H., Zhang, G.-M., Feng, Z.-H., et al. (2009). Regulation of HIF-1alpha and VEGF by miR-20b tunes tumor cells to adapt to the alteration of oxygen concentration. *PLoS One, 4*(10), e7629.

Li, Z., Hassan, M. Q., Jafferji, M., Aqeilan, R. I., Garzon, R., Croce, C. M., et al. (2009). Biological functions of miR-29b contribute to positive regulation of osteoblast differentiation. *The Journal of Biological Chemistry, 284*(23), 15676–15684.

Li, D., Li, X. I., Wang, A., Meisgen, F., Pivarcsi, A., Sonkoly, E., et al. (2015a). MicroRNA-31 promotes skin wound healing by enhancing keratinocyte proliferation and migration. *The Journal of Investigative Dermatology, 135*(6), 1676–1685.

Li, D., Wang, A., Liu, X., Meisgen, F., Grünler, J., Botusan, I. R., et al. (2015b). MicroRNA-132 enhances transition from inflammation to proliferation during wound healing. *The Journal of Clinical Investigation, 125*(8), 3008–3026.

Li, M., Ke, Q.-F., Tao, S.-C., Guo, S.-C., Rui, B.-Y., & Guo, Y.-P. (2016). Fabrication of hydroxyapatite/chitosan composite hydrogels loaded with exosomes derived from miR-126-3p overexpressed synovial mesenchymal stem cells for diabetic chronic wound healing. *Journal of Materials Chemistry. B: Materials for Biology and Medicine, 4*(42), 6830–6841.

Li, J.-Y., Cheng, B., Wang, X.-F., Wang, Z.-J., Zhang, H.-M., Liu, S.-F., et al. (2018a). Circulating MicroRNA-4739 may be a potential biomarker of critical limb ischemia in patients with diabetes. *BioMed Research International, 14*(2018), 4232794.

Li, H., Han, X., Zuo, K., Li, L., Liu, J., Yuan, X., et al. (2018b). miR-23b promotes cutaneous wound healing through inhibition of the inflammatory responses by targeting ASK1. *Acta Biochimica et Biophysica Sinica, 50*, 1104.

Li, Q., Xia, S., Yin, Y., Guo, Y., Chen, F., & Jin, P. (2018c). miR-5591-5p regulates the effect of ADSCs in repairing diabetic wound via targeting AGEs/AGER/JNK signaling axis. *Cell Death & Disease, 9*(5), 566.

Li, D., Peng, H., Qu, L., Sommar, P., Wang, A., Chu, T., et al. (2020). miR-19a/b and miR-20a promote wound healing by regulating the inflammatory response of keratinocytes. *The Journal of Investigative Dermatology, 16*, 659.

Liechty, C., Hu, J., Zhang, L., Liechty, K. W., & Xu, J. (2020). Role of microRNA-21 and its underlying mechanisms in inflammatory responses in diabetic wounds. *International Journal of Molecular Sciences, 21*(9).

Lin, C. J., Lan, Y. M., Ou, M. Q., Ji, L. Q., & Lin, S. D. (2019). Expression of miR-217 and HIF-1α/VEGF pathway in patients with diabetic foot ulcer and its effect on angiogenesis of diabetic foot ulcer rats. *Journal of Endocrinological Investigation, 11*, 1307.

Lucas, T., Schäfer, F., Müller, P., Eming, S. A., Heckel, A., & Dimmeler, S. (2017). Light-inducible antimiR-92a as a therapeutic strategy to promote skin repair in healing-impaired diabetic mice. *Nature Communications, 2*(8), 15162.

Lynch, S. E., Nixon, J. C., Colvin, R. B., & Antoniades, H. N. (1987). Role of platelet-derived growth factor in wound healing: Synergistic effects with other growth factors. *Proceedings of the National Academy of Sciences of the United States of America, 84*(21), 7696–7700.

Ma, Q., Dasgupta, C., Li, Y., Huang, L., & Zhang, L. (2019). MicroRNA-210 downregulates ISCU and induces mitochondrial dysfunction and neuronal death in neonatal hypoxic-ischemic brain injury. *Molecular Neurobiology, 56*(8), 5608–5625.

Maurer, B., Stanczyk, J., Jüngel, A., Akhmetshina, A., Trenkmann, M., Brock, M., et al. (2010). MicroRNA-29, a key regulator of collagen expression in systemic sclerosis. *Arthritis and Rheumatism, 62*(6), 1733–1743.

Minutti, C. M., Knipper, J. A., Allen, J. E., & Zaiss, D. M. W. (2017). Tissue-specific contribution of macrophages to wound healing. *Seminars in Cell & Developmental Biology, 61*, 3–11.

Mirza, R. E., Fang, M. M., Weinheimer-Haus, E. M., Ennis, W. J., & Koh, T. J. (2014). Sustained inflammasome activity in macrophages impairs wound healing in type 2 diabetic humans and mice. *Diabetes, 63*(3), 1103–1114.

Moura, J., Rodrigues, J., Gonçalves, M., Amaral, C., Lima, M., & Carvalho, E. (2016). Impaired T-cell differentiation in diabetic foot ulceration. *Cellular & Molecular Immunology, 14*(9), 758–769.

Moura, J., Sørensen, A., Leal, E. C., Svendsen, R., Carvalho, L., Willemoes, R. J., et al. (2019). microRNA-155 inhibition restores fibroblast growth factor 7 expression in diabetic skin and decreases wound inflammation. *Scientific Reports, 9*(1), 5836.

Mulholland, E. J., Dunne, N., & McCarthy, H. O. (2017). Microrna as therapeutic targets for chronic wound healing. *Molecular Therapy – Nucleic Acids, 15*(8), 46–55.

Nie, W., Huang, X., Zhao, L., Wang, T., Zhang, D., Xu, T., et al. (2023). Exosomal miR-17-92 derived from human mesenchymal stem cells promotes wound healing by enhancing angiogenesis and inhibiting endothelial cell ferroptosis. *Tissue & Cell, 83*, 102124.

Ozdemir, D., & Feinberg, M. W. (2018). MicroRNAs in diabetic wound healing: Pathophysiology and therapeutic opportunities. *Trends in Cardiovascular Medicine, 29*(3), 131–137.

Park, H., Huang, X., Lu, C., Cairo, M. S., & Zhou, X. (2015). MicroRNA-146a and microRNA-146b regulate human dendritic cell apoptosis and cytokine production by targeting TRAF6 and IRAK1 proteins. *The Journal of Biological Chemistry, 290*(5), 2831–2841.

Peiseler, M., & Kubes, P. (2019). More friend than foe: The emerging role of neutrophils in tissue repair. *The Journal of Clinical Investigation, 129*(7), 2629–2639.

Peña, O. A., & Martin, P. (2024). Cellular and molecular mechanisms of skin wound healing. *Nature Reviews. Molecular Cell Biology, 25*(8), 599–616.

Petkovic, M., Sørensen, A. E., Leal, E. C., Carvalho, E., & Dalgaard, L. T. (2020). Mechanistic actions of microRNAs in diabetic wound healing. *Cells, 9*(10).

Pin, A.-L., Houle, F., Guillonneau, M., Paquet, E. R., Simard, M. J., & Huot, J. (2012). miR-20a represses endothelial cell migration by targeting MKK3 and inhibiting p38 MAP kinase activation in response to VEGF. *Angiogenesis, 15*(4), 593–608.

Pottier, N., Maurin, T., Chevalier, B., Puisségur, M.-P., Lebrigand, K., Robbe-Sermesant, K., et al. (2009). Identification of keratinocyte growth factor as a target of microRNA-155 in lung fibroblasts: Implication in epithelial-mesenchymal interactions. *PLoS One, 4*(8), e6718.

Primo, M. N., Bak, R. O., Schibler, B., & Mikkelsen, J. G. (2012). Regulation of pro-inflammatory cytokines TNFα and IL24 by microRNA-203 in primary keratinocytes. *Cytokine, 60*(3), 741–748.

Ramirez, H. A., Pastar, I., Jozic, I., Stojadinovic, O., Stone, R. C., Ojeh, N., et al. (2018). Staphylococcus aureus triggers induction of miR-15B-5P to diminish DNA repair and deregulate inflammatory response in diabetic foot ulcers. *The Journal of Investigative Dermatology, 138*(5), 1187–1196.

Roy, S., & Sen, C. K. (2011). MiRNA in innate immune responses: Novel players in wound inflammation. *Physiological Genomics, 43*(10), 557–565.

Ruggeri, Z. M. (2007). The role of von Willebrand factor in thrombus formation. *Thrombosis Research, 120*(Suppl 1), S5–S9.

Rupaimoole, R., & Slack, F. J. (2017). MicroRNA therapeutics: Towards a new era for the management of cancer and other diseases. *Nature Reviews. Drug Discovery, 16*(3), 203–222.

Saleh, B., Dhaliwal, H. K., Portillo-Lara, R., Shirzaei Sani, E., Abdi, R., Amiji, M. M., et al. (2019). Local immunomodulation using an adhesive hydrogel loaded with miRNA-laden nanoparticles promotes wound healing. *Small, 15*(36), e1902232.

Sen, C. K. (2021). Human wound and its burden: Updated 2020 compendium of estimates. *Advances in Wound Care (New Rochelle), 10*(5), 281–292.

Seré, K. M., & Hackeng, T. M. (2003). Basic mechanisms of hemostasis. *Seminars in Vascular Medicine, 3*(1), 3–12.

Seyhan, A. A. (2024). Trials and tribulations of microrna therapeutics. *International Journal of Molecular Sciences, 25*(3).

Shanmugam, N., Reddy, M. A., & Natarajan, R. (2008). Distinct roles of heterogeneous nuclear ribonuclear protein K and microRNA-16 in cyclooxygenase-2 RNA stability induced by S100b, a ligand of the receptor for advanced glycation end products. *The Journal of Biological Chemistry, 283*(52), 36221–36233.

Sheedy, F. J., Palsson-McDermott, E., Hennessy, E. J., Martin, C., O'Leary, J. J., Ruan, Q., et al. (2010). Negative regulation of TLR4 via targeting of the proinflammatory tumor suppressor PDCD4 by the microRNA miR-21. *Nature Immunology, 11*(2), 141–147.

Sheng, W., Feng, Z., Song, Q., Niu, H., & Miao, G. (2016). Modulation of mesenchymal stem cells with miR-375 to improve their therapeutic outcome during scar formation. *American Journal of Translational Research, 8*(5), 2079–2087.

Short, W. D., Wang, X., & Keswani, S. G. (2022). The role of T lymphocytes in cutaneous scarring. *Advances in Wound Care (New Rochelle), 11*(3), 121–131.

Sîrbulescu, R. F., Boehm, C. K., Soon, E., Wilks, M. Q., Ilieş, I., Yuan, H., et al. (2017). Mature B cells accelerate wound healing after acute and chronic diabetic skin lesions. *Wound Repair and Regeneration, 25*(5), 774–791.

Siu, M. C., Voisey, J., Zang, T., & Cuttle, L. (2023). MicroRNAs involved in human skin burns, wound healing and scarring. *Wound Repair and Regeneration, 31*(4), 439–453.

Soliman, A. M., Das, S., Abd Ghafar, N., & Teoh, S. L. (2018). Role of microrna in proliferation phase of wound healing. *Frontiers in Genetics, 14*(9), 38.

Sundaram, G. M., Common, J. E. A., Gopal, F. E., Srikanta, S., Lakshman, K., Lunny, D. P., et al. (2013). "See-saw" expression of microRNA-198 and FSTL1 from a single transcript in wound healing. *Nature, 495*(7439), 103–106.

Taghdiri, M., & Mussolino, C. (2024). Viral and non-viral systems to deliver gene therapeutics to clinical targets. *International Journal of Molecular Sciences, 25*(13).

Tan, S., Li, R., Ding, K., Lobie, P. E., & Zhu, T. (2011). miR-198 inhibits migration and invasion of hepatocellular carcinoma cells by targeting the HGF/c-MET pathway. *FEBS Letters, 585*(14), 2229–2234.

Tanaka, K., Kim, S. E., Yano, H., Matsumoto, G., Ohuchida, R., Ishikura, Y., et al. (2016). MiR-142 is required for Staphylococcus aureus clearance at skin wound sites via small GTPase-mediated regulation of the neutrophil actin cytoskeleton. *The Journal of Investigative Dermatology, 137*(4), 931–940.

Tanno, H., Kawakami, K., Kanno, E., Suzuki, A., Takagi, N., Yamamoto, H., et al. (2017). Invariant NKT cells promote skin wound healing by preventing a prolonged neutrophilic inflammatory response. *Wound Repair and Regeneration, 25*(5), 805–815.

Tellechea, A., Leal, E. C., Kafanas, A., Auster, M. E., Kuchibhotla, S., Ostrovsky, Y., et al. (2016). Mast cells regulate wound healing in diabetes. *Diabetes, 65*(7), 2006–2019.

Teruel-Montoya, R., Rosendaal, F. R., & Martínez, C. (2015). MicroRNAs in haemostasis. *Journal of Thrombosis and Haemostasis, 13*(2), 170–181.

Testa, U., Pelosi, E., Castelli, G., & Labbaye, C. (2017). miR-146 and miR-155: Two key modulators of immune response and tumor development. *Noncoding RNA, 3*(3).

Tian, C., Ziegler, J. N., & Zucker, I. H. (2023). Extracellular vesicle micrornas in heart failure: Pathophysiological mediators and therapeutic targets. *Cells, 12*(17).

van Rooij, E., Sutherland, L. B., Thatcher, J. E., DiMaio, J. M., Naseem, R. H., Marshall, W. S., et al. (2008). Dysregulation of microRNAs after myocardial infarction reveals a role of miR-29 in cardiac fibrosis. *Proceedings of the National Academy of Sciences of the United States of America, 105*(35), 13027–13032.

Viticchiè, G., Lena, A. M., Cianfarani, F., Odorisio, T., Annicchiarico-Petruzzelli, M., Melino, G., et al. (2012). MicroRNA-203 contributes to skin re-epithelialization. *Cell Death & Disease, 3*(11), e435.

Wang, S., Aurora, A. B., Johnson, B. A., Qi, X., McAnally, J., Hill, J. A., et al. (2008). The endothelial-specific microRNA miR-126 governs vascular integrity and angiogenesis. *Developmental Cell, 15*(2), 261–271.

Wang, X. H., Qian, R. Z., Zhang, W., Chen, S. F., Jin, H. M., & Hu, R. M. (2009). MicroRNA-320 expression in myocardial microvascular endothelial cells and its relationship with insulin-like growth factor-1 in type 2 diabetic rats. *Clinical and Experimental Pharmacology & Physiology, 36*(2), 181–188.

Wang, B., Herman-Edelstein, M., Koh, P., Burns, W., Jandeleit-Dahm, K., Watson, A., et al. (2010). E-cadherin expression is regulated by miR-192/215 by a mechanism that is independent of the profibrotic effects of transforming growth factor-beta. *Diabetes, 59*(7), 1794–1802.

Wang, T., Feng, Y., Sun, H., Zhang, L., Hao, L., Shi, C., et al. (2012). miR-21 regulates skin wound healing by targeting multiple aspects of the healing process. *The American Journal of Pathology, 181*(6), 1911–1920.

Wang, Y., Hu, C., Cheng, J., Chen, B., Ke, Q., Lv, Z., et al. (2014a). MicroRNA-145 suppresses hepatocellular carcinoma by targeting IRS1 and its downstream Akt signaling. *Biochemical and Biophysical Research Communications, 446*(4), 1255–1260.

Wang, J.-M., Tao, J., Chen, D.-D., Cai, J.-J., Irani, K., Wang, Q., et al. (2014b). MicroRNA miR-27b rescues bone marrow-derived angiogenic cell function and accelerates wound healing in type 2 diabetes mellitus. *Arteriosclerosis, Thrombosis, and Vascular Biology, 34*(1), 99–109.

Wang, H., Li, X., Li, T., Wang, L., Wu, X., Liu, J., et al. (2019a). Multiple roles of microRNA-146a in immune responses and hepatocellular carcinoma. *Oncology Letters, 18*(5), 5033–5042.

Wang, C.-R., Zhu, H.-F., & Zhu, Y. (2019b). Knockout of MicroRNA-155 ameliorates the Th17/Th9 immune response and promotes wound healing. *Current Medical Science, 39*(6), 954–964.

Wang, L., Zheng, A., Zeng, N., Li, Z., Tang, L., Long, C., et al. (2023). Effects of intensive glycemic control on serum exosome miR-126-3p and miR-125b-1-3p levels and wound healing in patients with diabetic ulcers. *Evidence-Based Complementary and Alternative Medicine, 30*(2023), 2523245.

Wu, J., Li, X., Li, D., Ren, X., Li, Y., Herter, E. K., et al. (2019). MicroRNA-34 family enhances wound inflammation by targeting LGR4. *The Journal of Investigative Dermatology, 31*, 465.

Wu, N., Sun, H., Sun, Q., Zhang, F., Ma, L., Hu, Y., et al. (2023). Circulating microRNAs as diagnostic biomarkers for melanoma: A systematic review and meta-analysis. *BMC Cancer, 23*(1), 414.

Würdinger, T., Tannous, B. A., Saydam, O., Skog, J., Grau, S., Soutschek, J., et al. (2008). miR-296 regulates growth factor receptor overexpression in angiogenic endothelial cells. *Cancer Cell, 14*(5), 382–393.

Xia, W., Li, M., Jiang, X., Huang, X., Gu, S., Ye, J., et al. (2022). Young fibroblast-derived exosomal microRNA-125b transfers beneficial effects on aged cutaneous wound healing. *Journal of Nanobiotechnology, 20*(1), 144.

Xie, J., Wu, W., Zheng, L., Lin, X., Tai, Y., Wang, Y., et al. (2022). Roles of MicroRNA-21 in skin wound healing: A comprehensive review. *Frontiers in Pharmacology, 28*(13), 828627.

Xu, N., Brodin, P., Wei, T., Meisgen, F., Eidsmo, L., Nagy, N., et al. (2011). MiR-125b, a microRNA downregulated in psoriasis, modulates keratinocyte proliferation by targeting FGFR2. *The Journal of Investigative Dermatology, 131*(7), 1521–1529.

Xu, J., Wu, W., Zhang, L., Dorset-Martin, W., Morris, M. W., Mitchell, M. E., et al. (2012). The role of microRNA-146a in the pathogenesis of the diabetic wound-healing impairment: Correction with mesenchymal stem cell treatment. *Diabetes, 61*(11), 2906–2912.

Xu, W.-D., Feng, S.-Y., & Huang, A.-F. (2022a). Role of miR-155 in inflammatory autoimmune diseases: A comprehensive review. *Inflammation Research, 71*(12), 1501–1517.

Xu, M., Li, Y., Tang, Y., Zhao, X., Xie, D., & Chen, M. (2022b). Increased expression of miR-155 in peripheral blood and wound margin tissue of type 2 diabetes mellitus patients associated with diabetic foot ulcer. *Diabetes, Metabolic Syndrome and Obesity, 3*(15), 3415–3428.

Yan, Y., Wu, R., Bo, Y., Zhang, M., Chen, Y., Wang, X., et al. (2020). Induced pluripotent stem cells-derived microvesicles accelerate deep second-degree burn wound healing in mice through miR-16-5p-mediated promotion of keratinocytes migration. *Theranostics, 10*(22), 9970–9983.

Yang, X., Wang, J., Guo, S.-L., Fan, K.-J., Li, J., Wang, Y.-L., et al. (2011). miR-21 promotes keratinocyte migration and re-epithelialization during wound healing. *International Journal of Biological Sciences, 7*(5), 685–690.

Yang, J., Han, H., Zhao, Y., & Qin, H. (2015). Specific miRNA and its target in neutrophils after traumatic injury. *Acta Biochimica et Biophysica Sinica, 47*(9), 749–754.

Yang, Z.-G., Awan, F. M., Du, W. W., Zeng, Y., Lyu, J., Wu, D., et al. (2017). The circular RNA interacts with STAT3, increasing its nuclear translocation and wound repair by modulating Dnmt3a and miR-17 function. *Molecular Therapy, 25*(9), 2062–2074.

Yu, J., Ryan, D. G., Getsios, S., Oliveira-Fernandes, M., Fatima, A., & Lavker, R. M. (2008). MicroRNA-184 antagonizes microRNA-205 to maintain SHIP2 levels in epithelia. *Proceedings of the National Academy of Sciences of the United States of America, 105*(49), 19300–19305.

Yu, J., Peng, H., Ruan, Q., Fatima, A., Getsios, S., & Lavker, R. M. (2010). MicroRNA-205 promotes keratinocyte migration via the lipid phosphatase SHIP2. *The FASEB Journal, 24*(10), 3950–3959.

Zhang, S., Cheng, Z., Wang, Y., & Han, T. (2021a). The risks of miRNA therapeutics: In a drug target perspective. *Drug Design, Development and Therapy, 22*(15), 721–733.

Zhang, W., Qu, X., Zhu, Z., Wang, L., Qi, Q., Zhou, P., et al. (2021b). Inhibition of miR-139-5p by topical JTXK gel promotes healing of Staphylococcus aureus-infected skin wounds. *Cells and Development, 166*, 203658.

Zhang, X.-F., Wang, T., Wang, Z.-X., Huang, K.-P., Zhang, Y.-W., Wang, G.-L., et al. (2021c). Hypoxic ucMSC-secreted exosomal miR-125b promotes endothelial cell survival and migration during wound healing by targeting TP53INP1. *Molecular Therapy – Nucleic Acids, 3*(26), 347–359.

Zhou, S., Zhang, P., Liang, P., & Huang, X. (2014). The expression of miR-125b regulates angiogenesis during the recovery of heat-denatured HUVECs. *Burns, 41*(4), 803–811.

Zhou, W., Pal, A. S., Hsu, A. Y.-H., Gurol, T., Zhu, X., Wirbisky-Hershberger, S. E., et al. (2018). MicroRNA-223 suppresses the canonical NF-κB pathway in basal keratinocytes to dampen neutrophilic inflammation. *Cell Reports, 22*(7), 1810–1823.

# The Multifaceted Functions of MG53: Orchestrating Tissue Repair and Its Implications in Cancer and Human Pathologies

**Zhongguang Li, Fei Jiang, Isabella Melians, Matthew Bu, and Haichang Li**

**Abstract** Mitsugumin 53 (MG53), also known as TRIM72, is a member of the tripartite motif-containing (TRIM) protein family, which includes a conserved really interesting new gene (RING) domain, B-box domain, coiled-coil region, and C-terminal PRY- SplA and Ryanodine receptor (PRYSPRY) domain. It is predominantly expressed in skeletal and cardiac muscles in mice. Since its discovery in 2009, numerous studies have shown that MG53 plays a crucial role in cell membrane repair machinery. Its function in membrane repair involves facilitating the rapid resealing of damaged cellular membranes through interactions with phosphatidylserine (PS). In addition to membrane repair, MG53 regulates metabolic homeostasis and cellular signaling pathways utilizing its E3 ubiquitin ligase properties. Therapeutically, MG53 has been proven to protect tissues from injuries, including those in the heart, liver, lung, kidney, brain, skin, and muscle. Emerging evidence indicates that MG53 has antitumor effects in various cancers, such as lung cancer, pancreatic cancer, and colorectal cancer. MG53 may represent a potent adjuvant for treating therapy-resistant cancers. It is emerging as a versatile therapeutic target with applications in regenerative medicine and cancer, as well as related human diseases. The current review aims to summarize the latest literature concerning MG53 in light of its diverse functions in human diseases.

**Keywords** MG53 · Membrane repair · Regenerative medicine · Antitumor · Antivirus

---

Z. Li · F. Jiang · I. Melians · M. Bu
Department of Veterinary Biosciences, The Ohio State University, Columbus, OH, USA
e-mail: li.10025@osu.edu; jiang.2886@osu.edu; melians.1@buckeyemail.osu.edu; bu.99@buckeyemail.osu.edu

H. Li (✉)
Department of Veterinary Biosciences, The Ohio State University, Columbus, OH, USA

Department of Surgery, The Ohio State University, Columbus, OH, USA

The Ohio State University Comprehensive Cancer Center, Columbus, OH, USA
e-mail: li.3714@osu.edu



H. Das (ed.), *Tissue Repair and Regeneration*,
https://doi.org/10.1007/978-3-031-93677-7_2

## 1 Introduction

Effective membrane repair mechanisms are essential for maintaining cellular integrity, and their failure often exacerbates tissue damage. Defective cell membrane repair is a critical factor in the progression of diseases such as muscular dystrophy, cardiovascular disorders, and neurodegenerative conditions (Dias & Nylandsted, 2021; Ammendolia et al., 2021). Mitsugumin 53 (MG53) is a member of the tripartite motif-containing (TRIM) protein family and plays a crucial role in membrane repair, particularly in skeletal and cardiac muscles. The first identified function of MG53 was its facilitation of intracellular vesicles recruitment to sites of injury and promotion of membrane resealing via redox-sensitive oligomerization (Cai et al., 2009a).

As a member of the TRIM proteins, MG53 possesses a conserved really interesting new gene (RING) domain, a B-box domain, a coiled-coil region, and a C-terminal SplA and Ryanodine receptor (PRYSPRY) domain (Cai et al., 2009a). These structural features enable MG53 to function as an E3 ubiquitin ligase, mediating protein ubiquitination and regulating key cellular processes (Yi et al., 2013; Song et al., 2013). Initially characterized for its role in membrane repair in striated muscle cells, MG53 has since been implicated in a broad spectrum of physiological and pathological processes. Recombinant human MG53 (rhMG53) has also demonstrated therapeutic potential in regenerative medicine, as it can mitigate injury in cardiac and skeletal muscles, liver, lung, skin, brain, corneal, and renal tissues (Liu et al., 2015b; Han et al., 2022; Weisleder et al., 2012; Corona et al., 2014; Zhu et al., 2015; Jia et al., 2014; Duann et al., 2015; Li et al., 2015; Gushchina et al., 2017; Nagre et al., 2018; Guan et al., 2019; Jin et al., 2022; Gouchoe et al., 2024a, b). Toxicological evaluations confirmed rhMG53 safety with no treatment-emergent toxicities at supratherapeutic exposures, demonstrating favorable pharmacokinetic profiles and cross-species translatability from murine to canine species (Duann et al., 2015).

Additionally, MG53 plays a vital role in maintaining cellular homeostasis by modulating oxidative stress, autophagy, and inflammation (Gumpper-Fedus et al., 2022; Lijie et al., 2019; Wang et al., 2021, 2022b; Kenney et al., 2020). Emerging evidence suggests that MG53 can resist viral infections (Kenney et al., 2020; Sermersheim et al., 2020). Furthermore, MG53 has emerged as a tumor suppressor in cancers such as non-small cell lung cancer (NSCLC), and colorectal, gastric, and pancreatic cancers (Li et al., 2021a; Fang et al., 2023; Wang et al., 2024b).

This review aims to provide a comprehensive overview of MG53 research, with an emphasis on its discovery, molecular structure, diverse functions, therapeutic potential in tissue repair, antiviral defense, and cancer treatment. By explicating the mechanisms underlying MG53's multifunctionality, this review seeks to underscore its significance in translational medicine and future research opportunities.

## 2 Structure of MG53

The tripartite motif protein family comprises over 80 genes encoded in the human genome, all of which are characterized by the conserved RING-finger domain that is sequentially followed by one or two B-box domains and a coiled-coil domain (CCD) at their N-terminal region. The RING domain confers E3 ubiquitin protease activity; the RING and B-box domains contain conserved zinc finger motifs that bind two zinc atoms, while the coiled-coil domain forms antiparallel dimers that facilitate protein–protein interactions and serve as a platform for posttranslational modifications (Hatakeyama, 2017).

Genomic analyses categorize the human TRIM family into two groups based on differences in domain structure, genomic organization, and evolutionary properties. MG53, a member of group 2, is characterized by its carboxyl-terminal SplA and Ryanodine receptor (SPRY) domain, which includes conserved sequences such as those found in dual-specificity kinase SplA and ryanodine receptor (RyR). This group also presents a unique set of genes specific to each mammalian species, contributing to functional specificity (Sardiello et al., 2008).

MG53 is highly conserved across different species, including humans, mice, rats, and monkeys (Li et al., 2021b). It is abundantly expressed in skeletal and cardiac muscles, which can secrete MG53 when under stress and distribute it systemically through circulation, allowing it to reach various tissues and organs (Cai et al., 2009a; Lee et al., 2023; Whitson et al., 2021b; Shan et al., 2020). However, lower levels of MG53 are detected in lung epithelial cells, inner cortex of the kidney, tear film, corneal epithelium, and aqueous humor (Jia et al., 2014; Duann et al., 2015; Chandler et al., 2019). This extensive presence and conservation across species emphasize its essential role in many biological and physiological processes.

## 3 MG53-Mediated Membrane Repair

When cells are exposed to external stimuli—such as mechanical forces, chemicals, or drugs—the cell membrane is often disrupted; timely and effective intrinsic repair mechanisms are essential to maintain the integrity of the cell membrane. Cells require rapid repair of damaged plasma membranes to protect the cytoplasm from the external environment. In muscle tissue, this plasma membrane damage often occurs due to mechanical and metabolic stress. In 2009, Cai et al. first reported that MG53 is a key component of the plasma membrane repair mechanism in skeletal muscle, mediating the acute repair of membrane damage in an oxidation-dependent manner. Gene ablation of mouse MG53 ($mg53^{-/-}$) results in gradually worsening muscle problems and struggle with physical activity due to impaired cell membrane repair. Upon membrane injury, MG53 responds to oxidative stress by forming disulfide bonds that enable its oligomerization and movement of vesicles to membrane damage sites. At these injury sites, MG53 anchors to phosphatidylserine (PS) on the inner leaflet of the plasma membrane, assembling a repair complex that facilitates membrane restoration (Cai et al., 2009a).

Structural studies of MG53 via cryo-electron microscopy reveal that it exists as a homodimer, comprising a "body" and two "wings," and exhibiting an overall elongated shape (Ma et al., 2023; Park et al., 2023; Niu et al., 2022a). During membrane repair, MG53 binds to negatively charged phospholipids, such as PS and phosphatidylinositol, and oligomerizes into higher-order assemblies (Cai et al., 2009a; Ma et al., 2023). These assemblies form vertical columnar densities on lipid membranes, interconnected by unique horizontal bridge-like structures (Ma et al., 2023; Park et al., 2023). The PRYSPRY domain is positioned atop the columnar densities, while other domains (such as the coiled-coil domain [CCD], B-box, and RING) occupy the planar density above the lipid bilayer (Park et al., 2023). This assembly allows the positively charged surface of the PRYSPRY domain to closely interact with the lipid bilayer (Ma et al., 2023). The interactions between MG53 and PS are mediated by the PRYSPRY domain in conjunction with the coiled-coil domain (Ma et al., 2023; Park et al., 2023).

Functional experiments involving deletion mutants of MG53 during membrane repair have demonstrated that the absence of the RING domain does not impact membrane repair, whereas deletion of the B-box impairs MG53's membrane repair ability. In contrast, deletion of the SPRY domain eliminates membrane repair. These findings suggest that, although the RING domain is not essential for membrane repair, the B-box and PRYSPRY domains play critical roles (Ma et al., 2023). Furthermore, mutagenesis studies that convert all 16 cysteine residues of mouse MG53 to alanine revealed that only the C242A mutation completely negates MG53 oligomerization, demonstrating that C242 mediates oligomerization through disulfide bond formation formed under oxidative stress (Cai et al., 2009a). The K279A mutation in MG53 causes severe intracellular aggregation of the protein, resulting in a loss of positive charge at the lysine residue and significantly altering MG53 localization and function (Zhou et al., 2018).

During membrane repair, MG53 interacts with key regulators such as dysferlin-1, caveolin-3 (Cav3), polymerase I and transcript release factor (PTRF), and non-muscle myosin IIA (NM-IIA). Dysferlin-1 relies on MG53 to relocate to sites of membrane injury (Cai et al., 2009b). Cav3 modulates the membrane repair capacity of MG53; mutations such as Cav3 P104L or R26Q severely impair MG53 function (Cai et al., 2009b). PTRF binds to exposed cholesterol at injury sites, acting as a docking protein to recruit MG53 to the membrane and facilitate the formation of repair patches (Zhu et al., 2011). NM-IIA—a critical cytoskeletal motor protein—interacts with MG53 to transport MG53-containing vesicles to sites of membrane injury (Lin et al., 2012). Additionally, $Ca^{2+}$ and $Zn^{2+}$ ions interact with MG53, further promoting membrane repair (Cai et al., 2009a, 2015). These interactions underscore the interconnected roles that MG53 and its associated regulators play in the process of membrane repair.

# 4 Systemic Protective Effects of MG53 on Mammalian Organ Networks

Although MG53 is primarily expressed in skeletal muscle, MG53 is secreted from skeletal muscle and circulates throughout the body, reaching all tissues and organs to participate in tissue damage repair, primarily by mediating cell membrane repair, but also by activating cell survival pathways and reducing inflammation and cell death (Fig. 1).

## *4.1 MG53 in Skeletal Muscle Damage and Repair*

Although skeletal muscle integrity is critical for maintaining mobility, strength, and overall health, it is frequently challenged by mechanical stress, metabolic strain, and pathological conditions. The sarcolemmal membrane repair system plays a key role in defending against muscle injury by resealing disrupted membranes, preserving cellular homeostasis, and preventing further damage. Impairments in membrane repair mark various muscle disorders—including Duchenne muscular dystrophy (DMD) and inflammatory myopathies—emphasizing the need for efficient repair mechanisms to mitigate muscle degeneration and improve clinical outcomes (Dias & Nylandsted, 2021).

MG53 has emerged as a pivotal protein in the repair system, orchestrating the recruitment and fusion of repair vesicles at sites of sarcolemmal disruption. *Mg53*$^{-/-}$ mice exhibit impaired sarcolemmal repair, especially under mechanical stress,

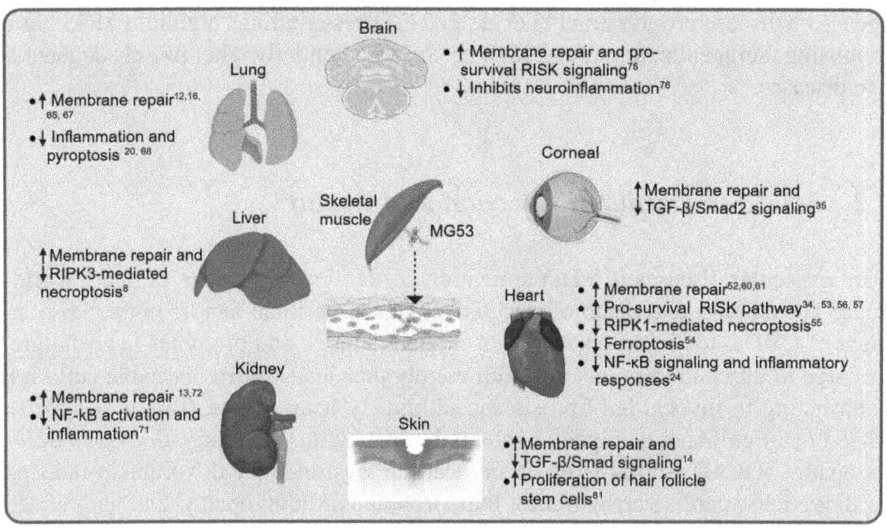


**Fig. 1** Tissue repair function of MG53 and potential mechanisms. (Created with BioRender.com)

emphasizing MG53's necessity in maintaining membrane integrity (Cai et al., 2009a). Stimuli such as exercise, remote ischemic preconditioning (RIPC), or insulin could promote the release of MG53 from skeletal muscle (Lee et al., 2023; Whitson et al., 2021b). This indicates that MG53 functions as a myokine that influences tissue repair and regeneration beyond skeletal muscle.

Preclinical studies using rhMG53 have highlighted its therapeutic value in mitigating skeletal muscle injury. For instance, rhMG53 injected via tail vein injection prevents acute and chronic muscle damage induced by eccentric-contraction in mice (Bian et al., 2019). *Mg53*$^{-/-}$ mice are more susceptible to ischemia/reperfusion (I/R) injury, while intravenous injection of rhMG53 protects muscles from such damage (Zhu et al., 2015).

DMD causes dysfunction and progressive damage to muscle structure with limited treatment options (Duan et al., 2021). Molecular complexes formed by MG53, dysferlin, and Cav3 are essential for repairing muscle membrane damage in DMD (Cai et al., 2009b). In a double-knockout model combining MG53 deficiency with the mdx DMD model, researchers found that the absence of MG53 exacerbates muscle pathology, reduces contractile function, increases fibrosis, and compromises membrane integrity in aged mice (Paleo et al., 2022). Intraperitoneal injection of rhMG53 can improve membrane repair and increase sarcolemmal membrane integrity in DMD mouse models (Weisleder et al., 2012; Gushchina et al., 2017). Furthermore, MG53 prevents inflammation by decreasing NLRP3 inflammasome activation and enhancing mitophagy in DMD mice (Wu et al., 2023).

Recombinant adeno-associated virus vectors that deliver MG53 systematically enhance membrane repair, improve pathology, and preserve muscle in δ-sarcoglycan deficient TO-2 hamsters, a model of muscular dystrophy (He et al., 2012). Systemic administration of rhMG53 in amyotrophic lateral sclerosis (ALS) mice protects against injury to the diaphragm muscle, preserves neuromuscular junction integrity, and slows disease progression (Yi et al., 2021). These findings highlight MG53 as a promising therapeutic target for DMD, ALS, and potentially other muscle degenerative diseases.

## 4.2 MG53 in Cardiac Protection and Repair

Cardiovascular diseases (CVDs) remain one of the leading causes of death worldwide. The damage and death of cardiomyocytes are critical factors in the development of CVDs. I/R injury exacerbates cardiac damage and promotes heart failure. Damage to unrepaired myocardial cell membranes leads to irreplaceable cell loss, contributing to myocardial fibrosis and ultimately heart failure (Liu et al., 2023). MG53 has significant therapeutic potential in managing I/R injury, diabetic cardiomyopathy (DCM), sepsis-induced myocardial dysfunction, doxorubicin-induced cardiotoxicity, cardiac arrhythmias, hypertrophic cardiomyopathy, and age-related heart failure due to its involvement in membrane repair, inhibition of inflammation and cell death, and modulation of redox signaling (Gumpper-Fedus et al., 2022; Cao et al., 2010; Wang et al., 2010, 2022a, 2024a; Zhang et al., 2011; Jiang et al., 2024; Lv et al., 2022; Feng et al., 2022; Liu et al., 2019).

MG53 facilitates the repair of sarcolemmal and intracellular membranes, which is crucial for cardiomyocyte survival during I/R injury. MG53 undergoes cholesterol-dependent translocation to injury sites, where it stabilizes compromised membranes and attenuates myocardial injury (Wang et al., 2010). Kohr et al. demonstrated that s-nitrosylation of MG53 at cysteine-144 is essential for its ability to mediate repair of cardiomyocyte membrane damage induced by I/R (Kohr et al., 2014). Furthermore, the mono-ADP-ribosylation of MG53, mediated by ADP-ribosyltransferase-1 and ADP-ribosyl arginine hydrolase-1, is critical for effective membrane repair and protection against I/R induced-myocardial injury (Ishiwata-Endo et al., 2018).

Ischemic preconditioning (IPC) and postischemic conditioning (PostC) protect against I/R-induced cardiac injury, and MG53 plays a critical role in both. I/R stress downregulates MG53 in the mouse heart (Cao et al., 2010). *Mg53*$^{-/-}$ mice exhibit increased susceptibility to I/R-induced cardiac injury and resistance to IPC and PostC-mediated cardioprotection (Cao et al., 2010; Wang et al., 2010; Zhang et al., 2011). MG53's cardioprotective effects rely on its interaction with Cav3 and the p85 subunit of phosphoinositide 3-kinase (p85-PI3K), which activates the reperfusion injury salvage kinase (RISK) pathway, including PI3K-Akt-GSK3β and ERK1/2 signaling (Shan et al., 2020; Zhang et al., 2011). IPC and oxidative stress stimulate cardiomyocyte secretion of MG53 via the $H_2O_2$-protein kinase-C-δ pathway, and extracellular MG53 contributes to IPC protection against cardiac I/R injury (Shan et al., 2020).

MG53 functions as an E3 ubiquitin ligase, which is essential for cardioprotection. By ubiquitinating and degrading RIPK1, it suppresses necroptosis, a programmed form of cell death implicated in myocardial injury (Wang et al., 2022a). Additionally, MG53 modulates the p53/SLC7A11/GPX4 pathway to reduce lipid peroxidation and ferroptosis, especially in drug-induced cardiotoxicity models such as doxorubicin-induced injury. MG53 ubiquitinates p53 degradation to promote the expression of the ferroptosis-resistant gene SLC7A11 (Jiang et al., 2024). Interaction between MG53 and p85α triggers ubiquitination and degradation of p85α, enhancing Akt phosphorylation and activating the Akt-GSK3β/Bcl-x(L) signaling pathway, which ultimately provides cardioprotection against I/R injury (Zhu et al., 2024).

Administration of rhMG53 has demonstrated significant cardioprotective effects in preclinical models, particularly in I/R injury, DCM, and age-related cardiac dysfunction (Li et al., 2021b; Wang et al., 2021; Shan et al., 2020; Feng et al., 2022). rhMG53 protects cardiomyocyte mitochondrial integrity during I/R-induced oxidative stress by binding to mitochondria-specific lipid cardiolipin (Gumpper-Fedus et al., 2022). Wang et al. found that MG53 expression decreases in human failing hearts, but rhMG53 treatment improves heart function and reduces inflammation in aged mice (Wang et al., 2021). rhMG53 is safe for repeated administration in aged mice. It suppresses the Nuclear factor kappa-light-chain-enhancer of activated B cells (NF-κB) signaling pathway, reducing inflammatory responses in age-related cardiac failure (Wang et al., 2021). Notably, specific MG53 mutants, such as S255A (rhMG53-S255A), retain cardioprotective properties while eliminating adverse effects related to E3 ligase activity, underscoring the therapeutic potential of modified MG53 variants (Lv et al., 2022). Another E3 ligase-dead mutant of MG53, rhMG53-C14A, maintains its membrane repair and cardioprotective capabilities while mitigating risks associated with the wild-type's E3 ligase activity (Feng et al., 2022). This mutant protects diabetic hearts from acute I/R injury and mitigates cardiometabolic damage caused by a high-fat diet (HFD). This E3 ligase activity mutant protein is a promising therapeutic candidate for patients with diabetes or metabolic syndrome who are at higher risk of ischemic heart diseases. Its refined mechanism balances therapeutic benefits with improved safety, making it suitable for long-term applications.

Cohort studies of human samples show a significant association between serum MG53 levels and both the presence and severity of stable coronary artery disease (CAD) and acute myocardial infarction (AMI). Thus, MG53 has the potential as a novel biomarker for diagnosing CAD and AMI and for identifying patients at high risk of developing CAD (Xie et al., 2020a, b).

MG53 represents a promising avenue for developing cardioprotective therapies. Its unique capacity to integrate membrane repair, mitochondrial preservation, mitigation of oxidative stress, inflammation, and regulation of cell death underscores its potential for treating cardiac and metabolic disorders. Future research should focus on optimizing MG53 delivery systems and exploring its long-term effects in clinical settings.

## 4.3 MG53 in Lung Injury and Repair

Acute and chronic lung injuries, including conditions like acute lung injury, ventilator-induced lung injury (VILI), and I/R injury, pose significant challenges in clinical medicine due to their complex pathophysiology and limited treatment options. Recent studies have highlighted MG53's pivotal roles in lung tissue repair, inflammation regulation, and oxidative stress mitigation, positioning it as a promising therapeutic target for various pulmonary diseases (Jia et al., 2014; Gouchoe et al., 2023, 2024a, b; Kim et al., 2014).

MG53 is expressed in lung tissue (Jia et al., 2014; Kim et al., 2014). *Mg53*$^{-/-}$ mice show increased susceptibility to I/R, overventilation, or nitrogen mustard-induced lung injury, emphasizing its indispensable role in pulmonary health (Jia et al., 2014; Gouchoe et al., 2024b; Li et al., 2022b). Lung injuries, whether from I/R events, mechanical ventilation, or environmental toxins, often disrupt cellular integrity and exacerbate inflammatory responses. MG53 mitigates these injuries through its membrane repair capabilities and regulation of inflammatory signaling pathways (Jia et al., 2014; Nagre et al., 2018; Kim et al., 2014; Li et al., 2022b).

Moreover, rhMG53 has demonstrated substantial therapeutic potential in various lung injury models, including those in mice, rat, and pigs, by leveraging its unique ability to repair cellular damage and modulate inflammatory responses (Jia et al., 2014; Nagre et al., 2018; Gouchoe et al., 2024a, b; Li et al., 2022b; Nagre et al.,

2016; Whitson et al., 2021a). It reduces oxidative stress and inflammation in nitrogen mustard-induced lung damage and lowers markers of apoptosis and inflammation in I/R injury, thereby improving lung compliance (Jia et al., 2014; Li et al., 2022b). In chronic lung diseases like asthma, rhMG53 reduces airway resistance and inflammatory cytokines via the NF-κB pathway. It also enhances graft outcomes in transplantation by alleviating I/R injury (Tan et al., 2023). Furthermore, rhMG53's rapid repair mechanisms make it highly effective in conditions like VILI and acute respiratory distress syndrome, offering targeted therapies to reduce inflammation and restore lung integrity (Jia et al., 2014; Nagre et al., 2018). These properties highlight rhMG53's broad applicability in treating acute, chronic, and transplant-related lung injuries.

In summary, MG53's ability to repair cellular damage, reduce inflammation, and improve lung function across diverse contexts makes it a promising therapeutic candidate for managing acute, chronic, and transplant-related pulmonary conditions. Future research should prioritize clinical trials, targeted delivery technologies, and expanding its therapeutic potential to systemic and multiorgan contexts.

## *4.4 MG53 in Kidney, Liver, Brain, and Corneal Injury and Repair*

*Mg53*$^{-/-}$ mice exhibit a renal pathologic phenotype, including tubulointerstitial damage, age-related renal fibrosis, disorganized microvilli, and a loss of brush border, underscoring the critical function of MG53 in maintaining kidney structure and function (Duann et al., 2015; Li et al., 2022a). These mice are more susceptible to I/R-induced acute kidney injury (AKI) and unilateral ureteral obstruction (UUO)-induced chronic kidney injury (CKI) (Duann et al., 2015; Li et al., 2022a). MG53 facilitates the repair of renal proximal tubular epithelial cells by translocating to injury sites, binding to PS, and reducing cell membrane damage and apoptosis (Li et al., 2022a). This protective mechanism has been demonstrated in models of I/R injury and nephrotoxin exposure (Li et al., 2022a; Liu et al., 2020). In UUO-induced CKI, MG53 reduces inflammation and fibrosis by modulating key pathways such as NF-κB and Transforming growth factor (TGF-β1), which suppress pro-inflammatory signaling and fibrotic remodeling (Li et al., 2022a). rhMG53 has shown significant therapeutic potential in rodent models of I/R-induced AKI, contrast-induced AKI, severe burn injury, and UUO-induced CKI (Duann et al., 2015; Li et al., 2022a; Liu et al., 2020; Wu et al., 2013). Given its ability to repair acute cell damage and modulate chronic disease pathways, MG53 represents a promising therapeutic target for both AKI and CKI.

MG53 is an essential membrane repair protein that protects hepatocytes from injury by repairing damaged cell membranes and reducing oxidative stress. It anchors to injury sites, often through interaction with dysferlin, to enhance localization and mitigate cell death in I/R injury models (Yao et al., 2017). In drug-induced liver injury (DILI), MG53 acts as an E3-ligase, reducing necroptosis pathways such as Receptor-interacting serine/threonine-protein kinase (RIPK3-mediated) Mixed lineage kinase domain like pseudokinase (MLKL) phosphorylation, which protects against liver necrosis and improves survival (Han et al., 2022). MG53 also prevents hepatocyte apoptosis by regulating mitochondrial permeability and the Bax/Bcl-2 ratio, thereby reducing caspase activation (Yao et al., 2017). Systemic administration of rhMG53 has shown effective results in both DILI and I/R contexts, highlighting its therapeutic potential for acute liver damage (Han et al., 2022; Yao et al., 2017).

MG53 plays a crucial role in brain injury repair and the modulation of neuroinflammation, by both repairing cell membrane damage and regulating inflammatory and oxidative stress pathways. MG53 facilitates membrane repair by forming a repair patch at injury sites and can cross the blood-brain barrier during injury to target damaged brain tissues (Yao et al., 2016). MG53 attenuates neuroinflammation by inhibiting pro-inflammatory pathways—such as TLR4/NF-κB and NLRP3 inflammasome activation—while activating the Nrf2/HO-1 signaling pathway to reduce oxidative stress and enhance neuronal survival in mouse and rat models (Yao et al., 2016; Ma et al., 2020; Zhu et al., 2023). MG53 also activates pro-survival pathways, including PI3K/Akt/GSK3β, and enhances the survival, migration, and efficacy of transplanted human umbilical cord mesenchymal stem cells, promoting neurogenesis and recovery in brain injury models (Guan et al., 2019; Cai et al., 2023).

rhMG53 protects against I/R-induced brain injury by promoting neuronal cell membrane repair, inhibiting neuronal cell apoptosis, and activating the pro-survival RISK signaling pathway (Yao et al., 2016). In the Yorkshire swine and mouse models, intravenously injected rhMG53 localized to injured brain areas, significantly reducing lesion size following traumatic brain injury. (Guan et al., 2019; Jin et al., 2022) rhMG53, in combination with stem cell therapies, has improved outcomes by inhibiting the NLRP3/Caspase-1/IL-1β axis in lipopolysaccharide-induced brain damage (Ma et al., 2020). Recently, a novel MG53/GMs/HA-Dex neural scaffold developed using gelatin microspheres, hyaluronic acid, and dextran loaded with MG53 protein, mitigated neuroinflammation, promoted neurogenesis and axon regeneration, and improved motor function in spinal cord injury (SCI) mice by regulating microglial polarization and inhibiting the Janus kinase 2/Signal transducer and activator of transcription 3 (JAK2/STAT3) pathway. This approach offers a promising strategy for SCI repair (Li et al., 2024). Exploring MG53's synergistic potential with other treatments, such as neuroprotective drugs or advanced stem cell therapies, could broaden its therapeutic applications.

Furthermore, *mg53*$^{-/-}$ mice result in impaired corneal wound healing, increased vascularization, and fibrotic remodeling following alkaline injury compared to wild-type littermates. Treatment with rhMG53 significantly improves healing capacity of *mg53*$^{-/-}$ corneas (Chandler et al., 2019). MG53 reduces fibrotic remodeling by inhibiting TGF-β/Smad2 signaling in corneal fibroblasts, lowering levels of pro-fibrotic markers such as Alpha Smooth Muscle Actin (α-SMA) and fibronectin (Chandler et al., 2019). In mouse and rat models of alkaline-induced corneal injury, topical rhMG53 significantly improved re-epithelialization, reduced fibrosis, and decreased vascularization (Chandler et al., 2019; Guo et al., 2021). These findings highlight the significant therapeutic potential of MG53 in corneal wound healing, offering a dual approach of promoting tissue repair and preventing fibrotic remodeling.

## *4.5 MG53 in Cutaneous Wound Healing*

Recent advances in the understanding of MG53 have significantly expanded its therapeutic potential, particularly in wound healing and tissue regeneration. Across multiple studies, MG53 has emerged as a critical player in membrane repair, angiogenesis, and fibrosis modulation, providing compelling insights into its multifunctional role in addressing complex medical challenges.

*Mg53*$^{-/-}$ mice exhibit abnormal skin structure, defective hair follicle morphology, abnormal hair follicle development, and dermal collagen overproduction, indicating MG53's essential role in maintaining dermal integrity (Li et al., 2015; Niu et al., 2022b). Furthermore, these mice are more susceptible to incisional dermal wounding, displaying delayed wound healing and abnormal scarring (Li et al., 2015). In contrast, *mg53*-overexpressing mice demonstrate elevated MG53 circulation levels and accelerate wound closure following punch-hole injury to ears (Bian et al., 2019). MG53 promotes wound healing and mitigates scarring by facilitating cell membrane repair, fibroblast migration, and the regulation of TGF-β signaling to limit fibrosis (Li et al., 2015). The Sundew-inspired hydrogel containing rhMG53 enhances wound healing by reducing scarring, promoting tissue regeneration, and supporting cellular activity. Its controlled degradation releases MG53 over time, making it ideal for diabetic wounds requiring prolonged treatment and reduced inflammation, while simultaneously overcoming MG53's short half-life (Li et al., 2017).

Several studies highlight MG53's therapeutic potential in diabetic wound healing. Wu et al. demonstrated that MG53 enhances the efficacy of bone marrow-derived mesenchymal stem cells by binding with Cav3 to activate the Endothelial Nitric Oxide Synthase/Nitric oxide (eNOS/NO) signaling pathway—promoting vascularization and re-epithelialization and addressing the impaired healing often associated with diabetes (Wu et al., 2024). Similarly, sustained delivery systems using rhMG53—such as the Reactive oxygen species (ROS)-scavenging hydrogel—significantly improve wound closure and skin regeneration in diabetic models (Niu et al., 2022b). In this system, rhMG53 sustains the release of rhMG53 and promotes healing of chronic cutaneous wounds and hair follicle development in *db/db* mice (Niu et al., 2022b). Mechanistically, rhMG53 protects hair follicle stem cells from oxidative stress-induced apoptosis and stimulates their differentiation into keratinocytes via modulation of Signal transducer and activator of transcription (STAT) and mitogen-activated protein kinase (MAPK) signaling (Niu et al., 2022b). This study supports the therapeutic potential of rhMG53 in chronic wound healing and hair follicle development, particularly under diabetic conditions (Niu et al., 2022b).

The versatility of MG53 in treating diverse tissue injuries—ranging from diabetic wounds to ocular and chronic wounds—positions it as a promising therapeutic target. However, challenges such as optimizing delivery systems, understanding long-term effects, and evaluating large-scale clinical applicability remain. Innovations in hydrogel formulations and combination therapies are likely to further enhance MG53's potential as a multifunctional agent in regenerative medicine.

# 5 MG53 in Insulin Resistance and Diabetes

Insulin resistance, type 2 diabetes (T2D), and metabolic syndrome are escalating global health concerns that involve abnormal regulation of the insulin signaling pathway. MG53 plays a crucial role in metabolic regulation; however, their precise function in metabolic disorders remains controversial.

Several studies suggest that MG53 promotes insulin resistance and metabolic disorders by ubiquitinating and degrading the insulin receptor and insulin receptor substrate-1 (IRS-1), impairing insulin signaling (Yi et al., 2013; Song et al., 2013). Song et al. observed that MG53 expression was significantly elevated in insulin-resistant mice induced by a high-fat diet (HFD) and in *db/db* diabetic mice (Song et al., 2013). Moreover, MG53 overexpression resulted in obesity, hypertension, and dyslipidemia (Song et al., 2013). Liu et al. further demonstrated that MG53 overexpression in the myocardium-activated peroxisome proliferator-activated receptor alpha, leads to lipid accumulation and cardiac dysfunction, thereby exacerbating DCM (Liu et al., 2015a). Additionally, Wu et al. proposed that MG53 may function as a myokine or cardiokine, secreted into circulation in response to high glucose and high insulin conditions, contributing to systemic insulin resistance by directly inhibiting insulin receptor function (Wu et al., 2019).

However, other studies challenge the detrimental role of MG53 in metabolic disorders. Ma et al. found that serum MG53 levels were reduced—not elevated—in a six-month HFD-induced metabolic syndrome mouse model, suggesting that MG53 might play a protective role in mitochondrial homeostasis and oxidative stress defense under diabetic conditions (Ma et al., 2015). Wang et al. observed that MG53 deficiency in knockout mice did not improve insulin sensitivity, while transgenic mice with sustained MG53 elevation in circulation exhibited no significant change in insulin resistance (Wang et al., 2020). Moreover, Bian et al. reported that prolonged elevation of serum MG53 did not exacerbate insulin resistance but enhanced tissue repair capacity, indicating its potential as a therapeutic target for diabetes-related tissue damage (Bian et al., 2019). Yi et al. showed that while MG53 mediates IRS-1 ubiquitination to regulate insulin signaling, its expression levels were not significantly altered in skeletal muscle samples from diabetic patients, further complicating the interpretation of MG53's role (Yi et al., 2013).

Clinical studies have also yielded conflicting insights into MG53's role in diabetes and metabolic syndrome. A 17-year prospective study by Bianchi et al. found that MG53 serum levels were elevated in T2D patients compared to healthy controls, with higher levels observed in women (Bianchi et al., 2023). However, these levels were not associated with cardiovascular disease risk or all-cause mortality, suggesting that MG53 may not serve as an independent predictor of cardiovascular events in T2D patients (Bianchi et al., 2023). Additionally, Philouze et al. reported that MG53 expression varied across different insulin-resistant and T2D populations, and in vitro experiments showed that neither MG53 knockdown nor overexpression significantly altered insulin signaling (Philouze et al., 2021). These findings imply that MG53's metabolic role may be influenced by individual variability, experimental conditions, and environmental factors—warranting further investigation to clarify its precise function in insulin resistance.

Notably, MG53's dual role in systemic metabolic regulation—as both an inhibitor of insulin signaling and a protector against oxidative stress and cellular damage—suggests that targeted therapeutic strategies must carefully balance its potential benefits and risks.

## 6 Antiviral Function of MG53

Recent studies have uncovered the significant involvement of MG53 in antiviral immune regulation. In a mouse model of influenza A virus (IAV) infection, systemic administration of rhMG53 demonstrated strong protective effects against IAV-induced lung damage (Kenney et al., 2020). MG53 mitigated lung tissue damage and enhanced survival rates by inhibiting NF-κB-mediated inflammatory responses, Gasdermin D (GSDMD-mediated pyroptosis), and reducing levels of pro-inflammatory cytokines such as type-I interferons β (IFN-β), IL-6, and IL-1β (Kenney et al., 2020). Additionally, MG53 regulates RyR-mediated calcium ion signaling, effectively suppressing excessive type-I interferons β (IFN-β) production induced by viral infection, thus preventing immune overactivation and tissue injury (Sermersheim et al., 2020). Interestingly, studies using *mg53*$^{-/-}$ mice demonstrate that the absence of MG53 results in increased morbidity and prolonged recovery following IAV infection, further underscoring MG53's protective role (Sermersheim et al., 2020). These findings suggest that MG53 primarily functions to suppress excessive inflammatory responses and facilitate tissue repair, rather than directly inhibiting viral replication.

Beyond its role in modulating host immune responses, MG53 has also demonstrated direct antiviral activity in certain infections. In the rabies virus model, MG53 induces K48-linked ubiquitination of the viral matrix protein, targeting it for proteasomal degradation, thereby reducing viral assembly and release (Sui et al., 2024). In the case of Coxsackievirus B3 (CVB3)-induced acute viral myocarditis, MG53 suppresses NOD-, LRR- and pyrin domain-containing protein 3 (NLRP3) inflammasome activation, decreases NF-κB signaling activity, and reduces cardiomyocyte pyroptosis, ultimately protecting cardiac function (Xue et al., 2024). Furthermore, MG53 overexpression has demonstrated strong protective effects against CVB3-induced myocarditis (Xue et al., 2024). These results highlight MG53's potential for therapeutic applications beyond its intrinsic cellular function.

Despite its promising antiviral potential, several challenges remain. While MG53 has been shown to effectively modulate host immune responses effectively, its broad-spectrum direct antiviral properties require further investigation. Additionally, its mode of action varies among different viral infections, potentially influenced by host cell types, viral species, and infection routes. This variation necessitates further research to clarify its generalizability. Future studies should aim to elucidate the precise molecular mechanisms by which MG53 exerts its antiviral effects, assess its role in other viral infections, and optimize MG53-based therapeutic strategies. Given its demonstrated efficacy in small-animal models, further preclinical and clinical trials are essential to evaluate MG53's feasibility as an antiviral and immune-modulating therapy.

## 7 Antitumor Effects of MG53

In addition to its role in tissue repair, recent studies have demonstrated that MG53 plays a crucial tumor-suppressive role in various cancers (Fig. 2), including colorectal cancer (CRC), hepatocellular carcinoma (HCC), pancreatic ductal adenocarcinoma (PDAC), non-small cell lung cancer (NSCLC), and tongue cancer (Li et al., 2021a; Fang et al., 2023; Wang et al., 2024b; Ma et al., 2022; Gupta et al., 2022; Yin et al., 2019). Several studies have shown that low expression of MG53 correlates with increased tumor invasiveness, reduced survival rates, and poorer prognosis,

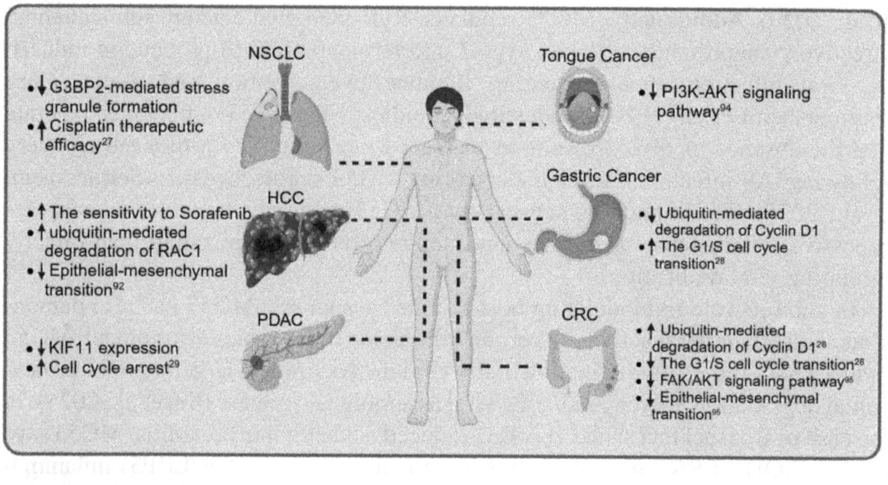


**Fig. 2** MG53 functions as a tumor suppressor in multiple cancers. (Created with BioRender.com)

suggesting that MG53 may serve as a potential biomarker for cancer prognosis (Fang et al., 2023; Wang et al., 2024b; Ma et al., 2022). Most excitingly, further studies suggest that rhMG53 could add synergy for chemotherapeutic reagents to induce death of cancer cells, and to suppress tumor growth of drug-resistant cancers in the xenograft model (Li et al., 2021a). Based on these findings, one can envision that rhMG53 protein can potentially be used as an adjuvant to enhance the therapy of NSCLCs.

The anticancer mechanisms of MG53 primarily involve ubiquitin-mediated protein degradation, regulation of signaling pathways, and inhibition of cell proliferation. In CRC, MG53 interacts with focal adhesion kinase (FAK), inhibiting the FAK/AKT (Protein kinase B) signaling pathway and reducing cell migration, invasion, and epithelial-mesenchymal transition—all of which are correlated with prolonged survival (Faleti et al., 2024). In HCC, MG53 suppresses the malignant phenotype through ubiquitin-mediated degradation of Ras-related C3 botulinum toxin substrate 1 (RAC1), enhancing the sensitivity of HCC cells to Sorafenib (Ma et al., 2022). In PDAC, MG53 functions as a transcriptional repressor of kinesin family member 11 (KIF11), leading to cell cycle arrest and inhibition of cancer cell proliferation (Wang et al., 2024b). In NSCLC, MG53 modulates G3BP2-mediated stress granule formation, reducing chemoresistance and enhancing cisplatin therapeutic efficacy (Li et al., 2021a). Furthermore, MG53 facilitates the ubiquitin-mediated degradation of Cyclin D1, blocking the G1/S cell cycle transition and effectively suppressing tumor growth in both gastric and colorectal cancers (Fang et al., 2023). In mouse models, *mg53*$^{-/-}$ mice exhibit accelerated tumor growth and increased metastasis in lung cancer, colorectal cancer, and pancreatic cancer models (Li et al., 2021a; Fang et al., 2023; Wang et al., 2024b).

Despite the experimental validation of MG53's anticancer effects, many research gaps remain. The precise molecular mechanisms underlying MG53's tumor-suppressive role across different cancer types remain incompletely understood—especially its involvement in tumor microenvironment regulation and immune modulation. Additionally, genetic mutations and crosstalk between signaling pathways may influence MG53's function in different cancer contexts, potentially leading to differential effects. Furthermore, enhancing MG53's stability and optimizing its in vivo delivery remain critical challenges for clinical translation. Future research should focus on developing precision therapeutic strategies that target MG53, which may offer novel treatment options for cancer patients and improve clinical outcomes.

## 8 Conclusions and Prospects

MG53 is a multifunctional protein that plays a crucial role in cell membrane repair, tissue regeneration, metabolic regulation, antitumor activity, and antiviral defense. It has demonstrated significant protective and reparative functions in multiple organs, including the heart, kidneys, lungs, liver, and brain, positioning it as a promising therapeutic target for acute and chronic tissue injuries.

In metabolic regulation, MG53 modulates insulin signaling through its E3 ubiquitin ligase activity, exhibiting dual effects—either promoting or inhibiting insulin sensitivity. This has important implications for the management of diabetes and metabolic syndrome. In cancer therapy, MG53 regulates cancer cell proliferation, apoptosis, and drug resistance, demonstrating therapeutic potential in various malignancies. Additionally, its membrane repair and immune-modulating properties may contribute to antiviral defense mechanisms.

Future research should focus on clarifying the molecular mechanisms of MG53 in metabolic disorders, cancer, and viral infections. Clarifying its dual role in insulin signaling and metabolic dysregulation will be essential to optimize its therapeutic application in diabetes and metabolic syndrome. Moreover, a comprehensive understanding of MG53's function in the tumor microenvironment is crucial to assess its feasibility as a cancer therapeutic target and to explore its potential synergy with chemotherapy and immunotherapy. The precise role of MG53 in antiviral defense and immune modulation also warrants investigation—specifically its function in viral infections and host immune responses—which could facilitate the development of MG53-based antiviral drugs or vaccines.

Overall, future research should prioritize optimizing MG53's therapeutic potential, developing drugs that selectively regulate MG53 function, and advancing its clinical translation for regenerative medicine, metabolic diseases, cancer, and infectious diseases.

**Acknowledgments** This work was supported by NIH grants (R21AR080628) to H.L. and partially supported by the start-up funding from The OSU Comprehensive Cancer Center to H.L.

## References

Ammendolia, D. A., Bement, W. M., & Brumell, J. H. (2021). Plasma membrane integrity: Implications for health and disease. *BMC Biology, 19*, 71. https://doi.org/10.1186/s12915-021-00972-y

Bian, Z., Wang, Q., Zhou, X., Tan, T., Park, K. H., Kramer, H. F., McDougal, A., Laping, N. J., Kumar, S., Adesanya, T. M. A., et al. (2019). Sustained elevation of MG53 in the bloodstream increases tissue regenerative capacity without compromising metabolic function. *Nature Communications, 10*, 4659. https://doi.org/10.1038/s41467-019-12483-0

Bianchi, C., Vaccaro, O., Distaso, M., Franzini, L., Raggi, F., & Solini, A. (2023). MG53 does not mark cardiovascular risk and all-cause mortality in subjects with type 2 diabetes: A prospective, observational study. *Diabetes Research and Clinical Practice, 204*, 110916. https://doi.org/10.1016/j.diabres.2023.110916

Cai, C., Masumiya, H., Weisleder, N., Matsuda, N., Nishi, M., Hwang, M., Ko, J. K., Lin, P., Thornton, A., Zhao, X., et al. (2009a). MG53 nucleates assembly of cell membrane repair machinery. *Nature Cell Biology, 11*, 56–64. https://doi.org/10.1038/ncb1812

Cai, C., Weisleder, N., Ko, J. K., Komazaki, S., Sunada, Y., Nishi, M., Takeshima, H., & Ma, J. (2009b). Membrane repair defects in muscular dystrophy are linked to altered interaction between MG53, caveolin-3, and dysferlin. *The Journal of Biological Chemistry, 284*, 15894–15902. https://doi.org/10.1074/jbc.M109.009589

Cai, C., Lin, P., Zhu, H., Ko, J. K., Hwang, M., Tan, T., Pan, Z., Korichneva, I., & Ma, J. (2015). Zinc binding to MG53 protein facilitates repair of injury to cell membranes. *The Journal of Biological Chemistry, 290*, 13830–13839. https://doi.org/10.1074/jbc.M114.620690

Cai, M., Gao, X., & Yu, S. (2023). Tripartite motif 72 inhibits apoptosis and mitochondrial dysfunction in neural stem cells induced by anesthetic sevoflurane by activating PI3K/AKT pathway. *The Chinese Journal of Physiology, 66*, 36–42. https://doi.org/10.4103/cjop.CJOP-D-22-00082

Cao, C. M., Zhang, Y., Weisleder, N., Ferrante, C., Wang, X., Lv, F., Zhang, Y., Song, R., Hwang, M., Jin, L., et al. (2010). MG53 constitutes a primary determinant of cardiac ischemic preconditioning. *Circulation, 121*, 2565–2574. https://doi.org/10.1161/CIRCULATIONAHA.110.954628

Chandler, H. L., Tan, T., Yang, C., Gemensky-Metzler, A. J., Wehrman, R. F., Jiang, Q., Peterson, C. M. W., Geng, B., Zhou, X., Wang, Q., et al. (2019). MG53 promotes corneal wound healing and mitigates fibrotic remodeling in rodents. *Communications Biology, 2*, 71. https://doi.org/10.1038/s42003-019-0316-7

Corona, B. T., Garg, K., Roe, J. L., Zhu, H., Park, K. H., Ma, J., & Walters, T. J. (2014). Effect of recombinant human MG53 protein on tourniquet-induced ischemia-reperfusion injury in rat muscle. *Muscle & Nerve, 49*, 919–921. https://doi.org/10.1002/mus.24160

Dias, C., & Nylandsted, J. (2021). Plasma membrane integrity in health and disease: Significance and therapeutic potential. *Cell Discovery, 7*, 4. https://doi.org/10.1038/s41421-020-00233-2

Duan, D., Goemans, N., Takeda, S., Mercuri, E., & Aartsma-Rus, A. (2021). Duchenne muscular dystrophy. *Nature Reviews. Disease Primers, 7*, 13. https://doi.org/10.1038/s41572-021-00248-3

Duann, P., Li, H., Lin, P., Tan, T., Wang, Z., Chen, K., Zhou, X., Gumpper, K., Zhu, H., Ludwig, T., et al. (2015). MG53-mediated cell membrane repair protects against acute kidney injury. *Science Translational Medicine, 7*, 279ra236. https://doi.org/10.1126/scitranslmed.3010755

Faleti, O. D., Gong, Y., Long, J., Luo, Q., Tan, H., Deng, S., Qiu, L., Lyu, X., Yao, J., & Wu, G. (2024). TRIM72 inhibits cell migration and epithelial-mesenchymal transition by attenuating FAK/akt signaling in colorectal cancer. *Heliyon, 10*, e37714. https://doi.org/10.1016/j.heliyon.2024.e37714

Fang, M., Wu, H. K., Pei, Y., Zhang, Y., Gao, X., He, Y., Chen, G., Lv, F., Jiang, P., Li, Y., et al. (2023). E3 ligase MG53 suppresses tumor growth by degrading cyclin D1. *Signal Transduction and Targeted Therapy, 8*, 263. https://doi.org/10.1038/s41392-023-01458-9

Feng, H., Shen, H., Robeson, M. J., Wu, Y. H., Wu, H. K., Chen, G. J., Zhang, S., Xie, P., Jin, L., He, Y., et al. (2022). MG53 E3 ligase-dead mutant protects diabetic hearts from acute ischemic/reperfusion injury and ameliorates diet-induced Cardiometabolic damage. *Diabetes, 71*, 298–314. https://doi.org/10.2337/db21-0322

Gouchoe, D. A., Whitson, B. A., & Zhu, H. (2023). The next frontier in lung transplantation: Protecting the endothelium and repairing organs for transplant utilizing MG53. *Clinical and Translational Discovery, 3*. https://doi.org/10.1002/ctd2.255

Gouchoe, D. A., Lee, Y. G., Kim, J. L., Zhang, Z., Marshall, J. M., Ganapathi, A., Zhu, H., Black, S. M., Ma, J., & Whitson, B. A. (2024a). Mitsugumin 53 mitigation of ischemia-reperfusion injury in a mouse model. *The Journal of Thoracic and Cardiovascular Surgery, 167*, e48–e58. https://doi.org/10.1016/j.jtcvs.2023.08.005

Gouchoe, D. A., Yi, T., Kim, J. L., Lee, Y. G., Black, S. M., Breuer, C., Ma, J., & Whitson, B. A. (2024b). MG53 mitigates warm ischemic lung injury in a murine model of transplantation. *The Journal of Thoracic and Cardiovascular Surgery, 168*, e13–e26. https://doi.org/10.1016/j.jtcvs.2023.10.056

Guan, F., Huang, T., Wang, X., Xing, Q., Gumpper, K., Li, P., Song, J., Tan, T., Yang, G. L., Zang, X., et al. (2019). The TRIM protein Mitsugumin 53 enhances survival and therapeutic efficacy of stem cells in murine traumatic brain injury. *Stem Cell Research & Therapy, 10*, 352. https://doi.org/10.1186/s13287-019-1433-4

Gumpper-Fedus, K., Park, K. H., Ma, H., Zhou, X., Bian, Z., Krishnamurthy, K., Sermersheim, M., Zhou, J., Tan, T., Li, L., et al. (2022). MG53 preserves mitochondrial integrity of cardiomyocytes during ischemia reperfusion-induced oxidative stress. *Redox Biology, 54*, 102357. https://doi.org/10.1016/j.redox.2022.102357

Guo, O., Ju, B., Shawver, M. H., Geng, B., Wei, S., Early, T., Yi, F., Tan, T., Chandler, H. L., Ma, J., & Zhu, H. (2021). Recombinant human MG53 protein protects against alkaline-induced corneal injuries in mice. *Military Medicine, 186*, 486–490. https://doi.org/10.1093/milmed/usaa357

Gupta, P., Li, H., Zhang, G. N., Barbuti, A. M., Yang, Y., Lin, P. H., Cai, C., Tan, T., Ma, J., & Chen, Z. S. (2022). MG53 inhibits cellular proliferation and tumor progression in colorectal carcinoma. *International Journal of Biological Sciences, 18*, 5221–5229. https://doi.org/10.7150/ijbs.67869

Gushchina, L. V., Bhattacharya, S., McElhanon, K. E., Choi, J. H., Manring, H., Beck, E. X., Alloush, J., & Weisleder, N. (2017). Treatment with recombinant human MG53 protein increases membrane integrity in a mouse model of limb girdle muscular dystrophy 2B. *Molecular Therapy, 25*, 2360–2371. https://doi.org/10.1016/j.ymthe.2017.06.025

Han, Y., Black, S., Gong, Z., Chen, Z., Ko, J. K., Zhou, Z., Xia, T., Fang, D., Yang, D., Gu, D., et al. (2022). Membrane-delimited signaling and cytosolic action of MG53 preserve hepatocyte integrity during drug-induced liver injury. *Journal of Hepatology, 76*, 558–567. https://doi.org/10.1016/j.jhep.2021.10.017

Hatakeyama, S. (2017). TRIM family proteins: Roles in autophagy, immunity, and carcinogenesis. *Trends in Biochemical Sciences, 42*, 297–311. https://doi.org/10.1016/j.tibs.2017.01.002

He, B., Tang, R. H., Weisleder, N., Xiao, B., Yuan, Z., Cai, C., Zhu, H., Lin, P., Qiao, C., Li, J., et al. (2012). Enhancing muscle membrane repair by gene delivery of MG53 ameliorates muscular dystrophy and heart failure in delta-Sarcoglycan-deficient hamsters. *Molecular Therapy, 20*, 727–735. https://doi.org/10.1038/mt.2012.5

Ishiwata-Endo, H., Kato, J., Tonouchi, A., Chung, Y. W., Sun, J., Stevens, L. A., Zhu, J., Aponte, A. M., Springer, D. A., San, H., et al. (2018). Role of a TRIM72 ADP-ribosylation cycle in myocardial injury and membrane repair. *JCI Insight, 3*. https://doi.org/10.1172/jci.insight.97898

Jia, Y., Chen, K., Lin, P., Lieber, G., Nishi, M., Yan, R., Wang, Z., Yao, Y., Li, Y., Whitson, B. A., et al. (2014). Treatment of acute lung injury by targeting MG53-mediated cell membrane repair. *Nature Communications, 5*, 4387. https://doi.org/10.1038/ncomms5387

Jiang, W., Yu, L., Mu, N., Zhang, Z., & Ma, H. (2024). MG53 inhibits ferroptosis by targeting the p53/SLC7A11/GPX4 pathway to alleviate doxorubicin-induced cardiotoxicity. *Free Radical Biology & Medicine, 223*, 224–236. https://doi.org/10.1016/j.freeradbiomed.2024.08.001

Jin, G., Ho, J. W., Keeney-Bonthrone, T. P., Ober, R. A., Liu, B., Chtraklin, K., Wang, X. L., Zhou, X., Cai, C., Li, Y., et al. (2022). Recombinant human MG53 protein attenuates brain lesion size in a large animal model of traumatic brain injury. *Journal of Trauma and Acute Care Surgery, 93*, 613–619. https://doi.org/10.1097/TA.0000000000003746

Kenney, A. D., Li, Z., Bian, Z., Zhou, X., Li, H., Whitson, B. A., Tan, T., Cai, C., Ma, J., & Yount, J. S. (2020). Recombinant MG53 protein protects mice from lethal influenza virus infection. *American Journal of Respiratory and Critical Care Medicine, 203*, 254. https://doi.org/10.1164/rccm.202007-2908LE

Kim, S. C., Kellett, T., Wang, S., Nishi, M., Nagre, N., Zhou, B., Flodby, P., Shilo, K., Ghadiali, S. N., Takeshima, H., et al. (2014). TRIM72 is required for effective repair of alveolar epithelial cell wounding. *American Journal of Physiology. Lung Cellular and Molecular Physiology, 307*, L449–L459. https://doi.org/10.1152/ajplung.00172.2014

Kohr, M. J., Evangelista, A. M., Ferlito, M., Steenbergen, C., & Murphy, E. (2014). S-nitrosylation of TRIM72 at cysteine 144 is critical for protection against oxidation-induced protein degradation and cell death. *Journal of Molecular and Cellular Cardiology, 69*, 67–74. https://doi.org/10.1016/j.yjmcc.2014.01.010

Lee, K. E., Kim, J., Park, K. H., Ma, J., & Zhu, H. (2023). Remote ischemic pre-conditioning (RIPC). *Methods in Molecular Biology, 2597*, 11–18. https://doi.org/10.1007/978-1-0716-2835-5_2

Li, H., Duann, P., Lin, P. H., Zhao, L., Fan, Z., Tan, T., Zhou, X., Sun, M., Fu, M., Orange, M., et al. (2015). Modulation of wound healing and scar formation by MG53 protein-mediated cell membrane repair. *The Journal of Biological Chemistry, 290*, 24592–24603. https://doi.org/10.1074/jbc.M115.680074

Li, M., Li, H., Li, X., Zhu, H., Xu, Z., Liu, L., Ma, J., & Zhang, M. (2017). A bioinspired alginate-gum Arabic hydrogel with micro-/nanoscale structures for controlled drug release in chronic wound healing. *ACS Applied Materials & Interfaces, 9*, 22160–22175. https://doi.org/10.1021/acsami.7b04428

Li, H., Lin, P. H., Gupta, P., Li, X., Zhao, S. L., Zhou, X., Li, Z., Wei, S., Xu, L., Han, R., et al. (2021a). MG53 suppresses tumor progression and stress granule formation by modulating G3BP2 activity in non-small cell lung cancer. *Molecular Cancer, 20*, 118. https://doi.org/10.1186/s12943-021-01418-3

Li, Z., Wang, L., Yue, H., Whitson, B. A., Haggard, E., Xu, X., & Ma, J. (2021b). MG53, a tissue repair protein with broad applications in regenerative medicine. *Cells, 10*. https://doi.org/10.3390/cells10010122

Li, H., Duann, P., Li, Z., Zhou, X., Ma, J., Rovin, B. H., & Lin, P. H. (2022a). The cell membrane repair protein MG53 modulates transcription factor NF-kappaB signaling to control kidney fibrosis. *Kidney International, 101*, 119–130. https://doi.org/10.1016/j.kint.2021.09.027

Li, H., Rosas, L., Li, Z., Bian, Z., Li, X., Choi, K., Cai, C., Zhou, X., Tan, T., Bergdall, V., et al. (2022b). MG53 attenuates nitrogen mustard-induced acute lung injury. *Journal of Cellular and Molecular Medicine, 26*, 1886–1895. https://doi.org/10.1111/jcmm.16917

Li, X., Ji, R., Duan, L., Hao, Z., Su, Y., Wang, H., Guan, F., & Ma, S. (2024). MG53/GMs/HA-Dex neural scaffold promotes the functional recovery of spinal cord injury by alleviating neuroinflammation. *International Journal of Biological Macromolecules, 267*, 131520. https://doi.org/10.1016/j.ijbiomac.2024.131520

Lijie, G., Yueyue, Z., Nan, Z., Ling, W., Xuan, W., & Weijie, Y. (2019). Mitsugumin 53 promotes mitochondrial autophagy through regulating Ambra1 expression in C2C12 myoblast cells. *Cell Biology International, 43*, 290–298. https://doi.org/10.1002/cbin.11097

Lin, P., Zhu, H., Cai, C., Wang, X., Cao, C., Xiao, R., Pan, Z., Weisleder, N., Takeshima, H., & Ma, J. (2012). Nonmuscle myosin IIA facilitates vesicle trafficking for MG53-mediated cell membrane repair. *The FASEB Journal, 26*, 1875–1883. https://doi.org/10.1096/fj.11-188599

Liu, F., Song, R., Feng, Y., Guo, J., Chen, Y., Zhang, Y., Chen, T., Wang, Y., Huang, Y., Li, C. Y., et al. (2015a). Upregulation of MG53 induces diabetic cardiomyopathy through transcriptional activation of peroxisome proliferation-activated receptor alpha. *Circulation, 131*, 795–804. https://doi.org/10.1161/CIRCULATIONAHA.114.012285

Liu, J., Zhu, H., Zheng, Y., Xu, Z., Li, L., Tan, T., Park, K. H., Hou, J., Zhang, C., Li, D., et al. (2015b). Cardioprotection of recombinant human MG53 protein in a porcine model of ischemia and reperfusion injury. *Journal of Molecular and Cellular Cardiology, 80*, 10–19. https://doi.org/10.1016/j.yjmcc.2014.12.010

Liu, W., Wang, G., Zhang, C., Ding, W., Cheng, W., Luo, Y., Wei, C., & Liu, J. (2019). MG53, a novel regulator of KChIP2 and Ito,f, plays a critical role in electrophysiological remodeling in cardiac hypertrophy. *Circulation, 139*, 2142–2156. https://doi.org/10.1161/CIRCULATIONAHA.118.029413

Liu, C., Hu, Y. H., Han, Y., Wang, Y. B., Zhang, Y., Zhang, X. Q., He, D. F., Ren, H. M., Liu, Y. K., Wang, H. Y., et al. (2020). MG53 protects against contrast-induced acute kidney injury by reducing cell membrane damage and apoptosis. *Acta Pharmacologica Sinica, 41*, 1457–1464. https://doi.org/10.1038/s41401-020-0420-8

Liu, Y., Li, L., Wang, Z., Zhang, J., & Zhou, Z. (2023). Myocardial ischemia-reperfusion injury, Molecular mechanisms and prevention. *Microvascular Research Journal, 149*, 104565. https://doi.org/10.1016/j.mvr.2023.104565

Lv, F., Wang, Y., Shan, D., Guo, S., Chen, G., Jin, L., Zheng, W., Feng, H., Zeng, X., Zhang, S., et al. (2022). Blocking MG53(S255) phosphorylation protects diabetic heart from ischemic injury. *Circulation Research, 131*, 962–976. https://doi.org/10.1161/CIRCRESAHA.122.321055

Ma, H., Liu, J., Bian, Z., Cui, Y., Zhou, X., Zhou, X., Zhang, B., Adesanya, T. M., Yi, F., Park, K. H., et al. (2015). Effect of metabolic syndrome on mitsugumin 53 expression and function. *PLoS One, 10*, e0124128. https://doi.org/10.1371/journal.pone.0124128

Ma, S., Wang, Y., Zhou, X., Li, Z., Zhang, Z., Wang, Y., Huang, T., Zhang, Y., Shi, J., & Guan, F. (2020). MG53 protects hUC-MSCs against inflammatory damage and synergistically enhances their efficacy in neuroinflammation injured brain through inhibiting NLRP3/Caspase-1/IL-1beta Axis. *ACS Chemical Neuroscience, 11*, 2590–2601. https://doi.org/10.1021/acschemneuro.0c00268

Ma, X., Ma, X., Zhu, L., Zhao, Y., Chen, M., Li, T., Lin, Y., Ma, D., Sun, C., & Han, L. (2022). The E3 ubiquitin ligase MG53 inhibits hepatocellular carcinoma by targeting RAC1 signaling. *Oncogene, 11*, 40. https://doi.org/10.1038/s41389-022-00414-6

Ma, Y., Ding, L., Li, Z., & Zhou, C. (2023). Structural basis for TRIM72 oligomerization during membrane damage repair. *Nature Communications, 14*, 1555. https://doi.org/10.1038/s41467-023-37198-1

Nagre, N., Wang, S., Kellett, T., Kanagasabai, R., Deng, J., Nishi, M., Shilo, K., Oeckler, R. A., Yalowich, J. C., Takeshima, H., et al. (2016). TRIM72 modulates caveolar endocytosis in repair of lung cells. *American Journal of Physiology. Lung Cellular and Molecular Physiology, 310*, L452–L464. https://doi.org/10.1152/ajplung.00089.2015

Nagre, N., Cong, X., Ji, H. L., Schreiber, J. M., Fu, H., Pepper, I., Warren, S., Sill, J. M., Hubmayr, R. D., & Zhao, X. (2018). Inhaled TRIM72 protein protects ventilation injury to the lung through injury-guided cell repair. *American Journal of Respiratory Cell and Molecular Biology, 59*, 635–647. https://doi.org/10.1165/rcmb.2017-0364OC

Niu, Y., Chen, G., Lv, F., Xiao, R. P., Hu, X., & Chen, L. (2022a). Cryo-EM structure of human MG53 homodimer. *The Biochemical Journal, 479*, 1909–1916. https://doi.org/10.1042/bcj20220385

Niu, H., Li, H., Guan, Y., Zhou, X., Li, Z., Zhao, S. L., Chen, P., Tan, T., Zhu, H., Bergdall, V., et al. (2022b). Sustained delivery of rhMG53 promotes diabetic wound healing and hair follicle development. *Bioactive Materials, 18*, 104–115. https://doi.org/10.1016/j.bioactmat.2022.03.017

Paleo, B. J., McElhanon, K. E., Bulgart, H. R., Banford, K. K., Beck, E. X., Sattler, K. M., Goines, B. N., Ratcliff, S. L., Crowe, K. E., & Weisleder, N. (2022). Reduced Sarcolemmal membrane repair exacerbates striated muscle pathology in a mouse model of Duchenne muscular dystrophy. *Cells, 11*. https://doi.org/10.3390/cells11091417

Park, S. H., Han, J., Jeong, B. C., Song, J. H., Jang, S. H., Jeong, H., Kim, B. H., Ko, Y. G., Park, Z. Y., Lee, K. E., et al. (2023). Structure and activation of the RING E3 ubiquitin ligase TRIM72 on the membrane. *Nature Structural & Molecular Biology, 30*, 1695–1706. https://doi.org/10.1038/s41594-023-01111-7

Philouze, C., Turban, S., Cremers, B., Caliez, A., Lamarche, G., Bernard, C., Provost, N., & Delerive, P. (2021). MG53 is not a critical regulator of insulin signaling pathway in skeletal muscle. *PLoS One, 16*, e0245179. https://doi.org/10.1371/journal.pone.0245179

Sardiello, M., Cairo, S., Fontanella, B., Ballabio, A., & Meroni, G. (2008). Genomic analysis of the TRIM family reveals two groups of genes with distinct evolutionary properties. *BMC Evolutionary Biology, 8*, 225. https://doi.org/10.1186/1471-2148-8-225

Sermersheim, M., Kenney, A. D., Lin, P. H., McMichael, T. M., Cai, C., Gumpper, K., Adesanya, T. M. A., Li, H., Zhou, X., Park, K. H., et al. (2020). MG53 suppresses interferon-beta and inflammation via regulation of ryanodine receptor-mediated intracellular calcium signaling. *Nature Communications, 11*, 3624. https://doi.org/10.1038/s41467-020-17177-6

Shan, D., Guo, S., Wu, H. K., Lv, F., Jin, L., Zhang, M., Xie, P., Wang, Y., Song, Y., Wu, F., et al. (2020). Cardiac ischemic preconditioning promotes MG53 secretion through H2O2-activated protein kinase C-delta signaling. *Circulation, 142*, 1077–1091. https://doi.org/10.1161/CIRCULATIONAHA.119.044998

Song, R., Peng, W., Zhang, Y., Lv, F., Wu, H. K., Guo, J., Cao, Y., Pi, Y., Zhang, X., Jin, L., et al. (2013). Central role of E3 ubiquitin ligase MG53 in insulin resistance and metabolic disorders. *Nature, 494*, 375–379. https://doi.org/10.1038/nature11834

Sui, B., Zheng, J., Fu, Z., Zhao, L., & Zhou, M. (2024). TRIM72 restricts lyssavirus infection by inducing K48-linked ubiquitination and proteasome degradation of the matrix protein. *PLoS Pathogens, 20*, e1011718. https://doi.org/10.1371/journal.ppat.1011718

Tan, S., Li, M., & Song, X. (2023). MG53 alleviates airway inflammatory responses by regulating nuclear factor-kappaB pathway in asthmatic mice. *Allergol Immunopathol (Madr), 51*, 175–181. https://doi.org/10.15586/aei.v51i4.880

Wang, X., Xie, W., Zhang, Y., Lin, P., Han, L., Han, P., Wang, Y., Chen, Z., Ji, G., Zheng, M., et al. (2010). Cardioprotection of ischemia/reperfusion injury by cholesterol-dependent MG53-mediated membrane repair. *Circulation Research, 107*, 76–83. https://doi.org/10.1161/CIRCRESAHA.109.215822

Wang, Q., Bian, Z., Jiang, Q., Wang, X., Zhou, X., Park, K. H., Hsueh, W., Whitson, B. A., Haggard, E., Li, H., et al. (2020). MG53 does not manifest the development of diabetes in db/db mice. *Diabetes, 69*, 1052–1064. https://doi.org/10.2337/db19-0807

Wang, X., Li, X., Ong, H., Tan, T., Park, K. H., Bian, Z., Zou, X., Haggard, E., Janssen, P. M., Merritt, R. E., et al. (2021). MG53 suppresses NF-kappaB activation to mitigate age-related heart failure. *JCI Insight, 6*. https://doi.org/10.1172/jci.insight.148375

Wang, Q., Park, K. H., Geng, B., Chen, P., Yang, C., Jiang, Q., Yi, F., Tan, T., Zhou, X., Bian, Z., et al. (2022a). MG53 inhibits necroptosis through ubiquitination-dependent RIPK1 degradation for cardiac protection following ischemia/reperfusion injury. *Frontiers in Cardiovascular Medicine, 9*, 868632. https://doi.org/10.3389/fcvm.2022.868632

Wang, C. C., Peng, H., Wang, Z., Yang, J., Hu, R. G., Li, C. Y., & Geng, W. J. (2022b). TRIM72-mediated degradation of the short form of p62/SQSTM1 rheostatically controls selective autophagy in human cells. *Military Medical Research, 9*, 35. https://doi.org/10.1186/s40779-022-00392-1

Wang, Y. F., An, Z. Y., Li, J. W., Dong, Z. K., & Jin, W. L. (2024a). MG53/TRIM72: Multi-organ repair protein and beyond. *Frontiers in Physiology, 15*, 1377025. https://doi.org/10.3389/fphys.2024.1377025

Wang, X. L., He, X., Gao, T., Zhou, X., Cruz-Monserrate, Z., Tsung, A., Ma, J., & Cai, C. (2024b). MG53 suppresses tumor growth via transcriptional inhibition of KIF11 in pancreatic cancer. *Translational Oncology, 50*, 102118. https://doi.org/10.1016/j.tranon.2024.102118

Weisleder, N., Takizawa, N., Lin, P., Wang, X., Cao, C., Zhang, Y., Tan, T., Ferrante, C., Zhu, H., Chen, P. J., et al. (2012). Recombinant MG53 protein modulates therapeutic cell membrane repair in treatment of muscular dystrophy. *Science Translational Medicine, 4*, 139ra185. https://doi.org/10.1126/scitranslmed.3003921

Whitson, B. A., Mulier, K., Li, H., Zhou, X., Cai, C., Black, S. M., Tan, T., Ma, J., & Beilman, G. J. (2021a). MG53 as a novel therapeutic protein to treat acute lung injury. *Military Medicine, 186*, 339–345. https://doi.org/10.1093/milmed/usaa313

Whitson, B. A., Tan, T., Gong, N., Zhu, H., & Ma, J. (2021b). Muscle multiorgan crosstalk with MG53 as a myokine for tissue repair and regeneration. *Current Opinion in Pharmacology, 59*, 26–32. https://doi.org/10.1016/j.coph.2021.04.005

Wu, Y., Huang, J., Liu, D., Tan, J., Peng, Y., Yang, J., Cui, Y., He, W., Luo, G., & Wu, J. (2013). Mitsugumin 53 protects the kidney from severe burn injury in mice. *Burns & Trauma, 1*, 128–133. https://doi.org/10.4103/2321-3868.123074

Wu, H. K., Zhang, Y., Cao, C. M., Hu, X., Fang, M., Yao, Y., Jin, L., Chen, G., Jiang, P., Zhang, S., et al. (2019). Glucose-sensitive Myokine/Cardiokine MG53 regulates systemic insulin response and metabolic homeostasis. *Circulation, 139*, 901–914. https://doi.org/10.1161/CIRCULATIONAHA.118.037216

Wu, M., Li, H., He, J., Liang, J., Liu, Y., & Zhang, W. (2023). TRIM72 alleviates muscle inflammation in mdx mice via promoting Mitophagy-mediated NLRP3 Inflammasome inactivation. *Oxidative Medicine and Cellular Longevity, 2023*, 8408574. https://doi.org/10.1155/2023/8408574

Wu, J., Feng, Y., Wang, Y., He, X., Chen, Z., Lan, D., Wu, X., Wen, J., Tsung, A., Wang, X., et al. (2024). MG53 binding to CAV3 facilitates activation of eNOS/NO signaling pathway to enhance the therapeutic benefits of bone marrow-derived mesenchymal stem cells in diabetic wound healing. *International Immunopharmacology, 136*, 112410. https://doi.org/10.1016/j.intimp.2024.112410

Xie, H., Wang, Y., Zhu, T., Feng, S., Yan, Z., Zhu, Z., Ni, J., Ni, J., Du, R., Zhu, J., et al. (2020a). Serum MG53/TRIM72 is associated with the presence and severity of coronary artery disease and acute myocardial infarction. *Frontiers in Physiology, 11*, 617845. https://doi.org/10.3389/fphys.2020.617845

Xie, H., Yan, Z., Feng, S., Zhu, T., Zhu, Z., Ni, J., Ni, J., Du, R., Zhu, J., Ding, F., et al. (2020b). Prognostic value of circulating MG53 levels in acute myocardial infarction. *Frontiers in Cardiovascular Medicine, 7*, 596107. https://doi.org/10.3389/fcvm.2020.596107

Xue, Y., Song, T., Ke, J., Lin, S., Zhang, J., Chen, Y., Wang, J., Fan, Q., & Chen, F. (2024). MG53 protects against Coxsackievirus B3-induced acute viral myocarditis in mice by inhibiting NLRP3 inflammasome-mediated pyroptosis via the NF-kappaB signaling pathway. *Biochemical Pharmacology, 223*, 116173. https://doi.org/10.1016/j.bcp.2024.116173

Yao, Y., Zhang, B., Zhu, H., Li, H., Han, Y., Chen, K., Wang, Z., Zeng, J., Liu, Y., Wang, X., et al. (2016). MG53 permeates through blood-brain barrier to protect ischemic brain injury. *Oncotarget, 7*, 22474–22485. https://doi.org/10.18632/oncotarget.7965

Yao, W., Li, H., Han, X., Chen, C., Zhang, Y., Tai, W. L., Xia, Z., & Hei, Z. (2017). MG53 anchored by dysferlin to cell membrane reduces hepatocyte apoptosis which induced by ischaemia/reperfusion injury in vivo and in vitro. *Journal of Cellular and Molecular Medicine, 21*, 2503–2513. https://doi.org/10.1111/jcmm.13171

Yi, J. S., Park, J. S., Ham, Y. M., Nguyen, N., Lee, N. R., Hong, J., Kim, B. W., Lee, H., Lee, C. S., Jeong, B. C., et al. (2013). MG53-induced IRS-1 ubiquitination negatively regulates skeletal myogenesis and insulin signalling. *Nature Communications, 4*, 2354. https://doi.org/10.1038/ncomms3354

Yi, J., Li, A., Li, X., Park, K., Zhou, X., Yi, F., Xiao, Y., Yoon, D., Tan, T., Ostrow, L. W., et al. (2021). MG53 preserves neuromuscular junction integrity and alleviates ALS disease progression. *Antioxidants (Basel), 10*. https://doi.org/10.3390/antiox10101522

Yin, W., Liu, Y., & Bian, Z. (2019). MG53 inhibits the progression of tongue cancer cells through regulating PI3K-AKT signaling pathway: Evidence from 3D cell culture and animal model. *Small, 15*, e1805492. https://doi.org/10.1002/smll.201805492

Zhang, Y., Lv, F., Jin, L., Peng, W., Song, R., Ma, J., Cao, C. M., & Xiao, R. P. (2011). MG53 participates in ischaemic postconditioning through the RISK signalling pathway. *Cardiovascular Research, 91*, 108–115. https://doi.org/10.1093/cvr/cvr029

Zhou, L., Middel, V., Reischl, M., Strahle, U., & Nienhaus, G. U. (2018). Distinct amino acid motifs carrying multiple positive charges regulate membrane targeting of dysferlin and MG53. *PLoS One, 13*, e0202052. https://doi.org/10.1371/journal.pone.0202052

Zhu, H., Lin, P., De, G., Choi, K. H., Takeshima, H., Weisleder, N., & Ma, J. (2011). Polymerase transcriptase release factor (PTRF) anchors MG53 protein to cell injury site for initiation of membrane repair. *The Journal of Biological Chemistry, 286*, 12820–12824. https://doi.org/10.1074/jbc.C111.221440

Zhu, H., Hou, J., Roe, J. L., Park, K. H., Tan, T., Zheng, Y., Li, L., Zhang, C., Liu, J., Liu, Z., et al. (2015). Amelioration of ischemia-reperfusion-induced muscle injury by the recombinant human MG53 protein. *Muscle & Nerve, 52*, 852–858. https://doi.org/10.1002/mus.24619

Zhu, X. Z., Wang, J. Q., & Wu, Y. H. (2023). MG53 ameliorates nerve injury induced neuropathic pain through the regulation of Nrf2/HO-1 signaling in rats. *Behavioural Brain Research, 449*, 114489. https://doi.org/10.1016/j.bbr.2023.114489

Zhu, K., Liu, Y., Dai, R., Wang, X., Li, J., Lin, Z., Du, L., Guo, J., Ju, Y., Zhu, W., et al. (2024). p85alpha deficiency alleviates ischemia-reperfusion injury by promoting cardiomyocyte survival. *Biochimica et Biophysica Acta – Molecular Basis of Disease, 1870*, 167318. https://doi.org/10.1016/j.bbadis.2024.167318

# Kidney Regeneration

# Matrix Proteins in Kidney Injury, Fibrosis, and Regeneration

**Pei-Hui Lin and Pu Duann**

**Abstract** The extracellular matrix (ECM) is a complex and dynamic cross-linked protein network that supports organ integrity and homeostasis through modulation of cellular functions within the organ and mediates crosstalk towards environmental stimuli. The dysregulated extracellular matrix is the hallmark of kidney injury, inflammation, and fibrosis. With the recent development of advanced proteomics, genetics, genomics, and epigenomics, studies of the interplay between ECM components and cellular and molecular mechanisms in kidney health and disease, particularly during the repair stage of kidney injury and fibrosis, are emerging. Better characterization of kidney matrisome proteins and their functions for organ cross-communication, in combination with stem cell treatment and novel extracellular vesicle delivery platform, will benefit not only biomarkers identification but also target therapeutics development. All these exciting advances will build new regenerative medicine to eradicate chronic kidney disease.

**Keywords** ECM · Inflammation · Cytokines/chemokines · Immune response · Myofibroblast · EMT · Fibrogenic signaling · Nanomedicine · Stem cell · Extracellular vesicles

## 1 Introduction

Kidney disease, which can be broadly classified as acute (acute kidney injury, AKI) or chronic (chronic kidney disease, CKD), has grown into a global health burden as currently there are no efficient therapies. CKD has a worldwide prevalence (8–15%) and is characterized by progressive loss of kidney function (Bowe et al., 2018; Kalantar-Zadeh et al., 2021). Clinical studies have shown that CKD serves as a

P.-H. Lin (✉) · P. Duann
Department of Internal Medicine, Division of Nephrology, The Ohio State University Wexner Medical Center, Columbus, OH, USA
e-mail: Pei-Hui.Lin@osumc.edu

 
H. Das (ed.), *Tissue Repair and Regeneration*,
https://doi.org/10.1007/978-3-031-93677-7_3

major risk factor for AKI, and episodes of AKI can accelerate the progression of CKD into a life-threatening end-stage renal failure, which has been regarded as an unhealing wound and could be related to renal cancer and requires life-saving kidney replacement treatment (Docs et al., 2023; Venkatachalam et al., 2015). Renal injury, inflammation, and subsequent fibrosis can be triggered by exogenous insults such as toxins, obstruction, and infection. Fibrosis driven by excessive accumulation of extracellular matrix (ECM) is the hallmark of CKD and the result of an uncontrolled wound-healing response to injury insults (Bülow & Boor, 2019; Huang et al., 2023; Panizo et al., 2021). Unlike well-studied skin wound healing that involves homeostasis, inflammation, extracellular matrix reorganization, and scar formation (Lin et al., 2017; Moretti et al., 2022), studies of the interplay between ECM components and cellular and molecular mechanisms in kidney fibrosis are emerging (Li et al., 2022; Moeller et al., 2021; Rayego-Mateos et al., 2021). As noted, the kidney ECM proteome, or matrisome, is altered in health, aging, and disease (Lausecker et al., 2022; Randles et al., 2021; Rende et al., 2023), which could also be affected by the genomics of fibrotic disease (Randles et al., 2015; Stone et al., 2020) or epigenetics (Kato & Natarajan, 2019; Rousselle et al., 2021; Xiang et al., 2022). Distinct morphological patterns of kidney fibrosis that appear in the nature and sub-compartments of nephrons suggest distinct pathogenic causes (Tampe et al., 2021).

In this chapter, we discuss (1) the major kidney matrisome compositions in physiology and disease, (2) the interplay between ECM components and cellular and molecular mechanisms in kidney fibrosis, and (3) ECM targeting combined with stem cell treatment as novel anti-fibrotic and regenerative therapies. We hope following a deeper mechanistic understanding of matrisome and cellular, molecular signaling in fibrogenesis will lead to the successful identification of biomarkers and novel anti-fibrotic drug targets that improve functional quality of life for kidney disease patients and decrease mortality from fibrotic diseases.

## 2 Major Kidney Matrisome Compositions in Physiology and Disease

The extracellular matrix network (matrisome) is critical in normal tissue architecture and provides a scaffold for cellular and molecular interaction. The ECM itself can regulate a range of cellular functions via its physical, mechanical, and molecular properties. In addition to structural proteins, ECM contains several cytokines, chemokines, and growth factors in immunomodulatory functions. These factors could be derived from various kidney resident cells, immune cells, and fibroblasts. Recent advances in mass-spectrometry-based proteomics and single cell transcriptomics (scRNAseq, snRNAseq) have overcome major technical challenges in studying kidney matrix, a heavily cross-linked insoluble and sometimes low-abundance scaffolds, and enable spatiotemporal precision in the characterization kidney matrisome composition in health, disease, aging, and also during mouse kidney development (Eckersley et al., 2023; Feng et al., 2019; Li et al., 2023a; Lipp et al., 2021; McCabe et al., 2023; Randles et al., 2021; Rende et al., 2023; Rinschen & Saez-Rodriguez, 2021).

The major kidney matriosome includes fibrillar collagen, elastin, fibronectin, laminin, glycoproteins, proteoglycans, glycosaminoglycans (GAGs, such as hyaluronan) (Rabelink et al., 2024), osteopontin, tenascin (TNC), and ECM-affiliated regulators/secreted factors and ECM remodeling enzymes (integrins, metalloproteinase, cross-linking enzymes) (Lausecker et al., 2022). In general, the matrix organizes into two distinct entities tubulointerstitial (TI) matrix or basement membrane (BM) matrix. The TI matrix serves as a tissue structural scaffold around tubules and mainly comprises fibrillar collagen I and III, elastin, laminin, and glycosaminoglycans (GAGs) (Lausecker et al., 2022). The basement membrane matrix mainly contains fibrillar collagen IV, laminin, and proteoglycan-protein perlecan with somewhat additional 200 more macromolecules to provide anchorage and adhesion platform for epithelial and endothelial cells (Lausecker et al., 2022; Randles et al., 2017). Glomerular basement membrane (GBM) is a specialized basement membrane in the kidney glomerulus, which mainly constitutes type IV collagen α3, 4, 5, laminin α5β2γ1, and perlecan. GBM provides podocyte anchorage and withstands tensional homeostasis during filtration mechanical force load to uniquely endow both size- and charge-selective blood filtration function in the glomerulus (Hobeika et al., 2017; Lausecker et al., 2022; Lennon et al., 2014; Naylor et al., 2021). Podocyte behavior such as morphology, adhesion, contraction, and interaction with the cytoskeletal network is heavily impacted by the composition and dynamic turnover of the GBM matrix to maintain its mechano-transduction and filtration function. Moreover, glomerular cell GBM crosstalk influences the composition and assembly of ECM and subsequent ECM-affiliated signaling (Byron et al., 2014).

CKD is histo-pathologically characterized by progressive ECM replacement of interstitial or glomerular tissues. The renal lesion starts with a fibrogenic niche integrating a myriad of diffusive and matrix-bound signaling molecules, including pro-inflammatory, profibrotic cytokines, growth factors secreted by diverse cell types, and progresses with a complex crosstalk of multiple cell types involving resident kidney cells, mesenchymal and immune cell populations. These interactions together drive dynamic changes in matrix composition, activation of fibrogenic macrophage, and trans-differentiation/activation of myofibroblast which actively secrete ECM proteins leading to changes in ECM fibrillar collagen networks, accompanied by excess GAG hyaluronan and increased tissue stiffness, loss of mechanical compliance, and altered ECM mechanical properties (Li et al., 2022). Upon successful tissue repair and clearance of injury cells and inflammatory signaling, which occurs more often in mild proximal tubular injury, activated myofibroblasts become quiescent or cleared by apoptosis to restore matrix homeostasis. On the contrary, prolonged stimulation of myofibroblasts and excessive fibrogenic ECM contribute to dynamic cell-responsive interactions between immune cells, fibroblasts, and renal parenchymal cells which form a complex feed-forward bidirectional circuit (further discussed below). Fibrotic ECM is well associated with the pathogenesis of different CKD etiologies, as exampled by diabetic nephropathy (Adeva-Andany & Carneiro-Freire, 2022; Hu et al., 2015), immunoglobulin A nephropathy (IgAN) (Paunas et al., 2019; Yuan et al., 2023), focal segmental glomerular sclerosis (FSGS) (Merchant et al., 2020), lupus nephritis (LN) (Zhou et al., 2022b), and Alport syndrome, a genetic disorder of collagen IV (Chakravarti et al., 2022; LeBleu et al., 2024). A comprehensive matrisome analysis to compare TI and various GBM dysregulations in disease and aging was performed recently (Lausecker et al., 2022).

In summary, matrisome homeostasis contributes greatly to kidney health and function. Better profiling of different CKD stages, from early to late, to cover the entire spectrum of fibrotic ECM changes is critical to understanding how ECM contributes to the onset, progression, resolution, and regeneration of kidney disease.

## 3 Interplay Between ECM Components and Cellular Molecular Mechanisms in Kidney Fibrosis

Kidney cellular injury resulting from genetic disorders, autoimmune diseases related to immune complexes accumulation, toxins, infections, ischemic/hypertensive nephropathy, or mechanical obstruction could all lead to kidney fibrosis. As such, kidney fibrogenesis is a highly dynamic and complex cellular event depending on the focal fibrogenic microenvironment, namely the specific cell type injury, immune cell infiltration, myofibroblast activation to deposit pathogenic ECMs, incomplete repair of tubular, and other parenchymal cells, leading to partial epithelial-mesenchymal transition (EMT), tubular atrophy, microvascular rarefaction, and glomerular sclerosis if it involves dysregulation of glomerular matrix, in which ECM has significant impact on all steps (Bülow & Boor, 2019; Lausecker et al., 2022; Li et al., 2022; Rayego-Mateos et al., 2021). We discuss this in more detail as follows.

### *3.1 Immune Cell-Fibroblast Circuit in ECM Modulation and Fibrosis*

Inflammation is an important component of fibrosis. Crosstalk of (myo)fibroblasts and immune cells, particularly macrophages, plays prominent roles in kidney injury and inflammation (Jiang et al., 2024; Tang et al., 2019a; Vlasschaert et al., 2022). In addition, resident fibroblasts transdifferentiate to become inflammatory myofibroblasts, serve as fibrosis drivers participating in CKD initiation and progression by orchestrating the formation of "tertiary lymphoid tissue (TLT)," an ectopic site of acquired immune reactions, to promote aberrant inflammation (Arai et al., 2021;

Sato & Yanagita, 2017). Diseased and aged kidneys have altered ECM and are often accompanied by chronic, non-resolving inflammation. Close contact and interaction lead to mutual interdependence between ECM and immune cells. Oxidatively stressed ECM, marked as lower glutathione peroxidase 3 (GPX3), could stimulate fibroblast activation (Li et al., 2023b). Whereas chemical and biophysical properties of ECM modulate immune cell survival, function, mobility, and localization; immune cells could alter ECM compositions and its mechanical properties as well (Bhattacharya & Ramachandran, 2023; Sutherland et al., 2023). For example, GAGs/heparan sulfate bind and present chemokines to direct immune cell chemotaxis, endowed with high affinity towards cytokines to regulate the bioavailability of cytokines, contribute to dynamic changes in ECM compositions, and serve an immune-modulation function on immune processes (Rabelink et al., 2024). In return, immune cells control ECM through mechanisms such as secretion of context-specific cytokines and chemokines, modulation of ECM components, and production of ECM remodeling proteinases. Particularly, subsets of macrophages participate in ECM turnover through mannose receptor-mediated uptake and degradation (Madsen et al., 2013). The bidirectional interplay between macrophage and fibroblast as a mechanism of pathogenesis was recently identified from human biopsies of transplanted kidneys (Setten et al., 2022).

Spatial and single-cell transcriptomics profiling deciphers the landscape of the cellular microenvironment of kidney fibrosis. Tracing the origin of human kidney myofibroblasts had identified distinct pericyte and fibroblast populations are the major source and Nkd2/NFκB signaling in kidney fibrosis (Kuppe et al., 2021). In a murine AKI-to-CKD transition model, a novel subset of macrophages enabling ECM remodeling was identified as persistent and profibrotic macrophages during the chronic stage (Zhang et al., 2024).

There are 12 myeloid cell subsets identified with distinct roles in kidney injury and repair as marked early proinflammatory monocytes (Arg1+), late-in injury (Ccr2+), and repair (MMP12+) (Conway et al., 2020). After kidney injury, the fibrotic scar-associated macrophage (SAMacs, marked as secreted phosphoprotein 1 (SPP1+) macrophage) expands and differentiates upon stimulation by CXCL4, a chemokine mainly secreted from platelets, to orchestrate fibroblast activation (Hoeft et al., 2023). In mouse models of CKD, macrophage assembles a fibrogenic niche enriched in ECM protein vitronectin to mediate fibroblast activation and proliferation, suggesting targeting such microenvironment is a viable therapeutic avenue (Peng et al., 2023). Interestingly, profibrotic renal tubule-basophils interaction contributes to kidney fibrosis as well. In a murine unilateral ureteral obstruction (UUO) kidney fibrotic model, subsets of profibrotic tubule cells secrete CXCL1 to recruit CXCR2+ basophils. These basophils then produce interleukin-6 (IL-6) and further cause differentiation of $T_H17$ cells to mediate downstream kidney fibrosis response, a phenomenon also validated in human CKD samples (Doke et al., 2022) (Fig. 1).

In summary, information derived from human multi-fibrotic organs and multi-omics studies identified several conserved fibrotic pathways involving SPP1+ macrophages (SAMacs) population, a close spatial proximity of SAMacs and fibroblasts in fibrotic foci, modulation of immune cell function by ECM, the key signaling

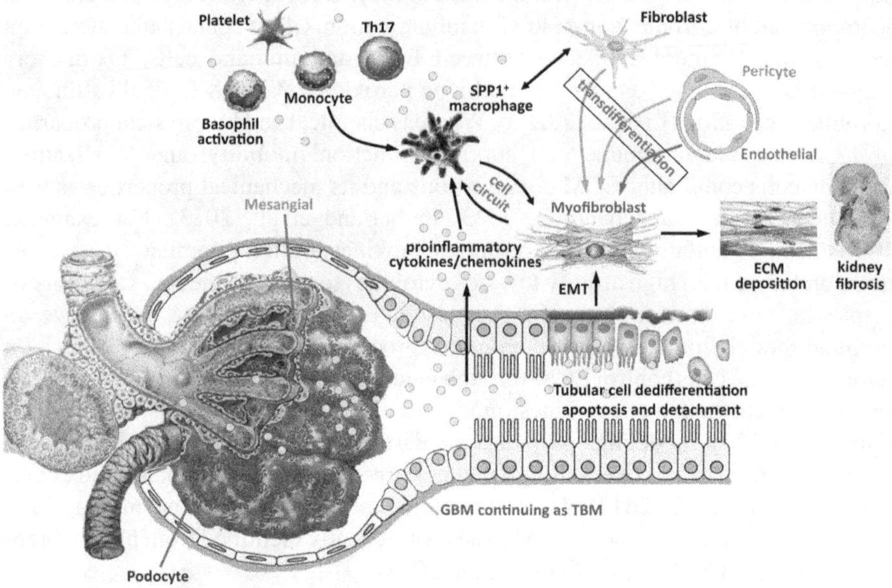


**Fig. 1** The dynamic interactions between immune cells, fibroblasts, and renal parenchymal cells in the formation of fibrotic niche, aberrant ECM, and kidney fibrosis. Kidney injury could cause glomerular damage, tubular cell death detachment, and dedifferentiation, and subsequently, accumulation of aberrant proinflammatory cytokines and chemokines. Circulating immune cells, particularly monocytes, are recruited and differentiate into SAMacs (SPP1+ macrophage) localized within the fibrogenic niche to establish cell circuit promoting myofibroblast trans-differentiation from fibroblasts, pericytes and endothelial cells, and epithelial-mesenchymal transition (EMT). Several inflammatory mediators and signaling pathways are involved in propagating fibrogenic pathways, causing dysregulated ECM and progressive kidney fibrosis

molecules (IL-1β, IL-17A, TGF-β, CXCL4…), the heterogeneities of both myeloid/macrophages populations, the ECM-producing fibroblasts, and capacity of fibroblasts in presenting antigen and activating T cells (Bhattacharya & Ramachandran, 2023; Sutherland et al., 2023).

## 3.2 *TGF-β Signaling in Dysregulation of ECM, EMT, and Activation of Myofibroblast*

Transforming growth factor-beta (TGF-β), a critical profibrotic growth factor in regulating ECM through promoting infiltration or proliferation of preexisting fibroblasts, myofibroblast activation, abnormal ECM deposition in the mesangium and tubulointerstitial, partial epithelial-mesenchymal transition (EMT) in tubular cells, and inhibition of collagenolysis, all promote pathological fibrosis (Higgins et al., 2021; Ma & Meng, 2019; Meng et al., 2016). The bioavailability of TGF-β itself is tightly regulated by its controlled release from its latent complex stored in ECM. In diabetic rats, injured glomeruli expressed high TGF-β to stimulate mesangial and glomerular matrix expansion, a hallmark of diabetic kidney disease (Lee, 2011; Ziyadeh, 2004). The disordered GBM remodeling includes high fibronectin, tenascin, and biglycan (Hu et al., 2015; Nakamura et al., 1992; Yamamoto et al., 1993). TGF-β is also the key regulator of tubulointerstitial (TI) fibrosis from renal injuries of all etiologies (Gewin & Zent, 2012). In CKD patients, necroptosis of injured tubular epithelial cells promotes excessive expression of TGF-β, collagen-1and fibronectin, and activation of fibroblast to myofibroblast (Lin et al., 2022). Mechanistically, TGF-β signaling impacts proximal tubule cell mitochondrial homeostasis, metabolic adaptation, and inflammation in CKD patients and animal models (Kayhan et al., 2023).

During TI fibrosis, TGF-β stimulates tubular EMT, a biological process in which tubular epithelial cells lose their phenotypes and new mesenchymal features are obtained, characterized by disassembly of tubular cell-cell junctions (loss of E-cadherin) and dramatic alterations in the actin cytoskeleton that facilitates cell-matrix adhesion (induction of α-SMA and vimentin) and stimulates migration (Hills & Squires, 2011; Tian et al., 2003). Interestingly, matrisome protein tenascin-C induces tubular cells undergoing partial EMT *in vitro* and contributes to AKI-to-CKD progression, highlighting a significant role of ECM in regulating tubular cell integrity (Zhu et al., 2020). Another important feature of kidney fibrosis is excessive proliferative fibroblast resulting from activation and trans-differentiation into ECM-producing myofibroblasts through cellular crosstalk of fibroblasts with injured proximal tubule and immune cells (Schuster et al., 2023; Yuan et al., 2019).

### 3.3 WNT/β-Catenin Pathways in ECM Regulation and Kidney Fibrosis

The WNT/β-catenin is a signaling cascade involved in embryogenesis, stem cell biology, cell polarity, and migration (Hayat et al., 2022). WNT signaling is generally suppressed in adulthood; however, transient induction of WNT/β-catenin signaling promotes repair and regeneration during acute kidney injury, whereas sustained (uncontrolled) WNT/β-catenin signaling promotes kidney fibrosis, podocyte damage, vascular calcification, and mineral bone disease associated with CKD (Schunk et al., 2021; Zhou et al., 2016). The renal tubular epithelium is the major source of WNT ligands in CKD. Moreover, WNT/β-catenin controls the expression of a wide variety of downstream mediators implicated in kidney fibrosis, such as fibronectin, plasminogen activator inhibitor (PAI)-1, matrix metalloproteinase (MMP)-7, and components of the renin-angiotensin system (He et al., 2012; Li et al., 2017). In a rat chronic allograft nephropathy model, SDF-1/CXCR4 induces tubular EMT via activation of WNT/β-catenin signaling (Tang et al., 2019b). Similar to TGF-β, many tubule-derived WNT signalings are mediators of fibroblast activation in CKD patients (Zhou et al., 2017). Exampled as a WNT-ECM crosstalk, biglycan, a small leucine-rich proteoglycan that is expressed de novo in the interstitial ECM of various CKD, directly interacts with WNT and its coreceptor low-density lipoprotein receptor-related protein 6 (LRP6) to regulate WNT availability for canonical WNT function in kidney fibrosis (Berendsen et al., 2011).

The detailed crosstalk in WNT signaling, ECM, and kidney fibrosis is still emerging. Not only tubular WNT/β-catenin is critical to kidney cell homeostasis, but activation of WNT/β-catenin signaling in macrophage promotes macrophage M2 polarization, highlighting the versatile function of cell type-specific WNT/*β*-catenin signaling in kidney fibrosis (Feng et al., 2018). Interestingly, macrophages supply a plethora of WNT ligands that support macrophage polarization, renal inflammation, and fibrosis after kidney injury (Tian et al., 2024). Although activation of canonical WNT signaling has been implicated in podocyte injury and proteinuria, however, podocyte-derived WNTs could promote non-canonical WNT/$Ca^{2+}$ signaling which protects podocytes from injury via suppressing β-catenin (Zhou et al., 2022a). More in-depth knowledge is required for cell type-specific WNT/β-catenin signaling in kidney homeostasis.

## *3.4 CCN2/CTGF and CKD Progression*

The cellular communication network 2/connective tissue growth factor (CCN2/CTGF) protein is an ECM matricellular protein that plays an important role in cell-cell communication responding to the cell microenvironment. CCN2 is implicated in the regulation of cell growth, apoptosis, EMT, EndMT transition, angiogenesis, inflammation, and fibrogenesis. Upregulation of CCN2 has been linked to various clinical fibrotic CKD (Ramazani et al., 2018). In the kidney, CCN2 binds to tubular epithelial cells and interacts with epidermal growth factor receptor (EGFR) to induce kidney fibrosis in a folic acid-induced AKI-to-CKD animal model (Rayego-Mateos et al., 2022; Rayego-Mateos et al., 2018; Toda et al., 2018).

Structurally, CCN2 is a tetra-modular molecule containing multiple binding sites with multiple binding partners. Full-length CCN2 is susceptible to proteolytic degradation and could be cleaved by a protease to release several different biological fragments from ECM, a concept of matrikine, to control its activities and downstream signaling (Kaasbøll et al., 2018). Different CCN2 modules exert different functions through their various binding partners, such as fibronectin (module I) and TGF-β (module 2) (Rayego-Mateos et al., 2021). The C-terminal CCN2 (module IV), released by MMP2, is implicated in the Th17 inflammatory response related to kidney inflammation and fibrosis (Rayego-Mateos et al., 2018; Rodrigues-Díez et al., 2013). High glucose stimulates interaction of CCN2 with insulin-like growth factor (IGF) to promote TGF-β expression, and ECM collagen deposition from kidney fibroblasts or mesangial cells (Lam et al., 2003). Research efforts have focused on anti-CCN2 antibodies as a therapeutic intervention for early human diabetes; future evaluations are needed to demonstrate whether CCN2 is an antifibrotic target in CKD.

### 3.5 ECM Stiffness in ECM Mechanosensing and CKD Progression

ECM stiffness affects cellular mechano-transduction, a signal mediating the conversion from mechanical signal to biochemical signal, and regulates many cellular activities. Higher ECM stiffness, as shown in denser cross-linkage of hyaluronan, impacts cytokine secretion by PBMC macrophages (Duann & Lin, 2021). Fibrotic ECM exhibits reduced mechanical compliance and increased tissue stiffness due to altered accumulation of fibrillar collagens and the cross-linking of the ECM components (Lausecker et al., 2022). Expression of cross-linking enzymes lysyl oxidase (LOX) and lysyl oxidase-like (LOXL) proteins promotes covalent cross-linking of type IV collagens and elastin in GBM (Pehrsson et al., 2021). LOXL proteins were shown upregulated in diabetic kidney fibrosis (Stangenberg et al., 2018). As such, changes in GBM affect glomerular stiffness and the underlying adhesion and cytoskeletal networks supporting podocyte function. Besides, changes in tubular base membrane stiffing also affect the growth of proximal tubule cells (Wang et al., 2022).

The mechano-transducer Yes-associated protein (YAP) and Transcriptional Coactivator with PDZ-binding Motif (TAZ) are key regulators of this process. In the kidney, YAP can be activated by ECM and transform fibroblast into myofibroblast; the interaction of YAP/TAZ and ECM forms a feed-forward loop resulting in kidney fibrosis (Liang et al., 2017). Moreover, increased stiffness in the fibrotic kidney leads to TGF-β activation which sensitizes Piezo1 calcium channel-mediated nuclear translocation and transcriptional activation of YAP to induce downstream gene expression involving cell directional migration along a matrix rigidity gradient. A process would promote further matrix secretion (Fu et al., 2021; Zhao et al., 2022). YAP/TAZ modulates podocytes' mechano-sensing activity and plays roles as a transcriptional regulator in diabetic kidney disease and focal segment glomerulosclerosis (FSGS) (Endlich et al., 2017; Ester et al., 2023). Recent studies have identified Piezo channels in mediating blood pressure (BP) control, hypertension pathogenesis, and hypertensive kidney damage (Nagase & Nagase, 2024). Targeting ECM stiffness could be a potential anti-fibrotic therapy and there are currently a number of clinical trials on this (Lampi & Reinhart-King, 2018).

### 3.6 ECM Remodeling with MMPs, ADAMs, and TIMPs in CKD Progression

Tissue injury and regeneration are often accompanied by ECM turnover and remodeling. Matrix proteinase metalloproteinases (MMPs) and disintegrin and metalloproteinases (ADAMs) are members of calcium-dependent, zinc-containing endopeptidases that function as soluble or membrane-bound proteases and play a pivotal role in ECM and tissue remodeling. These MMPs, ADAMs participate in tissue development, proliferation, and inflammatory responses, and are engaged in tissue remodeling after tissue injury or during the wound regeneration process (Lin et al., 2017; Rousselle et al., 2019). Dysregulation of MMPs, especially MMP-2, MMP-7, MMP-9, MMP-10, ADAM10, and ADAM17, has been implicated in AKI and CKD (Wozniak et al., 2021). The tissue inhibitors of metalloproteinases (TIMPs) inhibit the activity of MMPs and thus decrease ECM degradation. Various pathophysiological conditions have been linked with the imbalance of ECM synthesis and degradation (Lin et al., 2017; Wozniak et al., 2021). Abnormal high expression of TIMP-1 is associated with kidney fibrosis (Kökény et al., 2022). TIMP-3 is similarly upregulated in the kidney with age, and interestingly, loss of TIMP-3 enhances TI fibrosis and exacerbates diabetic nephropathy (Basu et al., 2012; Kassiri et al., 2009). These highlight the importance of homeostasis of these enzymes in keeping a healthy kidney matrix and function (Wozniak et al., 2021; Zakiyanov et al., 2021).

MMPs also cleave non-ECM substrates such as E-cadherin (an epithelium marker), cellular junction protein ZO1, FAS ligand, nephrin (a podocyte marker) to affect EMT, immune modulatory effect such as immune cells trafficking and recruitment, cell death pathways, tubular and podocyte biology. Moreover, ADAM10 and ADAM17 mediate signaling implicated in TGF-β, EGF, Klotho, Notch, IL-6, and chemotaxis, which are associated with various kidney diseases in glomerulonephritis, lupus nephritis, and kidney fibrosis (Wozniak et al., 2021; Zakiyanov et al., 2021). Some MMPs could be biomarkers for kidney diseases. For example, urinary MMP-7 levels correlate with CKD fibrosis in hypertensive patients (Sarangi et al., 2022), and serum MMP-7 is a biomarker of diabetic kidney disease from an unbiased single-cell transcriptomics and proteomics study (Hirohama et al., 2023). In animal studies, serum pro-MMP10 levels correlate with kidney podocyte function (Zuo et al., 2021). As MMPs, ADAMs, and TIMPs effect is context and therefore disease stage-dependent, therefore development of early intervention and novel therapeutics warrants complete characterization of these molecules.

## 4 ECM Targeting Combined with Stem Cell Treatment as Novel Anti-fibrotic and Regenerative Therapies

Kidney regeneration is hindered by a limited pool of intrinsic reparative cells. Despite extensive experimental data, the currently available strategies only alleviate symptoms or slow the progression of chronic kidney disease (CKD) but do not reverse fibrosis. Therefore, targeting the fibrotic remodeling of kidney tissue to combat CKD presents a promising therapeutic intervention. Recent reviews have examined both current and novel anti-fibrotic treatments, including several clinical trials for various types of CKD (Allinovi et al., 2018; Klinkhammer & Boor, 2023; Ruiz-Ortega et al., 2022). In summary, the approved anti-fibrotic therapies for CKD include drugs that modulate the renin-angiotensin-aldosterone system (RAAS), such as angiotensin-converting enzyme inhibitors (ACEIs), angiotensin II receptor blockers (ARBs), and statins, which likely influence TGF-β expression. Many therapies focus on key fibrogenic mediators at early stages to enhance effectiveness. Here, we highlight some innovative strategies and research trends, aside from those already mentioned in the referenced clinical trials, that have the potential to alter extracellular matrix (ECM) or related signaling in the progression of kidney fibrosis.

## 4.1 *ECM-Based Tissue Engineering Strategies with Decellularized Kidney Tissue Scaffolds*

Recent single-cell RNAseq studies revealed the complexity of human developing and adult kidneys (Schumacher et al., 2021). The kidney has a complex architecture with an intricate vascularization network, and parenchymal cells organized into a specific cortex and medullar compartments with more than 40 highly specialized cell types residing within. Each cell type is endowed with distinct cellular functions and needs a particular niche to maintain a unique cell phenotype. As such, it is still challenging to fabricate bioactive and dynamic ECM scaffolds for therapeutic application. Nevertheless, new developments in ECM-based tissue-engineered strategies have offered advances in bioengineered models of fibrosis, which have the potential to be scaled to implantable tissues for organ replacement in the future (Sobreiro-Almeida et al., 2021a, b). Emerging developments are undergoing to fabricate tunable biomaterials that mimic the 3D and fibrous nature of native ECMs which can be remodeled dynamically by cells to study ECM-cells interaction, cellular reprogramming kinetics, myofibroblast biology, mechanics, migration, cellular mechanosensing, and stem cell technologies, to improve the treatment of ECM-mediated diseases such as CKD.

Decellularization of kidney ECM scaffolds with an application of molecular tools, proteomics, and bioengineering offers a new mechanism in CKD fibrosis (Neishabouri et al., 2022; Yi et al., 2017; Zhang et al., 2022). Li et al. identified nine proteins that are differentially expressed in fibrotic UUO decellularized scaffolds and elucidated a new mechanism of kidney fibrosis, wherein reduced glutathione peroxidase 3 (GPX3) may amplify the nicotinamide adenine-dinucleotide phosphate (NADPH) oxidase 2/oxidative stress/p38 MAP kinase signaling pathway, resulting in fibrosis. Therefore, GPX3 serves as a countermeasure against hypoxia-induced ROS generation, representing a novel therapeutic strategy for CKD.

## 4.2 *Stem Cell Therapies in Bioengineered Scaffolds as Anti-fibrosis and Regeneration Treatment*

Application of bioengineered scaffolds or decellularized kidney matrices in combination with different stem cell therapies is a current trend in organ replacement or fibrosis treatment strategy which also highlights the importance of ECM scaffolds and the induction and interaction with embedding cells (Liu et al., 2021). These have been applied to efficiently generate kidney vasculature from human induced pluripotent stem cell (hiPSCs)-derived endothelial cells (hiPSC-ECs) with acellular rat kidney scaffolds. Successful generation of kidney vasculature with continuously distributed cells along the vessel wall, combined with compartments-specific endothelial cell features such as glomerular capillaries with fenestrae in the endothelium, but not in peritubular capillaries and small vessels (Ciampi et al., 2019). Scaffolds from a natural acellular entity or synthetic biomaterials and cell sheets could enhance the engraftment of exogenously-expanded mesenchymal stromal cells (MSC) for anti-fibrosis (Rockel et al., 2020). These include applications of bone marrow-derived mesenchymal stem cells (BM-MSC) or adipose-derived MSC (AD-MSC) to attenuate renal fibrosis following ischemia-reperfusion induced injury with benefits of diminished myofibroblast transition, decreased ECM accumulation, reduced MMP activity, and inhibited TGF-β/SMAD signaling in damaged kidney (Wang et al., 2023b; Zhou et al., 2020). An engineered BM-MSC cell sheets, retaining intrinsic cell adhesion proteins, alleviate kidney proteinuria and reduce profibrotic mediators in a rat chronic glomerulonephritis model (Wang et al., 2023a). Recent advanced applications include injectable hydrogel cross-links and decellularized porcine ECM to deliver and foster renal progenitor cell bioactivity to promote kidney regeneration (Quinteira et al., 2024).

## *4.3 Extracellular Vesicles as Nanomedicines to Treat Fibrotic Kidney Disease*

Small extracellular vesicles (EVs) are 30–100 nm-sized nanovesicles released by cells into ECM or circulation to affect the local tissue microenvironment or exert a remote effect, respectively. These EVs could be viewed as paracrines or as shuttering vehicles carrying proteins, mRNAs, microRNAs, or noncoding RNAs within lipid bilayer-enclosed vesicles for inter-cellular communication. Many MSCs used for tissue bioengineering shed EVs with anti-inflammation and anti-fibrosis effects. EVs therapeutics (nanomedicine) are based on the premise that EVs shed by stem cells exert similar biosignature which could serve as an alternative to stem cell therapies. EVs are naturally derived nanoparticles with innate biocompatibility, low immunogenicity, and high physiochemical stability which qualify them as potential acellular nanoscale therapeutics for a variety of diseases.

Depending on the origin of host cells, EVs have versatile effects on their specific target cells. Bioengineering EVs have been used in chronic skin wound healing to induce angiogenesis, promote re-epithelization, and reduce scar formation (Ni et al., 2023), and osteoblasts-derived EVs with hydrogel in bone regeneration (Man et al., 2022). EVs can be exploited as agents in renal regenerative medicine (Tang & Liu, 2019) and clinical biomarkers of kidney disease patients (Georgatzakou et al., 2020). Research shows that EVs regulate ECM (Patel et al., 2023). For example, the dual application of MSC-EVs and injectable collagen matrix provides sustained EV stability and a superior therapeutic effect on AKI (Tang & Liu, 2019). Mouse umbilical cord MSC paracrine alleviates kidney fibrosis in a diabetic nephropathy model (Li et al., 2020). Perhaps the biggest excitement (also a challenge) is its potential clinical application in kidney transplantation (Ashcroft et al., 2022). Moreover, advanced studies demonstrated that recombinant engineered multi-functional growth factors, such as angiogenic β-FGF, targeting ECM to reduce renal fibrosis, and facilitate recovery of AKI (Li et al., 2024). EVs thus serve as a versatile tool to modulate ECM and drug targets for kidney repair, fibrosis, and regeneration.

## 5 Conclusions and Future Perspectives

Current antifibrotic therapies aim at halting inflammation, stabilizing the fibrotic process, and relieving symptoms to improve patients' quality of life. However, efficient therapy to limit or reverse the progression of fibrotic CKD is still lacking. The development of efficient therapeutic drugs aiming at the prevention of the progressive loss of renal function requires an in-depth understanding of the complex microenvironment of CKD, ECM compositions, the underlying signaling and cellular crosstalk of matrix biology, cell biology, and immunology. In the past decade, the understanding of the pathophysiology of CKD fibrosis has significantly progressed due to multi-omics profiling of fibrotic tissues, including single-cell RNA-seq analysis, spatial transcriptomics analysis, ECM proteomics research, stem cell biology, tissue biomaterial bioengineering, and nanomedicine development. All these contribute to a better knowledge of the fibrotic niche, the communication between immune cells-fibroblasts, and the local microenvironmental crosstalk during the onset and progression of fibrosis, to help identification of new critical signaling networks for innovative strategies targeting fibrotic disease.

Although researchers already identified some conserved themes/molecular networks from recent studies on fibrotic niches derived from different fibrotic diseases, some targets appeared not specific enough to fully delineate the complexity of signaling networks. To adequately dissect such multicellular complex fibrogenic signaling within the niche, we will rely on the development of advanced molecular tools that enable specific spatial-temporally controlled expression of targeting molecules in a context and cell-type-dependent manner, with precise experimental and computer modeling of the interactions. We currently don't have too much knowledge on matrix remodeling enzyme activities and their association with activation of myofibroblasts or parenchymal cells, nor sufficient information on the synthesis and metabolism of matrikines in the regulation of onset and progression or resolution of CKD fibrotic niche. Furthermore, the impact of tissue crosstalk as exemplified by EVs derived from remote organs could affect inflammation and fibrosis of target organs. Additionally, the development of nanoparticles aims specifically to deliver targeted genetic messengers to modulate immune-fibroblast circuits. We are optimistic that knowledge gained from these studies will provide opportunities translated into novel biomarker identification, new target discovery, and the development of innovative targeted therapeutics for fibrotic CKD, which would maintain matrix homeostasis while inhibiting or reversing uncontrolled pathological ECM accumulation.

**Acknowledgments** Author Pei-Hui Lin is grateful to acknowledge the support of research grants from the National Institutes of Health for grants R01-DK106394, R01-AG062896, and R21-AI166273.

**Competing Interests** The authors declare no competing interests exist.

## References

Adeva-Andany, M. M., & Carneiro-Freire, N. (2022). Biochemical composition of the glomerular extracellular matrix in patients with diabetic kidney disease. *World Journal of Diabetes, 13*(7), 498–520. https://doi.org/10.4239/wjd.v13.i7.498

Allinovi, M., De Chiara, L., Angelotti, M. L., Becherucci, F., & Romagnani, P. (2018). Anti-fibrotic treatments: A review of clinical evidence. *Matrix Biology, 68-69*, 333–354. https://doi.org/10.1016/j.matbio.2018.02.017

Arai, H., Sato, Y., & Yanagita, M. (2021). Fibroblast heterogeneity and tertiary lymphoid tissues in the kidney. *Immunological Reviews, 302*(1), 196–210. https://doi.org/10.1111/imr.12969

Ashcroft, J., Leighton, P., Elliott, T. R., Hosgood, S. A., Nicholson, M. L., & Kosmoliaptsis, V. (2022). Extracellular vesicles in kidney transplantation: a state-of-the-art review. *Kidney International, 101*(3), 485–497. https://doi.org/10.1016/j.kint.2021.10.038

Basu, R., Lee, J., Wang, Z., Patel, V. B., Fan, D., Das, S. K., Liu, G. C., John, R., Scholey, J. W., Oudit, G. Y., & Kassiri, Z. (2012). Loss of TIMP3 selectively exacerbates diabetic nephropathy. *American Journal of Physiology. Renal Physiology, 303*(9), F1341–F1352. https://doi.org/10.1152/ajprenal.00349.2012

Berendsen, A. D., Fisher, L. W., Kilts, T. M., Owens, R. T., Robey, P. G., Gutkind, J. S., & Young, M. F. (2011). Modulation of canonical Wnt signaling by the extracellular matrix component biglycan. *Proceedings of the National Academy of Sciences of the United States of America, 108*(41), 17022–17027. https://doi.org/10.1073/pnas.1110629108

Bhattacharya, M., & Ramachandran, P. (2023). Immunology of human fibrosis. *Nature Immunology, 24*(9), 1423–1433. https://doi.org/10.1038/s41590-023-01551-9

Bowe, B., Xie, Y., Li, T., Mokdad, A. H., Xian, H., Yan, Y., Maddukuri, G., & Al-Aly, Z. (2018). Changes in the US burden of chronic kidney disease From 2002 to 2016: An analysis of The Global Burden of Disease Study. *JAMA Network Open, 1*(7), e184412. https://doi.org/10.1001/jamanetworkopen.2018.4412

Bülow, R. D., & Boor, P. (2019). Extracellular matrix in kidney fibrosis: More than just a scaffold. *The Journal of Histochemistry and Cytochemistry, 67*(9), 643–661. https://doi.org/10.1369/0022155419849388

Byron, A., Randles, M. J., Humphries, J. D., Mironov, A., Hamidi, H., Harris, S., Mathieson, P. W., Saleem, M. A., Satchell, S. C., Zent, R., Humphries, M. J., & Lennon, R. (2014). Glomerular cell cross-talk influences composition and assembly of extracellular matrix. *Journal of the American Society of Nephrology, 25*(5), 953–966. https://doi.org/10.1681/asn.2013070795

Chakravarti, S., Enzo, E., Monteiro, R., de Barros, M., Maffezzoni, M. B. R., & Pellegrini, G. (2022). Genetic disorders of the extracellular matrix: From cell and gene therapy to future applications in regenerative medicine. *Annual Review of Genomics and Human Genetics, 23*, 193–222. https://doi.org/10.1146/annurev-genom-083117-021702

Ciampi, O., Bonandrini, B., Derosas, M., Conti, S., Rizzo, P., Benedetti, V., Figliuzzi, M., Remuzzi, A., Benigni, A., Remuzzi, G., & Tomasoni, S. (2019). Engineering the vasculature of decellularized rat kidney scaffolds using human induced pluripotent stem cell-derived endothelial cells. *Scientific Reports, 9*(1), 8001. https://doi.org/10.1038/s41598-019-44393-y

Conway, B. R., O'Sullivan, E. D., Cairns, C., O'Sullivan, J., Simpson, D. J., Salzano, A., Connor, K., Ding, P., Humphries, D., Stewart, K., Teenan, O., Pius, R., Henderson, N. C., Bénézech, C., Ramachandran, P., Ferenbach, D., Hughes, J., Chandra, T., & Denby, L. (2020). Kidney single-cell atlas reveals myeloid heterogeneity in progression and regression of kidney disease. *Journal of the American Society of Nephrology, 31*(12), 2833–2854. https://doi.org/10.1681/asn.2020060806

Docs, J., Kovacs, G., & Peterfi, L. (2023). End-stage kidney disease: A never healing wound leading to another never healing wound, renal cancer. *Journal of Nephrology, 36*(6), 1673–1681. https://doi.org/10.1007/s40620-023-01694-w

Doke, T., Abedini, A., Aldridge, D. L., Yang, Y. W., Park, J., Hernandez, C. M., Balzer, M. S., Shrestra, R., Coppock, G., Rico, J. M. I., Han, S. Y., Kim, J., Xin, S., Piliponsky, A. M., Angelozzi, M., Lefebvre, V., Siracusa, M. C., Hunter, C. A., & Susztak, K. (2022). Single-cell analysis identifies the interaction of altered renal tubules with basophils orchestrating kidney fibrosis. *Nature Immunology, 23*(6), 947–959. https://doi.org/10.1038/s41590-022-01200-7

Duann, P., & Lin, P. H. (2021). Wound matrix stiffness imposes on macrophage activation. *Methods in Molecular Biology, 2193*, 111–120. https://doi.org/10.1007/978-1-0716-0845-6_11

Eckersley, A., Morais, M. R., Ozols, M., & Lennon, R. (2023). Peptide location fingerprinting identifies structural alterations within basement membrane components in ageing kidney. *Matrix Biology*. https://doi.org/10.1016/j.matbio.2023.07.001

Endlich, K., Kliewe, F., & Endlich, N. (2017). Stressed podocytes-mechanical forces, sensors, signaling and response. *Pflügers Archiv – European Journal of Physiology, 469*(7-8), 937–949. https://doi.org/10.1007/s00424-017-2025-8

Ester, L., Cabrita, I., Ventzke, M., Kieckhöfer, E., Christodoulou, M., Mandel, A. M., Diefenhardt, P., Fabretti, F., Benzing, T., Habbig, S., & Schermer, B. (2023). The role of the FSGS disease gene product and nuclear pore protein NUP205 in regulating nuclear localization and activity of transcriptional regulators YAP and TAZ. *Human Molecular Genetics*. https://doi.org/10.1093/hmg/ddad135

Feng, Y., Ren, J., Gui, Y., Wei, W., Shu, B., Lu, Q., Xue, X., Sun, X., He, W., Yang, J., & Dai, C. (2018). Wnt/β-catenin-promoted macrophage alternative activation contributes to kidney fibrosis. *Journal of the American Society of Nephrology, 29*(1), 182–193. https://doi.org/10.1681/asn.2017040391

Feng, D., Ngov, C., Henley, N., Boufaied, N., & Gerarduzzi, C. (2019). Characterization of matricellular protein expression signatures in mechanistically diverse mouse models of kidney injury. *Scientific Reports, 9*(1), 16736. https://doi.org/10.1038/s41598-019-52961-5

Fu, Y., Wan, P., Zhang, J., Li, X., Xing, J., Zou, Y., Wang, K., Peng, H., Zhu, Q., Cao, L., & Zhai, X. (2021). Targeting mechanosensitive Piezo1 alleviated renal fibrosis through p38MAPK-YAP pathway. *Frontiers in Cell and Development Biology, 9*, 741060. https://doi.org/10.3389/fcell.2021.741060

Georgatzakou, H. T., Pavlou, E. G., Papageorgiou, E. G., Papassideri, I. S., Kriebardis, A. G., & Antonelou, M. H. (2020). The multi-faced extracellular vesicles in the plasma of chronic kidney disease patients. *Frontiers in Cell and Development Biology, 8*, 227. https://doi.org/10.3389/fcell.2020.00227

Gewin, L., & Zent, R. (2012). How does TGF-β mediate tubulointerstitial fibrosis? *Seminars in Nephrology, 32*(3), 228–235. https://doi.org/10.1016/j.semnephrol.2012.04.001

Hayat, R., Manzoor, M., & Hussain, A. (2022). Wnt signaling pathway: A comprehensive review. *Cell Biology International, 46*(6), 863–877. https://doi.org/10.1002/cbin.11797

He, W., Tan, R. J., Li, Y., Wang, D., Nie, J., Hou, F. F., & Liu, Y. (2012). Matrix metalloproteinase-7 as a surrogate marker predicts renal Wnt/β-catenin activity in CKD. *Journal of the American Society of Nephrology, 23*(2), 294–304. https://doi.org/10.1681/asn.2011050490

Higgins, C. E., Tang, J., Higgins, S. P., Gifford, C. C., Mian, B. M., Jones, D. M., Zhang, W., Costello, A., Conti, D. J., Samarakoon, R., & Higgins, P. J. (2021). The genomic response to TGF-β1 dictates failed repair and progression of fibrotic disease in the obstructed kidney. *Frontiers in Cell and Development Biology, 9*, 678524. https://doi.org/10.3389/fcell.2021.678524

Hills, C. E., & Squires, P. E. (2011). The role of TGF-β and epithelial-to mesenchymal transition in diabetic nephropathy. *Cytokine & Growth Factor Reviews, 22*(3), 131–139. https://doi.org/10.1016/j.cytogfr.2011.06.002

Hirohama, D., Abedini, A., Moon, S., Surapaneni, A., Dillon, S. T., Vassalotti, A., Liu, H., Doke, T., Martinez, V., Md Dom, Z., Karihaloo, A., Palmer, M. B., Coresh, J., Grams, M. E., Niewczas, M. A., & Susztak, K. (2023). Unbiased human kidney tissue proteomics identifies matrix metalloproteinase 7 as a kidney disease biomarker. *Journal of the American Society of Nephrology, 34*(7), 1279–1291. https://doi.org/10.1681/asn.0000000000000141

Hobeika, L., Barati, M. T., Caster, D. J., McLeish, K. R., & Merchant, M. L. (2017). Characterization of glomerular extracellular matrix by proteomic analysis of laser-captured microdissected glomeruli. *Kidney International, 91*(2), 501–511. https://doi.org/10.1016/j.kint.2016.09.044

Hoeft, K., Schaefer, G. J. L., Kim, H., Schumacher, D., Bleckwehl, T., Long, Q., Klinkhammer, B. M., Peisker, F., Koch, L., Nagai, J., Halder, M., Ziegler, S., Liehn, E., Kuppe, C., Kranz, J., Menzel, S., Costa, I., Wahida, A., Boor, P., Schneider, R. K., Hayat, S., & Kramann, R. (2023). Platelet-instructed SPP1(+) macrophages drive myofibroblast activation in fibrosis in a CXCL4-dependent manner. *Cell Reports, 42*(2), 112131. https://doi.org/10.1016/j.celrep.2023.112131

Hu, C., Sun, L., Xiao, L., Han, Y., Fu, X., Xiong, X., Xu, X., Liu, Y., Yang, S., Liu, F., & Kanwar, Y. S. (2015). Insights into the mechanisms involved in the expression and regulation of extracellular matrix proteins in diabetic nephropathy. *Current Medicinal Chemistry, 22*(24), 2858–2870. https://doi.org/10.2174/0929867322666150625095407

Huang, R., Fu, P., & Ma, L. (2023). Kidney fibrosis: From mechanisms to therapeutic medicines. *Signal Transduction and Targeted Therapy, 8*(1), 129. https://doi.org/10.1038/s41392-023-01379-7

Jiang, Y., Cai, R., Huang, Y., Zhu, L., Xiao, L., Wang, C., & Wang, L. (2024). Macrophages in organ fibrosis: From pathogenesis to therapeutic targets. *Cell Death Discovery, 10*(1), 487. https://doi.org/10.1038/s41420-024-02247-1

Kaasbøll, O. J., Gadicherla, A. K., Wang, J. H., Monsen, V. T., Hagelin, E. M. V., Dong, M. Q., & Attramadal, H. (2018). Connective tissue growth factor (CCN2) is a matricellular preproprotein controlled by proteolytic activation. *The Journal of Biological Chemistry, 293*(46), 17953–17970. https://doi.org/10.1074/jbc.RA118.004559

Kalantar-Zadeh, K., Jafar, T. H., Nitsch, D., Neuen, B. L., & Perkovic, V. (2021). Chronic kidney disease. *Lancet, 398*(10302), 786–802. https://doi.org/10.1016/s0140-6736(21)00519-5

Kassiri, Z., Oudit, G. Y., Kandalam, V., Awad, A., Wang, X., Ziou, X., Maeda, N., Herzenberg, A. M., & Scholey, J. W. (2009). Loss of TIMP3 enhances interstitial nephritis and fibrosis. *Journal of the American Society of Nephrology, 20*(6), 1223–1235. https://doi.org/10.1681/asn.2008050492

Kato, M., & Natarajan, R. (2019). Epigenetics and epigenomics in diabetic kidney disease and metabolic memory. Nature reviews. *Nephrology, 15*(6), 327–345. https://doi.org/10.1038/s41581-019-0135-6

Kayhan, M., Vouillamoz, J., Rodriguez, D. G., Bugarski, M., Mitamura, Y., Gschwend, J., Schneider, C., Hall, A., Legouis, D., Akdis, C. A., Peter, L., Rehrauer, H., Gewin, L., Wenger, R. H., & Khodo, S. N. (2023). Intrinsic TGF-β signaling attenuates proximal tubule mitochondrial injury and inflammation in chronic kidney disease. *Nature Communications, 14*(1), 3236. https://doi.org/10.1038/s41467-023-39050-y

Klinkhammer, B. M., & Boor, P. (2023). Kidney fibrosis: Emerging diagnostic and therapeutic strategies. *Molecular Aspects of Medicine, 93*, 101206. https://doi.org/10.1016/j.mam.2023.101206

Kökény, G., Németh, Á., Kopp, J. B., Chen, W., Oler, A. J., Manzéger, A., Rosivall, L., & Mózes, M. M. (2022). Susceptibility to kidney fibrosis in mice is associated with early growth response-2 protein and tissue inhibitor of metalloproteinase-1 expression. *Kidney International, 102*(2), 337–354. https://doi.org/10.1016/j.kint.2022.03.029

Kuppe, C., Ibrahim, M. M., Kranz, J., Zhang, X., Ziegler, S., Perales-Patón, J., Jansen, J., Reimer, K. C., Smith, J. R., Dobie, R., Wilson-Kanamori, J. R., Halder, M., Xu, Y., Kabgani, N., Kaesler, N., Klaus, M., Gernhold, L., Puelles, V. G., Huber, T. B., Boor, P., Menzel, S., Hoogenboezem, R. M., Bindels, E. M. J., Steffens, J., Floege, J., Schneider, R. K., Saez-Rodriguez, J., Henderson, N. C., & Kramann, R. (2021). Decoding myofibroblast origins in human kidney fibrosis. *Nature, 589*(7841), 281–286. https://doi.org/10.1038/s41586-020-2941-1

Lam, S., van der Geest, R. N., Verhagen, N. A., van Nieuwenhoven, F. A., Blom, I. E., Aten, J., Goldschmeding, R., Daha, M. R., & van Kooten, C. (2003). Connective tissue growth factor and igf-I are produced by human renal fibroblasts and cooperate in the induction of collagen production by high glucose. *Diabetes, 52*(12), 2975–2983. https://doi.org/10.2337/diabetes.52.12.2975

Lampi, M. C., & Reinhart-King, C. A. (2018). Targeting extracellular matrix stiffness to attenuate disease: From molecular mechanisms to clinical trials. *Science Translational Medicine, 10*(422). https://doi.org/10.1126/scitranslmed.aao0475

Lausecker, F., Lennon, R., & Randles, M. J. (2022). The kidney matrisome in health, aging, and disease. *Kidney International, 102*(5), 1000–1012. https://doi.org/10.1016/j.kint.2022.06.029

LeBleu, V. S., Kanasaki, K., Lovisa, S., Alge, J. L., Kim, J., Chen, Y., Teng, Y., Gerami-Naini, B., Sugimoto, H., Kato, N., Revuelta, I., Grau, N., Sleeman, J. P., Taduri, G., Kizu, A., Rafii, S., Hochedlinger, K., Quaggin, S. E., & Kalluri, R. (2024). Genetic reprogramming with stem cells regenerates glomerular epithelial podocytes in Alport syndrome. *Life Science Alliance, 7*(6) https://doi.org/10.26508/lsa.202402664

Lee, H. S. (2011). Pathogenic role of TGF-β in the progression of podocyte diseases. *Histology and Histopathology, 26*(1), 107–116. https://doi.org/10.14670/hh-26.107

Lennon, R., Byron, A., Humphries, J. D., Randles, M. J., Carisey, A., Murphy, S., Knight, D., Brenchley, P. E., Zent, R., & Humphries, M. J. (2014). Global analysis reveals the complexity of the human glomerular extracellular matrix. *Journal of the American Society of Nephrology, 25*(5), 939–951. https://doi.org/10.1681/asn.2013030233

Li, Z., Zhou, L., Wang, Y., Miao, J., Hong, X., Hou, F. F., & Liu, Y. (2017). (Pro)renin receptor is an amplifier of Wnt/β-catenin signaling in kidney injury and fibrosis. *Journal of the American Society of Nephrology, 28*(8), 2393–2408. https://doi.org/10.1681/asn.2016070811

Li, H., Rong, P., Ma, X., Nie, W., Chen, Y., Zhang, J., Dong, Q., Yang, M., & Wang, W. (2020). Mouse umbilical cord mesenchymal stem cell paracrine alleviates renal fibrosis in diabetic nephropathy by reducing myofibroblast transdifferentiation and cell proliferation and upregulating MMPs in mesangial cells. *Journal Diabetes Research, 2020*, 3847171. https://doi.org/10.1155/2020/3847171

Li, L., Fu, H., & Liu, Y. (2022). The fibrogenic niche in kidney fibrosis: Components and mechanisms. *Nature Reviews. Nephrology, 18*(9), 545–557. https://doi.org/10.1038/s41581-022-00590-z

Li, L., He, M., Tang, X., Huang, J., Li, J., Hong, X., Fu, H., & Liu, Y. (2023a). Proteomic landscape of the extracellular matrix in the fibrotic kidney. *Kidney International, 103*(6), 1063–1076. https://doi.org/10.1016/j.kint.2023.01.021

Li, L., Lu, M., Peng, Y., Huang, J., Tang, X., Chen, J., Li, J., Hong, X., He, M., Fu, H., Liu, R., Hou, F. F., Zhou, L., & Liu, Y. (2023b). Oxidatively stressed extracellular microenvironment drives fibroblast activation and kidney fibrosis. *Redox Biology, 67*, 102868. https://doi.org/10.1016/j.redox.2023.102868

Li, X., Shi, C., Zhou, R., Chen, X., Xu, Q., Zhao, C., Ma, M., Ao, X., & Liu, Y. (2024). Modified EBP-bFGF targeting endogenous renal extracellular matrix protects against renal ischemia-reperfusion injury in rats. *Journal of Biomedical Materials Research. Part A, 112*(10), 1827–1839. https://doi.org/10.1002/jbm.a.37730

Liang, M., Yu, M., Xia, R., Song, K., Wang, J., Luo, J., Chen, G., & Cheng, J. (2017). Yap/Taz deletion in Gli(+) cell-derived myofibroblasts attenuates fibrosis. *Journal of the American Society of Nephrology, 28*(11), 3278–3290. https://doi.org/10.1681/asn.2015121354

Lin, P. H., Sermersheim, M., Li, H., Lee, P. H. U., Steinberg, S. M., & Ma, J. (2017). Zinc in wound healing modulation. *Nutrients, 10*(1). https://doi.org/10.3390/nu10010016

Lin, Z., Chen, A., Cui, H., Shang, R., Su, T., Li, X., Wang, K., Yang, J., Gao, K., Lv, J., Shen, J., Wang, S., Qi, Y., Guo, M., & Zhu, Y. (2022). Renal tubular epithelial cell necroptosis promotes tubulointerstitial fibrosis in patients with chronic kidney disease. *The FASEB Journal, 36*(12), e22625. https://doi.org/10.1096/fj.202200706RR

Lipp, S. N., Jacobson, K. R., Hains, D. S., Schwarderer, A. L., & Calve, S. (2021). 3D mapping reveals a complex and transient interstitial matrix during murine kidney development. *Journal of the American Society of Nephrology, 32*(7), 1649–1665. https://doi.org/10.1681/asn.2020081204

Liu, Y., Su, Y. Y., Yang, Q., & Zhou, T. (2021). Stem cells in the treatment of renal fibrosis: A review of preclinical and clinical studies of renal fibrosis pathogenesis. *Stem Cell Research & Therapy, 12*(1), 333. https://doi.org/10.1186/s13287-021-02391-w

Ma, T. T., & Meng, X. M. (2019). TGF-β/Smad and renal fibrosis. *Advances in Experimental Medicine and Biology, 1165*, 347–364. https://doi.org/10.1007/978-981-13-8871-2_16

Madsen, D. H., Leonard, D., Masedunskas, A., Moyer, A., Jürgensen, H. J., Peters, D. E., Amornphimoltham, P., Selvaraj, A., Yamada, S. S., Brenner, D. A., Burgdorf, S., Engelholm, L. H., Behrendt, N., Holmbeck, K., Weigert, R., & Bugge, T. H. (2013). M2-like macrophages are responsible for collagen degradation through a mannose receptor-mediated pathway. *The Journal of Cell Biology, 202*(6), 951–966. https://doi.org/10.1083/jcb.201301081

Man, K., Barroso, I. A., Brunet, M. Y., Peacock, B., Federici, A. S., Hoey, D. A., & Cox, S. C. (2022). Controlled release of epigenetically-enhanced extracellular vesicles from a GelMA/nanoclay composite hydrogel to promote bone repair. *International Journal of Molecular Sciences, 23*(2). https://doi.org/10.3390/ijms23020832

McCabe, M. C., Saviola, A. J., & Hansen, K. C. (2023). Mass spectrometry-based atlas of extracellular matrix proteins across 25 mouse organs. *Journal of Proteome Research, 22*(3), 790–801. https://doi.org/10.1021/acs.jproteome.2c00526

Meng, X. M., Nikolic-Paterson, D. J., & Lan, H. Y. (2016). TGF-β: The master regulator of fibrosis. Nature reviews. *Nephrology, 12*(6), 325–338. https://doi.org/10.1038/nrneph.2016.48

Merchant, M. L., Barati, M. T., Caster, D. J., Hata, J. L., Hobeika, L., Coventry, S., Brier, M. E., Wilkey, D. W., Li, M., Rood, I. M., Deegens, J. K., Wetzels, J. F., Larsen, C. P., Troost, J. P., Hodgin, J. B., Mariani, L. H., Kretzler, M., Klein, J. B., & McLeish, K. R. (2020). Proteomic analysis identifies distinct glomerular extracellular matrix in collapsing focal segmental glomerulosclerosis. *Journal of the American Society of Nephrology, 31*(8), 1883–1904. https://doi.org/10.1681/asn.2019070696

Moeller, M. J., Kramann, R., Lammers, T., Hoppe, B., Latz, E., Ludwig-Portugall, I., Boor, P., Floege, J., Kurts, C., Weiskirchen, R., & Ostendorf, T. (2021). New aspects of kidney fibrosis-from mechanisms of injury to modulation of disease. *Frontiers in Medicine, 8*, 814497. https://doi.org/10.3389/fmed.2021.814497

Moretti, L., Stalfort, J., Barker, T. H., & Abebayehu, D. (2022). The interplay of fibroblasts, the extracellular matrix, and inflammation in scar formation. *The Journal of Biological Chemistry, 298*(2), 101530. https://doi.org/10.1016/j.jbc.2021.101530

Nagase, T., & Nagase, M. (2024). Piezo ion channels: Long-sought-after mechanosensors mediating hypertension and hypertensive nephropathy. *Hypertension Research, 47*(10), 2786–2799. https://doi.org/10.1038/s41440-024-01820-6

Nakamura, T., Miller, D., Ruoslahti, E., & Border, W. A. (1992). Production of extracellular matrix by glomerular epithelial cells is regulated by transforming growth factor-beta 1. *Kidney International, 41*(5), 1213–1221. https://doi.org/10.1038/ki.1992.183

Naylor, R. W., Morais, M., & Lennon, R. (2021). Complexities of the glomerular basement membrane. Nature reviews. *Nephrology, 17*(2), 112–127. https://doi.org/10.1038/s41581-020-0329-y

Neishabouri, A., Soltani Khaboushan, A., Daghigh, F., Kajbafzadeh, A. M., & Majidi Zolbin, M. (2022). Decellularization in tissue engineering and regenerative medicine: Evaluation, modification, and application methods. *Frontiers in Bioengineering and Biotechnology, 10*, 805299. https://doi.org/10.3389/fbioe.2022.805299

Ni, H., Xi, J., Tang, J., Yan, Y., Chu, Y., & Zhou, J. (2023). Therapeutic potential of extracellular vesicles from different stem cells in chronic wound healing. *Stem Cell Reviews and Reports, 19*(6), 1596–1614. https://doi.org/10.1007/s12015-023-10540-2

Panizo, S., Martínez-Arias, L., Alonso-Montes, C., Cannata, P., Martín-Carro, B., Fernández-Martín, J. L., Naves-Díaz, M., Carrillo-López, N., & Cannata-Andía, J. B. (2021). Fibrosis in chronic kidney disease: Pathogenesis and consequences. *International Journal of Molecular Sciences, 22*(1). https://doi.org/10.3390/ijms22010408

Patel, N. J., Ashraf, A., & Chung, E. J. (2023). Extracellular vesicles as regulators of the extracellular matrix. *Bioengineering (Basel), 10*(2). https://doi.org/10.3390/bioengineering10020136

Paunas, F. T. I., Finne, K., Leh, S., Osman, T. A., Marti, H. P., Berven, F., & Vikse, B. E. (2019). Characterization of glomerular extracellular matrix in IgA nephropathy by proteomic analysis of laser-captured microdissected glomeruli. *BMC Nephrology, 20*(1), 410. https://doi.org/10.1186/s12882-019-1598-1

Pehrsson, M., Mortensen, J. H., Manon-Jensen, T., Bay-Jensen, A. C., Karsdal, M. A., & Davies, M. J. (2021). Enzymatic cross-linking of collagens in organ fibrosis – Resolution and assessment. *Expert Review of Molecular Diagnostics, 21*(10), 1049–1064. https://doi.org/10.1080/14737159.2021.1962711

Peng, Y., Li, L., Shang, J., Zhu, H., Liao, J., Hong, X., Hou, F. F., Fu, H., & Liu, Y. (2023). Macrophage promotes fibroblast activation and kidney fibrosis by assembling a vitronectin-enriched microenvironment. *Theranostics, 13*(11), 3897–3913. https://doi.org/10.7150/thno.85250

Quinteira, R., Gimondi, S., Monteiro, N. O., Sobreiro-Almeida, R., Lasagni, L., Romagnani, P., & Neves, N. M. (2024). Decellularized kidney extracellular matrix-based hydrogels for renal tissue engineering. *Acta Biomaterialia, 180*, 295–307. https://doi.org/10.1016/j.actbio.2024.04.026

Rabelink, T. J., Wang, G., van der Vlag, J., & van den Berg, B. M. (2024). The roles of hyaluronan in kidney development, physiology and disease. *Nature Reviews. Nephrology, 20*(12), 822–832. https://doi.org/10.1038/s41581-024-00883-5

Ramazani, Y., Knops, N., Elmonem, M. A., Nguyen, T. Q., Arcolino, F. O., van den Heuvel, L., Levtchenko, E., Kuypers, D., & Goldschmeding, R. (2018). Connective tissue growth factor (CTGF) from basics to clinics. *Matrix Biology, 68-69*, 44–66. https://doi.org/10.1016/j.matbio.2018.03.007

Randles, M. J., Woolf, A. S., Huang, J. L., Byron, A., Humphries, J. D., Price, K. L., Kolatsi-Joannou, M., Collinson, S., Denny, T., Knight, D., Mironov, A., Starborg, T., Korstanje, R., Humphries, M. J., Long, D. A., & Lennon, R. (2015). Genetic background is a key determinant of glomerular extracellular matrix composition and organization. *Journal of the American Society of Nephrology, 26*(12), 3021–3034. https://doi.org/10.1681/asn.2014040419

Randles, M. J., Humphries, M. J., & Lennon, R. (2017). Proteomic definitions of basement membrane composition in health and disease. *Matrix Biology, 57–58*, 12–28. https://doi.org/10.1016/j.matbio.2016.08.006

Randles, M. J., Lausecker, F., Kong, Q., Suleiman, H., Reid, G., Kolatsi-Joannou, M., Davenport, B., Tian, P., Falcone, S., Potter, P., Van Agtmael, T., Norman, J. T., Long, D. A., Humphries, M. J., Miner, J. H., & Lennon, R. (2021). Identification of an altered matrix signature in kidney aging and disease. *Journal of the American Society of Nephrology, 32*(7), 1713–1732. https://doi.org/10.1681/asn.2020101442

Rayego-Mateos, S., Morgado-Pascual, J. L., Rodrigues-Diez, R. R., Rodrigues-Diez, R., Falke, L. L., Mezzano, S., Ortiz, A., Egido, J., Goldschmeding, R., & Ruiz-Ortega, M. (2018). Connective tissue growth factor induces renal fibrosis via epidermal growth factor receptor activation. *The Journal of Pathology, 244*(2), 227–241. https://doi.org/10.1002/path.5007

Rayego-Mateos, S., Campillo, S., Rodrigues-Diez, R. R., Tejera-Muñoz, A., Marquez-Exposito, L., Goldschmeding, R., Rodríguez-Puyol, D., Calleros, L., & Ruiz-Ortega, M. (2021). Interplay between extracellular matrix components and cellular and molecular mechanisms in kidney fibrosis. *Clinical Science (London, England), 135*(16), 1999–2029. https://doi.org/10.1042/cs20201016

Rayego-Mateos, S., Morgado-Pascual, J. L., Lavoz, C., Rodrigues-Díez, R. R., Márquez-Expósito, L., Tejera-Muñoz, A., Tejedor-Santamaría, L., Rubio-Soto, I., Marchant, V., & Ruiz-Ortega, M. (2022). CCN2 binds to tubular epithelial cells in the kidney. *Biomolecules, 12*(2). https://doi.org/10.3390/biom12020252

Rende, U., Ahn, S. B., Adhikari, S., Moh, E. S. X., Pollock, C. A., Saad, S., & Guller, A. (2023). Deciphering the kidney matrisome: Identification and quantification of renal extracellular matrix proteins in healthy mice. *International Journal of Molecular Sciences, 24*(3). https://doi.org/10.3390/ijms24032827

Rinschen, M. M., & Saez-Rodriguez, J. (2021). The tissue proteome in the multi-omic landscape of kidney disease. Nature reviews. *Nephrology, 17*(3), 205–219. https://doi.org/10.1038/s41581-020-00348-5

Rockel, J. S., Rabani, R., & Viswanathan, S. (2020). Anti-fibrotic mechanisms of exogenously-expanded mesenchymal stromal cells for fibrotic diseases. *Seminars in Cell & Developmental Biology, 101*, 87–103. https://doi.org/10.1016/j.semcdb.2019.10.014

Rodrigues-Díez, R., Rodrigues-Díez, R. R., Rayego-Mateos, S., Suarez-Alvarez, B., Lavoz, C., Stark Aroeira, L., Sánchez-López, E., Orejudo, M., Alique, M., Lopez-Larrea, C., Ortiz, A., Egido, J., & Ruiz-Ortega, M. (2013). The C-terminal module IV of connective tissue growth factor is a novel immune modulator of the Th17 response. *Laboratory Investigation, 93*(7), 812–824. https://doi.org/10.1038/labinvest.2013.67

Rousselle, P., Montmasson, M., & Garnier, C. (2019). Extracellular matrix contribution to skin wound re-epithelialization. *Matrix Biology, 75-76*, 12–26. https://doi.org/10.1016/j.matbio.2018.01.002

Rousselle, T., Bardhi, E., Maluf, D. G., & Mas, V. R. (2021). Epigenetic modifications and the development of kidney graft fibrosis. *Current Opinion in Organ Transplantation, 26*(1), 1–9. https://doi.org/10.1097/mot.0000000000000839

Ruiz-Ortega, M., Lamas, S., & Ortiz, A. (2022). Antifibrotic agents for the management of CKD: A review. *American Journal of Kidney Diseases, 80*(2), 251–263. https://doi.org/10.1053/j.ajkd.2021.11.010

Sarangi, R., Tripathy, K. P., Bahinipati, J., Gupta, P., Pathak, M., Mahapatra, S., & Mohapatra, S. R. (2022). Urinary MMP-7: A predictive, noninvasive early marker for chronic kidney disease development in patients with hypertension. *Laboratoriums Medizin, 53*(4), 386–393. https://doi.org/10.1093/labmed/lmac003

Sato, Y., & Yanagita, M. (2017). Resident fibroblasts in the kidney: A major driver of fibrosis and inflammation. *Inflammation and Regeneration, 37*, 17. https://doi.org/10.1186/s41232-017-0048-3

Schumacher, A., Rookmaaker, M. B., Joles, J. A., Kramann, R., Nguyen, T. Q., van Griensven, M., & LaPointe, V. L. S. (2021). Defining the variety of cell types in developing and adult human kidneys by single-cell RNA sequencing. *NPJ Regenerative Medicine, 6*(1), 45. https://doi.org/10.1038/s41536-021-00156-w

Schunk, S. J., Floege, J., Fliser, D., & Speer, T. (2021). WNT-β-catenin signalling – A versatile player in kidney injury and repair. Nature reviews. *Nephrology, 17*(3), 172–184. https://doi.org/10.1038/s41581-020-00343-w

Schuster, R., Younesi, F., Ezzo, M., & Hinz, B. (2023). The role of myofibroblasts in physiological and pathological tissue repair. *Cold Spring Harbor Perspectives in Biology, 15*(1). https://doi.org/10.1101/cshperspect.a041231

Setten, E., Castagna, A., Nava-Sedeño, J. M., Weber, J., Carriero, R., Reppas, A., Volk, V., Schmitz, J., Gwinner, W., Hatzikirou, H., Feuerhake, F., & Locati, M. (2022). Understanding fibrosis pathogenesis via modeling macrophage-fibroblast interplay in immune-metabolic context. *Nature Communications, 13*(1), 6499. https://doi.org/10.1038/s41467-022-34241-5

Sobreiro-Almeida, R., Gómez-Florit, M., Quinteira, R., Reis, R. L., Gomes, M. E., & Neves, N. M. (2021a). Decellularized kidney extracellular matrix bioinks recapitulate renal 3D micro-environment in vitro. *Biofabrication, 13*(4). https://doi.org/10.1088/1758-5090/ac0fca

Sobreiro-Almeida, R., Quinteira, R., & Neves, N. M. (2021b). Renal regeneration: The role of extracellular matrix and current ECM-based tissue engineered strategies. *Advanced Healthcare Materials, 10*(14), e2100160. https://doi.org/10.1002/adhm.202100160

Stangenberg, S., Saad, S., Schilter, H. C., Zaky, A., Gill, A., Pollock, C. A., & Wong, M. G. (2018). Lysyl oxidase-like 2 inhibition ameliorates glomerulosclerosis and albuminuria in diabetic nephropathy. *Scientific Reports, 8*(1), 9423. https://doi.org/10.1038/s41598-018-27462-6

Stone, R. C., Chen, V., Burgess, J., Pannu, S., & Tomic-Canic, M. (2020). Genomics of human fibrotic diseases: Disordered wound healing response. *International Journal of Molecular Sciences, 21*(22). https://doi.org/10.3390/ijms21228590

Sutherland, T. E., Dyer, D. P., & Allen, J. E. (2023). The extracellular matrix and the immune system: A mutually dependent relationship. *Science, 379*(6633), eabp8964. https://doi.org/10.1126/science.abp8964

Tampe, D., Schridde, L., Korsten, P., Ströbel, P., Zeisberg, M., Hakroush, S., & Tampe, B. (2021). Different patterns of kidney fibrosis are indicative of injury to distinct renal compartments. *Cells, 10*(8). https://doi.org/10.3390/cells10082014

Tang, T. T., & Liu, B. C. (2019). Extracellular vesicles: Opportunities and challenges for the treatment of renal fibrosis. *Advances in Experimental Medicine and Biology, 1165*, 693–709. https://doi.org/10.1007/978-981-13-8871-2_34

Tang, P. M., Nikolic-Paterson, D. J., & Lan, H. Y. (2019a). Macrophages: versatile players in renal inflammation and fibrosis. Nature reviews. *Nephrology, 15*(3), 144–158. https://doi.org/10.1038/s41581-019-0110-2

Tang, H., Xu, Y., Zhang, Z., Zeng, S., Dong, W., Jiao, W., & Hu, X. (2019b). SDF-1/CXCR4 induces epithelial-mesenchymal transition through activation of the Wnt/β-catenin signaling pathway in rat chronic allograft nephropathy. *Molecular Medicine Reports, 19*(5), 3696–3706. https://doi.org/10.3892/mmr.2019.10045

Tian, Y. C., Fraser, D., Attisano, L., & Phillips, A. O. (2003). TGF-beta1-mediated alterations of renal proximal tubular epithelial cell phenotype. *American Journal of Physiology. Renal Physiology, 285*(1), F130–F142. https://doi.org/10.1152/ajprenal.00408.2002

Tian, Y., Chen, J., Huang, W., Ren, Q., Feng, J., Liao, J., Fu, H., Zhou, L., & Liu, Y. (2024). Myeloid-derived Wnts play an indispensable role in macrophage and fibroblast activation and kidney fibrosis. *International Journal of Biological Sciences, 20*(6), 2310–2322. https://doi.org/10.7150/ijbs.94166

Toda, N., Mukoyama, M., Yanagita, M., & Yokoi, H. (2018). CTGF in kidney fibrosis and glomerulonephritis. *Inflammation and Regeneration, 38*, 14. https://doi.org/10.1186/s41232-018-0070-0

Venkatachalam, M. A., Weinberg, J. M., Kriz, W., & Bidani, A. K. (2015). Failed tubule recovery, AKI-CKD transition, and kidney disease progression. *Journal of the American Society of Nephrology, 26*(8), 1765–1776. https://doi.org/10.1681/asn.2015010006

Vlasschaert, C., Moran, S. M., & Rauh, M. J. (2022). The myeloid-kidney interface in health and disease. *Clinical Journal of the American Society of Nephrology, 17*(2), 323–331. https://doi.org/10.2215/cjn.04120321

Wang, D., Sant, S., Lawless, C., & Ferrell, N. (2022). A kidney proximal tubule model to evaluate effects of basement membrane stiffening on renal tubular epithelial cells. *Integrative Biology, 14*(8-12), 171–183. https://doi.org/10.1093/intbio/zyac016

Wang, B., Kim, K., Tian, M., Kameishi, S., Zhuang, L., Okano, T., & Huang, Y. (2023a). Engineered bone marrow stem cell-sheets alleviate renal damage in a rat chronic glomerulonephritis model. *International Journal of Molecular Sciences, 24*(4). https://doi.org/10.3390/ijms24043711

Wang, H., Li, M., Fei, L., Xie, C., Ding, L., Zhu, C., Zeng, F., & Liu, N. (2023b). Bone marrow-derived mesenchymal stem cells transplantation attenuates renal fibrosis following acute kidney injury in rats by diminishing pericyte-myofibroblast transition and extracellular matrix augment. *Transplantation Proceedings, 55*(1), 225–234. https://doi.org/10.1016/j.transproceed.2022.12.002

Wozniak, J., Floege, J., Ostendorf, T., & Ludwig, A. (2021). Key metalloproteinase-mediated pathways in the kidney. Nature reviews. *Nephrology, 17*(8), 513–527. https://doi.org/10.1038/s41581-021-00415-5

Xiang, X., Zhu, J., Dong, G., & Dong, Z. (2022). Epigenetic regulation in kidney transplantation. *Frontiers in Immunology, 13*, 861498. https://doi.org/10.3389/fimmu.2022.861498

Yamamoto, T., Nakamura, T., Noble, N. A., Ruoslahti, E., & Border, W. A. (1993). Expression of transforming growth factor beta is elevated in human and experimental diabetic nephropathy. *Proceedings of the National Academy of Sciences of the United States of America, 90*(5), 1814–1818. https://doi.org/10.1073/pnas.90.5.1814

Yi, S., Ding, F., Gong, L., & Gu, X. (2017). Extracellular matrix scaffolds for tissue engineering and regenerative medicine. *Current Stem Cell Research & Therapy, 12*(3), 233–246. https://doi.org/10.2174/1574888x11666160905092513

Yuan, Q., Tan, R. J., & Liu, Y. (2019). Myofibroblast in kidney fibrosis: Origin, activation, and regulation. *Advances in Experimental Medicine and Biology, 1165*, 253–283. https://doi.org/10.1007/978-981-13-8871-2_12

Yuan, X., Su, Q., Wang, H., Shi, S., Liu, L., Lv, J., Wang, S., Zhu, L., & Zhang, H. (2023). Genetic variants of the COL4A3 , COL4A4 , and COL4A5 genes contribute to thinned glomerular basement membrane lesions in sporadic IgA nephropathy patients. *Journal of the American Society of Nephrology, 34*(1), 132–144. https://doi.org/10.1681/asn.2021111447

Zakiyanov, O., Kalousová, M., Zima, T., & Tesař, V. (2021). Matrix metalloproteinases and tissue inhibitors of matrix metalloproteinases in kidney disease. *Advances in Clinical Chemistry, 105*, 141–212. https://doi.org/10.1016/bs.acc.2021.02.003

Zhang, X., Chen, X., Hong, H., Hu, R., Liu, J., & Liu, C. (2022). Decellularized extracellular matrix scaffolds: Recent trends and emerging strategies in tissue engineering. *Bioactive Materials, 10*, 15–31. https://doi.org/10.1016/j.bioactmat.2021.09.014

Zhang, Y. L., Tang, T. T., Wang, B., Wen, Y., Feng, Y., Yin, Q., Jiang, W., Zhang, Y., Li, Z. L., Wu, M., Wu, Q. L., Song, J., Crowley, S. D., Lan, H. Y., Lv, L. L., & Liu, B. C. (2024). Identification of a novel ECM remodeling macrophage subset in AKI to CKD transition by integrative spatial and single-cell analysis. *Advanced Science, 11*(38), e2309752. https://doi.org/10.1002/advs.202309752

Zhao, X., Kong, Y., Liang, B., Xu, J., Lin, Y., Zhou, N., Li, J., Jiang, B., Cheng, J., Li, C., & Wang, W. (2022). Mechanosensitive Piezo1 channels mediate renal fibrosis. *JCI Insight, 7*(7). https://doi.org/10.1172/jci.insight.152330

Zhou, D., Tan, R. J., Fu, H., & Liu, Y. (2016). Wnt/β-catenin signaling in kidney injury and repair: A double-edged sword. *Laboratory Investigation, 96*(2), 156–167. https://doi.org/10.1038/labinvest.2015.153

Zhou, D., Fu, H., Zhang, L., Zhang, K., Min, Y., Xiao, L., Lin, L., Bastacky, S. I., & Liu, Y. (2017). Tubule-derived Wnts are required for fibroblast activation and kidney fibrosis. *Journal of the American Society of Nephrology, 28*(8), 2322–2336. https://doi.org/10.1681/asn.2016080902

Zhou, C., Zhou, L., Liu, J., Xu, L., Xu, Z., Chen, Z., Ge, Y., Zhao, F., Wu, R., Wang, X., Jiang, N., Mao, L., & Jia, R. (2020). Kidney extracellular matrix hydrogel enhances therapeutic potential of adipose-derived mesenchymal stem cells for renal ischemia reperfusion injury. *Acta Biomaterialia, 115*, 250–263. https://doi.org/10.1016/j.actbio.2020.07.056

Zhou, D., Wang, Y., Gui, Y., Fu, H., Zhou, S., Wang, Y., Bastacky, S. I., Stolz, D. B., & Liu, Y. (2022a). Non-canonical Wnt/calcium signaling is protective against podocyte injury and glomerulosclerosis. *Kidney International, 102*(1), 96–107. https://doi.org/10.1016/j.kint.2022.02.029

Zhou, X., Zhang, Y., & Wang, N. (2022b). Systematic identification of key extracellular proteins as the potential biomarkers in lupus nephritis. *Frontiers in Immunology, 13*, 915784. https://doi.org/10.3389/fimmu.2022.915784

Zhu, H., Liao, J., Zhou, X., Hong, X., Song, D., Hou, F. F., Liu, Y., & Fu, H. (2020). Tenascin-C promotes acute kidney injury to chronic kidney disease progression by impairing tubular integrity via αvβ6 integrin signaling. *Kidney International, 97*(5), 1017–1031. https://doi.org/10.1016/j.kint.2020.01.026

Ziyadeh, F. N. (2004). Mediators of diabetic renal disease: The case for tgf-Beta as the major mediator. *Journal of the American Society of Nephrology, 15*(*Suppl 1*), S55–S57. https://doi.org/10.1097/01.asn.0000093460.24823.5b

Zuo, Y., Wang, C., Sun, X., Hu, C., Liu, J., Hong, X., Shen, W., Nie, J., Hou, F. F., Zhou, L., & Liu, Y. (2021). Identification of matrix metalloproteinase-10 as a key mediator of podocyte injury and proteinuria. *Kidney International, 100*(4), 837–849. https://doi.org/10.1016/j.kint.2021.05.035

# Bone Regeneration

# Mechanisms of Bone Regeneration: An Update

**Jaganmay Sarkar, Vijay Bhaskar Reddy Konala, and Hiranmoy Das**

**Abstract** A significant global issue involves millions of individuals struggling with bone tissue defects. These defects, including serious conditions like osteonecrosis, arise from various factors such as fractures, tumors, trauma, and congenital disorders. Effective bone healing depends on the complex interaction between osteogenesis and the osteoimmune microenvironment. At the heart of bone regeneration are vital repair mechanisms, coordinating a series of physiological processes. These include the recruitment of inflammatory cells, the creation of vascular networks, and the development of callus tissue, followed by a later phase of remodeling. While conventional treatment options are routinely used, there has been a notable focus on alternative approaches, particularly autologous and allogeneic bone grafting. Unfortunately, the therapeutic results of these methods have often not met expectations due to limitations such as donor scarcity and graft rejection. In light of these challenges, researchers have been motivated to investigate innovative avenues for bone regeneration. This chapter begins an exploration of alternative therapeutic options, which include the use of stem and progenitor cells, the incorporation of cell-free scaffolds, and exosomes.

**Keywords** Bone regeneration · Stem cell · Oxidative stress · Cell-free scaffolds · Exosome

J. Sarkar
Department of Pharmaceutical Sciences, Jerry H. Hodge School of Pharmacy, Texas Tech University Health Sciences Center, Amarillo, TX, USA

Department of Integrated Drug Discovery, Yashraj Biotechnology Limited, Navi Mumbai, India

V. B. R. Konala
Department of Integrated Drug Discovery, Yashraj Biotechnology Limited, Navi Mumbai, India

H. Das (✉)
Department of Pharmaceutical Sciences, Jerry H. Hodge School of Pharmacy, Texas Tech University Health Sciences Center, Amarillo, TX, USA
e-mail: hiranmoy.das@ttuhsc.edu



H. Das (ed.), *Tissue Repair and Regeneration*,
https://doi.org/10.1007/978-3-031-93677-7_4

## 1 Introduction

Bone is a dynamic vascularized tissue whose natural composite controls its development, remodeling, and regeneration (Salhotra et al., 2020). Every year, a massive number of patients suffer from bone tissue defects and need treatment worldwide. Traffic accidents, congenital malformations, and other pathophysiological conditions like tumor resection can cause bone injuries and create challenges to healing itself. Previous studies have shown that the spontaneous healing ability of osteoblasts fails in the case of large bones, where the missing tissue is larger (Fabris et al., 2018; Rashid et al., 2020). Autologous grafting is the gold standard for small bone defects; however, this approach has some limitations because of the sources of graft. An alternative process is called allogeneic grafts, which may perhaps resolve the deficiency of autologous bone for transplantation, although it has a high risk of immune rejection and other medical and ethical concerns (Drujont et al., 2014; Hao et al., 2014; Spear et al., 1999).

With the unavailability of donors and insufficient supply of organs, along with other associated problems like immune rejection and ethical concerns about tissue replacement in patients, regenerative medicine comes in to gain prime focus nowadays. Stem cell therapy plays a major role in the field of regenerative medicine. Mesenchymal stromal cell (MSC) plays an important role in bone regeneration and formation by regulating the osteoblast, osteoid mineralization, extracellular matrix (ECM), and the formation of terminally differentiated osteocytes (Wang et al., 2013). However, stem cells may encounter challenges in addressing large tissue defects through injection alone.

Several alternative methods have been developed in recent years to address this shortcoming of stem cell therapy. Recent advances in bone tissue engineering (BTE) technology have allowed investigators to produce artificial bone by combining cells, biological factors, and biomaterials (Zhu et al., 2021). This artificial bone is advantageous because it does not show obvious complications, is biocompatible, has high modifiability, is osteoinductive, and has a low risk of infectivity (Henkel et al., 2013; Sheehy et al., 2019).

A noncellular 3D structure secreted by cells is called ECM. It provides support, integrity, and elasticity and maintains the homeostasis of bone tissue cells (Bonnans et al., 2014; Mouw et al., 2014). Bone ECM critically interacts with the osteoblast and osteoclast lineages of cells to control the development of new bone throughout regeneration. Several factors, such as an abundance of receptors, growth factors, and the pH of the local environment, control the ECM structure, which is subsequently involved in the development, function, and homeostasis of tissues and organs (Bonnans et al., 2014; Mouw et al., 2014).

In this review, we first summarize the hierarchical construction of the bone tissue and its microenvironment, which provides the physical, biochemical, and mechanical signals for bone tissue cells. Based on this, we brief the effect and application of ECM on the osteoblast lineages of cells, ECM-based scaffold, and delivery of bioactive factors based on the mode of action of the healing mechanism. We not only focus on the current BTE-induced biomimetic and bioinspired scaffold-based therapy in stem cell regenerative medicine but also discuss the limitations and improvements that need to be addressed.

## 2 Bone Microenvironment

Bone is a dynamic vascularized tissue that supports the physiological processes and structural integrity of the body. It also plays an important role in the body and acts as a protective barrier for vital organs. Bone is actively involved in the movement of body parts and the metabolic process and is a storehouse of several blood homeostasis substances, such as calcium, phosphate, and minerals (Marks & Popoff, 1988). Bone tissue is divided into the peripheral cortical part, the cortex, the internal spongy part, and bone marrow cavities, which are essential for hematopoiesis (Fukumoto & Martin, 2009; Marks & Popoff, 1988).

Bone homeostasis and rigidity were maintained by a heterogeneous population of cells, such as synovial fibroblast, macrophages, stem cells, osteoblast and osteoclast lineages of cells, and osteocytes with the involvement of ECM and minerals (Fig. 1). Bone cells continuously receive signals from the surrounding environment which guides bone for the remodeling process. Bone regeneration depends on the balance between bone resorption and bone deposition. Osteoblast originates from the MSC, which are the main contributors to bone formation and terminally differentiate into osteocytes (Ansari et al., 2021). After bone formation, majority of the

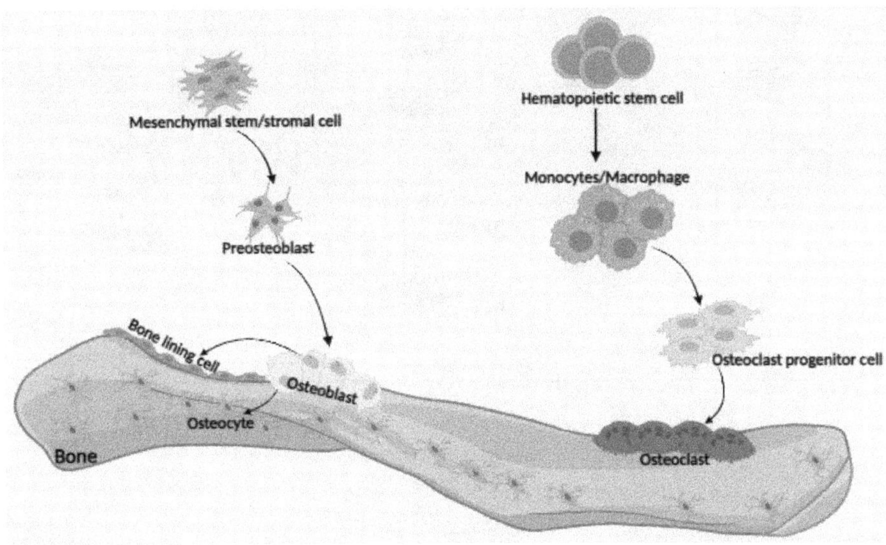


**Fig. 1 Schematic representation of bone homeostasis.** This image was created using BioRender.com

MSC undergoes apoptosis, and the rest of them become inactive and cover the surface of the quiescent bone (Franz-Odendaal et al., 2006; Marriott, 2013; Parfitt, 2001). In the presence of macrophage colony-stimulating factor (MCSF) and receptor activator of nuclear factor-kappa B ligand (RANKL) monocytes, and lineages of cells combine together to form multinucleated giant cells called osteoclast, which play a critical regulator for bone resorption (Sarkar et al., 2022). Osteocytes are an important player in bone turnover, which not only provides paracrine signaling to osteoblasts but also undergoes apoptosis to recruit osteoclasts (Lien et al., 2009).

## 3 Intracellular Signaling Pathways and Bone Regeneration

Several growth factors, cytokines, and hormones are released from the surrounding environment to produce autocrine, paracrine, and endocrine signaling into the bone to guide the remodeling process. Differentiation of osteogenesis is an important control measure for bone regeneration, where signaling pathways play a critical role. The signaling pathways control the transcriptional regulator which leads to differentiation, adhesion, and survival of the osteoblast during the bone repair and regeneration discussed below (Ponzetti & Rucci, 2021) (Fig. 2).

Transforming growth factor β (TGFβ) is an important factor for the regulation of cellular events like proliferation and differentiation, which also play an important role in osteoblast differentiation. Bone morphogenetic protein (BMP), activin, and TGFβ itself belong to the TGFβ superfamily that regulates osteoblast differentiation

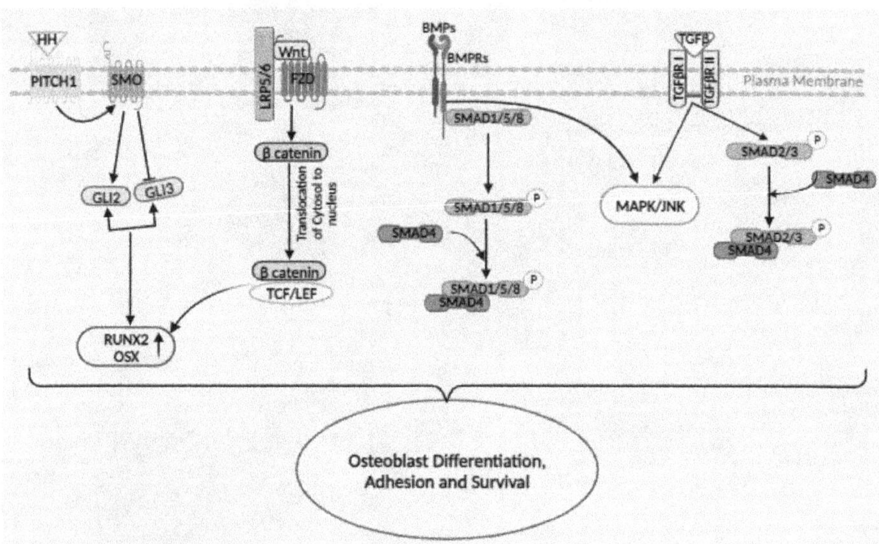


**Fig. 2 Signaling pathways involved in the regulation of osteoblast differentiation, adhesion, and survival.** This image was created using BioRender.com

(Rahman et al., 2015; Wang et al., 2014). BMPs bind to its serine/threonine kinase receptors called BMPR-I and BMPR-II to make a heterodimer complex to initiate the phosphorylation of SMAD1/5/8 and translocate the SMAD4 molecule to the nucleus to activate the transcription of osteogenic transcription factor. Studies have also shown that BMPs induce osteogenic properties via several non-canonical signaling pathways through MAPK/JNK (Salazar et al., 2016; Wang et al., 2014). A study by Bandyopadhyay et al. showed that BMP2 alone does not stop the differentiation of osteoblast; however, the combinational knockout of BMP2 along with BMP4 or BMP7 caused an anomalous osteoblast differentiation (Bandyopadhyay et al., 2006).

In 1980, Hedgehog was identified as a gene that is responsible for the development of anterior–posterior bodies in multiple species, including mammals (Ocbina & Anderson, 2008; Tickle & Towers, 2017). Two transmembrane proteins such as patched homolog 1 (PTCH1) and smoothened homolog (SMO), which is known as a receptors for Hedgehog (HH) signaling. Upon binding of HH with PTCH1 and SMO complex, SMO is released from the complex and activates the downstream signaling cascade to regulate the transcription factor GLI2 and GLI3, which is essential for normal skeletal development (Chinnaiya et al., 2014; Hojo et al., 2012). Although GLI1 does not directly take part in skeletal development, it synergistically involves GLI2 and GLI3 and upregulates osteogenesis-related genes like RUNX2 and OSX during bone formation (Hojo et al., 2012; Shi et al., 2015). In addition, the HH signaling pathways crosstalk with BMP and the Wnt/β-Catenin pathway to regulate osteoblast differentiation during bone regeneration (Chen et al., 2012; Mak et al., 2006).

The Wnt signaling pathway is another important factor for bone metabolism and homeostasis. Wnt signaling pathway elicits its effect through two different types of pathways called canonical and noncanonical pathways. In the canonical signaling pathway, where Wnt ligand binds with its Feizzled receptor (FZD) and lipoprotein receptor 5 and 6 (LPR5/6), and induces the translocation of β-Catenin from the cytosol to the nucleus, where it binds with T cell factor/Lymphoid enhancer-binding factor (TCF/LFF) and induce transcription of RUNX2 gene (Salhotra et al., 2020). Despite canonical pathways, noncanonical pathways also focus on bone cells, stimulate the replication of osteoprogenitors, and intracellular calcium release, induce osteoblastogenesis, and inhibit apoptosis of osteoblast cells (Baron & Kneissel, 2013; Bonewald & Johnson, 2008; Burgers & Williams, 2013; Macsai et al., 2008).

Notch signaling is another important pathway that crosstalks with the canonical Wnt/β-Catenin signaling pathway and inhibits osteoblast differentiation (Gao et al., 2021; Lin & Hankenson, 2011). Notch family has five receptors (Notch 1–5), which bind with a variety of ligands such as Jagged (JAG1 and JAG2), Delta-like (DIL1, DIL3, and DIL4) to initiate the signaling cascade and subsequently induce proteolytic cleavage of the γ-secretase complex in association with presenilin 1 or 2 (PS1 or PS2) (Pakvasa et al., 2021; Tu et al., 2012). Previous studies established the importance of Notch signaling in skeletal muscle development. In this study, the researcher used Prrx1Cre mice to reveal that inhibition of a few factors like PS1,

Notch 1, and Notch 2 in developing limb bud augmented mass of the trabecular bone (Hilton et al., 2008; Tu et al., 2012).

## 4 Oxidative Stress and Bone Regeneration

Oxidative stress occurs due to an imbalance of oxidant/antioxidant ratio and acts as a cellular alarm system during several pathophysiological conditions including bone defects. The oxidative stress promotes the activation of numerous enzymes and produces reactive oxygen species (ROS). ROS plays an important role in bone diseases by inducing osteoclast (OC) and inhibiting osteoblast (OB) differentiation (Liu et al., 2019; Maity et al., 2022; Zhang et al., 2023a).

Autophagy is another important factor involved in ROS-induced bone resorption, but the relationship between ROS signaling and autophagy is very complex. In our recent study we found that Epigallocatechin-3-gallate (EGCG), tea catechin acts as an antioxidant and inhibits intracellular and mitochondrial ROS production, which subsequently alters the autophagy-dependent mitochondrial function, inhibits the signaling pathway such as AKT, p38MAPK, and inhibited the epigenetic regulator such as SETD2, which is associated with the osteoclast differentiation using RAW264.7 cell line and murine bone marrow-derived primary monocytes (Sarkar et al., 2022). In another study, we showed that treatment of ferutinin (a phytoestrogen) treatment to the dental pulp-derived stem cell (DPSC) induces osteoblast differentiation by reducing intracellular and mitochondrial ROS generation and as well as inducing the production of intracellular $Ca^{2+}$ (Maity et al., 2022). Aucubin (AU), an iridoid glucoside, is well known for its anti-osteoporosis and anti-apoptosis activities. A study by Wang and his team using human bone marrow-derived mesenchymal stromal cells (hBM-MSC) showed that AU treatment increased osteogenic properties compared to untreated cells, and local injection of AU promoted bone regeneration in rat tibial fracture (Wang et al., 2022). AU treatment induces bone regeneration by regulating osteogenesis with the involvement partly via the BMP/SMAD signaling cascade and also inhibited oxidative stress damage via partly the Nrf2/HO1 signaling pathway (Wang et al., 2022).

Tartrate-resistant acid phosphatase (TRAP) which is an Fe-containing enzyme is reacting with hydrogen peroxide to generate ROS, which induces the fragmentation of collagen I during endocytosis (Halleen et al., 1999). A molecule released from biomaterials, called soluble silicate (Si), induces the bone regeneration process and inhibits bone resorption by regulating the OC formation (Zhou et al., 2016). In wastewater treatment and paper making, it is reported that Si may inhibit the free radical formation via Fenton reaction by reducing the availability of $Fe^{2+}$ (Cui et al., 2017; H. Hamäläinen et al., 2007). Therefore, Si may limit the ROS availability during the bone remodeling process via the Fenton reaction; however, the mechanism is not clear.

# 5 Stem Cell and Bone Regeneration

To overcome the problem of insufficient organ donors and associated ethical concerns, regenerative medicine gains prime focus in the field of bone grafting procedures. The self-renewal and multi-lineage differentiation properties of stem cells attract researchers' and clinicians' attention in the field of regenerative therapy during bone disorders. Based on the developmental hierarchy, stem cells were divided into embryonic stem cells (ESCs) and adult stem cells (ASCs). The inner cell mass of blastocyst consists of ESCs, which have a greater potency towards infinite proliferation and three germ layer differentiation, which helped them become ideal candidates for tissue engineering and regenerative medicine (Thomson et al., 1998; Whiting et al., 2015). However, the therapeutic application of ESCs has some controversies and limitations. Immune rejection or incompatibilities, along with the properties of formation of teratomas by administrative cells, hampered the derivation of ESCs and restricted their clinical application, creating an opportunity for ASCs for the therapeutic potential in regenerative medicine (Hmadcha et al., 2009; Tao et al., 2018).

MSCs belonging to the class of ASCs have the potential to differentiate into multiple lineages including bone, cartilage, and adipose tissue (Safarova et al., 2020). MSCs were isolated from the different types of tissues such as adipose issues, bone marrow, amniotic fluid, and synovial membrane; however, the preclinical application of MSC depends on its origin of isolation. MSCs derived from bone marrow and adipose tissues are commonly used in bone regeneration and formation, regulating the osteoblast, osteoid mineralization, extracellular matrix (ECM), and finally terminally differentiated osteocytes (Klingemann et al., 2008; Panepucci et al., 2004; Wang et al., 2013). Systemic administration involves intravenous (IV) and intra-arterial (IA) injection of cells, and the local administration involves direct injection of the cells into the regeneration site, which are the two main routes of transplantation of MSC. During the systemic administration, MSCs were concentrated in the lung and migrated to the inflamed tissues by sensing the chemokine CCL21 (De Becker & Riet, 2016; Eggenhofer et al., 2012; Fischer et al., 2009; Ma et al., 2014; Nystedt et al., 2013; Sasaki et al., 2008; Schrepfer et al., 2007). Thus, effective homing is essential for MSC transplantation. A previous study using an osteoporotic mice model showed that CXCR4 and core binding factor-1 (Cbfa1) increased the MSC homing, as well as induced bone regeneration and mass after 4 weeks of transplantation (Lien et al., 2009). Zwingenberger et al. showed that low doses of BMP-2 and stromal cell-derived factor 1 (SDF-1) induced the MSC recruitments in the inflated site, and subsequently increased the bone volume by increasing the ratio of osteoblast and osteoclast (Zwingenberger et al., 2014).

Gene therapy was originally used for the treatment of genetic diseases where one or more genes were missing or significantly altered. In the late 1990 s, researchers implemented the ex vivo gene therapy-based model for bone engineering by targeting cell types, molecular targets, and modes of gene transfer. Genetic modification of MSC increases the survival rate, homing efficiency, and differential potentials

(Nowakowski et al., 2016; Papadimitropoulos et al., 2015; Safarova et al., 2020; Wei et al., 2018). Genetic modification of MSC mainly focused on three aspects: (A) enhancing the activation of cytokine (BMPs, interleukin-4 (IL-4), IL-10, basic fibroblast growth factor (bFGF), and vascular endothelial growth factor (VEGF)) secretion to promote the survival and differentiation towards osteogenic lineages, which is vital for effective bone regeneration; (B) maintaining bone density and integrity by augmenting the expression of critical cell surface receptors, including CXCR4 for improved cell homing capabilities and RANK-Fc to hinder OC differentiation and activation; and (C) activating the transcriptional regulators and protein like Runx2, Cbfa1, and Sox11 and regulating the MSC migration, differentiation properties and facilitating their therapeutic effectiveness in bone regeneration.

In 2006, Yamanaka and Takahashi generated a cell line from the mouse fibroblast by inducing four factors such as OCT4, SOX2, c-MYC, and KLF4, called induced pluripotent cells (iPSCs) (Takahashi & Yamanaka, 2006). iPSCs have great potential for terminally differentiated cells and organoid formation, which have a huge application in drug discovery and disease model development (Sadar, 2020; Tsujimoto et al., 2020). In addition, iPSCs show almost similar morphological, functional, and genetic properties to ESCs in immunodeficient mice, which help iPSCs as an alternative therapeutic option in regenerative medicine (Lowry et al., 2008; Yu et al., 2007).

## 6 Biomaterials and Bone Regeneration

Growing incidents of bone defects were reported every day. This prompted scientists and/or clinicians to establish an alternative source for treatment by developing advanced bone transplant techniques that required bone substitute materials. A wide range of bioactive materials have an emerging future for bone tissue engineering with osteogenic induction ability.

MicroRNAs (miRNAs) are a class of noncoding RNAs that bind with the mRNA3′-UTR region to regulate the intracellular protein to induce osteogenic differentiation (Arfat et al., 2018; Tang et al., 2020). miRNA helped MSC (MSC) for osteogenic differentiation by regulating a myriad of signaling cascade pathways, Wnt/β-catenin, MAPK, BMPs/SMADs/RUNX2, PTEN/PI3K/AKT, and steering bioactive factors (Jia et al., 2019; Li et al., 2018; Shi et al., 2020; Sun et al., 2020). In addition, miRNAs also interfere with other noncoding RNAs like lncRNAs and cirRNAs, sponges of specific miRNAs to reduce the function of miRNAs to control osteogenic differentiation (Peng et al., 2019; Yuan et al., 2019). The immune response is an immediate action taken by the immune microenvironment upon bone tissue damage. A study by Zhang et al. (Zhang et al., 2023b) targets macrophages to regulate the immune microenvironment. In their study, they developed rapamycin-loaded virus-like hollow silica nanoparticles mediated intracellular delivery system to endorse autophagy-mediated M2 macrophage polarization to regulate bone regeneration by triggering osteogenic differentiation of mesenchymal stromal cells.

Considering the biocompatibility, mechanical properties, and osteoinductive activity of several biomaterials, titanium alloys have been widely used for orthopedic applications and substrates for extracellular vesicles (EV) manufacturing (Hanawa, 2019; Kaur & Singh, 2019; Ma et al., 2022). A study by Ma et al. assessed the impact of nanotomography on titanium surfaces and showed that nano topography enhanced the osteogenic differentiation potency of MSC-derived EV compared to polished titanium controls (Ma et al., 2022).

Nowadays 3D cell culture model gains prime focus in the field of regenerative medicine. 3D culture model replicates the proximity of cells native microenvironment, and it exhibits enhanced osteogenic capacity compared to the 2D culture system (Ham et al., 2019; Shen et al., 2013). Hydrogel is widely used for the production of EVs. Studies showed that when human periodontal ligament stem cells (hPDLSCs) were cultured in a collagen hydrogel-assisted 3D culture system give improved the yield of EV and osteoinductive potency compared to the 2D model for bone regeneration (Catoira et al., 2019; Yu et al., 2022). Previous studies have also shown that transplantation of MSC encapsulated in collagen sponge/hyaluronic acid-based hydrogel complex scaffold comprising VEGF and BMP-2 significantly increased BMD levels in canine maxillary alveolar bone defects (Kim et al., 2016). In the rat alveolar bone defect model, vesicles derived from the 3D model along with Matrigel augmented the bone healing process compared to the Matrigel alone group.

## 7 Exosomes and Bone Regeneration

Cells were secreted several extracellular vesicles, which were categorized as microvesicles, apoptotic bodies, and exosomes based on their size, origin of secretion, and biological function (De Jong et al., 2014; Mathivanan et al., 2010). Exosomes, a nanovesicles with a varied diameter of 40–100 nm, have gained prime attention of the researchers in treatment of various pathological conditions including bone diseases (Kalluri & LeBleu, 2020). In the context of a healthy bone, exosomes play a crucial role in communicating signals between the cell types engaged in maintaining bone homeostasis and also actively participating in the mineralization of the extracellular matrix. Macrophages are the important cell types for the bone defect mechanism. A study by Xiang et al. showed that serum exosome induces macrophage proliferation, decreases M1-related gene expression, augments angiogenesis in human umbilical vein endothelial cells (HUVEC), and stimulates bone repair through regulating vascular endothelial adhesion factor 1 (VCAM1) signaling cascade (Xiang et al., 2023). In addition, it is also established that exosomes derived from MSC can also be directly taken up by OB, promoting their proliferation by the production of GLUT3 and MAPK pathways-related proteins (Zhao et al., 2018).

## 8 Future Direction and Conclusion

The repair mechanism of bone is a complex event, which was controlled by several factors and cell types. Immune cells and pro/anti-inflammatory cytokines are some of such factors whose immunomodulatory role in bone regeneration is not clear to date. This will help researchers to identify a novel precise mechanism to understand the ongoing challenges of bone regeneration in the future and develop a new targeted therapy that improves the immune response to stimulate effective bone tissue regeneration.

In recent years, MSCs have gained prime focus in the field of cell-based therapies because of their multipotent, self-renewal properties. MSCs hold potential for bone restoration not only by exerting autocrine /paracrine signaling but it is reported that modified-MSC can deliver therapeutic molecules and genes in the defected bone area. However, more robust clinical studies are needed in the future to establish the effectiveness of MSC-based therapies in bone regeneration. Additionally, exploring new therapeutic approaches and combinations that influence the unique natures of MSC can lead to amended tactics for bone regeneration.

Finally, herein we conclude with an insight into the future perspective for designing next-generation scaffolds in combination with 3D printing, bone tissue engineering, and stem cells. This next-generation scaffold has a promising future in the field of regenerative medicine over the traditional method. However, further research should emphasize on refining the choice of raw materials, printing techniques, and scaffold designs to increase the clinical applicability and to mimic bone physiology and diseases is crucial. In addition, the choice between bone marrow-derived and peripheral blood-derived progenitor cells for these models remains an area of exploration, with a focus on optimizing cell extraction and differentiation from peripheral blood samples.

**Acknowledgments** We sincerely acknowledge Yashraj Biotechnology Limited, Navi Mumbai. India.

## References

Ansari, S., Ito, K., & Hofmann, S. (2021). Cell sources for human in vitro bone models. *Current Osteoporosis Reports, 19*(1), 88–100. https://doi.org/10.1007/s11914-020-00648-6

Arfat, Y., Basra, M. A. R., Shahzad, M., Majeed, K., Mahmood, N., & Munir, H. (2018). miR-208a-3p suppresses osteoblast differentiation and inhibits bone formation by targeting ACVR1. *Molecular Therapy – Nucleic Acids, 11*, 323–336. https://doi.org/10.1016/j.omtn.2017.11.009

Bandyopadhyay, A., Tsuji, K., Cox, K., Harfe, B. D., Rosen, V., & Tabin, C. J. (2006). Genetic analysis of the roles of BMP2, BMP4, and BMP7 in limb patterning and skeletogenesis. *PLoS Genetics, 2*(12), e216. https://doi.org/10.1371/journal.pgen.0020216

Baron, R., & Kneissel, M. (2013). WNT signaling in bone homeostasis and disease: From human mutations to treatments. *Nature Medicine, 19*(2), 179–192. https://doi.org/10.1038/nm.3074

Bonewald, L. F., & Johnson, M. L. (2008). Osteocytes, mechanosensing and Wnt signaling. *Bone, 42*(4), 606–615. https://doi.org/10.1016/j.bone.2007.12.224
Bonnans, C., Chou, J., & Werb, Z. (2014). Remodelling the extracellular matrix in development and disease. *Nature Reviews. Molecular Cell Biology, 15*(12), 786–801. https://doi.org/10.1038/nrm3904
Burgers, T. A., & Williams, B. O. (2013). Regulation of Wnt/β-catenin signaling within and from osteocytes. *Bone, 54*(2), 244–249. https://doi.org/10.1016/j.bone.2013.02.022
Catoira, M. C., Fusaro, L., Di Francesco, D., Ramella, M., & Boccafoschi, F. (2019). Overview of natural hydrogels for regenerative medicine applications. *Journal of Materials Science. Materials in Medicine, 30*(10), 115. https://doi.org/10.1007/s10856-019-6318-7
Chen, G., Deng, C., & Li, Y. P. (2012). TGF-β and BMP signaling in osteoblast differentiation and bone formation. *International Journal of Biological Sciences, 8*(2), 272–288. https://doi.org/10.7150/ijbs.2929
Chinnaiya, K., Tickle, C., & Towers, M. (2014). Sonic hedgehog-expressing cells in the developing limb measure time by an intrinsic cell cycle clock. *Nature Communications, 5*, 4230. https://doi.org/10.1038/ncomms5230
Cui, J., Wang, X., Zhang, J., Qiu, X., Wang, D., Zhao, Y., Xi, B., Alshawabkeh, A. N., & Mao, X. (2017). Disilicate-assisted iron electrolysis for sequential fenton-oxidation and coagulation of aqueous contaminants. *Environmental Science & Technology, 51*(14), 8077–8084. https://doi.org/10.1021/acs.est.7b01184
De Becker, A., & Riet, I. V. (2016). Homing and migration of mesenchymal stromal cells: How to improve the efficacy of cell therapy? *World Journal of Stem Cells, 8*(3), 73–87. https://doi.org/10.4252/wjsc.v8.i3.73
De Jong, O. G., Van Balkom, B. W., Schiffelers, R. M., Bouten, C. V., & Verhaar, M. C. (2014). Extracellular vesicles: potential roles in regenerative medicine. *Frontiers in Immunology, 5*, 608. https://doi.org/10.3389/fimmu.2014.00608
Drujont, L., Carretero-Iglesia, L., Bouchet-Delbos, L., Beriou, G., Merieau, E., Hill, M., Delneste, Y., Cuturi, M. C., & Louvet, C. (2014). Evaluation of the therapeutic potential of bone marrow-derived myeloid suppressor cell (MDSC) adoptive transfer in mouse models of autoimmunity and allograft rejection. *PLoS One, 9*(6), e100013. https://doi.org/10.1371/journal.pone.0100013
Eggenhofer, E., Benseler, V., Kroemer, A., Popp, F. C., Geissler, E. K., Schlitt, H. J., Baan, C. C., Dahlke, M. H., & Hoogduijn, M. J. (2012). Mesenchymal stem cells are short-lived and do not migrate beyond the lungs after intravenous infusion. *Frontiers in Immunology, 3*, 297. https://doi.org/10.3389/fimmu.2012.00297
Fabris, A., Faverani, L. P., Gomes-Ferreira, P. H. S., Polo, T. O. B., Santiago-Júnior, J. F., & Okamoto, R. (2018). Bone repair access of BoneCeramic™ in 5-mm defects: Study on rat calvaria. *Journal of Applied Oral Science, 26*, e20160531. https://doi.org/10.1590/1678-7757-2016-0531
Fischer, U. M., Harting, M. T., Jimenez, F., Monzon-Posadas, W. O., Xue, H., Savitz, S. I., Laine, G. A., & Cox, C. S., Jr. (2009). Pulmonary passage is a major obstacle for intravenous stem cell delivery: The pulmonary first-pass effect. *Stem Cells and Development, 18*(5), 683–692. https://doi.org/10.1089/scd.2008.0253
Franz-Odendaal, T. A., Hall, B. K., & Witten, P. E. (2006). Buried alive: How osteoblasts become osteocytes. *Developmental Dynamics, 235*(1), 176–190. https://doi.org/10.1002/dvdy.20603
Fukumoto, S., & Martin, T. J. (2009). Bone as an endocrine organ. *Trends in Endocrinology and Metabolism, 20*(5), 230–236. https://doi.org/10.1016/j.tem.2009.02.001
Gao, J., Fan, L., Zhao, L., & Su, Y. (2021). The interaction of notch and Wnt signaling pathways in vertebrate regeneration. *Cell Regeneration, 10*(1), 11. https://doi.org/10.1186/s13619-020-00072-2
Halleen, J. M., Räisänen, S., Salo, J. J., Reddy, S. V., Roodman, G. D., Hentunen, T. A., Lehenkari, P. P., Kaija, H., Vihko, P., & Väänänen, H. K. (1999). Intracellular fragmentation of bone resorption products by reactive oxygen species generated by osteoclastic tartrate-resistant

acid phosphatase. *The Journal of Biological Chemistry, 274*(33), 22907–22910. https://doi.org/10.1074/jbc.274.33.22907

Ham, J., Lever, L., Fox, M., & Reagan, M. R. (2019). In vitro 3D cultures to reproduce the bone marrow niche. *JBMR Plus, 3*(10), e10228. https://doi.org/10.1002/jbm4.10228

Hamäläinen, H., Aksela, R., Rautiainen, J., Sankari, M., Renvall, I., Paquet, R. (2007). Silicate-free peroxide bleaching of mechanical pulps: Efficiency of polymeric stabilizers. In *Proceedings TAPPI of international mechanical pulping conference*.

Hanawa, T. (2019). Titanium-tissue Interface reaction and its control with surface treatment. *Frontiers in Bioengineering and Biotechnology, 7*, 170. https://doi.org/10.3389/fbioe.2019.00170

Hao, C., Shintani, S., Shimizu, Y., Kondo, K., Ishii, M., Wu, H., & Murohara, T. (2014). Therapeutic angiogenesis by autologous adipose-derived regenerative cells: Comparison with bone marrow mononuclear cells. *American Journal of Physiology. Heart and Circulatory Physiology, 307*(6), H869–H879. https://doi.org/10.1152/ajpheart.00310.2014

Henkel, J., Woodruff, M. A., Epari, D. R., Steck, R., Glatt, V., Dickinson, I. C., Choong, P. F., Schuetz, M. A., & Hutmacher, D. W. (2013). Bone regeneration based on tissue engineering conceptions – A 21st century perspective. *Bone Research, 1*(3), 216–248. https://doi.org/10.4248/br201303002

Hilton, M. J., Tu, X., Wu, X., Bai, S., Zhao, H., Kobayashi, T., Kronenberg, H. M., Teitelbaum, S. L., Ross, F. P., Kopan, R., & Long, F. (2008). Notch signaling maintains bone marrow mesenchymal progenitors by suppressing osteoblast differentiation. *Nature Medicine, 14*(3), 306–314. https://doi.org/10.1038/nm1716

Hmadcha, A., Domínguez-Bendala, J., Wakeman, J., Arredouani, M., & Soria, B. (2009). The immune boundaries for stem cell based therapies: Problems and prospective solutions. *Journal of Cellular and Molecular Medicine, 13*(8a), 1464–1475. https://doi.org/10.1111/j.1582-4934.2009.00837.x

Hojo, H., Ohba, S., Yano, F., Saito, T., Ikeda, T., Nakajima, K., Komiyama, Y., Nakagata, N., Suzuki, K., Takato, T., Kawaguchi, H., & Chung, U. I. (2012). Gli1 protein participates in hedgehog-mediated specification of osteoblast lineage during endochondral ossification. *The Journal of Biological Chemistry, 287*(21), 17860–17869. https://doi.org/10.1074/jbc.M112.347716

Jia, B., Wang, Z., Sun, X., Chen, J., Zhao, J., & Qiu, X. (2019). Long noncoding RNA LINC00707 sponges miR-370-3p to promote osteogenesis of human bone marrow-derived mesenchymal stem cells through upregulating WNT2B. *Stem Cell Research & Therapy, 10*(1), 67. https://doi.org/10.1186/s13287-019-1161-9

Kalluri, R., & LeBleu, V. S. (2020). The biology, function, and biomedical applications of exosomes. *Science, 367*(6478). https://doi.org/10.1126/science.aau6977

Kaur, M., & Singh, K. (2019). Review on titanium and titanium based alloys as biomaterials for orthopaedic applications. *Materials Science & Engineering. C, Materials for Biological Applications, 102*, 844–862. https://doi.org/10.1016/j.msec.2019.04.064

Kim, S. K., Cho, T. H., Han, J. J., Kim, I. S., Park, Y., & Hwang, S. J. (2016). Comparative study of BMP-2 alone and combined with VEGF carried by hydrogel for maxillary alveolar bone regeneration. *Tissue Engineering and Regenerative Medicine, 13*(2), 171–181. https://doi.org/10.1007/s13770-015-0046-y

Klingemann, H., Matzilevich, D., & Marchand, J. (2008). Mesenchymal stem cells - sources and clinical applications. *Transfusion Medicine and Hemotherapy, 35*(4), 272–277. https://doi.org/10.1159/000142333

Li, H., Fan, J., Fan, L., Li, T., Yang, Y., Xu, H., Deng, L., Li, J., Li, T., Weng, X., Wang, S., & Chunhua Zhao, R. (2018). MiRNA-10b reciprocally stimulates osteogenesis and inhibits adipogenesis partly through the TGF-β/SMAD2 signaling pathway. *Aging and Disease, 9*(6), 1058–1073. https://doi.org/10.14336/ad.2018.0214

Lien, C. Y., Chih-Yuan Ho, K., Lee, O. K., Blunn, G. W., & Su, Y. (2009). Restoration of bone mass and strength in glucocorticoid-treated mice by systemic transplantation of CXCR4 and cbfa-1 co-expressing mesenchymal stem cells. *Journal of Bone and Mineral Research, 24*(5), 837–848. https://doi.org/10.1359/jbmr.081257

Lin, G. L., & Hankenson, K. D. (2011). Integration of BMP, Wnt, and notch signaling pathways in osteoblast differentiation. *Journal of Cellular Biochemistry, 112*(12), 3491–3501. https://doi.org/10.1002/jcb.23287

Liu, Y., Wang, C., Wang, G., Sun, Y., Deng, Z., Chen, L., Chen, K., Tickner, J., Kenny, J., Song, D., Zhang, Q., Wang, H., Chen, Z., Zhou, C., He, W., & Xu, J. (2019). Loureirin B suppresses RANKL-induced osteoclastogenesis and ovariectomized osteoporosis via attenuating NFATc1 and ROS activities. *Theranostics, 9*(16), 4648–4662. https://doi.org/10.7150/thno.35414

Lowry, W. E., Richter, L., Yachechko, R., Pyle, A. D., Tchieu, J., Sridharan, R., Clark, A. T., & Plath, K. (2008). Generation of human induced pluripotent stem cells from dermal fibroblasts. *Proceedings of the National Academy of Sciences of the United States of America, 105*(8), 2883–2888. https://doi.org/10.1073/pnas.0711983105

Ma, S., Xie, N., Li, W., Yuan, B., Shi, Y., & Wang, Y. (2014). Immunobiology of mesenchymal stem cells. *Cell Death and Differentiation, 21*(2), 216–225. https://doi.org/10.1038/cdd.2013.158

Ma, L., Li, G., Lei, J., Song, Y., Feng, X., Tan, L., Luo, R., Liao, Z., Shi, Y., Zhang, W., Liu, X., Sheng, W., Wu, S., & Yang, C. (2022). Nanotopography sequentially mediates human mesenchymal stem cell-derived small extracellular vesicles for enhancing Osteogenesis. *ACS Nano, 16*(1), 415–430. https://doi.org/10.1021/acsnano.1c07150

Macsai, C. E., Foster, B. K., & Xian, C. J. (2008). Roles of Wnt signalling in bone growth, remodelling, skeletal disorders and fracture repair. *Journal of Cellular Physiology, 215*(3), 578–587. https://doi.org/10.1002/jcp.21342

Maity, J., Barthels, D., Sarkar, J., Prateeksha, P., Deb, M., Rolph, D., & Das, H. (2022). Ferutinin induces osteoblast differentiation of DPSCs via induction of KLF2 and autophagy/mitophagy. *Cell Death & Disease, 13*(5), 452. https://doi.org/10.1038/s41419-022-04903-9

Mak, K. K., Chen, M. H., Day, T. F., Chuang, P. T., & Yang, Y. (2006). Wnt/beta-catenin signaling interacts differentially with Ihh signaling in controlling endochondral bone and synovial joint formation. *Development, 133*(18), 3695–3707. https://doi.org/10.1242/dev.02546

Marks, S. C., Jr., & Popoff, S. N. (1988). Bone cell biology: The regulation of development, structure, and function in the skeleton. *The American Journal of Anatomy, 183*(1), 1–44. https://doi.org/10.1002/aja.1001830102

Marriott, I. (2013). Apoptosis-associated uncoupling of bone formation and resorption in osteomyelitis. *Frontiers in Cellular and Infection Microbiology, 3*, 101. https://doi.org/10.3389/fcimb.2013.00101

Mathivanan, S., Ji, H., & Simpson, R. J. (2010). Exosomes: extracellular organelles important in intercellular communication. *Journal of Proteomics, 73*(10), 1907–1920. https://doi.org/10.1016/j.jprot.2010.06.006

Mouw, J. K., Ou, G., & Weaver, V. M. (2014). Extracellular matrix assembly: A multiscale deconstruction. *Nature Reviews. Molecular Cell Biology, 15*(12), 771–785. https://doi.org/10.1038/nrm3902

Nowakowski, A., Walczak, P., Lukomska, B., & Janowski, M. (2016). Genetic engineering of mesenchymal stem cells to induce their migration and survival. *Stem Cells International, 2016*, 4956063. https://doi.org/10.1155/2016/4956063

Nystedt, J., Anderson, H., Tikkanen, J., Pietilä, M., Hirvonen, T., Takalo, R., Heiskanen, A., Satomaa, T., Natunen, S., Lehtonen, S., Hakkarainen, T., Korhonen, M., Laitinen, S., Valmu, L., & Lehenkari, P. (2013). Cell surface structures influence lung clearance rate of systemically infused mesenchymal stromal cells. *Stem Cells, 31*(2), 317–326. https://doi.org/10.1002/stem.1271

Ocbina, P. J., & Anderson, K. V. (2008). Intraflagellar transport, cilia, and mammalian hedgehog signaling: Analysis in mouse embryonic fibroblasts. *Developmental Dynamics, 237*(8), 2030–2038. https://doi.org/10.1002/dvdy.21551

Pakvasa, M., Haravu, P., Boachie-Mensah, M., Jones, A., Coalson, E., Liao, J., Zeng, Z., Wu, D., Qin, K., Wu, X., Luo, H., Zhang, J., Zhang, M., He, F., Mao, Y., Zhang, Y., Niu, C.,

Wu, M., Zhao, X., Wang, H., Huang, L., Shi, D., Liu, Q., Ni, N., Fu, K., Lee, M. J., Wolf, J. M., Athiviraham, A., Ho, S. S., He, T. C., Hynes, K., Strelzow, J., El Dafrawy, M., & Reid, R. R. (2021). Notch signaling: Its essential roles in bone and craniofacial development. *Genes & Diseases, 8*(1), 8–24. https://doi.org/10.1016/j.gendis.2020.04.006

Panepucci, R. A., Siufi, J. L., Silva, W. A., Jr., Proto-Siquiera, R., Neder, L., Orellana, M., Rocha, V., Covas, D. T., & Zago, M. A. (2004). Comparison of gene expression of umbilical cord vein and bone marrow-derived mesenchymal stem cells. *Stem Cells, 22*(7), 1263–1278. https://doi.org/10.1634/stemcells.2004-0024

Papadimitropoulos, A., Scotti, C., Bourgine, P., Scherberich, A., & Martin, I. (2015). Engineered decellularized matrices to instruct bone regeneration processes. *Bone, 70*, 66–72. https://doi.org/10.1016/j.bone.2014.09.007

Parfitt, A. M. (2001). The bone remodeling compartment: A circulatory function for bone lining cells. *Journal of Bone and Mineral Research, 16*(9), 1583–1585. https://doi.org/10.1359/jbmr.2001.16.9.1583

Peng, W., Zhu, S., Chen, J., Wang, J., Rong, Q., & Chen, S. (2019). Hsa_circRNA_33287 promotes the osteogenic differentiation of maxillary sinus membrane stem cells via miR-214-3p/Runx3. *Biomedicine & Pharmacotherapy, 109*, 1709–1717. https://doi.org/10.1016/j.biopha.2018.10.159

Ponzetti, M., & Rucci, N. (2021). Osteoblast differentiation and signaling: Established concepts and emerging topics. *International Journal of Molecular Sciences, 22*(13). https://doi.org/10.3390/ijms22136651

Rahman, M. S., Akhtar, N., Jamil, H. M., Banik, R. S., & Asaduzzaman, S. M. (2015). TGF-β/BMP signaling and other molecular events: Regulation of osteoblastogenesis and bone formation. *Bone Research, 3*, 15005. https://doi.org/10.1038/boneres.2015.5

Rashid, M. S., Smith, R. D. J., Nagra, N., Wheway, K., Watkins, B., Snelling, S., Dakin, S. G., & Carr, A. J. (2020). Rotator cuff repair with biological graft augmentation causes adverse tissue outcomes. *Acta Orthopaedica, 91*(6), 782–788. https://doi.org/10.1080/17453674.2020.1793613

Sadar, M. D. (2020). Discovery of drugs that directly target the intrinsically disordered region of the androgen receptor. *Expert Opinion on Drug Discovery, 15*(5), 551–560. https://doi.org/10.1080/17460441.2020.1732920

Safarova, Y., Umbayev, B., Hortelano, G., & Askarova, S. (2020). Mesenchymal stem cells modifications for enhanced bone targeting and bone regeneration. *Regenerative Medicine, 15*(4), 1579–1594. https://doi.org/10.2217/rme-2019-0081

Salazar, V. S., Gamer, L. W., & Rosen, V. (2016). BMP signalling in skeletal development, disease and repair. *Nature Reviews. Endocrinology, 12*(4), 203–221. https://doi.org/10.1038/nrendo.2016.12

Salhotra, A., Shah, H. N., Levi, B., & Longaker, M. T. (2020). Mechanisms of bone development and repair. *Nature Reviews. Molecular Cell Biology, 21*(11), 696–711. https://doi.org/10.1038/s41580-020-00279-w

Sarkar, J., Das, M., Howlader, M. S. I., Prateeksha, P., Barthels, D., & Das, H. (2022). Epigallocatechin-3-gallate inhibits osteoclastic differentiation by modulating mitophagy and mitochondrial functions. *Cell Death & Disease, 13*(10), 908. https://doi.org/10.1038/s41419-022-05343-1

Sasaki, M., Abe, R., Fujita, Y., Ando, S., Inokuma, D., & Shimizu, H. (2008). Mesenchymal stem cells are recruited into wounded skin and contribute to wound repair by transdifferentiation into multiple skin cell type. *Journal of Immunology, 180*(4), 2581–2587. https://doi.org/10.4049/jimmunol.180.4.2581

Schrepfer, S., Deuse, T., Reichenspurner, H., Fischbein, M. P., Robbins, R. C., & Pelletier, M. P. (2007). Stem cell transplantation: The lung barrier. *Transplantation Proceedings, 39*(2), 573–576. https://doi.org/10.1016/j.transproceed.2006.12.019

Sheehy, E. J., Kelly, D. J., & O'Brien, F. J. (2019). Biomaterial-based endochondral bone regeneration: A shift from traditional tissue engineering paradigms to developmentally inspired strategies. *Materials Today Bio, 3*, 100009. https://doi.org/10.1016/j.mtbio.2019.100009

Shen, F. H., Werner, B. C., Liang, H., Shang, H., Yang, N., Li, X., Shimer, A. L., Balian, G., & Katz, A. J. (2013). Implications of adipose-derived stromal cells in a 3D culture system for osteogenic differentiation: An in vitro and in vivo investigation. *The Spine Journal, 13*(1), 32–43. https://doi.org/10.1016/j.spinee.2013.01.002

Shi, Y., Chen, J., Karner, C. M., & Long, F. (2015). Hedgehog signaling activates a positive feedback mechanism involving insulin-like growth factors to induce osteoblast differentiation. *Proceedings of the National Academy of Sciences of the United States of America, 112*(15), 4678–4683. https://doi.org/10.1073/pnas.1502301112

Shi, Z. L., Zhang, H., Fan, Z. Y., Ma, W., Song, Y. Z., Li, M., Li, T. Q., Cao, S. X., & Feng, G. J. (2020). Long noncoding RNA LINC00314 facilitates osteogenic differentiation of adipose-derived stem cells through the hsa-miR-129-5p/GRM5 axis via the Wnt signaling pathway. *Stem Cell Research & Therapy, 11*(1), 240. https://doi.org/10.1186/s13287-020-01754-z

Spear, M. A., Dupuy, D. E., Park, J. J., Halpern, E. F., & Spiro, I. J. (1999). Tolerance of autologous and allogeneic bone grafts to therapeutic radiation in humans. *International Journal of Radiation Oncology, Biology, Physics, 45*(5), 1275–1280. https://doi.org/10.1016/s0360-3016(99)00339-9

Sun, M., Hu, L., Wang, S., Huang, T., Zhang, M., Yang, M., Zhen, W., Yang, D., Lu, W., Guan, M., & Peng, S. (2020). Circulating MicroRNA-19b identified from osteoporotic vertebral compression fracture patients increases bone formation. *Journal of Bone and Mineral Research, 35*(2), 306–316. https://doi.org/10.1002/jbmr.3892

Takahashi, K., & Yamanaka, S. (2006). Induction of pluripotent stem cells from mouse embryonic and adult fibroblast cultures by defined factors. *Cell, 126*(4), 663–676. https://doi.org/10.1016/j.cell.2006.07.024

Tang, Z., Xu, T., Li, Y., Fei, W., Yang, G., & Hong, Y. (2020). Inhibition of CRY2 by STAT3/miRNA-7-5p promotes osteoblast differentiation through upregulation of CLOCK/BMAL1/P300 expression. *Molecular Therapy - Nucleic Acids, 19*, 865–876. https://doi.org/10.1016/j.omtn.2019.12.020

Tao, H., Chen, X., Wei, A., Song, X., Wang, W., Liang, L., Zhao, Q., Han, Z., Han, Z., Wang, X., & Li, Z. (2018). Comparison of teratoma formation between embryonic stem cells and parthenogenetic embryonic stem cells by molecular imaging. *Stem Cells International, 2018*, 7906531. https://doi.org/10.1155/2018/7906531

Thomson, J. A., Itskovitz-Eldor, J., Shapiro, S. S., Waknitz, M. A., Swiergiel, J. J., Marshall, V. S., & Jones, J. M. (1998). Embryonic stem cell lines derived from human blastocysts. *Science, 282*(5391), 1145–1147. https://doi.org/10.1126/science.282.5391.1145

Tickle, C., & Towers, M. (2017). Sonic hedgehog signaling in limb development. *Frontiers in Cell and Development Biology, 5*, 14. https://doi.org/10.3389/fcell.2017.00014

Tsujimoto, H., Kasahara, T., Sueta, S. I., Araoka, T., Sakamoto, S., Okada, C., Mae, S. I., Nakajima, T., Okamoto, N., Taura, D., Nasu, M., Shimizu, T., Ryosaka, M., Li, Z., Sone, M., Ikeya, M., Watanabe, A., & Osafune, K. (2020). A modular differentiation system maps multiple human kidney lineages from pluripotent stem cells. *Cell Reports, 31*(1), 107476. https://doi.org/10.1016/j.celrep.2020.03.040

Tu, X., Chen, J., Lim, J., Karner, C. M., Lee, S. Y., Heisig, J., Wiese, C., Surendran, K., Kopan, R., Gessler, M., & Long, F. (2012). Physiological notch signaling maintains bone homeostasis via RBPjk and hey upstream of NFATc1. *PLoS Genetics, 8*(3), e1002577. https://doi.org/10.1371/journal.pgen.1002577

Wang, X., Wang, Y., Gou, W., Lu, Q., Peng, J., & Lu, S. (2013). Role of mesenchymal stem cells in bone regeneration and fracture repair: A review. *International Orthopaedics, 37*(12), 2491–2498. https://doi.org/10.1007/s00264-013-2059-2

Wang, R. N., Green, J., Wang, Z., Deng, Y., Qiao, M., Peabody, M., Zhang, Q., Ye, J., Yan, Z., Denduluri, S., Idowu, O., Li, M., Shen, C., Hu, A., Haydon, R. C., Kang, R., Mok, J., Lee, M. J., Luu, H. L., & Shi, L. L. (2014). Bone morphogenetic protein (BMP) signaling in development and human diseases. *Genes & Diseases, 1*(1), 87–105. https://doi.org/10.1016/j.gendis.2014.07.005

Wang, K., Zhou, C., Li, L., Dai, C., Wang, Z., Zhang, W., Xu, J., Zhu, Y., & Pan, Z. (2022). Aucubin promotes bone-fracture healing via the dual effects of anti-oxidative damage and enhancing osteoblastogenesis of hBM-MSCs. *Stem Cell Research & Therapy, 13*(1), 424. https://doi.org/10.1186/s13287-022-03125-2

Wei, W., Huang, Y., Li, D., Gou, H. F., & Wang, W. (2018). Improved therapeutic potential of MSCs by genetic modification. *Gene Therapy, 25*(8), 538–547. https://doi.org/10.1038/s41434-018-0041-8

Whiting, P., Kerby, J., Coffey, P., da Cruz, L., & McKernan, R. (2015). Progressing a human embryonic stem-cell-based regenerative medicine therapy towards the clinic. *Philosophical Transactions of the Royal Society of London. Series B, Biological Sciences, 370*(1680), 20140375. https://doi.org/10.1098/rstb.2014.0375

Xiang, X., Pathak, J. L., Wu, W., Li, J., Huang, W., Wu, Q., Xin, M., Wu, Y., Huang, Y., Ge, L., & Zeng, S. (2023). Human serum-derived exosomes modulate macrophage inflammation to promote VCAM1-mediated angiogenesis and bone regeneration. *Journal of Cellular and Molecular Medicine, 27*(8), 1131–1143. https://doi.org/10.1111/jcmm.17727

Yu, J., Vodyanik, M. A., Smuga-Otto, K., Antosiewicz-Bourget, J., Frane, J. L., Tian, S., Nie, J., Jonsdottir, G. A., Ruotti, V., Stewart, R., Slukvin, I. I., & Thomson, J. A. (2007). Induced pluripotent stem cell lines derived from human somatic cells. *Science, 318*(5858), 1917–1920. https://doi.org/10.1126/science.1151526

Yu, W., Li, S., Guan, X., Zhang, N., Xie, X., Zhang, K., & Bai, Y. (2022). Higher yield and enhanced therapeutic effects of exosomes derived from MSCs in hydrogel-assisted 3D culture system for bone regeneration. *Biomaterials Advances, 133*, 112646. https://doi.org/10.1016/j.msec.2022.112646

Yuan, H., Xu, X., Feng, X., Zhu, E., Zhou, J., Wang, G., Tian, L., & Wang, B. (2019). A novel long noncoding RNA PGC1β-OT1 regulates adipocyte and osteoblast differentiation through antagonizing miR-148a-3p. *Cell Death and Differentiation, 26*(10), 2029–2045. https://doi.org/10.1038/s41418-019-0296-7

Zhang, C., Li, H., Li, J., Hu, J., Yang, K., & Tao, L. (2023a). Oxidative stress: A common pathological state in a high-risk population for osteoporosis. *Biomedicine & Pharmacotherapy, 163*, 114834. https://doi.org/10.1016/j.biopha.2023.114834

Zhang, Q., Xin, M., Yang, S., Wu, Q., Xiang, X., Wang, T., Zhong, W., Helder, M. N., Jaspers, R. T., Pathak, J. L., & Xiao, Y. (2023b). Silica nanocarrier-mediated intracellular delivery of rapamycin promotes autophagy-mediated M2 macrophage polarization to regulate bone regeneration. *Materials Today Bio, 20*, 100623. https://doi.org/10.1016/j.mtbio.2023.100623

Zhao, P., Xiao, L., Peng, J., Qian, Y. Q., & Huang, C. C. (2018). Exosomes derived from bone marrow mesenchymal stem cells improve osteoporosis through promoting osteoblast proliferation via MAPK pathway. *European Review for Medical and Pharmacological Sciences, 22*(12), 3962–3970. https://doi.org/10.26355/eurrev_201806_15280

Zhou, X., Moussa, F. M., Mankoci, S., Ustriyana, P., Zhang, N., Abdelmagid, S., Molenda, J., Murphy, W. L., Safadi, F. F., & Sahai, N. (2016). Orthosilicic acid, Si(OH)4, stimulates osteoblast differentiation in vitro by upregulating miR-146a to antagonize NF-κB activation. *Acta Biomaterialia, 39*, 192–202. https://doi.org/10.1016/j.actbio.2016.05.007

Zhu, G., Zhang, T., Chen, M., Yao, K., Huang, X., Zhang, B., Li, Y., Liu, J., Wang, Y., & Zhao, Z. (2021). Bone physiological microenvironment and healing mechanism: Basis for future bone-tissue engineering scaffolds. *Bioactive Materials, 6*(11), 4110–4140. https://doi.org/10.1016/j.bioactmat.2021.03.043

Zwingenberger, S., Yao, Z., Jacobi, A., Vater, C., Valladares, R. D., Li, C., Nich, C., Rao, A. J., Christman, J. E., Antonios, J. K., Gibon, E., Schambach, A., Maetzig, T., Goodman, S. B., & Stiehler, M. (2014). Enhancement of BMP-2 induced bone regeneration by SDF-1α mediated stem cell recruitment. *Tissue Engineering. Part A, 20*(3–4), 810–818. https://doi.org/10.1089/ten.TEA.2013.0222

# Osteoblastogenesis and Bone Regeneration

**Prathyusha Naidu and Hiranmoy Das**

**Abstract** Bone formation and regeneration are essential for maintaining and repairing the skeleton, with osteoblastogenesis as the key process where mesenchymal stem cells (MSCs) differentiate into osteoblasts. This complex process is regulated by various signaling pathways, which promote osteoblast differentiation, proliferation, and extracellular matrix mineralization in response to fractures or bone loss. This process requires a series of cellular events and is regulated by growth factors, cytokines, and mechanical stimuli. With large defects or pathological states, natural regeneration may be insufficient, requiring therapeutic interventions. Biomaterial scaffolds, gene therapy, stem cell therapy, and tissue engineering have greatly enhanced bone repair. The combination of MSC with bioactive agents has shown promise in promoting osteoblast formation and accelerating bone regeneration. Recent advancements in nanotechnology and 3D bioprinting have led to innovative methods for creating biomimetic scaffolds that support bone tissue formation through structural and biochemical cues. The research aims to optimize regenerative approaches, improve scaffold biocompatibility, and identify molecular targets that could greatly improve patient outcomes in orthopedics and reconstructive medicine. This chapter discusses microenvironments supporting osteogenesis, the long-term safety of new treatments, and the potential of targeting a transcription factor, and its downstream pathways to enhance bone regeneration.

**Keywords** Osteogenesis · Wnt/Β-catenin · BMP/SMAD · MSC · Inflammation · Osteoporosis

P. Naidu · H. Das (✉)
Department of Pharmaceutical Sciences, Jerry H. Hodge School of Pharmacy, Texas Tech University Health Sciences Center, Amarillo, TX, USA
e-mail: pratnaid@ttuhsc.edu; hiranmoy.das@ttuhsc.edu



H. Das (ed.), *Tissue Repair and Regeneration*,
https://doi.org/10.1007/978-3-031-93677-7_5

## 1 Introduction

Regeneration is a normal biological process through which organisms replace injured tissues, organs, or structures. Regeneration plays a crucial role in maintaining the integrity of living organisms, enabling them to heal from trauma or injury. Regeneration occurs at various levels in animals and human beings, depending on species, tissue, and level of injury (Elchaninov et al., 2021). Certain species, such as amphibians, enjoy an excellent capacity for regenerating full limbs or tails, while human beings and other mammal species enjoy only partial regenerating capacity, mostly wound healing or regenerating tissues like the liver (Chow & Barbul, 2014). A very interesting and complex form of regeneration found in human beings and animals includes bone regeneration, an essential form of regeneration crucial for the healing of skeletal trauma (Retzepi & Donos, 2010).

Bone regeneration means the natural bone tissue repair, whereby the body rebuilds the bone tissue structure due to fracture or injury. Bone tissue has an extraordinary capacity to heal, like other tissues, by regenerating bone tissue without leaving any trace, just like the tissue of the skin, except it does not develop any scar tissue (Patel & Wairkar, 2023). The bone regeneration commences with the development of clots over the fracture, by the subsequent mobility of bone-producing cells, referred to as osteoblasts, to the fracture region. These bone-producing cells develop bone matrices, later get hardened, and attach themselves to the original bone (Rodríguez-Merchán, 2021). While the body has an efficient capacity to regrow bone tissue in the majority, it gets complicated if there are enormous defects or diseases, such as bone cancer, or osteoporosis, whereby bone regeneration gets interrupted (Atala et al., 2010).

The importance of bone regeneration does not just relate to fracture healing. Bone plays a crucial role in the support of the body, enclosing vital organs, locomotion, and maintaining essential minerals such as calcium. If bone regeneration does not succeed, or if it gets interrupted, it could lead to severe problems that compromise mobility and general health (Ettinger, 2003). In bone fracture, if regeneration does not occur, or if it occurs partially, it could lead to nonunion, where the bone does not heal properly, possibly requiring an operation to stimulate healing (Wildemann et al., 2021).

In humans, bone regeneration impairment may result in an array of health issues, including chronic pain, deformity, or functional impairment. Diseases, such as an ailment known as osteoporosis, wherein bone weakening due to an imbalance between bone building and bone breakdown, are indicative of the need for bone regeneration so that bone integrity may be maintained over one's lifetime (Pignolo et al., 2021). Similarly, bone regeneration also plays an inevitable role during bone replacement processes, bone fusions, as well as bone surgeries. If bone regeneration does not occur well, bone unions with implants may not heal well, resulting in issues such as loosening of implants or further operation (Salgado et al., 2004).

For animals, bone regeneration is also crucial, particularly for those species that tend to get injured physically or indulge in dangerous behavior, such as some amphibians or reptiles that get injured by losing appendages or tails. These species' regenerative capabilities are very interesting areas of regenerative medicine study, as they could give us some clues toward developing human regenerative treatment (Rolph et al., 2020; Majidinia et al., 2018). Regenerative treatment has continued evolving, particularly in stem cell treatment, tissue engineering, and gene editing, three areas that can boost bone regeneration capabilities.

In particular, stem cell therapy can boost bone regeneration both in human beings and animals. Scientists have also conducted trials with various stem cell types, including mesenchymal stem cells (MSCs), that can form bone-producing cells. These stem cells have also exhibited promising indications of treating bone defects, fractures, and diseases, including osteoporosis (Knight & Hankenson, 2013; Maity et al., 2022). Besides, biocompatible scaffolds, typically coupled together with stem cells, provide support essential for bone tissue development when natural bone regeneration does not meet the demands. The approach can revolutionize treatment for individuals suffering from extensive bone defects, as well as those suffering from diseases that compromise normal bone healing (Knight & Hankenson, 2013).

Advancements in bone regeneration are crucial not only to further develop treatments for skeletal trauma but also to develop treatments for various bone diseases. The role played by bone toward general well-being cannot be overstated, as it serves as the foundation for locomotion, protects essential organs, and stores essential minerals. An example includes bone regeneration studies necessary for diseases like osteogenesis imperfecta, an inheritable bone frailty disorder. Bone regeneration treatments aimed at the genetic causes of these diseases could greatly boost the quality of life experienced by those suffering (Cole, 2002).

The importance of bone regeneration is also underscored by the use of regenerative medicine, whereby regenerative approaches such as tissue engineering, complemented by the utilization of 3D printing, are being exploited to create bone grafts or scaffolds. These technological innovations are making treatment processes faster and more personalized, which could lead to shorter time to heal and predictable outcomes (Arealis & Nikolaou, 2015). Tissue-engineered bone grafts could be particularly useful in individuals suffering from severe bone loss due to trauma, cancer, or congenital defects. This development emphasizes the necessity of bone regeneration both in the clinic and the lab.

## 2 Steps of Bone Regeneration

Bone regeneration involves an orderly, sequential series of events aimed at regaining the integrity and functioning of an injured bone. These processes, including inflammation, reparative, and remodeling, heal bone defects or fractures collectively. Regeneration occurs instantly upon injury, continuing over several months to several years.

(a) **Inflammatory Stage**

The initial phase of bone regeneration involves the inflammation phase, occurring instantly after bone injury. It lasts several days and plays an essential role in laying down the foundation for bone repair (Shrivats et al., 2014). Inflammation starts by causing a rupture of vessels in the injured tissue, resulting in the creation of a hematoma, or clot. The clot acts as a support for the arriving cells necessary for repair. It also serves to regulate bleeding, creating an environment that allows it to heal (Duda et al., 2023).

During the inflammatory phase, cytokines (i.e., tumor necrosis factor-α (TNF-α), interleukin-1β (IL-1β)) and growth factors (i.e., platelet-derived growth factor, transforming growth factor-beta) are released, recruiting immune cells such as neutrophils, macrophages, and lymphocytes (Knapik et al., 2014). These cells clear away the dead tissue, preventing infection, and facilitating healing (Einhorn & Lee, 2001). Macrophages, in particular, play an active role both during the inflammatory phase and the reparative phase by secreting cytokines that regulate entry into the next phase of healing (Henkel et al., 2013; Scharfetter-Kochanek et al., 1998).

Moreover, the inflammatory phase also starts fibrin mesh formation and initial synthesis of collagen, which shall provide the foundation for the ensuing bone repair (Hernández-Gil et al., 2006). Inflammation has an important role in initiating healing, while over-inflammation delays the same, besides causing problems like nonunion and infection (Giannoudis et al., 2015).

(b) **Reparative Stage**

The reparative phase follows, whereby there is the formation of new bone tissue, referred to as the callus, at the fracture site. It lasts several weeks to months, depending on the severity of the injury as well as the general well-being of the individual (Oryan et al., 2014). In this phase, the hematoma produced by the inflammatory phase gets gradually supplanted by the granulation tissue, which has an abundance of vascular elements as well as fibroblasts (Hankenson et al., 2015).

Chondrocytes and osteoblasts, two bone-healing central cell types, proliferate during the reparative phase. Chondrocytes develop cartilage, whereas osteoblasts lay down a collagen matrix, creating a soft callus over the fracture (Panetta et al., 2010). Over time, the cartilage gets ossified, wherein it gets replaced by bone tissue. The replacement is mediated by the effect of growth factors, bone morphogenetic proteins (BMPs), and vascular endothelial growth factor (VEGF) (Zhang et al., 2014).

Bone formation takes place primarily using intramembranous and endochondral ossification, varying by bone type and injury type. In the case of long bones, endochondral ossification prevails, whereby there first takes place an initial development of a cartilage template, subsequently slowly mineralizing into bone (Arvidson et al., 2011). Intramembranous ossification, through direct differentiation of the mesenchyme into bone tissue, prevails in the case of flat bones (Deschaseaux et al., 2009). This stage is also very responsive both internally (in terms of the presence of enough blood flow) and externally (in terms of the environment mechanically). An example is the non-stable fixation of the fracture, interference with the development of the soft callus, leading to nonunion or delay (Oryan et al., 2013). A balance, thus, has to be found between the mechanical stability and the biological factors if the reparative phase has any possibility of success.

(c) **Remodeling Stage**

The final stage of bone regeneration, remodeling, lasts several months to years. In this stage, the newly formed bone tissue gets matured and remodeled. In this remodeling, bone callus formed during the reparative phase gets slowly replaced by compact, well-ordered bone tissue. It is achieved by coordinated functions by both osteoclasts, responsible for bone breakdown, and osteoblasts, responsible for bone construction (Oryan et al., 2013).

Osteoclasts break down the soft callus produced during the reparative phase, while osteoblasts deposit fresh bone tissue replacing the resorbed matrix (Laha et al., 2022; Hernández-Gil et al., 2006). The bone regains its original form and strength by this method. The newly produced bone receives further mineralization and corticalization, giving a compact, strengthened bone structure (Isaksson, 2012). The bone remodeling occurs over many years, by which the bone adapts to the load it sustains, maximizing strength and functioning (Giannoudis et al., 2015).

One of the fundamental functions of bone remodeling involves the ability it has to regain the bone's functional capabilities. It plays a crucial role in bone system integrity. The bone, nonetheless, also has an active system of bone remodeling, whereby there is ongoing bone turnover over one's lifetime (Lieberman & Friedlaender, 2005). The bone remodeling time may also vary depending on various other factors, such as one's age, hormone balance, and dietary level, that may interfere with bone regeneration efficiency.

In addition to the initial processes of inflammation, reparative, and remodeling, bone healing also has other determinants that could regulate bone healing efficiency. These include the role played by stem cells during tissue regeneration as well as the mechanical environment of the injury site. Stem cells, that is, the MSC, play a crucial role during bone tissue regeneration by giving birth to osteoblasts, chondrocytes, and other bone tissue regeneration-involved cells (El-Jawhari et al., 2016). Besides, the bone load during bone regeneration can regulate the bone regeneration rate as well as bone regeneration result (Lieberman & Friedlaender, 2005).

## 3 Osteoblastogenesis and Bone Regeneration

Osteoblastogenesis refers to the differentiation and maturation of bone-producing cells, also referred to as osteoblasts. These osteoblasts are produced from MSC, and differentiation thereof plays a crucial role in normal bone development, maintenance, and repair. These cells play a crucial role in bone synthesis and mineralization, lending the bone structural support (Mutsaers & Walkley, 2014). Through osteoblastogenesis, the osteoblasts form collagen type I, osteocalcin, and other extracellular matrix proteins, which form the mineralized bone tissue. Any impairment or defect in the functioning or differentiation of the osteoblasts could lead to bone diseases such as osteoporosis, osteogenesis imperfecta, and fracture with retarded healing (Caetano-Lopes et al., 2007).

The process is initiated by committing MSC toward the osteogenic lineage, mediated by various molecular cascades. Wnt/β-catenin pathway, BMPs, and Notch signaling pathway are the cascades relevant here (Rolph et al., 2020; Zhou, 2011). Wnt signaling, in particular, plays a crucial role in initiating and progressing osteoblast differentiation by regulating gene expression involved in osteogenesis. BMPs, additionally, are excellent drivers of osteoblastic differentiation, facilitating the differentiation of MSC toward osteoblasts (Wozney, 1992). These cascades converge toward central transcriptional determinants Runx2 and Osterix (SP7), responsible for gene expression regulation of osteoblast-specific genes, further progressing toward the maturation of osteoblasts (Huang et al., 2007). Elucidation of these molecular processes offers insight into bone regeneration strategies in the clinic.

Bone remodeling and repair are fundamental processes that play an essential role in maintaining bone integrity and functionality. Bone remodeling, an active continuous process, involves the resorption of old bone by osteoclasts, coupled with the synthesis of fresh bone by osteoblasts (Hadjidakis & Androulakis, 2006). The coupled balance maintains bone homeostasis, allowing adaptation to stresses, as well as microdamage repair. Coordination between the functions of osteoblasts and the functions of osteoclasts is tightly coupled so as not to result in diseases, including osteoporosis, an overabundance of bone resorption, or osteopetrosis, by way of faulty bone resorption (Lemaire et al., 2004). Bone repair, on the other hand, occurs after trauma, injury, or fracture, starting with hematoma, followed by MSC, osteoprogenitor, and subsequent cartilaginous, later mineralized bone matrix (Deschaseaux et al., 2009). Regenerative capacity is mediated by coordination between the functions of osteoblasts, osteoclasts, and the surrounding extracellular matrix (ECM), both mechanically and biochemically.

The clinical application of bone regeneration has gained immense relevance in orthopedic as well as dental scenarios, where bone injury or bone loss comes to the forefront. In orthopedics, bone regeneration has an essential application in fracture management, bone defects, as well as diseases like osteoporosis (Chen et al., 2004). Therapeutic approaches such as bone grafting, application of osteoinductive materials like BMPs, and stem cell therapy attempt to enhance the natural repair by triggering osteoblastogenesis (Milner et al., 2011). In dental scenarios, bone regeneration has an essential application in successful implant placement, where dental implants require enough bone density and volume for anchorage (Egido-Moreno et al., 2021). Regenerating bone in the oral cavity, especially due to congenital defects or periodontal diseases, has also gained prominence as an area of study, utilizing approaches such as tissue engineering and gene therapy to induce differentiation of osteoblasts at the injury site (Pilipchuk et al., 2015).

In the quest for efficient bone regeneration approaches, it is necessary to comprehend the cellular and molecular processes underlying osteoblastogenesis, especially while formulating newer treatment approaches. Historically, the study of bone biology dates back to ancient civilizations, where early surgeons performed rudimentary bone-setting techniques (Buckwalter et al., 1995). However, modern advancements in bone biology research began in the late nineteenth century with the discovery of osteoblasts and osteoclasts. The introduction of electron microscopy in the mid-twentieth century allowed researchers to visualize osteoblast activity in greater detail (Chen et al., 2004). More recently, the focus has shifted toward understanding the molecular signals that control bone formation and regeneration. The advent of stem cell technology and tissue engineering has provided new avenues for accelerating bone regeneration, offering hope for patients suffering from severe bone injuries or degenerative bone diseases (Sui et al., 2019). Despite these advances, challenges remain in achieving efficient and sustainable bone regeneration, particularly in large bone defects or conditions involving impaired osteoblast function.

## 4 Cellular and Molecular Mechanisms of Osteoblastogenesis

### *4.1 MSC as Osteoblast Precursors*

MSCs are one of the well-known cell types involved in osteoblastogenesis, serving as precursors to the bone-producing osteoblasts. MSCs are located in many tissues, each containing varying benefits for differentiation into osteoblasts. Bone marrow has also been one of the best-studied sources, consisting of an abundant source of MSC that can develop into osteoblasts, chondrocytes, and adipocytes (Khan & Hardingham, 2012). Adipose tissue, also rich in MSC, has also proved to support osteogenic differentiation in vitro (Chen et al., 2016). In addition, dental pulp has also turned out to be an interesting source for MSC, with studies showing that MSC isolated from dental pulp display extensive osteogenic and neural capacity (Prateeksha et al., 2023b; Ito et al., 2011). Also, MSCs isolated from the periosteum have exhibited excellent osteogenic capacity, making them excellent targets for bone repair and regeneration (Ferretti et al., 2012).

The differentiation ability of MSC toward the lineage of osteoblasts is influenced by several factors, including microenvironmental stimuli and several signal transduction cascades. MSCs commit to the lineage of osteoblasts by activating lineage-specific transcriptional factors, including Runx2 and SP7 (Olmsted-Davis et al., 2003). Besides, the osteogenic ability of MSC from other sources also varies. Bone marrow-derived MSCs, for instance, are very efficient toward differentiation toward the lineage of osteoblasts, while adipose tissue-derived MSCs are less efficient, but this efficiency may be enhanced by utilizing several protocols of pre-differentiation (Ito et al., 2011). In addition, MSCs isolated from dental pulp and periosteum have also proved to play a crucial role in the generation of osteoblasts, commonly reporting greater efficiency toward tissue engineering and regenerative medicine (Mutsaers & Walkley, 2014) (Fig. 1).

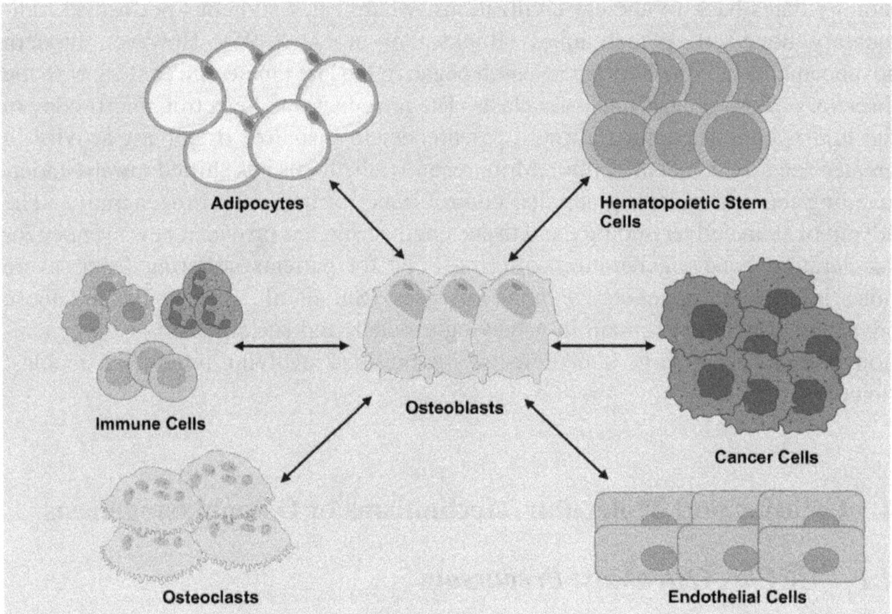


**Fig. 1** Osteoblast in bone marrow microenvironment

## *4.2 Key Signaling Pathways in Osteoblast Differentiation*

The regulation of differentiation of the osteoblasts includes an intricate system operated by several essential signaling routes. These include the Wnt/β-catenin pathway, BMPs, transforming growth factor-β (TGF-β) pathway, mitogen-activated protein kinase (MAPK) pathway, Hedgehog, and Notch pathway. These routes play an intricate, interconnected role, both effecting the pathway of osteogenesis and bone-producing osteoblasts (Fung Ling Chau et al., 2009; Prateeksha et al., 2023a).

The Wnt/β-catenin pathway has an active role in the differentiation of the osteoblasts and bone development. The pathway operates both via canonical and noncanonical processes. In the Wnt/β-catenin pathway, Wnt ligands attach to the Frizzled receptor, suppressing the degradation of β-catenin; thus, it accumulates in the cytoplasm and translocates to the nucleus. In the nucleus, β-catenin functions by activating those transcriptional factors essential for differentiation of the osteoblasts as well as mineralization, including Runx2, an essential master gene of osteogenesis (Lin & Hankenson, 2011). Wnt canonical signaling has also been shown to significantly enhance the differentiation of the osteoblasts by activating markers of osteogenesis (Yavropoulou & Yovos, 2007). In addition to the canonical pathway, there also exists noncanonical Wnt signaling, one that does not involve β-catenin, also responsible for differentiation of the osteoblasts, although less well known (Eijken et al., 2008).

BMPs including BMP2, BMP4, and BMP7 are key players in the regulation of osteogenesis. BMPs are also referred to as the TGF-β superfamily, known to be very powerful osteoinductive compounds. BMP2 has also been studied significantly due to its capacity to induce bone differentiation and bone generation (Yang et al., 2014). BMP4, as well as BMP7, also regulates the functions of the osteoblasts, while BMP7 has an active role specifically in bone repair and bone regeneration (Rahman et al., 2015). These BMPs achieve this by binding to the BMP receptors, triggering the intracellular signal cascades, including the Smad pathway, resulting in the induction of differentiation of the osteoblasts and bone matrix synthesis. BMPs play a crucial role both during embryonic bone generation and adult bone regeneration (Chen et al., 2004; Rahman et al., 2015).

The TGF-β signaling pathway plays a crucial role in early osteogenic differentiation. TGF-β exerts an effect on many cellular processes, including differentiation, proliferation, and apoptosis. In osteoblast differentiation, TGF-β signaling acts principally through Smad proteins, Smad2, and Smad3 that transduce the effect of TGF-β on gene expression (Eijken et al., 2008). TGF-β signaling plays a role in initial processes in the differentiation of osteoblasts by triggering the expression of crucial markers of osteogenic differentiation, such as Runx2 (Chen et al., 2012). The effect of TGF-β, however, is context-dependent, as TGF-β also inhibits bone formation under other contexts by triggering osteoclastogenesis, showing its bimodal role in bone biology (Li et al., 2022). MAPK signaling also plays an essential role in the differentiation of osteoblasts. MAPK pathway involves a network of serine/threonine kinases that regulate various cellular processes, including differentiation. The MAPK pathway, specifically the ERK1/2 and p38 MAPK branches, has also been shown to transfer the effects of mechanical load and growth factor on the differentiation and functioning of osteoblasts (Schindeler & Little, 2006). Activation of the pathway by ERK1/2 causes induction of osteoblast-specific transcriptional factors such as Runx2, further stimulating differentiation. Besides, p38, MAPK plays a role in transmitting stresses and inflammatory stimuli, perhaps altering osteoblast differentiation during pathological processes (Wu et al., 2013).

Hedgehog signaling is also one pathway that controls the differentiation of osteoblasts, although it has context-dependent functions. Activation by binding of Hedgehog ligands to the Patched receptor initiates Smoothened protein activation, ultimately leading to the release of Gli transcription factors. Indeed, the Hedgehog pathway has been found to regulate the differentiation of osteoblasts in some contexts, specifically bone mineralization regulation and maturation of osteoblasts (Kim et al., 2010). Contributions, however, of Hedgehog signaling specifically in the biology of osteoblasts remain under study, while it plays varying functions depending on tissue type or bone developmental stage (Cai et al., 2012).

Notch signaling also influences the differentiation of osteoblasts by controlling the choice of bone-producing cells' fate. Notch signaling is activated by Notch receptors' binding to Notch ligands, resulting in the Notch receptor becoming cleaved, and release of the Notch intracellular domain (NICD). NICD translocates to the nucleus, where it starts the transcription of differentiation-specific genes of osteoblasts (Cao et al., 2017). Notch signaling has also been shown to induce commitment of the MSC toward the osteoblastic lineage, while also preventing differentiation of osteoblasts under certain conditions (Ugarte et al., 2009). Notch signaling balance by other processes, such as Wnt and BMP, plays an essential role in the differentiation and functioning of osteoblasts (Luo et al., 2019).

## 4.3 Transcription Factors Regulating Osteoblastogenesis

Transcription factors, such as RUNX2, SP7, Dlx5, and Kruppel-like factors (KLFs), play crucial functions during bone development and differentiation of osteoblasts. RUNX2, in particular, has also been known as the master gene that plays the role of osteoblastogenesis, as well as osteogenesis. RUNX2 plays an initial role in bone development by initiating the differentiation of MSC into osteoblasts. RUNX2 also regulates gene expressions of other crucial gene expressions, such as alkaline phosphatase, and type I collagen, responsible for the generation of mineralized bone matrice (Komori, 2006). RUNX2 has also been found by studies to be increased by mechanical stresses, as well as by BMP signaling, further showing RUNX2 as an initial crucial regulator of differentiation of osteoblasts (Chan et al., 2021). RUNX2 has also been found in genetic studies conducted by studying mice, wherein genetic deactivation of RUNX2 has shown the complete absence of bone development, further showing RUNX2 as an essential gene responsible for bone development (Marie, 2008).

Osterix also plays an essential role in the differentiation of osteoblasts. While RUNX2 initiates differentiation of the osteoblasts, it is the later maturation of these cells that entails the essential role played by Osterix. RUNX2, alongside other cascades of signal, Wnt, and BMP, regulates the expression of Osterix (Zhang, 2010). It has also been noted that the lack of Osterix, as experienced by Osterix knockout mice, has an extensive reduction in bone development and ossification, devoid of matured osteoblasts, showing that terminal differentiation of the osteoblasts necessitates the presence of Osterix. Besides, it has also been noted that Osterix has an essential role in the regulation of mineralization by the osteoblasts, as it initiates gene expressions directly responsible for bone matrix deposition (Liu et al., 2020).

Dlx5, a homoeobox protein belonging to the gene family, has also been found to have a role in osteoblastogenesis. Dlx5 has also been found early during the differentiation of the osteoblast and has also cooperated with RUNX2 to facilitate commitment and differentiation of the osteoblast (Witkowska-Zimny et al., 2010). The gene *Dlx5* has an essential role during the development of the craniofacial skeleton and has a role in the regulation of bone-developing gene essential elements. Dlx5 has also been found to induce gene products relating to the osteoblast, such as osteocalcin, an essential gene involved in mineralization (Franceschi et al., 2007). Although less so than RUNX2, or Osterix, it still has an essential role in maintaining the regulation of the osteogenic lineage by binding other transcriptional elements as well as signal cascades.

## 5 KLFs and Functions

Kruppel-like factors (KLFs) are DNA-binding transcriptional factors that belong to the zinc finger family and perform a variety of roles in biological activities such as quiescence, proliferating, differing growth, maturation, and inflammation (Das et al., 2006). KLF family members are extremely conserved; at present, seventeen different members (KLF1 to KLF17) have been identified. The three Cys2/His2 zinc finger domains shared by every member of the aforementioned protein family enable them to attach to the shared DNA-binding sites of transcriptional sequences that are targeted (Feinberg et al., 2004).

The involvement of KLF2 in the physiological functioning of lymphocytes, particularly in terms of their existence, division, and trafficking, has received the greatest attention among all of the members (Jha & Das, 2017). KLF2 inhibits the expression of proinflammatory genes, such as those for attachment molecules like VCAM-1 and ICAM-1 and chemokines like MCP-1, to reduce inflammation in endothelial cells. It works by hooking up to gene promoters, preventing NF-B signaling, reducing the acquisition of immune cells, and preserving vascular integrity (SenBanerjee et al., 2004). KLF2 has a variety of functions in vascular growth, diseases, and inflammatory processes. By preventing the expression of proinflammatory genes, such as those encoding adhesion molecules and chemokines, it controls endothelial function, reducing immune cell recruitment and preserving vascular integrity (Lin et al., 2005; Novodvorsky & Chico, 2014; Turpaev, 2020). Additionally, KLF2 affects endothelial cells' ability to clot, which might affect cardiovascular health (Lin et al., 2005). KLF2 is implicated in the control of adipose tissue outside of the vasculature through its inhibition of adipogenesis and peroxisome proliferator-activated receptor expression (Banerjee et al., 2003). (Hansda et al., 2024).

## 6 KLF2 in Osteogenesis

More specifically, KLF2 plays a critical regulatory function in the differentiation of periodontal ligament stem cells into osteoblasts, which are bone-forming cells, throughout the course of osteogenesis. Osteoblast differentiation is the process by which precursor cells develop into adult osteoblasts, which are in the position of creating the bone matrix and assisting in the production of new bone. The *Runx2* gene is specifically targeted by the KLF2 to affect osteoblastogenesis (Hou et al., 2019). Runx2 is an essential transcription factor that regulates the expression of genes involved in osteoblast development and bone production(Komori et al., 1997). It has been demonstrated that KLF2 directly interacts with the Runx2 gene's regulatory regions, indicating that KLF2 regulates Runx2's expression through transcription. By boosting Runx2 expression and linking to it, KLF2 may also encourage the transcription of the osteoblastic differentiation marker genes Alp, Osx, and Ocn and enhance mineralization. Runx2 is regulated by factor KLF2, and Runx2

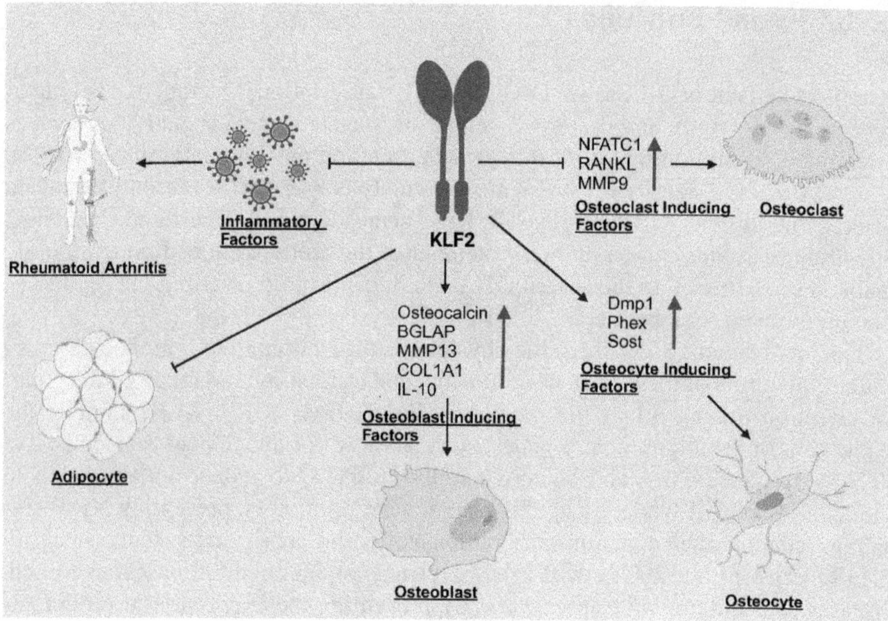


**Fig. 2** The schematic diagram of a transcription factor, KLF2, in the complex interaction with various bone cells and its regulatory role in bone marrow microenvironment

upregulation by KLF2 is enough to set off osteogenesis (Hou et al., 2019). By association with the protein IRF2BP2, KLF2 helps to promote the differentiation of osteoblasts. This association is a component of a wider regulatory axis that affects the differentiation of both osteoclasts and osteoblasts (Kim et al., 2019) (Fig. 2).

In addition to the role played by the transcriptional factors, epigenetic regulation also has an essential role to play in the differentiation and functioning of the osteoblasts. Epigenetic processes such as DNA methylation, modification of the histones, and noncoding RNAs are responsible for the precise regulation of gene expressions during the differentiation of the osteoblasts. DNA methyltransferases (DNMTs) such as DNMT1, DNMT3A, and DNMT3B have been shown to play essential functions in maintaining gene patterns responsible for the differentiation of the osteoblasts (Park-Min, 2017). The regulation of the expressions of these DNMTs occurs via various signal transduction cascades, including Wnt/β-catenin, BMP, and TGF-β (Adithya et al., 2022). Methylation of various CpG positions in RUNX2 and Osterix genes can enhance or suppress expressions, depending upon the level of development during the differentiation of the osteoblasts (Ling et al., 2017). Histone modifications, an epigenetic mode of regulation, also regulate the differentiation of the osteoblast. Acetylation and methylation of the histones regulate the accessibility of the chromatin, and, by extension, the gene expression of the osteoblast (Ghayor & Weber, 2016). For example, histone acetyltransferases (HATs) such as CREB-binding protein (CBP) and p300 have been shown to induce differentiation gene expression of the osteoblast by acetylating the histones and unwinding the chromatin (Xu et al., 2021). However, deacetylases (HDACs) downregulate the gene expression of the osteoblast by deacetylating the histones and compacting the chromatin (Barkhordarian et al., 2011). Also, histone methylation, specifically, at specified positions, such as H3K4me3, has been shown to mark the gene activation of the osteoblast, while the presence of the mark H3K27me3 signifies silencing of the non-osteoblast gene (Sharma et al., 2024).

Noncoding RNAs, including microRNAs (miRNAs) and long noncoding RNAs (lncRNAs), have emerged as crucial differentiation regulators of osteoblasts. MiRNAs are small, noncoding RNAs that regulate gene expressions by binding to the 3′ untranslated region (UTR) of target mRNAs, triggering the degradation of the mRNA or silencing the translation. MiR-204, miR-206, and miR-34, for instance, have been shown to regulate the differentiation of osteoblasts by targeting crucial transcriptional factors RUNX2 and Osterix (Lanzillotti et al., 2021). These miRNAs induce or inhibit osteoblastogenesis by altering the gene expressions responsible for differentiation. LncRNAs, on the other hand, have an intricate role in the differentiation of osteoblasts. LncRNAs regulate gene expression across various processes, including chromatin modification, transcriptional regulation, and posttranscriptional modification (Ko et al., 2020). An example is the lncRNA, H19, which is found to regulate RUNX2, other transcriptional factors, and differentiation of osteoblasts. LncRNAs, KCNQ1OT1, and MEG3 have also been found to interact with miRNAs and regulate the differentiation of osteoblasts by influencing signaling cascades, Wnt/β-catenin, and BMP (Yang et al., 2020).

## 7 Extracellular Matrix (ECM) and Bone Mineralization

ECM has an essential role in bone mineralization, both as a scaffold for bone form and as a depot for minerals necessary for bone strength. Bone ECM exists mostly as collagen and non-collagenous protein. Together, these elements facilitate bone mineralization, an active process essential for bone integrity and functionality. In this section, bone ECM composition, the functions of individual non-collagenous proteins, and mineralization, focusing specifically on the creation of hydroxyapatite, as well as alkaline phosphatase (ALP) functions, are covered.

## 8 Composition of Bone ECM

The bone ECM has an ordered protein network, the bulk of them being type I collagen. Collagen adds tensile strength to the bone matrix, mechanically strengthening it, as well as giving it a scaffold upon which minerals precipitate. Bone ECM also has a variety of non-collagenous proteins that perform several functions, including mineralization, cellular signaling, and regulation of bone remodeling. These are the proteins osteocalcin, osteopontin, bone sialoprotein (BSP), and fibronectin, performing discrete but interconnected functions during bone mineralization (Boskey, 1989).

Osteocalcin, a vitamin K-dependent protein, has emerged as one of the best-studied non-collagenous bone ECM proteins. Produced by osteoblasts, it plays an essential role in the regulation of mineralization. Hydroxyapatite crystals are bound by osteocalcin, thus allowing them to get incorporated into the ECM, and enhancing bone mineral depositions. The mechanically load-dependent expression of osteocalcin has been established, while the mineralization function has been found to relate to the capacity of osteocalcin to bind hydroxyapatite (Li et al., 2016).

Osteopontin, another essential non-collagen protein, plays an essential role in the regulation of ECM-cell interaction. It plays a role in osteoclast differentiation, attachment, and migration, essential processes involved in bone resorption, as well as bone remodeling. Through attachment to integrins located on the surface of the osteoclasts, it enhances attachment to the mineralized matrix, allowing bone tissue resorption (Giachelli & Steitz, 2000). Through this, it also plays an essential role in mineralization, together with the capacity to bind ions of calcium, playing an essential role in bone mineral homeostasis (Foster et al., 2018).

Bone sialoprotein (BSP) is an osteoblast-expressed glycoprotein that functions as an indicator of bone formation. It plays a crucial role during early mineralization, as it binds phosphate ions and calcium, facilitating the formation of hydroxyapatite. Bone sialoprotein has also been shown to nucleate the hydroxyapatite crystals, an essential function in the generation of an osseous mineralized matrix. Bone sialoprotein also associates with integrins, facilitating attachment of the osteoblasts to the ECM, allowing cellular functions necessary for bone generation (Gordon et al., 2007).

Fibronectin is a multifunctional protein that plays a role in cellular adhesion and migration. In bone ECM, it binds other ECM elements, including proteoglycans and collagen, to form an established scaffold for osteoblasts and osteoclasts. Fibronectin also regulates bone differentiation signal transduction by bone cells. It has also been shown that the binding of fibronectin to integrins influences the mineralized matrix laid down by osteoblasts, showing it plays a role in the mediation of the biochemical and biomechanical environment of bone mineralization (Bentmann et al., 2010).

## 9 Process of Bone Mineralization

Bone mineralization refers to the laying down of inorganic minerals, primarily calcium and phosphate, into the ECM. Hydroxyapatite (Ca10(PO4)6(OH)2), the bone's major mineral composition, exists as an ordered crystalline form that toughens and hardens the bone. Mineralization commences when the bone-producing osteoblasts release the ECM's organic materials, including collagen and non-collagenous protein. Calcium ions, together with phosphate ions, are deposited slowly while the matrix grows, forming hydroxyapatite crystals (Murshed, 2018).

The mineralization process also has several controlling factors, one being ALP enzyme activity. ALP is an enzyme that has an essential role in bone mineralization, produced by osteoblasts. It catalyzes the hydrolysis of the efficient mineralization

inhibitor, inorganic pyrophosphate. Enhancing the buildup of phosphate ions in the ECM increases the level of phosphate ions available, essential for the nucleation and crystallization of hydroxyapatite (Ruffoni et al., 2007). The level of enzyme activity has a close correlation with the differentiation of the osteoblasts as well as the ability of these cells to develop a mineralized matrix. High values of ALP activity are normally found alongside increased bone mineralization as well as an adult bone matrix (Bala et al., 2013).

The formation of hydroxyapatite is one crucial phase of bone mineralization. Hydroxyapatite crystallization takes place by precipitation of phosphate ions, together with precipitation of calcium ions, within the ECM. Hydroxyapatite crystallization involves the presence of non-collagenous proteins, such as osteocalcin, osteopontin, and bone sialoprotein, as these proteins can bind phosphate ions, together with calcium ions, and facilitate crystallization (Donnelly & Boskey, 2011). The role played by alkaline phosphatase during the development of hydroxyapatite is essential. Organic compounds are dephosphorylated by ALP, increasing the local phosphate ion level, which plays an essential role in developing hydroxyapatite (Bianco, 2020). It has been established through studies that there is greater mineralization of the matrix by those osteoblasts that have increased enzyme activity, further indicating the role played by the enzyme during bone mineralization (Boskey et al., 2006). In addition to playing the mineral depositing role, ALP also plays the role of maintaining differentiation of osteoblasts, so that mineralization occurs during the appropriate phase of bone development (Ruffoni et al., 2007).

## 10 Growth Factors and Cytokines in Osteoblastogenesis and Bone Regeneration

Bone regeneration, as well as osteoblastogenesis, is a complex process influenced by a multitude of cascades of signaling, including cytokines, as well as growth factors. Among them, BMPs, fibroblast growth factors (FGFs), insulin-like growth factors (IGFs), VEGF, proinflammatory cytokines, including tumor necrosis factor-alpha (TNF-$\alpha$), interleukin 6 (IL-6), and interleukin-1 beta (IL-1$\beta$), play crucial roles in osteogenesis, bone remodeling (Mundy et al., 1995).

## 11 BMPs and Their Therapeutic Applications

BMPs are one of the subsets of the TGF-$\beta$ superfamily and are one of the most essential bone formation growth factors (Wu et al., 2016). They play an essential role in the differentiation of MSC into bone-producing osteoblasts (97). Therapeutic application has shown that BMPs, including BMP-2, BMP-4, BMP-6, BMP-7, and BMP-9, have shown effectiveness in bone regeneration (Kempen et al., 2010). BMPs induce osteoblastogenesis by triggering intracellular signal cascades, including the Smad pathway, ultimately resulting in the gene transcription responsible for osteogenic differentiation (Toosi & Behravan, 2020). BMPs have gained extensive application studies due to their therapeutic application in bone tissue engineering, notably fracture treatment, spinal fusions, and bone defects (Bosetti et al., 2007).

## 12 Fibroblast Growth Factors (FGFs) and Bone Homeostasis

FGFs, especially FGF2, also known as basic FGF, play a crucial role in bone homeostasis and differentiation of osteoblasts. FGF2 has also been found to induce the proliferation of osteoprogenitor cells as well as induce the differentiation of osteoblasts (Charoenlarp et al., 2017). FGF2 also initiates several downstream signal cascades, including MAPK, and PI3K-Akt, that play a crucial role in osteoblastogenesis. FGF23, also an FGF, plays a crucial role in phosphate balance and mineral metabolism, indirectly playing a role in bone health (Safari et al., 2021). FGF23 has also been found to play a role in bone mineralization-related diseases, including osteomalacia, as well as rickets while playing an active role in bone regeneration has also been an active field (Charoenlarp et al., 2017).

## 13 Insulin-Like Growth Factor (IGF) and Osteogenic Signaling

IGFs, particularly IGF1 and IGF2, are essential bone growth and differentiation regulators of osteoblasts. IGF1 has been shown to facilitate the differentiation and proliferation of osteoblasts, as well as induce bone formation (Yu et al., 2012). It acts through the IGF1R and initiates PI3K-Akt and MAPK cascades, triggering osteogenesis (Al-Kharobi et al., 2014). IGF1 also acts together with the signal pathway of the hormone growth (GH), creating a synergistic effect on bone development (Feng et al., 2014). IGF2, on the other hand, has an increased role during fetal bone development, as well as has also been shown to induce differentiation of osteoblasts, as well as mineralization (Wang et al., 2012). Both IGFs have essential functions in bone remodeling by regulating the functioning of osteoblasts, as well as osteoclastogenesis, making them essential for bone regeneration (Yuan et al., 2019).

## 14 Vascular Endothelial Growth Factor (VEGF) in Bone Repair

VEGF is a crucial factor involved in the course of angiogenesis, fundamental to bone regeneration and repair. Bone tissue requires adequate vascularization necessary for the flow of nutrients, oxygen, and growth factors, fundamental to fracture

stem cells are easy enough to extract from pulled-out teeth, an advantage over other stem sources, as it does not involve an intrusive method. DPSCs have also found application in studies conducted on animals, applying them toward regenerating bone defects, showing promising results indicating increased bone regeneration when combined with scaffold materials (Qu et al., 2021). Similarly, PDLSCs, sourced from human tooth periodontal ligament, both have osteogenic and cementogenic differentiation capacity, making them very viable sources of bone regeneration in periodontal tissue (Lei et al., 2014). These stem cells have also found extensive application in periodontal tissue engineering, and it has been suggested that these stem cells play an active role not only in bone regeneration but also in periodontal ligament repair (Liu et al., 2009).

## 20 Induced Pluripotent Stem Cells (iPSCs) in Bone Regeneration

Induced pluripotent stem cells (iPSCs) are an evolving regenerative field. Fully differentiated somatic cells that are reprogrammed back into an embryonic state, iPSCs can develop into any form of cell, including an osteoblast. In comparison to adult stem cells, iPSCs offer an essentially limitless source of cells for regenerative application (Tang et al., 2014). In bone regeneration, iPSCs have been used in various studies both in vitro and in vivo to assess their ability to perform osteogenic differentiation. These cells have managed to be differentiated into an osteoblast under defined culture conditions and found to support bone healing in models (Teng et al., 2014). The advantage over other stem cells is the ability to customize them to an individual, ruling out the possibility of an immune response. While promising, iPSCs are still an area of intensive research, particularly the safety of application due to the possibility of tumor formation (teratoma) and genetic instability (Rana et al., 2019). Each type of stem cell has its advantages and drawbacks when it comes to bone regeneration. BMMSCs have had extensive research conducted on them and are the gold standard for bone tissue engineering. Their well-developed osteogenic ability, capacity to regulate the immune system, and accessibility from bone marrow make them an incredibly attractive candidate for clinical application (Illich et al., 2011).

ADSCs offer an advantage by being less intrusive when it comes to harvesting, while also displaying promising bone regeneration ability (Bastami et al., 2017). ADSCs, however, are still under investigation when it comes to whether it is safe and efficient over the long term. DPSCs and PDLSCs are promising when it comes to regenerating periodontal tissues and dental tissues but are less well-utilized compared to BMMSCs or ADSCs (Xie et al., 2016). iPSC offers an interesting option due to the ability it has to form pluripotent stem cells, as well as generate patient-specific cells, but the use of iPSC in the clinic still has limitations due to issues over whether it is safe, specifically whether it can induce tumorigenesis (Csobonyeiova et al., 2017).

### *20.1 Stem Cell Therapy for Bone Defects and Nonunion Fractures*

Stem cell therapy has also emerged prominently in bone defect treatment, as well as fracture nonunions, due to its capacity to induce bone regeneration and facilitate healing. Stem cells have proved to be effective, as recent studies have shown, by stimulating bone repair through induction of osteogenesis, angiogenesis, as well as synthesis of collagen (Killington et al., 2018). Stem cell transplant, an autologous approach wherein stem cells are harvested from the body of the patient, has found extensive use due to lowered risks of immune rejection. Allogeneic stem cell transplant, wherein stem cells are harvested from a donor, has also found use, but it has risks attached, including risks of immune incompatibility, as well as possible graft rejection (Shekkeris et al., 2012).

While both autologous and allogeneic stem cell therapy have seen success, both also have limitations. Autologous stem cells, while lessening the likelihood of immune rejection, have intrusive harvesting processes and are bound by the quality and quantity of stem cells that can be harvested. Allogeneic stem cells, while being bankable and more accessible, have the likelihood of triggering an immune reaction, especially in those suffering from underlying autoimmune diseases (Fayaz et al., 2011). In addition to these, there are also various impediments to the application of stem cell therapy toward bone regeneration. These include issues of stem cell source and scalability, ethical issues regarding the use of embryonic stem cells, and lack of ability toward stem cell differentiation and host tissue incorporation (Ehlen et al., 2024). The application of stem cell-based treatment toward bone defects and fracture must also address issues toward tumorigenicity as well as the integrity of transplanted stem cells over the long term. Besides, the economic cost of stem cell treatment, particularly for allogeneic transplants, still acts as an impediment to them being routinely practiced in the clinic (Killington et al., 2018). Despite these impediments, advancement toward stem cell biology as well as tissue engineering still makes stem cell-based treatment for bone defects and nonunion fracture more practical.

## 21 Biomaterials and Scaffold Strategies for Bone Regeneration

Bone regeneration has also greatly evolved by utilizing biomaterials and scaffold strategies. Collagen, hydroxyapatite, chitosan, and alginate, biomaterials of natural origin, have found extensive use due to compatibility as well as bone-healing ability. Collagen, one of the basic structural bone proteins, has found use in bone regeneration scaffolds due to its excellent ability to support attachment by cells as well as due to its ability to direct tissue development (Zhu et al., 2020). Hydroxyapatite, the mineral form of calcium phosphate, has found extensive use as bone repair biomaterial due to its similarity in composition to bone, facilitating bone development, as well as enhancing osteoconductivity (Zhu et al., 2020). Chitosan, the biopolymeric form of chitin, has found extensive use due to its multifaceted properties, including biodegradability, as well as antibacterial properties, making it an excellent choice for bone tissue engineering scaffolds (Battafarano et al., 2021). Alginate, the polysaccharide constituent found in algae, has found extensive use due to its ability to form a gel, as well as due to its ability to release bioactive compounds as well as cells in controlled release formulations (Iaquinta et al., 2019).

On the other hand, biomaterials produced from materials, such as bioceramics, bioactive glasses, and calcium phosphate cement, have emerged as prominent materials in bone tissue engineering due to their capacity to tailor mechanical properties and accommodate bone regeneration. Bioceramics, such as tricalcium phosphate, as well as calcium sulfate, are degradable, as well as osteoconductive, indicating that they allow bone conduction, and form a scaffold, enabling bone growth while they get slowly replaced by bone (Bentmann et al., 2010). Bioactive glasses, comprised of silica, phosphate, and calcium, release ions, enabling differentiation of the osteoblasts, as well as bone mineralization, so it has application in regenerative medicine (Tang et al., 2021). Calcium phosphate cement, mixed typically with other materials, provides an easy minimally invasive bone-filling method while displaying excellent osteoconductivity (Iaquinta et al., 2019).

The emergence of nanostructured scaffolds, as well as 3D printing, has greatly revolutionized the bone regeneration field by enabling the fabrication of scaffolds with precise geometrical, pore, and mechanical property control. These technologies facilitate the fabrication of scaffolds that accurately mimic the microarchitecture of bone, while also enabling increased cellular infiltration, as well as tissue integration (Polo-Corrales et al., 2014). Incorporation of nanomaterials, such as hydroxyapatite nanoparticles, as well as carbon nanotubes, into 3D-printed scaffolds has also shown promising toward increased mechanical strength, as well as functionalization, of these scaffolds, further enhancing their effectiveness toward bone repair (Battafarano et al., 2021). Decellularized bone matrices (DBMs) and extracellular vesicles (EVs) are newer bone regeneration strategies. Since naturally sourced from bone, DBMs carry the essential growth factors and extracellular matrices responsible for bone regeneration, as well as angiogenesis, constituting an original scaffold for bone regeneration (Zhou et al., 2021). In addition, nanoscale, cell-released extracellular vesicles have also emerged as an essential factor responsible for communications between cells and carry bioactive entities, including lipids, proteins, and RNAs, making them an interesting tool for enhancing bone scaffold regenerating capacity (Polo-Corrales et al., 2014).

# 22 Role of Mechanical and Biophysical Stimuli in Osteoblastogenesis

The role played by biophysical and mechanical stimuli during bone fracture repair and bone development has central importance. The bone-producing cells, referred to as osteoblasts, are responsive to various mechanical and biophysical stimuli, which govern their differentiation and functioning (Rosenberg, 2023). A case in point involves the relevance of mechanical loading, as well as mechanotransduction, whereby bone development relies upon them, as it initiates bone synthesis by the osteoblasts. In reaction to mechanical loading, the osteoblasts perform mechanotransduction, whereby force, whether compressive force or tension, gets transduced into biochemical stimuli, enhancing osteogenesis. It has been shown that it is mechanical loading that initiates cascades, including the Wnt/β-catenin pathway, as well as the MAPK pathway, both essential in bone development and differentiation of the osteoblasts (Aaron et al., 2006).

One prominent case study showing the effect of mechanical stimuli on the activity of osteoblasts includes the application of low-intensity pulsed ultrasound (LIPUS). LIPUS has been shown to induce cellular processes that facilitate bone regeneration and bone repair by activating osteoblasts. LIPUS has been shown by studies to enhance the level of crucial markers of bone generation, including alkaline phosphatase, as well as type I collagen, crucial for bone matrix synthesis. LIPUS has found extensive clinical application due to its capacity to heal fractures (Hao et al., 2021).

Besides, electromagnetic fields (EMFs), as well as piezoelectric materials, have also found application as possible stimulators of bone regeneration. Electromagnetic fields (EMFs) can regulate the functioning of osteoblasts by varying the electrical potentials across the membranes, as well as by activating cascades responsible for osteogenesis. Besides, when combined with piezoelectric materials, responsible for generating electric charges due to stresses, EMFs can synergize bone differentiation by osteoblasts, enhancing bone regeneration (Bai et al., 2024; Kang et al., 2014).

# 23 Hormonal and Nutritional Regulation of Osteoblast Differentiation

## 23.1 Hormonal Influences on Bone Formation

Osteoblast differentiation and bone formation are tightly regulated by various hormonal and dietary factors. Intermittent treatment by parathyroid hormone (PTH) plays an essential role in stimulating bone formation and the activity of osteoblasts. Intermittent treatment by PTH has also been shown to increase bone mass by stimulating various signal transduction cascades (Raisz & Bingham, 1972). Besides, estrogen, responsible for maintaining the skeletal system, has protective functions against bone diseases by targeting both osteoblasts and osteoclasts. Lack of estrogen, observed in postmenopausal women, leads to increased bone resorption, lowering bone density. The protective effect of estrogen against bone diseases lies in maintaining the balance between bone formation and bone resorption (Walsh, 2015). Vitamin D also plays an essential role in bone, primarily by maintaining the balance of calcium. Vitamin D increases the absorption of calcium by the intestine, necessary for bone mineralization and the functioning of osteoblasts. Vitamin D also contributes to aiding osteoblast differentiation and survival by regulating various gene expressions responsible for bone matrix synthesis (Kini & Nandeesh, 2012).

## 23.2 *Nutritional Factors in Osteogenesis*

Osteogenesis, or bone development, also has several nutritional factors. Among them, dietary minerals, such as calcium, phosphorus, and magnesium, are considered necessary elements for bone metabolism. Calcium, one of the prominent minerals contained in bone, has an active role both in bone construction and bone demolition. It has an active role in both bone-building (osteoblasts) and bone-breaking (osteoclasts) functioning (Yamaguchi, 2012). Phosphorus, typically found in phosphate form, also has an active role as one of the building stones of bone mineral hydroxyapatite, responsible for bone bulk. It, together with calcium, sustains bone mineral density. Magnesium, also an essential mineral, sustains bone health by playing an active role in bone-building (osteoblasts) functioning while also activating the bone to absorb calcium, ultimately sustaining bone mineralization (Trzeciakiewicz et al., 2009).

Beyond these traditional minerals, polyphenols and flavonoids have also been shown to be potential contributors to bone health. Resveratrol, a polyphenolic compound contained in wine, grapes, and other food, has proved able to boost bone building by stimulating bone building by the osteoblasts while inhibiting bone breakdown by the osteoclasts (Zambrano et al., 2014). Curcumin, the active constituent contained in turmeric, is also a polyphenol that has proved anti-inflammable and antioxidant functions, helpful in preventing bone loss due to diseases like osteoporosis (Chagas et al., 2012). Ellagic acid, present in many fruits, as well as berries, has also proved able to play a role in bone health by controlling oxidative stress and inflammation, both contributors to bone decline. These polyphenols, as well as flavonoids, when consumed as an equal balance, have synergist functions on bone metabolism, helpful in maintaining bone integrity and strength (De Amicis et al., 2024).

## 24 Pathological Conditions Affecting Osteoblastogenesis and Bone Regeneration

Osteoporosis and bone loss due to aging: In old people, osteoporosis indicates an imbalance between bone-building capacity and bone dissolution, resulting in bone mass reduction and increased fracture predisposition. Bone dissolution due to increased age has a close relation with reduced capacity for osteoblastogenesis, the creation of bone-building cells, and osteoblasts. Both the number and functioning capacity of the osteoblasts decline as individuals age, resulting in bone-building decline, and increased predisposition toward bone dissolution (Neve et al., 2011). Activation of the osteoclasts, together with lowered capacity for bone building, plays a crucial role in bone frailty seen in instances of osteoporosis (Valenti et al., 2016).

Delayed fracture healing, also known as nonunion, involves the failure of bone fracture to heal by the expected time. Delayed fracture healing, also known as nonunion, can result from various causes, including an impairment of the functioning capacity of osteoblasts, poor vascularization, and inflammatory reactions hindering bone healing. Delayed fracture healing has also been found to result due to an inadequate osteogenic capacity, wherein bone tissue regeneration employing functioning by osteoblasts takes a longer time, if it occurs, or does not occur (Dirckx et al., 2013). Nonunions have also been found to result due to alterations in molecular processes responsible for the differentiation of osteoblasts, including Wnt signaling, as well as BMP (bone morphogenetic protein) signaling. Also, underlying diseases, such as diabetes, or vascular insufficiency, may aggravate these by reducing functioning capacity by osteoblasts (Phan et al., 2004).

Diabetes, if uncontrolled, has previously been found to greatly interfere with bone regeneration. Both type 1 and type 2 diabetes can induce changes in the performance of osteoblasts, resulting in impairment of bone quality and bone capacity for healing. High blood glucose has previously been found, by studies, to interfere negatively with the differentiation of osteoblasts, due mostly to increased advanced glycation end products (AGEs), which interfere with the performance of osteoblasts, as well as bone mineralization deficits (Wu et al., 2016). The presence of diabetes also can alter the balance between osteoblasts and osteoclasts, resulting in bone resorption over bone formation, causing bone weakening, and slowing down healing (Picke et al., 2016).

Both periodontal disease and rheumatoid arthritis (RA) greatly compromise bone building, the process of osteogenesis. In RA, chronic inflammation releases cytokines TNF-α and IL-6, known to inhibit the differentiation of osteoblasts while enhancing the activity of osteoclasts, resulting in bone erosions (Zhou & Graves, 2022). In periodontal disease, the inflammatory process also plays an essential role in bone regeneration changes, periodontal pathogens, and inflammatory mediators prostaglandin E2 (PGE2) inhibiting the functions of osteoblasts while enhancing bone resorption. Both processes compromise bone regeneration and bone healing due to this imbalance between bone building and bone resorption (Komatsu & Takayanagi, 2022).

Cancer, specifically metastatic bone cancer, loses bone by secreting tumor-derived products, which interfere with the normal functioning of osteoblasts. Tumors, including those from cancers of the lung, prostate, and breasts, tend to induce osteolytic lesioning, whereby bone is lost faster, while minimal bone regeneration occurs (Shupp et al., 2018; Hansda & Das, 2025). In these cases, there is extensive impairment of osteoblastogenesis due to the release by the tumor of products such as RANKL (receptor activator of nuclear factor-κB ligand), activating differentiation and functioning of osteoclasts. Cancer treatment, including chemotherapy, interferes further by triggering cellular apoptosis or suppressing the proliferate functioning of osteoblasts, further enhancing bone loss as well as regeneration (Fornetti et al., 2018).

## 25 Advances in Bone Regenerative Medicine and Therapeutic Strategies

The exploration of gene therapy has seen notable advancement over recent years. Gene therapy aims to transfer gene constructs that allow bone regeneration or repair. A focal emphasis has been laid on the use of osteogenic genes, such as BMPs, vascular endothelial growth factor (VEGF), and fibroblast growth factors (FGFs), to allow bone regeneration (Oryan et al., 2014). A prominent method has been the use of CRISPR/Cas9 gene editing tooling to enhance the modification of osteogenic genes, representing an accurate gene editing tool that can optimize the differentiation and functioning of osteoblasts. Preclinical studies have indicated that CRISPR/Cas9 can allow targeted modification, enhancing bone repair processes (Chaparro & Linero, 2016). Transfer of these genetic constructs requires complex viral and non-viral gene transfer systems that allow maximum transfection efficiency along with safety in bone tissue. Adeno-associated viruses (AAVs) and lentivirus have routinely been researched, but there are still concerns regarding the generation of an immune response, alongside transgene expression. Alternative nonviral systems, such as lipid nanoparticles, alongside polymer-based systems, are being researched as less risky means by which gene transfer may be accomplished, delivering successful findings both in vitro and in vivo (Huang et al., 2021).

In addition to gene therapy, exosomes and extracellular vesicles also play central roles in bone regeneration. These minute, membrane-bound vesicles are naturally shed by cells and carry proteins, RNA, and lipids that govern the recipient cell behavior. Isolated exosomes from MSC have also shown promising ability to induce osteogenesis, as well as bone repair. Their ability to govern inflammation and cause stem cell recruitment to the bone defect site makes them an attractive choice over direct stem cell transplant (Ntege et al., 2020). Exosomes are particularly appealing due to their non-immunogenic property and ease of manipulation for targeted delivery, an option less invasive than traditional cell-based treatments (Mansouri et al., 2021).

The field of pharmaceutical and biological bone regeneration has also expanded, focusing on targeting small molecules that intervene in osteogenic pathways. Sclerostin-inhibiting small molecules and enhancing Wnt signaling have also been targeted for bone-promoting bone formation and increments in bone mass. Sclerostin-inhibiting pharmaceuticals, romosozumab, have also shown clinical relevance in treating bone diseases, including bone mineral density increments (Dantas et al., 2024). Antiresorptive drugs, denosumab, and bisphosphonates are also crucial in preventing bone resorption and fostering an environment that permits bone regeneration. These drugs achieve this by inhibiting the activity of osteoclasts, preventing bone mass decline, and supporting regeneration, especially in individuals suffering from bone loss due to old age, as well as those suffering from osteoporosis (Huang et al., 2021). Lastly, personalized treatment comes increasingly prominently into consideration in bone disorder treatment. By genetic profiling and individual-specific data, treatment can be tailor-made so that it best enhances bone regeneration and bone healing. Personalized regenerative treatment entails clinicians' capacity to select the optimum treatment option based on an individual's genetic predisposition, bone metabolism, and sensitivity to drugs (Shang et al., 2021). Personalization ensures that bone regenerating treatment, whether stem cell treatment, gene treatment, or pharmaceutical treatment, is maximally efficient and suited to the individual.

## 26 Clinical Applications and Future Perspectives

The management of bone defects and nonunions (also known as nonhealing fractures) has always posed an enormous clinical challenge, and treatment strategies continue to evolve. The traditional treatment approaches involve the application of autografts, allografts, and bone substitutes, aimed at stimulating bone regeneration by structurally giving support while stimulating osteogenesis (Arvidson et al., 2011). Growth factors, such as BMPs, have also been exploited to induce the differentiation of osteoblasts while stimulating bone regeneration. While these treatment approaches have had some success, there are still limitations due to concerns over donor site morbidity as well as the issue of immunogenicity when an allograft is applied. Besides, nonunions also tend to require further treatment approaches, including distraction osteogenesis, as well as the application of bioactive scaffolds aimed at stimulating local cellular processes (Vukicevic et al., 2014).

## 27 Challenges in Bone Tissue Engineering and Regeneration

Bone tissue engineering has also emerged as an interesting choice for bone defects and nonunion treatment, but there remain several obstacles preventing the transfer of data from the experimental environment to the clinic. Among the greatest challenges lies the generation of scaffolds that, apart from imitating bone's mechanical functions, also facilitate maximum attachment, proliferation, and differentiation of cells (Amini et al., 2012). Vascularization of engineered bone constructs also proves still an unresolved problem, as successful vascularization plays an essential role in bone regeneration (Moshiri et al., 2015). Stem cells, including MSC, have also shown promising performance, but their introduction into bone tissue proves problematic due to concerns over long-term survival, differentiation, and potential tumorigenicity. Overcoming these, therefore, requires innovative approaches that involve materials science, cell biology, and bioengineering to enhance bone regeneration processes (Moshiri et al., 2015).

## 28 Emerging Technologies: AI, Computational Techniques, and Bone Regeneration

Recent advancements in computer simulation, as well as artificial intelligence (AI), have also begun playing an ever-increasingly crucial role in bone regeneration. Optimal bone healing conditions, as well as treatment strategies, are also being forecast by machine-learning algorithms based on patient-specific parameters, such as comorbidities, fracture type, as well as age (Fan et al., 2024). AI-enhanced imaging has also significantly increased bone defect diagnosis, as well as bone health assessment, by enabling accurate, real-time bone health analysis (Akhtar, 2024). In bone tissue engineering, computer models are also being used to develop scaffold models that accurately mimic the bio-physical properties of natural bone, enhancing scaffold performance, as well as tissue integration. AI, together with computer simulation, has a vast scope to optimize bone regeneration therapy, as well as make it efficient, as well as personalized (Nosrati & Nosrati, 2023).

## 29 Future Directions

The future of studies on osteoblastogenesis depends on translating basic science findings into greater application in the clinic. Recent studies have illuminated the molecular processes underlying the differentiation and maturation of osteoblasts, including the role played by signaling molecules Wnt/β-catenin, Notch, and Hedgehog pathways (Dimitriou et al., 2011). Translational studies are aimed at harnessing these findings to develop newer pharmacological and gene therapies that are capable of stimulating osteoblastogenesis in those suffering from poor bone healing. Besides, newer biomaterials, functionalized hydrogels, and nanomaterials are being explored to facilitate greater differentiation of osteoblasts and bone regeneration. These advancements, together with personalized treatment approaches, have the potential to revolutionize bone defects and nonunion management, achieving greater efficiency, as well as customized treatment approaches, shortly (Dimitriou et al., 2011).

# 30 Conclusion

The field has witnessed vast progress over the last decade, driven by progress in the areas of osteoblastogenesis, stem cell biology, signal transduction, and biomaterials. Pivotal findings highlight the central role played by MSC as osteoblast precursors, the complex system of molecular signals that regulate osteogenesis, and the role played by the extracellular matrix (ECM) during bone mineralization. Progress, also, in tissue engineering, and scaffold technology, together with greater utilization of bioactive molecules, as well as growth factors, have significantly raised the possibility of successful bone regeneration therapy. In the context of ongoing clinical failures in treating bone defects, as well as nonunions, newer strategies continue to emerge, promising remedies that boost patient care. A central factor of bone regeneration lies in the ability to harness and regulate molecular processes underlying osteoblast differentiation and functioning. Wnt/β-catenin, BMP, and TGF-β signal transduction processes have emerged as central processes responsible for osteoblastogenesis, while RUNX2, alongside Osterix, are central transcriptional determinants of osteogenic differentiation. The extracellular matrix (ECM), comprised of collagen, alongside non-collagenous proteins, including osteopontin and osteocalcin, plays an essential role both in functioning osteoblasts and bone mineralization. The intricate coordination of ECM components, alongside cells, permits the generation of an ossein bone matrix, central to bone strength, alongside functionality. Besides, identification, alongside utilization, of growth factors, such as BMPs, IGFs, and VEGF, is also central to regenerative bone repair processes. In addition to cellular and molecular processes, the use of mechanical and biophysical stimuli has also emerged. Mechanoloading, mechanotransduction, LIPUS application, and electromagnetic field application have also proved to regulate the functions of osteoblasts, facilitate bone regeneration, and regulate bone repair. These approaches also highlight the importance not just of biochemical stimuli but also of physical stimuli toward maximizing bone repair, paving the way toward the coapplication of therapies that seek to enhance bone regeneration. The future of bone regenerative medicine is undoubtedly tied to further advancement of stem cell-based therapy, next-generation biomaterials, and gene modification. Stem cell therapy, particularly those employing autologous or allogeneic MSC, has shown success both in preclinical and clinical scenarios, while concerns regarding immune rejection, as well as the long-term survival of the cells, continue to plague them. Gene therapy, including the use of CRISPR/Cas9 gene modification of osteogenic genes, can resolve these concerns by enhancing osteoblast differentiation directly at the genetic level. Besides, biomaterial innovations, such as bioactive scaffolds, decellularized matrices, and nanostructured materials, offer the possibility of more efficient bone regeneration by creating an environment optimum for cellular attachment and incorporation. Looking ahead, bone regeneration therapy has an upcoming future relying on computer-aided strategies complemented by AI. AI can play an active role in customized treatment, leveraging data from an individual patient to project the best treatment protocols, as well as track bone regeneration progress in real time. Computer models also can generate scaffolds with precise mechanical and biological specifications tailored to individual patients, maximizing the success rate of regenerative treatment. AI-enabled diagnostics and tracking also can revolutionize bone defects diagnosis and management, enabling early treatment, as well as maximizing treatment efficiency. Despite these promising gains, there remain several issues confronting the transfer of findings from the laboratory environment to successful clinic application. Among these are concerns over immune compatibility, scaffold incorporation, and vascularization of engineered bone tissue, still waiting for innovative solutions. In addition, bone regeneration demands an interdisciplinary approach, bringing together bioengineering, stem cell biology, and molecular medicine, so successful as well as safe therapies may emerge.

## References

Aaron, R. K., Ciombor, D. M., Wang, S., & Simon, B. (2006). Clinical biophysics: The promotion of skeletal repair by physical forces. *Annals of the New York Academy of Sciences, 1068*(1), 513–531.

Adithya, S. P., Balagangadharan, K., & Selvamurugan, N. (2022). Epigenetic modifications of histones during osteoblast differentiation. *Biochimica et Biophysica Acta (BBA)-Gene Regulatory Mechanisms, 1865*(1), 194780.

Akhtar, Z. (2024). Exploring biomedical engineering (BME): Advances within accelerated computing and regenerative medicine for a computational and medical science perspective exploration analysis. *Journal of Emergency Medicine: Open Access, 2*, 1–23.

Al-Kharobi, H., El-Gendy, R., Devine, D., & Beattie, J. (2014). The role of the insulin-like growth factor (IGF) axis in osteogenic and odontogenic differentiation. *Cellular and Molecular Life Sciences, 71*, 1469–1476.

Amini, A. R., Laurencin, C. T., & Nukavarapu, S. P. (2012). Bone tissue engineering: Recent advances and challenges. *Critical Reviews™ in Biomedical Engineering, 40*(5), 363.

Arealis, G., & Nikolaou, V. S. (2015). Bone printing: New frontiers in the treatment of bone defects. *Injury, 46*, S20–S22.

Arvidson, K., Abdallah, B., Applegate, L., Baldini, N., Cenni, E., Gomez-Barrena, E., Granchi, D., Kassem, M., Konttinen, Y., & Mustafa, K. (2011). Bone regeneration and stem cells. *Journal of Cellular and Molecular Medicine, 15*(4), 718–746.

Atala, A., Irvine, D. J., Moses, M., & Shaunak, S. (2010). Wound healing versus regeneration: Role of the tissue environment in regenerative medicine. *MRS Bulletin, 35*(8), 597–606.

Bai, Y., Li, X., Wu, K., Heng, B. C., Zhang, X., & Deng, X. (2024). Biophysical stimuli for promoting bone repair and regeneration. *Medical Review*.

Bala, Y., Farlay, D., & Boivin, G. (2013). Bone mineralization: From tissue to crystal in normal and pathological contexts. *Osteoporosis International, 24*, 2153–2166.

Banerjee, S. S., Feinberg, M. W., Watanabe, M., Gray, S., Haspel, R. L., Denkinger, D. J., Kawahara, R., Hauner, H., & Jain, M. K. (2003). The Krüppel-like factor KLF2 inhibits peroxisome proliferator-activated receptor-γ expression and adipogenesis. *Journal of Biological Chemistry, 278*(4), 2581–2584.

Barba, M., Di Taranto, G., & Lattanzi, W. (2017). Adipose-derived stem cell therapies for bone regeneration. *Expert Opinion on Biological Therapy, 17*(6), 677–689.

Barkhordarian, A., Sison, J., Cayabyab, R., Mahanian, N., & Chiappelli, F. (2011). Epigenetic regulation of osteogenesis: Human embryonic palatal mesenchymal cells. *Bioinformation, 5*(7), 278.

Bastami, F., Nazeman, P., Moslemi, H., Rezai Rad, M., Sharifi, K., & Khojasteh, A. (2017). Induced pluripotent stem cells as a new getaway for bone tissue engineering: A systematic review. *Cell Proliferation, 50*(2), e12321.

Battafarano, G., Rossi, M., De Martino, V., Marampon, F., Borro, L., Secinaro, A., & Del Fattore, A. (2021). Strategies for bone regeneration: From graft to tissue engineering. *International Journal of Molecular Sciences, 22*(3), 1128.

Bentmann, A., Kawelke, N., Moss, D., Zentgraf, H., Bala, Y., Berger, I., Gasser, J. A., & Nakchbandi, I. A. (2010). Circulating fibronectin affects bone matrix, whereas osteoblast fibronectin modulates osteoblast function. *Journal of Bone and Mineral Research, 25*(4), 706–715.

Bianco, P. (2020). Structure and mineralization of bone. In *Calcification in biological systems* (pp. 243–268). CRC Press.

Bosetti, M., Boccafoschi, F., Leigheb, M., & Cannas, M. F. (2007). Effect of different growth factors on human osteoblasts activities: A possible application in bone regeneration for tissue engineering. *Biomolecular Engineering, 24*(6), 613–618.

Boskey, A. L. (1989). Noncollagenous matrix proteins and their role in mineralization. *Bone and Mineral, 6*(2), 111–123.

Boskey, A. L., Seibel, M., Robins, S., & Bilezikian, J. (2006). Mineralization, structure and function of bone. *Dynamics of Bone and Cartilage Metabolism: Principles and Clinical Applications, 2*, 201–209.

Buckwalter, J., Glimcher, M., Cooper, R., & Recker, R. (1995). Bone biology. *Journal of Bone and Joint Surgery, 77*(8), 1276–1289.

Caetano-Lopes, J., Canhão, H., & Fonseca, J. E. (2007). Osteoblasts and bone formation. *Acta Reumatológica Portuguesa, 32*(2), 103–110.

Cai, J. Q., Huang, Y. Z., Chen, X. H., Xie, H. L., Zhu, H. M., Tang, L., Yang, Z. M., Huang, Y. C., & Deng, L. (2012). Sonic hedgehog enhances the proliferation and osteogenic differentiation of bone marrow-derived mesenchymal stem cells. *Cell Biology International, 36*(4), 349–355.

Cao, J., Wei, Y., Lian, J., Yang, L., Zhang, X., Xie, J., Liu, Q., Luo, J., He, B., & Tang, M. (2017). Notch signaling pathway promotes osteogenic differentiation of mesenchymal stem cells by enhancing BMP9/Smad signaling. *International Journal of Molecular Medicine, 40*(2), 378–388.

Cha, Y., Jeon, M., Lee, H.-S., Kim, S., Kim, S.-O., Lee, J.-H., & Song, J. S. (2015). Effects of in vitro osteogenic induction on in vivo tissue regeneration by dental pulp and periodontal ligament stem cells. *Journal of Endodontics, 41*(9), 1462–1468.

Chagas, C. E. A., Roque, J. P., Peters, B. S. E., Lazaretti-Castro, M., & Martini, L. A. (2012). Do patients with osteogenesis imperfecta need individualized nutritional support? *Nutrition, 28*(2), 138–142.

Chan, W. C. W., Tan, Z., To, M. K. T., & Chan, D. (2021). Regulation and role of transcription factors in osteogenesis. *International Journal of Molecular Sciences, 22*(11), 5445.

Chaparro, O., & Linero, I. (2016). Regenerative medicine: a new paradigm in bone regeneration. *Advanced Techniques in Bone Regeneration, 12*, 253–274.

Charoenlarp, P., Rajendran, A. K., & Iseki, S. (2017). Role of fibroblast growth factors in bone regeneration. *Inflammation and Regeneration, 37*, 1–7.

Chen, D., Zhao, M., & Mundy, G. R. (2004). Bone morphogenetic proteins. *Growth Factors, 22*(4), 233–241. https://doi.org/10.1080/08977190412331279890

Chen, G., Deng, C., & Li, Y.-P. (2012). TGF-β and BMP signaling in osteoblast differentiation and bone formation. *International Journal of Biological Sciences, 8*(2), 272.

Chen, Q., Shou, P., Zheng, C., Jiang, M., Cao, G., Yang, Q., Cao, J., Xie, N., Velletri, T., & Zhang, X. (2016). Fate decision of mesenchymal stem cells: Adipocytes or osteoblasts? *Cell Death & Differentiation, 23*(7), 1128–1139.

Chow, O., & Barbul, A. (2014). Immunonutrition: Role in wound healing and tissue regeneration. *Advances in Wound Care, 3*(1), 46–53.

Cole, W. G. (2002). Advances in osteogenesis imperfecta. *Clinical Orthopaedics and Related Research®, 401*, 6–16.

Csobonyeiova, M., Polak, S., Zamborsky, R., & Danisovic, L. (2017). iPS cell technologies and their prospect for bone regeneration and disease modeling: A mini review. *Journal of Advanced Research, 8*(4), 321–327.

Dantas, L. R., Ortis, G. B., Suss, P. H., & Tuon, F. F. (2024). Advances in regenerative and reconstructive medicine in the prevention and treatment of bone infections. *Biology, 13*(8), 605.

Das, H., Kumar, A., Lin, Z., Patino, W. D., Hwang, P. M., Feinberg, M. W., Majumder, P. K., & Jain, M. K. (2006). Kruppel-like factor 2 (KLF2) regulates proinflammatory activation of monocytes. *Proceedings of the National Academy of Sciences, 103*(17), 6653–6658.

De Amicis, R., Bertoli, S., Bruno, A., De Carlo, G., Battezzati, A., Foppiani, A., Leone, A., & Lo Mauro, A. (2024). Short-term effects of Mediterranean diet on nutritional status in adults affected by Osteogenesis Imperfecta: A pilot study. *Orphanet Journal of Rare Diseases, 19*(1), 94.

Deschaseaux, F., Sensébé, L., & Heymann, D. (2009). Mechanisms of bone repair and regeneration. *Trends in Molecular Medicine, 15*(9), 417–429.

Dimitriou, R., Jones, E., McGonagle, D., & Giannoudis, P. V. (2011). Bone regeneration: Current concepts and future directions. *BMC Medicine, 9*, 1–10.

Dirckx, N., Van Hul, M., & Maes, C. (2013). Osteoblast recruitment to sites of bone formation in skeletal development, homeostasis, and regeneration. *Birth Defects Research Part C: Embryo Today: Reviews, 99*(3), 170–191.

Donnelly, E., & Boskey, A. L. (2011). Mineralization. In *Vitamin D* (pp. 381–401). Elsevier.

Duda, G. N., Geissler, S., Checa, S., Tsitsilonis, S., Petersen, A., & Schmidt-Bleek, K. (2023). The decisive early phase of bone regeneration. *Nature Reviews Rheumatology, 19*(2), 78–95.

Egido-Moreno, S., Valls-Roca-Umbert, J., Céspedes-Sánchez, J. M., López-López, J., & Velasco-Ortega, E. (2021). Clinical efficacy of mesenchymal stem cells in bone regeneration in oral implantology. Systematic review and meta-analysis. *International Journal of Environmental Research and Public Health, 18*(3), 894.

Ehlen, Q. T., Costello, J. P., Mirsky, N. A., Slavin, B. V., Parra, M., Ptashnik, A., Nayak, V. V., Coelho, P. G., & Witek, L. (2024). Treatment of bone defects and nonunion via novel delivery mechanisms, growth factors, and stem cells: A review. *ACS Biomaterials Science & Engineering, 10*(12), 7314–7336.

Eijken, M., Meijer, I. M., Westbroek, I., Koedam, M., Chiba, H., Uitterlinden, A., Pols, H., & Van Leeuwen, J. (2008). Wnt signaling acts and is regulated in a human osteoblast differentiation dependent manner. *Journal of Cellular Biochemistry, 104*(2), 568–579.

Einhorn, T. A., & Lee, C. A. (2001). Bone regeneration: New findings and potential clinical applications. *JAAOS-Journal of the American Academy of Orthopaedic Surgeons, 9*(3), 157–165.

El-Jawhari, J. J., Jones, E., & Giannoudis, P. V. (2016). The roles of immune cells in bone healing; what we know, do not know and future perspectives. *Injury, 47*(11), 2399–2406.

Elchaninov, A., Sukhikh, G., & Fatkhudinov, T. (2021). Evolution of regeneration in animals: A tangled story. *Frontiers in Ecology and Evolution, 9*, 621686.

Ettinger, M. P. (2003). Aging bone and osteoporosis: Strategies for preventing fractures in the elderly. *Archives of Internal Medicine, 163*(18), 2237–2246.

Fan, J., Xu, J., Wen, X., Sun, L., Xiu, Y., Zhang, Z., Liu, T., Zhang, D., Wang, P., & Xing, D. (2024). The future of bone regeneration: Artificial intelligence in biomaterials discovery. *Materials Today Communications, 109982*, 109982.

Fayaz, H. C., Giannoudis, P. V., Vrahas, M. S., Smith, R. M., Moran, C., Pape, H. C., Krettek, C., & Jupiter, J. B. (2011). The role of stem cells in fracture healing and nonunion. *International Orthopaedics, 35*, 1587–1597.

Feinberg, M. W., Lin, Z., Fisch, S., & Jain, M. K. (2004). An emerging role for Krüppel-like factors in vascular biology. *Trends in Cardiovascular Medicine, 14*(6), 241–246.
Feng, X., Huang, D., Lu, X., Feng, G., Xing, J., Lu, J., Xu, K., Xia, W., Meng, Y., & Tao, T. (2014). Insulin-like growth factor 1 can promote proliferation and osteogenic differentiation of human dental pulp stem cells via m TOR pathway. *Development, Growth & Differentiation, 56*(9), 615–624.
Ferretti, C., Borsari, V., Falconi, M., Gigante, A., Lazzarini, R., Fini, M., Di Primio, R., & Mattioli-Belmonte, M. (2012). Human periosteum-derived stem cells for tissue engineering applications: The role of VEGF. *Stem Cell Reviews and Reports, 8*, 882–890.
Fornetti, J., Welm, A. L., & Stewart, S. A. (2018). Understanding the bone in cancer metastasis. *Journal of Bone and Mineral Research, 33*(12), 2099–2113.
Foster, B., Ao, M., Salmon, C., Chavez, M., Kolli, T., Tran, A., Chu, E., Kantovitz, K., Yadav, M., & Narisawa, S. (2018). Osteopontin regulates dentin and alveolar bone development and mineralization. *Bone, 107*, 196–207.
Franceschi, R. T., Ge, C., Xiao, G., Roca, H., & Jiang, D. (2007). Transcriptional regulation of osteoblasts. *Annals of the New York Academy of Sciences, 1116*(1), 196–207.
Fung Ling Chau, J., Fook Leong, W., & Li, B. (2009). Signaling pathways governing osteoblast proliferation, differentiation and function. *Histology and Histopathology.*
Ghayor, C., & Weber, F. E. (2016). Epigenetic regulation of bone remodeling and its impacts in osteoporosis. *International Journal of Molecular Sciences, 17*(9), 1446.
Giachelli, C. M., & Steitz, S. (2000). Osteopontin: A versatile regulator of inflammation and biomineralization. *Matrix Biology, 19*(7), 615–622.
Giannoudis, P. V., Hak, D., Sanders, D., Donohoe, E., Tosounidis, T., & Bahney, C. (2015). Inflammation, bone healing, and anti-inflammatory drugs: An update. *Journal of Orthopaedic Trauma, 29*, S6–S9.
Gordon, J. A., Tye, C. E., Sampaio, A. V., Underhill, T. M., Hunter, G. K., & Goldberg, H. A. (2007). Bone sialoprotein expression enhances osteoblast differentiation and matrix mineralization in vitro. *Bone, 41*(3), 462–473.
Grottkau, B. E., & Lin, Y. (2013). Osteogenesis of adipose-derived stem cells. *Bone Research, 1*(1), 133–145.
Hadjidakis, D. J., & Androulakis, I. I. (2006). Bone remodeling. *Annals of the New York Academy of Sciences, 1092*(1), 385–396.
Hankenson, K., Gagne, K., & Shaughnessy, M. (2015). Extracellular signaling molecules to promote fracture healing and bone regeneration. *Advanced Drug Delivery Reviews, 94*, 3–12.
Hansda, S., & Das, H. (2025). Insights into cancer-associated thrombosis leading towards ischemic stroke. *Biology, 14*(1), 50. https://www.mdpi.com/2079-7737/14/1/50
Hansda, S., Prateeksha, P., & Das, H. (2024). Krüppel-like factor 2 (KLF2), a potential target for neuroregeneration. *Neural Regeneration Research, 19*(11), 2327–2328. https://doi.org/10.4103/nrr.Nrr-d-23-01758
Hao, Z., Xu, Z., Wang, X., Wang, Y., Li, H., Chen, T., Hu, Y., Chen, R., Huang, K., & Chen, C. (2021). Biophysical stimuli as the fourth pillar of bone tissue engineering. *Frontiers in Cell and Developmental Biology, 9*, 790050.
Henkel, J., Woodruff, M. A., Epari, D. R., Steck, R., Glatt, V., Dickinson, I. C., Choong, P. F., Schuetz, M. A., & Hutmacher, D. W. (2013). Bone regeneration based on tissue engineering conceptions—A 21st century perspective. *Bone Research, 1*(1), 216–248.
Hernández-Gil, I., Gracia, M. A. A., Pingarrón, M., & Jerez, L. (2006). Physiological bases of bone regeneration II. The remodeling process. *Medicina Oral, Patología Oral y Cirugía Bucal, 11*, E151–E215.
Hou, Z., Wang, Z., Tao, Y., Bai, J., Yu, B., Shen, J., Sun, H., Xiao, L., Xu, Y., & Zhou, J. (2019). KLF2 regulates osteoblast differentiation by targeting of Runx2. *Laboratory Investigation, 99*(2), 271–280.

Howlader, M. S. I., Prateeksha, P., Hansda, S., Naidu, P., Das, M., Barthels, D., & Das, H. (2024). Corrigendum to "secretory products of DPSC mitigate inflammatory effects in microglial cells by targeting MAPK pathway" Biomed. Pharmacother. 170(2024) 115971. *Biomedicine & Pharmacotherapy, 177*, 117061. https://doi.org/10.1016/j.biopha.2024.117061

Hu, K., & Olsen, B. R. (2016). The roles of vascular endothelial growth factor in bone repair and regeneration. *Bone, 91*, 30–38.

Hu, K., & Olsen, B. R. (2017). Vascular endothelial growth factor control mechanisms in skeletal growth and repair. *Developmental Dynamics, 246*(4), 227–234.

Huang, W., Yang, S., Shao, J., & Li, Y.-P. (2007). Signaling and transcriptional regulation in osteoblast commitment and differentiation. *Frontiers in Bioscience: a Journal and Virtual Library, 12*, 3068.

Huang, Z., Tian, Z., Zhu, M., Wu, C., & Zhu, Y. (2021). Recent advances in biomaterial scaffolds for integrative tumor therapy and bone regeneration. *Advanced Therapeutics, 4*(3), 2000212.

Iaquinta, M. R., Mazzoni, E., Manfrini, M., D'Agostino, A., Trevisiol, L., Nocini, R., Trombelli, L., Barbanti-Brodano, G., Martini, F., & Tognon, M. (2019). Innovative biomaterials for bone regrowth. *International Journal of Molecular Sciences, 20*(3), 618.

Illich, D. J., Demir, N., Stojković, M., Scheer, M., Rothamel, D., Neugebauer, J., Hescheler, J., & Zöller, J. E. (2011). Concise review: Induced pluripotent stem cells and lineage reprogramming: Prospects for bone regeneration. *Stem Cells, 29*(4), 555–563.

Isaksson, H. (2012). Recent advances in mechanobiological modeling of bone regeneration. *Mechanics Research Communications, 42*, 22–31.

Ito, K., Yamada, Y., Nakamura, S., & Ueda, M. (2011). Osteogenic potential of effective bone engineering using dental pulp stem cells, bone marrow stem cells, and periosteal cells for osseointegration of dental implants. *International Journal of Oral & Maxillofacial Implants, 26*(5), 947.

Panetta, N. J., Gupta, D. M., & Longaker, M. T. (2010). Bone regeneration and repair. *Current Stem Cell Research & Therapy, 5*(2), 122–128.

Jafarian, M., Eslaminejad, M. B., Khojasteh, A., Abbas, F. M., Dehghan, M. M., Hassanizadeh, R., & Houshmand, B. (2008). Marrow-derived mesenchymal stem cells-directed bone regeneration in the dog mandible: A comparison between biphasic calcium phosphate and natural bone mineral. *Oral Surgery, Oral Medicine, Oral Pathology, Oral Radiology, and Endodontology, 105*(5), e14–e24.

Jha, P., & Das, H. (2017). KLF2 in regulation of NF-κB-mediated immune cell function and inflammation. *International Journal of Molecular Sciences, 18*(11), 2383.

Kang, K. S., Hong, J. M., Jeong, Y. H., Seol, Y.-J., Yong, W.-J., Rhie, J.-W., & Cho, D.-W. (2014). Combined effect of three types of biophysical stimuli for bone regeneration. *Tissue Engineering Part A, 20*(11–12), 1767–1777.

Kempen, D. H., Creemers, L. B., Alblas, J., Lu, L., Verbout, A. J., Yaszemski, M. J., & Dhert, W. J. (2010). Growth factor interactions in bone regeneration. *Tissue Engineering Part B: Reviews, 16*(6), 551–566.

Khan, W. S., & Hardingham, T. E. (2012). Mesenchymal stem cells, sources of cells and differentiation potential. *Journal of Stem Cells, 7*(2), 75.

Killington, K., Mafi, R., Mafi, P., & Khan, W. S. (2018). A systematic review of clinical studies investigating mesenchymal stem cells for fracture non-union and bone defects. *Current Stem Cell Research & Therapy, 13*(4), 284–291.

Kim, I., Kim, J. H., Kim, K., Seong, S., & Kim, N. (2019). The IRF2BP2-KLF2 axis regulates osteoclast and osteoblast differentiation. *BMB Reports, 52*(7), 469.

Kim, W. K., Meliton, V., Bourquard, N., Hahn, T. J., & Parhami, F. (2010). Hedgehog signaling and osteogenic differentiation in multipotent bone marrow stromal cells are inhibited by oxidative stress. *Journal of Cellular Biochemistry, 111*(5), 1199–1209.

Kini, U., & Nandeesh, B. (2012). Physiology of bone formation, remodeling, and metabolism. *Radionuclide and Hybrid Bone Imaging*, 29–57.

Knapik, D. M., Perera, P., Nam, J., Blazek, A. D., Rath, B., Leblebicioglu, B., Das, H., Wu, L. C., Hewett, T. E., & Agarwal, S. K., Jr. (2014). Mechanosignaling in bone health, trauma and inflammation. *Antioxidants & Redox Signaling, 20*(6), 970–985.

Knight, M. N., & Hankenson, K. D. (2013). Mesenchymal stem cells in bone regeneration. *Advances in Wound Care, 2*(6), 306–316.

Ko, N.-Y., Chen, L.-R., & Chen, K.-H. (2020). The role of micro RNA and long-non-coding RNA in osteoporosis. *International Journal of Molecular Sciences, 21*(14), 4886.

Komatsu, N., & Takayanagi, H. (2022). Mechanisms of joint destruction in rheumatoid arthritis—Immune cell–fibroblast–bone interactions. *Nature Reviews Rheumatology, 18*(7), 415–429.

Komori, T. (2006). Regulation of osteoblast differentiation by transcription factors. *Journal of Cellular Biochemistry, 99*(5), 1233–1239.

Komori, T., Yagi, H., Nomura, S., Yamaguchi, A., Sasaki, K., Deguchi, K., Shimizu, Y., Bronson, R., Gao, Y.-H., & Inada, M. (1997). Targeted disruption of Cbfa1 results in a complete lack of bone formation owing to maturational arrest of osteoblasts. *Cell, 89*(5), 755–764.

Laha, D., Sarkar, J., Maity, J., Pramanik, A., Howlader, M. S. I., Barthels, D., & Das, H. (2022). Polyphenolic compounds inhibit osteoclast differentiation while reducing autophagy through limiting ROS and the mitochondrial membrane potential. *Biomolecules, 12*(9), 1220. https://www.mdpi.com/2218-273X/12/9/1220

Lanzillotti, C., De Mattei, M., Mazziotta, C., Taraballi, F., Rotondo, J. C., Tognon, M., & Martini, F. (2021). Long non-coding RNAs and microRNAs interplay in osteogenic differentiation of mesenchymal stem cells. *Frontiers in Cell and Developmental Biology, 9*, 646032.

Lei, G., Wang, Y., Yu, Y., Li, Z., Lu, J., Ge, X., Li, N., Manduca, A. G. C., & Yu, J. (2020). Dentin-derived inorganic minerals promote the Osteogenesis of bone marrow-derived mesenchymal stem cells: Potential applications for bone regeneration. *Stem Cells International, 2020*(1), 8889731.

Lei, M., Li, K., Li, B., Gao, L.-N., Chen, F.-M., & Jin, Y. (2014). Mesenchymal stem cell characteristics of dental pulp and periodontal ligament stem cells after in vivo transplantation. *Biomaterials, 35*(24), 6332–6343.

Lemaire, V., Tobin, F. L., Greller, L. D., Cho, C. R., & Suva, L. J. (2004). Modeling the interactions between osteoblast and osteoclast activities in bone remodeling. *Journal of Theoretical Biology, 229*(3), 293–309. https://doi.org/10.1016/j.jtbi.2004.03.023

Li, J., Ge, L., Zhao, Y., Zhai, Y., Rao, N., Yuan, X., Yang, J., Li, J., & Yu, S. (2022). TGF-β2 and TGF-β1 differentially regulate the odontogenic and osteogenic differentiation of mesenchymal stem cells. *Archives of Oral Biology, 135*, 105357.

Li, J., Zhang, H., Yang, C., Li, Y., & Dai, Z. (2016). An overview of osteocalcin progress. *Journal of Bone and Mineral Metabolism, 34*, 367–379.

Lieberman, J. R., & Friedlaender, G. E. (2005). *Bone regeneration and repair: Biology and clinical applications*. Springer.

Lin, G. L., & Hankenson, K. D. (2011). Integration of BMP, Wnt, and notch signaling pathways in osteoblast differentiation. *Journal of Cellular Biochemistry, 112*(12), 3491–3501.

Lin, Z., Kumar, A., SenBanerjee, S., Staniszewski, K., Parmar, K., Vaughan, D. E., Gimbrone, M. A., Jr., Balasubramanian, V., García-Cardeña, G., & Jain, M. K. (2005). Kruppel-like factor 2 (KLF2) regulates endothelial thrombotic function. *Circulation Research, 96*(5), e48–e57.

Ling, M., Huang, P., Islam, S., Heruth, D. P., Li, X., Zhang, L. Q., Li, D.-Y., Hu, Z., & Ye, S. Q. (2017). Epigenetic regulation of Runx2 transcription and osteoblast differentiation by nicotinamide phosphoribosyltransferase. *Cell & Bioscience, 7*, 1–10.

Liu, L., Ling, J., Wei, X., Wu, L., & Xiao, Y. (2009). Stem cell regulatory gene expression in human adult dental pulp and periodontal ligament cells undergoing odontogenic/osteogenic differentiation. *Journal of Endodontics, 35*(10), 1368–1376.

Liu, Q., Li, M., Wang, S., Xiao, Z., Xiong, Y., & Wang, G. (2020). Recent advances of osterix transcription factor in osteoblast differentiation and bone formation. *Frontiers in Cell and Developmental Biology, 8*, 601224.

Luo, Z., Shang, X., Zhang, H., Wang, G., Massey, P. A., Barton, S. R., Kevil, C. G., & Dong, Y. (2019). Notch signaling in osteogenesis, osteoclastogenesis, and angiogenesis. *The American Journal of Pathology, 189*(8), 1495–1500.

Maity, J., Barthels, D., Sarkar, J., Prateeksha, P., Deb, M., Rolph, D., & Das, H. (2022). Ferutinin induces osteoblast differentiation of DPSCs via induction of KLF2 and autophagy/mitophagy. *Cell Death & Disease, 13*(5), 452.

Majidinia, M., Sadeghpour, A., & Yousefi, B. (2018). The roles of signaling pathways in bone repair and regeneration. *Journal of Cellular Physiology, 233*(4), 2937–2948.

Manolagas, S. C. (1995). Role of cytokines in bone resorption. *Bone, 17*(2), S63–S67.

Mansouri, V., Beheshtizadeh, N., Gharibshahian, M., Sabouri, L., Varzandeh, M., & Rezaei, N. (2021). Recent advances in regenerative medicine strategies for cancer treatment. *Biomedicine & Pharmacotherapy, 141*, 111875.

Marie, P. J. (2008). Transcription factors controlling osteoblastogenesis. *Archives of Biochemistry and Biophysics, 473*(2), 98–105.

Marie, P. J., & Fromigué, O. (2006). Osteogenic differentiation of human marrow-derived mesenchymal stem cells. *Regenerative Medicine, 1*(4), 539–548.

Milner, P. I., Clegg, P. D., & Stewart, M. C. (2011). Stem cell–based therapies for bone repair. *Veterinary Clinics: Equine Practice, 27*(2), 299–314.

Moshiri, A., Oryan, A., & Shahrezaee, M. (2015). An overview on bone tissue engineering and regenerative medicine: Current challenges, future directions and strategies. *Journal of Sports Medicine and Doping Studies, 5*, e144.

Mundy, G., Boyce, B., Hughes, D., Wright, K., Bonewald, L., Dallas, S., Harris, S., Ghosh-Choudhury, N., Chen, D., & Dunstan, C. (1995). The effects of cytokines and growth factors on osteoblastic cells. *Bone, 17*(2), S71–S75.

Murshed, M. (2018). Mechanism of bone mineralization. *Cold Spring Harbor Perspectives in Medicine, 8*(12), a031229.

Mutsaers, A. J., & Walkley, C. R. (2014). Cells of origin in osteosarcoma: Mesenchymal stem cells or osteoblast committed cells? *Bone, 62*, 56–63.

Neve, A., Corrado, A., & Cantatore, F. P. (2011). Osteoblast physiology in normal and pathological conditions. *Cell and Tissue Research, 343*, 289–302.

Nosrati, H., & Nosrati, M. (2023). Artificial intelligence in regenerative medicine: Applications and implications. *Biomimetics, 8*(5), 442.

Novodvorsky, P., & Chico, T. J. (2014). The role of the transcription factor KLF2 in vascular development and disease. *Progress in Molecular Biology and Translational Science, 124*, 155–188.

Ntege, E. H., Sunami, H., & Shimizu, Y. (2020). Advances in regenerative therapy: A review of the literature and future directions. *Regenerative Therapy, 14*, 136–153.

Olmsted-Davis, E. A., Gugala, Z., Camargo, F., Gannon, F. H., Jackson, K., Kienstra, K. A., Shine, H. D., Lindsey, R. W., Hirschi, K. K., & Goodell, M. A. (2003). Primitive adult hematopoietic stem cells can function as osteoblast precursors. *Proceedings of the National Academy of Sciences, 100*(26), 15877–15882.

Oryan, A., Alidadi, S., & Moshiri, A. (2013). Current concerns regarding healing of bone defects. *Hard Tissue, 2*(2), 1–12.

Oryan, A., Alidadi, S., Moshiri, A., & Maffulli, N. (2014). Bone regenerative medicine: Classic options, novel strategies, and future directions. *Journal of Orthopaedic Surgery and Research, 9*, 1–27.

Park-Min, K. H. (2017). Epigenetic regulation of bone cells. *Connective Tissue Research, 58*(1), 76–89.

Patel, D., & Wairkar, S. (2023). Bone regeneration in osteoporosis: Opportunities and challenges. *Drug Delivery and Translational Research, 13*(2), 419–432.

Phan, T., Xu, J., & Zheng, M. (2004). Interaction between osteoblast and osteoclast: Impact in bone disease. *Histology and Histopathology, 19*(4), 1325–1344.

Picke, A.-K., Alaguero, I. G., Campbell, G., Glüer, C.-C., Salbach-Hirsch, J., Rauner, M., Hofbauer, L., & Hofbauer, C. (2016). Bone defect regeneration and cortical bone parameters of type 2 diabetic rats are improved by insulin therapy. *Bone, 82*, 108–115.

Pignolo, R. J., Law, S. F., & Chandra, A. (2021). Bone aging, cellular senescence, and osteoporosis. *Journal of Bone and Mineral Research Plus, 5*(4), e10488.

Pilipchuk, S. P., Plonka, A. B., Monje, A., Taut, A. D., Lanis, A., Kang, B., & Giannobile, W. V. (2015). Tissue engineering for bone regeneration and osseointegration in the oral cavity. *Dental Materials, 31*(4), 317–338.

Polo-Corrales, L., Latorre-Esteves, M., & Ramirez-Vick, J. E. (2014). Scaffold design for bone regeneration. *Journal of Nanoscience and Nanotechnology, 14*(1), 15–56.

Prateeksha, P., Howlader, M. S. I., Hansda, S., Naidu, P., Das, M., Abo-Aziza, F., & Das, H. (2023a). Secretome of dental pulp-derived stem cells reduces inflammation and proliferation of glioblastoma cells by deactivating Mapk-Akt pathway. *Disease Research, 3*(2), 74–86. https://doi.org/10.54457/dr.202302006

Prateeksha, P., Naidu, P., Das, M., Barthels, D., & Das, H. (2023b). KLF2 regulates neural differentiation of dental pulp-derived stem cells by modulating autophagy and Mitophagy. *Stem Cell Reviews and Reports, 19*(8), 2886–2900. https://doi.org/10.1007/s12015-023-10607-0

Qu, G., Li, Y., Chen, L., Chen, Q., Zou, D., Yang, C., & Zhou, Q. (2021). Comparison of osteogenic differentiation potential of human dental-derived stem cells isolated from dental pulp, periodontal ligament, dental follicle, and alveolar bone. *Stem Cells International, 2021*(1), 6631905.

Rahman, M. S., Akhtar, N., Jamil, H. M., Banik, R. S., & Asaduzzaman, S. M. (2015). TGF-β/BMP signaling and other molecular events: Regulation of osteoblastogenesis and bone formation. *Bone Research, 3*(1), 1–20.

Raisz, L., & Bingham, P. (1972). Effect of hormones on bone development. *Annual Review of Pharmacology, 12*(1), 337–352.

Rana, D., Kumar, S., Webster, T. J., & Ramalingam, M. (2019). Impact of induced pluripotent stem cells in bone repair and regeneration. *Current Osteoporosis Reports, 17*, 226–234.

Retzepi, M., & Donos, N. (2010). Guided bone regeneration: Biological principle and therapeutic applications. *Clinical Oral Implants Research, 21*(6), 567–576.

Rodríguez-Merchán, E. C. (2021). A review of recent developments in the molecular mechanisms of bone healing. *International Journal of Molecular Sciences, 22*(2), 767.

Rolph, D. N., Deb, M., Kanji, S., Greene, C. J., Das, M., Joseph, M., Aggarwal, R., Leblebicioglu, B., & Das, H. (2020). Ferutinin directs dental pulp-derived stem cells towards the osteogenic lineage by epigenetically regulating canonical Wnt signaling. *Biochimica et Biophysica Acta (BBA)-Molecular Basis of Disease, 1866*(4), 165314.

Roodman, G. D. (1993). Role of cytokines in the regulation of bone resorption. *Calcified Tissue International, 53*(Suppl 1), S94–S98.

Rosenberg, N. (2023). The theoretical context of biophysical stimulation of osteoblasts. In *Biophysical osteoblast stimulation for bone grafting and regeneration: From basic science to clinical applications* (pp. 3–12). Springer.

Ruffoni, D., Fratzl, P., Roschger, P., Klaushofer, K., & Weinkamer, R. (2007). The bone mineralization density distribution as a fingerprint of the mineralization process. *Bone, 40*(5), 1308–1319.

Shekkeris, S. K., Jaiswal, P. K., & Khan, W. S. (2012). Clinical applications of mesenchymal stem cells in the treatment of fracture non-union and bone defects. *Current Stem Cell Research & Therapy, 7*(2), 127–133.

Safari, B., Davaran, S., & Aghanejad, A. (2021). Osteogenic potential of the growth factors and bioactive molecules in bone regeneration. *International Journal of Biological Macromolecules, 175*, 544–557.

Salgado, A. J., Coutinho, O. P., & Reis, R. L. (2004). Bone tissue engineering: State of the art and future trends. *Macromolecular Bioscience, 4*(8), 743–765.

Scharfetter-Kochanek, K., Klein, P., & Krieg, T. (1998). Cellular and molecular mechanisms of tissue repair. *Basic Research in Cardiology, 93*, s001–s003.

Schett, G. (2011). Effects of inflammatory and anti-inflammatory cytokines on the bone. *European Journal of Clinical Investigation, 41*(12), 1361–1366.

Schindeler, A., & Little, D. G. (2006). Ras-MAPK signaling in osteogenic differentiation: Friend or foe? *Journal of Bone and Mineral Research, 21*(9), 1331–1338.

SenBanerjee, S., Lin, Z., Atkins, G. B., Greif, D. M., Rao, R. M., Kumar, A., Feinberg, M. W., Chen, Z., Simon, D. I., & Luscinskas, F. W. (2004). KLF2 is a novel transcriptional regulator of endothelial proinflammatory activation. *The Journal of Experimental Medicine, 199*(10), 1305–1315.

Shang, F., Yu, Y., Liu, S., Ming, L., Zhang, Y., Zhou, Z., Zhao, J., & Jin, Y. (2021). Advancing application of mesenchymal stem cell-based bone tissue regeneration. *Bioactive Materials, 6*(3), 666–683.

Sharma, G., Sultana, A., Abdullah, K. M., Pothuraju, R., Nasser, M. W., Batra, S. K., & Siddiqui, J. A. (2024). Epigenetic regulation of bone remodeling and bone metastasis. *Seminars in Cell & Developmental Biology, 154*, 275.

Shrivats, A. R., Alvarez, P., Schutte, L., & Hollinger, J. O. (2014). Bone regeneration. In *Principles of tissue engineering* (pp. 1201–1221). Elsevier.

Shupp, A. B., Kolb, A. D., Mukhopadhyay, D., & Bussard, K. M. (2018). Cancer metastases to bone: Concepts, mechanisms, and interactions with bone osteoblasts. *Cancers, 10*(6), 182.

Street, J., Bao, M., deGuzman, L., Bunting, S., Peale, F. V., Jr., Ferrara, N., Steinmetz, H., Hoeffel, J., Cleland, J. L., & Daugherty, A. (2002). Vascular endothelial growth factor stimulates bone repair by promoting angiogenesis and bone turnover. *Proceedings of the National Academy of Sciences, 99*(15), 9656–9661.

Sui, B.-D., Hu, C.-H., Liu, A.-Q., Zheng, C.-X., Xuan, K., & Jin, Y. (2019). Stem cell-based bone regeneration in diseased microenvironments: Challenges and solutions. *Biomaterials, 196*, 18–30.

Takeuchi, R., Katagiri, W., Endo, S., & Kobayashi, T. (2019). Exosomes from conditioned media of bone marrow-derived mesenchymal stem cells promote bone regeneration by enhancing angiogenesis. *PLoS One, 14*(11), e0225472.

Tang, G., Liu, Z., Liu, Y., Yu, J., Wang, X., Tan, Z., & Ye, X. (2021). Recent trends in the development of bone regenerative biomaterials. *Frontiers in Cell and Developmental Biology, 9*, 665813.

Tang, M., Chen, W., Liu, J., Weir, M. D., Cheng, L., & Xu, H. H. (2014). Human induced pluripotent stem cell-derived mesenchymal stem cell seeding on calcium phosphate scaffold for bone regeneration. *Tissue Engineering Part A, 20*(7–8), 1295–1305.

Teng, S., Liu, C., Krettek, C., & Jagodzinski, M. (2014). The application of induced pluripotent stem cells for bone regeneration: Current progress and prospects. *Tissue Engineering Part B: Reviews, 20*(4), 328–339.

Toosi, S., & Behravan, J. (2020). Osteogenesis and bone remodeling: A focus on growth factors and bioactive peptides. *BioFactors, 46*(3), 326–340.

Trzeciakiewicz, A., Habauzit, V., & Horcajada, M.-N. (2009). When nutrition interacts with osteoblast function: Molecular mechanisms of polyphenols. *Nutrition Research Reviews, 22*(1), 68–81.

Turpaev, K. (2020). Transcription factor KLF2 and its role in the regulation of inflammatory processes. *Biochemistry (Moscow), 85*(1), 54–67.

Ugarte, F., Ryser, M., Thieme, S., Fierro, F. A., Navratiel, K., Bornhäuser, M., & Brenner, S. (2009). Notch signaling enhances osteogenic differentiation while inhibiting adipogenesis in primary human bone marrow stromal cells. *Experimental Hematology, 37*(7), 867–875. e861.

Valenti, M. T., Dalle Carbonare, L., & Mottes, M. (2016). Osteogenic differentiation in healthy and pathological conditions. *International Journal of Molecular Sciences, 18*(1), 41.

Vukicevic, S., Oppermann, H., Verbanac, D., Jankolija, M., Popek, I., Curak, J., Brkljacic, J., Pauk, M., Erjavec, I., & Francetic, I. (2014). The clinical use of bone morphogenetic proteins revisited: A novel biocompatible carrier device OSTEOGROW for bone healing. *International Orthopaedics, 38*, 635–647.

Walsh, J. S. (2015). Normal bone physiology, remodelling and its hormonal regulation. *Surgery (Oxford), 33*(1), 1–6.

Wang, S., Mu, J., Fan, Z., Yu, Y., Yan, M., Lei, G., Tang, C., Wang, Z., Zheng, Y., & Yu, J. (2012). Insulin-like growth factor 1 can promote the osteogenic differentiation and osteogenesis of stem cells from apical papilla. *Stem Cell Research, 8*(3), 346–356.

Wildemann, B., Ignatius, A., Leung, F., Taitsman, L. A., Smith, R. M., Pesántez, R., Stoddart, M. J., Richards, R. G., & Jupiter, J. B. (2021). Non-union bone fractures. *Nature Reviews Disease Primers, 7*(1), 57.

Witkowska-Zimny, M., Wrobel, E., & Przybylski, J. (2010). The most important transcriptional factors of osteoblastogenesis. *Advances in Cell Biology, 2*, 17–28.

Wozney, J. M. (1992). The bone morphogenetic protein family and osteogenesis. *Molecular Reproduction and Development, 32*(2), 160–167.

Wu, M., Chen, G., & Li, Y.-P. (2016). TGF-β and BMP signaling in osteoblast, skeletal development, and bone formation, homeostasis and disease. *Bone Research, 4*(1), 1–21.

Wu, Y., Yang, Y., Yang, P., Gu, Y., Zhao, Z., Tan, L., Zhao, L., Tang, T., & Li, Y. (2013). The osteogenic differentiation of PDLSCs is mediated through MEK/ERK and p38 MAPK signalling under hypoxia. *Archives of Oral Biology, 58*(10), 1357–1368.

Xie, J., Peng, C., Zhao, Q., Wang, X., Yuan, H., Yang, L., Li, K., Lou, X., & Zhang, Y. (2016). Osteogenic differentiation and bone regeneration of iPSC-MSCs supported by a biomimetic nanofibrous scaffold. *Acta Biomaterialia, 29*, 365–379.

Xu, F., Li, W., Yang, X., Na, L., Chen, L., & Liu, G. (2021). The roles of epigenetics regulation in bone metabolism and osteoporosis. *Frontiers in Cell and Developmental Biology, 8*, 619301.

Yamaguchi, M. (2012). Nutritional factors and bone homeostasis: Synergistic effect with zinc and genistein in osteogenesis. *Molecular and Cellular Biochemistry, 366*, 201–221.

Yang, J., Shi, P., Tu, M., Wang, Y., Liu, M., Fan, F., & Du, M. (2014). Bone morphogenetic proteins: Relationship between molecular structure and their osteogenic activity. *Food Science and Human Wellness, 3*(3–4), 127–135.

Yang, Y., Yujiao, W., Fang, W., Linhui, Y., Ziqi, G., Zhichen, W., Zirui, W., & Shengwang, W. (2020). The roles of miRNA, lncRNA and circRNA in the development of osteoporosis. *Biological Research, 53, 40.*

Yavropoulou, M. P., & Yovos, J. G. (2007). The role of the Wnt signaling pathway in osteoblast commitment and differentiation. *Hormones-Athens, 6*(4), 279.

Yu, Y., Mu, J., Fan, Z., Lei, G., Yan, M., Wang, S., Tang, C., Wang, Z., Yu, J., & Zhang, G. (2012). Insulin-like growth factor 1 enhances the proliferation and osteogenic differentiation of human periodontal ligament stem cells via ERK and JNK MAPK pathways. *Histochemistry and Cell Biology, 137*, 513–525.

Yuan, Y., Duan, R., Wu, B., Huang, W., Zhang, X., Qu, M., Liu, T., & Yu, X. (2019). Gene expression profiles and bioinformatics analysis of insulin-like growth factor-1 promotion of osteogenic differentiation. *Molecular Genetics & Genomic Medicine, 7*(10), e00921.

Zambrano, M. B., Brizola, E. S., Refosco, L., Giugliani, R., & Felix, T. M. (2014). Anthropometry, nutritional status, and dietary intake in pediatric patients with osteogenesis imperfecta. *Journal of the American College of Nutrition, 33*(1), 18–25.

Zhang, C. (2010). Transcriptional regulation of bone formation by the osteoblast-specific transcription factor Osx. *Journal of Orthopaedic Surgery and Research, 5*, 1–8.

Zhang, W., Zhu, C., Wu, Y., Ye, D., Wang, S., Zou, D., Zhang, X., Kaplan, D. L., & Jiang, X. (2014). VEGF and BMP-2 promote bone regeneration by facilitating bone marrow stem cell homing and differentiation. *European Cells & Materials, 27*, 1–11.; discussion 11-12. https://doi.org/10.22203/ecm.v027a01

Zhou, J., Zhang, Z., Joseph, J., Zhang, X., Ferdows, B. E., Patel, D. N., Chen, W., Banfi, G., Molinaro, R., & Cosco, D. (2021). Biomaterials and nanomedicine for bone regeneration: Progress and future prospects. *Exploration, 1*(2), 20210011.

Zhou, M., & Graves, D. T. (2022). Impact of the host response and osteoblast lineage cells on periodontal disease. *Frontiers in Immunology, 13*, 998244.

Zhou, S. (2011). TGF-β regulates β-catenin signaling and osteoblast differentiation in human mesenchymal stem cells. *Journal of Cellular Biochemistry, 112*(6), 1651–1660.

Zhu, L., Luo, D., & Liu, Y. (2020). Effect of the nano/microscale structure of biomaterial scaffolds on bone regeneration. *International Journal of Oral Science, 12*(1), 6.

# Multifaceted Role of Osteoclasts in Bone Regeneration

**Md Sariful Islam Howlader and Hiranmoy Das**

**Abstract** Osteoclasts (OC), traditionally regarded as the primary cells responsible for bone resorption, are essential for skeletal remodeling, mineral homeostasis, and structural integrity. Studies have shown that OC actively participate in bone regeneration by interacting with osteoblasts (OB), mesenchymal stem cells (MSC), and immune cells to coordinate bone resorption and formation. This chapter focuses on exploring the dual functions of OC in bone health, starting with an analysis of their origin, differentiation, and the molecular mechanisms involved in bone resorption. It then investigates the signaling pathways and coupling factors, such as TGF-β, BMPs, and IGFs, that connect OC-mediated resorption to subsequent OB-driven bone formation. Additionally, the interactions between OC and immune cells within the bone microenvironment and age-related changes in OC activity are examined to highlight OC's roles in various pathological conditions, including osteoporosis, arthritis, and bone metastasis. Emerging therapeutic approaches, such as anti-resorptive drugs, combination therapies, and novel treatments designed to modulate OC function to enhance regeneration while avoiding inhibition of bone resorption, are also discussed. The potential for personalized medicine using biomarkers, gene editing, and pharmacogenomic strategies is further emphasized as a promising direction for optimizing treatment efficacy and safety. By shedding light on the multifaceted roles of OC, this chapter highlights their significance in maintaining bone homeostasis and provides insights for the development of innovative therapies that balance resorptive and regenerative functions, ultimately improving outcomes for patients with bone-related diseases.

**Keywords** Osteoclasts (OC) · Osteoblasts (OB) · Bone resorption · Bone regeneration · Skeletal remodeling · Mesenchymal stem cells (MSC) · Osteoporosis · Arthritis · Bone homeostasis · Regenerative medicine

M. S. I. Howlader · H. Das (✉)
Department of Pharmaceutical Sciences, Jerry H. Hodge School of Pharmacy, Texas Tech University Health Sciences Center, Amarillo, TX, USA
e-mail: mdhowlad@ttuhsc.edu; hiranmoy.das@ttuhsc.edu

 
H. Das (ed.), *Tissue Repair and Regeneration*,
https://doi.org/10.1007/978-3-031-93677-7_6

# 1 Introduction

Bone remodeling is a continuous process involving the resorption of old bones by osteoclasts (OC) and the formation of new bone by osteoblasts (OB). This tightly regulated cycle maintains skeletal integrity and supports metabolic functions by releasing calcium and phosphorus as needed. Each remodeling phase consists of distinct cellular activities: the initial phase of bone resorption mediated by OC, followed by a reversal phase, and the final formation phase, where OB deposits a new matrix and mineralizes the bone. This coupling between bone resorption and formation is crucial for maintaining bone mass and quality throughout life (Florencio-Silva et al., 2015).

The balance between OC and OB activity is under the control of systemic and local factors, including parathyroid hormone (PTH), vitamin D, estrogen, and various cytokines. The RANK/RANKL/OPG signaling pathway is particularly critical in this regulation; RANKL (receptor activator of nuclear factor κB ligand) binds to its receptor RANK on OC precursors, promoting OC differentiation and activity. Osteoprotegerin (OPG), a decoy receptor produced by OB, inhibits OC formation by competing with RANK for RANKL binding, thus balancing bone resorption and formation (Boyle et al., 2003; Raggatt & Partridge, 2010).

Bone homeostasis ensures the preservation of bone strength, structural integrity, and mineral storage. This dynamic process involves a precise balance between bone formation by OB and bone resorption by OC. Dysregulation of bone remodeling leads to various pathological conditions: osteoporosis, where resorption outpaces formation, resulting in reduced bone density; and osteopetrosis, where impaired resorption causes excessive bone density but structural fragility. Additionally, inflammation and aging affect bone remodeling by promoting OC activity and suppressing OB function (Henriksen et al., 2011; Raisz, 2005).

Calcium homeostasis is another critical aspect of bone health, with bones acting as a reservoir for this essential mineral. The release of calcium into the bloodstream through OC-mediated resorption ensures the availability of calcium for cellular functions throughout the body. Conversely, OB activity facilitates calcium uptake into the skeleton, maintaining systemic calcium balance (Florencio-Silva et al., 2015).

Historically, osteoclasts were first described by German anatomist Wilhelm His in the nineteenth century, who noted their distinctive multinucleated morphology and their role in bone resorption. Early research focused on their resorptive function, categorizing them as bone-destroying cells that degrade bone matrix and liberate calcium (Teitelbaum, 2000). The later discovery of the RANK/RANKL/OPG system and the role of cytokines like macrophage colony-stimulating factors (M-CSF) marked a turning point in OC research, providing insight into their differentiation and regulation (Boyle et al., 2003).

More recent studies have highlighted the role of OC beyond resorption. Research into OC-OB interactions revealed that OC plays an active role in bone formation by releasing signaling molecules, such as TGF-β, that stimulate OB recruitment and differentiation. These findings shifted the perception of OC from mere bone restorers to multifunctional cells that also participate in bone regeneration (Henriksen et al., 2011).

This chapter aims to elucidate the multifaceted role of OC in bone remodeling, focusing on their dual functions in bone resorption and regeneration. Traditionally, therapeutic approaches have focused on inhibiting OC activity to treat bone diseases characterized by excessive bone loss, such as osteoporosis (Das et al., 2019; Deb et al., 2021). However, this inhibition often disrupts the coupling process, reducing bone formation as well. Understanding the regenerative roles of OC, including their interactions with OB and mesenchymal stem cells (MSC), is essential for developing therapies that can selectively modulate OC activity to enhance bone regeneration while minimizing bone resorption (Zaidi, 2007).

By exploring both OC's resorptive and regenerative roles, this chapter highlights new perspectives and therapeutic potentials for treating bone diseases that extend beyond conventional anti-resorptive strategies.

## 2 Osteoclast Biology and Bone Resorption

Osteoclasts (OC) are specialized, multinucleated cells essential for bone resorption and play a crucial role in bone remodeling. These cells originate from hematopoietic stem cells (HSCs), specifically from the monocyte/macrophage lineage. Their differentiation and activation are tightly regulated through interactions with specific signaling pathways and transcription factors, which coordinate their functions in both physiological and pathological conditions (Maity et al., 2022; Rolph & Das, 2020). Understanding the biology of osteoclasts is critical for developing therapeutic strategies for bone diseases characterized by excessive bone resorption or inadequate remodeling.

### *2.1 Origin and Differentiation of Osteoclasts*

Osteoclastogenesis, the process of osteoclast formation, begins with monocyte/macrophage precursors that differentiate into mature osteoclasts under the influence of receptor activator of nuclear factor κB ligand (RANKL) and macrophage colony-stimulating factor (M-CSF). M-CSF, binding to its receptor c-Fms on osteoclast precursors, promotes their proliferation and survival. Concurrently, RANKL, expressed on osteoblasts and stromal cells, interacts with RANK on osteoclast precursors, activating downstream signaling cascades that drive differentiation (Boyle et al., 2003). This interaction triggers nuclear factor κB (NF-κB) and mitogen-activated protein kinase (MAPK) pathways, essential for initiating osteoclastogenesis (Teitelbaum, 2000).

Key transcription factors play critical roles in this process. NFATc1 (nuclear factor of activated T cells cytoplasmic 1) is the master regulator of osteoclast differentiation. Activated by RANKL signaling, NFATc1 induces the expression of genes responsible for the osteoclastic phenotype, including cathepsin K and tartrate-resistant acid phosphatase (TRAP) (Henriksen et al., 2011). Another important factor, c-Fos, collaborates with NFATc1 to ensure proper maturation of osteoclasts. These transcriptional programs integrate external signals into a coordinated cellular response, culminating in the formation of mature, functional osteoclasts.

## 2.2 *Morphology and Functional Characteristics of Osteoclasts*

Osteoclasts exhibit unique structural features optimized for their bone-resorbing function. A hallmark of osteoclast morphology is multinucleation, achieved through the fusion of mononuclear precursors. This multinucleated state increases their resorptive efficiency, allowing them to degrade extensive areas of the bone matrix effectively (Zaidi, 2007).

The resorption zone, comprising the sealing zone and ruffled border, is critical for osteoclast function. The sealing zone, a specialized actin-rich structure, firmly anchors the osteoclast to the bone matrix, creating a closed microenvironment for resorption. The ruffled border, characterized by extensive plasma membrane infoldings, increases the surface area for the secretion of resorptive enzymes and protons. These structures are essential for the degradation of both inorganic and organic components of bone (Takayanagi, 2007).

## 2.3 *Molecular Mechanisms of Bone Resorption*

Osteoclasts degrade the bone matrix through a combination of acidification and enzymatic activity. The acidification process begins with the secretion of protons by vacuolar H + -ATPase pumps located on the ruffled border. This creates an acidic microenvironment that dissolves the inorganic hydroxyapatite crystals of bone (Henriksen et al., 2011).

In addition to acidification, osteoclasts secrete proteolytic enzymes, such as TRAP and cathepsin K, to degrade the organic matrix. TRAP plays a key role in breaking down the collagen matrix, while cathepsin K is a potent protease that specifically targets type I collagen, the main organic component of bone (Teitelbaum, 2000). The combined actions of acidification and enzymatic degradation result in the formation of resorption pits, known as Howship's lacunae, which are characteristic features of osteoclast activity. These pits mark areas of bone resorption, releasing calcium and phosphate into the bloodstream (Florencio-Silva et al., 2015).

## 2.4 *Pathological Conditions Linked to Osteoclast Dysfunction*

The dysregulation of osteoclast activity is associated with various bone disorders characterized by either excessive bone resorption or impaired remodeling.

### 2.4.1 Osteoporosis

Osteoporosis is a condition in which the rate of bone resorption exceeds that of bone formation, leading to reduced bone mass and increased fracture risk. Overactive osteoclasts play a central role in this imbalance, highlighting the need for therapeutic interventions targeting osteoclast activity. Anti-resorptive drugs, such as bisphosphonates and RANKL inhibitors like denosumab, effectively suppress osteoclast-mediated bone resorption, improving bone density and reducing fracture risk (Raisz, 2005).

### 2.4.2 Rheumatoid Arthritis

Rheumatoid arthritis (RA) is an inflammatory disease characterized by localized bone erosion. Pro-inflammatory cytokines such as tumor necrosis factor-alpha (TNF-α) and interleukin-1 (IL-1) promote osteoclastogenesis, resulting in focal bone loss and joint damage. Excessive osteoclast activity around inflamed joints contributes to the severity of RA. Targeting inflammatory cytokines and osteoclast activity, through agents like TNF inhibitors or RANKL inhibitors, is an effective strategy for mitigating bone destruction in RA (Martin & Sims, 2005).

Rheumatoid arthritis (RA) is a chronic autoimmune disease characterized by persistent inflammation that primarily affects synovial joints, leading to progressive joint destruction and disability. The pathogenesis of RA involves immune cell infiltration, synovial hyperplasia, and excessive production of pro-inflammatory cytokines such as tumor necrosis factor-alpha (TNF-α), interleukin-1β (IL-1β), and interleukin-6 (IL-6) (McInnes & Schett, 2011). These inflammatory mediators activate synovial fibroblasts and osteoclasts, resulting in bone erosion and cartilage degradation (Firestein & McInnes, 2017).

As the disease progresses, synovial membrane inflammation leads to pannus formation—an abnormal tissue growth that invades and damages the cartilage and bone. The breakdown of cartilage, primarily mediated by matrix metalloproteinases (MMPs), leads to loss of joint cushioning and pain (Smolen et al., 2016). Concurrently, osteoclast-driven bone resorption erodes the subchondral bone, further exacerbating joint deformities and dysfunction. The narrowing of the joint space due to cartilage loss contributes to stiffness and restricted mobility, significantly impairing quality of life (Scott et al., 2010).

Current therapeutic strategies target the inflammatory pathways using disease-modifying antirheumatic drugs (DMARDs) and biologics, such as TNF inhibitors

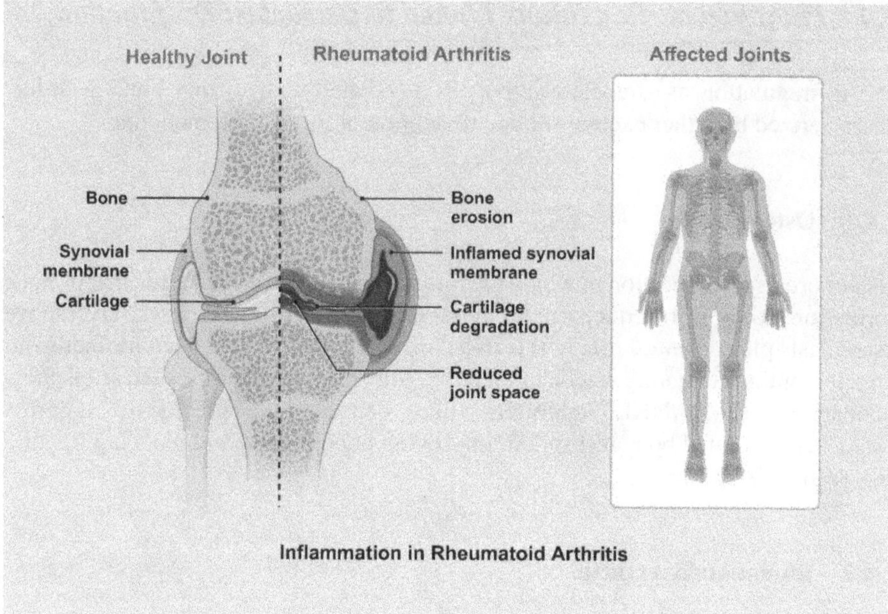


**Fig. 1** Comparison of a healthy joint and a joint affected by rheumatoid arthritis (RA). The illustration compares a normal joint with a joint affected by rheumatoid arthritis. The healthy joint maintains structural integrity with smooth cartilage, an intact synovial membrane, and sufficient joint space for movement. In contrast, RA leads to several pathological changes: (1) bone erosion due to excessive osteoclast activity, (2) inflammation of the synovial membrane, resulting in swelling and pain, (3) cartilage degradation, impairing joint function, and (4) joint space reduction, leading to stiffness and restricted movement

and Janus kinase (JAK) inhibitors, to slow disease progression and preserve joint function (Smolen et al., 2010). However, effective disease management requires early diagnosis and intervention to prevent irreversible joint damage (Fig. 1).

## Stepwise Pathogenesis of Rheumatoid Arthritis

Rheumatoid arthritis is a complex autoimmune disorder driven by immune dysregulation, chronic inflammation, and excessive osteoclast-mediated bone resorption. Targeted therapies, including TNF inhibitors, IL-6 receptor antagonists, and JAK inhibitors, aim to disrupt these pathological mechanisms and preserve joint function. However, early diagnosis and intervention remain crucial for preventing irreversible joint damage (Fig. 2).

### Step 1: Synovial Inflammation and Immune Cell Infiltration

Rheumatoid arthritis begins with an aberrant immune response in genetically predisposed individuals, triggered by environmental factors such as infections or

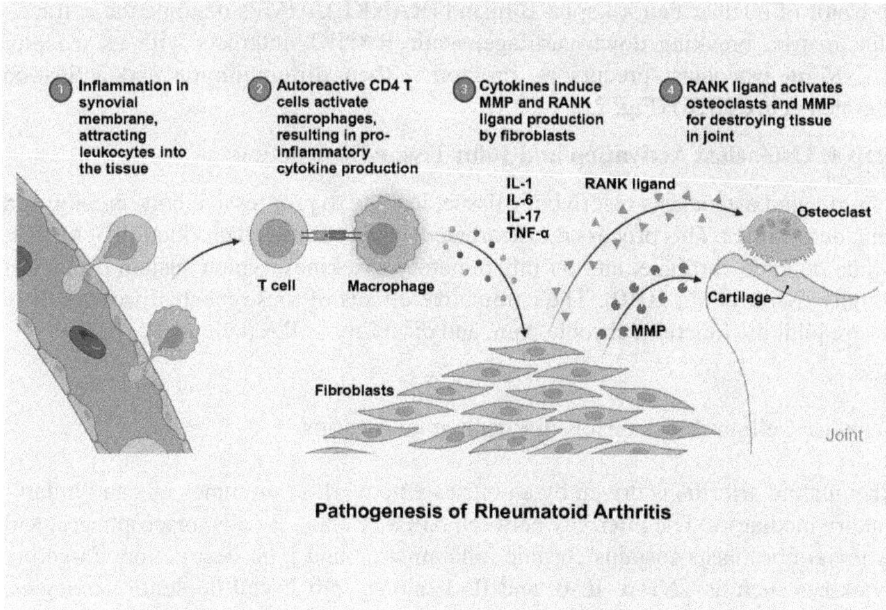


**Fig. 2** Pathogenesis of rheumatoid arthritis (RA). The stepwise progression of rheumatoid arthritis, beginning with synovial inflammation and immune cell infiltration. (1) Inflammatory signals recruit leukocytes to the synovial membrane, initiating immune activation. (2) Autoreactive CD4$^+$ T cells stimulate macrophages, leading to the release of pro-inflammatory cytokines such as TNF-α, IL-1β, and IL-6. (3) These cytokines trigger fibroblasts to produce matrix metalloproteinases (MMPs) and receptor activator of nuclear factor kappa-B ligand (RANKL), promoting tissue degradation. (4) RANKL activates osteoclasts, while MMPs degrade the extracellular matrix, resulting in bone erosion, cartilage destruction, and joint deformities

smoking. The synovial membrane becomes inflamed, leading to the recruitment of leukocytes, including macrophages, dendritic cells, B cells, and autoreactive CD4$^+$ T cells (McInnes & Schett, 2011). This results in synovial hyperplasia and the formation of pannus—an aggressive fibrovascular tissue that invades joint structures (Fig. 2).

### Step 2: CD4$^+$ T Cell Activation and Pro-Inflammatory Cytokine Release

Activated CD4$^+$ T cells, particularly Th1 and Th17 subsets, drive RA progression by secreting inflammatory cytokines such as TNF-α, IL-1β, IL-6, and IL-17 (Firestein & McInnes, 2017). These cytokines enhance synovial inflammation and stimulate macrophages, which further amplify cytokine production, fueling chronic inflammation (Fig. 2).

### Step 3: Fibroblast Activation and MMP/RANKL Induction

Cytokines released by T cells and macrophages activate synovial fibroblasts, prompting them to produce matrix metalloproteinases (MMPs) and receptor activator of nuclear factor kappa-B ligand (RANKL). MMPs degrade the extracellular matrix, breaking down cartilage, while RANKL interacts with its receptor RANK on osteoclast precursors, promoting their differentiation and activation (Smolen et al., 2016) (Fig. 2).

**Step 4: Osteoclast Activation and Joint Tissue Destruction**

Activated osteoclasts resorb bone tissue, leading to progressive bone erosion and joint deformities. This process is exacerbated by the continued production of MMPs, which degrade cartilage, and by inflammatory cytokines, which sustain osteoclast activity (Scott et al., 2010). The cumulative effects of these mechanisms result in severe joint dysfunction, chronic pain, and disability in RA patients (Fig. 2).

### Stepwise Cellular Mechanisms in Rheumatoid Arthritis

Rheumatoid arthritis is driven by an intricate network of immune cells and inflammatory mediators. The interplay between APCs, T cells, B cells, macrophages, and synovial fibroblasts sustains chronic inflammation and joint destruction. Targeting cytokines such as TNF-α, IL-6, and IL-1, along with B cell depletion strategies, remain a cornerstone of RA therapy, but novel approaches targeting cellular interactions hold promise for future treatments (Fig. 3).

**Step 1: Antigen Presentation and T Cell Activation**

The pathogenesis of RA begins with the activation of antigen-presenting cells (APCs), such as dendritic cells, which present autoantigens to naïve $CD4^+$ T cells via major histocompatibility complex (MHC) class II molecules. This interaction, facilitated by costimulatory molecules (e.g., CD80/CD86 binding to CD28 on T cells), initiates T cell activation and the secretion of pro-inflammatory cytokines (McInnes & Schett, 2011) (Fig. 3).

**Step 2: B Cell Activation and Autoantibody Production**

Activated $CD4^+$ T cells release IL-4, IL-6, and IL-10, which stimulate B cells to differentiate into plasma cells. Plasma cells then produce rheumatoid factor (RF) and anti-citrullinated protein antibodies (ACPAs) (Firestein & McInnes, 2017). These autoantibodies form immune complexes that deposit in the synovium, triggering complement activation and further inflammatory responses, leading to synovial hyperplasia and pannus formation (Fig. 3).

**Step 3: Macrophage Activation and Cytokine Release**

In parallel, activated $CD4^+$ T cells secrete IFN-γ, which enhances macrophage activation. Activated macrophages produce key pro-inflammatory cytokines, including tumor necrosis factor-alpha (TNF-α), interleukin-1 (IL-1), and interleukin-6 (IL-6) (Smolen et al., 2016). These cytokines perpetuate chronic inflammation, recruit additional immune cells, and contribute to joint destruction (Fig. 3).

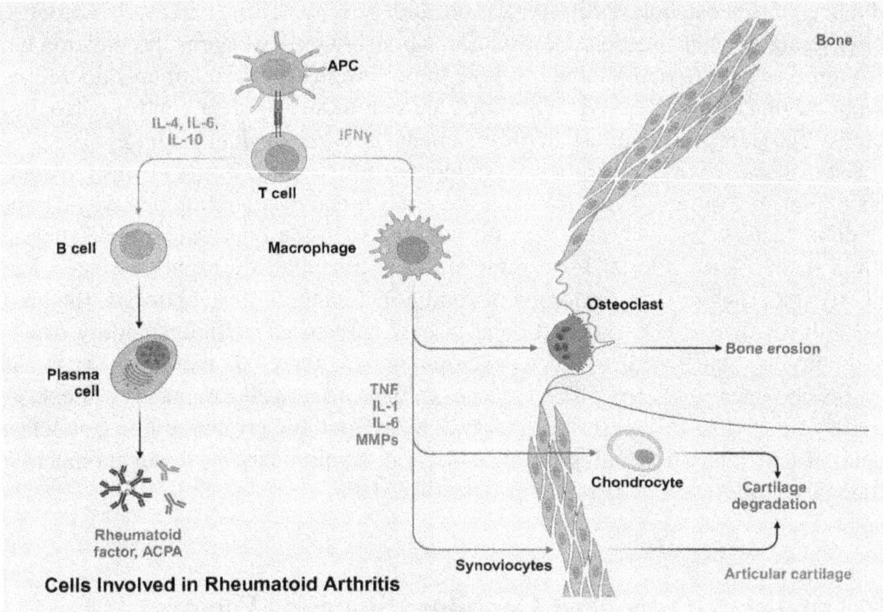


**Fig. 3** Cellular mechanisms in rheumatoid arthritis (RA)The immune cell interactions that drive the pathogenesis of rheumatoid arthritis. (1) Antigen-presenting cells (APCs) activate CD4⁺ T cells, which secrete pro-inflammatory cytokines IL-4, IL-6, and IL-10, stimulating B cell activation. (2) Activated B cells differentiate into plasma cells, which produce rheumatoid factor (RF) and anti-citrullinated protein antibodies (ACPAs), leading to immune complex formation and chronic inflammation. (3) In parallel, APCs activate CD4⁺ T cells to secrete IFN-γ, which promotes macrophage activation. (4) Activated macrophages produce TNF-α, IL-1, IL-6, and matrix metalloproteinases (MMPs), contributing to synovial inflammation, cartilage degradation, and bone erosion

### Step 4: Cartilage and Bone Destruction via MMPs and Osteoclast Activation

Pro-inflammatory cytokines stimulate synovial fibroblasts to release matrix metalloproteinases (MMPs), enzymes that degrade cartilage by breaking down extracellular matrix proteins (Scott et al., 2010). Additionally, TNF-α and IL-1 promote the expression of receptor activator of nuclear factor kappa-B ligand (RANKL) by synovial fibroblasts. RANKL binds to its receptor RANK on osteoclast precursors, driving osteoclast differentiation and bone resorption, ultimately leading to bone erosion (McInnes & Schett, 2011) (Fig. 3).

### 2.4.3 Cancer-Induced Bone Disease

Certain cancers, such as multiple myeloma and metastatic breast cancer, stimulate osteoclast activity, leading to cancer-induced bone disease. Tumor cells produce factors, such as parathyroid hormone-related protein (PTHrP) and interleukins, which enhance osteoclast differentiation and activity. This increased osteoclast activity causes extensive bone destruction, which not only weakens the skeleton but also releases growth factors that promote tumor progression. Treatments, including bisphosphonates and RANKL inhibitors, are commonly used to control osteoclast activity and mitigate bone loss in these patients (Henriksen et al., 2011).

Osteoclasts are integral to bone remodeling, functioning as the primary mediators of bone resorption. Their activities are finely tuned by signaling pathways, transcription factors, and interactions with other cells in the bone microenvironment. While their role in normal bone remodeling is essential, dysregulated osteoclast activity underlies several pathological conditions, including osteoporosis, rheumatoid arthritis, and cancer-induced bone disease. Advances in understanding osteoclast biology have led to the development of targeted therapies, such as bisphosphonates and RANKL inhibitors, that effectively manage osteoclast-mediated bone loss (Glass II & Karsenty, 2006). Continued research into osteoclast function and regulation holds promise for the development of novel therapeutic strategies to address a wide range of bone disorders.

## 3 Osteoclast-Osteoblast Coupling: Signaling Factors in Bone Regeneration

Bone remodeling is a dynamic and continuous process that maintains skeletal integrity and mineral homeostasis. It relies on the coordinated actions of osteoclasts (OC) and osteoblasts (OB), which execute bone resorption and formation, respectively. Traditionally, OC has been viewed solely as bone-resorbing cells that degrade the mineralized bone matrix, releasing calcium and phosphate into circulation. However, emerging research has highlighted their pivotal role in orchestrating bone regeneration by interacting with OB, mesenchymal stem cells (MSC), and immune cells. These interactions enable OC to serve as signaling hubs that couple resorption to subsequent bone formation. This chapter delves into the mechanisms underlying osteoclast-osteoblast coupling, explores the signaling pathways and cellular interactions involved in bone regeneration, and discusses the therapeutic implications for managing bone-related disorders.

### *3.1 Osteoclast-Osteoblast Coupling*

Osteoclasts are integral not only to bone degradation but also to the initiation of subsequent bone formation. During bone resorption, OC releases signaling molecules stored in the bone matrix, which act on OB to stimulate new bone deposition.

#### 3.1.1 Key Factors Released by Osteoclasts

During the resorption process, OC release several growth factors from the degraded bone matrix, including transforming growth factor-beta (TGF-β), bone morphogenetic proteins (BMPs), and insulin-like growth factors (IGFs). TGF-β plays a critical role in coupling bone resorption and formation by enhancing the recruitment and differentiation of OB at resorption sites (Bonewald, 2011, Bonewald, 2017). BMPs, particularly BMP-2 and BMP-7, are key mediators of MSC differentiation into OB, thereby ensuring the availability of new bone-forming cells (Florencio-Silva et al., 2015). Additionally, IGFs stimulate OB proliferation and matrix production, promoting the regenerative phase following osteoclastic activity. Together, these factors create a microenvironment that promotes efficient bone formation, tightly coupled to the resorptive actions of OC.

#### 3.1.2 Mechanism of Osteoclast-Osteoblast Coupling

Osteoclast-osteoblast coupling involves the spatial and temporal release of signaling molecules into the resorption pit, where OB are recruited to initiate bone formation. This process is marked by a "reversal phase," during which the bone surface is cleaned and primed by OB precursors before new bone deposition begins (Kim et al., 2020; Marie, 2015; Tripuwabhrut, 2014). The factors released by OC during this phase not only recruit OB but also stimulate their differentiation and functional activity, ensuring the continuity of bone remodeling. By coupling resorption with formation, OC maintain skeletal homeostasis and prevent pathological bone loss.

#### 3.1.3 Therapeutic Implications

The signaling pathways involved in osteoclast-osteoblast coupling represent promising therapeutic targets for bone-related disorders. Enhancing OB activity through TGF-β or BMP-based therapies offers the potential to promote bone regeneration in conditions characterized by impaired remodeling. Furthermore, combination therapies that pair anti-resorptive agents, such as bisphosphonates, with anabolic treatments, like parathyroid hormone analogs, may improve bone density and quality by balancing resorption and formation.

### *3.2 OC Interaction with Mesenchymal Stem Cells (MSC)*

In addition to interacting with OB, OC influences MSC recruitment and differentiation, playing a central role in maintaining the cellular pool required for sustained bone regeneration.

### 3.2.1 Recruitment of MSC to Resorption Sites

Osteoclasts release sphingosine-1-phosphate (S1P), a lipid mediator that acts as a chemotactic signal to recruit MSC to sites of bone resorption. Once at these sites, MSC differentiate into OB, contributing to new bone formation (Raggatt & Partridge, 2010). Additionally, OC activates the Wnt/β-catenin signaling pathway, which is critical for MSC migration, proliferation, and osteogenic differentiation (Glass II & Karsenty, 2006; Henriksen et al., 2011; Houschyar et al., 2019). These interactions underscore the multifaceted role of OC in orchestrating cellular processes essential for bone regeneration.

### 3.2.2 MSC Differentiation and Bone Formation

Upon recruitment to resorption sites, MSC are exposed to BMPs, IGFs, and other factors released by OC, which drive their differentiation into OB. This ensures a continuous supply of bone-forming cells to replace resorbed bone. The interaction between OC and MSC highlights the importance of crosstalk among bone cells in maintaining the dynamic equilibrium of bone remodeling.

### 3.2.3 Therapeutic Potential of MSC Recruitment

Targeting pathways that regulate MSC recruitment and differentiation, such as S1P and Wnt, holds significant therapeutic potential for enhancing bone regeneration. These strategies may be particularly effective in conditions like osteoporosis, where the regenerative capacity of bone is compromised.

## *3.3 Osteoclast-Immune Cell Crosstalk and Inflammatory Response in Bone Regeneration*

Osteoclasts interact with immune cells, including T cells and macrophages, modulating inflammatory responses that are critical for bone remodeling and regeneration.

### 3.3.1 Crosstalk with T Cells and Macrophages

T cells contribute to OC differentiation by producing RANKL, a key regulator of osteoclastogenesis. While this promotes resorption during inflammation, T cells can also influence OB activity, balancing resorption with regeneration. Similarly, macrophages produce cytokines such as TNF-α and IL-1, which enhance OC activity during inflammation but can also support OB differentiation under specific conditions (Henriksen et al., 2011). These interactions underscore the dual role of immune cells in bone remodeling.

### 3.3.2 Regenerative Impact of Immune Cell Interaction

The interplay between OC and immune cells regulates the balance between resorption and formation during injury and inflammation. Cytokines and growth factors released during immune responses create a microenvironment that supports bone regeneration while preventing excessive resorption.

### 3.3.3 Implications for Inflammatory Bone Disorders

Aberrant interactions between OC and immune cells contribute to diseases such as rheumatoid arthritis and periodontitis, where excessive bone resorption occurs. Modulating these interactions through targeted therapies could enhance regenerative outcomes while mitigating bone loss.

## *3.4 Signaling Pathways Involved in Osteoclast-Mediated Regeneration*

Several signaling pathways coordinate the regenerative roles of OC, coupling bone resorption with formation.

### 3.4.1 RANKL/OPG Pathway

The RANKL/OPG axis is central to OC differentiation and activity. RANKL binds to RANK on OC precursors to promote their maturation, while OPG, a decoy receptor, inhibits this interaction, regulating OC activity (Boyce, 2013). Therapeutic modulation of the RANKL/OPG balance offers a means to enhance regeneration while controlling resorption.

### 3.4.2 Wnt/β-Catenin Pathway

The Wnt pathway, activated during OC activity, promotes MSC differentiation into OB and couples' resorption to regeneration. Therapeutic interventions targeting this pathway have shown promise in enhancing bone formation (Glass II & Karsenty, 2006).

### 3.4.3 Notch Signaling Pathway

Notch signaling, which is activated during OC differentiation, coordinates the activities of OC and OB to ensure synchronized remodeling. Therapeutic targeting of this pathway could enhance regenerative functions while mitigating excessive resorption.

### 3.4.4 Additional Pathways

Other pathways, such as NF-κB and BMP/TGF-β, play crucial roles in osteoclast-mediated regeneration. NF-κB influences OC activity during inflammation, while BMP and TGF-β pathways drive MSC and OB differentiation, linking resorption to regeneration (Teitelbaum, 2000).

Osteoclasts, once regarded solely as bone-resorbing cells, are now recognized as pivotal regulators of bone regeneration. Through interactions with OB, MSC, and immune cells, OC couple bone resorption with formation, maintaining skeletal integrity. Understanding these interactions and the signaling pathways involved provides valuable insights into bone biology and offers new avenues for therapeutic intervention. By harnessing the regenerative potential of OC while controlling their resorptive activity, it may be possible to develop effective treatments for conditions such as osteoporosis, rheumatoid arthritis, and other bone-related diseases.

## 4 Osteoclast Function Across Lifespan and Pathological Conditions

Bone is a dynamic and metabolically active tissue that undergoes continuous remodeling to maintain structural integrity and mineral homeostasis. Osteoclasts (OC), derived from hematopoietic stem cells (HSCs), are specialized multinucleated cells responsible for bone resorption. For decades, OC have been considered the primary agents of bone degradation, facilitating the release of calcium and phosphate for systemic mineral balance. However, emerging research highlights their multifaceted role in bone health, where OC not only contribute to bone resorption but also play critical roles in regeneration by coupling with osteoblasts (OB) and interacting with mesenchymal stem cells (MSC) and immune cells. Dysregulation of OC activity, whether due to aging, hormonal imbalances, or pathological conditions, disrupts bone remodeling and contributes to diseases such as osteoporosis, osteopetrosis, Paget's disease, and cancer-induced bone loss. This chapter explores OC function across the lifespan, its role in metabolic bone disorders, and therapeutic strategies to address these challenges.

## *4.1 Age-Related Changes in Osteoclast and Osteoblast Activity*

### 4.1.1 OC and OB Balance in Youth and Adulthood

During youth and early adulthood, bone remodeling is characterized by a delicate balance between OC-mediated resorption and OB-driven bone formation. This efficient coupling mechanism allows the skeleton to adapt to mechanical stress, repair microdamage, and maintain peak bone mass. Hormones such as growth hormone (GH), insulin-like growth factor-1 (IGF-1), and sex steroids (estrogen and testosterone) play pivotal roles in regulating this balance. Estrogen inhibits OC activity by reducing receptor activator of nuclear factor kappa-B ligand (RANKL) expression and increasing osteoprotegerin (OPG) production. OPG serves as a decoy receptor for RANKL, preventing it from binding to RANK on OC precursors and thus inhibiting OC differentiation (Boyle et al., 2003) (Khosla & Monroe, 2018; Khosla et al., 2012). This tightly regulated process ensures that bone resorption is matched by formation, resulting in optimal skeletal health and structural integrity.

### 4.1.2 Aging and OC Overactivity

With advancing age, the balance between OC and OB activity shifts, leading to increased bone resorption relative to formation. This imbalance is attributed to hormonal changes, reduced osteogenic potential of MSC, and increased osteoclastogenesis. In postmenopausal women, estrogen deficiency amplifies RANKL expression, enhances OC activity, and accelerates bone resorption. Similarly, in aging men, declining testosterone and GH levels impair OB function and matrix production, exacerbating bone loss (Khosla et al., 2012). Aging also impairs MSC function, favoring adipogenesis over osteogenesis, which further reduces bone formation (Florencio-Silva et al., 2015; Prateeksha et al., 2023). This age-related shift in remodeling dynamics increases susceptibility to fractures and skeletal fragility, particularly in trabecular bone, which is more metabolically active.

### 4.1.3 Hormonal Influence on OC Activity with Age

Hormonal regulation is central to OC and OB activity across the lifespan. Estrogen deficiency in postmenopausal women is a major driver of increased OC activity, as it reduces the inhibitory effect on RANKL and enhances bone resorption. Testosterone and GH also contribute to skeletal health, with their decline leading to reduced OB activity and impaired bone matrix deposition. Parathyroid hormone (PTH), which regulates calcium homeostasis, plays a dual role in bone remodeling. Intermittent PTH administration can stimulate OB activity and enhance bone formation, whereas chronic elevations in PTH, as seen in hyperparathyroidism, increase OC activity and cortical bone resorption (Raisz, 2005). Collectively, hormonal changes in aging create a pro-resorptive environment, heightening the risk of osteoporosis and related fractures.

## *4.2 OC Function in Metabolic Bone Disorders*

### 4.2.1 Osteoporosis

Osteoporosis is characterized by reduced bone mass and microarchitectural deterioration, resulting in increased fracture risk. The condition arises from excessive OC activity coupled with insufficient OB function, leading to net bone loss. Estrogen deficiency in postmenopausal women is a key driver of osteoporosis, as it enhances RANKL expression and OC-mediated resorption, particularly in trabecular bone (Raisz, 2005). Therapeutic strategies primarily focus on reducing OC activity using anti-resorptive agents such as bisphosphonates and RANKL inhibitors like denosumab. These therapies effectively reduce fracture risk but may suppress OB activity due to disrupted coupling, leading to adverse effects such as atypical femoral fractures and osteonecrosis of the jaw (Zaidi, 2007).

### 4.2.2 Paget's Disease of Bone

Paget's disease is a localized bone remodeling disorder characterized by excessive OC activity followed by disorganized OB-driven bone formation. This results in structurally abnormal and weakened bone prone to deformities and fractures. Genetic mutations, particularly in the SQSTM1 gene, and environmental factors, such as viral infections, contribute to the disease's pathogenesis (Roodman, 2004). Bisphosphonates remain the first-line treatment, effectively suppressing OC activity and alleviating symptoms. Emerging therapies targeting RANKL inhibition or immune modulation offer additional benefits by addressing the underlying drivers of OC hyperactivity (Boyle et al., 2003).

### 4.2.3 Cancer-Induced Bone Disease

In cancers such as breast, prostate, and multiple myeloma, tumor cells secrete factors like parathyroid hormone-related protein (PTHrP), which upregulates RANKL expression and enhances OC activity. This creates a vicious cycle of bone resorption and tumor growth, as OC release growth factors like transforming growth factor-beta (TGF-β) from the bone matrix, which further stimulates tumor proliferation (Roodman, 2004). Denosumab, a RANKL inhibitor, and bisphosphonates are the mainstay therapies for managing cancer-induced bone disease, reducing skeletal complications and alleviating pain.

### 4.2.4 Rheumatoid Arthritis and Inflammatory Bone Diseases

Rheumatoid arthritis (RA) and other inflammatory bone diseases are characterized by elevated pro-inflammatory cytokines, such as tumor necrosis factor-alpha (TNF-$\alpha$) and interleukin-1 (IL-1), which drive OC differentiation and activity. This leads to localized bone erosion and joint deformities. Anti-inflammatory therapies, including TNF-$\alpha$ inhibitors, not only reduce inflammation but also indirectly suppress OC activity, offering dual benefits for managing bone loss and inflammation (Sarkar et al., 2022; Takayanagi, 2007; Teitelbaum, 2000).

## *4.3 Cancer and Bone Metastasis: OC-Driven Bone Resorption and Osteolysis*

### 4.3.1 Role of OC in Bone Metastasis

Bone metastases are driven by tumor-OC interactions, where tumor cells secrete factors such as PTHrP to upregulate RANKL and enhance OC activity. This leads to extensive bone resorption and the release of matrix-embedded growth factors like TGF-$\beta$, which promote tumor growth and survival (Boyle et al., 2003; Roodman, 2004). The resulting osteolysis contributes to the progression of skeletal-related complications, including pain and fractures.

### 4.3.2 Impact on Bone Integrity and Fracture Risk

OC-driven bone resorption in metastatic lesions weakens bone integrity, increasing the risk of pathological fractures and hypercalcemia. This significantly reduces patient quality of life and complicates cancer management. Understanding the molecular mechanisms underlying tumor-OC crosstalk is critical for developing effective therapeutic strategies (Roodman, 2004).

### 4.3.3 Therapeutic Approaches for Cancer-Related Bone Disease

Therapeutic options for cancer-induced bone disease include anti-resorptive agents such as denosumab and bisphosphonates, which target OC activity to reduce bone resorption and alleviate symptoms. Combination therapies, incorporating chemotherapy and radiotherapy with anti-resorptive treatments, provide a comprehensive approach to managing skeletal complications. Novel agents targeting specific OC pathways, such as cathepsin K inhibitors and immune-modulating therapies, hold promise for enhancing treatment efficacy and improving patient outcomes (Howlader et al., 2024; Teitelbaum, 2000).

Osteoclasts play a central role in bone remodeling and regeneration, acting not only as bone-resorbing cells but also as key regulators of skeletal health. Their interactions with OB, MSC, and immune cells underscore their dual functionality in coupling bone resorption with formation. Dysregulated OC activity in aging and pathological conditions highlights the need for targeted therapeutic strategies that preserve their regenerative potential while mitigating excessive resorption. Advances in OC-targeted therapies, including combination treatments and novel agents, offer promising avenues for managing bone diseases and improving patient outcomes.

# 5 Therapeutic Strategies Targeting Osteoclasts

Osteoclasts (OC), derived from hematopoietic stem cells (HSCs), are specialized multinucleated cells responsible for bone resorption, a process essential for skeletal remodeling and maintaining calcium balance (Boyle et al., 2003). Dysregulation of OC activity contributes to a range of bone-related disorders, including osteoporosis, Paget's disease, and cancer-induced bone loss, making OC a key therapeutic target. This chapter explores anti-resorptive therapies, the challenges associated with these treatments, and emerging therapeutic approaches aimed at selectively modulating OC function while preserving bone formation. By examining these strategies, we aim to highlight their clinical potential and implications for improving bone health.

## *5.1 Anti-resorptive Therapies Targeting Osteoclasts*

Anti-resorptive therapies aim to reduce OC-mediated bone resorption to maintain skeletal integrity and prevent bone loss in disorders like osteoporosis. These therapies primarily include bisphosphonates, RANKL inhibitors, and selective estrogen receptor modulators (SERMs).

### 5.1.1 Bisphosphonates

Bisphosphonates are widely used anti-resorptive agents that bind to hydroxyapatite in bone mineral surfaces. Once internalized by OC during resorption, these compounds inhibit farnesyl pyrophosphate synthase, a key enzyme in the mevalonate pathway, disrupting OC function and inducing apoptosis (Rodan & Martin, 2000). Common bisphosphonates include alendronate, primarily used for osteoporosis, and zoledronic acid, a potent agent effective against metastatic bone disease and Paget's disease. Bisphosphonates significantly reduce fracture risk and slow bone loss; however, prolonged use is associated with adverse effects such as osteonecrosis of the jaw (ONJ) and atypical femoral fractures due to excessive suppression of bone turnover (Raisz, 2005). These risks underscore the importance of balancing the duration and intensity of treatment.

### 5.1.2 Denosumab (RANKL Inhibitor)

Denosumab, a monoclonal antibody targeting RANKL, prevents its interaction with RANK receptors on OC precursors, thereby inhibiting OC differentiation and activation (Boyle et al., 2003; Laha et al., 2022; Sarkar et al., 2022). Approved for postmenopausal osteoporosis and cancer-induced bone disease, denosumab has demonstrated superior efficacy in reducing vertebral, non-vertebral, and hip fracture risks. Despite its benefits, denosumab presents unique challenges, such as rebound bone loss and increased fracture risk upon discontinuation, necessitating follow-up therapies to sustain bone density. Additionally, hypocalcemia is a common side effect due to reduced calcium release from bone.

### 5.1.3 Selective Estrogen Receptor Modulators (SERMs)

SERMs, such as raloxifene, mimic estrogen's bone-protective effects by binding to estrogen receptors on OC, reducing their activity. These agents are effective in maintaining bone density and preventing fractures in postmenopausal women, with the added benefit of reducing breast cancer risk. However, SERMs have a modest effect on fracture prevention compared to bisphosphonates and denosumab and carry risks such as venous thromboembolism, limiting their widespread use (Khosla & Monroe, 2018; Khosla et al., 2012).

### 5.1.4 Advanced Drug Delivery Systems for Rheumatoid Arthritis Treatment

Nanotechnology-based drug delivery systems improve RA treatments' efficacy, specificity, and safety. Intravenous, oral, transdermal, and intraarticular drug delivery strategies provide targeted action, reducing systemic toxicity and enhancing patient compliance. Developing polymeric nanoparticles, dendrimers, lipid-based formulations, and micelles represents a promising direction for improving RA management (Fig. 4).

**Step 1: Intravenous Drug Delivery Using PAMAM Dendrimer**

- Poly(amidoamine) (PAMAM) dendrimers are nanoscale, highly branched polymeric carriers that enhance drug solubility and controlled release.
- **Methotrexate (MTX)**, a disease-modifying antirheumatic drug (DMARD), is conjugated with PAMAM dendrimers to improve systemic circulation, targeting inflamed joints with higher specificity (Khandare et al., 2012).

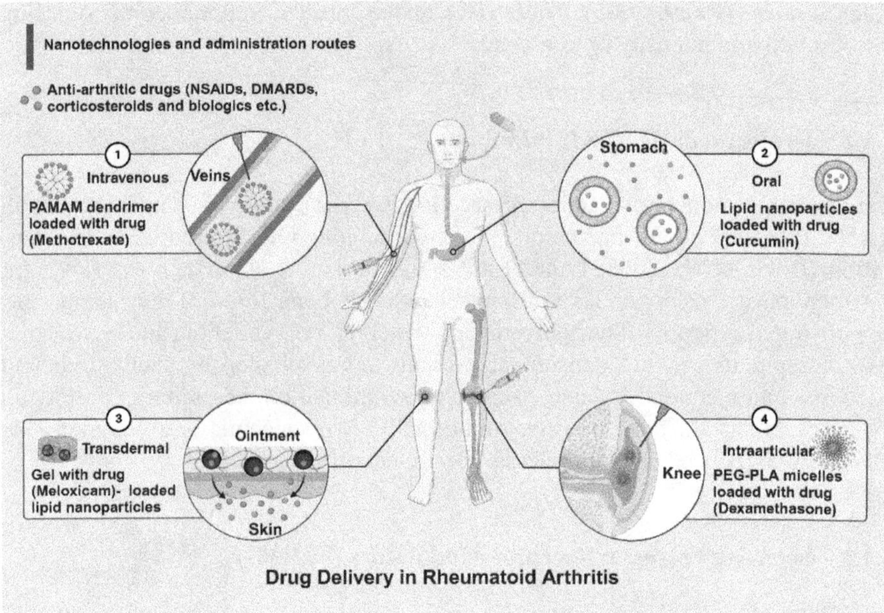


**Fig. 4** **Drug delivery strategies for rheumatoid arthritis (RA)**. This figure illustrates various drug delivery methods for rheumatoid arthritis, including (1) **intravenous delivery**: poly(amidoamine) (PAMAM) dendrimer-loaded methotrexate for systemic circulation and targeted joint inflammation; (2) **oral delivery**: lipid nanoparticles encapsulating curcumin for improved bioavailability and systemic anti-inflammatory effects; (3) **transdermal delivery**: meloxicam-loaded lipid nanoparticles in gel form for localized pain and inflammation relief through skin penetration; (4) **intraarticular delivery**: PEG-PLA micelles carrying dexamethasone for direct delivery to the inflamed joint, reducing systemic side effects. These advanced drug delivery systems aim to enhance therapeutic efficacy while minimizing adverse effects compared to conventional treatments such as NSAIDs, DMARDs, corticosteroids, and biologics

- This delivery system enhances MTX retention in inflamed synovial tissue and reduces systemic toxicity, leading to better therapeutic outcomes in RA patients.

### Step 2: Oral Drug Delivery Using Lipid Nanoparticles

- Lipid nanoparticles enhance the oral bioavailability of hydrophobic drugs like **curcumin**, a natural anti-inflammatory and antioxidant agent.
- Curcumin-loaded lipid nanoparticles protect the drug from degradation in the gastrointestinal tract, improve absorption, and provide sustained release in systemic circulation (Ravindranath & Chandrasekhara, 1980).
- This strategy helps mitigate oxidative stress and inflammation in RA while reducing gastrointestinal irritation compared to conventional NSAIDs.

**Step 3: Transdermal Drug Delivery Using Lipid Nanoparticle-Based Gel**

- **Meloxicam**, a selective COX-2 inhibitor, is incorporated into lipid nanoparticles and formulated as a **topical gel**.
- This approach allows drug penetration through the skin, directly targeting inflamed joints with minimal systemic exposure (Attia et al., 2019).
- Transdermal delivery avoids gastrointestinal side effects associated with oral NSAIDs and offers prolonged local pain relief.

**Step 4: Intraarticular Drug Delivery Using PEG-PLA Micelles**

- **Dexamethasone**, a potent corticosteroid, is encapsulated within **PEG-PLA (polyethylene glycol-polylactic acid) micelles** for **intraarticular injection**.
- This system ensures localized drug retention in the joint, reducing systemic corticosteroid-related adverse effects (Beloqui et al., 2016).
- PEG-PLA micelles enhance drug stability and prolong anti-inflammatory activity in the joint space.

## *5.2 Challenges and Side Effects of Anti-resorptive Therapies*

While anti-resorptive therapies effectively reduce bone loss, they are associated with significant challenges. By suppressing OC activity, these therapies also impair the natural coupling between OC and osteoblasts (OB), reducing bone formation. This disruption in bone remodeling can lead to an accumulation of microdamage, resulting in compromised bone quality and adaptive remodeling. Long-term use of bisphosphonates and denosumab has been linked to ONJ and atypical femoral fractures, particularly in patients undergoing high-dose therapy (Zaidi, 2007). Additionally, rebound bone loss after denosumab discontinuation highlights the need for sequential therapies to maintain treatment benefits.

## *5.3 Emerging Therapeutic Approaches Targeting Osteoclast Function*

Emerging therapies aim to modulate OC activity more selectively, preserving their resorptive function while enhancing bone regeneration.

### 5.3.1 Therapies Targeting OC-OB Coupling Mechanisms

Targeting the coupling mechanisms between OC and OB offers a promising approach to maintaining bone homeostasis. Therapies enhancing the Wnt/β-catenin signaling pathway, such as sclerostin inhibitors like romosozumab, stimulate OB activity while moderating OC resorption. Romosozumab has demonstrated clinical potential in increasing bone formation and improving bone density while reducing fracture risk (Howlader et al., 2024; Kim et al., 2020).

### 5.3.2 MSC Recruitment and Differentiation Modulation

Mesenchymal stem cells (MSC) are crucial for bone regeneration, as they differentiate into OB that form new bone. Therapies targeting sphingosine-1-phosphate (S1P) signaling enhance MSC recruitment to bone resorption sites, promoting OB differentiation and coupling bone resorption with regeneration. Anabolic combination therapies, such as pairing parathyroid hormone (PTH) analogs with anti-resorptives, have shown potential in optimizing bone regeneration by simultaneously enhancing OB activity and moderating OC function (Marie, 2015).

### 5.3.3 Selective Modulators of Osteoclast Activity

Cathepsin K, an enzyme secreted by OC for bone matrix degradation, has emerged as a therapeutic target. Inhibitors like odanacatib selectively reduce OC-mediated resorption without completely suppressing bone remodeling. However, clinical trials of odanacatib revealed adverse cardiovascular effects, highlighting the need for safer alternatives.

# 6 Novel Therapeutic Approaches for Bone Regeneration

## *6.1 Anabolic and Combination Therapies*

The need to balance bone resorption and formation in therapeutic strategies for bone disorders has led to exploring anabolic and combination therapies. Anabolic therapies, which stimulate bone formation, provide an important counterpoint to traditional anti-resorptive agents that primarily inhibit bone resorption. By combining these two approaches, therapeutic strategies can more effectively address bone diseases characterized by both bone loss and impaired bone formation.

I. **Parathyroid Hormone (PTH) Analogs**
   PTH analogs, such as teriparatide, are the only FDA-approved anabolic agents that stimulate bone formation by promoting osteoblast (OB) activity. Intermittent administration of PTH analogs has enhanced OB function, stimulated collagen synthesis, and facilitated new bone formation. Studies show that PTH therapy can significantly increase bone mass and improve bone microarchitecture, especially in patients with osteoporosis (Kim et al., 2020). However, PTH treatment also requires close monitoring to mitigate side effects, such as hypercalcemia and, with long-term use, the potential risk of osteosarcoma.

II. **Sclerostin Inhibitors**
Sclerostin, a glycoprotein produced by osteocytes, acts as an inhibitor of OB activity through the Wnt signaling pathway. Sclerostin inhibitors, such as romosozumab, enhance OB activity by blocking sclerostin, thereby promoting bone formation and reducing bone resorption. Clinical trials have demonstrated that romosozumab can increase bone mineral density and reduce fracture risk (Saag et al., 2017). This dual-action effect is beneficial for bone regeneration and makes sclerostin inhibitors a promising choice for patients with high fracture risk.

III. **Combination Therapy: Anabolic and Anti-resorptive**
Combination therapies using anti-resorptive agents (e.g., bisphosphonates, denosumab) with anabolics (e.g., PTH analogs, sclerostin inhibitors) have shown potential to further enhance bone density gains while controlling excessive resorption. Studies combining teriparatide and denosumab in osteoporosis patients reported greater increases in bone mineral density compared to either agent alone (Takayanagi, 2007; Teitelbaum, 2000). Sclerostin inhibitors combined with anti-resorptives may offer similar benefits, providing a balanced approach to bone regeneration. The combination of anabolic and anti-resorptive agents aims to maximize bone formation while limiting bone resorption, making it a powerful therapeutic strategy for individuals with severe osteoporosis or high fracture risk.

## *6.2 Stem Cell-Based Approaches*

Stem cell-based therapies offer an innovative approach to bone regeneration by harnessing the ability of mesenchymal stem cells (MSC) to differentiate into OB. MSC are pluripotent cells capable of self-renewal and differentiation into bone-forming cells, making them promising candidates for bone tissue engineering. Enhanced through osteoclast (OC) signaling, MSC can be recruited and stimulated to form new bone tissue, supporting the restoration of bone mass in areas where remodeling is needed.

I. **OC-Mediated MSC Recruitment and Differentiation**
OC release several signaling molecules, including sphingosine-1-phosphate (S1P) and transforming growth factor-β (TGF-β), that can attract MSC to resorption sites, promoting their differentiation into OB (Bandy et al., 2023; Greene et al., 2022; Howlader et al., 2024; Laha et al., 2022; Prateeksha et al., 2023; Sarkar et al., 2022). Leveraging OC signaling to enhance MSC recruitment could facilitate endogenous bone repair in patients with bone loss or fractures. MSC respond to these signaling molecules by migrating to the area of bone resorption, where they contribute to bone formation, filling in resorption pits and maintaining skeletal integrity.

II. **Ex Vivo MSC Expansion and Implantation**
Ex vivo expansion of MSC, followed by their reimplantation into patients, is a promising approach for enhancing bone regeneration. This technique involves isolating MSC from a patient's bone marrow, expanding them in culture, and then implanting them at the desired bone site. Studies have demonstrated that MSC can improve bone healing in patients with bone defects or fractures (Grayson et al., 2010). Enhanced by growth factors or OC signaling molecules, ex vivo expanded MSC could be strategically used to treat conditions like non-union fractures or bone defects following trauma.

III. **Genetically Modified MSC**
To further enhance MSC efficacy, researchers are exploring genetic modifications that can boost MSC differentiation and bone-forming potential. Modifications to increase the expression of osteogenic markers like Runx2 or BMP-2 have shown promise in promoting MSC differentiation into OB, accelerating bone formation (Kim et al., 2020; Marie, 2015; Tripuwabhrut, 2014). Genetically modified MSC may offer a targeted approach to regenerate bone in patients with impaired bone healing, though more studies are needed to ensure their safety and efficacy.

## 6.3 *Biomaterials and Scaffolds in Bone Repair*

Bioengineered scaffolds and biomimetic materials represent a rapidly evolving area in bone regeneration, supporting bone repair by providing structural support and enhancing OC-OB dynamics. These materials can be designed to mimic the extracellular matrix (ECM) of bone, promoting cell adhesion, proliferation, and differentiation, while also allowing for the controlled release of growth factors and cytokines.

I. **Scaffolds Supporting OC and OB Dynamics**
Scaffolds play a crucial role in maintaining the balance between OC and OB activity, supporting bone regeneration by providing a three-dimensional structure for bone cells to grow and differentiate. By mimicking the composition and structure of the bone ECM, scaffolds can encourage OB to deposit new bone matrix, while also maintaining sufficient porosity to facilitate nutrient and oxygen exchange (Kim et al., 2020). Scaffolds that incorporate materials like collagen, hydroxyapatite, and bioactive glass can provide mechanical strength and enhance OB function, supporting new bone formation.

II. **Bioactive Scaffolds with Growth Factors**
Bioactive scaffolds can be embedded with growth factors, such as BMP-2, TGF-β, and IGFs, which are released gradually to stimulate OB differentiation and bone formation. These growth factors can also encourage MSC to migrate to the scaffold, where they differentiate into OB and contribute to the bone regeneration process. Research on BMP-2-infused scaffolds has shown promis-

ing results in patients with critical-sized bone defects, improving bone density and repair outcomes (Zaidi, 2007).

III. **Biomimetic Materials and Smart Scaffolds**
   Advance in materials science have led to the development of biomimetic and "smart" scaffolds that respond to the bone microenvironment. These scaffolds can be designed to release growth factors or anti-inflammatory agents in response to changes in pH, temperature, or other biochemical signals within the bone tissue. For example, scaffolds that release anti-inflammatory agents in response to elevated cytokine levels could be beneficial in treating inflammatory bone diseases, where inflammation often inhibits OB activity (Liu et al., 2023). Smart scaffolds capable of modulating the bone microenvironment offer a more adaptive approach to bone regeneration, enhancing the healing process while reducing the risk of complications.

The field of bone regeneration has advanced significantly with the development of novel therapeutic approaches that combine anti-resorptive and anabolic agents, leverage MSC-based therapies, and employ innovative biomaterials and scaffolds. Anabolic and combination therapies offer balanced approaches to enhance bone formation while preventing excessive resorption, making them particularly valuable in treating osteoporosis and severe bone loss. Stem cell-based therapies, enhanced by OC signaling, provide a promising avenue for regenerating bone tissue, especially in cases of nonunion fractures or large bone defects (Kim et al., 2020). Bioengineered scaffolds and biomimetic materials represent an exciting frontier, offering structural support and modulating the bone microenvironment to optimize OC-OB dynamics and promote bone regeneration.

The integration of these approaches, along with advancements in personalized medicine and biomarker development, suggests a future where bone regeneration treatments can be highly customized to meet the unique needs of each patient. By targeting specific pathways and cell interactions, these therapies have the potential to address not only osteoporosis and fractures but also more complex bone diseases. Future research focusing on refining these therapeutic strategies, exploring combinatory approaches, and enhancing patient-specific customization will likely drive the next wave of innovation in bone health, improving outcomes and quality of life for patients with bone-related disorders.

## 7 Role of Osteoclasts in the Bone Microenvironment

Osteoclasts (OC) are specialized bone-resorbing cells essential for maintaining skeletal integrity and calcium balance. Traditionally known for their role in bone resorption, recent studies have expanded our understanding of OC as key regulators within the bone microenvironment. This chapter examines the multifaceted interactions of OC with the extracellular matrix (ECM), their coupling mechanisms with osteoblasts (OB), and the influence of immune cell interactions through osteoimmunology. Understanding these relationships is crucial for developing targeted therapies for bone remodeling disorders, including osteoporosis, arthritis, and metastatic bone disease.

- **Osteoclast Function in Bone Remodeling**
  Osteoclasts are large, multinucleated cells derived from the hematopoietic stem cell (HSC) lineage, specifically the monocyte/macrophage lineage. Their primary function is bone resorption, a process necessary for skeletal remodeling and calcium homeostasis (Bonewald, 2011, 2017; Boyce, 2013). Through the secretion of acidic compounds and proteolytic enzymes, OC degrade the bone matrix, facilitating calcium release into the bloodstream and maintaining bone integrity (Takayanagi, 2007; Teitelbaum, 2000).
- **Importance of the Bone Microenvironment**
  The bone microenvironment comprises various cell types, including osteoblasts, osteocytes, MSC, and immune cells, embedded within a complex extracellular matrix (ECM). This environment is essential for supporting cellular interactions and signaling necessary for bone remodeling and homeostasis (Sarkar et al., 2022; Schini et al., 2023; Sims & Martin, 2014). This chapter focuses on how OC interact within the bone microenvironment, particularly through ECM adhesion, crosstalk with OB, and immune cell interactions, which regulate their activity in health and disease.

## *7.1 Extracellular Matrix (ECM) and Cell Adhesion*

### 7.1.1 Role of the ECM in Bone Homeostasis

- **ECM Composition and Structure**
  The ECM in bone primarily consists of collagen type I fibers, proteoglycans, glycoproteins, and hydroxyapatite crystals, which provide structural stability and serve as an anchor for cellular adhesion (Florencio-Silva et al., 2015). The ECM also stores various growth factors, such as TGF-β and IGF, which are released during bone resorption, promoting osteoblast recruitment and coupling resorption with new bone formation.
- **Bone Resorption and ECM Remodeling**
  OC initiate bone resorption by secreting hydrochloric acid and proteolytic enzymes, including cathepsin K, to break down the mineral matrix and decompose organic components. This process releases growth factors embedded in the ECM, essential for bone remodeling (Zaidi, 2007).

### 7.1.2 Osteoclast-ECM Adhesion Mechanisms

- **Role of Integrins**
  Integrins, especially αvβ3, are transmembrane receptors critical for OC adhesion to the bone surface. This adhesion enables OC to form a sealing zone, isolating the resorptive area and creating an acidic environment for matrix breakdown (Helfrich, 2003).
  - **αvβ3 Integrin in Bone Resorption**
  - αvβ3 integrin binds to ECM proteins, including osteopontin and vitronectin, anchoring OC to the bone matrix. This interaction is essential for forming a resorption lacuna, where matrix degradation occurs (Grässel et al., 2021).
- **Formation of the Sealing Zone and Ruffled Border**
  The sealing zone confines the acidic environment needed for matrix degradation. OC podosomes, actin-rich structures, form a ring around this area, maintaining cell adhesion and enabling enzyme localization to the resorption site (Destaing et al., 2003).

### 7.1.3 ECM-Derived Signals and OC Regulation

- **Matrix-Derived Signaling Molecules**
  Growth factors stored in the ECM, such as TGF-β and IGF, are released upon matrix resorption. These factors stimulate osteoblast recruitment and differentiation, linking bone formation with resorption(Boyle et al., 2003).
- **Therapeutic Potential**
  Inhibiting αvβ3 integrin in OC has shown potential in moderating bone resorption without disrupting overall bone turnover, offering therapeutic options for conditions involving excessive OC activity (Boyle et al., 2003).

## *7.2 Crosstalk with Other Bone Cells and Signaling Molecules*

### 7.2.1 Osteoclast-Osteoblast Coupling

- **Concept of Bone Remodeling Coupling**
  Coupling refers to the sequential process in which OC-mediated bone resorption signals OB to initiate bone formation in resorbed areas, maintaining bone mass and quality (Sims & Martin, 2014).

- **Release of Coupling Factors During Resorption**
  - **Transforming growth factor-β (TGF-β)**
    TGF-β, stored in the bone matrix, is released during resorption. It stimulates OB differentiation and promotes matrix formation (Mundy, 1996).
  - **Insulin-like growth factors (IGFs)**
    IGFs, also released from the bone matrix, enhance OB function, further supporting bone regeneration post-resorption (Raisz, 2005).
  - **Wnt signaling pathway**
    OC-derived factors, such as sphingosine-1-phosphate (S1P), influence Wnt signaling in OB, aiding in OB recruitment and matrix deposition(Glass II & Karsenty, 2006; Houschyar et al., 2019).

### 7.2.2 Role of Osteocytes in OC Regulation

- **Osteocyte-Secreted Regulatory Factors**
  Osteocytes, embedded within the bone matrix, regulate OC activity through RANKL and sclerostin secretion. They adapt to changes in mechanical load or microdamage, coordinating OC and OB activity to ensure balanced remodeling (Bonewald, 2011, 2017; Boyce, 2013; Boyle et al., 2003).
- **RANK/RANKL/OPG Axis in OC Activation**
  Osteocytes produce RANKL, which binds to RANK on OC precursors, promoting OC differentiation and activation. OPG acts as a decoy receptor for RANKL, modulating OC function and maintaining bone integrity (Karsenty & Wagner, 2002; Khosla & Monroe, 2018; Khosla et al., 2012).
- **Sclerostin's Role in OC Modulation**
  Sclerostin, produced by osteocytes, inhibits OB function and indirectly influences OC by reducing bone formation, thereby impacting the bone remodeling cycle (Florencio-Silva et al., 2015).

## *7.3 Immune System Interactions and Osteoimmunology*

### 7.3.1 Overview of Osteoimmunology

Immune Cells in Bone Remodeling

Osteoimmunology explores the interactions between immune and bone cells, recognizing that immune cells can influence OC activity through cytokine secretion, especially during inflammation (Takayanagi, 2007; Teitelbaum, 2000; Tognana, 2008; Tripuwabhrut, 2014). This interaction is crucial for adapting OC activity to both bone remodeling and immune challenges.

T Cell-Osteoclast Interactions

- **T Cell Contribution to RANKL Production**
  Activated T cells produce RANKL, a major stimulator of OC formation and activity, particularly during inflammation or infection (Takayanagi, 2007).
- **Pro-inflammatory Cytokines and OC Activity**
  - **Tumor Necrosis Factor-Alpha (TNF-α)**
    TNF-α enhances OC differentiation and resorptive activity, linking chronic inflammation to accelerated bone loss, as seen in rheumatoid arthritis (Karsenty & Wagner, 2002; Khosla & Monroe, 2018; Khosla et al., 2012).
  - **Interleukin-1 (IL-1) and Interleukin-6 (IL-6)**
    These cytokines promote OC formation and resorption, contributing to bone loss in chronic inflammatory conditions (Boyce, 2013; Boyle et al., 2003).
- **Therapeutic Targeting of Immune-OC Interactions**
  Modulating inflammatory cytokines in conditions like rheumatoid arthritis has shown promise in controlling OC activity and reducing bone erosion (Takayanagi, 2007; Teitelbaum, 2000).

Osteoclast-Macrophage Interactions in Bone Healing

- **OC Precursors and Macrophage Lineage**
  OC share a precursor lineage with macrophages, which play a vital role in immune response and inflammation modulation. Macrophage colony-stimulating factor (M-CSF) is essential for OC differentiation and survival (Henriksen et al., 2011; Henriksen et al., 2009; Houschyar et al., 2019).
- **Macrophages in Bone Repair**
  Macrophages contribute to bone healing by recruiting OC precursors and modulating inflammation, which is essential for fracture repair and regeneration (Sims & Martin, 2014).
- **Therapeutic Implications**
  Understanding immune-OC interactions provides insights into developing therapies that balance bone remodeling while mitigating inflammation-related bone loss (Takayanagi, 2007).

## 8 Future Directions and Research Needs for Osteoclast Function and Regenerative Applications

Osteoclasts (OC) have traditionally been studied as cells responsible for bone resorption, crucial in maintaining bone integrity and mineral homeostasis. However, emerging research reveals that OC play significant roles beyond resorption, including contributions to bone regeneration and interactions with osteoblasts (OB), mesenchymal stem cells (MSC), and immune cells. This chapter explores research gaps in understanding OC, recent advancements in OC-targeted therapies, and the potential for personalized medicine. By enhancing our knowledge of OC function, we can unlock new therapeutic avenues for treating bone disorders, such as osteoporosis, arthritis, and bone regeneration post-injury.

## *8.1 Research Gaps in OC Function and Regeneration*

### 8.1.1 Understanding the Regenerative Signaling of OC

Despite significant advances in osteoclast biology, the full regenerative potential of OC remains incompletely understood. Traditional models of bone remodeling emphasize OC primarily as bone-resorbing cells, which release calcium and phosphate from bone for mineral homeostasis. However, recent studies show OC also produce and release factors such as transforming growth factor-beta (TGF-β), insulin-like growth factors (IGFs), and bone morphogenetic proteins (BMPs), all of which stimulate osteoblastic differentiation and function (Boyce, 2013; Boyle et al., 2003; Martin & Sims, 2005; Mundy, 1996). Current research is needed to elucidate the complex signaling pathways involved in these regenerative roles and how they may differ under normal physiological conditions versus pathologies, such as osteoporosis or arthritis.

Further exploration into OC-derived signaling molecules, including sphingosine-1-phosphate (S1P) and Wnt pathway activators, is necessary to determine how these molecules recruit and stimulate MSC. Understanding this regenerative signaling could provide insight into therapies that activate OC in controlled ways to promote bone regeneration without excessive resorption (Martin & Sims, 2005; Mundy, 1996).

### 8.1.2 Elucidating Crosstalk with Immune Cells

The role of OC in immunomodulation within the bone microenvironment is another area with research gaps. OC interact with immune cells such as T cells and macrophages through osteoimmunology pathways, notably the RANK/RANKL/OPG axis. Immune cells, especially T cells, secrete RANKL during inflammation, promoting OC differentiation and activity. However, the reciprocal impact of OC activity on immune responses and inflammation is not yet fully understood. New research should examine whether OC could be modulated to mitigate inflammation, offering dual benefits of bone preservation and inflammation control in diseases like rheumatoid arthritis and periodontitis (Khosla & Monroe, 2018; Khosla et al., 2012).

### 8.1.3 Mechanisms of OC Coupling with OB

The interdependency of OC and OB in bone homeostasis highlights the need for further investigation into OC-OB coupling mechanisms. Studies show that coupling factors released by OC, such as TGF-β and IGFs, recruit OB to sites of resorption to initiate new bone formation, a process essential for maintaining bone density and integrity (Sims & Martin, 2014; Takayanagi, 2007). Understanding the dynamics of this coupling, particularly in conditions where resorption exceeds formation, is crucial for developing treatments that can prevent bone loss while enhancing regeneration.

### 8.1.4 OC Function in Aging and Bone Diseases

Aging is associated with increased OC activity and decreased MSC-derived osteogenesis, often leading to diseases like osteoporosis. The mechanisms driving this shift in OC function with age, and the resultant imbalance in bone remodeling, remain poorly understood. Further research is needed to understand how the aging bone microenvironment influences OC and OB interactions, particularly in conditions where resorption becomes pathologically dominant (Khosla & Monroe, 2018; Khosla et al., 2012; Zaidi, 2007).

## 8.2 *Advances in OC-Targeted Therapies*

### 8.2.1 Gene Editing and Molecular Targeting

Advancements in gene editing technologies, such as CRISPR-Cas9, offer promising tools for precisely targeting OC-related pathways. For instance, targeting genes involved in OC differentiation, such as RANK or cathepsin K, could help modulate OC activity without impairing the entire bone remodeling process. Specific inhibitors of the RANK/RANKL pathway, which directly influences OC formation, are already in use (e.g., denosumab), but precision gene-editing tools offer the potential to selectively modulate gene expression in response to local bone turnover needs (Roodman, 2004; Teitelbaum, 2000).

### 8.2.2 Pharmacogenomic Approaches to Optimize OC Activity

Pharmacogenomics, which focuses on how genes affect drug response, offers a pathway to individualized OC-modulating treatments. For example, the genetic polymorphisms in RANK, RANKL, or OPG genes among patients with osteoporosis suggest that certain individuals might respond better to targeted OC therapies (Bandy et al., 2023; Barthels et al., 2023; Raisz, 2005). Pharmacogenomic profiling could help clinicians identify patients most likely to benefit from OC-modulating drugs, minimizing side effects while maximizing efficacy in bone regeneration.

### 8.2.3 Innovative Molecules for Selective OC Modulation

Novel OC-targeted agents, such as sclerostin inhibitors and cathepsin K inhibitors, show potential for enhancing bone regeneration while minimizing the adverse effects of excessive OC suppression. Sclerostin inhibitors, for example, work by increasing bone formation via OB while indirectly affecting OC function. These agents offer an approach to rebalance bone remodeling by shifting the focus toward enhancing regeneration rather than merely inhibiting resorption (Henriksen et al., 2011; Henriksen et al., 2009; Houschyar et al., 2019; Howlader et al., 2024).

### 8.2.4 Combination Therapies: Anti-resorptive and Anabolic Agents

Combination therapies that use both anti-resorptive drugs and anabolic agents, such as parathyroid hormone (PTH) analogs, are emerging as a strategy to stimulate bone formation while controlling resorption. This approach aligns with recent findings suggesting that OC play a regenerative role that should be preserved rather than completely inhibited (Boyce, 2013; Boyle et al., 2003). Combining therapies can help retain the beneficial coupling of bone resorption and formation while reducing the risk of excessive bone loss seen with high-dose anti-resorptive agents alone.

## *8.3 Toward Personalized Medicine in Bone Disease Treatment*

### 8.3.1 Integrating OC Activity Profiles with Clinical Interventions

Personalized medicine aims to tailor therapies to individual patients, based on specific biomarkers and genetic factors. The wide variability in OC activity among individuals underscores the importance of personalized approaches in bone disease treatment. For instance, age, sex, hormonal status, and genetic background influence OC activity, and these factors could be integrated into a clinical decision-making process for individualized treatment plans (Khosla & Monroe, 2018; Khosla et al., 2012).

### 8.3.2 Biomarkers for OC Activity

Identifying biomarkers of OC activity, such as serum levels of tartrate-resistant acid phosphatase (TRAP) and cathepsin K, can provide clinicians with insights into disease progression and treatment efficacy. Biomarkers related to RANKL and OPG levels in blood can serve as indicators of OC differentiation and activation status, allowing for real-time monitoring of OC activity during treatment. Advances in biomarker technology could lead to more precise treatment adjustments based on dynamic changes in OC function (Bonewald, 2017; Prateeksha et al., 2023; Raggatt & Partridge, 2010).

### 8.3.3 Long-Term Monitoring of OC-Related Therapies

For chronic bone diseases like osteoporosis, long-term monitoring of OC function is critical for ensuring that therapies remain effective and safe. Techniques like bone turnover markers (BTMs) and imaging technologies, such as high-resolution micro-computed tomography, offer noninvasive methods to assess OC activity over time. This approach can help optimize dosing regimens, especially for anti-resorptive therapies, and prevent adverse effects associated with excessive suppression of bone turnover (Saag et al., 2017).

As research expands our understanding of the multifaceted role of osteoclasts in bone regeneration, new therapeutic possibilities emerge. Investigating OC-regenerative signaling, understanding immune interactions, and leveraging gene-editing technologies hold promise for advancing OC-targeted therapies. Personalized medicine and biomarker-driven approaches offer the potential to refine treatments, optimizing OC function for bone health without disrupting homeostasis. By addressing the current research gaps, we can develop interventions that harness osteoclasts' full regenerative potential, providing more effective treatments for osteoporosis, arthritis, and other bone-related diseases.

# 9 Summary

## *9.1 Recap of Key Points*

Osteoclasts (OC) have long been understood as cells specialized in bone resorption, a function essential for bone remodeling, skeletal integrity, and mineral homeostasis. This traditional view primarily emphasized their role in degrading bone matrix and releasing calcium into the bloodstream, thus maintaining systemic calcium levels. Recent research, however, has broadened our understanding of OC, revealing their contributions to bone regeneration through signaling and cellular interactions within the bone microenvironment. OC actively participate in a feedback loop that fosters bone formation, not only through their direct activity on the bone matrix but also through interactions with osteoblasts (OB), mesenchymal stem cells (MSC), and immune cells. These multifaceted roles of OC have redefined them from purely resorptive cells to vital regulators of bone health, coupling bone breakdown with renewal.

This dual functionality of OC underscores the need for a balanced approach in therapeutic strategies. While anti-resorptive therapies, such as bisphosphonates and RANKL inhibitors, are effective in curbing OC-mediated bone degradation in conditions like osteoporosis, they often disrupt the coupling mechanism, inadvertently suppressing OB activity and reducing bone formation. This realization has led to the exploration of combination therapies and novel therapeutic approaches that target both resorption and regeneration. Emerging research into OC signaling pathways, gene-editing tools, and pharmacogenomic profiling suggests that it may be possible to selectively modulate OC activity, preserving their resorptive functions while enhancing their regenerative potential.

## 9.2 *Implications for Bone Health*

The multifaceted roles of OC are integral to maintaining skeletal health, as they not only control bone resorption but also regulate OB differentiation and MSC recruitment, essential for new bone formation. OC initiates the remodeling process, setting the stage for OB-driven bone regeneration by releasing growth factors, such as transforming growth factor-beta (TGF-β), bone morphogenetic proteins (BMPs), and insulin-like growth factors (IGFs), which are embedded within the bone matrix and released upon resorption. These signaling molecules, in turn, stimulate OB recruitment and activity, linking the phases of resorption and formation.

OC interactions with MSC and immune cells further support bone regeneration, especially in response to bone damage or inflammation. OC release sphingosine-1-phosphate (S1P) and engages the Wnt/β-catenin pathway, both of which facilitate MSC migration and OB differentiation. Additionally, osteoimmunology reveals that immune cells, such as T cells and macrophages, communicate with OC in response to inflammation, influencing OC differentiation and activity. This crosstalk is essential for a balanced immune response that controls bone degradation in inflammatory conditions like rheumatoid arthritis, where elevated pro-inflammatory cytokines stimulate OC activity and bone resorption.

The dual role of OC in both resorption and regeneration have practical implications for bone health management, particularly in age-related and inflammatory bone diseases. In osteoporosis, for example, postmenopausal estrogen deficiency heightens RANKL expression, driving excessive OC activity and bone resorption. Traditional anti-resorptive therapies, though effective, often lead to decreased OB activity due to the disrupted coupling mechanism, highlighting the need for a nuanced therapeutic approach. Conversely, conditions like osteopetrosis, characterized by deficient OC activity, emphasize the importance of OC in ensuring balanced bone turnover and skeletal integrity.

## 9.3 Outlook

The expanded understanding of OC functions presents promising opportunities for future therapeutic advancements. Potential therapies targeting OC-OB coupling, MSC recruitment, and immune modulation offer new avenues for managing bone diseases. Several emerging areas of research could hold significant promises for future therapies that preserve the resorptive function of OC while enhancing their regenerative capabilities:

I. **Selective Modulation of OC Activity**: Advances in gene-editing technologies, like CRISPR-Cas9, allow for precise targeting of genes involved in OC differentiation and activity, such as RANK or cathepsin K. Such tools provide the potential for selective OC modulation, offering therapeutic benefits without widespread suppression of bone turnover. Additionally, pharmacogenomic profiling could identify patients who are most likely to benefit from specific OC-targeted therapies, thus paving the way for personalized treatments.
II. **Combined Anti-resorptive and Anabolic Therapies**: Combining anti-resorptive agents with anabolic therapies could allow for comprehensive bone health management. For instance, parathyroid hormone (PTH) analogs, which stimulate OB activity, combined with anti-resorptive drugs, may help balance bone resorption and formation. Sclerostin inhibitors, which promote OB activity while indirectly modulating OC function, are another promising therapeutic direction. This balanced approach may preserve skeletal integrity while reducing the risk of fractures and improving patient outcomes in osteoporosis and other bone degenerative diseases.
III. **Stem Cell-Based and Biomimetic Therapies**: The potential to use MSC, especially those enhanced by OC signaling, for bone regeneration is an exciting frontier. These therapies could leverage OC signaling to recruit MSC to damaged bone sites, facilitating OB differentiation and bone regeneration. Biomimetic scaffolds and materials that mimic the extracellular matrix (ECM) may further enhance the regenerative process, supporting the dynamic interactions between OC, OB, and MSC within the bone microenvironment. These scaffolds could be used in clinical settings to repair and (Takayanagi, 2007; Teitelbaum, 2000; Tognana, 2008; Tripuwabhrut, 2014) regenerate bone in cases of significant loss or trauma, providing structural support while promoting cellular activity.
IV. **Osteoimmunology and Immune Modulation**: The field of osteoimmunology has provided insights into how immune cells influence OC activity through cytokine signaling, especially in inflammatory bone diseases. Future therapies that modulate immune responses could offer dual benefits of controlling inflammation and promoting bone regeneration. For example, immune-targeted treatments, such as TNF-$\alpha$ inhibitors, may reduce OC activity and preserve bone mass in inflammatory conditions like rheumatoid arthritis. Additionally, selective modulation of immune-OC interactions could help maintain bone

health in diseases where inflammation plays a critical role in bone degradation.

V. **Monitoring and Biomarker Development**: Real-time monitoring of OC activity through biomarkers and imaging technologies could improve the efficacy and safety of OC-targeted therapies. Biomarkers, such as serum levels of tartrate-resistant acid phosphatase (TRAP) and cathepsin K, along with the RANKL/OPG ratio, can provide insights into disease progression and treatment efficacy. Coupled with imaging technologies, such as micro-computed tomography, these tools could help clinicians monitor OC activity over time, adjust treatment regimens, and prevent adverse effects from long-term anti-resorptive therapies.

## 10 Conclusion

The multifaceted roles of OC in bone resorption and regeneration highlight their importance as dynamic contributors to bone health. Their interactions with OB, MSC, and immune cells emphasize the complexity of bone remodeling, demonstrating that OC do more than degrade bone—actively participate in its renewal. Understanding these roles paves the way for innovative therapies that strike a balance between resorptive and regenerative functions, moving beyond conventional anti-resorptive treatments to more holistic approaches that preserve skeletal integrity.

Future research and therapeutic development will likely focus on selective modulation of OC activity, combining anti-resorptive and anabolic therapies, and integrating personalized medicine approaches to optimize treatment for each patient. By addressing the regenerative potential of OC while controlling their resorptive activity, it may be possible to develop more effective and balanced treatments for osteoporosis, arthritis, and other bone-related diseases. Through these advancements, the future of bone health holds the promise of enhanced patient outcomes and improved quality of life for individuals facing bone-related disorders.

## References

Attia, M. F., Anton, N., Wallyn, J., Omran, Z., & Vandamme, T. F. (2019). An overview of active and passive targeting strategies to improve the nanocarriers efficiency to tumour sites. *Journal of Pharmacy and Pharmacology, 71*(8), 1185–1198.

Bandy, R., Shahi, S., Quagraine, N., Shahbazi Nia, S., Howlader, M. S. I., Srivenugopal, K., Stephan, C., Das, H., Mikelis, C. M., & German, N. A. (2023). Mechanistic aspects of biphenyl urea-based analogues in triple-negative breast cancer cell lines. *ACS Pharmacology & Translational Science, 7*(1), 120–136.

Barthels, D., Prateeksha, P., Nozohouri, S., Villalba, H., Zhang, Y., Sharma, S., Anderson, S., Howlader, M. S. I., Nambiar, A., & Abbruscato, T. J. (2023). Dental pulp-derived stem cells

preserve astrocyte health during induced gliosis by modulating mitochondrial activity and functions. *Cellular and Molecular Neurobiology, 43*(5), 2105–2127.
Beloqui, A., Solinís, M. Á., Rodríguez-Gascón, A., Almeida, A. J., & Préat, V. (2016). Nanostructured lipid carriers: Promising drug delivery systems for future clinics. *Nanomedicine: Nanotechnology, Biology and Medicine, 12*(1), 143–161.
Bonewald, L. F. (2011). The amazing osteocyte. *Journal of Bone and Mineral Research, 26*(2), 229–238.
Bonewald, L. F. (2017). The role of the osteocyte in bone and nonbone disease. *Endocrinology and Metabolism Clinics, 46*(1), 1–18.
Boyce, B. F. (2013). Advances in osteoclast biology reveal potential new drug targets and new roles for osteoclasts. *Journal of Bone and Mineral Research, 28*(4), 711–722.
Boyle, W. J., Simonet, W. S., & Lacey, D. L. (2003). Osteoclast differentiation and activation. *Nature, 423*(6937), 337–342.
Das, M., Deb, M., Laha, D., Joseph, M., Kanji, S., Aggarwal, R., Iwenofu, O. H., Pompili, V. J., Jarjour, W., & Das, H. (2019). Myeloid krüppel-like factor 2 critically regulates K/BxN serum-induced arthritis. *Cells, 8*(8), 908.
Deb, M., Laha, D., Maity, J., & Das, H. (2021). SETD2-mediated epigenetic regulation of noncanonical Wnt5A during osteoclastogenesis. *Clinical Epigenetics, 13*, 1–14.
Destaing, O., Saltel, F., Géminard, J.-C., Jurdic, P., & Bard, F. (2003). Podosomes display actin turnover and dynamic self-organization in osteoclasts expressing actin-green fluorescent protein. *Molecular Biology of the Cell, 14*(2), 407–416.
Firestein, G. S., & McInnes, I. B. (2017). Immunopathogenesis of rheumatoid arthritis. *Immunity, 46*(2), 183–196.
Florencio-Silva, R., de Slva Sasso, G. R., Sasso-Cerri, E., Simões, M. J., & Cerri, P. S. (2015). Biology of bone tissue: Structure, function, and factors that influence bone cells. *BioMed Research International, 2015*(1), 421746.
Glass, D. A., II, & Karsenty, G. (2006). Molecular bases of the regulation of bone remodeling by the canonical Wnt signaling pathway. *Current Topics in Developmental Biology, 73*, 43–84.
Grässel, S., Zaucke, F., & Madry, H. (2021). Osteoarthritis: Novel molecular mechanisms increase our understanding of the disease pathology. *Journal of Clinical Medicine, 10*(9), 1938.
Grayson, W. L., Fröhlich, M., Yeager, K., Bhumiratana, S., Chan, M. E., Cannizzaro, C., Wan, L. Q., Liu, X. S., Guo, X. E., & Vunjak-Novakovic, G. (2010). Engineering anatomically shaped human bone grafts. *Proceedings of the National Academy of Sciences, 107*(8), 3299–3304.
Greene, C. J., Anderson, S., Barthels, D., Howlader, M. S. I., Kanji, S., Sarkar, J., & Das, H. (2022). DPSC products accelerate wound healing in diabetic mice through induction of SMAD molecules. *Cells, 11*(15), 2409.
Helfrich, M. H. (2003). Osteoclast diseases. *Microscopy Research and Technique, 61*(6), 514–532.
Henriksen, K., Neutzsky-Wulff, A. V., Bonewald, L. F., & Karsdal, M. A. (2009). Local communication on and within bone controls bone remodeling. *Bone, 44*(6), 1026–1033.
Henriksen, K., Bollerslev, J., Everts, V., & Karsdal, M. (2011). Osteoclast activity and subtypes as a function of physiology and pathology—Implications for future treatments of osteoporosis. *Endocrine Reviews, 32*(1), 31–63.
Houschyar, K. S., Tapking, C., Borrelli, M. R., Popp, D., Duscher, D., Maan, Z. N., Chelliah, M. P., Li, J., Harati, K., & Wallner, C. (2019). Wnt pathway in bone repair and regeneration–what do we know so far. *Frontiers in Cell and Developmental Biology, 6*, 170.
Howlader, M. S. I., Prateeksha, P., Hansda, S., Naidu, P., Das, M., Barthels, D., & Das, H. (2024). Corrigendum to "secretory products of DPSC mitigate inflammatory effects in microglial cells by targeting MAPK pathway" Biomed. Pharmacother. 170 (2024) 115971. *Biomedicine & Pharmacotherapy= Biomedecine & Pharmacotherapie, 177*, 117061–117061.
Karsenty, G., & Wagner, E. F. (2002). Reaching a genetic and molecular understanding of skeletal development. *Developmental Cell, 2*(4), 389–406.
Khandare, J., Calderón, M., Dagia, N. M., & Haag, R. (2012). Multifunctional dendritic polymers in nanomedicine: Opportunities and challenges. *Chemical Society Reviews, 41*(7), 2824–2848.

Khosla, S., & Monroe, D. G. (2018). Regulation of bone metabolism by sex steroids. *Cold Spring Harbor Perspectives in Medicine, 8*(1), a031211.
Khosla, S., Oursler, M. J., & Monroe, D. G. (2012). Estrogen and the skeleton. *Trends in Endocrinology & Metabolism, 23*(11), 576–581.
Kim, J.-M., Lin, C., Stavre, Z., Greenblatt, M. B., & Shim, J.-H. (2020). Osteoblast-osteoclast communication and bone homeostasis. *Cells, 9*(9), 2073.
Laha, D., Sarkar, J., Maity, J., Pramanik, A., Howlader, M. S. I., Barthels, D., & Das, H. (2022). Polyphenolic compounds inhibit osteoclast differentiation while reducing autophagy through limiting ROS and the mitochondrial membrane potential. *Biomolecules, 12*(9), 1220.
Liu, S., Yao, S., Yang, H., Liu, S., & Wang, Y. (2023). Autophagy: Regulator of cell death. *Cell Death & Disease, 14*(10), 648.
Maity, J., Barthels, D., Sarkar, J., Prateeksha, P., Deb, M., Rolph, D., & Das, H. (2022). Ferutinin induces osteoblast differentiation of DPSCs via induction of KLF2 and autophagy/mitophagy. *Cell Death & Disease, 13*(5), 452.
Marie, P. J. (2015). Osteoblast dysfunctions in bone diseases: From cellular and molecular mechanisms to therapeutic strategies. *Cellular and Molecular Life Sciences, 72*(7), 1347–1361.
Martin, T. J., & Sims, N. A. (2005). Osteoclast-derived activity in the coupling of bone formation to resorption. *Trends in Molecular Medicine, 11*(2), 76–81.
McInnes, I. B., & Schett, G. (2011). The pathogenesis of rheumatoid arthritis. *New England Journal of Medicine, 365*(23), 2205–2219.
Mundy, G. R. (1996). Regulation of bone formation by bone morphogenetic proteins and other growth factors. *Clinical Orthopaedics and Related Research®, 324*, 24–27.
Prateeksha, P., Howlader, M. S. I., Hansda, S., Naidu, P., Das, M., Abo-Aziza, F., & Das, H. (2023). Secretome of dental pulp-derived stem cells reduces inflammation and proliferation of glioblastoma cells by deactivating Mapk-Akt pathway. *Diseases & Research, 3*(2), 74.
Raggatt, L. J., & Partridge, N. C. (2010). Cellular and molecular mechanisms of bone remodeling. *Journal of Biological Chemistry, 285*(33), 25103–25108.
Raisz, L. G. (2005). Pathogenesis of osteoporosis: Concepts, conflicts, and prospects. *The Journal of Clinical Investigation, 115*(12), 3318–3325.
Ravindranath, V., & Chandrasekhara, N. (1980). Absorption and tissue distribution of curcumin in rats. *Toxicology, 16*(3), 259–265.
Rodan, G. A., & Martin, T. J. (2000). Therapeutic approaches to bone diseases. *Science, 289*(5484), 1508–1514.
Rolph, D., & Das, H. (2020). Transcriptional regulation of osteoclastogenesis: The emerging role of KLF2. *Frontiers in Immunology, 11*, 937.
Roodman, G. D. (2004). Mechanisms of bone metastasis. *New England Journal of Medicine, 350*(16), 1655–1664.
Saag, K. G., Petersen, J., Brandi, M. L., Karaplis, A. C., Lorentzon, M., Thomas, T., Maddox, J., Fan, M., Meisner, P. D., & Grauer, A. (2017). Romosozumab or alendronate for fracture prevention in women with osteoporosis. *New England Journal of Medicine, 377*(15), 1417–1427.
Sarkar, J., Das, M., Howlader, M. S. I., Prateeksha, P., Barthels, D., & Das, H. (2022). Epigallocatechin-3-gallate inhibits osteoclastic differentiation by modulating mitophagy and mitochondrial functions. *Cell Death & Disease, 13*(10), 908.
Schini, M., Vilaca, T., Gossiel, F., Salam, S., & Eastell, R. (2023). Bone turnover markers: Basic biology to clinical applications. *Endocrine Reviews, 44*(3), 417–473.
Scott, D. L., Wolfe, F., & Huizinga, T. W. (2010). Rheumatoid arthritis. *Lancet, 376*, 1094–1108.
Sims, N. A., & Martin, T. J. (2014). Coupling the activities of bone formation and resorption: A multitude of signals within the basic multicellular unit. *BoneKEy Reports, 3*, 481.
Smolen, J., Aletaha, D., & McInnes, I. B. (2016). Rheumatoid arthritis. *Lancet, 388*(10055), 2023–2038.
Smolen, J. S., Landewé, R., Breedveld, F. C., Dougados, M., Emery, P., Gaujoux-Viala, C., Gorter, S., Knevel, R., Nam, J., & Schoels, M. (2010). EULAR recommendations for the management

of rheumatoid arthritis with synthetic and biological disease-modifying antirheumatic drugs. *Annals of the Rheumatic Diseases, 69*(6), 964–975.
Takayanagi, H. (2007). Osteoimmunology: Shared mechanisms and crosstalk between the immune and bone systems. *Nature Reviews Immunology, 7*(4), 292–304.
Teitelbaum, S. L. (2000). Bone resorption by osteoclasts. *Science, 289*(5484), 1504–1508.
Tognana, E. (2008). Scaffolds based on hyaluronan derivatives in biomedical applications. In *Natural-based polymers for biomedical applications* (pp. 291–314). Elsevier.
Tripuwabhrut, P. (2014). Inflammatory responses of immune cells and osteoblasts in orthodontically-induced bone remodelling and root resorption: in vitro and in vivo studies. Thesis Published on 11 07 2024, https://hdl.handle.net/1956/8759
Zaidi, M. (2007). Skeletal remodeling in health and disease. *Nature Medicine, 13*(7), 791–801.

# Dental Tissue Regeneration

# Tissue Regeneration for Periodontal and Alveolar Bone Defects

**Yun-Han Ma Becky and Binnaz Leblebicioglu**

**Abstract** Tissue regeneration around periodontally involved teeth and at edentulous sites with alveolar bony defects has been one of the fastest-developing research fields in dental biomaterials and tissue engineering. The requirement for three different tissue types around a root with an avascular surface characteristic, bacterial biofilm contamination, and an anatomy causing limited access have been major challenges in research progression, in addition to difficulties in designing well-controlled randomized clinical studies to evaluate treatment outcomes. Tooth loss itself causes significant alveolar bone resorption. Regenerating lost bone and ridge volume are critical to rehabilitate dentition to its pristine functional and physiological capacities. This chapter aims to provide a review of the periodontal anatomy and bone defect types in relation to periodontal tissue regeneration. Root surface conditioning and Guided Tissue Regeneration topics are covered as part of the historical development of regenerative treatment modalities. Available biomaterials and their limitations are discussed. Biologics are considered as potential products to overcome these limitations. The second part of the chapter reviews chronological changes in alveolar ridge dimensions and related mechanism(s) following tooth loss. Regenerative procedures indicatable to prevent and/or restore post-extraction tissue loss and their limitations are explored. A summary on current trends in biologics applied around teeth as well as at alveolar defect sites is provided. In preparation of this chapter, a simple plain written format is chosen for readers with no dental/periodontal background.

**Keywords** Bone regeneration · Guided tissue regeneration · Periodontium · Alveolar resorption · Tooth loss · Biomaterial · Bone graft · Biologics · Treatment outcomes · Alveolar ridge augmentation

---

Y.-H. Ma Becky · B. Leblebicioglu (✉)
Division of Periodontology, College of Dentistry, The Ohio State University, Columbus, OH, USA
e-mail: ma.2614@buckeyemail.osu.edu; Leblebicioglu.1@osu.edu



H. Das (ed.), *Tissue Repair and Regeneration*,
https://doi.org/10.1007/978-3-031-93677-7_7

# 1 Periodontal Anatomy and Regeneration

Tissue regeneration methods are frequently employed within the oral cavity, and various surgical procedures have been developed. In general, the volume and the type of tissue deficiency, and not the biomaterials per se, are the main determinants differentially affecting regenerative outcomes. Thus, understanding the site-specific anatomical factors is imperative. The periodontal apparatus consists of cementum, periodontal ligament, alveolar bone, and gingiva. Periodontal diseases cause destruction to these structures and lead to clinical attachment loss. The ramifications include loss of periodontal support, tooth mobility, gingival inflammation and bleeding, and eventually loss of teeth. Efforts have been made to salvage the teeth and slow down the disease progression. Non-surgical periodontal therapy focuses on eliminating the etiology and reinforcing oral hygiene, but it has some limitations especially at sites with severe loss in supporting tissues. Surgical periodontal therapy further improves clinical treatment outcomes by gaining access to the root surfaces and bone defects. Among the surgical approaches, regenerative surgical modalities present the best opportunities to turn the clock backwards. However, they are also the most complicated in terms of technique, material as well as case selection: in general, they are considered less predictable compared to other surgical modalities that require elimination of the defects (e.g., resective surgeries). Therefore, it is of paramount importance to understand the mechanisms behind the periodontal regeneration concept.

# 2 Bone Defect Types

Predictable outcomes of any periodontal surgeries mainly depend on existing bony defect shape and morphology. The importance of bone defect structural differences around teeth was first discussed in relation to osseous resective surgery treatment. This surgical manipulation basically requires removing the defect by reducing bony walls. In 1964, Ochsenbein and Bohannan (Ochsenbein & Bohannan, 1964) classified the bony defects around teeth (craters) into four classes based on the depth of osseous concavity and the thickness of buccal and lingual walls:

1. Class I: 2–3 mm osseous concavity surrounded with thick bony walls
2. Class II: 4–5 mm concavity with a wide orifice and thinner walls (the type most encountered)
3. Class III: 6–7 mm concavity with a sharp drop into a broad, flat base
4. Class IV: Variable depth with extremely thin walls (the least common type)

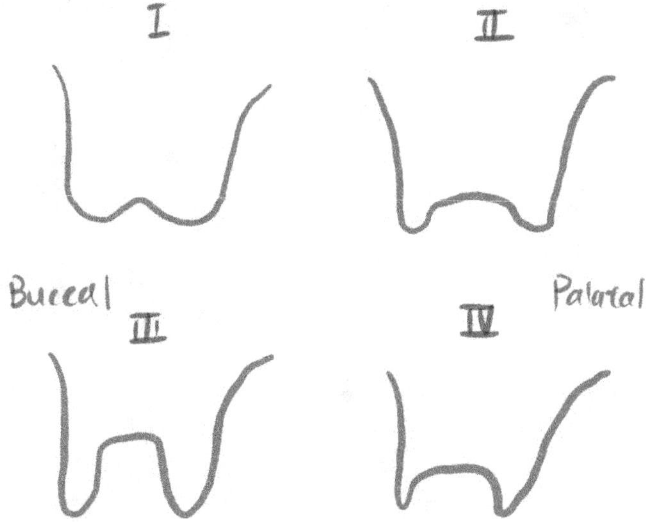


This classification was later improved by Ochsenbein (1986), by further classifying the bony craters into classes A, B, and C:

1. Class A: Shallow crater—1 to 2 mm
2. Class B: Medium crater—3 to 4 mm
3. Class C: Deep crater—5 mm or more

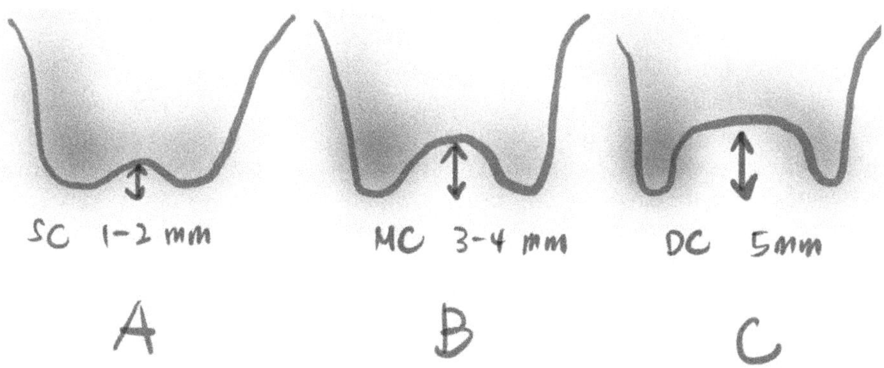


The impact of bony defect morphology on regeneration was highlighted at the American Academy of Periodontology (AAP)'s Workshop in 2015 (Reynolds et al., 2015). It is well documented that the predictability of treatment outcomes of regenerative surgery is differentially affected by the bony defect morphologies: infrabony (subcrestal) defects that have a vertical depth of at least 3 mm respond most favorably to regenerative procedures. Additionally, a defect surrounded with intact bony walls (three-wall defect, Fig. 1) is advantageous since regenerative outcomes

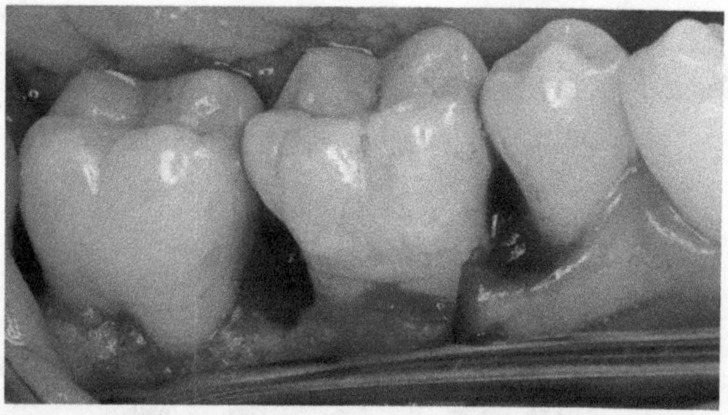

**Fig. 1** Clinical photo of a three-wall periodontal bone defect between mandibular molars (distal of the 1st molar), with buccal, distal and lingual walls present

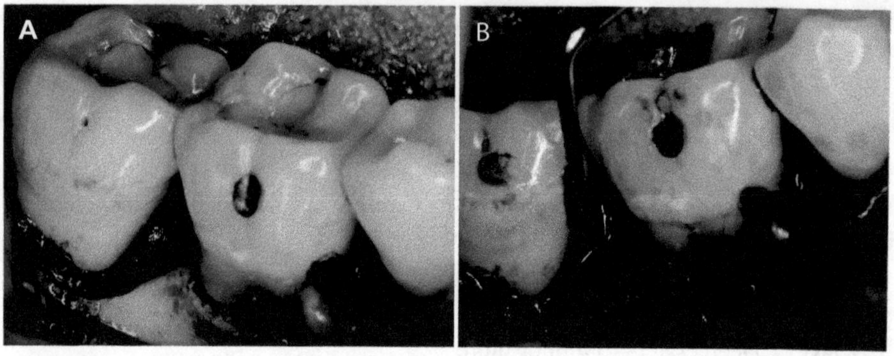



**Fig. 2** (**a** and **b**) Clinical photos of a two-wall periodontal bone defect between mandibular molars. The buccal wall is absent and the defect is less containable, which would require a more rigid membrane on top of bone grafts for space maintenance

become less predictable with defects that are surrounded with fewer bony walls (two- or one-wall defects, Fig. 2). Periodontal tissue loss due to bone resorption that occurs with no walls (horizontal bone loss) cannot be treated through regenerative procedures since the predictability of regenerating any tissue would be very low. Thus, the latter cases are generally treated through resective tissue contouring procedures instead of regeneration. Recently, Hasegawa et al. proposed a new classification of tooth position in the alveolar bone housing (Hasegawa et al., 2021). The significance of this new classification is that it focuses on the progression of the regenerative process and the factors that may influence the prognosis of teeth after regenerative surgery. Tooth positions were divided into two groups:

1. Grade I: The whole tooth is centrally positioned in the existing bone housing.
2. Grade II: A part of the tooth is exposed out of the existing bone housing. Further divided into two subgroups: subgroup A, where the alveolar bone housing is broader than the tooth; subgroup B, where the alveolar bone housing is narrower than the tooth.

In 2021, Nibali et al. published a systematic review of healing by defect morphology following regenerative surgeries (2021). It was concluded that deeper bony defects around teeth with narrower angles and a higher number of walls present improved clinical attachment level and radiographic bone gain at 1-year post-regenerative surgery. This finding corroborates the previous statement that three-wall defects respond more predictably than all other types of defects. Additional factors in relation to periodontal bony defect morphology, such as its extension from interdental space to other tooth surfaces, need to be explored in relation to regenerative outcomes.

## 3 Guided Tissue Regeneration: Historic Development and Biological Concepts

Regeneration of periodontal tissues lost to disease is the ultimate goal of surgical treatment when bone loss creates an intrabony defect. By definition guided tissue regeneration (GTR) is a dental surgical procedure that utilizes barrier membranes to direct new periodontal tissue growth in areas with insufficient gingiva or bone. The function of the barrier membrane is to stop epithelial cell migration into the defect and maintain the defect space so that undifferentiated mesenchymal cells located in remaining periodontal ligament apparatus at the base of the defect will have a chance to migrate into the defect, differentiate, and regenerate new attachment; ideally this should include new cementum, new bone, and new periodontal ligament attachment between these two newly forming tissues. In 1982, Nyman et al. published the first case report of GTR by simply applying a Millipore filter as a barrier membrane around a tooth with severe supporting tissue loss (1982b). Bone grafts and substitutes entered into dental practice mainly to help maintain the space, especially for bony defects with wide orifices. By 1985, the protection of soft tissue through minimally invasive surgery approaches were initiated in relation to GTR since the importance of primary wound closure and stability as well as the need for blood supply to the biomaterials were critical for better predictable regenerative and esthetic outcomes. In 1985, Takei et al. described the papilla preservation flap technique for the first time (1985). Ten years later, Cortellini et al. described the modified papilla preservation technique (MPPT) as a new approach for interproximal regenerative procedures (Cortellini et al., 1995b). In 1999, the same group published another technique called the simplified papilla preservation flap (SPPF) in order to obtain tissue regeneration together with good esthetic outcomes (Cortellini et al., 1999).

As the surgical techniques evolve together with biomaterials, the predictability and success of GTR still mostly depend on the fundamental biological concepts behind wound healing. The general principles of healing involve clot formation followed by an early stage of inflammation. Within 3 days, the late inflammatory stage starts when macrophages migrate into the wound area and secrete growth factors as well as cytokines (Polimeni et al., 2006). Fibroblasts, endothelial cells, and smooth muscle cells are attracted by the signals and therefore proliferation and migration take place, which leads to the formation of granulation tissue. Approximately 1 week following the wounding, the provisional extracellular matrix formed by the initial clot and fibrin is replaced by a collagen-rich matrix synthesized by fibroblasts. After their mission is completed, some fibroblasts undergo transformation into myofibroblasts and express alpha-smooth muscle actin, which is responsible for wound contraction. The abovementioned incisional wound healing phases are summarized in Fig. 3. The maturation and remodeling of granulation tissue contribute to the regeneration or repair (scar formation) depending upon two crucial factors: the availability of the cell types needed and the presence or absence of cues and signals necessary to recruit and stimulate these cells (Grzesik & Narayanan, 2002).

The wound healing following GTR around a root is complex in that there are several local factors other than bony defect type differentially affecting regenerative outcome. A mucoperiosteal flap is adapted to a mechanically and/or chemically cleaned but avascular root surface (Polimeni et al., 2006); the source of cells for the new tissue regeneration is limited to surrounding alveolar bone and leftover periodontal ligament apparatus localized at the bottom of the bony defect.

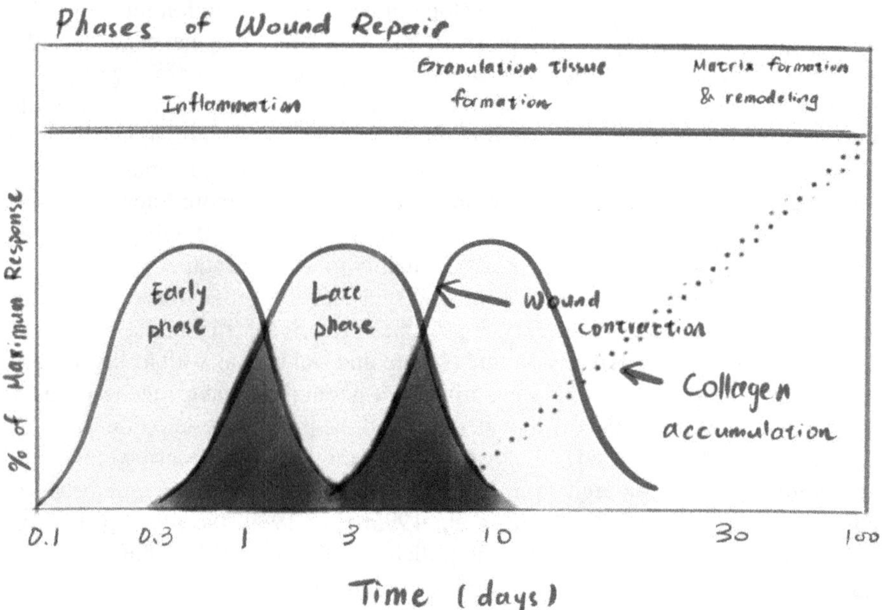


**Fig. 3** Phases of wound healing (incisional wounds), including an early (within hours) and a late (within days) phase of inflammation dominated by polymorphonuclear neutrophils and macrophages, respectively. The magnitude of wound contraction parallels the phase of granulation tissue formation. Collagen accumulation is first observed during the phase of granulation tissue formation, continuing through the phase of matrix formation and remodeling (Polimeni et al., 2006). (Adapted from Polimeni et al. 2006)

As early as 1976, Melcher postulated that periodontal structures are subdivided into four categories (gingiva, periodontal ligament, cementum, and bone) and the nature of the new attachment following periodontal surgery is determined by the cells repopulating the root surface (1976). The classic studies conducted by Karring and Nyman et al. provided the first evidence that periodontal regeneration can be obtained at a periodontitis-affected tooth (Karring et al., 1993; Karring et al., 1980; Nyman et al., 1980, 1982a, b). The same group proved that it was the cells from the periodontal ligament that have the capacity to regenerate periodontal attachment, but not those from alveolar bone and gingival connective tissue (Karring et al., 1993). Furthermore, Linghorne and O'Connell suggested that the main factor in the postoperative formation of a new long junctional epithelium is a lack of mechanical stability. This concept was further elucidated by Hiatt et al., who demonstrated the fundamental role of the fibrin clot in preventing apical migration of the gingival epithelium (Hiatt et al., 1968). Polson and Proye also corroborated this viewpoint by using a monkey model and showed that the unimpeded absorption, adhesion and maturation of the fibrin clot could favor connective tissue attachment over long junctional epithelium in the healing phase (Polson & Proye, 1983). On the other hand, Hiatt et al. investigated into the tensile strength of the tooth-gingival flap interface following reconstructive surgery over canine teeth in dogs (Hiatt et al., 1968). They discovered that the relatively small periodontal wound might not reach functional integrity until 2 weeks post-surgery. The tensile strength increases from 340 g at 5–7 days post-surgery to >1700 g at 2 weeks. These data suggest that wound integrity during the early healing phase rests primarily upon the stabilization from suturing. Consequently, the choice of suture material, placement techniques and removal timing, as well as the post-operative protocols should be dictated by such observations to protect the wound healing from disturbance (Leknes et al., 2005; Selvig et al., 1998).

Susin et al. further explored the biological principles that could unleash the innate potential of the periodontium to achieve clinically meaningful periodontal regeneration (Susin et al., 2015). Space provision, angiogenesis, wound stability and wound healing through primary intention are the four main principles that clinicians should bear in mind (Haney et al., 1993; Kim et al., 2004; Sigurdsson et al., 1994; Wikesjö et al., 2003). As technology advances and dental biomaterial research develops, we are in an era when more and more tools are in our armamentarium to confront challenges posed by periodontal disease and attachment loss. However, neglect of the aforementioned factors in the surgical planning and intervention phase will lead to compromised clinical and biological outcomes.

## 4 Regenerative Biomaterials

In 1982, Nyman et al. first introduced the concept of GTR by using a Millipore filter to prevent the migration of dentogingival epithelium alongside of the root surface and thus, providing time for the undifferentiated mesenchymal cells originating from the remaining periodontal ligament apparatus to differentiate and migrate into the bony defect (1982). The significance of a barrier membrane as a tool to prevent epithelial growth was discussed in the literature. Initial trials implemented a nonresorbable polytetrafluoroethylene (PTFE) membrane which was followed by resorbable collagen based options. Later on, Cortellini and Tonetti conducted controlled clinical trials using titanium reinforced membranes and bioresorbable membranes to achieve periodontal regeneration of human intrabony defects (Cortellini et al., 1995a; Cortellini et al., 1996). Furthermore, Falk et al. investigated into the clinical and radiographic findings of 203 consecutive GTR cases using a bioabsorbable matrix barrier (Falk et al., 1997). Today, resorbable barrier membranes are preferred to regenerate periodontal infrabony defects, to prevent the necessity of a second surgery to remove the material.

Autogenous bone graft has been introduced as the gold standard due to its osteogenic capacity. Nabers and O'Leary were the pioneers, discussing autogenous bone graft application in a case report (Nabers & O'Leary, 1967). Later on, Bowers et al. demonstrated that the healing of infrabony defects with the placement of decalcified freeze-dried bone allograft (DFDBA) in either a submerged (removal of the tooth crown and submerging the root by suturing the gingival flap on it) or non-submerged environment formed significantly more new attachment apparatus, new cementum, new connective tissue, and new bone than non-grafted sites (Bowers et al., 1989a, b). Today, allografts and xenografts are mostly preferred over autogenous bone graft since the latter generally requires a second wound and the bone material obtained from such a site is very limited. Unfortunately, currently available allografts and xenografts are mainly osteoconductive and not osteogenic. Thus, they mainly serve as a scaffold for cell chemotaxis and attachment. Autogenous bone and some allografts may have osteoinductive properties due to their growth factor content. Currently, biologics and/or autogenous bone grafts are mixed with allografts and xenografts to induce their osteoconductive and osteoinductive capacities.

## 5 Current Trends: Biologics

Growth factors serve as signaling molecules and recruit the necessary cells to complete the wound healing and regeneration process. The rationale of incorporating such molecules into the wound is mainly to regulate crucial cellular events involved in tissue repair or regeneration, including DNA synthesis, cell replication, chemotaxis, differentiation, matrix synthesis, and tissue vascularization (Giannobile & Somerman, 2003). They are often considered for the treatment of periodontal and alveolar bone defects through regeneration by mixing them with actual bone graft materials, since they have been reported to improve osteoinductive and osteoconductive properties of bone graft materials. The current trend of biologics application in regenerative approaches is reviewed in this section, including the widely used enamel matrix derivative (EMD) and recombinant human platelet-derived growth factor-BB (rhPDGF-BB), as both already have long-term data (5+ years) pertaining to the use for the treatment of periodontal infrabony defects (Avila-Ortiz et al., 2022).

## 5.1 *Enamel Matrix Derivative (EMD)*

It has been 25 years since Rasperini published case reports to demonstrate periodontal regeneration with the application of EMD in 1999 (Rasperini et al., 1999). EMD is rich in amelogenins and, when directly applied to root surfaces, these proteins are absorbed into the hydroxyapatite and collagen fibers of the root surface. This, in turn, induces cementum formation followed by periodontal regeneration (Kao et al., 2009; Sculean et al., 2003, 2008). EMD alone or in combination with graft materials provides improved clinical outcomes and long-term clinical stability comparable with GTR (Heijl et al., 1997; Rasperini et al., 2005; Sculean et al., 2007, 2008). In a systematic review on periodontal regeneration of infrabony defects from the AAP Regeneration Workshop (Kao et al., 2015), it was concluded that EMD is generally comparable to demineralized freeze-dried bone allograft and GTR, and superior to open flap debridement alone in improving clinical supporting tissue levels.

## 5.2 *Recombinant Human Platelet-Derived Growth Factor-BB (rhPDGF-BB)*

PDGF was first shown to promote bone regeneration by Lynch et al. (1989) Mechanistic studies have demonstrated that periodontal ligament (PDL) fibroblasts, osteoblasts, bone marrow derived stem cells, and pericytes express multiple receptors ($\alpha$, $\beta$, $\chi$, $\delta$) for PDGFs, which enhances the proliferation and chemotaxis of these cells (Cho et al., 1995; Park et al., 1995). Among the various isoforms of PDGF (PDGF-AA, PDGF-AB, and PDGF-BB), rhPDGF-BB has been reported as being the most potent in terms of its effects on the mitogenic and chemotactic responses of PDL fibroblasts(Boyan et al., 1994). rhPDGF-BB was also cited as one of the biologics with therapeutic effects comparable with other regenerative approaches (Kao et al., 2015). The unique advantages of this growth factor are that its concentration stays consistent at the delivered site, and it does not need a barrier membrane. Clinical application of rhPDGF-BB has been shown to be effective in regenerating bone, periodontal ligament and cementum (Camelo et al., 2003; Nevins et al., 2003). In a prospective, masked randomized clinical trial (RCT) evaluating the safety and efficacy of rhPDGF-BB, it was demonstrated that although a decrease in probing depths and an increase in clinical attachment levels around the tooth root were not significantly superior to bone grafted (control) defects, there was significant improvement in bone formation and accelerated wound healing (Nevins et al.,

2005). Follow-up studies showed that the improved bone fill and linear bone growth continued to improve over 36 months (McGuire et al., 2006; Nevins et al., 2013). Another multicenter RCT confirmed not only improved bone fill but also significant decrease in periodontal probing depths and an increase in clinical attachment levels after 6 months (Jayakumar et al., 2011). Additionally, the AAP 2022 best evidence consensus statement on the use of biologics in clinical practice concluded that rhPDGF-BB and platelet-rich fibrin (PRF) are associated with superior clinical and radiographic outcomes compared to EMD and platelet-rich plasma (PRP). Moreover, xenogeneic bone grafts with rhPDGF-BB or PRF are the best combination therapy to maintain the stability of the gingival margin following regenerative treatment of periodontal infrabony defects (Avila-Ortiz et al., 2022). In a follow-up best evidence systematic review and network meta-analysis, rhPDGF-BB demonstrated the largest effect size for CAL gain, PD reduction, less gingival recession and better radiographic linear bone gain compared to EMD and PRP (Tavelli et al., 2022). The significant beneficial effects on gingival recession have been interpreted as a possible result of enhanced angiogenic properties (Nasirzade et al., 2020; Steed, 2006; Tavelli et al., 2020; Zhang et al., 2020).

## 6 Tooth Loss and Changes in Alveolar Ridge Dimensions

Tooth loss is inevitable in the presence of severe periodontal tissue breakdown due to disease and/or trauma. Once the tooth is lost, the bundle bone which is a part of the periodontium and constitutes the bone housing of the tooth will start to resorb, since its existence relies on the presence of the periodontal ligament and the root surface that provides support. The resulting atrophic alveolar ridge creates a major challenge in reconstructing the lost dentition, especially with implant supported fixed type dental restorations, as sufficient alveolar bone ridge volume is a prerequisite for dental implant consideration (Fig. 4).

Due to jaw bone physiology, maxillary bone generally presents less density with thin cortical plate compared to mandibular bone. Therefore, post-tooth extraction alveolar ridge dimensional changes are more prominent in the maxilla than the mandible. Additionally, the bone resorption becomes noticeable in the horizontal plane immediately after the tooth extraction at both maxilla and mandible and, continues in the vertical dimensions, creating complex ridge defects challenging to regenerative through reconstructive procedures.

Trombelli et al. evaluated the healing of human extraction sockets utilizing semi-quantitative analysis of tissues and cell populations over a 6-month period (Trombelli et al., 2008). The healing was divided into the early (2–4 weeks), intermediate (6–8 weeks), and late (12–24 weeks) phases. It was revealed that granulation tissue was present in large amounts in the early healing phase and then replaced with provisional matrix and woven bone at the interval between the early and intermediate phases. The presence of osteoblasts peaked at 6–8 weeks and remained stable thereafter. It was also observed that while a provisional connective tissue consistently

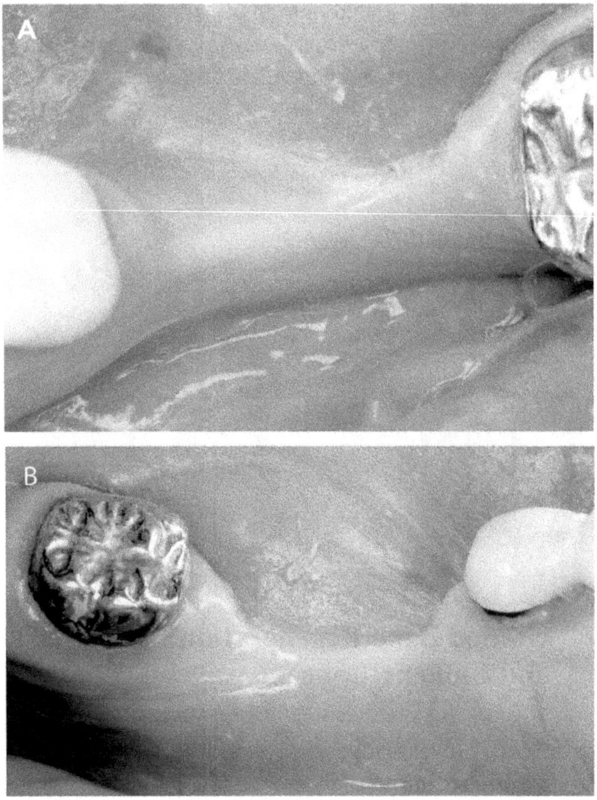


**Fig. 4** (**A** and **B**) Clinical photos of severe atrophic mandibular alveolar ridge of a 75-year-old female patient who was seeking dental implants as restorative dental treatment. The insufficient height and width of bone will require extensive bone augmentation surgical procedures and a long healing time.

forms within the first weeks of healing, it is much less predictable to determine the initiation of mineralized bone formation.

Heggeler et al. reported ridge reductions of 2.6–4.5 mm and 0.4–3.9 mm in the horizontal and vertical planes, respectively, within naturally healed alveolar bone sockets (Ten Heggeler et al., 2011). Post-extraction horizontal bone loss is usually more observable, predominantly involving the buccal aspect (30%) rather than the lingual (10%) aspect (Cardaropoli et al., 2003). Similar to periodontal bone defects, alveolar ridge defects have been described through several classifications. Among those, Seibert's classification is the most commonly used among clinicians (Seibert, 1983), and it includes the following groups:

Class I: Bone loss affecting the buccal and lingual (horizontal) aspects of the edentulous area of the ridge

Class II: Bone loss affecting the occlusal and cervical (vertical) aspects of the edentulous area of the ridge

Class III: A combination of both class 1 and class 2 bone loss, representing a complex ridge defect.

Class III is the most challenging alveolar ridge defect type to regenerate due to difficulties in primary wound closure, angiogenesis into the graft material, stability of the bone graft and barrier membrane, and space maintenance.

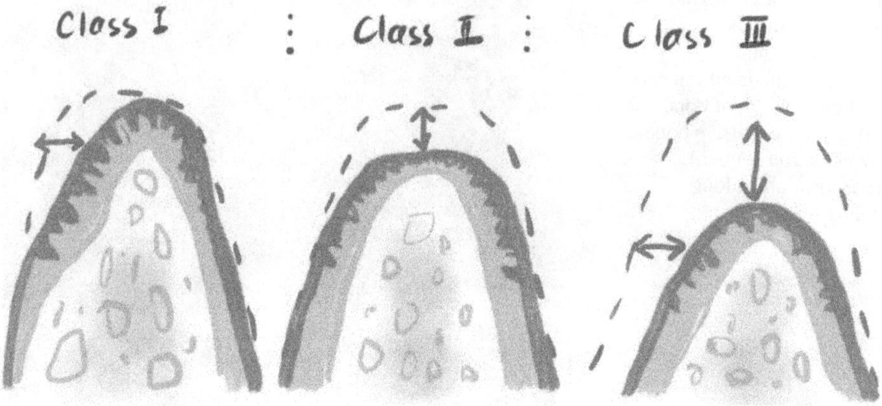


## 7 Guided Bone Regeneration: Historic Development and Biologic Concepts

In order to prevent ridge resorption following tooth loss, a ridge preservation technique with an intrasocket bone graft placement sealed with a barrier membrane immediately following tooth extraction was proposed. Iasella et al. conducted a randomized, controlled, blinded, clinical investigation to compare the post-extraction dimensional changes following tooth extraction plus ridge preservation versus extraction alone (Iasella et al., 2003). Freeze-dried allograft and a collagen membrane were used and compared to an alveolar bone socket that was left to heal without grafting. The histologic analysis revealed more bone in the ridge preservation group, and the alveolar bone ridge dimensional changes between the two groups were statistically significant (Iasella et al., 2003). These results supported the idea of ridge preservation to prevent post-extraction bone loss, especially in areas where loss of ridge height would impact and compromise the esthetics.

The bone graft of choice for Alveolar Ridge Preservation (ARP) is primarily allografts such as free-dried bone allografts (FDBA). In the clinical setting, the extraction socket, which normally has 4 intact bony walls, is filled with bone grafts and then covered with a collagen plug or barrier membrane. If the socket walls are all intact, a collagen plug is placed and sutured to seal the grafted site (Fig. 5). On the other hand, if the buccal or lingual bone wall is absent, a more rigid barrier membrane such as non-resorbable polytetrafluoroethylene (PTFE) membrane with better space-maintaining properties would be selected to cover the defect and occlude epithelial cell migration into the socket (Fig. 6).

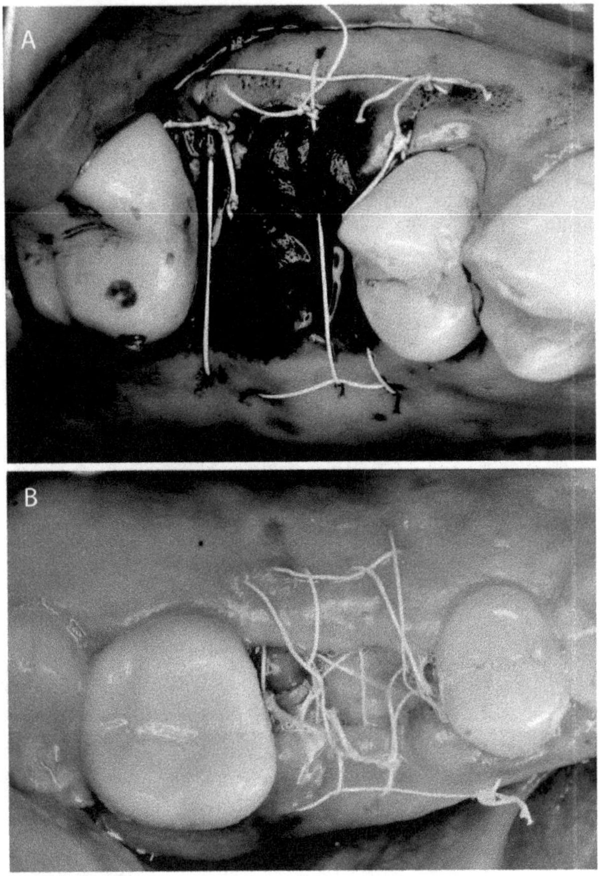


**Fig. 5** (**A** and **B**) Clinical photos of alveolar ridge preservation with collagen plug. (**a**) Immediate post-OP, (**b**) 2-week postoperative healing

The results of alveolar ridge preservation (ARP) are also determined by anatomic location. Leblebicioglu et al. investigated the outcomes following ARP in the posterior maxilla and mandible. It was found that buccal plate thickness and tooth root length were determinants of outcomes. Moreover, the initial ridge width was positively correlated with ridge width loss (Leblebicioglu et al., 2013).

As ARP became a well-accepted treatment option, the effect of different modalities of ARP as compared to tooth extraction alone was critically analyzed in the systematic review written by Avila-Ortiz et al. (2019) The pooled quantitative analysis revealed that ARP via socket grafting prevented horizontal, vertical mid-buccal and vertical mid-lingual bone resorption compared to extraction alone. However, current evidence is not strong enough to support any specific ARP approach compared to other available surgical modalities.

Similar to periodontal ridge defects, alveolar ridge defect type more than biomaterial characteristics plays an important role in decision making. In general, alveolar ridges that are deficient in the horizontal plane can be successfully treated with the application of particulate allogenic or xenogenic grafts covered with a resorbable

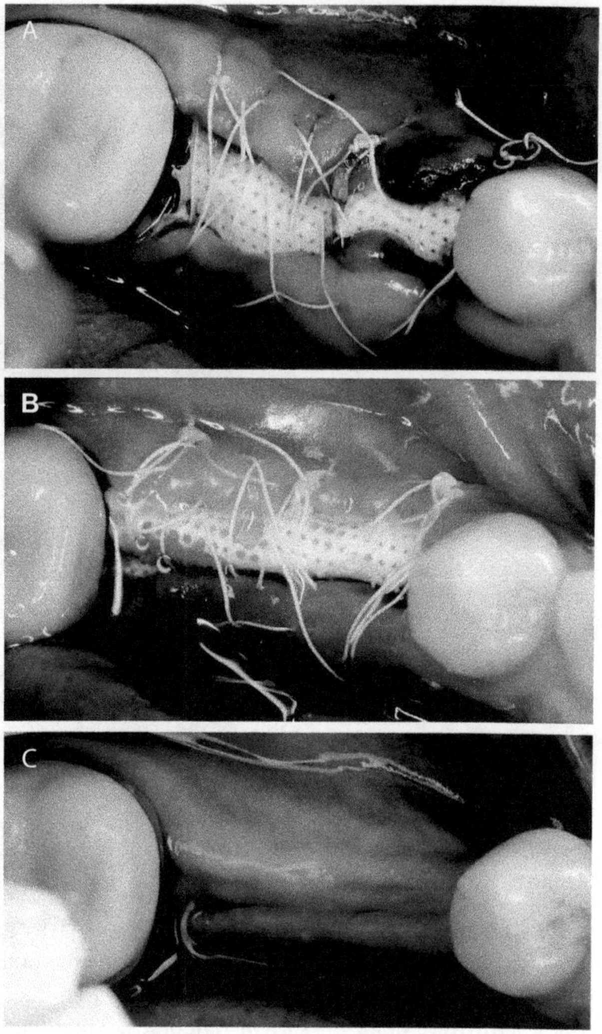


**Fig. 6** (**a**–**c**) Clinical photos of alveolar ridge preservation with non-resorbable PTFE membrane. (**a**) Immediate post-OP, (**b**) 4-week postoperative healing, and (**c**) 4-month postoperative healing

barrier membrane (Buser et al., 1993). As the ridge deficiency changes to more tissue loss in the vertical, or in both the horizontal and vertical dimensions, materials of choice would change from particulate bone graft to block bone; from a resorbable barrier membrane to a more rigid and possibly non-resorbable one (Urban et al., 2017). Additionally, the "sandwich" bone augmentation technique was proposed. Biomaterials with different strengths in properties were applied layer by layer for the optimal augmentation outcomes. A pilot case study reported using autograft, DFDBA, hydroxyapatite, and collagen membrane to treat the implant dehiscence defects (Wang et al., 2004). On the same note, a prospective case series study investigated the GBR with a two-layer composite graft and reported stable bone contours with low risk of mucosal recession over 10 years obsevation period (Chappuis et al., 2018).

There is a strong interest in developing and applying digital technology and customized 3D printing to engineer scaffold as a bone graft substitute, space filler and simultaneous drug delivery tool for biologics. Similarly, custom made 3D printing to engineered titanium alloy barrier membranes are already in clinical practice (Shi et al., 2022). They are indicated in treating complex ridge defects with both horizontal and vertical dimensional deficiencies.

## 8 Conclusion

The ultimate goal of periodontal therapy through GTR is the regeneration of functional and esthetic periodontal attachment apparatus. Current evidence and ongoing scientific research are bringing forth a better understanding of the innate potential of the periodontium to achieve clinically meaningful regeneration. It is critical to perform prudent case selection along with suitable surgical approaches in order to optimize the healing outcomes after regenerative therapy. While ARP is an effective tool to prevent post-extraction alveolar bone loss, GBR allows to treat already existing ridge defects. General principles of wound healing together with primary closure, angiogenesis, stability and space maintenance (P, A, S, S) are key factors for successful regenerative outcomes around teeth or at edentulous sites (Wang & Boyapati, 2006). Recent development in biologics and digital technology, specifically, in 3D printing has been promising.

## References

Avila-Ortiz, G., Chambrone, L., & Vignoletti, F. (2019). Effect of alveolar ridge preservation interventions following tooth extraction: A systematic review and meta-analysis. *Journal of Clinical Periodontology, 46*(Suppl 21), 195–223. https://doi.org/10.1111/jcpe.13057

Avila-Ortiz, G., Ambruster, J., Barootchi, S., Chambrone, L., Chen, C. Y., Dixon, D. R., Geisinger, M. L., Giannobile, W. V., Goss, K., Gunsolley, J. C., Heard, R. H., Kim, D. M., Mandelaris, G. A., Monje, A., Nevins, M. L., Palaiologou-Gallis, A., Rosen, P. S., Scheyer, E. T., Suarez-Lopez Del Amo, F., Tavelli, L., Velasquez, D., Wang, H. L., & Mealey, B. L. (2022). American Academy of periodontology best evidence consensus statement on the use of biologics in clinical practice. *Journal of Periodontology, 93*(12), 1763–1770. https://doi.org/10.1002/jper.22-0361

Bowers, G. M., Chadroff, B., Carnevale, R., Mellonig, J., Corio, R., Emerson, J., Stevens, M., & Romberg, E. (1989a). Histologic evaluation of new attachment apparatus formation in humans. Part II. *Journal of Periodontology, 60*(12), 675–682. https://doi.org/10.1902/jop.1989.60.12.675

Bowers, G. M., Chadroff, B., Carnevale, R., Mellonig, J., Corio, R., Emerson, J., Stevens, M., & Romberg, E. (1989b). Histologic evaluation of new attachment apparatus formation in humans. Part III. *Journal of Periodontology, 60*(12), 683–693. https://doi.org/10.1902/jop.1989.60.12.683

Boyan, L. A., Bhargava, G., Nishimura, F., Orman, R., Price, R., & Terranova, V. P. (1994). Mitogenic and chemotactic responses of human periodontal ligament cells to the different isoforms of platelet-derived growth factor. *Journal of Dental Research, 73*(10), 1593–1600. https://doi.org/10.1177/00220345940730100301

Buser, D., Dula, K., Belser, U., Hirt, H. P., & Berthold, H. (1993). Localized ridge augmentation using guided bone regeneration. 1. Surgical procedure in the maxilla. *The International Journal of Periodontics & Restorative Dentistry, 13*(1), 29–45.

Camelo, M., Nevins, M. L., Schenk, R. K., Lynch, S. E., & Nevins, M. (2003). Periodontal regeneration in human class II furcations using purified recombinant human platelet-derived growth factor-BB (rhPDGF-BB) with bone allograft. *The International Journal of Periodontics & Restorative Dentistry, 23*(3), 213–225.

Cardaropoli, G., Araujo, M., & Lindhe, J. (2003). Dynamics of bone tissue formation in tooth extraction sites. An experimental study in dogs. *Journal of Clinical Periodontology, 30*(9), 809–818. https://doi.org/10.1034/j.1600-051x.2003.00366.x

Chappuis, V., Rahman, L., Buser, R., Janner, S. F. M., Belser, U. C., & Buser, D. (2018). Effectiveness of contour augmentation with guided bone regeneration: 10-year results. *Journal of Dental Research, 97*(3), 266–274. https://doi.org/10.1177/0022034517737755

Cho, M. I., Lin, W. L., & Genco, R. J. (1995). Platelet-derived growth factor-modulated guided tissue regenerative therapy. *Journal of Periodontology, 66*(6), 522–530. https://doi.org/10.1902/jop.1995.66.6.522

Cortellini, P., Pini Prato, G., & Tonetti, M. S. (1995a). Periodontal regeneration of human intrabony defects with titanium reinforced membranes. A controlled clinical trial. *Journal of Periodontology, 66*(9), 797–803. https://doi.org/10.1902/jop.1995.66.9.797

Cortellini, P., Prato, G. P., & Tonetti, M. S. (1995b). The modified papilla preservation technique. A new surgical approach for interproximal regenerative procedures. *Journal of Periodontology, 66*(4), 261–266. https://doi.org/10.1902/jop.1995.66.4.261

Cortellini, P., Pini Prato, G., & Tonetti, M. S. (1996). Periodontal regeneration of human intrabony defects with bioresorbable membranes. A controlled clinical trial. *Journal of Periodontology, 67*(3), 217–223. https://doi.org/10.1902/jop.1996.67.3.217

Cortellini, P., Prato, G. P., & Tonetti, M. S. (1999). The simplified papilla preservation flap. A novel surgical approach for the management of soft tissues in regenerative procedures. *The International Journal of Periodontics & Restorative Dentistry, 19*(6), 589–599.

Falk, H., Laurell, L., Ravald, N., Teiwik, A., & Persson, R. (1997). Guided tissue regeneration therapy of 203 consecutively treated intrabony defects using a bioabsorbable matrix barrier. Clinical and radiographic findings. *Journal of Periodontology, 68*(6), 571–581. https://doi.org/10.1902/jop.1997.68.6.571

Giannobile, W. V., & Somerman, M. J. (2003). Growth and Amelogenin-like factors in periodontal wound healing. A systematic review. *Annals of Periodontology, 8*(1), 193–204. https://doi.org/10.1902/annals.2003.8.1.193

Grzesik, W. J., & Narayanan, A. S. (2002). Cementum and periodontal wound healing and regeneration. *Critical Reviews in Oral Biology and Medicine, 13*(6), 474–484. https://doi.org/10.1177/154411130201300605

Haney, J. M., Nilvéus, R. E., McMillan, P. J., & Wikesjö, U. M. (1993). Periodontal repair in dogs: Expanded polytetrafluoroethylene barrier membranes support wound stabilization and enhance bone regeneration. *Journal of Periodontology, 64*(9), 883–890. https://doi.org/10.1902/jop.1993.64.9.883

Hasegawa, Y., Takayama, T., & Iwano, Y. (2021). Clinical classification of tooth position in the alveolar bone housing with periodontal defects. *Journal of Dental Sciences, 16*(2), 795–798. https://doi.org/10.1016/j.jds.2020.11.011

Heijl, L., Heden, G., Svärdström, G., & Ostgren, A. (1997). Enamel matrix derivative (EMDOGAIN) in the treatment of intrabony periodontal defects. *Journal of Clinical Periodontology, 24*(9 Pt 2), 705–714. https://doi.org/10.1111/j.1600-051x.1997.tb00253.x

Hiatt, W. H., Stallard, R. E., Butler, E. D., & Badgett, B. (1968). Repair following mucoperiosteal flap surgery with full gingival retention. *Journal of Periodontology, 39*(1), 11–16. https://doi.org/10.1902/jop.1968.39.1.11

Iasella, J. M., Greenwell, H., Miller, R. L., Hill, M., Drisko, C., Bohra, A. A., & Scheetz, J. P. (2003). Ridge preservation with freeze-dried bone allograft and a collagen membrane compared to extraction alone for implant site development: A clinical and histologic study in humans. *Journal of Periodontology, 74*(7), 990–999. https://doi.org/10.1902/jop.2003.74.7.990

Jayakumar, A., Rajababu, P., Rohini, S., Butchibabu, K., Naveen, A., Reddy, P. K., Vidyasagar, S., Satyanarayana, D., & Pavan Kumar, S. (2011). Multi-Centre, randomized clinical trial on the efficacy and safety of recombinant human platelet-derived growth factor with β-tricalcium phosphate in human intra-osseous periodontal defects. *Journal of Clinical Periodontology, 38*(2), 163–172. https://doi.org/10.1111/j.1600-051X.2010.01639.x

Kao, R. T., Murakami, S., & Beirne, O. R. (2009). The use of biologic mediators and tissue engineering in dentistry. *Periodontology 2000, 2000*(50), 127–153. https://doi.org/10.1111/j.1600-0757.2008.00287.x

Kao, R. T., Nares, S., & Reynolds, M. A. (2015). Periodontal regeneration – Intrabony defects: A systematic review from the AAP regeneration workshop. *Journal of Periodontology, 86*(2 Suppl), S77–S104. https://doi.org/10.1902/jop.2015.130685

Karring, T., Nyman, S., & Lindhe, J. (1980). Healing following implantation of periodontitis affected roots into bone tissue. *Journal of Clinical Periodontology, 7*(2), 96–105. https://doi.org/10.1111/j.1600-051x.1980.tb01952.x

Karring, T., Nyman, S., Gottlow, J., & Laurell, L. (1993). Development of the biological concept of guided tissue regeneration–animal and human studies. Periodontology 2000 2000, 1, 26–35.

Kim, C. S., Choi, S. H., Chai, J. K., Cho, K. S., Moon, I. S., Wikesjö, U. M., & Kim, C. K. (2004). Periodontal repair in surgically created intrabony defects in dogs: Influence of the number of bone walls on healing response. *Journal of Periodontology, 75*(2), 229–235. https://doi.org/10.1902/jop.2004.75.2.229

Leblebicioglu, B., Salas, M., Ort, Y., Johnson, A., Yildiz, V. O., Kim, D. G., Agarwal, S., & Tatakis, D. N. (2013). Determinants of alveolar ridge preservation differ by anatomic location. *Journal of Clinical Periodontology, 40*(4), 387–395. https://doi.org/10.1111/jcpe.12065

Leknes, K. N., Selvig, K. A., Bøe, O. E., & Wikesjö, U. M. (2005). Tissue reactions to sutures in the presence and absence of anti-infective therapy. *Journal of Clinical Periodontology, 32*(2), 130–138. https://doi.org/10.1111/j.1600-051X.2005.00647.x

Lynch, S. E., Williams, R. C., Polson, A. M., Howell, T. H., Reddy, M. S., Zappa, U. E., & Antoniades, H. N. (1989). A combination of platelet-derived and insulin-like growth factors enhances periodontal regeneration. *Journal of Clinical Periodontology, 16*(8), 545–548. https://doi.org/10.1111/j.1600-051x.1989.tb02334.x

McGuire, M. K., Kao, R. T., Nevins, M., & Lynch, S. E. (2006). rhPDGF-BB promotes healing of periodontal defects: 24-month clinical and radiographic observations. *The International Journal of Periodontics & Restorative Dentistry, 26*(3), 223–231.

Melcher, A. H. (1976). On the repair potential of periodontal tissues. *Journal of Periodontology, 47*(5), 256–260. https://doi.org/10.1902/jop.1976.47.5.256

Nabers, C. L., & O'Leary, T. J. (1967). Autogenous bone grafts: Case report. *Periodontics, 5*(5), 251–253. https://www.ncbi.nlm.nih.gov/pubmed/5234057

Nasirzade, J., Kargarpour, Z., Hasannia, S., Strauss, F. J., & Gruber, R. (2020). Platelet-rich fibrin elicits an anti-inflammatory response in macrophages in vitro. *Journal of Periodontology, 91*(2), 244–252. https://doi.org/10.1002/jper.19-0216

Nevins, M., Camelo, M., Nevins, M. L., Schenk, R. K., & Lynch, S. E. (2003). Periodontal regeneration in humans using recombinant human platelet-derived growth factor-BB (rhPDGF-BB) and allogenic bone. *Journal of Periodontology, 74*(9), 1282–1292. https://doi.org/10.1902/jop.2003.74.9.1282

Nevins, M., Giannobile, W. V., McGuire, M. K., Kao, R. T., Mellonig, J. T., Hinrichs, J. E., McAllister, B. S., Murphy, K. S., McClain, P. K., Nevins, M. L., Paquette, D. W., Han, T. J., Reddy, M. S., Lavin, P. T., Genco, R. J., & Lynch, S. E. (2005). Platelet-derived growth factor stimulates bone fill and rate of attachment level gain: Results of a large multicenter randomized controlled trial. *Journal of Periodontology, 76*(12), 2205–2215. https://doi.org/10.1902/jop.2005.76.12.2205

Nevins, M., Kao, R. T., McGuire, M. K., McClain, P. K., Hinrichs, J. E., McAllister, B. S., Reddy, M. S., Nevins, M. L., Genco, R. J., Lynch, S. E., & Giannobile, W. V. (2013). Platelet-derived growth factor promotes periodontal regeneration in localized osseous defects: 36-month extension results from a randomized, controlled, double-masked clinical trial. *Journal of Periodontology, 84*(4), 456–464. https://doi.org/10.1902/jop.2012.120141

Nibali, L., Sultan, D., Arena, C., Pelekos, G., Lin, G.-H., & Tonetti, M. (2021). Periodontal infrabony defects: Systematic review of healing by defect morphology following regenerative surgery. *Journal of Clinical Periodontology, 48*(1), 101–114. https://doi.org/10.1111/jcpe.13381

Nyman, S., Karring, T., Lindhe, J., & Plantén, S. (1980). Healing following implantation of periodontitis-affected roots into gingival connective tissue. *Journal of Clinical Periodontology, 7*(5), 394–401. https://doi.org/10.1111/j.1600-051x.1980.tb02012.x

Nyman, S., Gottlow, J., Karring, T., & Lindhe, J. (1982a). The regenerative potential of the periodontal ligament. An experimental study in the monkey. *Journal of Clinical Periodontology, 9*(3), 257–265. https://doi.org/10.1111/j.1600-051x.1982.tb02065.x

Nyman, S., Lindhe, J., Karring, T., & Rylander, H. (1982b). New attachment following surgical treatment of human periodontal disease. *Journal of Clinical Periodontology, 9*(4), 290–296. https://doi.org/10.1111/j.1600-051x.1982.tb02095.x

Ochsenbein, C. (1986). A primer for osseous surgery. *The International Journal of Periodontics & Restorative Dentistry, 6*(1), 8–47.

Ochsenbein, C., & Bohannan, H. M. (1964). The palatal approach to osseous surgery II. Clinical application. *The Journal of Periodontology, 35*(1), 54–68. https://doi.org/10.1902/jop.1964.35.1.54

Park, J. B., Matsuura, M., Han, K. Y., Norderyd, O., Lin, W. L., Genco, R. J., & Cho, M. I. (1995). Periodontal regeneration in class III furcation defects of beagle dogs using guided tissue regenerative therapy with platelet-derived growth factor. *Journal of Periodontology, 66*(6), 462–477. https://doi.org/10.1902/jop.1995.66.6.462

Polimeni, G., Xiropaidis, A. V., & Wikesjö, U. M. (2006). Biology and principles of periodontal wound healing/regeneration. *Periodontology 2000, 2000*(41), 30–47. https://doi.org/10.1111/j.1600-0757.2006.00157.x

Polson, A. M., & Proye, M. P. (1983). Fibrin linkage: A precursor for new attachment. *Journal of Periodontology, 54*(3), 141–147. https://doi.org/10.1902/jop.1983.54.3.141

Rasperini, G., Ricci, G., & Silvestri, M. (1999). Surgical technique for treatment of infrabony defects with enamel matrix derivative (Emdogain): 3 case reports. *The International Journal of Periodontics & Restorative Dentistry, 19*(6), 578–587.

Rasperini, G., Silvestri, M., & Ricci, G. (2005). Long-term clinical observation of treatment of infrabony defects with enamel matrix derivative (Emdogain): Surgical reentry. *The International Journal of Periodontics & Restorative Dentistry, 25*(2), 121–127.

Reynolds, M. A., Kao, R. T., Camargo, P. M., Caton, J. G., Clem, D. S., Fiorellini, J. P., Geisinger, M. L., Mills, M. P., Nares, S., & Nevins, M. L. (2015). Periodontal regeneration – Intrabony defects: A consensus report from the AAP regeneration workshop. *Journal of Periodontology, 86*(2 Suppl), S105–S107. https://doi.org/10.1902/jop.2015.140378

Sculean, A., Junker, R., Donos, N., Windisch, P., Brecx, M., & Dünker, N. (2003). Immunohistochemical evaluation of matrix molecules associated with wound healing following treatment with an enamel matrix protein derivative in humans. *Clinical Oral Investigations, 7*(3), 167–174. https://doi.org/10.1007/s00784-003-0212-9

Sculean, A., Schwarz, F., Chiantella, G. C., Donos, N., Arweiler, N. B., Brecx, M., & Becker, J. (2007). Five-year results of a prospective, randomized, controlled study evaluating treatment of intra-bony defects with a natural bone mineral and GTR. *Journal of Clinical Periodontology, 34*(1), 72–77. https://doi.org/10.1111/j.1600-051X.2006.01007.x

Sculean, A., Kiss, A., Miliauskaite, A., Schwarz, F., Arweiler, N. B., & Hannig, M. (2008). Ten-year results following treatment of intra-bony defects with enamel matrix proteins and guided tissue regeneration. *Journal of Clinical Periodontology, 35*(9), 817–824. https://doi.org/10.1111/j.1600-051X.2008.01295.x

Seibert, J. S. (1983). Reconstruction of deformed, partially edentulous ridges, using full thickness onlay grafts. Part I. Technique and wound healing. *The Compendium of Continuing Education in Dentistry(Lawrenceville), 4*(5), 437–453.

Selvig, K. A., Biagiotti, G. R., Leknes, K. N., & Wikesjö, U. M. (1998). Oral tissue reactions to suture materials. *The International Journal of Periodontics & Restorative Dentistry, 18*(5), 474–487.

Shi, Y., Liu, J., Du, M., Zhang, S., Liu, Y., Yang, H., Shi, R., Guo, Y., Song, F., Zhao, Y., & Lan, J. (2022). Customized barrier membrane (titanium alloy, poly ether-ether ketone and unsintered hydroxyapatite/poly-l-Lactide) for guided bone regeneration [review]. *Frontiers in Bioengineering and Biotechnology, 10*. https://doi.org/10.3389/fbioe.2022.916967

Sigurdsson, T. J., Hardwick, R., Bogle, G. C., & Wikesjö, U. M. (1994). Periodontal repair in dogs: Space provision by reinforced ePTFE membranes enhances bone and cementum regeneration in large supraalveolar defects. *Journal of Periodontology, 65*(4), 350–356. https://doi.org/10.1902/jop.1994.65.4.350

Steed, D. L. (2006). Clinical evaluation of recombinant human platelet-derived growth factor for the treatment of lower extremity ulcers. *Plastic and Reconstructive Surgery, 117*(7 Suppl) 143S–149S; discussion 150S–151S. https://doi.org/10.1097/01.prs.0000222526.21512.4c

Susin, C., Fiorini, T., Lee, J., De Stefano, J. A., Dickinson, D. P., & Wikesjö, U. M. (2015). Wound healing following surgical and regenerative periodontal therapy. *Periodontology 2000, 68*(1), 83–98. https://doi.org/10.1111/prd.12057

Takei, H. H., Han, T. J., Carranza, F. A., Jr., Kenney, E. B., & Lekovic, V. (1985). Flap technique for periodontal bone implants. Papilla preservation technique. *Journal of Periodontology, 56*(4), 204–210. https://doi.org/10.1902/jop.1985.56.4.204

Tavelli, L., McGuire, M. K., Zucchelli, G., Rasperini, G., Feinberg, S. E., Wang, H. L., & Giannobile, W. V. (2020). Biologics-based regenerative technologies for periodontal soft tissue engineering. *Journal of Periodontology, 91*(2), 147–154. https://doi.org/10.1002/jper.19-0352

Tavelli, L., Chen, C. J., Barootchi, S., & Kim, D. M. (2022). Efficacy of biologics for the treatment of periodontal infrabony defects: An American Academy of periodontology best evidence systematic review and network meta-analysis. *Journal of Periodontology, 93*(12), 1803–1826. https://doi.org/10.1002/jper.22-0120

Ten Heggeler, J. M., Slot, D. E., & Van der Weijden, G. A. (2011). Effect of socket preservation therapies following tooth extraction in non-molar regions in humans: A systematic review. *Clinical Oral Implants Research, 22*(8), 779–788. https://doi.org/10.1111/j.1600-0501.2010.02064.x

Trombelli, L., Farina, R., Marzola, A., Bozzi, L., Liljenberg, B., & Lindhe, J. (2008). Modeling and remodeling of human extraction sockets. *Journal of Clinical Periodontology, 35*(7), 630–639. https://doi.org/10.1111/j.1600-051X.2008.01246.x

Urban, I. A., Monje, A., Lozada, J., & Wang, H. L. (2017). Principles for vertical ridge augmentation in the atrophic posterior mandible: A technical review. *The International Journal of Periodontics & Restorative Dentistry, 37*(5), 639–645. https://doi.org/10.11607/prd.3200

Wang, H. L., & Boyapati, L. (2006). "PASS" principles for predictable bone regeneration. *Implant Dentistry, 15*(1), 8–17. https://doi.org/10.1097/01.id.0000204762.39826.0f

Wang, H. L., Misch, C., & Neiva, R. F. (2004). "Sandwich" bone augmentation technique: Rationale and report of pilot cases. *The International Journal of Periodontics & Restorative Dentistry, 24*(3), 232–245.

Wikesjö, U. M., Lim, W. H., Thomson, R. C., & Hardwick, W. R. (2003). Periodontal repair in dogs: Gingival tissue occlusion, a critical requirement for GTR? *Journal of Clinical Periodontology, 30*(7), 655–664. https://doi.org/10.1034/j.1600-051x.2003.00362.x

Zhang, J., Yin, C., Zhao, Q., Zhao, Z., Wang, J., Miron, R. J., & Zhang, Y. (2020). Anti-inflammation effects of injectable platelet-rich fibrin via macrophages and dendritic cells. *Journal of Biomedical Materials Research Part A, 108*(1), 61–68. https://doi.org/10.1002/jbm.a.36792

# Corneal Tissue Regeneration

# Mechanisms of Corneal Epithelial Homeostasis, Repair, and Regeneration

**Matilda F. Chan and Finn D. Wolfreys**

**Abstract** The maintenance of the cornea during homeostasis and the reparative processes that occur after injury involves a complex and dynamic set of cellular and molecular events. This chapter presents an overview of the normal physiological processes that occur to maintain the cornea in its steady state and the cellular and molecular processes that become activated following injury to help restore the cornea back to its natural state. Furthermore, specific diseases of the corneal epithelium are discussed in the context of these processes.

**Keywords** Corneal epithelium · Corneal wound healing · Limbal stem cells · Corneal epithelial homeostasis · Corneal epithelial regeneration

## 1 Introduction

The maintenance of the corneal epithelium during homeostasis and the reparative processes that occur after injury involves a complex and dynamic set of cellular and molecular events. Key cellular and molecular factors include stem cells, growth factors, matrix proteins, immune cells, and corneal nerves. Here, we describe the normal physiologic processes that occur to maintain the cornea in its homeostatic state and the processes that are activated following trauma to promote repair and regeneration. We additionally describe how the dysfunction of these molecular processes can lead to corneal epithelial diseases.

---

M. F. Chan (✉)
Department of Ophthalmology, University of California, San Francisco, CA, USA

Francis I. Proctor Foundation, University of California, San Francisco, CA, USA
e-mail: matilda.chan@ucsf.edu

F. D. Wolfreys
Department of Ophthalmology, University of California, San Francisco, CA, USA
e-mail: finn.wolfreys@ucsf.edu

 
H. Das (ed.), *Tissue Repair and Regeneration*,
https://doi.org/10.1007/978-3-031-93677-7_8

## 2 Corneal Anatomy and Epithelium Development

The cornea, one of the main components of the ocular surface, has important refractive and barrier functions. The cornea bends light rays through the pupil and lens and focuses them onto the retina. The transparent human cornea has an average refractive index of 1.3375 and a central anterior radius of curvature of 7.8 mm making it responsible for three-fourth of the total refractive power of the human eye (Eghrari et al., 2015).

The cornea has a relatively simple anatomy and consists of three major specialized cellular layers: epithelium, stroma, and endothelium (Fig. 1). The corneal epithelium is the outermost layer of the cornea and serves as the major protective barrier to environmental challenges and pathogens. The corneal epithelium consists of five to seven layers of stratified, squamous, nonkeratinized epithelial cells and originates from the surface ectoderm (Hassell & Birk, 2010; Sridhar, 2018). During development, the surface ectoderm in the head region thickens and gives rise to bilateral lens placodes that invaginate to form lens pits. Once the lens pits separate from the overlying ectoderm, the remaining ectoderm gives rise to the presumptive corneal epithelium. During embryogenesis, the corneal epithelium expands from a two-layered structure into a five to seven cell-layered epithelium (Walker et al., 2021).

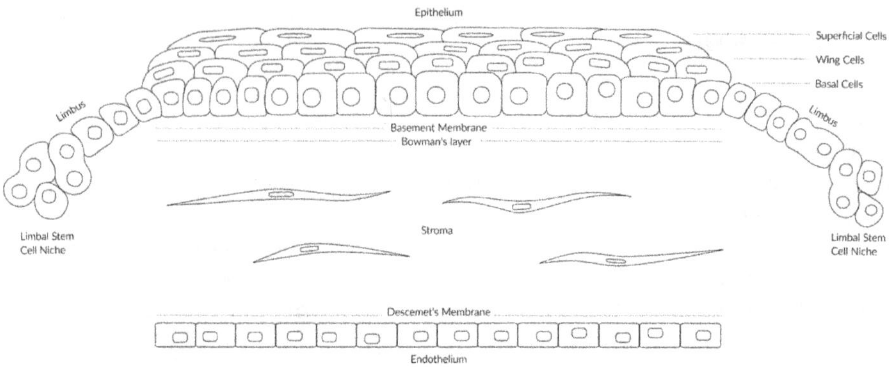


**Fig. 1** A simplified representation of the cornea's five major layers. The epithelium sits on a basement membrane distinct from Bowman's layer, which divides the epithelium from the corneal stroma, the largest of the layers. Division of stem cells in the limbal stem cell niche produces transit-amplifying cells, which migrate from the periphery and establish a basal cell layer. The XYZ hypothesis posits that the upper layers of epithelial cells, distinguished by a flatter morphology, are supplied by the upward movement of cells from the basal layer which differentiate as they move. Superficial cells are, at the same time, removed in part by blinking

## 3 Maintenance of Corneal Epithelial Homeostasis

The junctional zone between the cornea and the conjunctiva is the limbus (Fig. 1). The limbus harbors a specialized population of limbal epithelial stem/progenitor cells (LESCs) that function to maintain corneal homeostasis and drive reparative processes (Bonnet et al., 2021; Di Girolamo et al., 2015; Tseng, 1989). During homeostasis, LESCs are typically stationary, undifferentiated, and slow-cycling (Bonnet et al., 2022). LESCs are able to divide symmetrically and produce two daughter stem cells or divide asymmetrically to produce one daughter stem cell to replenish the stem cell pool and a transit amplifying cell (TAC). TACs are progenitor cells that are still undifferentiated but have finite proliferative potential. TACs are motile and function to replace terminally differentiated cells (TDCs) (Di Girolamo et al., 2015; Tseng, 1989). Corneal epithelial cells repopulate and maintain homeostasis through the stratification, proliferation, and differentiation of cells under the regulation of bone morphogenic protein-6 (BMP-6) (Tiwari et al., 2020). The posterior layer of the corneal epithelium consists of the basal epithelial cells. In mice, basal limbal epithelial cells express Keratin-14 (K14) and can divide and differentiate. As these cells proliferate, mature, and migrate anteriorly to repopulate the cornea, they lose K14 expression and gain K3/K12 expression (Di Girolamo & Park, 2023; Park et al., 2019). During homeostasis, the corneal epithelium is continuously replenished according to the XYZ hypothesis, in which desquamating superficial cells (Z) are replenished by proliferating basal cells (X) that migrate in a centripetal direction (Y) (Thoft & Friend, 1983).

A major function of the cornea is to serve as a structural barrier between traumatic insults from the external environment and intraocular contents. The corneal epithelium barrier function is achieved through its architectural components of adherens junctions between cells and three types of epithelial cells (superficial, wing, and basal). Superficial epithelial cells have tight junctions, suprabasal epithelial cells have desmosomes, and gap junctions and hemidesmosomes containing $\alpha6\beta4$ integrin are found between basal epithelial cells (Di Girolamo & Park, 2023).

## 4 The Process of Corneal Epithelial Wound Healing

It is perhaps the balance of barrier function and transparency that most clearly motivates the events that follow wounding: vulnerability to infection favors rapid wound closure, whereas vision

depends on corneal transparency, and thus the carefully organized structure of the corneal epithelium. Corneal epithelial wound healing requires that cells migrate rapidly to cover the wound site, proliferate to replace lost cells, and subsequently reestablish a highly specialized structure. Although the process is highly choreographed, establishing mechanistic causality in corneal wound healing is, in general, difficult. Here we cover the major processes observed so that the reader may better contextualize mechanisms of healing, and its dysfunction presented here and in the wider literature.

## 4.1 The Wounded Cornea Is Covered by a Single Sheet of Cells

Upon wounding, and after apoptosis has seen damaged cells removed, proteases release these remaining corneal epithelial cells from the epithelial structure to which they are usually anchored (Morimoto et al., 1993). A single sheet of cells migrates rapidly, and without cell division, from the wound periphery covering the exposed cornea (Fig. 2) (Crosson et al., 1986). These cells will form the basal layer of the repaired epithelium whose division, under normal circumstances, establishes its distinct layers. A sliding model posits that analogous to a landslide, cells from the surrounding epithelial layers move in sheets (Hanna, 1966), whereas a rolling model suggests cells from the upper epithelial layers roll into the wound (Krawczyk, 1971). However, like cells of the limbus, and unlike those of the central epithelium, the migrating basal layer expresses K14. Supported by lineage tracing, this has motivated an alternative model in which pressure from increased cell division at the limbus drives existing basal cells into the wound area (Park et al., 2019). In all cases, however, replacement of the lost basal layer is integral to wound healing, both to repair the damaged matrix on which they sit and to reestablish the upper layers whose population they supply.

## 4.2 A Temporary Basement Membrane Facilitates Migration

Migrating corneal epithelial cells must traverse the potentially damaged, or nonexistent, basement membrane which usually separates the basal cell layer and Bowman's layer. Perhaps due to the speed at which they must advance, as they move, they deposit a temporary basement membrane distinguished by its high fibronectin content (Fig. 3) (Fujikawa et al., 1984; Suda et al., 1981). Fibrin, hypothesized to be deposited via the tears, is another conspicuous addition to the temporary basement membrane, while laminin and collagen IV, major constituents of the permanent epithelial basement membrane, are not detected under the migrating cell front (Fujikawa et al., 1984; Suzuki et al., 2000). To better attach to the fibronectin-rich temporary membrane, migrating epithelial cells upregulate the expression of a number of integrins (Murakami et al., 1992). Coverage of the wound site coincides with the disappearance of fibronectin from the basement membrane and the reappearance of collagen IV and laminin (Suda et al., 1981; Suzuki et al., 2000). Together, these observations support a model of basement membrane repair that proceeds by rapid coverage by a new basal cell layer, followed by remodeling of the matrix below.

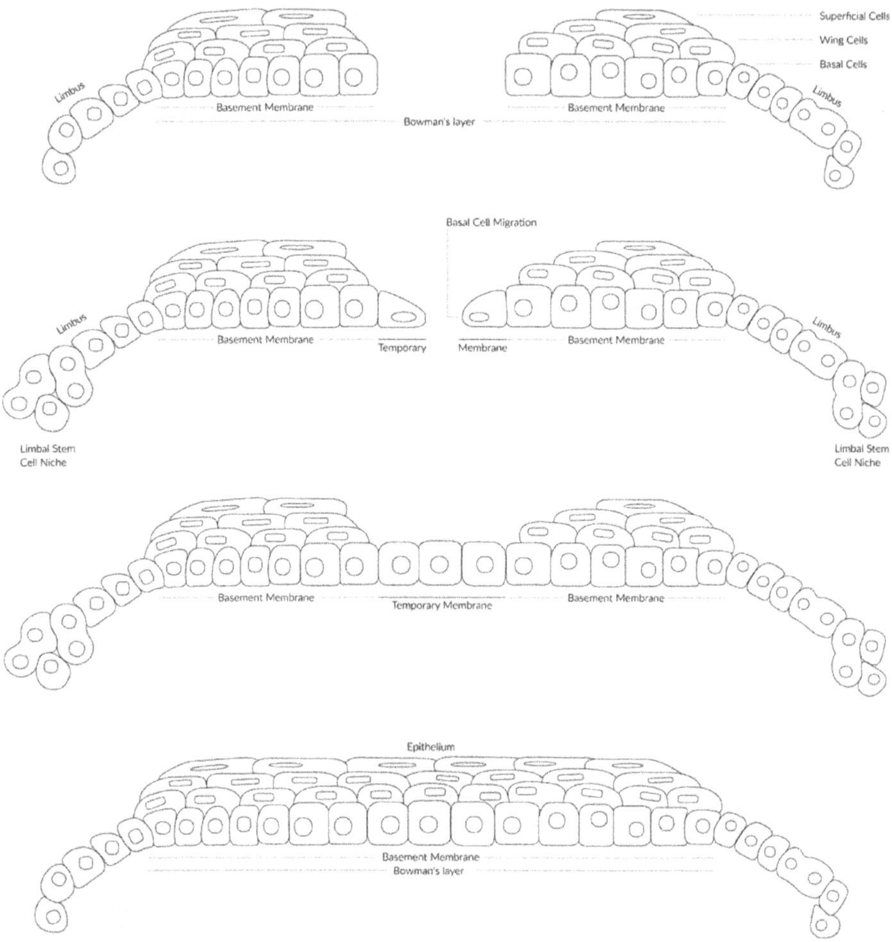


**Fig. 2** Major processes in the healing of the wounded epithelium. The initial migration of a single sheet of cells covers the wound. Although models differ in their description of the identity of these cells, it appears that reestablishing a basal layer is a prerequisite for subsequent reepithelization. A temporary basement membrane is synthesized as the new basal layer migrates and later is replaced by a permanent membrane, while the upper layers of the epithelium are reestablished and cell-cell contacts reformed. A significant increase in stem cell division in the limbus supplies replacement cells for the new epithelium

## *4.3 Cell Division in the Limbus Supplies New Corneal Epithelial Cells*

Establishing a new basal layer appears to be a precondition for the replacement of the upper epithelial layers (Fig. 2). In the healthy cornea, the XYZ hypothesis posits that the cornea's stratified cell layers, in part, by an equilibrium of basal epithelial

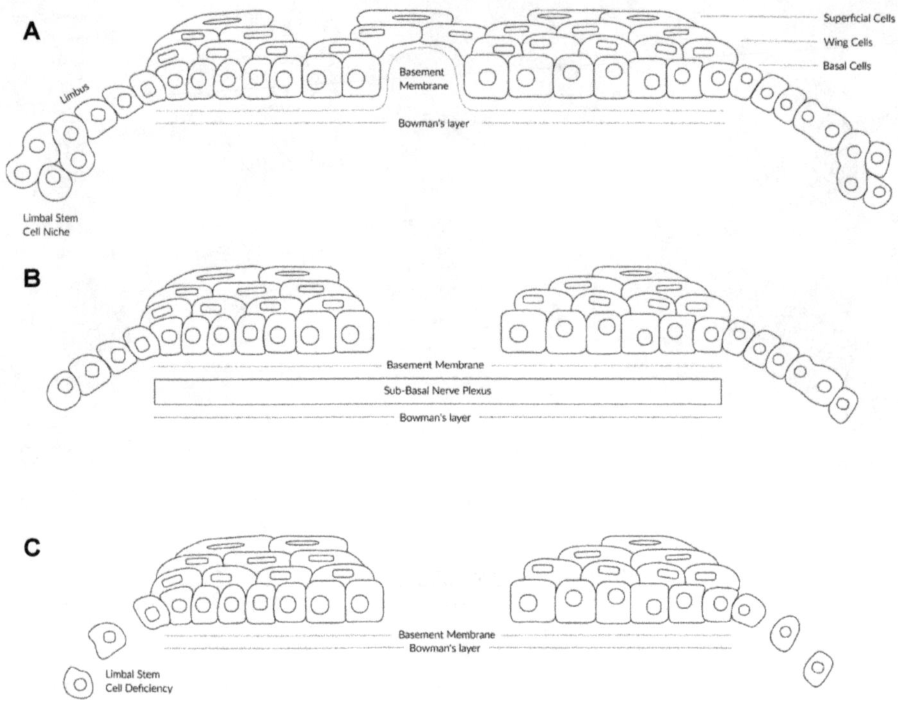


**Fig. 3** Corneal diseases result from damaged or abnormal cellular components of the corneal epithelium. The corneal epithelial basement membrane can be irregular due to prior injury or abnormal corneal dystrophies, which results in an abnormal adherence to its overlying epithelial cells (**a**). Corneal nerve loss or damage can result from various causes, including diabetes mellitus, and this leads to impaired corneal wound repair and persistent corneal epithelial defects (**b**). The stem cells in the corneal limbus are important in maintaining corneal homeostasis as well as reestablishing all layers of the corneal epithelium. Loss or damage of the limbal corneal epithelial stem cells can also result in impaired corneal wound repair and persistent corneal epithelial defects (**c**)

cell division, X, centripetal movement of cells from the cornea's periphery, Y, and the steady loss of differentiated cells at the surface of the cornea, Z (Thoft & Friend, 1983). Although this hypothesis has seen a challenge in corneal homeostasis (Majo et al., 2008), the supply of new cells from the limbus in the wounded cornea is evidenced by a marked increase in their usually slow rate of division (Park et al. 2019). Lineage tracing experiments lend further support to "pressure" from the division at the limbus being the significant force in reestablishing a basal cell layer in the wounded cornea (Park et al., 2019).

## 4.4 *Corneal Epithelial Cells Must Break and Reestablish Contacts*

The corneal epithelium relies on complex, carefully organized contacts with both other cells and the extracellular matrix (ECM). Tight junctions characterize the upper corneal epithelium and underpin its barrier function, whereas the wing cells below maintain a connection to each other through desmosomes. Cells of the basal layer form gap junctions, while adherens junctions are a feature common to all layers of the epithelium (Suzuki et al., 2000). During healing, however, some of these cell connections are broken: in the migrating cell front, gap junctions, and desmosomes, as identified by the markers connexin43 and desmoglein, disappear, whereas tight and adherens junctions, as identified by occludin and E-cadherin, appear to remain largely intact (Suzuki et al., 2000). Sometime after the basal layer is reestablished, concomitant with the reappearance of laminin in the basement membrane, gap junctions and desmosomes are reestablished (McCartney & Cantu-Crouch, 1992; Suzuki et al., 2000). Tight junctions appear maintained in the migration of corneal epithelial cells, while the gap junctions associated with the basal layer are not, perhaps further reinforcing the constraint of barrier maintenance on wound healing, as well as the importance of rapidly reestablishing a basal layer.

## 4.5 *Signaling and Immune Interaction Choreographs Epithelial Wound Healing*

For the corneal epithelium to heal, a single cell layer must migrate while synthesizing a temporary basement membrane. Limbal stem cells must replace lost epithelial cells by dramatically increasing their rate of division. The cells which have filled in the wound area must reestablish the matrix and contacts that make the cornea a transparent barrier. Coordination is achieved by signals from the wounded epithelium itself, keratocytes of the stroma below, infiltrating immune cells, corneal nerves, and the extracellular matrix (Ljubimov & Saghizadeh, 2015). A plethora of growth factors, cytokines, and other molecules have demonstrated roles in, or at least influence over, wound healing. However, how their actions are integrated to support the process of healing remains an important area of research.

Despite this, common functions are observed among the factors that influence corneal wound healing. For instance, members of the epidermal growth factor family tend to enhance proliferation and migration (Wilson et al., 1994) with their ablation generally detrimental to corneal wound healing (Mann et al., 1993; Yoshioka et al., 2010). Likewise, keratinocyte growth factor (KGF), hepatocyte growth factor (HGF), and insulin-like growth factor (IGF) also generally stimulate the proliferation and migration of corneal epithelial cells, with some exerting influence on the deposition of extracellular matrix and differentiation of fibroblasts (Ljubimov & Saghizadeh, 2015). Cytokines stored in epithelial cells, such as interleukin-1 (IL-1), mediate communication with the immune cells that infiltrate the usually immune-privileged cornea upon wounding; however, this inflammatory response, although necessary for healing, can also stimulate corneal scarring (Stapleton et al., 2008).

Wound healing is necessarily collaborative, and the effect of a signal is not merely a function of its identity or concentration but it also depends on the context in which it is applied. Transforming growth factor- β (TGF-β) accelerates corneal wound healing in rabbits, an effect that is amplified by co-administration with fibronectin (Er & Uzmez, 1998). However, in vitro, TGF-β can antagonize the aforementioned proliferative effects of growth factors such as EGF, KGF, HGF, and IGF (Honma et al., 1997; Mishima et al., 1992). Cooperativity is further observed between the neuropeptide substance P, which causes IGF to enhance wound healing when it would otherwise be inactive alone (Nakamura et al., 1997; Nishida et al., 1996). These representative processes underscore that understanding the wound healing process, and its dysfunction, is not simply a matter of understanding individual actors, but the context of their actions and the constraints of the aforementioned processes that they influence.

## 5 Diseases of the Corneal Epithelium and Associated Cellular Mechanisms

The following sections describe clinical conditions in which corneal epithelial healing is disrupted and highlight the complexity and importance of matrix proteins, corneal nerves, immune cells, and limbal stem cells in the corneal epithelial healing and regenerative processes. A common theme in clinical conditions is the close interplay between these various reparative and regenerative factors.

### *5.1 Epithelial Recurrent Erosions and Epithelial Basement Membrane Corneal Dystrophy (EBMCD)*

Epithelial recurrent erosions are a clinical condition in which corneal epithelial healing is disrupted, and patients experience repeated spontaneous corneal epithelial defects. They typically occur in patients with either a history of a prior corneal abrasion or a history of a corneal dystrophy, such as epithelial basement membrane corneal dystrophy (EBMCD). EBMCD has alternatively been named map-dot-fingerprint dystrophy and anterior basement membrane dystrophy based on clinical findings. Slit lamp findings of the cornea in EBMCD include "map"-like changes characterized as amorphous or geographic gray zones, "dots" that appear as small, irregular putty-like grayish white opacities, or "fingerprints" seen as small clusters of curved, parallel lines. Light microscopy of EBMCD shows that excessive basement membrane material intervening between distorted epithelium and the intact Bowman layer form redundant sheets that appear as maps and fingerprint lines or blebs that appear as dots (Weiss et al., 2015).

Patients with epithelial recurrent erosions often have a history of waking up with sharp eye pain associated with conjunctival redness, light sensitivity, blurred vision, and tearing. On examination, an epithelial defect, or recently healed epithelial defect, is usually noted depending on the size of the epithelial wound and the length of time that has passed from the start of symptoms. In patients with a healed epithelial defect, loose epithelium may still be noted. For some patients, the frequency of recurrent erosions can be as often as several episodes per month.

Although the mouse cornea lacks a thick Bowman's layer under its epithelial basement membrane (EBM), mouse models of corneal epithelial recurrent erosions have given insight into pathophysiologic mechanisms (Le et al., 2023; Pal-Ghosh et al., 2004, 2008, 2016). A mouse model of recurrent erosions that uses a dull blade to generate a single corneal debridement wound results in recurrent epithelial erosions in mice that are noted within 2 weeks of injury (Pal-Ghosh et al., 2004; Stepp et al., 2014). This recurrent erosion mouse model has demonstrated important cell-to-cell and cell-to-matrix disruptions that lead to recurrent epithelial erosions (Miller et al., 2019). It was found that high matrix metalloproteinase-9 (MMP-9) levels and excessive MMP-9 activity localize at recurrent erosion sites, and this leads to the abnormal cleavage of β integrins, which are integral membrane components of hemidesmosomes (Pal-Ghosh et al., 2011). Additionally, α3β1 integrins, which mediate corneal epithelial adhesion to the basement membrane through focal adhesions, have impaired relocalization to the basement membrane zone following injury in recurrent erosion corneas.(Pal-Ghosh et al., 2004).

This recurrent erosion mouse model has also demonstrated an abnormal inflammatory response to corneal epithelial injury. In C57BL/6 mice injured with corneal debridement, numerous Arg1+ CD206+ macrophages are associated with the occurrence of corneal erosions, both of which are features of M2 macrophages (Le et al., 2023). This finding suggests that tissue memory derived from prior injury helps to protect the cornea from rupturing under conditions when the cornea is subjected to recurrent epithelial defects (Le et al., 2023).

Treatment for recurrent corneal erosions include aggressive lubrication to keep the ocular surface moist, hypertonic saline to dehydrate the corneal epithelium causing it to better adhere to the underlying stroma, bandage contact lens to stabilize and protect the corneal epithelium, and anterior stromal micro-puncture to incite a fibrotic response between the epithelium and the underlying anterior stroma of the cornea to increase their adherence to each other (Miller et al., 2019). For recurrent erosions recalcitrant to these treatments, oral doxycycline has been used as supplemental therapy as an inhibitor of matrix metalloproteinase-9 (MMP-9), and for its anti-inflammatory effects (Wang et al., 2008).

## 5.2 Persistent Epithelial Defect Due to Neurotrophic Keratopathy

The cornea is one of the most densely innervated tissues in the body with a nerve fiber density of approximately 46 mm/mm$^2$ at the basal epithelium layer (Marfurt et al., 2010). Corneal sensory nerves originate from the ophthalmic division of the trigeminal nerve system (cranial nerve V), travel in the nasociliary nerve and its long ciliary nerve branches, and branch into nerve fibers that run parallel between the basal epithelium and Bowman's layer and form the subbasal nerve plexus (SBNP) in the cornea (Patel et al., 2021). The superficial nerve terminals (SNT) arising from the SBNP terminate as free nerve endings within the apical layers of the epithelium (Wu et al., 2022).

Corneal nerves protect the cornea from injury and maintain corneal homeostasis by regulating tear production, modulating blink reflexes, and releasing numerous neurotropic substances (Nishida, 2005). Substance P (SP) and calcitonin gene-related peptide (CGRP) are examples of neurotropic substances released by corneal sensory nerves (Nishida, 2005), and vasoactive intestinal polypeptide (VIP) is mostly produced from autonomic nerves (He & Bazan, 2016). These corneal neuropeptides have been demonstrated to promote corneal epithelial healing. A clinical trial demonstrated that eye drops containing SP and insulin-like growth factor (IGF)-derived peptides induce a rapid re-epithelialization of persistent epithelial defects (Yamada et al., 2008). CGRP has been shown to enhance epithelial wound healing in vitro (Mikulec & Tanelian, 1996) and regeneration of CGRP+ nerves expressing CGRP after corneal injury accelerated corneal epithelial healing (Cortina et al., 2012). Furthermore, corneal innervation has been shown to play a critical role in maintaining limbal stem cells and/or the stem cell niche. Mice with denervated corneas through the electrocoagulation of the ophthalmic branch of the trigeminal nerve demonstrated reduced numbers of corneolimbal stem/progenitor cells (Ueno et al., 2012).

Diabetes mellitus (DM) is a chronic, metabolic disease characterized by elevated levels of blood glucose and roughly 422 million people worldwide have diabetes. DM can damage several ocular tissues, including the cornea. Loss of corneal nerves associated with DM is called diabetic neurotrophic keratopathy and clinical manifestations include superficial punctate keratopathy and persistent epithelial defects associated with delayed wound healing. A rat DM model of corneal epithelial healing found many genes to be differentially expressed (Bettahi et al., 2014). Genome-wide cDNA microarray analysis revealed 1888 differentially expressed genes in the healing epithelia of normal versus DM rat corneas, and TGF-β signaling as a major altered pathway (Bettahi et al., 2014). A rat DM model also helped demonstrate epithelial structural changes that delay epithelial repair (Yu et al., 2022). Streptozocin (STZ)-induced rats had impaired reformation of cell–cell junctions of corneal epithelia following injury, as well as altered AKT (a serine/threonine kinase that is a major downstream target and effector protein of phosphatidylinositol 3-kinase (PI3K)) signaling (Yin et al., 2011). Additionally, the delayed epithelial wound healing in diabetic neurotrophic keratopathy has been attributed to corneal epithelial stem cell deficiency and the decreased expression of several putative stem cell markers (Saghizadeh et al., 2013; Ueno et al., 2014). Adenoviral gene therapy targeting corneal limbal stem cells successfully accelerated wound healing in human diabetic corneas (Kramerov et al., 2016).

Several growth factors have been shown to successfully promote corneal epithelial wound healing, including EGF, TGFβ, PDGF, CNTF, HGF, and NGF (Yu et al., 2022). The U.S. Food and Drug Administration (FDA) has approved the first drug, Oxervate (cenegermin), for the treatment of neurotrophic keratitis. Cenegermin is a recombinant human NGF (rhNGF) produced in *Escherichia coli* (Pflugfelder et al., 2020). A multicenter, randomized, double-masked, vehicle-controlled trial showed that cenegermin treatment resulted in higher rates of corneal healing compared with vehicle control in patients with neurotrophic keratopathy associated with nonhealing corneal defects (Pflugfelder et al., 2020). Patients with neurotrophic keratitis treated with at least 4 weeks of rhNGF showed a significant increase in corneal nerve densities after treatment by in vivo confocal microscopy (IVCM) (Balbuena-Pareja et al., 2023).

## *5.3 Limbal Stem Cell Deficiency (LSCD)*

Limbal epithelial stem/progenitor cells (LESCs) function to maintain corneal homeostasis and drive reparative processes (Bonnet et al., 2021; Di Girolamo et al., 2015; Tseng, 1989). Genetic causes of limbal stem cell deficiency include aniridia, Peter's anomaly, ectrodactyly-ectodermal-dysplasia-clefting syndrome, keratitis-ichthyosis-deafness (KID) syndrome, xeroderma pigmentosum, dominantly inherited keratitis, Turner syndrome, and dyskeratosis congenita (Moshirfar et al., 2023). However, the majority of cases of limbal stem cell deficiency are acquired and result from contact lens wear, ocular surgery, chemical/thermal burns, topical medication toxicity, pterygia, ocular infections, and ocular surface tumors (Moshirfar et al., 2023). For severe cases of LSCD, several surgical treatment options are available, and the approach taken depends on how many clock hours of limbal tissue are diseased and whether the LSCD is unilateral or bilateral. Direct limbal stem cell transplantation was the initial surgical therapeutic approach and is performed using either autologous or allogeneic donor tissue. Cultivated limbal stem cell transplantation using ex vivo cultivated limbal epithelial cells for transplantation was then developed to reduce the size and amount of the donor limbal tissue. The cell sources of LSCs have expanded to include induced pluripotent stem cells (iPSCs) derived from the patient's own somatic cells, cultivated autologous oral mucosal epithelium, and mesenchymal stem cell (MSCs) (Genna et al., 2025).

Several signaling pathways are involved in regulating limbal epithelial stem cells. The Notch signaling pathway is a regulator of self-renewal and differentiation in several tissues and cell types. Notch receptors N1, N2, and target ligands HES1 and HEY1 are expressed in the human corneal epithelium and help regulate the proliferation and differentiation abilities of LESCs (Gonzalez et al., 2019; Ma et al., 2007). In adult tissue, Notch activation may occur sporadically when regeneration of the corneal surface is needed during normal homeostasis or wound healing (Gonzalez et al., 2019). The Wnt/β-catenin signaling pathway comprises a family of proteins that play critical roles in embryonic development and adult tissue homeostasis. LESCs express multiple Wnt ligands, and different Wnt ligands can cooperate and synergize for stem cell proliferation (Alok et al., 2017). This knowledge about the signaling pathways can be exploited to assist with corneal epithelial regeneration. For example, a small-molecule Wnt mimic was shown to promote the proliferation and repopulation of the residual LSCs in partial LSCD through Wnt activation (Bonnet et al., 2023).

## 6 Conclusions

The corneal epithelium functions as a main refractive component of the eye, as well as a major protective barrier to environmental and external insults. Here, we have reviewed various interactions between multiple cellular and molecular factors, including stem cells, growth factors, matrix proteins, immune cells, cytokines, and corneal nerves, which occur to maintain the cornea in its homeostatic state. Following injury, these cellular processes are accelerated to facilitate rapid wound closure, and additional processes are activated to regenerate the corneal layers and restore the cornea back to its original state. Corneal epithelial disease can occur when there is a major loss or dysfunction of some of these key factors, including basement membrane proteins, corneal nerves, and limbal stem cells. By targeting a specific regenerative signaling pathway, a medication that promotes corneal repair has been successfully developed. As our understanding of the cellular and regenerative processes of corneal epithelial repair continues to expand, more promising therapeutic targets to treat corneal epithelial disease are likely to be uncovered.

**Financial Support** This study was supported by funds from the National Institutes of Health (R01EY032161 and R21EY035294 to MFC, R21EY035829 to MFC and FDW, and NIH-NEI P30 EY002162—Core Grant for Vision Research), Larry L. Hillblom Foundation (to FDW), Research to Prevent Blindness (an unrestricted grant to the UCSF Department of Ophthalmology), and All May See Foundation.

## References

Alok, A., Lei, Z., Jagannathan, N. S., Kaur, S., Harmston, N., Rozen, S. G., Tucker-Kellogg, L., & Virshup, D. M. (2017). Wnt proteins synergize to activate beta-catenin signaling. *Journal of Cell Science, 130*(9), 1532–1544. https://doi.org/10.1242/jcs.198093

Balbuena-Pareja, A., Bogen, C. S., Cox, S. M., & Hamrah, P. (2023). Effect of recombinant human nerve growth factor treatment on corneal nerve regeneration in patients with neurotrophic keratopathy. *Frontiers in Neuroscience, 17*, 1210179. https://doi.org/10.3389/fnins.2023.1210179

Bettahi, I., Sun, H., Gao, N., Wang, F., Mi, X., Chen, W., Liu, Z., & Yu, F. S. (2014). Genome-wide transcriptional analysis of differentially expressed genes in diabetic, healing corneal epithelial cells: Hyperglycemia-suppressed TGFbeta3 expression contributes to the delay of epithelial wound healing in diabetic corneas. *Diabetes, 63*(2), 715–727. https://doi.org/10.2337/db13-1260

Bonnet, C., Gonzalez, S., Roberts, J. S., Robertson, S. Y. T., Ruiz, M., Zheng, J., & Deng, S. X. (2021). Human limbal epithelial stem cell regulation, bioengineering and function. *Progress in Retinal and Eye Research, 85*, 100956. https://doi.org/10.1016/j.preteyeres.2021.100956

Bonnet, C., Chauhan, T., Encampira Luna, E., Le, Q., Tseng, C. H., & Deng, S. X. (2022). Cell morphology as an in vivo parameter for the diagnosis of limbal stem cell deficiency. *Cornea, 41*(8), 995–1001. https://doi.org/10.1097/ICO.0000000000002955

Bonnet, C., Gonzalez, S., Deng, S. X., & Zheng, J. J. (2023). Wnt activation as a potential therapeutic approach to treat partial limbal stem cell deficiency. *Scientific Reports, 13*(1), 15670. https://doi.org/10.1038/s41598-023-42794-8

Cortina, M. S., He, J., Li, N., Bazan, N. G., & Bazan, H. E. (2012). Recovery of corneal sensitivity, calcitonin gene-related peptide-positive nerves, and increased wound healing induced by pigment epithelial-derived factor plus docosahexaenoic acid after experimental surgery. *Archives of Ophthalmology, 130*(1), 76–83. https://doi.org/10.1001/archophthalmol.2011.287

Crosson, C. E., Klyce, S. D., & Beuerman, R. W. (1986). Epithelial wound closure in the rabbit cornea. A biphasic process. *Investigative Ophthalmology & Visual Science, 27*(4), 464–473. https://www.ncbi.nlm.nih.gov/pubmed/3957565

Di Girolamo, N., & Park, M. (2023). Cell identity changes in ocular surface epithelia. *Progress in Retinal and Eye Research, 95*, 101148. https://doi.org/10.1016/j.preteyeres.2022.101148

Di Girolamo, N., Bobba, S., Raviraj, V., Delic, N. C., Slapetova, I., Nicovich, P. R., Halliday, G. M., Wakefield, D., Whan, R., & Lyons, J. G. (2015). Tracing the fate of limbal epithelial progenitor cells in the murine cornea. *Stem Cells, 33*(1), 157–169. https://doi.org/10.1002/stem.1769

Eghrari, A. O., Riazuddin, S. A., & Gottsch, J. D. (2015). Overview of the cornea: Structure, function, and development. *Progress in Molecular Biology and Translational Science, 134*, 7–23. https://doi.org/10.1016/bs.pmbts.2015.04.001

Er, H., & Uzmez, E. (1998). Effects of transforming growth factor-beta 2, interleukin 6 and fibronectin on corneal epithelial wound healing. *European Journal of Ophthalmology, 8*(4), 224–229. https://doi.org/10.1177/112067219800800404

Fujikawa, L. S., Foster, C. S., Gipson, I. K., & Colvin, R. B. (1984). Basement membrane components in healing rabbit corneal epithelial wounds: Immunofluorescence and ultrastructural studies. *The Journal of Cell Biology, 98*(1), 128–138. https://doi.org/10.1083/jcb.98.1.128

Genna, V. G., Maurizi, E., Rama, P., & Pellegrini, G. (2025). Biology and medicine on ocular surface restoration: Advancements and limits of limbal stem cell deficiency treatments. *The Ocular Surface, 35*, 57–67. https://doi.org/10.1016/j.jtos.2024.11.005

Gonzalez, S., Uhm, H., & Deng, S. X. (2019). Notch inhibition prevents differentiation of human limbal stem/progenitor cells in vitro. *Scientific Reports, 9*(1), 10373. https://doi.org/10.1038/s41598-019-46793-6

Hanna, C. (1966). Proliferation and migration of epithelial cells during corneal wound repair in the rabbit and the rat. *American Journal of Ophthalmology, 61*(1), 55–63. https://www.ncbi.nlm.nih.gov/pubmed/5904378

Hassell, J. R., & Birk, D. E. (2010). The molecular basis of corneal transparency [research support, N.I.H., Extramural review]. *Experimental Eye Research, 91*(3), 326–335. https://doi.org/10.1016/j.exer.2010.06.021

He, J., & Bazan, H. E. (2016). Neuroanatomy and neurochemistry of mouse cornea. *Investigative Ophthalmology & Visual Science, 57*(2), 664–674. https://doi.org/10.1167/iovs.15-18019

Honma, Y., Nishida, K., Sotozono, C., & Kinoshita, S. (1997). Effect of transforming growth factor-beta1 and -beta2 on in vitro rabbit corneal epithelial cell proliferation promoted by epidermal growth factor, keratinocyte growth factor, or hepatocyte growth factor. *Experimental Eye Research, 65*(3), 391–396. https://doi.org/10.1006/exer.1997.0338

Kramerov, A. A., Saghizadeh, M., & Ljubimov, A. V. (2016). Adenoviral gene therapy for diabetic keratopathy: Effects on wound healing and stem cell marker expression in human organ-cultured corneas and Limbal epithelial cells. *JoVE: Journal of Visualized Experiments, 110*, e54058. https://doi.org/10.3791/54058

Krawczyk, W. S. (1971). A pattern of epidermal cell migration during wound healing. *The Journal of Cell Biology, 49*(2), 247–263. https://doi.org/10.1083/jcb.49.2.247

Le, P. M., Pal-Ghosh, S., Menko, A. S., & Stepp, M. A. (2023). Immune cells localize to sites of corneal erosions in C57BL/6 mice. *Biomolecules, 13*(7). https://doi.org/10.3390/biom13071059

Ljubimov, A. V., & Saghizadeh, M. (2015). Progress in corneal wound healing. *Progress in Retinal and Eye Research, 49*, 17–45. https://doi.org/10.1016/j.preteyeres.2015.07.002

Ma, A., Boulton, M., Zhao, B., Connon, C., Cai, J., & Albon, J. (2007). A role for notch signaling in human corneal epithelial cell differentiation and proliferation. *Investigative Ophthalmology & Visual Science, 48*(8), 3576–3585. https://doi.org/10.1167/iovs.06-1373

Majo, F., Rochat, A., Nicolas, M., Jaoude, G. A., & Barrandon, Y. (2008). Oligopotent stem cells are distributed throughout the mammalian ocular surface. *Nature, 456*(7219), 250–254. https://doi.org/10.1038/nature07406

Mann, G. B., Fowler, K. J., Gabriel, A., Nice, E. C., Williams, R. L., & Dunn, A. R. (1993). Mice with a null mutation of the TGF alpha gene have abnormal skin architecture, wavy hair, and curly whiskers and often develop corneal inflammation. *Cell, 73*(2), 249–261. https://doi.org/10.1016/0092-8674(93)90227-h

Marfurt, C. F., Cox, J., Deek, S., & Dvorscak, L. (2010). Anatomy of the human corneal innervation. *Experimental Eye Research, 90*(4), 478–492. https://doi.org/10.1016/j.exer.2009.12.010

McCartney, M. D., & Cantu-Crouch, D. (1992). Rabbit corneal epithelial wound repair: Tight junction reformation. *Current Eye Research, 11*(1), 15–24. https://doi.org/10.3109/02713689209069163

Mikulec, A. A., & Tanelian, D. L. (1996). CGRP increases the rate of corneal re-epithelialization in an in vitro whole mount preparation. *Journal of Ocular Pharmacology and Therapeutics, 12*(4), 417–423. https://doi.org/10.1089/jop.1996.12.417

Miller, D. D., Hasan, S. A., Simmons, N. L., & Stewart, M. W. (2019). Recurrent corneal erosion: A comprehensive review. *Clinical Ophthalmology, 13*, 325–335. https://doi.org/10.2147/OPTH.S157430

Mishima, H., Nakamura, M., Murakami, J., Nishida, T., & Otori, T. (1992). Transforming growth factor-beta modulates effects of epidermal growth factor on corneal epithelial cells. *Current Eye Research, 11*(7), 691–696. https://doi.org/10.3109/02713689209000742

Morimoto, K., Mishima, H., Nishida, T., & Otori, T. (1993). Role of urokinase type plasminogen activator (u-PA) in corneal epithelial migration. *Thrombosis and Haemostasis, 69*(4), 387–391. https://www.ncbi.nlm.nih.gov/pubmed/8497852

Moshirfar, M., Masud, M., Harvey, D. H., Payne, C., Bruce, E., Ronquillo, Y. C., & Hoopes, P. C. (2023). The multifold etiologies of Limbal stem cell deficiency: A comprehensive review on the etiologies and additional treatment options for Limbal stem cell deficiency. *Journal of Clinical Medicine, 12*(13). https://doi.org/10.3390/jcm12134418

Murakami, J., Nishida, T., & Otori, T. (1992). Coordinated appearance of beta 1 integrins and fibronectin during corneal wound healing. *The Journal of Laboratory and Clinical Medicine, 120*(1), 86–93. https://www.ncbi.nlm.nih.gov/pubmed/1377217

Nakamura, M., Ofuji, K., Chikama, T., & Nishida, T. (1997). Combined effects of substance P and insulin-like growth factor-1 on corneal epithelial wound closure of rabbit in vivo. *Current Eye Research, 16*(3), 275–278. https://doi.org/10.1076/ceyr.16.3.275.15409

Nishida, T. (2005). Neurotrophic mediators and corneal wound healing. *The Ocular Surface, 3*(4), 194–202. https://doi.org/10.1016/s1542-0124(12)70206-9

Nishida, T., Nakamura, M., Ofuji, K., Reid, T. W., Mannis, M. J., & Murphy, C. J. (1996). Synergistic effects of substance P with insulin-like growth factor-1 on epithelial migration

of the cornea. *Journal of Cellular Physiology, 169*(1), 159–166. https://doi.org/10.1002/(SICI)1097-4652(199610)169:1<159::AID-JCP16>3.0.CO;2-8
Pal-Ghosh, S., Pajoohesh-Ganji, A., Brown, M., & Stepp, M. A. (2004). A mouse model for the study of recurrent corneal epithelial erosions: alpha9beta1 integrin implicated in progression of the disease. *Investigative Ophthalmology & Visual Science, 45*(6), 1775–1788. https://doi.org/10.1167/iovs.03-1194
Pal-Ghosh, S., Tadvalkar, G., Jurjus, R. A., Zieske, J. D., & Stepp, M. A. (2008). BALB/c and C57BL6 mouse strains vary in their ability to heal corneal epithelial debridement wounds. *Experimental Eye Research, 87*(5), 478–486. https://doi.org/10.1016/j.exer.2008.08.013
Pal-Ghosh, S., Blanco, T., Tadvalkar, G., Pajoohesh-Ganji, A., Parthasarathy, A., Zieske, J. D., & Stepp, M. A. (2011). MMP9 cleavage of the beta4 integrin ectodomain leads to recurrent epithelial erosions in mice. *Journal of Cell Science, 124*(Pt 15), 2666–2675. https://doi.org/10.1242/jcs.085480
Pal-Ghosh, S., Pajoohesh-Ganji, A., Tadvalkar, G., Kyne, B. M., Guo, X., Zieske, J. D., & Stepp, M. A. (2016). Topical Mitomycin-C enhances subbasal nerve regeneration and reduces erosion frequency in the debridement wounded mouse cornea. *Experimental Eye Research, 146*, 361–369. https://doi.org/10.1016/j.exer.2015.08.023
Park, M., Richardson, A., Pandzic, E., Lobo, E. P., Whan, R., Watson, S. L., Lyons, J. G., Wakefield, D., & Di Girolamo, N. (2019). Visualizing the contribution of keratin-14(+) limbal epithelial precursors in corneal wound healing. *Stem Cell Reports, 12*(1), 14–28. https://doi.org/10.1016/j.stemcr.2018.11.014
Patel, S., Hwang, J., Mehra, D., & Galor, A. (2021). Corneal nerve abnormalities in ocular and systemic diseases. *Experimental Eye Research, 202*, 108284. https://doi.org/10.1016/j.exer.2020.108284
Pflugfelder, S. C., Massaro-Giordano, M., Perez, V. L., Hamrah, P., Deng, S. X., Espandar, L., Foster, C. S., Affeldt, J., Seedor, J. A., Afshari, N. A., Chao, W., Allegretti, M., Mantelli, F., & Dana, R. (2020). Topical recombinant human nerve growth factor (Cenegermin) for neurotrophic keratopathy: A multicenter randomized vehicle-controlled pivotal trial. *Ophthalmology, 127*(1), 14–26. https://doi.org/10.1016/j.ophtha.2019.08.020
Saghizadeh, M., Epifantseva, I., Hemmati, D. M., Ghiam, C. A., Brunken, W. J., & Ljubimov, A. V. (2013). Enhanced wound healing, kinase and stem cell marker expression in diabetic organ-cultured human corneas upon MMP-10 and cathepsin F gene silencing. *Investigative Ophthalmology & Visual Science, 54*(13), 8172–8180. https://doi.org/10.1167/iovs.13-13233
Sridhar, M. S. (2018). Anatomy of cornea and ocular surface. *Indian Journal of Ophthalmology, 66*(2), 190–194. https://doi.org/10.4103/ijo.IJO_646_17
Stapleton, W. M., Chaurasia, S. S., Medeiros, F. W., Mohan, R. R., Sinha, S., & Wilson, S. E. (2008). Topical interleukin-1 receptor antagonist inhibits inflammatory cell infiltration into the cornea. *Experimental Eye Research, 86*(5), 753–757. https://doi.org/10.1016/j.exer.2008.02.001
Stepp, M. A., Zieske, J. D., Trinkaus-Randall, V., Kyne, B. M., Pal-Ghosh, S., Tadvalkar, G., & Pajoohesh-Ganji, A. (2014). Wounding the cornea to learn how it heals. *Experimental Eye Research, 121*, 178–193. https://doi.org/10.1016/j.exer.2014.02.007
Suda, T., Nishida, T., Ohashi, Y., Nakagawa, S., & Manabe, R. (1981). Fibronectin appears at the site of corneal stromal wound in rabbits. *Current Eye Research, 1*(9), 553–556. https://doi.org/10.3109/02713688109069181
Suzuki, K., Tanaka, T., Enoki, M., & Nishida, T. (2000). Coordinated reassembly of the basement membrane and junctional proteins during corneal epithelial wound healing. *Investigative Ophthalmology & Visual Science, 41*(9), 2495–2500. https://www.ncbi.nlm.nih.gov/pubmed/10937559
Thoft, R. A., & Friend, J. (1983). The X, Y, Z hypothesis of corneal epithelial maintenance. *Investigative Ophthalmology & Visual Science, 24*(10), 1442–1443. https://www.ncbi.nlm.nih.gov/pubmed/6618809
Tiwari, A., Swamynathan, S., Campbell, G., Jhanji, V., & Swamynathan, S. K. (2020). BMP6 regulates corneal epithelial cell stratification by coordinating their proliferation and differentiation

and is upregulated in Pterygium. *Investigative Ophthalmology & Visual Science, 61*(10), 46. https://doi.org/10.1167/iovs.61.10.46

Tseng, S. C. (1989). Concept and application of limbal stem cells. *Eye (London, England), 3*(Pt 2), 141–157. https://doi.org/10.1038/eye.1989.22

Ueno, H., Ferrari, G., Hattori, T., Saban, D. R., Katikireddy, K. R., Chauhan, S. K., & Dana, R. (2012). Dependence of corneal stem/progenitor cells on ocular surface innervation. *Investigative Ophthalmology & Visual Science, 53*(2), 867–872. https://doi.org/10.1167/iovs.11-8438

Ueno, H., Hattori, T., Kumagai, Y., Suzuki, N., Ueno, S., & Takagi, H. (2014). Alterations in the corneal nerve and stem/progenitor cells in diabetes: Preventive effects of insulin-like growth factor-1 treatment. *International Journal of Endocrinology, 2014*, 312401. https://doi.org/10.1155/2014/312401

Walker, H., Taiyab, A., Deschamps, P., Williams, T., & West-Mays, J. A. (2021). Conditional deletion of AP-2beta in the periocular mesenchyme of mice alters corneal epithelial cell fate and stratification. *International Journal of Molecular Sciences, 22*(16). https://doi.org/10.3390/ijms22168730

Wang, L., Tsang, H., & Coroneo, M. (2008). Treatment of recurrent corneal erosion syndrome using the combination of oral doxycycline and topical corticosteroid. *Clinical & Experimental Ophthalmology, 36*(1), 8–12. https://doi.org/10.1111/j.1442-9071.2007.01648.x

Weiss, J. S., Moller, H. U., Aldave, A. J., Seitz, B., Bredrup, C., Kivela, T., Munier, F. L., Rapuano, C. J., Nischal, K. K., Kim, E. K., Sutphin, J., Busin, M., Labbe, A., Kenyon, K. R., Kinoshita, S., & Lisch, W. (2015). IC3D classification of corneal dystrophies–edition 2. *Cornea, 34*(2), 117–159. https://doi.org/10.1097/ICO.0000000000000307

Wilson, S. E., He, Y. G., Weng, J., Zieske, J. D., Jester, J. V., & Schultz, G. S. (1994). Effect of epidermal growth factor, hepatocyte growth factor, and keratinocyte growth factor, on proliferation, motility and differentiation of human corneal epithelial cells. *Experimental Eye Research, 59*(6), 665–678. https://doi.org/10.1006/exer.1994.1152

Wu, M., Hill, L. J., Downie, L. E., & Chinnery, H. R. (2022). Neuroimmune crosstalk in the cornea: The role of immune cells in corneal nerve maintenance during homeostasis and inflammation. *Progress in Retinal and Eye Research, 91*, 101105. https://doi.org/10.1016/j.preteyeres.2022.101105

Yamada, N., Matsuda, R., Morishige, N., Yanai, R., Chikama, T. I., Nishida, T., Ishimitsu, T., & Kamiya, A. (2008). Open clinical study of eye-drops containing tetrapeptides derived from substance P and insulin-like growth factor-1 for treatment of persistent corneal epithelial defects associated with neurotrophic keratopathy. *The British Journal of Ophthalmology, 92*(7), 896–900. https://doi.org/10.1136/bjo.2007.130013

Yin, J., Huang, J., Chen, C., Gao, N., Wang, F., & Yu, F. S. (2011). Corneal complications in streptozocin-induced type I diabetic rats. *Investigative Ophthalmology & Visual Science, 52*(9), 6589–6596. https://doi.org/10.1167/iovs.11-7709

Yoshioka, R., Shiraishi, A., Kobayashi, T., Morita, S., Hayashi, Y., Higashiyama, S., & Ohashi, Y. (2010). Corneal epithelial wound healing impaired in keratinocyte-specific HB-EGF-deficient mice in vivo and in vitro. *Investigative Ophthalmology & Visual Science, 51*(11), 5630–5639. https://doi.org/10.1167/iovs.10-5158

Yu, F. X., Lee, P. S. Y., Yang, L., Gao, N., Zhang, Y., Ljubimov, A. V., Yang, E., Zhou, Q., & Xie, L. (2022). The impact of sensory neuropathy and inflammation on epithelial wound healing in diabetic corneas. *Progress in Retinal and Eye Research, 89*, 101039. https://doi.org/10.1016/j.preteyeres.2021.101039

# Corneal Epithelial Wound Healing: Cellular Mechanisms and Therapeutic Approaches

**Joshua Ong, Chanon Thanitcul, Navid Fotovat-Ahmadi, Christian Reinhardt, and Vinay Kumar Aakalu**

**Abstract** The corneal epithelium is a unique structure that protects and covers the anterior portion of the cornea. Although only approximately 50–60 μm in thickness, maintenance of corneal epithelial health is of utmost importance for corneal clarity and the visual pathway. The corneal epithelium is also one of the unique structures in the human body that is constantly being renewed by limbal epithelial stem cells. These mechanisms also play a role in corneal epithelial wound healing. In this chapter, we review the anatomy of the cornea, particularly the cornea epithelium and its sublayers. We discuss the mechanism of corneal epithelial wound healing and its phases which are characterized by the lag phase, migration phase, proliferation phase, and adhesion phase. We then discuss factors that affect corneal epithelial wound healing including genetic and molecular factors, local factors (e.g., hypoxia and inflammation), systemic factors (e.g., aging, diabetes), and environmental factors (e.g., ultraviolet light exposure). We then conclude by reviewing therapeutic interventions to optimize corneal epithelial healing including pharmacologic, non-surgical, and surgical approaches.

**Keywords** Corneal epithelium · Wound healing · Ocular surface · Limbal stem cells · Growth factors

## 1 Introduction

The corneal epithelium is the most external layer of the cornea, and it can sustain various mechanical, chemical, and pathological damage. A process of corneal epithelial wound healing ensues once the epithelium is disrupted, and this process is a

J. Ong · C. Thanitcul · N. Fotovat-Ahmadi · C. Reinhardt · V. K. Aakalu (✉)
Department of Ophthalmology and Visual Sciences, Kellogg Eye Center, University of Michigan, Ann Arbor, MI, USA
e-mail: ongjo@med.umich.edu; thanitcu@med.umich.edu; nfotovat@med.umich.edu; caureinh@umich.edu; vaakalu@med.umich.edu

 
H. Das (ed.), *Tissue Repair and Regeneration*,
https://doi.org/10.1007/978-3-031-93677-7_9

complex interplay involving cell migration, proliferation, and differentiation (Ljubimov & Saghizadeh, 2015). Various growth factors, cytokines, and extracellular matrix (ECM) remodeling occur in this process (Yu et al., 2010). When the cornea is damaged, it may involve other corneal layers, including the stroma and the endothelium. Though the wound healing process also occurs in the stroma and the endothelium, the process of epithelial wound healing is distinct because it heals primarily from movement and differentiation of limbal stem cells and does not undergo cell transformation to different cell types, unlike wound healing processes in the stroma and the endothelium (Ljubimov & Saghizadeh, 2015).

While the cornea plays a crucial role in eyesight, it lies vulnerable to the external environment, leaving itself particularly susceptible to injury (Ljubimov & Saghizadeh, 2015). Every year, it is estimated that 107,000 people are blinded by chemical injury, and various other injuries to the cornea often can occur from infectious or traumatic processes (Allen et al., 2024). Corneal epithelial wound healing is critical in maintaining the structural integrity of the cornea. The goal of many treatments is to allow the epithelium's self-regenerative mechanisms to proliferate without disruption. Disruption of this healing can result in persistent corneal epithelial defects (PCED), where the healing of corneal surface defects is disrupted. There are an estimated 73,434–99,465 PCED cases per year in the United States. PCEDs can take weeks or months to heal but are susceptible to reoccurrence. PCED can lead to corneal ulceration, perforation, and loss of vision (Wirostko et al., 2015). Ultimately, corneal epithelial wound healing is essential to the proper functioning of the visual system.

The corneal epithelium undergoes four phases in wound healing: lag, migration, proliferation, and adhesion phase. Briefly, in the lag phase, cells that are damaged by the insult undergo apoptosis (Estil et al., 2002). This phase also involves cellular reorganization to promote migration of the epithelial cells at the wound edge (Kuwabara et al., 1976). During the migration phase, cells at the wound margin spread into a single layer and utilize fibronectin and cytoskeletal contractile mechanisms to move centrally (Ashby et al., n.d.). In the proliferation phase, there is a marked increase in proliferation at the limbus and the peripheral cornea, and the daughter cells spread centrally and superficially to restore epithelial cell density (Ashby et al., n.d.). Lastly, during the attachment phase, a reattachment of the newly regenerated epithelium to the underlying stroma is made via hemidesmosomes (Ashby et al., n.d.). These phases will be discussed more in-depth throughout the chapter.

In this chapter, we provide an overview of the anatomy and structure of the corneal epithelium. We then discuss the phases of corneal epithelium wound healing and factors that affect this process. We then review the pharmacologic, nonsurgical, and surgical approaches that have been developed to improve the process of corneal epithelium wound healing.

## 2 Overview of Corneal Structures

As the first structure light encounters, the cornea, along with its tear film, plays a crucial role in bending light to ensure it is focused correctly on the retina. This role wouldn't be possible without the cornea's remarkable transparency. In the following section, we will discuss the layers of the cornea and how they contribute to its function.

### 2.1 *Tear Film*

The tear film, the outermost layer of the eye, is essential for maintaining a healthy and functional cornea. It provides hydration and lubrication, supports oxygen delivery, and contains protective substances such as immunoglobulins, lysozymes, lactoferrin, and α and β defensins. It also ensures a smooth interface for light to pass through (Sridhar, 2018).

This film is made up of three interconnected layers. The outermost layer, consisting of oily secretions formed by meibomian gland secretions, prevents evaporation and creates a smooth surface for light refraction. Beneath this lies the aqueous layer, produced mainly by the lacrimal gland located in the upper outer corner of the eye. The aqueous solution is released through ducts near the upper eyelid and plays a critical role in hydration, as well as fighting off bacteria and safeguarding the cornea. Mucins from conjunctival goblet cells and epithelial cells help to anchor the tear film to the eye and support debris removal and immune defense (Sridhar, 2018; American Academy of Ophthalmology, 2023). Beneath the tear film lies the first anatomical layer of the cornea, the epithelium.

### 2.2 *Corneal Epithelium*

The corneal epithelium is the delicate, 50–60 μm thick outermost layer of the cornea. It is composed of stratified squamous nonkeratinized epithelial cells. As the top layer of the cornea, it serves the function of protecting the eye, and it does so with its superficial layer of cells that form a protective barrier through tight junctions. The apical surface features many microvilli and microplates that attach and stabilize the tear film. It has many desmosomes that hold adjacent cells together and hemidesmosomes anchor the basal epithelial cells to the basal lamina below. The epithelium sits on a basement membrane which hooks it to the bowman's layer with the aid of anchoring fibrils (type VII collagen) and anchoring plaques (type VI collagen) (ClinicalKey, n.d.).

## 2.3 Bowman's Layer

Bowman's layer is a thin (8–12 μm), acellular structure composed of randomly arranged collagen fibrils (primarily type I and V) as well as proteoglycans (Sridhar, 2018). Its posterior boundary merges with the collagen lamellae of the stroma, while its anterior surfaces support the epithelial basement membrane (Wilson, 2020). Its main function is to maintain corneal shape (Sridhar, 2018).

## 2.4 Stroma

The stroma, the thickest layer of the cornea, constitutes 80–85% of its overall thickness. This layer is essential for both structural integrity and transparency of the cornea. Its transparency allows for light transmission. It can remain transparent from the arrangement of thin collagen lamellae, stacked at right angles to adjacent layers. The structure is regulated by glycosaminoglycans (GAGs) such as keratan sulfate. Keratocytes, the predominant cells in the stroma, are spread between these lamellae, playing an important role in maintaining the extracellular matrix by producing collagen, GAGs, and enzymes. The stroma's central region, particularly near the pupil, features densely packed lamellae for added strength, ensuring both optical clarity and mechanical stability (Sridhar, 2018). Beyond the stroma, the next layer contributing to structural and functional integrity is Descemet's membrane.

## 2.5 Descemet's Membrane

Descemet's membrane serves as the basement membrane for the endothelium of the cornea. It is made up of two main parts—an anterior banded region that contains type VIII collagen and a posterior nonbanded region made up of type IV collagen. On its periphery, it transitions to the trabecular meshwork, which is important for draining the aqueous humor in the eye. The trabecular meshwork has thickened edges that mark its anterior limit called the Schwalbe lines (ClinicalKey, n.d.).

## 2.6 Corneal Endothelium

The corneal endothelium plays a critical role in maintaining corneal hydration and transparency, compensating for fluid loss through evaporation. This simple squamous epithelial layer lies on the posterior surface of the cornea. Its polygonal cells are highly specialized, actively transporting fluid from the stroma into the aqueous humor to preserve clarity (ClinicalKey, n.d.).

## 3 Anatomy and Stratified Structures of the Corneal Epithelium

The corneal epithelium consists of three sublayers of cells, including superficial cells (outer two three cell layers), wing cells (middle two to three layers), and basal cells (inner one layer) (Paley et al., 2022). Basal cells have the capability to proliferate and migrate superficially as well as centrally, and they can also differentiate into superficial cells and wing cells (Paley et al., 2022). The proliferative process happens most actively at the limbus, where the immortal progenitor stem cells reside (Paley et al., 2022).

### 3.1 *Basal Layer*

The basal layer of the corneal epithelium consists of a single layer of columnar cells anchored to the basement membrane through hemidesmosomes. These junctions along with anchoring filaments secure the basal cells to the basement membrane (ClinicalKey, n.d.). They are also connected across their lateral membranes with desmosomes (Sridhar, 2018).

Basal cells act as the stem cells of the epithelium and are the only layer capable of mitosis. As new cells are produced, older basal cells are gradually pushed upward. During this migration, they first change into wing cells, then differentiate further into superficial cells. At the surface, these cells lose their attachments and are shed into the tear film, a natural process of epithelial renewal (eOphtha, n.d.). The process of cell movement towards the surface epithelium is called the XYZ hypothesis (Sridhar, 2018).

### 3.2 *Wing Cells*

The wing cells form the middle layer of the corneal epithelium and are arranged in two to three layers. They are named for their characteristic wing-like shape, with elongated processes extending outward from their central body. These cells represent an intermediate stage between the basal and superficial layers (Sridhar, 2018; eOphtha, n.d.). They are connected together by desmosomes. This allows for cohesion as the cells migrate towards superficial cells (Sridhar, 2018).

### *3.3 Superficial Cells*

The superficial layer of cells makes up the top two to three layers of the corneal epithelium. It consists of flat cells that are multisided. Each cell has a diameter of about 40–60 μ along with a thickness of 2–6 μ. These cells have many microvilli and microplicae on their superficial surface, which serve to increase the surface area of the epithelium (Sridhar, 2018) and bind stronger to the tear film. This strong binding of tear film is achieved with the help of glycolax, a glue-like substance that binds the tear film (eOphtha, n.d.). The superficial cells are held together by desmosomes as well as tight junctions, which are only found in the superficial layer. They create a seal that acts as a protective barrier for the cornea, regulating the movement of substances between cells (Sridhar, 2018).

## 4 Limbal Stem Cells

The limbus is the transitional zone at the point where the cornea meets with the whiteness of the sclera. It serves several important functions including nourishment of the peripheral cornea, detecting and regulating immune responses, and contributing to corneal wound healing (ClinicalKey, n.d.). This last function is facilitated by the basal layer of the limbus, which is the storage site of the limbal stem cells (LSCs) (Ebrahimi et al., 2009). More specifically they are located in the Palisades of Vogt. These ridge-like structures provide a protective environment to the limbal stem cells and also supply them with nutrients from blood vessels (Alio et al., 2019).

## 5 Limbal Stem Cells in Corneal Epithelial Wound Healing

When corneal injury is detected, LSCs are activated, resulting in the activation of transient-amplifying cells (TACs). TACs serve to amplify stem cell division through their own division and differentiation (Saghizadeh et al., 2017; Ebrahimi et al., 2009). The activation of this LSC-TAC interplay is not fully understood, but it has been shown that certain growth factors, cytokines, and chemotactic molecules play a role in this activation. This includes keratinocyte growth factor (KGF), ciliary neurotrophic factor (CNTF), and the c-met receptor (Ljubimov & Saghizadeh, 2015). While the LSCs remain in the basal layer of the corneal limbal region, the TACs themselves migrate toward the wounded epithelia (Notara et al., 2018). The TACs, then, differentiate into suprabasal postmitotic cells, which then differentiate into terminally differentiated cells (TDCs) (Ebrahimi et al., 2009). The healing takes place through four main phases: lag, migration, proliferation, and adhesion (Estil et al., 2002). These phases are greatly detailed in the next section.

# 6 Limbal Stem Cell Deficiency

Limbal stem cell deficiency (LSCD) is a condition in which there is an insufficient or total loss of limbal stem cells (LSCs). Damage to limbal stem cells can lead to LSCD. This can result from genetic dysfunction in the LSC microenvironment (primary) or direct damage to the LSCs (secondary) (Ebrahimi et al., 2009; Le et al., 2018). LSCD causes the limbus/LSC barrier between the corneal epithelium and conjunctival epithelium to be disrupted. This results in the abnormal regeneration of the corneal epithelium, where the conjunctival epithelial cells centripetally migrate and replace the corneal epithelial cells. The fibrovascular conjunctiva causes neovascularization as well as a loss in transparency (conjunctivalization), leading to corneal opacity and vision loss (Saghizadeh et al., 2017; Ebrahimi et al., 2009; Le et al., 2018; Di Girolamo, 2015; Ramos et al., 2015). LSCD includes direct damage to either the LSC microenvironment, the LSCs themselves or both. This is commonly caused by chemical/thermal injuries, chronic inflammation, contact-lens-induced-keratopathy, Stevens-Johnson syndrome, graft-versus-host disease, and severe microbial infection (Ebrahimi et al., 2009; Le et al., 2018). There are also various congenital conditions that can cause LSCD including aniridia, xeroderma pigmentosum, and keratitis associated with various systemic diseases. The clinical characteristics of these diseases that can cause LSCD are diverse; covering each disease is outside the scope of this specific chapter but many of these congenital or pathologic causes are discussed in the literature (Saghizadeh et al., 2017; Ebrahimi et al., 2009; Le et al., 2018; Di Girolamo, 2015; Ramos et al., 2015).

# 7 Phases of Corneal Epithelial Wound Healing

Once the corneal epithelium is injured, the disrupted epithelial cells as well as the underlying basement membrane are responsible for releasing various chemical signals to promote the process of wound healing (Paley et al., 2022). Notable signals include epidermal growth factor (EGF) (Hongo et al., 1992), nerve growth factor (NGF) (Lambiase et al., 2000), alpha-enolase (Chung et al., 1995), fibronectin (Gipson et al., 1993), vinculin (Zieske et al., 1989), and integrin (Murakami et al., 1992). The process of corneal epithelial wound healing is characterized by four phases: the lag phase, migration phase, proliferation phase, and adhesion phase. (Ghafar et al., 2021) In this section, we discuss about each phase and the various factors that affect its impact on corneal epithelium wound healing.

## *7.1 Lag Phase*

The lag phase is the initial phase of corneal epithelial wound healing. It is seen as a preparation phase for the epithelium prior to the upcoming repair process. When the corneal epithelium is injured, the affected epithelial cells undergo apoptosis and are shed into the tear film and lost as tears. Their apoptosis triggers many pathways that begin the healing of the corneal epithelium. Healing begins with the release of IL-1 and TNF-α, as well as EGF and PDGF. These then cause the activation of various intracellular signaling pathways, namely ERK, MAP kinase, and NF-κB (Ljubimov & Saghizadeh, 2015). These pathways then cause the rest of the lag phase to take place.

Fibronectin creates a temporary scaffold over the damaged surfaces to allow for cellular access to the damaged area. The hemidesmosome attachments of basal epithelial cells break apart in the injured region. This leads to the dispersion of α6β4 integrin to the surrounding regions. Since α6β4's main function is to create a bond between the basal cells and their associated basement membrane, loss of such leads to a weakened junction.

The basal cell membrane also expresses integrin α5β1 and αvβ6 receptors, which play an important role in the formation of adhesions with actin and vinculin. A parallel process happens where adherents and gap junctions disassemble. Also, it leads to weakened cellular connections and sets the stage for epithelial migration in the next phase (Ghafar et al., 2021).

## *7.2 Migration Phase*

The migration phase starts as the basal layer in the regions that are undamaged flatten and moves towards the damaged area to create a cast over the injury. They are able to achieve this movement by creating finger-like projections, called filopodia, and broad sheet-like extensions, called lamellipodia (Lee, 2016). The leading cells move towards the damaged epithelium and drag the remainder of the epithelial cells behind them. The system is able to move via cytoskeleton contractile mechanisms. As the cells migrate towards the damaged area behind them, they form a zonula occludens. This helps to restore the function of the epithelial barrier as the migration is actively going on. The migrating cells secrete laminin, which serves two functions:

1. Remodeling the basement membrane (BM): Laminin helps rebuild the BM as the cells move forward.
2. Anchoring and propulsion: Laminin in the BM binds to integrin αvβ6 on the migrating cells' membrane. This interaction provides a temporary foothold that allows the cells to propel themselves forward, similar to a climbing mechanism (Ghafar et al., 2021).

This cycle repeats until a single layer of cells fully seals the defect, a process that generally takes 24–36 hours (Lee, 2016). Various factors can affect the migration phase for which we discuss in subsequent sections.

### 7.3 *Proliferation Phase*

Twenty-four hours following the injury, the limbal epithelial stem cells begin to divide and proliferate increasing up to sevenfold (Ghafar et al., 2021). They go on to produce transient-amplifying cells (TACs). The TACs have two main functions:

- Rapid division: TACs divide rapidly via mitosis and produce a high number of cells.
- Migration: TACS migrate into the wounded area to fill in all the damaged region (Lee, 2016).

Although TACs have a high proliferative capacity, they have a limit on the total number of divisions they can do. These cells are then converted to basal cells, wing cells, and eventually superficial cells. However, the exact mechanism of this conversion and differentiation is still not fully understood (Ghafar et al., 2021). Various factors play a role in activation and/or regulation of the proliferation phase including epidermal growth factor (EGFR), keratinocyte growth factor (KGF), and transforming growth factor-beta (TGF-β) (Ruan et al., 2021). We discuss various mediators and pathways in subsequent sections that can affect the proliferation phase.

### 7.4 *Adhesion Phase*

The adhesion phase is the final phase of corneal wound healing. Initially, hemidesmosomes reform and aid in anchoring the epithelial layer to the underlying basement membrane and stroma. Anchoring fibrils link to plaques to ensure a strong connection. This process usually takes a few days but could be prolonged to months if the underlying basement membrane was also damaged in the initial injury (Lee, 2016). Poor corneal epithelial adhesion can lead to recurrent corneal erosions, a common syndrome characterized by eye pain upon waking up. These recurrent erosions are often experienced after trauma or in patients with various corneal dystrophies such as epithelial basement membrane dystrophy (Fig. 1) (Ramamurthi et al., 2006).

## 8 Pathways and Factors Affecting Corneal Epithelial Wound Healing

Even though each epithelial wound undergoes the same phases, there are several factors that can influence the wound-healing process. These factors range from molecular modification of the gene and local environment, such as inflammation

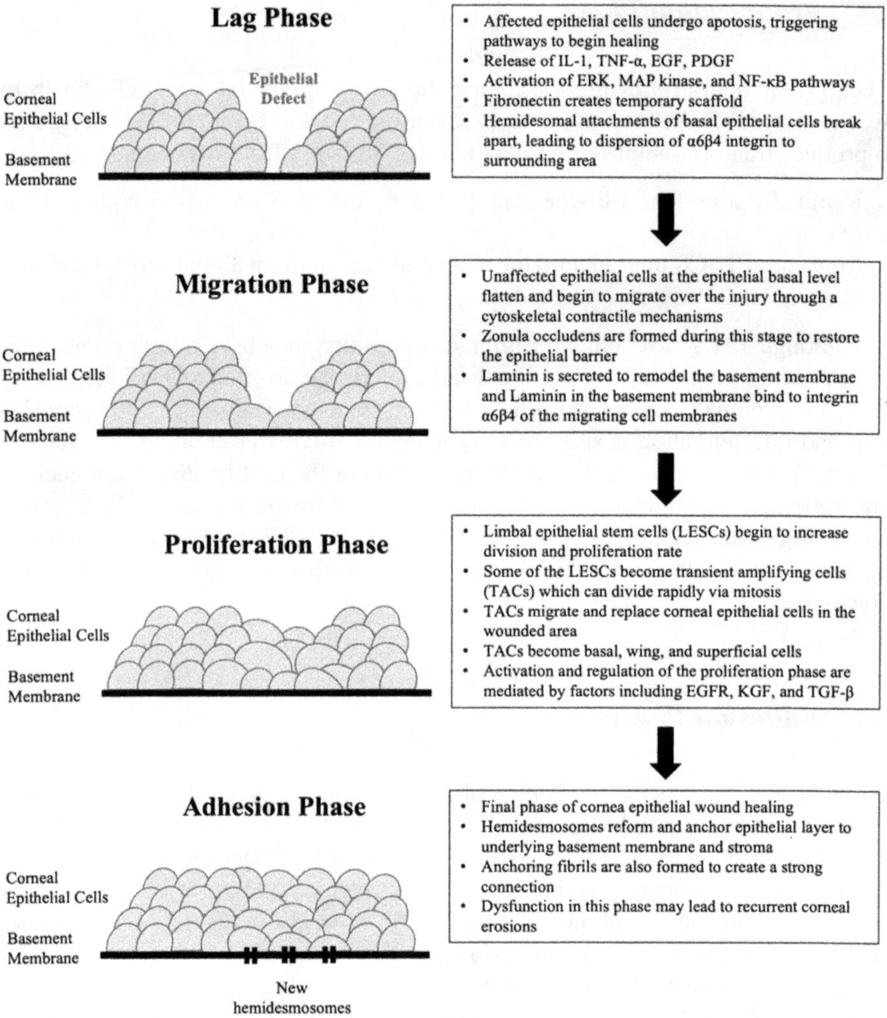


**Fig. 1** Stages of corneal epithelial wound healing and the summary of various mechanisms involved. *IL-1* interleukin-1, *TNF-α* tumor necrosis factor-alpha, *EGF* epidermal growth factor, *PDGF* platelet-derived growth factor, *ERK* extracellular signal-regulated kinase, *MAP kinase* mitogen-activated protein kinase, *NF-κB* nuclear factor kappa B, *EGFR* epidermal growth factor receptor, *KGF* keratinocyte growth factor, *TGF-β* transforming growth factor-beta

and hypoxia, to systemic comorbidities, such as diabetes and aging. In addition, larger environmental factors such as ultraviolet (UV) exposure and dryness are also implicated in the epithelial wound healing process.

## 8.1 *Cell Signaling Pathways*

Cell signaling pathways play an important role in the healing of the corneal epithelium. They manage processes like cell growth, migration, and repair. The following section will go into more detail about how some of them work. The four main signaling molecules are EGF (epidermal growth factor), KGF-1 (keratinocyte growth factor-1), HGF (hepatocyte growth factor), and TGF-β (transforming growth factor-beta) (Liu & Kao, 2015).

## 8.2 *Epidermal Growth Factor (EGF)*

EGF exists in many different types of bodily fluids, including tears. EGF acts almost immediately after corneal epithelium damage. This is due to EGF already existing in many parts of the body before damage. They then act to promote cellular proliferation. EGF is always the baseline molecule that signals when the epithelium is injured. The following two will then become activated in an as-needed matter, depending on the amount of damage (Liu & Kao, 2015).

Its specific mechanism of action is EGF binds to the EGFR which induces a cell signal that in turn activates the Src/PI3K/AKT pathway. This pathway then goes on to activate metalloproteinases (specifically MMP-9) in the extracellular matrix. MMP-9 then activates a positive feedback loop by mediating ectodomain shedding. Ectodomain shedding causes proteins like EGF and HB-EGF to be released from the cell surface by being cut free. This allows them to activate more receptors like EGFR and promote cell growth and repair. MMPs also activate proliferating cell nuclear antigens which help in the proliferation of cells (Chen et al., 2022).

## 8.3 *Keratinocyte Growth Factor-1 (KGF-1)*

KGF-1 originates a few layers away from the epithelium in the corneal stroma. However, KGF-1 receptors are in the corneal epithelium. KGF-1 signals via paracrine signaling. Similar to EGF it acts to promote cellular proliferation (Liu & Kao, 2015).

## 8.4 *Hepatocyte Growth Factor (HGF)*

HGF has a lot of similarities to KGF-1. It also exists in the corneal stroma and acts via paracrine signaling. HGF also acts on cellular proliferation. The way you can differentiate between the three lies in their speed of action as well as where they act.

EGF acts more generally in all layers of the epithelium. However, studies suggest that KGF-1 proliferates limbal and conjunctival epithelial cells and HGF proliferates central and peripheral corneal epithelial cells (Liu & Kao, 2015).

## 8.5 *Transforming Growth Factor-Beta (TGF-β)*

TGF-β acts differently than the other three cell signals. While they all play a role in proliferation of corneal epithelial cells, TGF-β plays more of an inhibitory role. It is more of a family of proteins that all fall under the general category of TGF-β. The different isoforms are TGF-β1, −β2, and -β3. They all play a function in antagonizing EGF's role on corneal epithelial cells. TGF-β1, and -β2 also play roles in inhibiting the proliferative effects of KGF-1 and HGF (Liu & Kao, 2015).

## 8.6 *Genetics and Molecular Factors*

A specific transcription factor ZEB1 has recently been identified to play a crucial role in epithelial wound healing process. Based on a mouse model, it was found that an increase in Zeb1 expression is associated with augmentation of apoptosis, proliferation, as well as cell mobility (Zhang et al., 2023). Another molecular factor involved in the healing process is RNA modification. It has been found that the RNA methylase NOP2/Sun domain family member 2 (NSUN2), which catalyzes 5-methylcytosine (m (Wirostko et al., 2015)C) modification of mRNAs, may play a critical role (Luo et al., 2023). NSUN2 expression is significantly increased in the corneal epithelial wound healing process, and a lack of NSUN2 function significantly delays the healing process (Luo et al., 2023). This has important implications, since gene therapy that can deliver genetic material to augment NSUN2 expression may help augment the corneal epithelial wound healing process (Luo et al., 2023).

## 8.7 *Local Factors: Hypoxia and Inflammation*

Given that the cornea is an avascular structure, its function relies heavily on atmospheric air as well as the aqueous humor (Pang et al., 2021). Hypoxia, which refers to the state where oxygen is insufficient for the tissue to maintain homeostasis (Bhutta et al., 2024), can have detrimental effects on the cornea as well as the epithelium healing process. Hypoxia has multiple etiologies including contact lens use, inflammation, and systemic diseases. Prolonged contact lens use physically limits oxygen permeability. Inflamed tissues have higher oxygen demand than normal tissues at baseline, creating a state of hypoxia. Systemic diseases can also reduce the amount of oxygen delivered via the aqueous humor (Bhutta et al., 2024). Once hypoxia develops, the corneal epithelium undergoes apoptosis (Ghafar et al., 2021). Exposure to hypoxia for 24–48 hours also attenuates the signaling cascade between the epithelial cells and corneal nerves, which is a crucial process in augmenting the wound healing process (Lee et al., 2014). Specifically, it was found that hypoxia impairs the cellular response to adenosine triphosphate (ATP) and glutamate which are typically released by damaged cells.

While pro-inflammatory cytokines are necessary to properly initiate the wound healing process, excessive inflammation, commonly seen during infection, can delay corneal wound healing. For instance, it has been demonstrated that reducing the level of toll-like receptor 9 (TLR9), a pattern recognition receptor implicated in pseudomonas aeruginosa keratitis, leads to better outcomes in wound healing (Huang et al., 2005). Neurotrophic keratitis is also another factor that can impact corneal epithelial wound healing, as trophic factors are critical for the various phases, as well as the overall health of the cornea (Versura et al., 2018). While an in-depth discussion on the complex pathophysiology of neurotrophic keratitis is out of the scope of this chapter, it is an important disease that can affect the epithelial wound healing process.

## 8.8 *Systemic Factors: Aging and Diabetes*

It has been known that age-related changes can alter the structural and functional properties of the cornea, which in turn affect the ability of the cornea to refract light, repair damage, and maintain proper barrier function (Yu et al., 2021). It has been found in a mouse model that aged mice can express more than 150 genes that are different compared to young mice. Several pathways are downregulated with age including biosynthesis, assembly, and degradation (Le et al., 2018). It has also been found that aged mice, irrespective of sex, have a delayed rate of re-epithelialization in the first 48 hours, which was hypothesized to be due to an altered transcriptome (Abu-Romman et al., 2024).

Diabetes, which is a major global health burden (GBD, 2023), has been implicated in the process of corneal epithelial wound healing. This is because the epithelium relies on sensory nerve endings to properly release neuropeptides to reduce inflammation and promote wound healing (Yu et al., 2022). Since diabetes commonly causes sensory neuropathy, it impairs the ability of the sensory nerves to properly respond to an injury and thus delays regeneration and the wound healing process (Yu et al., 2022). In addition to altering sensory nerves, it has been found that diabetes can affect the stiffness of the biomaterial substrates in the corneal regeneration process. This has important implications as the stiffness is known to modulate cell behavior such as migration, proliferation, and morphology (Ghafar et al., 2021). It has been found that the corneal epithelium is softer in diabetic cornea. Cells cultured on softer materials are found to express lower levels of proliferative markers such as extracellular signal-regulated kinase (pERK) and Ki67 (Ghafar et al., 2021) The collagen anchoring fibrils in diabetic corneas are also found to be abnormally shallow, which may lead to reduced adhesion and impaired barrier function (Azar et al., 1989; Gekka et al., 2004).

## 8.9 Environmental Factors: Dry Eye Disease and UV Exposure

Dry eye disease is a common condition estimated to affect 10–20% of the population over the age of 40 (Britten-Jones et al., 2024). It is characterized by tear film instability, hyperosmolarity, ocular surface inflammation, damage, and possible neurosensory abnormalities (Zemanová, 2021). Dry eye disease can delay corneal epithelial wound healing by altering the normal molecular signaling pathway involved in the wound healing process. One of the cytokines that are released is TNF-α, and it influences the expression of growth factors including EGF, KGF, HGF, and TGF (Yu et al., 2010). It has been shown that TNF-α is increased in dry eye, and consequently impedes the migration of epithelial cells (Okada et al., 2007). In addition, dry eye can affect wound healing via the phosphatidylinositol 3 kinase (PI3K/AKT) signaling pathway, which is responsible for apoptosis, proliferation, metabolism, and protein synthesis (Chen et al., 2022). Dry eye disease can also lead to hyperosmolarity and increased reactive oxygen species (ROS) accumulation, which in turn suppresses the PI3K/AKT pathway (Chen et al., 2022). Hyperglycemia secondary to diabetes can also delay epithelial wound healing by directly inhibiting the PI3K/AKT pathway (Chen et al., 2022). Lastly, dry eye disease can damage the type IV collagen scaffold, which is an integral component of the epithelial basement membrane, and therefore results in slowing the healing process (Baratta et al., 2022).

Exposure to UV radiation is an important environmental factor that can affect epithelial wound healing. It has been found that even mild irradiations with UV light can increase intracellular ROS. Excessive accumulation of ROS can lead to damage to proteins, membranes, as well as DNA (Zinflou & Rochette, 2017). Intense UV irradiation can also overwhelm the antioxidant mechanisms, promote excessive apoptosis, and increase inflammation, which in turn deter wound healing (Zernii et al., 2018).

## 8.10 Antimicrobial Peptides and Corneal Wound Healing

The ocular surface is constantly exposed to many different bacteria, fungi, viruses, and parasites that can be pathogenic. Additionally, the ocular surface consists of its own complex interplay of microorganisms that maintain homeostasis. To maintain this niche corneal environment, antimicrobial peptides (AMPs)—the body's natural antibiotic and innate immune effector—are deployed from various areas of the eye

(McIntosh et al., 2005; Gallo & Hooper, 2012; McDermott, 2009). At the ocular surface, various families of AMPs are found including α- and β-defensins, cathelicidins, and other lesser-known types of AMPs (McDermott, 2009). α-defensins—HNP-1, HNP-2, and HNP-3—have been found to be released into the ocular surface from infiltrating neutrophils (McDermott, 2009; Haynes et al., 1999; Eshac et al., 2021). β-defensins—hBD-1, hBD-2, and hBD-3—are expressed by corneal and conjunctival epithelial cells (McDermott, 2009; Haynes et al., 1999; Eshac et al., 2021). LL-37 is expressed by both corneal and conjunctival epithelial cells and is the only AMP from the cathelicidin family (McDermott, 2009; Eshac et al., 2021; Gordon et al., 2005). Some histatins—Hist-1—are found in the tear film and multiple histatins have been shown to enhance corneal epithelial wound healing (McDermott, 2009; Kalmodia et al., 2019; Shah et al., 2020; Son et al., 2024; Shah et al., 2017; Son et al., 2021). Other AMPs such as thymosin β4 (Tβ4)—expressed by corneal and conjunctival cells—and lacritin—the human tear glycoprotein—have been shown to exhibit important antimicrobial wound healing roles (McDermott, 2009; Eshac et al., 2021). Overall, these AMPs play important roles in maintaining homeostasis in the corneal environment and protecting the ocular surface from the many pathogenic organisms it encounters.

## 9 Therapeutic Interventions to Optimize Corneal Epithelial Healing

Several therapeutic interventions have been investigated to optimize epithelial wound healing. They can be divided broadly into pharmacologic approach, nonsurgical approach, and surgical approach.

### *9.1 Molecular and Pharmacologic Approaches*

Given that the process of epithelial wound healing involves multiple inflammatory markers and signaling molecules, these have been targets of investigation. Rho kinases (ROCK) regulate wound healing by negatively modulating cell migration and cell adhesion (Yin & Yu, 2008). Rho kinase inhibitors have been explored extensively for the corneal endothelium; however, it has also been shown to have effects on the corneal epithelium as well. Rho kinase inhibitors Y-27673 have been investigated in corneal epithelium and have been shown to promote basal wound healing and enhanced cell migration as well as adhesion to matrices (Yin & Yu, 2008).

Another molecule involved in regulating wound healing is ascorbic acid. It has been shown in a mouse model that ascorbic acid promotes stemness of corneal epithelial progenitor cells by increasing clone formation ability and expression of stemness markers (Chen et al., 2017). Another molecule, trehalose, which is a natural nonreducing disaccharide, has been shown to help preserve morphological features of epithelium among a small group of patients undergoing laser subepithelial keratomileusis (LASEK) (Aragona et al., 2014). Specifically, epithelium treated with trehalose showed larger areas and clearer cytoplasm in basal cells and higher epithelial heights (Aragona et al., 2014).

Fibronectin, which is present at the site of epithelial defects, has been shown to stimulate epithelial wound healing in vitro and in an animal model (Nishida, 2010). Eye drops containing fibronectin have been reported to be an effective treatment of persistent corneal epithelial defects (Nishida, 2010). Eye drops containing substance P and insulin-like growth factor-1 (IGF-1) are also effective in treating epithelial wounds in neurotrophic keratopathy. This is because substance P-derived peptides and IGF-1-derived peptides upregulate the expression of integrins in epithelial cells and therefore promote cell attachment and cell migration (Nishida, 2010). Specifically, the C-domain of IGF-1, which includes a four amino-acid sequence (serine-serine-serine-arginine), was identified to play a role in the crucial synergistic mechanism with substance P (Nishida, 2010). Exposure of corneal epithelial cells to substance P and IGF-1 increased the expression of mRNAs for alpha-5 and beta-1 integrins (Nishida, 2010). In addition, substance P and IGF-1 synergistically stimulate the phosphorylation of paxillin and focal adhesion kinase, which are components of the integrin signaling pathway (Nishida, 2010). Interestingly, the use of insulin, even at a concentration 1000 times that at which IGF is found to be effective, did not promote epithelial cell migration (Nishida, 2010).

Coenzyme Q10, an organic molecule that is found largely in mitochondria, has been shown to enhance epithelial wound healing (Asena et al., 2020). Coenzyme Q10 acts as an antioxidant and free radical scavenger, and thus improves the viability of epithelial cells and protects them against oxidative damage to mitochondrial membranes (Asena et al., 2020). Oral tetracycline and doxycycline have also been shown to promote epithelial healing in patients with persistent epithelial defects (American Academy of Ophthalmology, 2014).

Among molecules involved in the signaling cascade of epithelial wound healing, EGF has been shown to be an effective treatment of epithelial defects in randomized, double-blind, placebo-controlled, multicenter clinical trials (Pastor & Calonge, 1992). This trial showed that the number of epithelial defects completely healed at 1, 2, and 3 days was significantly higher among the EGF-treated group compared to the placebo-treated group (Pastor & Calonge, 1992).

Gene therapy with the use of microRNA (miRNA) has been investigated extensively in the past decades as a potential approach to promote epithelial wound healing. A study found that the use of miRNA mimics, or antagomir, can promote the normalization of epithelial wound healing in human diabetic organ-cultured corneas by enhancing cell migration (Winkler et al., 2014). More recently, an injection of adipose tissue-derived stem cells modified to release IGF-1 has shown promising preclinical results in the ability to promote recovery from an alkaline-burn injury (Fagoonee et al., 2023). It has been proposed that cells carrying IGF-modified RNA can benefit repairs of corneal epithelial and limbal cells (Fagoonee et al., 2023).

## 9.2 Nonsurgical Approaches

Nonsurgical approach to enhance corneal epithelial wound healing is often the first-line treatment of epithelial defects. Nonsurgical treatments focus on providing adequate lubrication to the epithelium and preventing further damage to the epithelium. These treatments include lubrication with preservative-free artificial tears, placing a punctual plug to increase tear retention, and placing a bandage contact lens to prevent further trauma to the epithelium (American Academy of Ophthalmology, 2014). Scleral contact lenses can also be used to enhance hydration (American Academy of Ophthalmology, 2014). For larger epithelial defects, an amniotic membrane graft may be used. Amniotic membrane graft has been shown to promote epithelial healing by increasing cell migration and proliferation. The degree of increment is associated with an increasing concentration of amniotic membrane suspension (Choi et al., 2009). It is also known to have anti-inflammatory and antiscarring properties (Rauz & Saw, 2010).

## 9.3 Surgical Approaches

One of the most straightforward methods to optimize a persistent epithelial defect is tarsorrhaphy. This can be either a partial or complete closure of the eyelid to protect the epithelial and allow it to heal itself. Tarsorrhaphy may also reduce the risk of infection associated with the use of bandage contact lenses (American Academy of Ophthalmology, 2014). However, a more advanced surgical method that has been gaining significant attention is limbal stem cell transplant. Since the corneal epithelial wound is healed by stem cells located at the limbus, patients who lack these cells will result in limbal stem cell deficiency (LSCD), which in turn leads to persistent epithelial defects and potential conjunctivalization of the cornea (Fernandez-Buenaga et al., 2018).

In limbal stem cell transplantation, the stem cell source can be obtained via autograft, allograft, or culture (Fernandez-Buenaga et al., 2018). In cultivated limbal epithelial transplantation (CLET), a small amount of donor limbal epithelium can be co-cultured with fibroblasts, resulting in an expansion of cultivated limbal epithelium, which can then be transplanted (Pellegrini et al., 1997). A newer technique, called simple limbal epithelial transplantation (SLET), involves harvesting donor epithelial cells and transplanting them directly onto the human amniotic membrane that is placed over the ocular surface without the need for culturing epithelial cells (Sangwan et al., 2012). This simplifies the logistics involved in culturing epithelial cells and the resources needed for the transplant (Sangwan et al., 2012). Another newer method, which is particularly advantageous when the epithelial defect is extensive and bilateral, involves obtaining mucosal epithelium from a nonocular origin. This is known as cultivated oral mucosal epithelial transplant (COMET). Transplantation with this method has been shown to demonstrate stability after 1 year postoperatively with sustained epithelialization, reduced conjunctivalization, and improved vision (Prabhasawat et al., 2016).

Most recently, an approach was developed such that the source of stem cells does not necessarily need to be from the epithelium. Instead, induced pluripotent stem cells (iPS) are harvested from the blood cells of a healthy donor and reprogrammed to transform into corneal epithelial cells (Mallapaty, 2024).

Preliminary clinical results show that most patients who received these transplants have improvements of their vision. However, larger clinical trials are needed to assess the efficacy of this approach.

| Technique | Description | References |
|---|---|---|
| *Pharmacologic* | | |
| Rho kinase inhibitors | Enhance cell migration and adhesion to matrices | Yin and Yu (2008) |
| Ascorbic acid | Promote stemness of epithelial progenitor cells | Chen et al. (2017) |
| Trehalose | Preserve morphological features of epithelium, clearer cytoplasm, higher epithelial heights | Aragona et al. (2014) |
| Fibronectin | Stimulate epithelial wound healing in vitro and animal model | Nishida (2010) |
| Substance P and IGF-1 | Upregulate expression of integrins in epithelial cells, promote attachment and migration | Nishida (2010) |
| Coenzyme Q10 | Improve viability of epithelial cells by protecting against oxidative damages | Asena et al. (2020) |
| Tetracycline/ doxycycline | Promote epithelial healing in persistent epithelial defect | American Academy of Ophthalmology (2014) |
| EGF | Show efficacy in randomized, double-blind multicenter trial | Pastor and Calonge (1992) |
| microRNA | Promote normalization of epithelial cell wound healing in diabetic organ-cultured by enhancing cell migration | Winkler et al. (2014) |
| Injection of adipose-derived stem cells | Promote recovery from alkaline-burn injury | Fagoonee et al. (2023) |
| *Nonsurgical* | | |
| Artificial tears | Provide adequate lubrication | American Academy of Ophthalmology (2014) |
| Punctal plug | Increase tear retention | American Academy of Ophthalmology (2014) |
| Bandage contact lens | Prevent further trauma to epithelium | American Academy of Ophthalmology (2014) |
| Scleral contact lens | Enhance hydration | American Academy of Ophthalmology (2014) |

| Technique | Description | References |
|---|---|---|
| Amniotic membrane graft | Increase cell migration and proliferation, provide anti-inflammatory and antiscarring properties | Choi et al. (2009) |
| *Surgical* | | |
| Tarsorrhaphy | Partial or complete closure of eyelid to protect epithelium | American Academy of Ophthalmology (2014) |
| CLET | Donor limbal epithelium co-cultured with fibroblast prior to transplant | Pellegrini et al. (1997) |
| SLET | Donor epithelium transplanted on amniotic membrane without culturing | Sangwan et al. (2012) |
| COMET | Epithelium for transplant obtained from nonocular origin (mucosa) | Prabhasawat et al. (2016) |
| IPS-derived epithelium transplant | Induced pluripotent stem cells reprogramed to transform to epithelial cells | Mallapaty (2024) |

*CLET* cultivated limbal epithelial transplantation, *COMET* cultivated oral mucosal epithelial transplant, *EGF* epidermal growth factor, *IGF* insulin-like growth factor, *IPS* induced pluripotent stem cells, *SLET* simple limbal epithelial transplantation

## 10 Conclusion

The corneal epithelium is a complex structure that is vital to clearing the cornea. A diverse multitude of molecular factors play a role in maintaining the health of the corneal epithelium and its wound-healing capabilities. Each phase in the wound healing process plays a critical in ensuring a healthy corneal epithelium, and studies have shown the various external factors that can affect corneal epithelium wound healing. Research has also shown a myriad of diverse methods to improve wound healing processes, including pharmacologic interventions, nonsurgical interventions (e.g., artificial tears, bandage contact lenses), and surgical interventions (e.g., cultivated limbal epithelial transplantation). Ultimately, future research may be targeted toward further personalized treatment and understanding the long-term efficacy of current treatments.

**Conflict of Interest** VKA is an equity owner of ViSo Therapeutics Inc.

**Funding Statement** JO, CT, and VKA are supported by an Unrestricted Grant from Research to Prevent Blindness, New York, NY. VKA has support from The National Institutes of Health, National Eye Institute (R01EY029409), and National Institute of Neurological Disorders and Stroke (R01NS124784).

## References

Abu-Romman, A., et al. (2024). Age-related differences in the mouse corneal epithelial transcriptome and their impact on corneal wound healing. *Investigative Ophthalmology & Visual Science, 65*, 21.

Alio, J. L., et al. (Eds.). (2019). *Corneal regeneration: Therapy and surgery*. Springer. https://doi.org/10.1007/978-3-030-01304-2

Allen, N. E., Crawford, A. Z., McGhee, C. N. J., & Meyer, J. J. (2024). Chemical eye injuries: A 10 year retrospective review of acute presentations and clinical outcomes in Auckland, New Zealand. *Scientific Reports, 14*, 8264.

American Academy of Ophthalmology. (2014). *Healing a persistent corneal epithelial defect*. https://www.aao.org/eyenet/article/healing-persistent-corneal-epithelial-defect

American Academy of Ophthalmology. (2023). *Facts about tears*. https://www.aao.org/eye-health/tips-prevention/facts-about-tears

Aragona, P., et al. (2014). Protective effects of Trehalose on the corneal epithelial cells. *Scientific World Journal, 2014*, 717835.

Asena, L., Alkayid, H., & Altınörs, D. D. (2020). Corneal epithelial wound healing and management strategies. In M. A. Shiffman & M. Low (Eds.), *Plastic and thoracic surgery, orthopedics and ophthalmology* (pp. 91–102). Springer. https://doi.org/10.1007/15695_2017_65

Ashby, B. D., Garrett, Q. & Willcox, M. D. Corneal Injuries and Wound Healing Review of Processes and Therapies.

Azar, D. T., Spurr-Michaud, S. J., Tisdale, A. S., & Gipson, I. K. (1989). Decreased penetration of anchoring fibrils into the diabetic stroma. A morphometric analysis. *Archives of Ophthalmology Chic Ill, 1960*(107), 1520–1523.

Baratta, R. O., Schlumpf, E., Buono, B. J. D., DeLorey, S., & Calkins, D. J. (2022). Corneal collagen as a potential therapeutic target in dry eye disease. *Survey of Ophthalmology, 67*, 60–67.

Bhutta, B. S., Alghoula, F. & Berim, I. Hypoxia. in *StatPearls* (StatPearls Publishing, Treasure Island (FL), 2024).

Britten-Jones, A. C., et al. (2024). Epidemiology and risk factors of dry eye disease: Considerations for clinical management. *Medicina (Kaunas, Lithuania), 60*, 1458.

Chen, J., et al. (2017). Ascorbic acid promotes the Stemness of corneal epithelial stem/progenitor cells and accelerates epithelial wound healing in the cornea. *Stem Cells Translational Medicine, 6*, 1356–1365.

Chen, K., et al. (2022). The role of the PI3K/AKT signalling pathway in the corneal epithelium: Recent updates. *Cell Death & Disease, 13*, 1–13.

Choi, J. A., et al. (2009). Effects of amniotic membrane suspension in human corneal wound healing in vitro. *Molecular Vision, 15*, 2230–2238.

Chung, E. H., DeGregorio, P. G., Wasson, M., & Zieske, J. D. (1995). Epithelial regeneration after limbus-to-limbus debridement. Expression of alpha-enolase in stem and transient amplifying cells. *Investigative Ophthalmology & Visual Science, 36*, 1336–1343.

ClinicalKey. (n.d.). *Anatomy of the eye and orbit*. ClinicalKey. https://www-clinicalkey-com.proxy.lib.umich.edu/#!/content/book/3-s2.0-B9780702079931000011?scrollTo=%23hl0002559

Di Girolamo, N. (2015). Moving epithelia: Tracking the fate of mammalian limbal epithelial stem cells. *Progress in Retinal and Eye Research, 48*, 203–225.

Ebrahimi, M., Taghi-Abadi, E., & Baharvand, H. (2009). Limbal stem cells in review. *Journal of Ophthalmic and Vision Research, 4*, 40–58.

eOphtha. (n.d.). *Anatomy of cornea*. https://www.eophtha.com/posts/anatomy-of-cornea.

Eshac, Y., Redfern, R. L., & Aakalu, V. K. (2021). The role of endogenous antimicrobial peptides in modulating innate immunity of the ocular surface in dry eye diseases. *International Journal of Molecular Sciences, 22*, 721.

Estil, S., et al. (2002). Pilot study on the time course of apoptosis in the regenerating corneal epithelium. *Acta Ophthalmologica Scandinavica, 80*, 517–523.

Fagoonee, S., Saccu, G., & Bussolati, B. (2023). Innovative stem cell-based strategies for corneal wound healing: A step forward. *Molecular Therapy, 31*, 2307–2308.
Fernandez-Buenaga, R., Aiello, F., Zaher, S. S., Grixti, A., & Ahmad, S. (2018). Twenty years of limbal epithelial therapy: An update on managing limbal stem cell deficiency. *BMJ Open Ophthalmol., 3*, e000164.
Gallo, R. L., & Hooper, L. V. (2012). Epithelial antimicrobial defence of the skin and intestine. *Nature Reviews. Immunology, 12*, 503–516.
GBD. (2023). 2021 Diabetes Collaborators. Global, regional, and national burden of diabetes from 1990 to 2021, with projections of prevalence to 2050: a systematic analysis for the Global Burden of Disease Study 2021. *The Lancet London, 402*, 203–234.
Gekka, M., et al. (2004). Corneal epithelial barrier function in diabetic patients. *Cornea, 23*, 35–37.
Ghafar, N. A., Jalil, N. A. A., & Kamarudin, T. A. (2021). Wound healing of the corneal epithelium: A review. *Asian biomedicine : Research, Reviews and News, 15*, 199–212.
Gipson, I. K., Watanabe, H., & Zieske, J. D. (1993). Corneal wound healing and fibronectin. *International Ophthalmology Clinics, 33*, 149–163.
Gordon, Y. J., et al. (2005). Human cathelicidin (LL-37), a multifunctional peptide, is expressed by ocular surface epithelia and has potent antibacterial and antiviral activity. *Current Eye Research, 30*, 385–394.
Haynes, R. J., Tighe, P. J., & Dua, H. S. (1999). Antimicrobial defensin peptides of the human ocular surface. *The British Journal of Ophthalmology, 83*, 737–741.
Hongo, M., Itoi, M., Yamaguchi, N., & Imanishi, J. (1992). Distribution of epidermal growth factor (EGF) receptors in rabbit corneal epithelial cells, keratocytes and endothelial cells, and the changes induced by transforming growth factor-beta 1. *Experimental Eye Research, 54*, 9–16.
Huang, X., Barrett, R. P., McClellan, S. A., & Hazlett, L. D. (2005). Silencing toll-like receptor-9 in Pseudomonas aeruginosa keratitis. *Investigative Ophthalmology & Visual Science, 46*, 4209–4216.
Kalmodia, S., et al. (2019). Presence of Histatin-1 in human tears and association with aqueous deficient dry eye diagnosis: A preliminary study. *Scientific Reports, 9*, 10304.
Kuwabara, T., Perkins, D. G., & Cogan, D. G. (1976). Sliding of the epithelium in experimental corneal wounds. *Investigative Ophthalmology, 15*, 4–14.
Lambiase, A., et al. (2000). Nerve growth factor promotes corneal healing: Structural, biochemical, and molecular analyses of rat and human corneas. *Investigative Ophthalmology & Visual Science, 41*, 1063–1069.
Le, Q., Xu, J., & Deng, S. X. (2018). The diagnosis of limbal stem cell deficiency. *The Ocular Surface, 16*, 58–69.
Lee, T. (2016). The ins and outs of corneal wound healing. *Review of Optometry*.
Lee, A., et al. (2014). Hypoxia-induced changes in ca(2+) mobilization and protein phosphorylation implicated in impaired wound healing. *American Journal of Physiology. Cell Physiology, 306*, C972–C985.
Liu, C.-Y., & Kao, W. W.-Y. (2015). Chapter five – Corneal epithelial wound healing. In J. F. Hejtmancik & J. M. Nickerson (Eds.), *Progress in molecular biology and translational science* (Vol. 134, pp. 61–71). Academic.
Ljubimov, A. V., & Saghizadeh, M. (2015). Progress in corneal wound healing. *Progress in Retinal and Eye Research, 49*, 17–45.
Luo, G., et al. (2023). The RNA m5C Methylase NSUN2 modulates corneal epithelial wound healing. *Investigative Ophthalmology & Visual Science, 64*, 5.
Mallapaty, S. (2024). World-first stem-cell treatment restores vision in people. *Nature, 635*, 533–534.
McDermott, A. M. (2009). The role of antimicrobial peptides at the ocular surface. *Ophthalmic Research, 41*, 60–75.
McIntosh, R. S., et al. (2005). The spectrum of antimicrobial peptide expression at the ocular surface. *Investigative Ophthalmology & Visual Science, 46*, 1379–1385.

Murakami, J., Nishida, T., & Otori, T. (1992). Coordinated appearance of beta 1 integrins and fibronectin during corneal wound healing. *The Journal of Laboratory and Clinical Medicine, 120*, 86–93.

Nishida, T. (2010). Translational research in corneal epithelial wound healing. *Eye & Contact Lens, 36*, 300.

Notara, M., Lentzsch, A., Coroneo, M., & Cursiefen, C. (2018). The role of Limbal epithelial stem cells in regulating corneal (lymph)angiogenic privilege and the micromilieu of the Limbal niche following UV exposure. *Stem Cells International, 2018*, 8620172.

Okada, Y., et al. (2007). TNFalpha suppression of corneal epithelium migration. *Molecular Vision, 13*, 1428–1435.

Paley, G. L. *et al.* Corneal wound healing, recurrent corneal erosions, and persistent epithelial defects. in Albert and Jakobiec's principles and practice of ophthalmology (eds. Albert, D. M., Miller, J. W., Azar, D. T. & Young, L. H.) 331–360 (Springer, , 2022). doi:https://doi.org/10.1007/978-3-030-42634-7_212.

Pang, K., Lennikov, A., & Yang, M. (2021). Hypoxia adaptation in the cornea: Current animal models and underlying mechanisms. *Animal Models and Experimental Medicine, 4*, 300–310.

Pastor, J. C., & Calonge, M. (1992). Epidermal growth factor and corneal wound healing. A multicenter study. *Cornea, 11*, 311–314.

Pellegrini, G., et al. (1997). Long-term restoration of damaged corneal surfaces with autologous cultivated corneal epithelium. *The Lancet, 349*, 990–993.

Prabhasawat, P., et al. (2016). Long-term result of autologous cultivated oral mucosal epithelial transplantation for severe ocular surface disease. *Cell and Tissue Banking, 17*, 491–503.

Ramamurthi, S., Rahman, M. Q., Dutton, G. N., & Ramaesh, K. (2006). Pathogenesis, clinical features and management of recurrent corneal erosions. *Eye, 20*, 635–644.

Ramos, T., Scott, D., & Ahmad, S. (2015). An update on ocular surface epithelial stem cells: Cornea and conjunctiva. *Stem Cells International, 2015*, 601731.

Rauz, S., & Saw, V. P. (2010). Serum eye drops, amniotic membrane and limbal epithelial stem cells—Tools in the treatment of ocular surface disease. *Cell and Tissue Banking, 11*, 13–27.

Ruan, Y., Jiang, S., Musayeva, A., Pfeiffer, N., & Gericke, A. (2021). Corneal epithelial stem cells-physiology, pathophysiology and therapeutic options. *Cells, 10*, 2302.

Saghizadeh, M., Kramerov, A. A., Svendsen, C. N., & Ljubimov, A. V. (2017). Concise review: Stem cells for corneal wound healing. *Stem cells (Dayton, Ohio), 35*, 2105–2114.

Sangwan, V. S., Basu, S., MacNeil, S., & Balasubramanian, D. (2012). Simple limbal epithelial transplantation (SLET): A novel surgical technique for the treatment of unilateral limbal stem cell deficiency. *The British Journal of Ophthalmology, 96*, 931–934.

Shah, D., Ali, M., Shukla, D., Jain, S., & Aakalu, V. K. (2017). Effects of histatin-1 peptide on human corneal epithelial cells. *PLoS One, 12*, e0178030.

Shah, D., et al. (2020). Wound healing properties of Histatin-5 and identification of a functional domain required for Histatin-5-induced cell migration. *Molecular Therapy – Methods & Clinical Development, 17*, 709–716.

Son, K.-N., et al. (2021). Histatin-1 is an endogenous ligand of the sigma-2 receptor. *The FEBS Journal, 288*, 6815–6827.

Son, K.-N., et al. (2024). Identifying the crucial binding domain of histatin-1 to recombinant TMEM97 in activating chemotactic migration in human corneal epithelial cells. *Biochemical and Biophysical Research Communications, 739*, 150991.

Sridhar, M. S. (2018). Anatomy of cornea and ocular surface. *Indian Journal of Ophthalmology, 66*, 190–194.

Versura, P., Giannaccare, G., Pellegrini, M., Sebastiani, S., & Campos, E. C. (2018). Neurotrophic keratitis: Current challenges and future prospects. *Eye Brain, 10*, 37–45.

Wilson, S. E. (2020). Bowman's layer in the cornea– Structure and function and regeneration. *Experimental Eye Research, 195*, 108033.

Winkler, M. A., Dib, C., Ljubimov, A. V., & Saghizadeh, M. (2014). Targeting miR-146a to treat delayed wound healing in human diabetic organ-cultured corneas. *PLoS One, 9*, e114692.

Wirostko, B., Rafii, M., Sullivan, D. A., Morelli, J., & Ding, J. (2015). Novel therapy to treat corneal epithelial defects: A hypothesis with growth hormone. *The Ocular Surface, 13*, 204–212.e1.

Yin, J., & Yu, F.-S. X. (2008). Rho kinases regulate corneal epithelial wound healing. *American Journal of Physiology-Cell Physiology, 295*, C378–C387.

Yu, F.-S. X., Yin, J., Xu, K., & Huang, J. (2010). Growth factors and corneal epithelial wound healing. *Brain Research Bulletin, 81*, 229–235.

Yu, Q. et al. (2021). Canonical NF-κB signaling maintains corneal epithelial integrity and prevents corneal aging via retinoic acid. *eLife, 10*, e67315.

Yu, F. X., et al. (2022). The impact of sensory neuropathy and inflammation on epithelial wound healing in diabetic corneas. *Progress in Retinal and Eye Research, 89*, 101039.

Zemanová, M. (2021). Dry eye disease: A review. *Ceská a slovenská oftalmologie: casopis Ceské oftalmologické spolecnosti a Slovenské oftalmologické spolecnosti, 77*, 107–119.

Zernii, E. Y., et al. (2018). Mitochondria-targeted antioxidant SKQ1 protects cornea from oxidative damage induced by ultraviolet irradiation and mechanical injury. *BMC Ophthalmology, 18*, 336.

Zhang, Y., et al. (2023). Zeb1 facilitates corneal epithelial wound healing by maintaining corneal epithelial cell viability and mobility. *Communications Biology, 6*, 1–12.

Zieske, J. D., Bukusoglu, G., & Gipson, I. K. (1989). Enhancement of vinculin synthesis by migrating stratified squamous epithelium. *The Journal of Cell Biology, 109*, 571–576.

Zinflou, C., & Rochette, P. J. (2017). Ultraviolet A-induced oxidation in cornea: Characterization of the early oxidation-related events. *Free Radical Biology & Medicine, 108*, 118–128.

# Stem Cells for Corneal Endothelial Disease

**Elena Bellina, Farshad Abedi, Mahsa Rashidi, Nuran Aly, and Ali R. Djalilian**

**Abstract** The corneal endothelium is crucial in maintaining corneal transparency via regulating corneal hydration. Corneal endothelial injury and dysfunction are two of the main causes of impaired vision, secondary to corneal haze and opacification. Human corneal endothelium has very limited regeneration capacity. Different methods of corneal endothelial transplantation are currently the mainstay of the treatment of corneal endothelial dysfunction. Due to the limited availability of donor corneal transplant tissues, cell and cell-based treatments are currently an important area of research to help improve the function or regeneration of corneal endothelium without transplant surgeries. Different types of stem cells and acellular stem cell-based treatments such as extracellular vesicles (EVs) and secretomes are good therapeutic candidates due to their regenerative and immunomodulatory properties. Umbilical cord and embryonic stem cells showed encouraging results for their ability to differentiate into corneal endothelial cell (CEC)-like cells and improve corneal endothelial function. Induced pluripotent stem cells (iPSCs), created by reprogramming adult somatic cells, were also successfully differentiated into CEC-like cells and improved corneal endothelial function in animal experimental models. Mesenchymal stromal cells (MSCs), their secreted factors (secretome), and extracellular vesicles (EVs) yielded encouraging outcomes in improving corneal endothelial regeneration and function via their paracrine immunomodulatory and regenerative properties. Clinical trials are necessary to translate these promising experimental results into treatments for human corneal endothelial disease.

E. Bellina
Humanitas University, Milan, Italy

F. Abedi · A. R. Djalilian (✉)
University of Illinois, Chicago, IL, USA
e-mail: fabedi3@uic.edu; adjalili@uic.edu

M. Rashidi
University of Melbourne, Melbourne, Australia
e-mail: MRashidi@alumni.unimelb.edu.au

N. Aly
Cairo University, Giza, Egypt

 
H. Das (ed.), *Tissue Repair and Regeneration*,
https://doi.org/10.1007/978-3-031-93677-7_10

**Keywords** Corneal endothelium · Regeneration/repair · Stem cells · Stromal cells · MSCs

## 1 Introduction to Corneal Endothelial Disease and Stem Cell-Based Therapies Rho-Associated Protein Kinase Inhibitor

The transparent cornea is crucial for focusing light and maintaining clear vision. The corneal endothelium, the innermost layer of the cornea, maintains corneal transparency by pumping excess fluid out of the cornea and its barrier function. Damage or dysfunction of the corneal endothelium can arise from a variety of causes, including age-related degeneration, genetic disorders like Fuchs' endothelial dystrophy, trauma, infections, and surgical complications. Dysfunction of the corneal endothelium leads to corneal edema, opacification, and, ultimately, vision loss (Feizi, 2018; Van Den Bogerd et al., 2019).

Treatment of corneal endothelial diseases continues to be challenging due to the very limited regenerative capacity of human corneal endothelial cells (*CECs*) (Fu et al., 2024; Shen et al., 2017).

Corneal transplantation has been the gold standard for treatment, including penetrating keratoplasty (*PK*) and more refined techniques such as Descemet's stripping endothelial keratoplasty (*DSEK*), Descemet's membrane endothelial transfer (*DMET*), and Descemet's membrane endothelial keratoplasty (*DMEK*). These surgeries depend on the availability of healthy corneal tissue from donors to replace the damaged endothelium. However, the global shortage of donor corneas and the risks associated with transplant surgeries (e.g., graft rejection, infection, and the need for long-term immunosuppressive therapy) underscore the need for alternative therapies that can regenerate or replace the corneal endothelium without relying on donor tissue (Feizi, 2018; Moshirfar et al., 2024; Price et al., 2021).

In recent years, stem cell-based therapies have emerged as a promising solution. Stem cells can differentiate into specialized cell types or secrete trophic factors, making them ideal for regenerating damaged tissues. These cells offer the potential to generate functional corneal endothelial-like cells in vitro, which can then be transplanted to restore corneal clarity and function (Bartakova et al., 2014; Chen et al., 2021; Ng et al., 2023).

The following chapter provides an overview of the stem cell types that have been examined for corneal endothelial regeneration to date, their unique properties and mechanisms, and the extensive body of research that supports their potential application, providing a comprehensive understanding of future opportunities and challenges in developing stem cell-based therapies for corneal endothelial diseases.

## 2 Stem Cell Types and Their Application in Corneal Endothelial Repair

Stem cell research for corneal endothelial injuries encompasses a diverse array of cell types, each offering different advantages and challenges. These include but are not limited to umbilical cord stem cells, induced pluripotent stem cells (iPSCs), embryonic stem cells (ESCs), and mesenchymal stromal cells (MSCs).

## 3 Umbilical Cord Stem Cells

Human umbilical cord blood MSCs (hUCB-MSCs) have multipotent differentiation capabilities but are particularly known for their strong paracrine activity. Joyce et al. showed that hUCB-MSCs can differentiate into corneal endothelial-like cells under specific conditions. In an ex vivo model, hUCB-MSCs attached to the injured corneal endothelial cells showed relative expression of the genes, suggestive of a more human corneal endothelial cell-like (CEC-like) phenotype. UCB MSCs treated with lens epithelial cell-conditioned medium formed closely opposed elongated cells in parallel sheets (Joyce et al., 2012).

## 4 Embryonic Stem Cells

Embryonic stem cells (ESCs) are pluripotent cells with the potential to generate CEC-like cells. Derived from the inner cell mass of blastocysts, ESCs are highly versatile and capable of differentiating into any cell type. When directed into a corneal endothelial lineage, ESCs exhibit some properties of the native CECs. Despite their high differentiation potential, the use of ESCs carries ethical concerns and immune rejection risks. Regulatory and societal barriers further hinder their widespread adoption. Song et al. described a stepwise protocol to guide hESCs toward a CEC-like lineage (Song et al., 2016). The initial step in differentiation involved the induction of neural crest cells (NCCs), which are precursors to corneal endothelial cells. Small molecules like SB431542 (a transforming growth factor-β (TGF-beta) receptor inhibitor) and CHIR99021 (a GSK-3 inhibitor) were used to push the hESCs toward the neural crest lineage. After neural crest induction, cells were further directed towards a periocular mesenchyme (POM) stage, which is a precursor to corneal endothelial cells. Retinoic acid was added to promote the formation of structures similar to the corneal endothelium and helped guide cells. Transcriptomic analysis using RNA sequencing revealed that the gene expression profiles of the hESC-derived CEC-like cells closely resembled those of adult CECs, indicating successful differentiation into mature, endothelial-like cells (Song et al., 2016). The hESC-derived CECs transplanted into rabbit models with corneal endothelial dysfunction yielded significant improvements in corneal transparency and thickness. However, since hESCs are not autologous, there is a risk of immune rejection, which may necessitate the use of immunosuppressive therapy and may raise ethical considerations (Song et al., 2016).

## 4.1 Induced Pluripotent Stem Cells

Another cell therapy involves the use of induced pluripotent stem cells (iPSCs) created by reprogramming adult somatic cells, such as fibroblasts or blood cells. The iPSCs have the potential to differentiate into any cell type, including CEC-like cells. Differentiation of iPSCs into CECs involves a combination of growth factors, such as TGF-β, bone morphogenic protein (BMP) inhibitors, and Rho-associated protein kinase (ROCK) inhibitors, which guide their development into endothelial cells. The iPSC-derived CECs demonstrate key properties of native cells, including hexagonal morphology, tight junction formation, and pump function, showing the advantage of their autologous nature and minimizing immune rejection. However, iPSCs face significant hurdles, including the risk of tumorigenicity from undifferentiated cells, high production costs, and the complexity of differentiation protocols (Grönroos et al., 2021).

Ali et al. investigated the feasibility of the application of pluripotent stem cell-derived corneal endothelial cells (PSC-CECs). (Ali et al., 2021) By employing a robust differentiation protocol, pluripotent stem cells (PSCs) were treated with specific combinations of growth factors (e.g., TGF-β, BMP inhibitors) to promote lineage-specific differentiation. ROCK inhibitors were applied during and after differentiation for optimization of the process. The differentiated cells were subjected to gene and protein expression analyses, showing markers such as Vascular Endothelial (VE)-cadherin, ZO-1 and Na+/K+-ATPase, CD31, and Tie-2 that are characteristic of endothelial cells, excellent cellular viability, tight intercellular junctions, and proper pump function, which are indicative of functional corneal endothelial cells (Chatterjee et al., 2015; Ng et al., 2023; Sun et al., 2021; Zhao & Afshari, 2016; Zhou et al., 2023). When transplanted into animal models of corneal endothelial dysfunction (e.g., rabbits and monkeys), the PSC-CECs restored corneal clarity and thickness, indicating successful functionality in vivo (Hatou et al., 2021).

Another study used human induced pluripotent stem cells (*hiPSCs*) to generate neural crest-derived cells (*NCCs*) (Gong et al., 2021). Neural crest cells are multipotent cells that arise during embryonic development and can differentiate into various cell types, including corneal endothelial cells. By directing hiPSCs to differentiate into NCCs, the study aimed to replicate the natural development of corneal endothelial cells. Results were monitored for cell survival, migration, and functionality, including their ability to restore normal corneal transparency and function. They showed that hiPSC-derived NCCs could differentiate into CEC-like cells and integrate into the corneal tissue, adhered to the endothelium, and helped to reduce corneal thickness following intracameral injection to rabbit eyes after removal of the corneal endothelium.

A novel therapeutic strategy for corneal endothelial dysfunction explored the combination of human pluripotent stem cell-derived corneal endothelial precursors (*hPSC-CEPs*) with nicotinamide (*NAM*) (Li et al., 2022). NAM plays a vital role in protecting transplanted cells from premature aging and endothelial-mesenchymal transition (*EMT*), the processes triggered by the TGF-β-rich environment in the eye that can impair cell functionality. In rabbits, intracameral injection of hPSC-CEPs stimulated endogenous regeneration, enabling the host's cells to repair the cornea. However, in nonhuman primates, where corneal regeneration is limited, hPSC-CEPs integrated into the tissue and restored corneal clarity and function. Over 3 years, the NAM-supported cells maintained corneal transparency and prevented edema, showing their effectiveness as a long-term treatment. This innovative approach provides strong evidence for the simultaneous use of hPSC-CEPs and nicotinamide as a potential therapeutic strategy highlighting the dual benefits (Li et al., 2022).

## 5 Mesenchymal Stem Cells, Their Exosomes, and Secretome

Another increasingly recognized approach involves the use of mesenchymal stem cells (*MSCs*), which have gained significant attention due to their paracrine effects and differentiation potential. MSCs are multipotent cells isolated from various sources, including bone marrow, adipose tissue, umbilical cord (Wharton's jelly), dental pulp, and even corneal stroma. Unlike iPSCs or ESCs, MSCs primarily exert their effects through the secretion of growth factors and cytokines that reduce inflammation, promote cell survival, and enhance endogenous repair mechanisms. One critical distinction about MSCs is their dual role in therapeutic applications. While they can differentiate into mesodermal-lineage cells (e.g., osteoblasts, chondrocytes, and adipocytes) under the right conditions, their primary mechanism of action in most treatments, including corneal repair, is their paracrine activity. This trophic effect involves the secretion of bioactive factors, such as cytokines, growth factors, and extracellular vesicles (exosomes), which modulate the immune response, reduce inflammation, and promote tissue repair. For corneal endothelial repair, MSCs' secretion of such factors enhances the survival and proliferation of existing CECs rather than directly replacing the injured cells (Kobal et al., 2024; Zhang et al., 2015). Additionally, human mesenchymal stem cells (*hMSCs*) exhibit low immunogenicity, as they express low levels of major histocompatibility complex (*MHC*) class II molecules and lack costimulatory signals. This property reduces the risk of adverse immune responses and minimizes the need for long-term immunosuppressive therapy (Castro-Manrreza & Montesinos, 2015; Soleimani et al., 2023). Furthermore, MSCs have immunomodulatory properties, which may reduce the risk of immune rejection following transplantation (Joyce et al., 2012; Ye et al., 2022).

Several studies reported the potential applications of MSCs in corneal repair. For example, adipose-derived mesenchymal stem cells (*ADMSCs*) effectively promoted corneal regeneration in preclinical animal models (Alió Del Barrio et al., 2022; Marta et al., 2021). MSCs release paracrine factors such as VEGF and TGF-β, which help reduce inflammation, fibrosis, and cell death in damaged corneal tissues. Preclinical animal studies demonstrated that ADMSC-derived endothelial cells improved corneal clarity, reduced edema, and restored endothelial function (Alió Del Barrio et al., 2022; Marta et al., 2021). MSCs from Wharton's jelly, the gelatinous substance in the umbilical cord, demonstrate high regenerative potential. They can differentiate into CEC-like cells under specific conditions, expressing key markers and exhibiting functional properties. Feiertag et al. aimed to harness this potential to treat corneal diseases by differentiating these stem cells into CEC-like cells that exhibited typical endothelial morphology and expressed key markers associated with the corneal endothelium (Feiertag et al., 2020).

MSC-derived extracellular vesicles (*MSC-EVs*), also known as exosomes, have emerged as an acellular therapy for corneal endothelial damage, by safeguarding human CECs from cell death induced by endoplasmic reticulum (ER) stress (Buono et al., 2021). Research shows that MSC-EVs, which contain bioactive molecules like proteins, lipids, and RNAs, reduce ER stress-induced apoptosis in CECs by modulating signaling pathways, enhancing mitochondrial function, and reducing oxidative damage, suggesting that MSC-EVs could serve as a noninvasive alternative therapy for corneal endothelial injuries (Buono et al., 2021).

The condition media, the so-called secretome, of MSCs also contain trophic factors and chemokines that harbor regenerative properties. Djalilian et al. investigated the effect of corneal mesenchymal stromal cell secretome introduced to the anterior chamber of porcine eyes following phacoemulsification injury to corneal endothelial cells.(Rouhbakhshzaeri et al., 2019) They used staining with trypan blue to highlight the areas of endothelial cell loss. They showed that corneal MSC secretome significantly reduced endothelial cell loss (Fig. 1). These results were

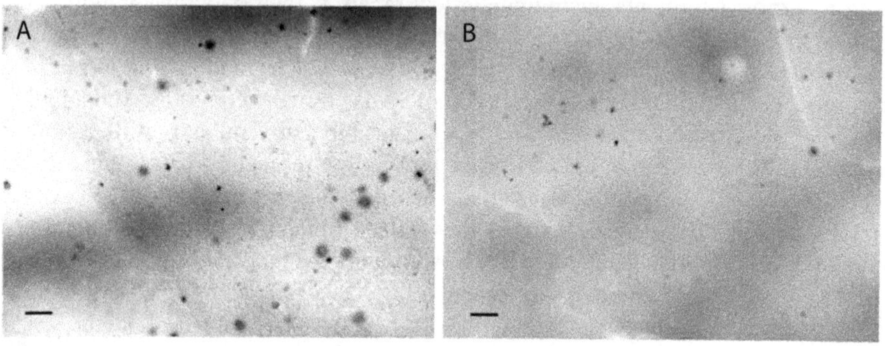


**Fig. 1** Trypan blue staining of porcine corneas with phacoemulsification corneal endothelial injury following incubation with unconditioned media (A) and corneal MSC-conditioned media (B). The blue stains represent areas of dead corneal endothelial cells. Porcine corneas treated with MSC-conditioned media (secretome) showed smaller and fewer areas of corneal endothelial cell loss [30]

validated in another ex vivo porcine model with corneal endothelial injury secondary to an oxidative injury.(Rabiee et al., 2020).

Rahimiyan et al. evaluated the effect of conditioned medium (*CM*) from stem cells derived from the apical papilla (*SCAPs*) on the proliferation of rat corneal endothelial cells (*rCECs*) (Rahimiyan et al., 2024). SCAPs are multipotent stem cells found in the apical papilla, a tissue located at the tip of the developing teeth. These cells have been shown to possess regenerative potential and can be induced to differentiate into various cell types. Exposure to a SCAP-conditioned medium significantly increased the proliferation of rat corneal endothelial cells. This suggests that the soluble factors in the secretome of SCAPs may promote the growth of corneal endothelial cells, which could be useful in regenerating the corneal endothelium.

## 6 Future Directions

In vitro and in vivo experiments on lab animals are encouraging that stem cell-based therapies have great potential as a therapeutic option for corneal endothelial disease. While more animal experiments are required to further explore and optimize the best choices of stem cells, routes of delivery, and other conditions to maximize the therapeutic effects, the next step is to translate the promising findings from animal experiments into treatments for human corneal endothelial disease. Phase 1 and phase 2 clinical trials are required to investigate the safety and efficacy of stem cell-based treatments such as MSCs, their secretome, and exosomes for corneal endothelial disease.

## References

Ali, M., Khan, S. Y., Gottsch, J. D., Hutchinson, E. K., Khan, A., & Riazuddin, S. A. (2021). Pluripotent stem cell-derived corneal endothelial cells as an alternative to donor corneal endothelium in keratoplasty. *Stem Cell Reports, 16*(9), 2320–2335. https://doi.org/10.1016/j.stemcr.2021.07.008

Alió Del Barrio, J. L., De la Mata, A., De Miguel, M. P., Arnalich-Montiel, F., Nieto-Miguel, T., El Zarif, M., Cadenas-Martín, M., López-Paniagua, M., Galindo, S., Calonge, M., & Alió, J. L. (2022). Corneal regeneration using adipose-derived mesenchymal stem cells. *Cells, 11*(16). https://doi.org/10.3390/cells11162549

Bartakova, A., Kunzevitzky, N. J., & Goldberg, J. L. (2014). Regenerative cell therapy for corneal endothelium. *Current Ophthalmology Reports, 2*(3), 81–90. https://doi.org/10.1007/s40135-014-0043-7

Buono, L., Scalabrin, S., De Iuliis, M., Tanzi, A., Grange, C., Tapparo, M., Nuzzi, R., & Bussolati, B. (2021). Mesenchymal stem cell-derived extracellular vesicles protect human corneal endothelial cells from endoplasmic reticulum stress-mediated apoptosis. *International Journal of Molecular Sciences, 22*(9). https://doi.org/10.3390/ijms22094930

Castro-Manrreza, M. E., & Montesinos, J. J. (2015). Immunoregulation by mesenchymal stem cells: Biological aspects and clinical applications. *Journal of Immunology Research, 2015*, 1–20. https://doi.org/10.1155/2015/394917

Chatterjee, I., Li, F., Kohler, E. E., Rehman, J., Malik, A. B., & Wary, K. K. (2015). Induced pluripotent stem (iPS) cell culture methods and induction of differentiation into endothelial cells. In *In* (Vol. 1357, pp. 311–327). Springer. https://doi.org/10.1007/7651_2015_203

Chen, J., Ou, Q., Wang, Z., Liu, Y., Hu, S., Liu, Y., Tian, H., Xu, J., Gao, F., Lu, L., Jin, C., Xu, G.-T., & Cui, H.-P. (2021). Small-molecule induction promotes corneal endothelial cell differentiation from human iPS cells. *Frontiers in Bioengineering and Biotechnology, 9*. https://doi.org/10.3389/fbioe.2021.788987

Feiertag, E., Maassen, J., Mueller, A., Harnisch, E., Skazik-Voogt, C., Engelmann, K., Schrage, N., & Gutermuth, A. (2020). From cord to eye: Wharton jelly-derived stem cells differentiate into corneal endothelial-like cells. *Cornea, 39*(7), 877–885. https://doi.org/10.1097/ico.0000000000002319

Feizi, S. (2018). Corneal endothelial cell dysfunction: Etiologies and management. *Therapeutic Advances in Ophthalmology, 10*, 251584141881580. https://doi.org/10.1177/2515841418815802

Fu, Y. K., Lin, M., Hung, K.-H., Yeh, L.-K., & Tan, H.-Y. (2024). Excessive corneal endothelial single cell loss following endothelial injuries. *Cold Spring Harbor Laboratory*. https://doi.org/10.1101/2024.09.09.24313319

Gong, Y., Duan, H., Wang, X., Zhao, C., Li, W., Dong, C., Li, Z., & Zhou, Q. (2021). Transplantation of human induced pluripotent stem cell-derived neural crest cells for corneal endothelial regeneration. *Stem Cell Research & Therapy, 12*(1), 214. https://doi.org/10.1186/s13287-021-02267-z

Grönroos, P., Ilmarinen, T., & Skottman, H. (2021). Directed differentiation of human pluripotent stem cells towards corneal endothelial-like cells under defined conditions. *Cells, 10*(2). https://doi.org/10.3390/cells10020331

Hatou, S., Sayano, T., Higa, K., Inagaki, E., Okano, Y., Sato, Y., Okano, H., Tsubota, K., & Shimmura, S. (2021). Transplantation of iPSC-derived corneal endothelial substitutes in a monkey corneal edema model. *Stem Cell Research, 55*, 102497. https://doi.org/10.1016/j.scr.2021.102497

Joyce, N. C., Harris, D. L., Markov, V., Zhang, Z., & Saitta, B. (2012). Potential of human umbilical cord blood mesenchymal stem cells to heal damaged corneal endothelium. *Molecular Vision, 18*, 547–564.

Kobal, N., Marzidovšek, M., Schollmayer, P., Maličev, E., Hawlina, M., & Marzidovšek, Z. L. (2024). Molecular and cellular mechanisms of the therapeutic effect of mesenchymal stem cells and extracellular vesicles in corneal regeneration. *International Journal of Molecular Sciences, 25*(20), 11121. https://doi.org/10.3390/ijms252011121

Li, Z., Duan, H., Jia, Y., Zhao, C., Li, W., Wang, X., Gong, Y., Dong, C., Ma, B., Dou, S., Zhang, B., Li, D., Cao, Y., Xie, L., Zhou, Q., & Shi, W. (2022). Long-term corneal recovery by simultaneous delivery of hPSC-derived corneal endothelial precursors and nicotinamide. *The Journal of Clinical Investigation, 132*(1). https://doi.org/10.1172/jci146658

Marta, C. M., Adrian, M., Jorge, F. D., Francisco, A. M., & De Miguel, M. P. (2021). Improvement of an effective protocol for directed differentiation of human adipose tissue-derived adult mesenchymal stem cells to corneal endothelial cells. *International Journal of Molecular Sciences, 22*(21). https://doi.org/10.3390/ijms222111982

Moshirfar, M., Thomson, A. C., & Ronquillo, Y. (2024). Corneal endothelial transplantation. In *StatPearls*. StatPearls Publishing. Copyright © 2024, StatPearls Publishing LLC.

Ng, X. Y., Peh, G. S. L., Yam, G. H.-F., Tay, H. G., & Mehta, J. S. (2023). Corneal endothelial-like cells derived from induced pluripotent stem cells for cell therapy. *International Journal of Molecular Sciences, 24*(15), 12433. https://doi.org/10.3390/ijms241512433

Price, M. O., Feng, M. T., & Price, F. W. J. (2021). Endothelial keratoplasty update 2020. *Cornea, 40*(5), 541–547. https://doi.org/10.1097/ico.0000000000002565

Rabiee, B., Bheri, A., Farooq, A. V., Yazdanpanah, G., Jabbehdari, S., Ghassemi, M., & Djalilian, A. R. (2020). The effect of mesenchymal stem cell secretome on corneal endothelial cell preservation in an oxidative injury model. *Cornea, 39*(11), 1426–1430.

Rahimiyan, K., Nasr Esfahani, M. H., & Karamali, F. (2024). The proliferative effects of stem cells from apical papilla-conditioned medium on rat corneal endothelial cells. *Wound Repair and Regeneration, 32*(3), 292–300. https://doi.org/10.1111/wrr.13161

Rouhbakhshzaeri, M., Rabiee, B., Azar, N., Ghahari, E., Putra, I., Eslani, M., & Djalilian, A. R. (2019). New ex vivo model of corneal endothelial phacoemulsification injury and rescue therapy with mesenchymal stromal cell secretome. *Journal of Cataract & Refractive Surgery, 45*(3), 361–366.

Shen, L., Sun, P., Zhang, C., Yang, L., Du, L., & Wu, X. (2017). Therapy of corneal endothelial dysfunction with corneal endothelial cell-like cells derived from skin-derived precursors. *Scientific Reports, 7*(1), 13400. https://doi.org/10.1038/s41598-017-13787-1

Soleimani, M., Mirshahi, R., Cheraqpour, K., Baharnoori, S. M., Massoumi, H., Chow, C., Shahjahan, S., Momenaei, B., Ashraf, M. J., Koganti, R., Ghassemi, M., Anwar, K. N., Jalilian, E., & Djalilian, A. R. (2023). Intrastromal versus subconjunctival injection of mesenchymal stem/stromal cells for promoting corneal repair. *The Ocular Surface, 30*, 187–195. https://doi.org/10.1016/j.jtos.2023.09.008

Song, Q., Yuan, S., An, Q., Chen, Y., Mao, F. F., Liu, Y., Liu, Q., & Fan, G. (2016). Directed differentiation of human embryonic stem cells to corneal endothelial cell-like cells: A transcriptomic analysis. *Experimental Eye Research, 151*, 107–114.

Sun, B., Bikkuzin, T., Li, X., Shi, Y., & Zhang, H. (2021). Human-induced pluripotent stem cells-derived corneal endothelial-like cells promote corneal transparency in a rabbit model of bullous keratopathy. *Stem Cells and Development, 30*(17), 856–864. https://doi.org/10.1089/scd.2020.0205

Van Den Bogerd, B., Zakaria, N., Adam, B., Matthyssen, S., Koppen, C., & Ní Dhubhghaill, S. (2019). Corneal endothelial cells over the past decade: Are we missing the mark(er)? *Translational Vision Science & Technology, 8*(6), 13. https://doi.org/10.1167/tvst.8.6.13

Ye, E. A., Chung, H. S., Park, Y., Sunwoo, J. H., Lee, W., Kim, J., Tchah, H., Lee, H., & Kim, J. Y. (2022). Induction of corneal endothelial-like cells from mesenchymal stem cells of the umbilical cord. *International Journal of Molecular Sciences, 23*(23). https://doi.org/10.3390/ijms232315408

Zhang, L., Coulson-Thomas, V. J., Ferreira, T. G., & Kao, W. W. Y. (2015). Mesenchymal stem cells for treating ocular surface diseases. *BMC Ophthalmology, 15*(S1), 155. https://doi.org/10.1186/s12886-015-0138-4

Zhao, J. J., & Afshari, N. A. (2016). Generation of human corneal endothelial cells via in vitro ocular lineage restriction of pluripotent stem cells. *Investigative Opthalmology & Visual Science, 57*(15), 6878. https://doi.org/10.1167/iovs.16-20024

Zhou, Q., Li, Z., & Duan, H. (2023). iPSC-derived corneal endothelial cells. *Handbook of Experimental Pharmacology, 281*, 257–276. https://doi.org/10.1007/164_2023_644

# Neural Regeneration

# Mechanisms and Challenges of Stem Cell-Mediated Neuroregeneration

**Abraham Al-Ahmad**

**Abstract** Neurological disorders, particularly neurodegenerative diseases, pose significant therapeutic challenges due to the limited regenerative capacity of the central nervous system (CNS) and the restrictive nature of brain barriers that obstruct drug delivery. Stem cell-based therapies present a promising approach for neuronal regeneration, neuroprotection, and disease modification. This chapter explores the developmental origins of neurons and glial cells, establishing a foundation for understanding regenerative strategies. It also includes a thorough review of clinical trials examining stem cell therapies for amyotrophic lateral sclerosis (ALS), Alzheimer's disease, Parkinson's disease, Huntington's disease, stroke, and macular degeneration. While numerous phase I/II trials affirm the safety of mesenchymal stem cells (MSCs), neural progenitor cells, and neurons derived from embryonic stem cells, their long-term effectiveness remains uncertain. Key challenges—including immune rejection, limited stem cell survival, and the challenging blood-brain barrier—hinder clinical translation. Nonetheless, advancements in stem cell engineering, biomaterials, and targeted delivery methods provide promising strategies to address these obstacles. This chapter underscores the need for further research and optimization to fully realize the therapeutic potential of stem cell-based treatments for neurological diseases.

**Keywords** Stem cell therapy · Neurodegenerative diseases · Neuronal regeneration · Blood-brain barrier · Clinical translation

A. Al-Ahmad (✉)
Department of Pharmaceutical Sciences, Jerry H. Hodge School of Pharmacy, Texas Tech University Health Sciences Center, Amarillo, TX, USA
e-mail: Abraham.Al-Ahmad@ttuhsc.edu

 
H. Das (ed.), *Tissue Repair and Regeneration*,
https://doi.org/10.1007/978-3-031-93677-7_11

# 1 Introduction

The nervous system is unique within the human body because it contains neurons, a cell type characterized by two important features: its excitatory function (which occurs via electrical or electrochemical coupling) and its postmitotic function (mature neurons are postmitotic cells).

These features of neurons represent a major challenge in treating neurological diseases: the former results in brain barriers (which limit the penetration of therapeutics inside the central nervous system), while the latter is hindered by the inability to regenerate and replace neurons lost due to neurodegenerative diseases.

The latter constitutes an important aspect of regenerative medicine focused on neurological diseases: repopulating cerebral areas with stem cells that can restore partially or fully the cellular functions lost due to neurodegenerative diseases. The premise and potential of such interventions are encouraging; however, the challenges related to both the materials and the environment remain significant today.

In this chapter, we will explore the perspective of stem cell therapy for neurological diseases. We will include a section on ontology, a review of the literature on clinical trials using stem cell-based therapies for selected neurological conditions, and a discussion of the challenges identified in these approaches. Additionally, this chapter provides a review of current literature on stem cell therapies aimed at macular degeneration, recognizing that the eyes are a natural extension of the central nervous system.

# 2 Ontology of the Nervous System

It is important to understand the developmental origin of neurons and non-neuron cells as found in the nervous system before we discuss the opportunities of regenerative medicine for neurological diseases. Neuron origins can be traced from the primordial embryonic layer, from a subpopulation that differentiates into neurectoderm during the gastrulation phase. Such neurectoderm leads to the rise of the neural tube, which will become the future brain (encephalon) and spinal cord in the developing fetus. Interestingly, during the neurulation stage, at the molecular level, the differentiation of such tube involves a gradient of various signaling molecules (morphogens) that will dictate the differentiation of the neural tube based on its rostrocaudal and dorsoventral pattern. The differentiation of the neural tube in its dorsoventral section involves two opposite gradients: a gradient represented by Sonic hedgehog (Shh) originating ventrally, counteracting a dorsal gradient producing both bone morphogen proteins (BMPs) and Wnt. On the other hand, the differentiation of the neural tube in its rostrocaudal pattern is driven by the presence of fibroblast growth factor (FGF) in its extremities, with a gradient of retinoic acid (RA) originating from the middle section of the neural tube and expanding towards both the rostral and the caudal extremities.

In the developing primitive brain, the neuroepithelial cells lining the neural tube quickly differentiate into the radial glia (RG), which will rapidly develop cellular elongation within the periphery towards the region that will become the future cortex. Such RG will undergo active cell divisions early during the gestation (fourth to fifth week of human gestation), which will result in the formation of immature and migrating neurons (neuroblasts) by providing these cells with a "cellular" rail to guide them in their migration, and ultimately populate the outmost layers of the future cortical region, populating the cortical layers from the most distal to the proximal layers (with migration achieved by 28th week of gestation). On the other hand, this RG also gives rise to the neural progenitor cells (NPCs) and to the outer RG cells (which will all. The RG cells in proximity to the ventricular surface will progressively evolve and differentiate towards an ependymocyte lineage (and delineate the future ventricular system with the ependyma). Both the proximal and distal RG will also contribute to the gliogenesis leading to the formation of astrocyte lineage (starting in the 28th week of gestation) and ultimately to the oligodendrocyte lineage (starting by the 32nd week of gestation).

NPCs provide a common ancestry for most of the cell types found in the central nervous system, including neurons and glial cells (except for the microglia and Schwann cells, which have distinct embryological origins). Notably, the differentiation of NPCs into the different cell types is not a synchronous process, as the different cell types will appear at different stages of brain development: first, neurons, followed by astrocytes, and finally by oligodendrocytes.

In human adults, the presence of neural stem cells (as a reservoir for the generation of new neurons) has been restricted to specific locations inside the brain, in areas referred to as "neurovascular niches" (areas marked by a rich neural stem cell population surrounded by angiogenic blood vessels) located in the olfactory bulb and the hippocampus. These niches serve as a reservoir to produce olfactory neurons and hippocampal neurons, playing an important role in the reception of olfactory stimuli, as well as memory formation and retention.

## 3 Use of Stem Cell-Based Therapies in Neurological Diseases

The differentiation of neuroblasts into the different types of neurons, based on the type of neurotransmitters or neuropeptides excreted, involves various signaling molecules that will direct them into the right lineage. Such directed differentiation is important when it comes to engineering stem cell-based therapies for certain diseases, in which the implantation of the right types of neurons in a location is needed. To maintain consistency, we will focus on the current literature addressing stem cell therapy as cellular therapy (direct administration of stem cells) rather than discussing stem cell-derived products (e.g., exosomes, biologics). We will not discuss neurological diseases indirectly targeted by stem cell therapies (e.g., autoimmune neurological diseases).

## 3.1 *Amyotrophic Lateral Sclerosis*

Amyotrophic lateral sclerosis (ALS) is a neurodegenerative disease that primarily affects motoneurons in the spinal cord. The etiology of the disease remains unclear, although two forms of the disease (sporadic and familial) have been documented over the years.

As of today, the use of stem cells in ALS has been primarily focused on disease modeling and drug discovery by using patient-derived induced pluripotent stem cells (iPSCs) (Ito et al., 2023; Morimoto et al., 2023). However, several clinical trials have been reported in the literature, with a primary end goal to demonstrate safety in patients (phase I/IIa clinical trials). One clinical trial used modified allogeneic neural progenitor cells (CNS10-NPC) designed to overexpress glial-derived neurotrophic factor (GDNF) (Baloh et al., 2022) in 18 patients. Such cells were injected into the dorsal horn of the spinal cord (with the hope of these cells migrating into the ventral horn, which houses motoneurons). This clinical trial concluded that the transplant appeared well-tolerated (neuropathic pain was reported by 50% of the patients), and there were no signs of immune rejection (a common issue reported with stem cell transplants in the central nervous system, which appears diminished using immunosuppressive drugs). Although the safety profile looked promising, there were also clinical outcomes that raised the concerns of the ability of such intervention to succeed in phase II clinical trials: patients were able to identify the injection site, the limited ability of such progenitor cells to migrate into the ventral horn and also no differences in the number of motoneurons (ChAT+ motor neurons) in the treated side from the untreated side.

Another phase I/IIa study, using embryonic stem cell-derived astrocytes (AstroRx®), was recently reported (Gotkine et al., 2023). Again, this study was exploratory in nature, and the primary outcome was assessing safety, yet the authors reported a possible benefit in the first 3 months (such results have to be tempered to the small patient size ($N = 10$), the early intervention and also the possible need to inject cells every 3 months.

Finally, the latest trial (phase I/IIa) reported in the literature was based on the use of multilineage-differentiating stress-enduring (MUSE) cells (CL2020). Unlike other trials using cells differentiated into a single lineage (e.g., neural progenitor cells and astrocytes), these cells originated from a subpopulation of bone marrow mesenchymal stem cells. The authors concluded that there was a very modest slow-down of the progression of the disease, which has to be further investigated in a larger clinical trial.

## 3.2 *Alzheimer's Disease*

Alzheimer's disease is a progressive neurodegenerative disease characterized by the progressive loss of neurons from the hippocampus which ultimately extends to the cerebral cortex. The etiology of Alzheimer's disease remains unknown, but it is

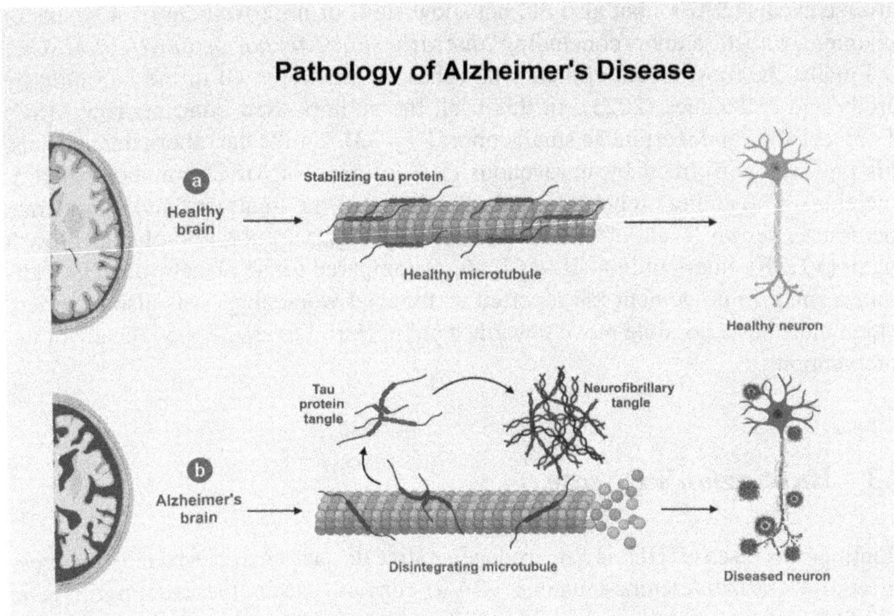


**Fig. 1** Healthy brain and Alzheimer's disease pathology. (**a**) In a healthy state, neurons maintain structural integrity with stable microtubules, supported by properly functioning tau proteins. (**b**) In Alzheimer's disease, tau proteins become hyperphosphorylated, leading to neurofibrillary tangles that disrupt the stability of microtubules. The figure illustrates the transition from a normal neuronal structure to the disrupted state that is characteristic of Alzheimer's disease

often accompanied by the presence of amyloid beta (Aβ)-rich senile plaques in the brain parenchyma and around brain blood vessels, as well as the presence of neurofibrillary tangles enriched in hyperphosphorylated Tau protein (a microtubule-associated protein).

The number of clinical trials using stem cells in targeting Alzheimer's disease remains limited in the literature. The first documented clinical trial is from Duma and colleagues (2019) (Fig. 1).

In this study, the authors performed an intracerebroventricular (ICV) injection of cells obtained from the stromal vascular fraction (SVF) obtained from the patient's adipose tissue. The SVF composition is usually a mixture of various cell types, which include adipose-derived stem cells (ADSCs), hematopoietic stem cells (HSCs), vascular cells (e.g., endothelial cells, pericytes), immune cells (e.g., T cells, macrophage), and other cells (Gentile et al., 2012). The authors reported an overall good tolerance to the treatment; however, the outcomes remained limited, with the authors reporting that two patients in their small cohort ($N = 10$) did "subjectively better" at 15- and 20-month follow-up visits.

Another clinical trial (phase I) based on ICV injections, this time using human umbilical cord mesenchymal stem cells (hUCB-MSCs), by Kim and colleagues, was reported (2021). Again, the intervention was not accompanied by serious adverse events (SAEs), but also did not show signs of improvement by 4 weeks of treatment, with the authors concluding "*due to the short life span of the hUCB-MSCs.*"

Finally, the most recent clinical trial (phase I) was reported in the literature by Brody and colleagues (2023). In this trial, the authors used bone marrow MSCs (Lomecel-B©) and recruited a small cohort ($N = 33$). Unlike the other clinical trials, this one was performed by intravenous (IV) infusion of MSCs, rather than ICV injections. The authors reported good tolerability of the treatment and a moderate increase in serum levels of various biomarkers (e.g., vascular endothelial growth factor (VEGF), interleukin-4 (IL-4), IL-6…) compared to the placebo arm. In addition, a small improvement (as reported as reduced worsening) was also reported, which indicates a possible move towards a phase II trial to assess the efficacy of the intervention.

## 3.3 Huntington's Disease

Huntington's disease (HD) is a neurodegenerative disease characterized by the presence of Cytidine-Adenine-Guanine (CAG) repeats within the Huntingtin gene (HTT) above the normal value (<26 CAG repeats). Such an abnormally high number of CAG repeats results in the generation of a polyglutamine domain, which provides a gain-of-function through the ability of the mutated Huntingtin protein (mHTT) to aggregate together and result in neurotoxicity. As of today, there are no reported clinical trials using stem cells in HD patients.

## 3.4 Macular Degeneration

Macular degeneration (MD) (age-related macular degeneration (AMD) is a condition associated with a progressive loss of the central vision in juvenile patients (Stargardt Macular Degeneration, STGD1) (Fujinami et al., 2015), or in patients aged 50 and older (Stefansson et al., 2011). AMD is usually divided into two subtypes: the "dry" AMD (most common form) in which the blood vessels under the macula become atrophied and lead to the deposition of drusen. The "wet" AMD, on the other hand, is characterized by the presence of an abnormal angiogenesis, usually leading to the formation of unstable and leaky blood vessels.

The first clinical trial (phase I) reported in the literature was published by Schwartz and colleagues (2015), using human embryonic stem cell (hESC)-derived retinal pigmented epithelial cells (RPEs) in patients with AMD ($N = 9$) with RPE transplanted. The authors reported that the transplantation to the patients appeared well-tolerated and even reported slight improvement in terms of vision. A second clinical trial (phase I), using the same approach (hESC-derived RPEs), was published around the same time by Song and colleagues (2015) in two patients with AMD. Although the short-term follow-up showed encouraging signs of safety (and a slight improvement in the vision of the patient), one AMD patient had to be removed from the trial due to the development of choroid neovascularization (CNV) at 33 weeks postinjection. Interestingly, the reported use of a special bioengineered material (ultrathin perylene membrane) seems to alleviate the risk of neovascularization associated with hESC-RPE (Kashani et al., 2021).

The latest clinical trial (phase I) using hESC-RPEs has been published by Li and colleagues (2021) in patients with STGD1, in a small subset of patients ($N = 7$), over 5 years. The outcome in terms of safety was considered adequate. Early results in terms of efficacy suggested visual improvements in three of the patients in the short-term (<6 months), but the 5-year follow-up suggested a worsening of the visual activity in the operated eye compared to the fellow (untreated) eye. A similar outcome in terms of safety for STGD1 patients was reported later by Brant Fernandes and colleagues (2023).

## 3.5 *Parkinson's Disease*

Parkinson's disease (PD) is a neurodegenerative disease targeting a subset of neurons of the substantia nigra (dopaminergic neurons), resulting in the onset of α-synuclein accumulation and aggregation inside such cells (as Lewy bodies) and ultimately death of such neurons. The clinical presentation of PD usually involves the progressive onset of tremors in various parts of the body, muscle stiffness, reduced movement speed, and impaired balance and coordination (Fig. 2). As of today, there is no cure for the disease, with only interventions aimed at relieving the symptoms and improving the patient's quality of life. The peculiar nature of the disease made it a strong candidate for targeting with stem cell therapies.

The first clinical trial (non-randomized open-label study) reported in the literature is from Venkataramana and colleagues (2010), in which seven patients ($N = 7$) were selected in which bone marrow MSCs (BM-MSCs) were injected in the subventricular zone (SVZ). The patients were followed up to 36 months posttransplantation. The authors claimed that the patients overall supported well the transplants, did not display any specific SAEs, and even alleged some improvements in patients. Similar descriptions were made in another study (also using BM-MSCs, but this time autologous administration) (Brazzini et al., 2010), as well as the use of fetal RPEs (Yin et al., 2012).

Later, the approach has transitioned to the use of neural progenitor cells (NPCs) in place of MSCs (Jiang et al., 2024; Lige & Zengmin, 2016; Madrazo et al., 2019). All these studies were registered as phase I clinical trials and reported evidence of safety as well as alleged improvement in patient outcomes.

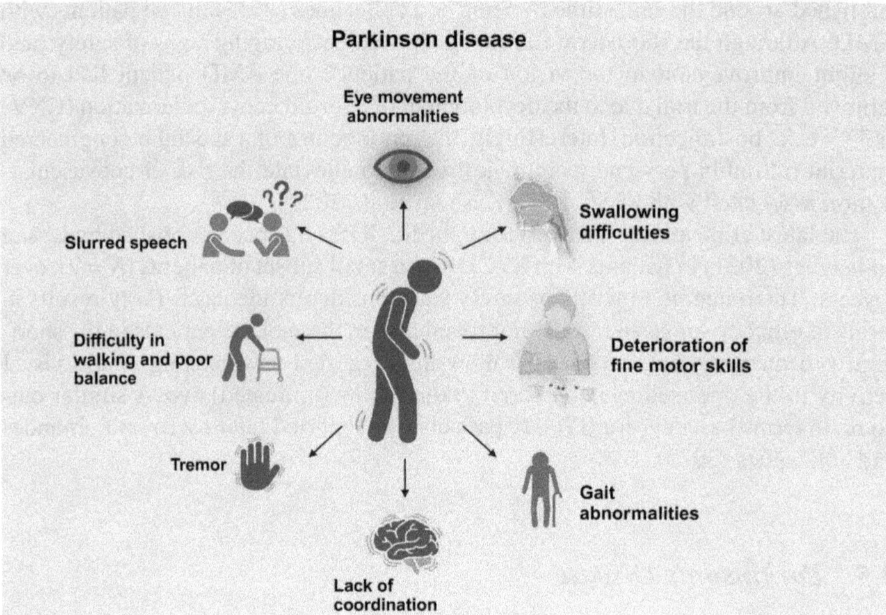


**Fig. 2** Neurological and motor impairments in Parkinson's disease. Parkinson's disease is marked by the loss of dopaminergic neurons in the substantia nigra, resulting in various motor and non-motor functional abnormalities. The figure depicts key manifestations, including eye movement abnormalities, difficulties with swallowing, and slurred speech, as well as the decline of fine motor skills, walking difficulties, poor balance, tremors, gait abnormalities, and lack of coordination

## 3.6 Stroke

Stroke is the fifth leading cause of death in the United States and a major cause of disability in adults. Most of the stroke events are characterized as "ischemic stroke" resulting in the occlusion of one or several blood vessels by a blood clot, while "hemorrhagic stroke" (lesser in incidence, but greater in fatality rate) is characterized by the rupture of a blood vessel within the brain, resulting in blood leak within the brain parenchyma.

In both cases, such events lead to neuron starvation for oxygen and nutrients, which results in massive cell depolarization and glutamate (a major neurotransmitter in mammalian brains) release. Such a massive release of glutamate results in "excitotoxicity" by activating various glutamate receptors, including receptor channels such as NMDA and AMPA receptors, a significant increase in intracellular calcium level, and activation of the calpains pathways, ultimately leading to apoptosis onset and neuron cell death.

It is commonly accepted that the recovery phase following such an insult modestly involves neuronal regeneration within the ischemic core while remodeling in the penumbra region surrounding such core could contribute to the restoration of some function within weeks of the injury. It is also assumed that such recovery

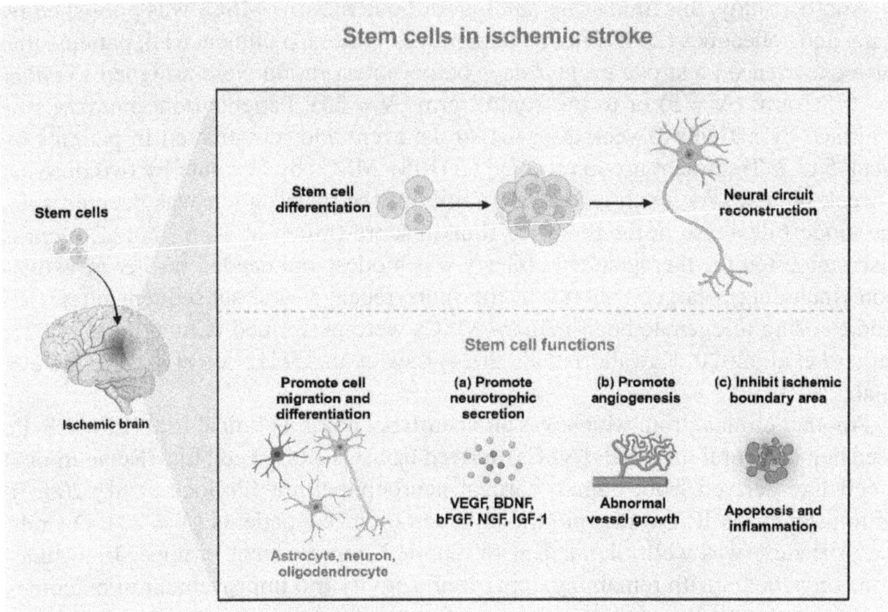


**Fig. 3** Stem cells are critical in recovery from ischemic stroke. The figure illustrates how stem cells contribute to neuroprotection and repair by (**a**) secreting neurotrophic factors such as vascular endothelial growth factor (VEGF), brain-derived neurotrophic factor (BDNF), basic fibroblast growth factor (bFGF), nerve growth factor (NGF), and insulin-like growth factor-1 (IGF-1), which support neuronal survival and regeneration; (**b**) promoting angiogenesis to restore blood supply and facilitate the formation of functional vasculature; and (**c**) inhibiting apoptosis and inflammation in the ischemic boundary zone, thereby reducing secondary neuronal damage

process by 6 months poststroke event will arrive at a halt. Therefore, the idea of transplanting immature neurons or neural progenitor cells to help recover cerebral functions lost at the ischemic core appears tempting (Fig. 3).

The first clinical study reported in the literature is the pilot study by Kondziolka and colleagues (2000). In this open-label clinical trial, 12 patients ($N = 12$) who had a stroke event in the past (mean time...55 months) were enrolled. They utilized differentiated (using 10 μM retinoic acid) NT2/D1 human carcinoma cell line (Pleasure & Lee, 1993), a cell line that harbors neural progenitor cell features upon induction with RA treatment, injected within the injury and followed patients for up to 52 weeks postinjection.

All these patients showed signs of safety, with no SAEs reported over 52 weeks. Out of the 12 patients, 6 reported improved scores (European Stroke Scale (ESS)) compared to baseline, three remained unchanged, and three showed worsened scores at 6 months post-intervention. Assessment of the region's cellular repopulation (using positron-emission tomography) at 6 months noted a modest increase in $^{18}$F-deoxyglucose ($^{18}$FDG) in the transplant site or the ipsilateral site in about half of the patients.

Another study, this time using autologous bone marrow MSCs was published by Bang and colleagues (2005). In this randomized phase I/II clinical trial, patients that have experienced a stroke event 7 days before intervention were assigned to either the MSC arm ($N = 5$) or to the control arm ($N = 25$). Patients bone marrow was aspirated by patients 1 week after the stroke event and reimplanted in patients by week 5 to 7. Treatment group received CD105+ MSCs by IV route by two doses at 1-week interval. As much as the tolerability of the intervention was deemed safe, the modest decrease in the modified Rankin score (mRS) at 3-, 6- and 12-months visit suggested the therapeutical efficacy was modest and needed further investigation (including a larger cohort). In the more recent years, subsequent phase I/II studies using allogeneic bone-marrow MSCs were performed (Chung et al., 2021; Jaillard et al., 2020; Kawabori et al., 2024; Law et al., 2021; Lee et al., 2022; Levy et al., 2019).

Another clinical trial, which was an open-label phase I clinical trial (PISCES-1), used human neural stem cells (NSCs) derived from CTX0E03 cell line (ReNeurons® a cell line derived from human cortical neuroepithelium (Pollock et al., 2006)), administered via ICV route into the putamen of stroke patients ($N = 23$). Overall, the procedure was well-tolerated, with reported improvement in upper limb functions on patients with remaining upper limb activity (no improvement in outcomes in patients that have lost their upper limb functions).

Notably, there is a phase I open-label clinical trial (PASSIoN) performed on an infant patient population targeting pediatric acute ischemic stroke (PAIS) (Baak et al., 2022), in which bone marrow MSCs were administered nasally within seven days of diagnosis, with a 3-month follow-up of the cohort. The procedure was deemed safe by the authors, but the authors have been cautious to conclude on efficacy at this stage.

A clinical trial involving IV infusion of bone marrow mononuclear cells (MNCs) in a phase I clinical trial was reported and showed safety with modest efficacy (reduction in mRS by 1 point) (Vahidy et al., 2019).

A phase I clinical study using autologous adipose MSCs was also reported in the literature (Chiu et al., 2022), quickly followed by a phase II study from another group (de Celis-Ruiz et al., 2022). In this double-blinded, placebo-controlled, randomized clinical trial, 20 patients ($N = 20$) were enrolled. Cells were administered via IV infusion to patients as a single treatment in patients within 2 weeks of their stroke event. Overall, the study showed the safety of the treatment but fell short of showing efficacy of the treatment, with no statistical differences noted between treatment and placebo group after 18 months of treatment. A limitation of such short outcome was the very small number size of the cohort ($N = 4$ in treatment group, $N = 9$ in placebo group), which could have underpowered the statistical analysis.

A phase I/II randomized placebo-controlled study (J-REPAIR) using dental pulp mesenchymal stem cells (DPSCs) has also been reported in the literature (Suda et al., 2022). In this study, patients received a single IV infusion of JTR-161 DPSCs within 48 hours of ischemic stroke events and were followed for up to 12 months. The study published (with the clinical trial (NCT04608838, labeled as "completed" in the clinicaltrials.gov website) fell short of providing published results by the time of this literature review was performed.

A clinical trial using Muse cells was reported by Niizuma and colleagues (2023). In this phase II placebo-controlled randomized clinical trial (treatment $N = 27$, placebo $N = 10$), MSCs derived from the CL2020 cell line were administered via IV infusion in patients (with an average poststroke intervention of 19–20 days), with a primary endpoint for safety at week 12 and week 52 post-intervention. The CL2020-treated group showed no additional adverse events compared to the placebo arm, while the authors reported a decrease of 1 point and 2 points in the mRS score at 12 and 52 weeks, respectively. Notably, the upper limbs' function was significantly improved in the treatment group compared to placebo, but no differences were observed in the lower limbs.

One recent study has been reporting the use of umbilical cord blood stem (UCB) cells (COBIS2) (Laskowitz et al., 2024). This phase II, double-blinded, placebo-controlled randomized clinical trial study, injected UCB cells via IV infusion. Patients were monitored for the first 24 hours and were followed by 30- and 90-day post-intervention. The results suggested a positive effect associated with treatment in regard to the mRS score, and interestingly, the data also suggested that a delayed intervention (9–11 days latency between stroke event and intervention) favored such improvement.

Another study (phase II, placebo-controlled randomized clinical trial) using UCB has been recently published by another group (Lin et al., 2024). Although this study showed safety, it failed to show statistically meaningful efficacy, possibly due to the relatively low cohort ($N = 14$ per group).

## 3.7 Conclusion

Neurological diseases, in particular neurodegenerative diseases, remain extremely challenging diseases to treat. Despite the important investment made in the research of therapeutics targeting these diseases, the prognosis and the treatment remain at most symptomatic and only minimally impact the progression of such diseases.

The possible use of regenerative medicine, especially using stem cells, has been raising promising prospects for these diseases. Yet, this chapter also highlighted that despite showing signs of safety, the paucity of phase II clinical trials in the last 25 years raises the shortfall of such promising techniques.

It is important to note that most of the clinical trials enumerated in this chapter heavily rely on the use of mesenchymal stem cells, which were obtained in their majority from bone marrow, followed from adipose tissues.

One possible issue is the risk of immune rejection, as stem cells can still raise an immune reaction (and therefore a risk of graft rejection) despite being administered in an immune privileged tissue (e.g., brain), as several studies have reported the use of immunomodulators concomitantly with the intervention. The limited survival of these cells may be beneficial in the short-term for safety but may also have contributed to the modest therapeutic effect documented.

The other issue that we could highlight is the administration route utilized for these cells. The use of the intravenous (IV) route appeared to be the preferred route, as it is much less invasive than the intracerebroventricular (ICV) route. The presence of a blood-brain barrier (BBB) and blood-spinal cord barrier (BSCB) will greatly limit the ability of stem cells administered via IV route, as the presence of such barriers does not allow the diapedesis of such cells due to their size and by the absence of adhesion molecules needed for cell-cell adhesion.

In conclusion, despite 20 years of research and effort aimed at treating neurological diseases with stem cells, the efficacy of such an approach remains limited despite the track record of safety.

## References

Baak, L. M., Wagenaar, N., van der Aa, N. E., Groenendaal, F., Dudink, J., Tataranno, M. L., Mahamuud, U., Verhage, C. H., Eijsermans, R., Smit, L. S., Jellema, R. K., de Haan, T. R., Ter Horst, H. J., de Boode, W. P., Steggerda, S. J., Prins, H. J., de Haar, C. G., de Vries, L. S., van Bel, F., Heijnen, C. J., Nijboer, C. H., & Benders, M. (2022). Feasibility and safety of intranasally administered mesenchymal stromal cells after perinatal arterial ischaemic stroke in The Netherlands (PASSIoN): A first-in-human, open-label intervention study. *Lancet Neurology, 21*(6), 528–536. https://doi.org/10.1016/S1474-4422(22)00117-X

Baloh, R. H., Johnson, J. P., Avalos, P., Allred, P., Svendsen, S., Gowing, G., Roxas, K., Wu, A., Donahue, B., Osborne, S., Lawless, G., Shelley, B., Wheeler, K., Prina, C., Fine, D., Kendra-Romito, T., Stokes, H., Manoukian, V., Muthukumaran, A., Garcia, L., Banuelos, M. G., Godoy, M., Bresee, C., Yu, H., Drazin, D., Ross, L., Naruse, R., Babu, H., Macklin, E. A., Vo, A., Elsayegh, A., Tourtellotte, W., Maya, M., Burford, M., Diaz, F., Patil, C. G., Lewis, R. A., & Svendsen, C. N. (2022). Transplantation of human neural progenitor cells secreting GDNF into the spinal cord of patients with ALS: A phase 1/2a trial. *Nature Medicine, 28*(9), 1813–1822. https://doi.org/10.1038/s41591-022-01956-3

Bang, O. Y., Lee, J. S., Lee, P. H., & Lee, G. (2005). Autologous mesenchymal stem cell transplantation in stroke patients. *Annals of Neurology, 57*(6), 874–882. https://doi.org/10.1002/ana.20501

Brant Fernandes, R. A., Lojudice, F. H., Zago Ribeiro, L., Santos da Cruz, N. F., Polizelli, M. U., Cristovam, P. C., Innocenti, F., Morimoto, L., Magalhaes, O., Jr., Ferraz Sallum, J. M., Penha, F. M., Sogayar, M. C., Belfort, R., Jr., & Maia, M. (2023). Transplantation of subretinal stem cell-derived retinal pigment epithelium for Stargardt disease: A phase I clinical trial. *Retina, 43*(2), 263–274. https://doi.org/10.1097/IAE.0000000000003655

Brazzini, A., Cantella, R., De la Cruz, A., Yupanqui, J., Leon, C., Jorquiera, T., Brazzini, M., Ortega, M., & Saenz, L. N. (2010). Intraarterial autologous implantation of adult stem cells for patients with Parkinson disease. *Journal of Vascular and Interventional Radiology, 21*(4), 443–451. https://doi.org/10.1016/j.jvir.2010.01.008

Brody, M., Agronin, M., Herskowitz, B. J., Bookheimer, S. Y., Small, G. W., Hitchinson, B., Ramdas, K., Wishard, T., McInerney, K. F., Vellas, B., Sierra, F., Jiang, Z., McClain-Moss, L., Perez, C., Fuquay, A., Rodriguez, S., Hare, J. M., Oliva, A. A., Jr., & Baumel, B. (2023). Results and insights from a phase I clinical trial of Lomecel-B for Alzheimer's disease. *Alzheimers Dement, 19*(1), 261–273. https://doi.org/10.1002/alz.12651

Chiu, T. L., Baskaran, R., Tsai, S. T., Huang, C. Y., Chuang, M. H., Syu, W. S., Harn, H. J., Lin, Y. C., Chen, C. H., Huang, P. C., Wang, Y. F., Chuang, C. H., Lin, P. C., & Lin, S. Z. (2022). Intracerebral transplantation of autologous adipose-derived stem cells for chronic ischemic stroke: A phase I study. *Journal of Tissue Engineering and Regenerative Medicine, 16*(1), 3–13. https://doi.org/10.1002/term.3256

Chung, J. W., Chang, W. H., Bang, O. Y., Moon, G. J., Kim, S. J., Kim, S. K., Lee, J. S., Sohn, S. I., Kim, Y. H., & Collaborators, S.-. (2021). Efficacy and safety of intravenous mesenchymal stem cells for ischemic stroke. *Neurology, 96*(7), e1012–e1023. https://doi.org/10.1212/WNL.0000000000011440

de Celis-Ruiz, E., Fuentes, B., Alonso de Lecinana, M., Gutierrez-Fernandez, M., Borobia, A. M., Gutierrez-Zuniga, R., Ruiz-Ares, G., Otero-Ortega, L., Laso-Garcia, F., Gomez-de Frutos, M. C., & Diez-Tejedor, E. (2022). Final results of allogeneic adipose tissue-derived mesenchymal stem cells in acute ischemic stroke (AMASCIS): A phase II, randomized, double-blind, placebo-controlled, single-center, pilot clinical trial. *Cell Transplant, 31*, 9636897221083863. https://doi.org/10.1177/09636897221083863

Duma, C., Kopyov, O., Kopyov, A., Berman, M., Lander, E., Elam, M., Arata, M., Weiland, D., Cannell, R., Caraway, C., Berman, S., Scord, K., Stemler, L., Chung, K., Khoudari, S., McRory, R., Duma, C., Farmer, S., Bravo, A., Yassa, C., Sanathara, A., Singh, E., & Rapaport, B. (2019). Human intracerebroventricular (ICV) injection of autologous, non-engineered, adipose-derived stromal vascular fraction (ADSVF) for neurodegenerative disorders: Results of a 3-year phase 1 study of 113 injections in 31 patients. *Molecular Biology Reports, 46*(5), 5257–5272. https://doi.org/10.1007/s11033-019-04983-5

Fujinami, K., Zernant, J., Chana, R. K., Wright, G. A., Tsunoda, K., Ozawa, Y., Tsubota, K., Robson, A. G., Holder, G. E., Allikmets, R., Michaelides, M., & Moore, A. T. (2015). Clinical and molecular characteristics of childhood-onset Stargardt disease. *Ophthalmology, 122*(2), 326–334. https://doi.org/10.1016/j.ophtha.2014.08.012

Gentile, P., Orlandi, A., Scioli, M. G., Di Pasquali, C., Bocchini, I., & Cervelli, V. (2012). Concise review: Adipose-derived stromal vascular fraction cells and platelet-rich plasma: Basic and clinical implications for tissue engineering therapies in regenerative surgery. *Stem Cells Translational Medicine, 1*(3), 230–236. https://doi.org/10.5966/sctm.2011-0054

Gotkine, M., Caraco, Y., Lerner, Y., Blotnick, S., Wanounou, M., Slutsky, S. G., Chebath, J., Kuperstein, G., Estrin, E., Ben-Hur, T., Hasson, A., Molakandov, K., Sonnenfeld, T., Stark, Y., Revel, A., Revel, M., & Izrael, M. (2023). Safety and efficacy of first-in-man intrathecal injection of human astrocytes (AstroRx(R)) in ALS patients: Phase I/IIa clinical trial results. *Journal of Translational Medicine, 21*(1), 122. https://doi.org/10.1186/s12967-023-03903-3

Ito, D., Morimoto, S., Takahashi, S., Okada, K., Nakahara, J., & Okano, H. (2023). Maiden voyage: Induced pluripotent stem cell-based drug screening for amyotrophic lateral sclerosis. *Brain, 146*(1), 13–19. https://doi.org/10.1093/brain/awac306

Jaillard, A., Hommel, M., Moisan, A., Zeffiro, T. A., Favre-Wiki, I. M., Barbieux-Guillot, M., Vadot, W., Marcel, S., Lamalle, L., Grand, S., & Detante, O. (2020). Autologous mesenchymal stem cells improve motor recovery in subacute ischemic stroke: A randomized clinical trial. *Translational Stroke Research, 11*(5), 910–923. https://doi.org/10.1007/s12975-020-00787-z

Jiang, S., Wang, H., Yang, C., Feng, F., Xu, D., Zhang, M., Xie, M., Cui, R., Zhu, Z., Jia, C., Liu, L., Wang, L., Yang, X., Yang, Y., Hao, H., Liu, Z., Wu, Z., Leng, L., Li, X., Sun, X., Zhao, X., Xu, J., Zhang, Y., Wan, X., Bao, X., & Wang, R. (2024). Phase 1 study of safety and preliminary efficacy of intranasal transplantation of human neural stem cells (ANGE-S003) in Parkinson's disease. *Journal of Neurology, Neurosurgery, and Psychiatry, 95*(12), 1102–1111. https://doi.org/10.1136/jnnp-2023-332921

Kashani, A. H., Lebkowski, J. S., Rahhal, F. M., Avery, R. L., Salehi-Had, H., Chen, S., Chan, C., Palejwala, N., Ingram, A., Dang, W., Lin, C. M., Mitra, D., Pennington, B. O., Hinman, C., Faynus, M. A., Bailey, J. K., Mohan, S., Rao, N., Johnson, L. V., Clegg, D. O., Hinton, D. R., & Humayun, M. S. (2021). One-year follow-up in a phase 1/2a clinical trial of an allogeneic RPE cell bioengineered implant for advanced dry age-related macular degeneration. *Translational Vision Science & Technology, 10*(10), 13. https://doi.org/10.1167/tvst.10.10.13

Kawabori, M., Kuroda, S., Shichinohe, H., Kahata, K., Shiratori, S., Ikeda, S., Harada, T., Hirata, K., Tha, K. K., Aragaki, M., Terasaka, S., Ito, Y. M., Nishimoto, N., Ohnishi, S., Yabe, I., Kudo, K., Houkin, K., & Fujimura, M. (2024). Intracerebral transplantation of MRI-trackable autologous bone marrow stromal cells for patients with subacute ischemic stroke. *Med, 5*(5), 432–444 e434. https://doi.org/10.1016/j.medj.2024.02.009

Kim, H. J., Cho, K. R., Jang, H., Lee, N. K., Jung, Y. H., Kim, J. P., Lee, J. I., Chang, J. W., Park, S., Kim, S. T., Moon, S. W., Seo, S. W., Choi, S. J., & Na, D. L. (2021). Intracerebroventricular injection of human umbilical cord blood mesenchymal stem cells in patients with Alzheimer's disease dementia: A phase I clinical trial. *Alzheimer's Research & Therapy, 13*(1), 154. https://doi.org/10.1186/s13195-021-00897-2

Kondziolka, D., Wechsler, L., Goldstein, S., Meltzer, C., Thulborn, K. R., Gebel, J., Jannetta, P., DeCesare, S., Elder, E. M., McGrogan, M., Reitman, M. A., & Bynum, L. (2000). Transplantation of cultured human neuronal cells for patients with stroke. *Neurology, 55*(4), 565–569. https://doi.org/10.1212/wnl.55.4.565

Laskowitz, D. T., Troy, J., Poehlein, E., Bennett, E. R., Shpall, E. J., Wingard, J. R., Freed, B., Belagaje, S. R., Khanna, A., Jones, W., Volpi, J. J., Marrotte, E., & Kurtzberg, J. (2024). A randomized, placebo-controlled, phase II trial of intravenous allogeneic non-HLA matched, unrelated donor, cord blood infusion for ischemic stroke. *Stem Cells Translational Medicine, 13*(2), 125–136. https://doi.org/10.1093/stcltm/szad080

Law, Z. K., Tan, H. J., Chin, S. P., Wong, C. Y., Wan Yahya, W. N. N., Muda, A. S., Zakaria, R., Ariff, M. I., Ismail, N. A., Cheong, S. K., Abdul Wahid, S. F., & Mohamed Ibrahim, N. (2021). The effects of intravenous infusion of autologous mesenchymal stromal cells in patients with subacute middle cerebral artery infarct: A phase 2 randomized controlled trial on safety, tolerability and efficacy. *Cytotherapy, 23*(9), 833–840. https://doi.org/10.1016/j.jcyt.2021.03.005

Lee, J., Chang, W. H., Chung, J. W., Kim, S. J., Kim, S. K., Lee, J. S., Sohn, S. I., Kim, Y. H., Bang, O. Y., & Collaborators, S.-. (2022). Efficacy of intravenous mesenchymal stem cells for motor recovery after ischemic stroke: A neuroimaging study. *Stroke, 53*(1), 20–28. https://doi.org/10.1161/STROKEAHA.121.034505

Levy, M. L., Crawford, J. R., Dib, N., Verkh, L., Tankovich, N., & Cramer, S. C. (2019). Phase I/II study of safety and preliminary efficacy of intravenous allogeneic mesenchymal stem cells in chronic stroke. *Stroke, 50*(10), 2835–2841. https://doi.org/10.1161/STROKEAHA.119.026318

Li, S. Y., Liu, Y., Wang, L., Wang, F., Zhao, T. T., Li, Q. Y., Xu, H. W., Meng, X. H., Hao, J., Zhou, Q., Wang, L., & Yin, Z. Q. (2021). A phase I clinical trial of human embryonic stem cell-derived retinal pigment epithelial cells for early-stage Stargardt macular degeneration: 5-years' follow-up. *Cell Proliferation, 54*(9), e13100. https://doi.org/10.1111/cpr.13100

Lige, L., & Zengmin, T. (2016). Transplantation of neural precursor cells in the treatment of Parkinson disease: An efficacy and safety analysis. *Turkish Neurosurgery, 26*(3), 378–383. https://doi.org/10.5137/1019-5149.JTN.10747-14.4

Lin, H. S., Sung, P. H., Huang, S. H., Lin, W. C., Chiang, J. Y., Ma, M. C., Chen, Y. L., Chen, K. H., Lee, F. Y., Ko, S. F., & Yip, H. K. (2024). Long term outcomes of intracarotid arterial transfusion of circulatory-derived autologous CD34 + cells for acute ischemic stroke patients-A randomized, open-label, controlled phase II clinical trial. *Stem Cell Research & Therapy, 15*(1), 443. https://doi.org/10.1186/s13287-024-04021-7

Madrazo, I., Kopyov, O., Avila-Rodriguez, M. A., Ostrosky, F., Carrasco, H., Kopyov, A., Avendano-Estrada, A., Jimenez, F., Magallon, E., Zamorano, C., Gonzalez, G., Valenzuela, T., Carrillo, R., Palma, F., Rivera, R., Franco-Bourland, R. E., & Guizar-Sahagun, G. (2019). Transplantation of human neural progenitor cells (NPC) into Putamina of parkinsonian patients: A case series study, safety and efficacy four years after surgery. *Cell Transplantation, 28*(3), 269–285. https://doi.org/10.1177/0963689718820271

Morimoto, S., Takahashi, S., Ito, D., Date, Y., Okada, K., Kato, C., Nakamura, S., Ozawa, F., Chyi, C. M., Nishiyama, A., Suzuki, N., Fujimori, K., Kondo, T., Takao, M., Hirai, M., Kabe,

Y., Suematsu, M., Jinzaki, M., Aoki, M., Fujiki, Y., Sato, Y., Suzuki, N., Nakahara, J., Open-Access, P. R., & A. L. S. C. T. C., & Okano, H. (2023). Phase 1/2a clinical trial in ALS with ropinirole, a drug candidate identified by iPSC drug discovery. *Cell Stem Cell, 30*(6), 766–780 e769. https://doi.org/10.1016/j.stem.2023.04.017

Niizuma, K., Osawa, S. I., Endo, H., Izumi, S. I., Ataka, K., Hirakawa, A., Iwano, M., & Tominaga, T. (2023). Randomized placebo-controlled trial of CL2020, an allogenic muse cell-based product, in subacute ischemic stroke. *Journal of Cerebral Blood Flow and Metabolism, 43*(12), 2029–2039. https://doi.org/10.1177/0271678X231202594

Pleasure, S. J., & Lee, V. M. (1993). NTera 2 cells: A human cell line which displays characteristics expected of a human committed neuronal progenitor cell. *Journal of Neuroscience Research, 35*(6), 585–602. https://doi.org/10.1002/jnr.490350603

Pollock, K., Stroemer, P., Patel, S., Stevanato, L., Hope, A., Miljan, E., Dong, Z., Hodges, H., Price, J., & Sinden, J. D. (2006). A conditionally immortal clonal stem cell line from human cortical neuroepithelium for the treatment of ischemic stroke. *Experimental Neurology, 199*(1), 143–155. https://doi.org/10.1016/j.expneurol.2005.12.011

Schwartz, S. D., Regillo, C. D., Lam, B. L., Eliott, D., Rosenfeld, P. J., Gregori, N. Z., Hubschman, J. P., Davis, J. L., Heilwell, G., Spirn, M., Maguire, J., Gay, R., Bateman, J., Ostrick, R. M., Morris, D., Vincent, M., Anglade, E., Del Priore, L. V., & Lanza, R. (2015). Human embryonic stem cell-derived retinal pigment epithelium in patients with age-related macular degeneration and Stargardt's macular dystrophy: Follow-up of two open-label phase 1/2 studies. *Lancet, 385*(9967), 509–516. https://doi.org/10.1016/S0140-6736(14)61376-3

Song, W. K., Park, K. M., Kim, H. J., Lee, J. H., Choi, J., Chong, S. Y., Shim, S. H., Del Priore, L. V., & Lanza, R. (2015). Treatment of macular degeneration using embryonic stem cell-derived retinal pigment epithelium: Preliminary results in Asian patients. *Stem Cell Reports, 4*(5), 860–872. https://doi.org/10.1016/j.stemcr.2015.04.005

Stefansson, E., Geirsdottir, A., & Sigurdsson, H. (2011). Metabolic physiology in age related macular degeneration. *Progress in Retinal and Eye Research, 30*(1), 72–80. https://doi.org/10.1016/j.preteyeres.2010.09.003

Suda, S., Nito, C., Ihara, M., Iguchi, Y., Urabe, T., Matsumaru, Y., Sakai, N., Kimura, K., & group, J. R. t. (2022). Randomised placebo-controlled multicentre trial to evaluate the efficacy and safety of JTR-161, allogeneic human dental pulp stem cells, in patients with acute Ischaemic stRoke (J-REPAIR). *BMJ Open, 12*(5), e054269. https://doi.org/10.1136/bmjopen-2021-054269

Vahidy, F. S., Haque, M. E., Rahbar, M. H., Zhu, H., Rowan, P., Aisiku, I. P., Lee, D. A., Juneja, H. S., Alderman, S., Barreto, A. D., Suarez, J. I., Bambhroliya, A., Hasan, K. M., Kassam, M. R., Aronowski, J., Gee, A., Cox, C. S., Jr., Grotta, J. C., & Savitz, S. I. (2019). Intravenous bone marrow mononuclear cells for acute ischemic stroke: Safety, feasibility, and effect size from a phase I clinical trial. *Stem Cells, 37*(11), 1481–1491. https://doi.org/10.1002/stem.3080

Venkataramana, N. K., Kumar, S. K., Balaraju, S., Radhakrishnan, R. C., Bansal, A., Dixit, A., Rao, D. K., Das, M., Jan, M., Gupta, P. K., & Totey, S. M. (2010). Open-labeled study of unilateral autologous bone-marrow-derived mesenchymal stem cell transplantation in Parkinson's disease. *Translational Research, 155*(2), 62–70. https://doi.org/10.1016/j.trsl.2009.07.006

Yin, F., Tian, Z. M., Liu, S., Zhao, Q. J., Wang, R. M., Shen, L., Wieman, J., & Yan, Y. (2012). Transplantation of human retinal pigment epithelium cells in the treatment for Parkinson disease. *CNS Neuroscience & Therapeutics, 18*(12), 1012–1020. https://doi.org/10.1111/cns.12025

# Effects of Alcohol and Opioids on Neuroregeneration

**Elmira Zolali, Lance R. McMahon, and Samuel Obeng**

**Abstract** Neuroregeneration is essential for restoring neural function and repairing injuries in the nervous system. The sub-ventricular and sub-granular zones in the brain are key regions for neural stem cell activity, supporting neurogenesis and synaptic plasticity. These processes are influenced by various factors, including hormones, neurotrophic factors, and neurotransmitters. It has been shown that drugs of abuse, such as alcohol and opioids, disrupt neurotransmission, and alter neurotrophic signaling, thereby impairing neuroregeneration through the inhibition of neurogenesis and induction of neural damage. However, prolonged abstinence from these substances has been shown to promote neuroregeneration and restore neural function. Additionally, several US Food and Drug Administration (FDA)-approved pharmacotherapies for substance use disorders facilitate neural recovery and enhance neurogenesis, although some studies report conflicting results. This chapter will aid in propelling more effective therapeutic strategies for substance use disorders that reduce drug craving and withdrawal symptoms as well as enhance neuroregeneration.

**Keywords** Neural cells · Neurogenesis · Plasticity · Hormones · Neurotrophic factors · Neurotransmitters · Drugs · Abuse

## 1 Introduction

Neuroregeneration is the regrowth or repair of existing neurons or differentiation and maturation of neural stem cells into new neurons. This physiological process requires different mechanisms and regulatory factors to support and restore neural cells (Cope & Gould, 2019). Notably, two brain regions serve as reservoirs for

E. Zolali · L. R. McMahon · S. Obeng (✉)
Department of Pharmaceutical Sciences, Jerry H. Hodge School of Pharmacy, Texas Tech University Health Sciences Center, Amarillo, TX, USA
e-mail: Elmira.Zolali@ttuhsc.edu; Lance.McMahon@ttuhsc.edu; Samuel.Obeng@ttuhsc.edu



H. Das (ed.), *Tissue Repair and Regeneration*,
https://doi.org/10.1007/978-3-031-93677-7_12

neural stem cells; the first is the sub-granular zone (SGZ) located in the hippocampus, and the other is the sub-ventricular zone (SVZ) located in the lateral ventricles (Alvarez-Buylla & Garcıa-Verdugo, 2002). Both cells in the central nervous system (CNS) and the peripheral nervous system (PNS) can regenerate, although the extent of neuroregeneration in the PNS is higher than in the CNS. The PNS has both intrinsic and extrinsic factors that facilitate neuroregeneration, whereas the CNS has limited intrinsic and extrinsic factors. In addition, the CNS has factors that antagonize regeneration and factors that facilitate glial inhibition, which result in restricting neuroregeneration (Nagappan et al., 2020).

Glial cells in the CNS can be converted into new neurons under the influence of factors like fibroblast growth factor (FGF), epidermal growth factor (EGF), brain derived neurotrophic factor (BDNF), and vascular endothelial growth factor (VEGF). These factors can induce glial cell-reprogramming which leads to the formation of functional neurons (Wakhloo et al., 2022). However, intrinsic factors such as bone morphogenetic protein (BMP), phosphatase and tensin homolog (PTEN), and suppressor of cytokine signaling 3 (SOCS3), as well as extrinsic factors like inhibitory molecules from myelin debris inhibit neuroregeneration in the CNS (Mira et al., 2010; Nagappan et al., 2020). Therefore, the loss of the intrinsic ability of neurons to regenerate during development and the presence of suppressor factors diminish neuroregeneration in the CNS (Nagappan et al., 2020).

After brain injury, reactive glial cells form glial scars which limit brain damage. Changes in signaling pathways and concentration of physiological factors like cytokines and growth factors play important roles in glial cell-reprogramming and glial scar formation (Wang et al., 2021a). The ability of the brain to adapt and form new neural connections after injury is referred to as neuroplasticity, which exists in four different forms, including long-term potentiation (LTP), long-term depression (LTD), synaptogenesis, and neurogenesis. This process depends on the distance between the damaged and healthy axon, changes in the existing synaptic strength, and modification of the gamma-aminobutyric acid (GABA) interneurons (Nagappan et al., 2020). After injury, other regions of the brain become more reactive to compensate for the lost function of the damaged area. Consequently, this neuromodulation can help the brain restore its function (Grefkes & Ward, 2014).

Another important terminology related to the repair of brain injury is neurogenesis, which refers to the production of new neurons. This physiological process involves the proliferation and differentiation of neural stem cells into neurons, the maturation of new neurons, and the migration of neurons into existing neural networks (Van Praag et al., 2002). Several factors regulate neurogenesis, including hormones, humoral factors, stress proteins, and neurotransmitters (Bobkova et al., 2020). Thyroid hormones and corticosteroids are among the most critical regulators of this process (Bobkova et al., 2020). Additionally, stressors such as injury can trigger the production of stress proteins, which in turn influence neurogenesis. One key stress protein involved in this regulation is heat shock protein 70 (HSP70), which not only supports neurogenesis but also protects against neuronal death (Bobkova et al., 2020). The inhibitory and excitatory regulators, GABA and glutamate, influence neurogenesis. For example, GABA receptor activation can inhibit dormant neural stem cell proliferation and induce the differentiation of neural progenitors (Crowther & Song, 2014). Injection of N-methyl-D-aspartate (NMDA) in the dentate gyrus of an adult rat decreased cell proliferation, whereas NMDA receptor antagonism increased the rate of production of new cells (Cameron et al., 1995). Similarly, serotonin and dopamine have also been shown to contribute to the regulation of neurogenesis (Park & Enikolopov, 2010). Because of the diversity of serotonin receptors in the dentate gyrus, serotonin has both pro- and anti-neurogenesis effects (Crowther & Song, 2014). Nevertheless, selective serotonin reuptake inhibitors like fluoxetine have been shown to induce the function, proliferation of neuronal cells, and neurogenesis in the hippocampus of an adult rat (Malberg et al., 2000). Activation of the dopamine D2 receptor has been shown to stimulate neurogenesis in the SVZ (O'Keeffe et al., 2009). Moreover, in a study by Takamura et al., they show that activation of dopamine D1-like receptors induced neurogenesis, whereas activation of dopamine D2-like receptors did not change the proliferation and survival of neural precursor cells in the dentate gyrus of an adult rat (Takamura et al., 2014).

Dopamine is the primary neurotransmitter associated with the drug reward system in the CNS. All addictive drugs increase dopamine concentrations in the ventral tegmental area (VTA) and the nucleus accumbens (NAc), either directly or indirectly, thereby influencing the reward process (Wise, 2008). The common classes of abused substances include alcohol, opioids, and stimulants (Volkow et al., 2019). Alcohol modulates a broad range of receptors and indirectly increases dopamine concentrations in the NAc (Wiers et al., 2016). Opioids, such as heroin, morphine, and fentanyl, act as agonists at mu opioid receptors (MOR), which result in increased dopamine release in the VTA (Macey et al., 2006). Cocaine, a stimulant, inhibits the dopamine transporter (DAT), thereby preventing dopamine reuptake from the synaptic space and increasing extracellular dopamine concentrations. Amphetamines on the other hand reverse the activity of DAT and deplete vesicular dopamine stores, leading to an accumulation of dopamine in the synaptic space (Kahlig & Galli, 2003). In this chapter, we focus on the effects of alcohol and opioids on neuroregeneration.

## 2 Alcohol Use Disorder (AUD)

### *2.1 Statistics*

Alcohol is the most abused substance, and according to recent data from the National Institute on Alcohol Abuse and Alcoholism (NIAAA), 224.3 million individuals aged 12 and older in the United States—representing 79.1% of this age group—reported consuming alcohol at some point in their lifetime (NIAAA, 2025). In 2021, alcohol-impaired driving fatalities reached 13,384, marking a 14% increase from 2020 and accounting for 31% of all traffic-related fatalities in the United States (Stewart, 2023).

## 2.2 Effects of Alcohol on the CNS

Alcohol is a CNS depressant which activates inhibitory signaling pathways, such as GABA neurotransmission, and inhibits excitatory ion channels, such as glutamate receptors (Woodward, 2009). Chronic alcohol consumption has been shown to reduce the volume of both white and gray matter in the brain, particularly in cortical and subcortical regions (Mackey et al., 2019). The most prevalent cognitive impairments observed in individuals with AUD include motor skill disruptions, impaired decision-making and judgment, and behavioral changes like reduced motivation and increased anxiety. Studies suggest that neurobehavioral deficits may improve following abstinence from alcohol (Oscar-Berman & Marinković, 2007).

## 2.3 Effects of Alcohol on Neurons

Beyond behavioral changes, prolonged alcohol consumption contributes to neuronal loss and impaired brain function, either directly through the toxic effects of acetaldehyde or indirectly due to thiamine (vitamin B1) deficiency (Ridley et al., 2013). Thiamine is a crucial cofactor for proper neuronal function. Alcohol not only reduces thiamine absorption in the gastrointestinal tract but also disrupts cellular utilization, making chronic thiamine deficiency a key factor in the development of alcoholic neuropathy (Nutt et al., 2021). For instance, thiamine deficiency has been shown to impair the activity of Purkinje cells in the cerebellum, which in turn affects cortical function (Sullivan et al., 2005). The most significant clinical disorder associated with thiamine deficiency is Wernicke-Korsakoff syndrome, which consists of two distinct phases. Wernicke encephalopathy which occurs during the acute phase of alcohol consumption may be reversible with treatment, whereas Korsakoff's syndrome is a chronic condition that results in irreversible brain damage (Nutt et al., 2021). Additionally, alcohol-induced neurotoxicity can occur due to the production of acetaldehyde, a metabolite that directly damages cells, disrupts DNA function, and contributes to neuronal degeneration (Takeuchi & Saito, 2005).

## 2.4 Effects of Alcohol on Neuroregeneration

In the context of neuroregeneration, chronic alcohol intake can lead to neuronal loss and altered neurogenesis, particularly in the frontal cortex and cerebellum. For example, animal studies have shown that alcohol consumption over just four days can decrease neurogenesis and increase neurodegeneration in the brain (Crews & Nixon, 2009). Furthermore, in vitro studies suggest that alcohol's neurodegenerative effects are due to its inhibition of neural stem cell regeneration (Crews et al., 2004).

Another mechanism of alcohol-related damage is neuroinflammation. Alcohol consumption induces the release of intestinal inflammatory biomarkers, such as the nuclear protein high-mobility group box 1 (HMGB1), which subsequently activates toll-like receptor 4 (TLR4) and other pro-inflammatory cytokines. These cytokines can cross the blood-brain barrier (BBB) and trigger inflammatory signaling pathways in the brain—pathways that are typically inactive in a healthy brain (Ferrier et al., 2006). In particular, TLR4 activation leads to the release of interleukin-1β (IL-1β) and the phosphorylation of NMDA receptors, resulting in increased glutamate concentrations, greater susceptibility to seizures, and glutamate-induced excitotoxicity (Ward et al., 2009). Microglial activation following alcohol exposure triggers inflammatory pathways and neuronal apoptosis, mediated by nuclear factor kappa B (NF-κB), a key inflammatory regulator (Guadagno et al., 2013). Another cytokine, transforming growth factor-beta1 (TGF-β1), has also been shown to play a crucial role in neuronal apoptosis (Crews et al., 2015; Kuhn & Sarkar, 2008; Subbanna et al., 2018). Additionally, activated microglia secrete reactive oxygen species (ROS), which damage cell membranes and lead to apoptotic neuronal death (Boyadjieva & Sarkar, 2013). Chronic alcohol exposure also induces the neuroimmune gene cyclooxygenase-2 (COX-2) in various brain regions, a process linked to TLR4 activation (Alfonso-Loeches et al., 2010). Thus, cytokines, glutamate, and ROS all contribute to microglial-dependent neuronal apoptosis (Socodato et al., 2015).

## 2.5 Neuroregeneration Following Abstinence from Alcohol

It has been shown that the brain has the capacity to regenerate damaged neuronal cells after alcohol abstinence. For instance, neuroimaging studies indicate that gray matter volume increases in various brain regions within a month of abstinence (O'Neill et al., 2001). Interestingly, magnetic resonance imaging (MRI) scans of alcohol-dependent patients reveal that gray matter recovery begins within days after abstinence. However, the extent of recovery varies across brain regions; for instance, regions associated with improvements in learning and memory such as the insula, cingulate gyrus, cerebellum, anterior frontal gyrus, and hippocampus recover quickly (van Eijk et al., 2013). At the cellular level, it has been shown that cell proliferation begins within a day of abstinence. By the second day, cell development occurs in both gray and white matter as well as in several non-cortical areas, with most newly generated cells being microglia. These microglia can remain active for months and relocate to different brain regions (Crews & Nixon, 2009; Nixon et al., 2008). One potential explanation for cell proliferation after abstinence is degeneration-induced regeneration and brain plasticity (Crews & Nixon, 2009). Another contributing factor is CREB's (cyclic AMP-responsive element binding protein) role in neuronal maintenance and plasticity. Chronic ethanol exposure impairs CREB, a cellular regulatory and transcription protein that plays a crucial role in protecting neurons from apoptosis and promoting neuronal survival and plasticity (Crews et al., 2015; Mantamadiotis et al., 2002; Zou & Crews, 2006). CREB regulates the expression of BDNF, which is crucial for cell survival, neurogenesis, and neuroregeneration (Lonze & Ginty, 2002). During alcohol intoxication, CREB's function is inhibited, but during abstinence, its reactivation—along with elevated BDNF levels—enhances cell growth, plasticity, and neurogenesis (Crews & Nixon, 2009; Köhler et al., 2013).

## 2.6 *Effects of Drugs Used to Treat AUD on Neuroregeneration*

The US Food and Drug Administration (FDA) has approved three drugs with distinct mechanisms of action for AUD; these include acamprosate, disulfiram, and naltrexone (Cohen et al., 2022).

## 2.7 *Acamprosate*

Since NMDA receptors are upregulated during alcohol consumption, acamprosate reduces neuronal damage by modulating GABA and glutamate neurotransmission (Popp & Lovinger, 2000). A study by Doeppner et al. demonstrated that acamprosate administration in a post-ischemic brain injury model increased neuroregeneration and prevented secondary cell death of newly formed neurons. They also found that acamprosate reduced oxidative stress and inflammatory responses while strengthening BBB integrity (Doeppner et al., 2015). Similarly, Choi et al. found that acamprosate protected against neuronal cell death by reducing glial activation and delaying neuronal loss. Additionally, animals treated with acamprosate exhibited improved neurological and cognitive functions (Choi et al., 2019).

## 2.8 *Disulfiram*

Disulfiram is a fat-soluble compound that easily crosses the BBB after oral administration. Its mechanism of action involves inhibiting acetaldehyde dehydrogenase, an enzyme critical for alcohol metabolism (Lanz et al., 2023). Some studies suggest that disulfiram's toxic metabolites increase ROS levels, leading to cell cytotoxicity and neurological side effects via microglial activation (Kast et al., 2014). A study by Wood et al. on human lymphoblasts found that disulfiram's neurotoxic effects may be associated with reduced levels of structural glycerophospholipids and sphingolipids (Wood et al., 2014). However, other studies indicate that disulfiram possesses antioxidant properties and can protect against ROS generation and inflammation (Deng et al., 2020; Kanai et al., 2011). A study by Xu et al. demonstrated that disulfiram inhibits inflammation, mitigates neurological dysfunction, reduces glial cell activation, and reduces neuronal cell death after intracranial hemorrhage. These protective effects were attributed to the inhibition of caspase-1, TLR4, and NF-κB signaling pathways (Xu et al., 2024).

### 2.9 Naltrexone

Naltrexone is an antagonist at opioid receptors which is selective for the MOR over the kappa opioid receptor (KOR) and the delta opioid receptor (DOR) (Peng et al., 2007). In individuals with AUD, naltrexone administration reduces dopamine release in the nucleus accumbens, thereby diminishing alcohol's reinforcing effects (Anton, 2008). Research suggests that naltrexone induces neuroprotective effects through downregulating apoptotic proteins such as Bad and Bax, thereby reducing neural cell death (San-Emeterio & Hurlé, 2006). Additionally, naltrexone inhibits the TLR4 signaling pathway and suppresses microglial activation, which in turn reduces hippocampal neuronal death (Grace et al., 2015). Wang et al. demonstrated that naltrexone protects neural cells against inflammation and neurodegeneration by reducing glutamate-induced excitotoxicity and microglia activation (Wang et al., 2021b). In support of these findings, Kučić et al. demonstrated that naltrexone exerts neuroprotective effects by deactivating microglia via modulation of the mechanistic target of rapamycin (mTOR) signaling pathway, which regulates cell growth and survival (Kučić et al., 2021).

## 3 Opioid Use Disorder (OUD)

### 3.1 Statistics

Opioids are the leading cause of death due to drug overdose. The United States has the highest prevalence of opioid use, and according to recent reports from the National Institute on Drug Abuse (NIDA), opioid overdoses were involved in more than 81,000 deaths in 2022, with the vast majority of these fatalities due to synthetic opioids like fentanyl (NIDA, 2024).

### 3.2 Effects of Opioids on the CNS

Opioids are potent analgesic compounds used for pain management but have a high potential for abuse and addiction. These compounds bind to opioid receptors which include the MOR, KOR, and DOR. Opioids increase dopamine concentrations in the VTA, thereby inducing reward pathways, resulting in reinforcing effects (Taylor

& Samet, 2022). Additionally, opioid receptors decrease neurotransmitter release and alter postsynaptic excitation by inhibiting voltage-gated calcium channels and activating potassium channels, influencing neurotransmission and regulating neuroplasticity (Reeves et al., 2022). Opioids can induce respiratory depression through the activation of MORs in the medullary preBötzinger Complex (Baertsch et al., 2021). The most abused opioids include heroin, oxycodone, morphine, and fentanyl. Fentanyl, a potent synthetic opioid, is approximately 100 times more potent than morphine and its abuse has dramatically increased over the past decade, becoming a major cause of overdose deaths (Barletta et al., 2025). A significant concern related to fentanyl is its coadministration with xylazine (an alpha 2 adrenergic receptor agonist), which has become a prevalent compound in illicit drug supplies in the United States (Young et al., 2025).

## 3.3 *Effects of Opioids on Neuroregeneration*

Long-term opioid use significantly decreases gray and white matter volumes and compromises their integrity. However, long-term abstinence from opioids may partially restore affected brain regions (Wollman et al., 2017; Wollman et al., 2015). Notably, opioid-induced brain hypoxia, resulting from respiratory depression, can lead to ischemic neuronal damage and neuronal cell loss. Such damage may reduce the number of neurons in the globus pallidus and alter brain metabolic activity, potentially leading to mental and motor dysfunction (Andersen & Skullerud, 1999; Solis et al., 2017; Solis et al., 2018). Opioids influence neural plasticity by affecting LTP and neurogenesis in the hippocampus. They can also impact neural stem cells, altering their differentiation and proliferation (Zhang et al., 2016). For instance, Eisch et al. demonstrated that chronic morphine exposure in animals inhibited neurogenesis, reduced cell proliferation, and decreased the number of newly formed neurons, negatively impacting learning, memory, and cognition (Eisch et al., 2000). Similarly, it has been shown that chronic morphine exposure significantly reduces neuronal cell proliferation in the SGZ (Arguello et al., 2009; Kahn et al., 2005). Interestingly, a week after morphine abstinence, cell proliferation rebounded, returning to normal levels after two weeks (Kahn et al., 2005). Moreover, Harburg et al. examined MOR's role in neurogenesis using MOR knockout (KO) mice, and they found that while cell proliferation was unaffected, the persistent reduction in MOR enhanced granule cell survival and improved hippocampal neurogenesis (Harburg et al., 2007).

Similar to long-term effects of opioids on neurogenesis, an acute administration of morphine in an in vitro assay was shown to decrease cell proliferation, increase the activity of the pro-apoptotic enzyme caspase-3 in neural progenitor cells and astrocytes, and alter neuronal differentiation. These acute effects of morphine were reversed by naloxone (Willner et al., 2014). On the other hand, Xu et al. found that morphine promoted cell proliferation, albeit this was limited to astrocytes where neural progenitor cells preferentially differentiated into astrocytes in the presence of morphine (Xu et al., 2014).

## 3.4 Mechanisms of Opioids on Regulation of Neurogenesis and Neurodegeneration

Morphine regulates neurogenesis through an increase in Notch1 expression and the downregulation of Prospero homeobox protein 1 (Prox1) (Xu et al., 2014) (Hu et al., 2020). Notch is a regulator of neural development and can inhibit neurogenesis, whereas Prox1 is necessary for progenitor cell proliferation and survival (Hitoshi et al., 2002). That is, although morphine can increase cell proliferation, this is limited to astrocytes; morphine inhibits the differentiation of neural precursor cells and the survival of neurons (Hu et al., 2020).

In a study in mice, Tramullas et al. showed that chronic administration of heroin up-regulated pro-apoptotic proteins Bad, Fas, and FasL in the cortex and hippocampus without affecting the levels of anti-apoptotic proteins Bcl-2 and Bcl-XL (Tramullas et al., 2008). This resulted in neuronal apoptosis, i.e., neurodegeneration. Morphine has also been shown to induce neuronal apoptosis within the spinal cord dorsal horn which is mediated through the regulation of cyclic adenosine monophosphate (cAMP) and protein kinase A (PKA) pathways, and the induction of caspase-3 and mitogen-activated protein kinase (MAPK) expression (Lim et al., 2005). Additionally, prolonged morphine administration in neonatal rats was also shown to cause neuronal apoptosis in the cortex and amygdala, but not in the hippocampus (Bajic et al., 2013). In summary, studies show that opioids can induce neuronal apoptosis and decrease neuronal cell proliferation and survival in the SGZ, which negatively affects memory, cognition, and learning.

## 3.5 Neuroregeneration Following Abstinence from Opioids

Clinical studies have shown that the concentration of nerve growth factor (NGF) increased in serum samples of opiate-dependent patients after withdrawal from heroin compared to healthy subjects, while the level of VEGF did not change (Heberlein et al., 2012). Like BDNF, NGF is a neurotrophic factor that plays an essential role in the growth and survival of neurons and prevents neuronal apoptosis (Airaksinen & Saarma, 2002). In a study in male Swiss Webster mice evaluating the withdrawal effects of morphine and oxycodone, the authors reported that withdrawal from morphine or oxycodone did not change VEGF levels. However, there was a significant decrease in NGF expression during oxycodone withdrawal, but not in morphine withdrawal, suggesting that oxycodone may have greater adverse effects on the CNS compared with morphine (Bhalla et al., 2015) (Fig. 1).

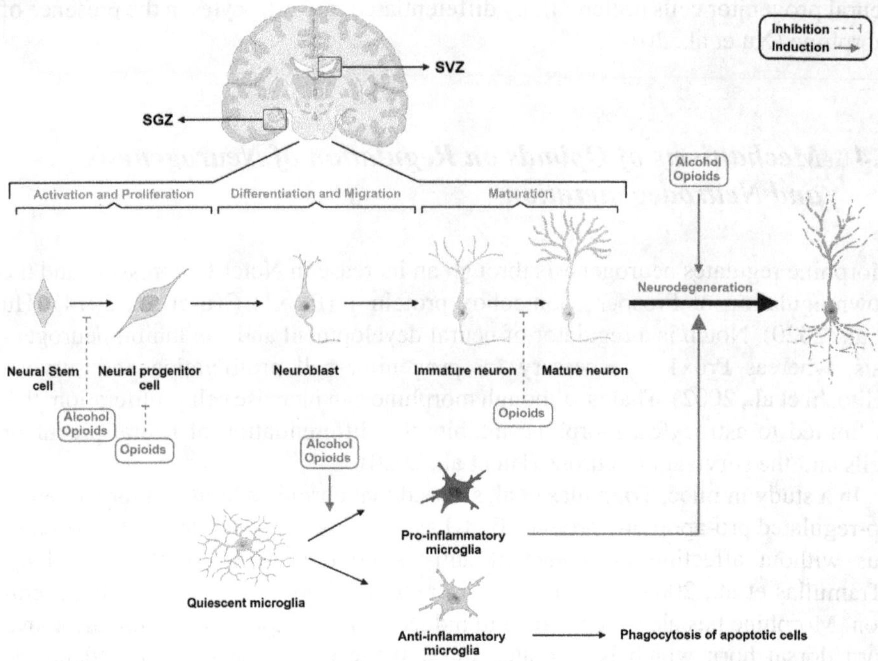


**Fig. 1** Graphical representation of neurogenesis, cellular stages, and effects of alcohol and opioids on neurogenesis and neurodegeneration. SVZ: sub-ventricular zone, SGZ: sub-granular zone. Figure created with BioRender.com

## 3.6 *Effects of Drugs Used to Treat OUD on Neuroregeneration*

There are three FDA-approved agents for treating opioid use disorder: methadone, naltrexone, and buprenorphine (Taylor & Samet, 2022). Methadone is a long-acting full MOR agonist that reduces craving and withdrawal symptoms. Buprenorphine is a partial MOR agonist and, like methadone, can be used to treat withdrawal symptoms. Naltrexone is a MOR antagonist and, unlike methadone and buprenorphine, is not effective in treating withdrawal symptoms and does not have reinforcing effects (Taylor & Samet, 2022).

## 3.7 *Methadone*

A clinical study investigating inflammatory and neurotrophic markers in OUD found a negative correlation between memory function and inflammatory biomarkers. Specifically, in patients treated with methadone, there was a significant improvement in inflammatory status, which can directly affect neural plasticity, neurogenesis, and neuroprotection. Furthermore, methadone decreased the compensatory elevation of BDNF and TGF-β1 in response to neuronal injuries (Wang et al., 2018). However, animal studies evaluating methadone's effects on hippocampal neurogenesis showed no alteration in parameters related to neurogenesis in the hippocampus of adults (Sankararaman et al., 2012). Despite its benefits, methadone has been shown to cause cognitive impairments. For instance, Wu et al. demonstrated that methadone negatively affects neural function, development, and maturation (Wu et al., 2020). Recent studies revealed that methadone can trigger neuronal apoptosis, reduce synaptic density, impair myelination, and induce the proinflammatory activation of microglia and astrocytes (De Gregorio et al., 2024). Additionally, preliminary research in heroin-addicted patients suggested that long-term methadone administration might impair white matter integrity (Wang et al., 2011). This finding raises concerns about the long-term neurological consequences of methadone maintenance therapy.

## 3.8 *Buprenorphine*

In vitro studies on oligodendrocytes showed that low-dose buprenorphine increased the levels of myelin basic proteins, which are markers of mature oligodendrocytes. In contrast, high doses of buprenorphine had opposing effects, inhibiting oligodendrocyte differentiation and myelin synthesis (Eschenroeder et al., 2012). Moreover, animal studies on perinatal exposure to buprenorphine revealed a decrease in the number of myelinated axons in the corpus callosum and a disproportionately thinner myelin sheath, which could affect axon-glial interaction in the nervous system (Sanchez et al., 2008). These findings suggest that early-life exposure to buprenorphine may have long-term consequences on neural connectivity. Similarly, buprenorphine's effect on hippocampal neurogenesis showed a reduction in neural cell proliferation in the hippocampus of adults, but interestingly, acute exposure transiently increased the survival of newly formed neurons (Pettit et al., 2012). In addition, Wu et al. also showed that prenatal exposure to buprenorphine decreased stem/progenitor cell proliferation, impaired BDNF expression and signaling, and consequently reduced neurogenesis (Wu et al., 2014). More recently, a study using human brain spheroids and animal models showed that prenatal buprenorphine exposure could alter interneuron migration into the developing cortex and disrupt cortical network development, potentially leading to long-term developmental consequences (Nieto-Estévez et al., 2022).

## 3.9 *Naltrexone*

Naltrexone is used to treat both OUD and AUD. It has been shown that the activity of dopamine neurons in the VTA is higher in morphine-dependent animals, and administration of naltrexone decreases the tonic and burst firing rate of neurons in the VTA (Georges et al., 2006). Furthermore, Eisch et al. demonstrated that naltrexone administration in morphine-dependent animals abolished the morphine-induced decrease in cell proliferation in the dentate gyrus granule cell layer (Eisch et al., 2000). These results suggest that naltrexone may counteract opioid-induced neurogenic deficits. Naltrexone has been shown to inhibit microglial-induced ROS production and the TLR-4 signaling pathway, reducing neural damage (Wang et al., 2016). On the other hand, chronic morphine exposure has been shown to increase CCL5, a neuroprotective chemokine from astrocytes. Interestingly, morphine withdrawal and naltrexone administration decreased CCL5 levels in the frontal cortex, induced microglia proliferation, and increased proinflammatory biomarkers such as TNF-α and IL-1β (Campbell et al., 2013).

## 4 Conclusion

Neurogenesis is a fundamental process in neural recovery, essential for maintaining brain function and adaptability following injury. The detrimental effects of alcohol and opioids on neurogenesis, synaptic plasticity, and neural integrity underscore the profound impact of substance abuse on brain health. These substances disrupt neurotransmitter systems, signaling pathways, and glial cell activity, leading to neurodegeneration and impaired neural repair mechanisms. While abstinence has been shown to promote neurogenic recovery, particularly in the hippocampus, the effects of pharmacological treatments for substance use disorders on neurogenesis remain inconsistent. Although these treatments effectively manage withdrawal symptoms and reduce craving, their influence on neuroregeneration varies, with some studies reporting paradoxical outcomes. Thus, these findings emphasize the need for future research to develop therapeutic strategies that not only address substance dependence but also enhance neurogenesis and protect against neurodegeneration, ultimately improving long-term neural recovery and cognitive function, as well as reducing drug craving and dependence.

**Competing Interests** The authors declare no competing financial and/or nonfinancial interests in relation to the content of the chapter.

## References

National Institute on Drug Abuse (2024). *Drug Overdose Deaths: Facts and Figures*. https://nida.nih.gov/research-topics/trends-statistics/overdose-death-rates#Fig5

Airaksinen, M. S., & Saarma, M. (2002). The GDNF family: Signalling, biological functions and therapeutic value. *Nature Reviews Neurosciencen̉, 3*(5), 383–394.

National Institute on Alcohol Abuse and Alcoholism (2025). Alcohol facts and statistics. National Institutes of Health. https://www.niaaa.nih.gov/alcohols-effects-health/alcohol-topics/alcohol-facts-and-statistics

Alfonso-Loeches, S., Pascual-Lucas, M., Blanco, A. M., Sanchez-Vera, I., & Guerri, C. (2010). Pivotal role of TLR4 receptors in alcohol-induced neuroinflammation and brain damage. *Journal of Neuroscience, 30*(24), 8285–8295.

Alvarez-Buylla, A., & Garcıa-Verdugo, J. M. (2002). Neurogenesis in adult subventricular zone. *Journal of Neuroscience, 22*(3), 629–634.

Andersen, S. N., & Skullerud, K. (1999). Hypoxic/ischaemic brain damage, especially pallidal lesions, in heroin addicts. *Forensic Science International, 102*(1), 51–59.

Anton, R. F. (2008). Naltrexone for the management of alcohol dependence. *New England Journal of Medicine, 359*(7), 715–721.

Arguello, A. A., Fischer, S. J., Schonborn, J. R., Markus, R. W., Brekken, R. A., & Eisch, A. J. (2009). Effect of chronic morphine on the dentate gyrus neurogenic microenvironment. *Neuroscience, 159*(3), 1003–1010.

Baertsch, N. A., Bush, N. E., Burgraff, N. J., & Ramirez, J. M. (2021). Dual mechanisms of opioid-induced respiratory depression in the inspiratory rhythm-generating network. *eLife, 10*. https://doi.org/10.7554/eLife.67523

Bajic, D., Commons, K. G., & Soriano, S. G. (2013). Morphine-enhanced apoptosis in selective brain regions of neonatal rats. *International Journal of Developmental Neuroscience, 31*(4), 258–266.

Barletta, C., Di Natale, V., Esposito, M., Chisari, M., Cocimano, G., Di Mauro, L., Salerno, M., & Sessa, F. (2025). The rise of fentanyl: Molecular aspects and forensic investigations. *International Journal of Molecular Sciences, 26*(2), 444.

Bhalla, S., Pais, G., Tapia, M., & Gulati, A. (2015). Endothelin ETA receptor antagonist reverses naloxone-precipitated opioid withdrawal in mice. *Canadian Journal of Physiology and Pharmacology, 93*(11), 935–944.

Bobkova, N., Poltavtseva, R., Leonov, S., & Sukhikh, G. (2020). Neuroregeneration: Regulation in neurodegenerative diseases and aging. *Biochemistry (Moscow), 85*, 108–130.

Boyadjieva, N. I., & Sarkar, D. K. (2013). Cyclic adenosine monophosphate and brain-derived neurotrophic factor decreased oxidative stress and apoptosis in developing hypothalamic neuronal cells: Role of microglia. *Alcoholism: Clinical and Experimental Research, 37*(8), 1370–1379.

Cameron, H. A., McEwen, B. S., & Gould, E. (1995). Regulation of adult neurogenesis by excitatory input and NMDA receptor activation in the dentate gyrus. *Journal of Neuroscience, 15*(6), 4687–4692.

Campbell, L. A., Avdoshina, V., Rozzi, S., & Mocchetti, I. (2013). CCL5 and cytokine expression in the rat brain: Differential modulation by chronic morphine and morphine withdrawal. *Brain, Behavior, and Immunity, 34*, 130–140.

Choi, B. Y., Lee, S. H., Choi, H. C., Lee, S.-K., Yoon, H. S., Park, J. B., Chung, W. S., & Suh, S. W. (2019). Alcohol dependence treating agent, acamprosate, prevents traumatic brain injury-induced neuron death through vesicular zinc depletion. *Translational Research, 207*, 1–18.

Cohen, S. M., Alexander, R. S., & Holt, S. R. (2022). The spectrum of alcohol use: Epidemiology, diagnosis, and treatment. *Medical Clinics, 106*(1), 43–60.

Cope, E. C., & Gould, E. (2019). Adult neurogenesis, glia, and the extracellular matrix. *Cell Stem Cell, 24*(5), 690–705.

Crews, F. T., Collins, M. A., Dlugos, C., Littleton, J., Wilkins, L., Neafsey, E. J., Pentney, R., Snell, L. D., Tabakoff, B., & Zou, J. (2004). Alcohol-induced neurodegeneration: When, where and why? *Alcoholism: Clinical and Experimental Research, 28*(2), 350–364.

Crews, F. T., & Nixon, K. (2009). Mechanisms of neurodegeneration and regeneration in alcoholism. *Alcohol & Alcoholism, 44*(2), 115–127.

Crews, F. T., Sarkar, D. K., Qin, L., Zou, J., Boyadjieva, N., & Vetreno, R. P. (2015). Neuroimmune function and the consequences of alcohol exposure. *Alcohol Research: Current Reviews, 37*(2), 331.

Crowther, A. J., & Song, J. (2014). Activity-dependent signaling mechanisms regulating adult hippocampal neural stem cells and their progeny. *Neuroscience Bulletin, 30*, 542–556.

De Gregorio, C., Gallardo, J., Berríos-Cárcamo, P., Handy, Á., Santapau, D., González-Madrid, A., Ezquer, M., Morales, P., Luarte, A., & Corvalán, D. (2024). Methadone directly impairs central nervous system cells in vitro. *Scientific Reports, 14*(1), 16978.

Deng, W., Yang, Z., Yue, H., Ou, Y., Hu, W., & Sun, P. (2020). Disulfiram suppresses NLRP3 inflammasome activation to treat peritoneal and gouty inflammation. *Free Radical Biology & Medicine, 152*, 8–17. https://doi.org/10.1016/j.freeradbiomed.2020.03.007

Doeppner, T. R., Pehlke, J. R., Kaltwasser, B., Schlechter, J., Kilic, E., Bähr, M., & Hermann, D. M. (2015). The indirect NMDAR antagonist acamprosate induces postischemic neurologic recovery associated with sustained neuroprotection and neuroregeneration. *Journal of Cerebral Blood Flow & Metabolism, 35*(12), 2089–2097.

Eisch, A. J., Barrot, M., Schad, C. A., Self, D. W., & Nestler, E. J. (2000). Opiates inhibit neurogenesis in the adult rat hippocampus. *Proceedings of the National Academy of Sciences, 97*(13), 7579–7584.

Eschenroeder, A. C., Vestal-Laborde, A. A., Sanchez, E. S., Robinson, S. E., & Sato-Bigbee, C. (2012). Oligodendrocyte responses to buprenorphine uncover novel and opposing roles of μ-opioid-and nociceptin/orphanin FQ receptors in cell development: Implications for drug addiction treatment during pregnancy. *Glia, 60*(1), 125–136.

Ferrier, L., Bérard, F., Debrauwer, L., Chabo, C., Langella, P., Buéno, L., & Fioramonti, J. (2006). Impairment of the intestinal barrier by ethanol involves enteric microflora and mast cell activation in rodents. *The American Journal of Pathology, 168*(4), 1148–1154.

Georges, F., Le Moine, C., & Aston-Jones, G. (2006). No effect of morphine on ventral tegmental dopamine neurons during withdrawal. *Journal of Neuroscience, 26*(21), 5720–5726.

Grace, P. M., Shimizu, K., Strand, K. A., Rice, K. C., Deng, G., Watkins, L. R., & Herson, P. S. (2015). (+)-naltrexone is neuroprotective and promotes alternative activation in the mouse hippocampus after cardiac arrest/cardiopulmonary resuscitation. *Brain, Behavior, and Immunity, 48*, 115–122.

Grefkes, C., & Ward, N. S. (2014). Cortical reorganization after stroke: How much and how functional? *The Neuroscientist, 20*(1), 56–70.

Guadagno, J., Xu, X., Karajgikar, M., Brown, A., & Cregan, S. (2013). Microglia-derived TNFα induces apoptosis in neural precursor cells via transcriptional activation of the Bcl-2 family member puma. *Cell Death & Disease, 4*(3), e538–e538.

Harburg, G. C., Hall, F. S., Harrist, A. V., Sora, I., Uhl, G. R., & Eisch, A. J. (2007). Knockout of the mu opioid receptor enhances the survival of adult-generated hippocampal granule cell neurons. *Neuroscience, 144*(1), 77–87.

Heberlein, A., Dürsteler-MacFarland, K. M., Frieling, H., Gröschl, M., Lenz, B., Bönsch, D., Kornhuber, J., Wiesbeck, G. A., Bleich, S., & Hillemacher, T. (2012). Association of nerve growth factor and vascular endothelial growth factor A with psychometric measurements of opiate dependence: Results of a pilot study in patients participating in a structured diamorphine maintenance program. *European Addiction Research, 18*(5), 213–219.

Hitoshi, S., Alexson, T., Tropepe, V., Donoviel, D., Elia, A. J., Nye, J. S., Conlon, R. A., Mak, T. W., Bernstein, A., & Van Der Kooy, D. (2002). Notch pathway molecules are essential for the maintenance, but not the generation, of mammalian neural stem cells. *Genes & Development, 16*(7), 846–858.

Hu, J., Cui, F., & Zhang, X. (2020). Morphine modulates hippocampal neurogenesis and contextual memory extinction via miR-34c/Notch1 pathway in male ICR mice. *Open Life Sciences, 15*(1), 51–59.

Kahlig, K. M., & Galli, A. (2003). Regulation of dopamine transporter function and plasma membrane expression by dopamine, amphetamine, and cocaine. European Journal of Pharmacology, 479(1–3), 153–158, 153.

Kahn, L., Alonso, G., Normand, E., & Manzoni, O. J. (2005). Repeated morphine treatment alters polysialylated neural cell adhesion molecule, glutamate decarboxylase-67 expression and cell proliferation in the adult rat hippocampus. *European Journal of Neuroscience, 21*(2), 493–500.

Kanai, K., Itoh, N., Ito, Y., Nagai, N., Hori, Y., Chikazawa, S., Hoshi, F., & Higuchi, S. (2011). Anti-inflammatory potency of oral disulfiram compared with dexamethasone on endotoxin-induced uveitis in rats. *Journal of Veterinary Medical Science, 73*(4), 517–520.

Kast, R. E., Karpel-Massler, G., & Halatsch, M.-E. (2014). CUSP9* treatment protocol for recurrent glioblastoma: Aprepitant, artesunate, auranofin, captopril, celecoxib, disulfiram, itraconazole, ritonavir, sertraline augmenting continuous low dose temozolomide. *Oncotarget, 5*(18), 8052.

Köhler, S., Klimke, S., Hellweg, R., & Lang, U. E. (2013). Serum brain-derived neurotrophic factor and nerve growth factor concentrations change after alcohol withdrawal: Preliminary data of a case-control comparison. *European Addiction Research, 19*(2), 98–104.

Kučić, N., Rački, V., Šverko, R., Vidović, T., Grahovac, I., & Mršić-Pelčić, J. (2021). Immunometabolic modulatory role of naltrexone in BV-2 microglia cells. *International Journal of Molecular Sciences, 22*(16), 8429.

Kuhn, P., & Sarkar, D. K. (2008). Ethanol induces apoptotic death of β-endorphin neurons in the rat hypothalamus by a TGF-β1-dependent mechanism. *Alcoholism: Clinical and Experimental Research, 32*(4), 706–714.

Lanz, J., Biniaz-Harris, N., Kuvaldina, M., Jain, S., Lewis, K., & Fallon, B. A. (2023). Disulfiram: Mechanisms, applications, and challenges. *Antibiotics, 12*(3), 524.

Lim, G., Wang, S., Lim, J.-A., & Mao, J. (2005). Activity of adenylyl cyclase and protein kinase A contributes to morphine-induced spinal apoptosis. *Neuroscience Letters, 389*(2), 104–108.

Lonze, B. E., & Ginty, D. D. (2002). Function and regulation of CREB family transcription factors in the nervous system. *Neuron, 35*(4), 605–623.

Macey, T. A., Lowe, J. D., & Chavkin, C. (2006). Mu opioid receptor activation of ERK1/2 is GRK3 and arrestin dependent in striatal neurons. *Journal of Biological Chemistry, 281*(45), 34515–34524.

Mackey, S., Allgaier, N., Chaarani, B., Spechler, P., Orr, C., Bunn, J., Allen, N. B., Alia-Klein, N., Batalla, A., & Blaine, S. (2019). Mega-analysis of gray matter volume in substance dependence: General and substance-specific regional effects. *American Journal of Psychiatry, 176*(2), 119–128.

Malberg, J. E., Eisch, A. J., Nestler, E. J., & Duman, R. S. (2000). Chronic antidepressant treatment increases neurogenesis in adult rat hippocampus. *Journal of Neuroscience, 20*(24), 9104–9110.

Mantamadiotis, T., Lemberger, T., Bleckmann, S. C., Kern, H., Kretz, O., Villalba, A. M., Tronche, F., Kellendonk, C., Gau, D., & Kapfhammer, J. (2002). Disruption of CREB function in brain leads to neurodegeneration. *Nature Genetics, 31*(1), 47–54.

Mira, H., Andreu, Z., Suh, H., Lie, D. C., Jessberger, S., Consiglio, A., San Emeterio, J., Hortigüela, R., Marqués-Torrejón, M. Á., & Nakashima, K. (2010). Signaling through BMPR-IA regulates quiescence and long-term activity of neural stem cells in the adult hippocampus. *Cell Stem Cell, 7*(1), 78–89.

Nagappan, P. G., Chen, H., & Wang, D.-Y. (2020). Neuroregeneration and plasticity: A review of the physiological mechanisms for achieving functional recovery postinjury. *Military Medical Research, 7*, 1–16.

Nieto-Estévez, V., Donegan, J. J., McMahon, C. L., Elam, H. B., Chavera, T. A., Varma, P., Berg, K. A., Lodge, D. J., & Hsieh, J. (2022). Buprenorphine exposure alters the development and migration of interneurons in the cortex. *Frontiers in Molecular Neuroscience, 15*, 889922.

Nixon, K., Kim, D., Potts, E., He, J., & Crews, F. (2008). Distinct cell proliferation events during abstinence after alcohol dependence: Microglia proliferation precedes neurogenesis. *Neurobiology of Disease, 31*(2), 218–229.

Nutt, D., Hayes, A., Fonville, L., Zafar, R., Palmer, E. O., Paterson, L., & Lingford-Hughes, A. (2021). Alcohol and the brain. *Nutrients, 13*(11), 3938.

O'Neill, J., Cardenas, V. A., & Meyerhoff, D. J. (2001). Effects of abstinence on the brain: Quantitative magnetic resonance imaging and magnetic resonance spectroscopic imaging in chronic alcohol abuse. *Alcoholism: Clinical and Experimental Research, 25*(11), 1673–1682.

O'Keeffe, G. C., Barker, R. A., & Caldwell, M. A. (2009). Dopaminergic modulation of neurogenesis in the subventricular zone of the adult brain. *Cell Cycle, 8*(18), 2888–2894.

Oscar-Berman, M., & Marinković, K. (2007). Alcohol: Effects on neurobehavioral functions and the brain. *Neuropsychology Review, 17*, 239–257.

Park, J.-H., & Enikolopov, G. (2010). Transient elevation of adult hippocampal neurogenesis after dopamine depletion. Experimental Neurology, 222(2), 267–276.

Peng, X., Knapp, B. I., Bidlack, J. M., & Neumeyer, J. L. (2007). Pharmacological properties of bivalent ligands containing butorphan linked to nalbuphine, naltrexone, and naloxone at μ, δ, and κ opioid receptors. *Journal of Medicinal Chemistry, 50*(9), 2254–2258.

Pettit, A., Desroches, R., & Bennett, S. (2012). The opiate analgesic buprenorphine decreases proliferation of adult hippocampal neuroblasts and increases survival of their progeny. *Neuroscience, 200*, 211–222.

Popp, R. L., & Lovinger, D. M. (2000). Interaction of acamprosate with ethanol and spermine on NMDA receptors in primary cultured neurons. *European Journal of Pharmacology, 394*(2–3), 221–231.

Reeves, K. C., Shah, N., Muñoz, B., & Atwood, B. K. (2022). Opioid receptor-mediated regulation of neurotransmission in the brain. *Frontiers in Molecular Neuroscience, 15*, 919773.

Ridley, N. J., Draper, B., & Withall, A. (2013). Alcohol-related dementia: An update of the evidence. *Alzheimer's Research & Therapy, 5*, 1–8.

San-Emeterio, E. P., & Hurlé, M. A. (2006). Modulation of brain apoptosis-related proteins by the opioid antagonist naltrexone in mice. *Neuroscience Letters, 403*(3), 276–279.

Sanchez, E. S., Bigbee, J. W., Fobbs, W., Robinson, S. E., & Sato-Bigbee, C. (2008). Opioid addiction and pregnancy: Perinatal exposure to buprenorphine affects myelination in the developing brain. *Glia, 56*(9), 1017–1027.

Sankararaman, A., Masiulis, I., Richardson, D. R., Andersen, J. M., Mørland, J., & Eisch, A. J. (2012). Methadone does not alter key parameters of adult hippocampal neurogenesis in the heroin-naïve rat. *Neuroscience Letters, 516*(1), 99–104.

Socodato, R., Portugal, C. C., Canedo, T., Domith, I., Oliveira, N. A., Paes-de-Carvalho, R., Relvas, J. B., & Cossenza, M. (2015). C-Src deactivation by the polyphenol 3-O-caffeoylquinic acid abrogates reactive oxygen species-mediated glutamate release from microglia and neuronal excitotoxicity. *Free Radical Biology & Medicine, 79*, 45–55. https://doi.org/10.1016/j.freeradbiomed.2014.11.019

Solis, E., Cameron-Burr, K. T., Shaham, Y., & Kiyatkin, E. A. (2017). Intravenous heroin induces rapid brain hypoxia and hyperglycemia that precede brain metabolic response. *eneuro, 4*(3), ENEURO.0151.

Solis, E., Jr., Afzal, A., & Kiyatkin, E. A. (2018). Changes in brain oxygen and glucose induced by oxycodone: Relationships with brain temperature and peripheral vascular tone. *Neuropharmacology, 133*, 481–490.

Stewart, T. (2023). *Overview of motor vehicle traffic crashes in*, 2021. No. DOT HS 813 435. United States. Department of Transportation. National Highway Traffic Safety Administration, 2023.

Subbanna, S., Nagre, N. N., Shivakumar, M., Joshi, V., Psychoyos, D., Kutlar, A., Umapathy, N. S., & Basavarajappa, B. S. (2018). CB1R-mediated activation of Caspase-3 causes epigenetic and neurobehavioral abnormalities in postnatal ethanol-exposed mice. *Frontiers in Molecular Neuroscience, 11*, 45. https://doi.org/10.3389/fnmol.2018.00045

Sullivan, E. V., Deshmukh, A., De Rosa, E., Rosenbloom, M. J., & Pfefferbaum, A. (2005). Striatal and forebrain nuclei volumes: Contribution to motor function and working memory deficits in alcoholism. *Biological Psychiatry, 57*(7), 768–776.

Takamura, N., Nakagawa, S., Masuda, T., Boku, S., Kato, A., Song, N., An, Y., Kitaichi, Y., Inoue, T., & Koyama, T. (2014). The effect of dopamine on adult hippocampal neurogenesis. *Progress in Neuro-Psychopharmacology and Biological Psychiatry, 50*, 116–124.

Takeuchi, M., & Saito, T. (2005). Cytotoxicity of acetaldehyde-derived advanced glycation end-products (AA-AGE) in alcoholic-induced neuronal degeneration. *Alcoholism: Clinical and Experimental Research, 29*, 220S–224S.

Taylor, J. L., & Samet, J. H. (2022). Opioid use disorder. *Annals of Internal Medicine, 175*(1), ITC1-ITC16.
Tramullas, M., Martínez-Cué, C., & Hurlé, M. A. (2008). Chronic administration of heroin to mice produces up-regulation of brain apoptosis-related proteins and impairs spatial learning and memory. *Neuropharmacology, 54*(4), 640–652.
van Eijk, J., Demirakca, T., Frischknecht, U., Hermann, D., Mann, K., & Ende, G. (2013). Rapid partial regeneration of brain volume during the first 14 days of abstinence from alcohol. *Alcoholism: Clinical and Experimental Research, 37*(1), 67–74.
Van Praag, H., Schinder, A. F., Christie, B. R., Toni, N., Palmer, T. D., & Gage, F. H. (2002). Functional neurogenesis in the adult hippocampus. *Nature, 415*(6875), 1030–1034.
Volkow, N. D., Michaelides, M., & Baler, R. (2019). The neuroscience of drug reward and addiction. *Physiological Reviews, 99*(4), 2115–2140.
Wakhloo, D., Oberhauser, J., Madira, A., & Mahajani, S. (2022). From cradle to grave: Neurogenesis, neuroregeneration and neurodegeneration in Alzheimer's and Parkinson's diseases. *Neural Regeneration Research, 17*(12), 2606–2614.
Wang, F., Cheng, L., & Zhang, X. (2021a). Reprogramming glial cells into functional neurons for neuro-regeneration: Challenges and promise. *Neuroscience Bulletin, 37*(11), 1625–1636.
Wang, T.-Y., Lee, S.-Y., Chang, Y.-H., Chen, S.-L., Chen, P. S., Chu, C.-H., Huang, S.-Y., Tzeng, N.-S., Lee, I. H., & Chen, K. C. (2018). Correlation of cytokines, BDNF levels, and memory function in patients with opioid use disorder undergoing methadone maintenance treatment. *Drug and Alcohol Dependence, 191*, 6–13.
Wang, X., Zhang, Y., Peng, Y., Hutchinson, M. R., Rice, K., Yin, H., & Watkins, L. (2016). Pharmacological characterization of the opioid inactive isomers (+)-naltrexone and (+)-naloxone as antagonists of toll-like receptor 4. *British Journal of Pharmacology, 173*(5), 856–869.
Wang, Y., Li, W., Li, Q., Yang, W., Zhu, J., & Wang, W. (2011). White matter impairment in heroin addicts undergoing methadone maintenance treatment and prolonged abstinence: A preliminary DTI study. *Neuroscience Letters, 494*(1), 49–53.
Wang, Y. S., Hung, T. W., Bae, E. K., Wu, K. J., Hsieh, W., & Yu, S. J. (2021b). Naltrexone is neuroprotective against traumatic brain injury in mu opioid receptor knockout mice. *CNS Neuroscience & Therapeutics, 27*(7), 831–841.
Ward, R. J., Lallemand, F., & De Witte, P. (2009). Biochemical and neurotransmitter changes implicated in alcohol-induced brain damage in chronic or 'binge drinking'alcohol abuse. *Alcohol & Alcoholism, 44*(2), 128–135.
Wiers, C. E., Cabrera, E., Skarda, E., Volkow, N. D., & Wang, G.-J. (2016). PET imaging for addiction medicine: From neural mechanisms to clinical considerations. *Progress in Brain Research, 224*, 175–201.
Willner, D., Cohen-Yeshurun, A., Avidan, A., Ozersky, V., Shohami, E., & Leker, R. R. (2014). Short term morphine exposure in vitro alters proliferation and differentiation of neural progenitor cells and promotes apoptosis via mu receptors. *PLoS One, 9*(7), e103043.
Wise, R. A. (2008). Dopamine and reward: The anhedonia hypothesis 30 years on. *Neurotoxicity Research, 14*, 169–183.
Wollman, S. C., Alhassoon, O. M., Hall, M. G., Stern, M. J., Connors, E. J., Kimmel, C. L., Allen, K. E., Stephan, R. A., & Radua, J. (2017). Gray matter abnormalities in opioid-dependent patients: A neuroimaging meta-analysis. *The American Journal of Drug and Alcohol Abuse, 43*(5), 505–517. https://doi.org/10.1080/00952990.2016.1245312
Wollman, S. C., Alhassoon, O. M., Stern, M. J., Hall, M. G., Rompogren, J., Kimmel, C. L., & Perez-Figueroa, A. M. (2015). White matter abnormalities in long-term heroin users: A preliminary neuroimaging meta-analysis. *The American Journal of Drug and Alcohol Abuse, 41*(2), 133–138.
Wood, P. L., Alexander, R. S., & Felty, C. C. (2014). Disulfiram neurotoxicity: Decrements in ethanolamine, serine, and inositol glycerophospholipids. *Journal of Molecular Pathophysiology, 3*, 33–37.
Woodward, J. (2009). The pharmacology of alcohol. *Principles of addiction medicine*, 85–97.

Wu, C.-C., Hung, C.-J., Shen, C.-H., Chen, W.-Y., Chang, C.-Y., Pan, H.-C., Liao, S.-L., & Chen, C.-J. (2014). Prenatal buprenorphine exposure decreases neurogenesis in rats. *Toxicology Letters, 225*(1), 92–101.

Wu, W., Yao, H., Dwivedi, I., Negraes, P. D., Zhao, H. W., Wang, J., Trujillo, C. A., Muotri, A. R., & Haddad, G. G. (2020). Methadone suppresses neuronal function and maturation in human cortical organoids. *Frontiers in Neuroscience, 14*, 593248.

Xu, C., Jiang, F., Mao, Y., Wei, W., Song, J., Jia, F., Du, X., Zhong, D., & Li, G. (2024). Disulfiram attenuates cell and tissue damage and blood–brain barrier dysfunction after intracranial haemorrhage by inhibiting the classical pyroptosis pathway. *Scientific Reports, 14*(1), 21860.

Xu, C., Zhang, Y., Zheng, H., Loh, H. H., & Law, P.-Y. (2014). Morphine modulates mouse hippocampal progenitor cell lineages by upregulating miR-181a level. *Stem Cells, 32*(11), 2961–2972.

Young, B. L., Gqamana, P. P., Qiu, Y., Nacca, N. E., & Zhang, Y. V. (2025). The prevalence of Xylazine in patient urine samples that were positive for fentanyl in Western New York. *The Journal of Applied Laboratory Medicine, jfae158*, 572.

Zhang, Y., Loh, H. H., & Law, P.-Y. (2016). Effect of opioid on adult hippocampal neurogenesis. *The Scientific World Journal, 2016*(1), 2601264.

Zou, J., & Crews, F. (2006). CREB and NF-κB transcription factors regulate sensitivity to excitotoxic and oxidative stress induced neuronal cell death. *Cellular and Molecular Neurobiology, 26*, 383–403.

# Cancer and Regeneration

# Regeneration and Cancer: Molecular Drivers and Therapeutic Implications

**Durga Puro and Surajit Hansda**

**Abstract** Regeneration is a highly complex process involving cellular differentiation and proliferation, orchestrated by important signaling pathways such as Wnt/β--catenin, mitogen-activated protein kinase/ extracellular signal-related kinase (MAPK/ERK), Notch, and Hedgehog. These pathways play fundamental roles in both physiological tissue repair and pathological conditions, including cancer. Dysregulation of these signaling mechanisms can lead to aberrant growth and malignancy, supporting the concept that cancer represents a dysregulated form of regeneration. Furthermore, factors such as microenvironment, inflammation, and epigenetic modifications significantly influence both regeneration and cancer progression, revealing shared molecular pathways that may serve as therapeutic targets. Insights derived from highly regenerative organisms provide a valuable framework for developing novel cancer therapies aimed at restoring controlled cellular behavior and enhancing tissue repair. This chapter comprehensively examines the intersection between regenerative biology and tumorigenesis, emphasizing the potential to leverage regenerative mechanisms for cancer treatment strategies.

**Keywords** Regeneration · Cancer · Wound healing · Tissue repair · Inflammation

## 1 Introduction

Regeneration generally refers to the precisely orchestrated processes of restoring cells, tissues, and organs that are lost or damaged, whether in the form of structure or function. This restructuring process requires linking and recreating missing parts while ensuring seamless integration between newly formed and existing tissues to bring about functional and structural changes. When regeneration involves cell

D. Puro (✉) · S. Hansda
Department of Pharmaceutical Sciences, Jerry H. Hodge School of Pharmacy, Texas Tech University Health Sciences Center, Amarillo, TX, USA
e-mail: durpuro@ttuhsc.edu; shansda@ttuhsc.edu

 
H. Das (ed.), *Tissue Repair and Regeneration*,
https://doi.org/10.1007/978-3-031-93677-7_13

proliferation (epimorphosis), it relies on regulatory signals that control the cell cycle to produce enough cells for complete repair. The cells are precisely directed to specific areas, and once regeneration concludes, signals are necessary to indicate completion and halt the process. Without such cues, ongoing regenerative activity could disrupt the body's internal balance (Oviedo & Beane, 2009).

Regeneration is facilitated through diverse mechanisms that vary depending on the organism, organ type, and cellular context. In amphibians such as adult newts, regeneration is primarily mediated by postmitotic differentiated cells that reenter the S-phase of the cell cycle, undergoing dedifferentiation to regain proliferative capacity. In contrast, planarians rely on the proliferation of resident adult somatic stem cells as the principal mechanism of regeneration. Similarly, in mammals, dedicated stem cell populations support tissue homeostasis and regeneration in highly proliferative organs such as the skin and intestine, where continuous stem cell-driven cell turnover is essential for maintaining tissue integrity (Fig. 1). Conversely, in quiescent tissues such as the liver and pancreas, regeneration is suggested to occur through alternative mechanisms that do not primarily rely on classical stem cell activity. These mechanisms include activation of dormant stem cells, transdifferentiation (conversion of one differentiated cell type into another), metaplasia (adaptive transformation of one tissue type into another), and compensatory proliferation of mature cells to restore tissue function following injury. These distinct regenerative strategies emphasize the contextual nature of tissue repair, governed by intrinsic cellular plasticity and the microenvironmental signals regulating regenerative capacity across different species and organ systems (Charni et al., 2017).

Though it seems to be paradoxical, regeneration has the dual potential to both contribute to the onset of abnormal growth and act as a mechanism to mitigate or rectify such abnormalities. Malignant tumor formation may arise from disrupted or incomplete regenerative processes; conversely, regeneration can play a regulatory role in suppressing the unchecked proliferation of malignant cells. Disruption of normal regenerative processes is marked by deficits in tissue formation, morphogenesis, and the completion of repair mechanisms despite intact damage-sensing and cell proliferation pathways. This highlights the importance of elucidating the signaling pathways that regulate the later stages of regeneration, as they may reveal therapeutic targets for suppressing abnormal cell proliferation and, ultimately, halting cancer progression (Riss et al., 2006; Schäfer & Werner, 2008).

On the other hand, cellular proliferation induced during regeneration, when integrated with morphogenetic processes, has the potential not only to suppress abnormal growth but also to reverse malignancies and restore morphostasis—phenomena predominantly observed in organisms with exceptional regenerative capacities. Intriguingly, the biological mechanisms driving epimorphic regeneration can yield profoundly contrasting outcomes, ranging from tissue destruction in certain contexts to the complete restoration of form and function in others (Brockes, 1998; White & Zon, 2008).

Cancers have often been described as wounds that fail to heal, and the intricate relationships between carcinogenesis, inflammation, and local tissue repair (wound healing) have been extensively studied (Fig. 1). In this chapter, we aim to elucidate

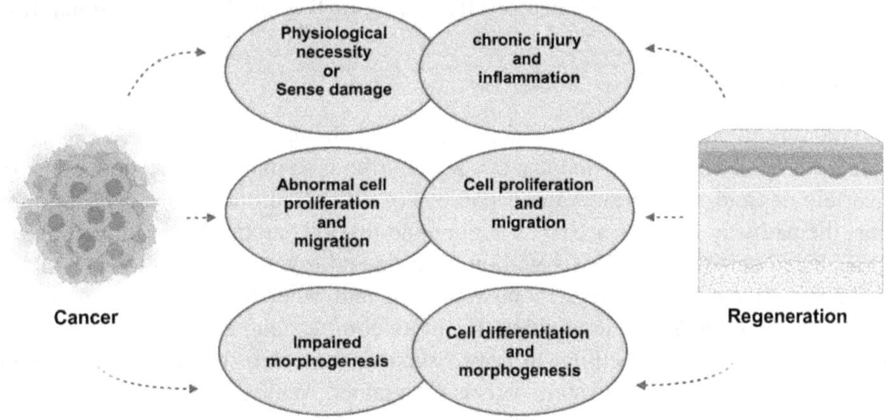


**Fig. 1** Events involved in cancer and regeneration

the overlapping signaling pathways shared by cancer, wound healing, and regeneration and their therapeutic implications.

## 2 Molecular Pathways in Regeneration and Tissue Repair

Regeneration is a widely observed phenomenon in the animal kingdom, though the extent of regenerative capacity varies significantly among species. Invertebrates, such as hydra and planarians, possess the remarkable ability to regenerate their entire body. This regeneration in invertebrates occurs mainly through two processes: epimorphic and morphallactic regeneration. In epimorphic regeneration, terminally differentiated cells dedifferentiate (reenter the cell cycle) and form an undifferentiated mass of cells that subsequently develop into the regenerated appendage or organ. In contrast, morphallactic regeneration involves the repatterning of pre-existing tissues with minimal new cell growth (Galliot et al., 2008; Gilbert, 2000).

In vertebrates, regenerative potential is generally more pronounced during embryonic or larval stages but declines significantly as development progresses. While many adult vertebrates, excluding mammals, retain the ability to fully regenerate specific organs or appendages after injury, the regenerative capacity in adult mammals is comparatively limited (Gao et al., 2021). In mammals, regeneration is typically restricted to specific tissues, such as the liver, bones, skeletal muscle, and, to some extent, skin. However, other tissues, like the heart and sensory hair cells, show minimal capacity for structural or functional recovery following injury (Galliot et al., 2008; Okamoto & Watanabe, 2004; Vujic et al., 2020; C. Zhang et al., 2024). Several signaling pathways play a fundamental role in tissue and organ regeneration across both vertebrates and invertebrates. These pathways regulate cellular proliferation, differentiation, migration, and tissue remodeling, contributing to effective regeneration. Below, we discuss a few of the most crucial regeneration pathways.

- Wnt/*β-Catenin Pathway*

The Wnt/β-catenin pathway plays a critical role in embryonic and postnatal development in vertebrates and insects. Its key role is to maintain the stem cells. In invertebrates and select vertebrates capable of regeneration, such as urodele amphibians, the pathway remains active throughout adulthood, supporting stem cell maintenance and self-renewal for tissue homeostasis and regeneration following injury (Volckaert & De Langhe, 2015). In mammals and other non-regenerating vertebrates, the pathway's role in adulthood is more complex and context-dependent. In organs with continuous cellular turnover, such as the colon, where a well-defined population of tissue-resident stem cells is maintained, Wnt signaling remains active and regulates stem cell homeostasis. However, in non-regenerating organs, such as the heart in higher vertebrates, the pathway is typically quiescent and is only activated in response to injury (Bastakoty & Young, 2016).

Wnt signaling comprises canonical and noncanonical pathways. Noncanonical Wnt pathways include the Wnt/$Ca^{2+}$ and planar cell polarity pathways. The noncanonical pathway functions independently of β-catenin and the T cell factor/lymphoid enhancer-binding factor (TCF/LEF) complex. However, the canonical Wnt pathway, also known as the Wnt/β-catenin pathway, involves the nuclear translocation of β-catenin and the activation of target genes via TCF/LEF transcription factors. While canonical pathways primarily regulate cell proliferation, noncanonical Wnt signaling governs cell polarity and migration. These pathways interact through a network of mutual regulation, ensuring coordinated cellular responses.

Wnt signaling plays a crucial role in developing and renewing small intestinal epithelial tissue, particularly by promoting the differentiation of Paneth cells at the crypt base. Beyond the intestine, Wnt signaling is integral to liver metabolism and regeneration, lung tissue repair and homeostasis, hair follicle renewal, hematopoietic system development, and osteoblast maturation and activity. Through these diverse functions, the pathway maintains tissue integrity and regeneration across multiple organ systems (Liu et al., 2022).

- *MAPK/ERK Pathway*

The mitogen-activated protein kinase (MAPK) pathway is a significant signal transduction cascade that regulates various physiological processes, including cell differentiation, proliferation, gene expression, metabolism, and survival. This family of kinases is activated through a phosphorylation cascade and requires an extracellular stimulus for initiation leading to four distinct signaling cascades, including the extracellular signal-related kinases (ERK1/2), Jun amino-terminal kinases (JNK1/2/3), p38-MAPK, and ERK5. The extracellular signal-regulated kinase (ERK1/2) cascade was the first MAPK pathway to be characterized. In this pathway, ERK1/2 is phosphorylated by MAPK kinase 1 (MEK1), thereby activating downstream MAPK signaling. (Cervellini et al., 2018) The ERK1/2 pathway promotes cell proliferation and differentiation, while the JNK and p38 pathways regulate stress responses and tissue repair.

Various studies done in vertebrates and invertebrates have shown that the MAPK/ERK pathway is being initiated in various organs during regeneration promoting various processes such as cell cycle progression, dedifferentiation, osteoblast proliferation, cardiomyocyte proliferation, and hepatocyte proliferation. Moreover, MAPK signaling is essential for neural regeneration, myelination, and skeletal muscle repair, as well as inflammatory responses in wound healing (Wen et al., 2022).

- *Notch Pathway*

The Notch signaling pathway regulates cell fate determination, differentiation, and proliferation in both vertebrates and invertebrates. Notch signaling has been extensively reported to play a critical role in tissue regeneration and developmental patterning in various organisms, including sea urchins. In arthropods, this highly conserved pathway is integral to sequential segmentation, as observed in diverse taxa such as spiders, myriapods, insects, and crustaceans.(Li et al., 2024).

This pathway is activated when Notch receptors interact with ligands (Jagged/Delta-like) on neighboring cells, triggering the release of the Notch intracellular domain (NICD), which enters the nucleus to influence gene expression. The canonical Notch signaling pathway regulates gene expression through its target gene families, primarily the hairy/enhancer of split (HES) and hairy/enhancer of split-related with YRPW motif (HEY) families (Slowik & Bermingham-McDonogh, 2014). It regulates the development and function of the cardiovascular and skeletal systems through its downstream target genes, particularly members of the HEY family, which are involved in cell fate determination and tissue morphogenesis. It maintains neural stem cells in the central nervous system and promotes muscle regeneration by regulating satellite cell activity. It is also involved in cardiac regeneration, intestinal stem cell renewal, and epidermal repair, playing a fundamental role in tissue homeostasis (Zhou et al., 2022).

- *Hedgehog Pathway*

The Hedgehog (Hh) signaling pathway plays a fundamental role in regenerative processes. In amphibian limb regeneration, studies have shown that while the dedifferentiation of mature cell lineages occurs independently of Hh signaling, the proliferation and migration of dedifferentiated cells are critically regulated by this pathway (Singh et al., 2012).

It is mediated by three key ligands—Sonic Hedgehog (Shh), Indian Hedgehog (Ihh), and Desert Hedgehog (Dhh). Hedgehog signaling is activated when Hh ligands bind to the Patched (Ptch) receptor, leading to activation of the smoothened (Smo) protein and downstream Gli transcription factors. This pathway is vital for neuronal regeneration, skeletal muscle repair, epidermal regeneration, and bone and cartilage formation (Jing et al., 2023). The Hedgehog (Hh) signaling pathway has been demonstrated to facilitate axonal regeneration and neurogenesis in both the central and peripheral nervous systems. Furthermore, it plays a critical role in the regeneration of various tissues, including the lens, limbs, vasculature, prostate, retina, facial nerve, bile duct, lung, and bone following injury. Additionally, Hh signaling is essential for the maintenance and regulation of adult stem cell populations in the skin and brain (Petrova & Joyner, 2014).

- *Hippo Pathway*

The Hippo signaling pathway is a key regulator of organ size, tissue homeostasis, and stem cell activity. In mammals, the Hippo signaling pathway is regulated by various stimuli, including mechanical signals, cell shape, extracellular matrix (ECM) stiffness, cell polarity, metabolism, and cell-cell interactions. The core Hippo pathway includes kinases MST1/2 and LATS1/2, which inhibit the transcriptional coactivators YAP/TAZ. When Hippo signaling is suppressed, YAP1 and TAZ remain unphosphorylated and translocate into the nucleus, where they bind to transcription factors TEAD1-4. This interaction promotes the expression of genes involved in cell proliferation, cancer progression, stemness, and epithelial-mesenchymal transition (EMT) (Riley et al., 2022).

The Hippo pathway is believed to have evolved from unicellular organisms. In mammals, its structure and function are largely preserved. Studies in mice show that deleting Hippo pathway genes or increasing the activity of its downstream effectors, especially in the liver, often leads to significant organ enlargement. In humans, this pathway is particularly important for liver regeneration, as YAP activation promotes hepatocyte proliferation. It also plays a role in cardiac regeneration in neonatal mammals and regulates epidermal stem cell activation during wound healing (Nguyen-Lefebvre et al., 2021).

- *PI3K/Akt/mTOR Signaling Pathway*

The PI3K/Akt/mTOR pathway is a central regulator of cell survival, metabolism, and tissue regeneration. It is activated by growth factors binding to receptor tyrosine kinases, leading to phosphatidylinositol 3-kinase (PI3K) activation and the production of PIP3; this pathway leads to the phosphorylation of Akt (also known as protein kinase B), which subsequently activates the mechanistic target of rapamycin (mTOR). Downstream signaling through mTOR complex 1 (mTORC1) stimulates protein synthesis by phosphorylating key effectors such as p70S6 kinase (S6K) and the eukaryotic initiation factor 4E-binding protein 1 (4E-BP1), thereby promoting cell growth and proliferation. mTOR signaling also regulates autophagy through the inhibition of the ULK1 complex. Additionally, Akt activation inhibits pro-apoptotic proteins like BAD and activates the transcription factor FOXO, modulating cell survival, glucose metabolism, and oxidative stress responses (Y. He et al., 2021).

PI3K signaling plays a critical role in various regenerative processes across species. In *Hydra*, PI3K is essential for head formation by activating early head-specific gene expression. In amphibians, such as *Xenopus laevis* froglets, PI3K facilitates limb regeneration by regulating the re-entry of mesodermal cells into the cell cycle, enabling blastema formation. In mammalian models, including mice, PI3K activation promotes sensory axon regeneration and enhances hepatic regenerative responses (Zheng et al., 2021). Additionally, the PI3K/Akt/mTOR pathway enhances muscle regeneration by stimulating satellite cell proliferation and protein synthesis, supports neuronal survival and axonal regrowth in the nervous system, promotes angiogenesis for vascular repair, and regulates osteoblast and chondrocyte activity for bone and cartilage regeneration. The pathway's downstream targets orchestrate a wide range of cellular processes, highlighting its pivotal role in tissue homeostasis, repair, and regeneration (J. S. L. Yu & Cui, 2016).

- *TGF-β Signaling Pathway*

The transforming growth factor-β (TGF-β) signaling pathway plays a dual role in regeneration by promoting tissue repair while also contributing to fibrosis. TGF-β signaling is mediated by Suppressor of Mothers against Decapentaplegic (SMAD)-dependent and non-SMAD pathways and regulates fibroblast activation, extracellular matrix deposition, and immune responses.

TGF-β signaling is initiated upon ligand binding to its serine/threonine kinase receptors, leading to receptor-mediated phosphorylation of downstream Smad2/3 transcription factors. Once phosphorylated, pSmad2/3 associates with Smad4, forming a heteromeric complex that translocates into the nucleus. Within the nucleus, this complex interacts with co-activators or co-repressors to modulate the transcription of target genes, thereby regulating key cellular processes such as proliferation, differentiation, apoptosis, and extracellular matrix remodeling (Nakamura et al., 2021). It is crucial for cartilage and bone regeneration, liver repair, and wound healing. However, excessive TGF-β activity can lead to pathological fibrosis, limiting regenerative potential in organs like the heart and lungs (Sader & Roy, 2022).

## 3 Cancer

Cancer remains a debilitating disease that, despite extensive research and therapeutic advancements, remains challenging to cure. Most treatment strategies are developed based on insights into its pathogenesis through the analysis of gene and protein expression in clinically obtained specimens from cancer patients. However, the differences between in vivo and in vitro models often fail to fully replicate the complex dynamics of the tumor microenvironment. Cancer tumors have been described as "wounds that do not heal," arising from persistent tissue insult that leads to uncontrolled cellular proliferation (Riss et al., 2006; Schäfer & Werner, 2008).

Both tissue regeneration and tumorigenesis are highly dynamic and adaptive, regulated by signals from the host and the tissue microenvironment. A complex network of molecular cues governs response to injury, facilitating regeneration and wound repair. Both processes involve key cellular mechanisms such as proliferation, survival, and migration, which are modulated by growth factors, cytokines, inflammatory mediators, and angiogenic signals (Coussens & Werb, 2002) (Hansda & Ghosh, 2022). These pro-proliferative and pro-survival signals originate from various cellular and extracellular components within the wound and tumor microenvironment. As a result, wounds and cancers exhibit significant phenotypic similarities in cellular behavior, signaling pathways, and gene expression profiles.

Investigating these similarities between wounds and cancers can provide valuable insights into the mechanisms underlying the malignant phenotype. Additionally, identifying the key differences, particularly those associated with "impaired healing" may elucidate the loss of regulatory control in cancer. This understanding could serve as the foundation for the development of novel diagnostic and therapeutic targets (Riss et al., 2006; Schäfer & Werner, 2008).

## 4 Molecular Drivers in Cancer and Regeneration

Human cancers exhibit significant similarities to both developmental processes and tissue regeneration, not only at the cellular level but also at genetic, genomic, and epigenetic levels. Dysregulation of key regenerative signaling pathways plays a critical role in facilitating oncogenesis and promoting cancer progression (White & Zon, 2008).

At the genetic level, oncogene expression does not inherently disrupt the normal morphogenetic potential of the cells. Oncogenes, rather than being exclusive cancer-causing genes, are normal regulatory genes essential for organismal development and, potentially, cellular immortality. The characteristics commonly associated with cancer, except for the inhibition of redifferentiation, reflect the reactivation of embryonic developmental pathways and self-renewal mechanisms (Baramiya & Baranov, 2020).

External stimuli, including chronic inflammation and recurrent insults, frequently initiate tumorigenesis, creating a microenvironment conducive to malignant transformation. Notably, several molecular drivers are shared between cancer and regeneration, underscoring the long-standing notion that cancer represents a dysregulated form of regenerative growth. This convergence is particularly evident in the activation of pathways such as PI3K/Akt/mTOR, *Wnt*/β-catenin, Hippo/YAP, and TGF-β, which orchestrate both tissue repair and oncogenic progression, thereby blurring the distinction between physiological regeneration and pathological proliferation.

- *Cellular Proliferation and Plasticity*

Cellular plasticity refers to the capacity of cells to undergo reversible phenotypic transitions in response to intrinsic and extrinsic cues. A prominent example is observed during early metazoan embryogenesis, where epithelial cells acquire mesenchymal characteristics through a process known as epithelial-to-mesenchymal transition (EMT). EMT facilitates the transformation of epithelial cells into a motile, invasive mesenchymal state, enabling migration and invasion into adjacent tissues. This transition is characterized by the loss of apical-basal polarity, the downregulation of cell-cell adhesion molecules (e.g., E-cadherin), and the upregulation of mesenchymal markers (e.g., N-cadherin and vimentin) (Nieto, 2013).

The dynamic nature of EMT is crucial for embryonic development, as it allows cells to migrate into the interior of the embryo, traverse long distances, and contribute to the formation of internal organs. Importantly, EMT is inherently reversible; upon reaching their destination, cells typically undergo a mesenchymal-to-epithelial transition (MET), restoring epithelial characteristics to facilitate tissue organization, proliferation, and differentiation into specialized structures. This bidirectional plasticity is fundamental to developmental processes and plays a pivotal role in wound healing, fibrosis, and cancer progression.(Qin et al., 2020) Several signaling pathways and transcription factors orchestrate the mesenchymal-to-epithelial transition (MET), and their dysregulation can lead to the acquisition of a mesenchymal phenotype, which may be reversible or irreversible depending on the context. Key molecular regulators of MET include TGF-β, *Wnt*, Notch, and receptor tyrosine kinases (RTKs).

TGF-β signaling, which is often aberrantly activated in cancer, plays a pivotal role in regulating both epithelial-to-mesenchymal transition (EMT) and MET. TGF-β can induce EMT by promoting the loss of epithelial markers and the acquisition of mesenchymal characteristics. It can also facilitate MET under certain conditions, contributing to the reversion of the mesenchymal phenotype during processes such as metastasis and tissue repair (Bhat et al., 2024).

An important regulator of the PI3K signaling pathway, PTEN is among the most frequently mutated tumor suppressor genes in human cancer. Loss of PTEN function leads to hyperactivation of PI3K/Akt/mTOR signaling, promoting uncontrolled cell proliferation, survival, and invasion. In planarians, PTEN deficiency results in tissue disorganization, disruption of basement membrane integrity, and aberrant epithelial-mesenchymal interactions, leading to the emergence of invasive, ectopic outgrowths akin to neoblasts. This phenotype parallels chemically induced tumor-like growths in planarians, suggesting that molecular dysregulation of PI3K signaling can drive abnormal stem cell proliferation, disrupt epithelial-mesenchymal crosstalk, and facilitate cellular infiltration—hallmarks of cancer progression in vertebrates. These findings highlight the evolutionary conservation of PI3K/PTEN signaling in regulating cellular plasticity and underscore its dual role in both regeneration and oncogenesis (Fig. 2) (Oviedo & Beane, 2009).

In multiple stem cell compartments, including the hematopoietic and enteric systems, *Wnt*/β-catenin signaling plays a pivotal role in regulating progenitor cell proliferation and maintaining multipotency. The Hedgehog (Hh) and Notch pathways serve as critical modulators of enteric stem cell homeostasis by balancing proliferative potential, restricting excessive cell division, and promoting lineage-specific differentiation.

Disruptions in these signaling pathways can lead to uncontrolled stem cell expansion. Specifically, loss-of-function mutations in Hh or Notch signaling, or gain-of-function mutations in *Wnt*/β-catenin signaling, result in the loss of regulatory control over crypt stem cell proliferation. This dysregulation leads to an unchecked expansion of the stem cell compartment, where mutant progenitor cells continuously generate daughter cells harboring the same genetic alterations.

The selective advantage conferred by these mutations drives the clonal expansion of self-renewing, aberrantly proliferating progenitor cells. Over time, this enlarges the pool of cells predisposed to additional genetic or epigenetic alterations,

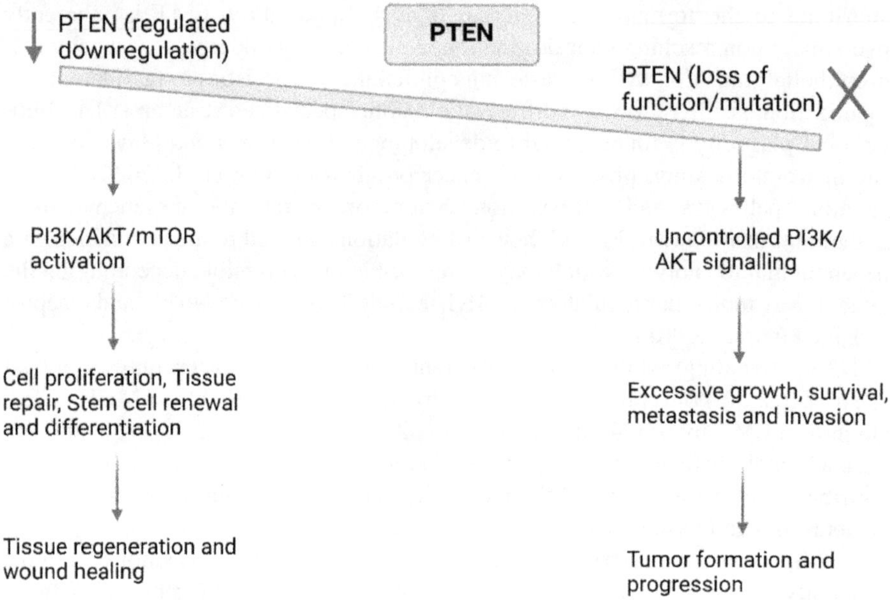


**Fig. 2** Dual role of PTEN in cancer and regeneration

thereby increasing susceptibility to further oncogenic mutations and malignant transformation. Such cumulative effects contribute to the early stages of tumorigenesis, highlighting the critical role of tightly regulated signaling networks in maintaining stem cell equilibrium and preventing neoplastic progression (Tataria et al., 2006).

- *The Tumor Microenvironment and Regeneration Microenvironment*

Tissue maintenance and regeneration in mammals are governed by highly complex and dynamic microenvironmental systems, collectively termed stem cell niches. These specialized regulatory domains comprise stem and progenitor cells, diverse somatic cell populations, extracellular matrix (ECM) components, and a network of signaling mediators, all of which engage in mutual interactions to maintain tissue integrity and function. Under normal adult homeostasis, stem cell niches predominantly remain quiescent, preserving a delicate balance between self-renewal and differentiation. (Howlader et al., 2024; Prateeksha et al., 2023) However, upon tissue injury or pathological insult, these niches undergo extensive remodeling, often reverting to embryonic or early postnatal regulatory states. This shift establishes a pro-regenerative microenvironment characterized by the activation, proliferation, and lineage commitment of stem and progenitor cells, ultimately leading to the restoration of functional tissue architecture. In parallel, immune cells and angiogenic factors play pivotal roles in modulating this regenerative milieu, ensuring coordinated tissue formation and functional restoration. Macrophages, T cells, and cytokines regulate inflammatory responses and cellular crosstalk, while angiogenic signals facilitate vascular remodeling and oxygenation, essential for effective tissue regeneration.

In the tumor microenvironment (TME), a similarly intricate and dynamic cellular milieu is present, comprising a diverse array of cell types that collectively influence tumor progression and immune evasion. These specialized regulatory domains consist of stem and progenitor cells, cancer-associated fibroblasts (CAFs), immune cells such as tumor-associated macrophages (TAMs) and tumor-infiltrating lymphocytes (TILs), endothelial cells, pericytes, extracellular matrix (ECM) components, and a complex network of signaling mediators, all of which engage in reciprocal interactions to sustain tissue integrity in regeneration or drive malignant transformation in cancer (De Visser & Joyce, 2023).

During tissue regeneration, macrophages initially adopt an M1 pro-inflammatory phenotype, facilitating pathogen clearance, debris phagocytosis, and cytokine secretion to recruit monocytes. As repair progresses, they transition to the M2 anti-inflammatory phenotype, promoting neutrophil apoptosis, inflammation resolution, and tissue restoration. A balanced M1-to-M2 switch is crucial for proper regeneration. In the tumor microenvironment (TME), cancer cells manipulate macrophage polarization. Early-stage tumors exhibit M1-like macrophages, which suppress tumor growth by recruiting CD8+ T and NK cells through antigen presentation and chemokine signaling (CXCL9, CXCL10, CXCL11). However, as cancer progresses, tumor cells induce a dominant M2-like phenotype, driving immune suppression, angiogenesis, and tumor progression (Boutilier & Elsawa, 2021).

- *Epigenetic Regulation*

Epigenetic regulation serves as a critical interface between environmental stimuli and the genome, enabling cells to dynamically modulate gene expression in response to external cues. Epigenetic alterations such as DNA methylation, histone modifications, and non-coding RNA-mediated regulation may arise when cells within tissues are subjected to persistent or intense environmental signals, leading to modifications. These changes, without altering the underlying genetic sequence, can influence cellular function, adaptation, and even long-term phenotypic outcomes (Aspriţoiu et al., 2021).

DNA methylation occurs through two primary mechanisms: maintenance methylation and de novo methylation. DNA methyltransferase 1 (DNMT1) ensures the precise transmission of CpG methylation patterns during cell division, while DNMT3 enzymes mediate de novo methylation by modifying previously unmethylated cytosines, thereby influencing gene expression. These enzymes are crucial for embryonic development and cellular differentiation, exhibiting high expression in undifferentiated embryonic stem cells and downregulation upon differentiation. Embryonic stem cells deficient in DNMTs retain pluripotency but fail to differentiate unless DNMT activity is restored. Moreover, embryonic stem cells lacking all DNMTs show upregulated expression of tissue-specific transcription factors and signaling molecules, highlighting the role of epigenetic modifications in repressing differentiation and maintaining stem cell identity.(Aguilar & Gardiner, 2015) A study conducted by Liu et al. demonstrated the fundamental role of DNA methylation in liver regeneration following tissue injury. The research elucidated the mechanistic involvement of DNA methylation at the p53 locus during both the dedifferentiation and redifferentiation phases of biliary-mediated liver regeneration. These findings underscore the significance of epigenetic modifications, particularly DNA methylation and p53 regulation, as key determinants of the liver's regenerative capacity (J. He et al., 2022). Cancer cells frequently exhibit hypermethylation within the promoter regions of CpG island-associated tumor suppressor genes, including glutathione S-transferase pi (GSTP1), retinoblastoma (RB1), and E-cadherin (CDH1). This aberrant DNA methylation leads to transcriptional silencing of these critical regulatory genes, resulting in loss of function and contributing to tumorigenesis. Additionally, the combined effects of hypermethylation and global hypomethylation reshape the epigenomic landscape of cancer cells, often exerting a more significant impact on gene regulation than mutations within coding regions, which relatively occur at lower frequencies (McCabe et al., 2009).

Chromatin remodeling is the dynamic alteration of chromatin architecture to regulate gene expression by facilitating or restricting access to transcriptional machinery. In eukaryotes, ATP-dependent chromatin remodeling complexes are classified into four groups based on their ATPase and protein subunits: switching defective/sucrose nonfermenting (SWI/SNF), imitation switch (ISWI), chromodomain helicase DNA binding (CHD), and inositol requiring 80 (INO80) (Tang et al., 2017). Chromatin structure and its regulation are critical for maintaining both stem cell quiescence and cellular plasticity. In regenerative contexts, such as the blastema, tissue-specific stem cells must balance high proliferative capacity with the ability to differentiate into appropriate cellular lineages as needed. This requires chromatin to remain in an open conformation, allowing cells to rapidly respond to signals that regulate the expression of pluripotency-associated or differentiation-associated genes. For instance, DNase I hypersensitivity assays have demonstrated that intestinal progenitor cells maintain an open chromatin state, enabling cell fate transitions. Additionally, epigenetic regulation, which is well-established in cell cycle control and differentiation, plays a crucial role in governing quiescence, facilitating either its maintenance or release in response to environmental cues (Martinez-Redondo & Izpisua Belmonte, 2020).

The regulation of gene expression is governed by the intricate interplay between transcription factors and chromatin-remodeling proteins. Disruptions in this regulatory network can lead to inappropriate gene activation or silencing, resulting in cellular dysfunction and disease. A strong link between cancer and chromatin remodeling is evidenced by studies on rhabdoid tumors, an aggressive pediatric malignancy affecting the brain and soft tissues. The majority of these tumors harbor mutations in hSNF5/INI1, a core subunit of the SWI/SNF chromatin remodeling complex. Notably, hSNF5/INI1 mutations are also observed in other pediatric cancers, including choroid plexus carcinoma, medulloblastoma, central primitive neuroectodermal tumors, and certain forms of acute leukemia. Additionally, mutations in BRG1, the ATPase subunit of the SWI/SNF complex, have been identified in multiple cancer cell lines, suggesting its role as a potential oncogenic target, further supporting its involvement in tumorigenesis. Mice with BRG1+/− exhibit an increased predisposition to tumor formation, reinforcing the significance of chromatin remodeling dysfunction in cancer development (Davis & Brachmann, 2003; Neely, 2002; Roberts & Orkin, 2004).

## 5 Inflammation in Cancer and Regeneration

Inflammation is an important defense mechanism that allows the body to heal from injuries and fight off infections by sending immune cells to the affected area, essentially acting as a vital part of the body's healing process; however, when uncontrolled, chronic inflammation can contribute to various diseases. Inflammation is a central player linking regeneration and cancer, acting as a mediator of both tissue repair and tumorigenesis. In regeneration, acute inflammation initiates healing by recruiting immune cells and releasing cytokines like interleulin-6 (IL-6), tumor necrosis factor-α (TNF-α), and TGF-β, which activate stem cells and promote tissue remodeling (Hansda & Das, 2025). Conversely, chronic inflammation creates a microenvironment conducive to cancer by inducing oxidative stress, DNA damage, and angiogenesis, while sustaining proliferative signaling through pathways like nuclear factor kappa B (NF-κB), STAT3, and COX-2. For example, regenerative processes in liver injury and chronic hepatitis show how persistent inflammation can evolve into hepatocellular carcinoma. Similarly, in the intestine, inflammation associated with colitis drives tumor development, emphasizing the shared pathways between wound healing and cancer progression.

In regeneration, NF-κB is essential for coordinating inflammatory responses and promoting cell proliferation and survival. It is activated in response to tissue injury, recruiting immune cells, stimulating the production of cytokines, and enhancing the regenerative capacity of stem or progenitor cells. This is particularly evident in tissues like the liver, where NF-κB activation in hepatocytes and Kupffer cells is necessary for compensatory growth following partial hepatectomy. However, the same pro-survival and proliferative signals that facilitate regeneration also contribute to oncogenesis when dysregulated. In cancer, chronic NF-κB activation is associated with tumor initiation, progression, and resistance to therapy. Many tumors exploit NF-κB signaling to evade apoptosis, sustain inflammation, and enhance angiogenesis, creating a microenvironment conducive to malignancy. Additionally, NF-κB-mediated immune modulation can suppress anti-tumor immunity, further promoting tumor survival. Thus, while NF-κB is indispensable for regeneration by orchestrating cellular repair and growth, its persistent activation in cancer leads to unchecked proliferation and therapy resistance (Chen et al., 2001; Park & Hong, 2016). In regeneration, STAT3 is activated in response to signals such as interleukin-6 (IL-6) and other cytokines, promoting cell proliferation, survival, and tissue repair. This is particularly evident in the liver, where STAT3 activation following injury or partial hepatectomy facilitates hepatocyte proliferation and enhances the regenerative process. Similarly, in skin and muscle regeneration, STAT3 supports stem cell function and modulates inflammation to create a favorable microenvironment for tissue repair. However, the same regenerative properties of STAT3 are co-opted by cancer cells to drive tumorigenesis. Persistent STAT3 activation in cancer contributes to uncontrolled cell proliferation, evasion of apoptosis, and tumor immune evasion. STAT3 also plays a key role in promoting angiogenesis by upregulating VEGF, thereby ensuring a sustained supply of nutrients to growing tumors. Additionally, STAT3-mediated immune suppression via regulatory T cells (Tregs) and myeloid-derived suppressor cells (MDSCs) allow tumors to escape immune surveillance. Chronic inflammation, often associated with cancer, further fuels STAT3 activity, establishing a feed-forward loop that sustains tumor progression. Thus, while STAT3 is indispensable for regeneration by facilitating tissue repair and homeostasis, its aberrant activation in cancer leads to malignant transformation, making it a critical target for anti-cancer therapies aimed at disrupting its signaling while preserving its regenerative functions (Bareyre et al., 2011; Erratum, 2022).

Targeting inflammatory mediators provides therapeutic opportunities. Anti-inflammatory agents like NSAIDs or IL-1β inhibitors have shown potential in preventing chronic inflammation-driven cancers while preserving regenerative capacity. These insights stress the importance of balancing pro- and anti-inflammatory mechanisms to optimize outcomes in both regeneration and cancer treatment (Coussens & Werb, 2002; Eming et al., 2017; Grivennikov et al., 2010).

## 6 Biomarkers in Cancer and Regeneration

Studying biomarkers of cancer and regeneration together is crucial for understanding shared molecular mechanisms and their divergent roles in tissue repair and malignancy. Regeneration involves controlled cell growth and differentiation, processes often hijacked in cancer. Regenerative processes involve tightly regulated cellular mechanisms that prevent uncontrolled proliferation. Biomarkers such as p53, Wnt/β-catenin, and BMP signaling pathways play dual roles in regeneration and cancer. For example, Wnt/β-catenin is crucial for tissue repair, but its dysregulation leads to tumorigenesis. Sox2, a marker of stemness, is another key regulator of cellular plasticity in regeneration and is implicated in cancer development when overexpressed (Liu et al., 2022; S. Zhang et al., 2020). Cancer biomarkers include HER2 in breast cancer, KRAS mutations in lung and colorectal cancers, and PD-L1, which modulate immune evasion. These markers help identify aggressive tumor types, monitor treatment responses, and guide targeted therapy. Some biomarkers, like TGF-β, have overlapping roles in regeneration and cancer, promoting tissue repair in normal settings but driving metastasis in cancers (Das et al., 2023; Principe et al., 2014). Targeting NF-κB signaling as a biomarker remains a major focus in cancer research, aiming to inhibit its oncogenic effects while preserving its regenerative functions. Integrating knowledge from these fields can lead to innovative therapies that promote regeneration while preventing or treating cancer, ultimately advancing precision medicine and tissue engineering.

## 7 Regeneration Models in Cancer

Regeneration models are pivotal in studying cancer mechanisms, evaluating therapies, and understanding treatment resistance. There are some notable models used in both clinical and preclinical settings, such as organoids, mouse models, and syngeneic models. In vitro 2D cell culture is widely used for studying cancer biology and drug screening. They are simple and cost-effective but lack the complexity of the tumor microenvironment. Co-culture system models also involve cancer cells interacting with stromal or immune cells, providing insights into tumor-immune interactions. Patient-derived xenografts (PDXs) are widely used for testing systemic therapies and predicting clinical outcomes, especially for precision oncology. Tumors from patients are implanted into immunodeficient mice, retaining tumor histology and genetics. Three-dimensional (3D) organoids are derived from patient tumors; these 3D in vitro cultures mimic the tumor microenvironment, providing a scalable model for drug testing and translational research. Another model is genetically engineered mouse models (GEMMs), which allow precise genetic manipulations to study specific oncogenes and tumor suppressors in vivo. Syngeneic models use mouse tumors implanted in immunocompetent mice to study immune-oncology therapies. Translational research bridges preclinical insights into human trials, leveraging models like humanized PDXs to assess immunotherapies. In zebrafish models, transparent larvae are used for visualizing tumor growth, angiogenesis, and drug effects in real time. (Yu et al., 2024).

Axolotls and planarians have both regeneration ability with cancer insight that they are used as model. Axolotls can regenerate limbs, spinal cords, heart tissues, and even parts of the brain. Their robust regenerative capabilities make them ideal for studying cancer, particularly mechanisms of cell cycle control, immune response, and tumor suppression. On cancer insights axolotls exhibit low rates of spontaneous tumors, suggesting unique anti-tumor properties in their regenerative pathways. Their ability to maintain cellular plasticity without forming malignancies is of significant interest in cancer biology. Planarians possess abundant neoblasts, totipotent stem cells responsible for their remarkable regeneration of any body part. These cells allow for the study of stem cell regulation, apoptosis, and differentiation—key areas in cancer research. Furthermore, under specific conditions, planarian regeneration can lead to uncontrolled growth, providing insights into tumor formation and mechanisms that prevent cancer in regenerating organisms (McCusker & Gardiner, 2011; Reddien, 2018). These models are invaluable for advancing cancer research, although they are resource-intensive.

## 8 Therapeutic Implications in Cancer and Regeneration

The study of regeneration and cancer offers critical therapeutic implications. Regenerative models, such as axolotls and planarians, reveal mechanisms for maintaining controlled cell proliferation, differentiation, and immune regulation, all of which are crucial for preventing tumorigenesis. These models provide insights into how tissues can repair themselves without triggering malignant transformations, highlighting pathways like *Wnt*, BMP, and p53 signaling (McCusker & Gardiner, 2011; Reddien, 2018). In cancer therapy, understanding these pathways can lead to innovative treatments that suppress tumor growth while promoting healthy tissue repair, such as through stem cell therapies or immune modulation. Regenerative medicine approaches in the treatment of cancer, such as stem cell therapies, may counteract the side effects of chemotherapy by restoring damaged tissues and reducing fibrosis.

Other therapeutic approaches, such as exosome-based strategies, are actively being investigated. Exosomes, small membrane-bound vesicles, mediate the intracellular transfer of molecules, antibodies, and therapeutic agents through endocytosis or membrane fusion. As key components of both the stem cell niche and the tumor microenvironment, exosomes are being explored for their role in targeted drug delivery, intercellular communication, and tumor modulation. By influencing critical signaling pathways such as Hedgehog (Hh), TGF-β, and Wnt, exosome-based therapies present a promising avenue for regenerative medicine and cancer treatment (Hao et al., 2022).

Furthermore, studies on organisms with high regenerative potential can inform therapies designed to re-establish controlled cellular behavior and inhibit tumor growth. For example, manipulating epigenetic regulation and immune pathways from regenerative models may enhance precision oncology strategies. Moreover, regeneration research aids in developing strategies to mitigate the damage caused by cancer treatments, including chemotherapy and radiation, by enhancing tissue repair and minimizing fibrosis (Baddour et al., 2012).

## 9 Future Perspective

The intersection of regeneration and cancer research has vast potential for future exploration. Understanding molecular drivers like Wnt signaling, p53 regulation, and epigenetic modifications in regenerative models (e.g., axolotls and planarians) can shed light on tumor suppression and controlled proliferation. These insights may lead to novel cancer therapies that mimic regenerative processes to repair tissues without promoting malignancy. Additionally, advancements in regenerative medicine, such as 3D bioprinting, stem cell therapies, and organoid systems, can be optimized to reduce treatment side effects. Integration with precision medicine could personalize therapies to target cancer cells while promoting tissue repair. Harnessing immune modulation observed in regenerative organisms may further improve immune-oncology therapies. Finally, interdisciplinary approaches involving bioinformatics and CRISPR-based genome editing could enhance our ability to decode regeneration mechanisms and translate them into innovative cancer therapies. Collaboration between regenerative biology and oncology promises transformative breakthroughs in cancer treatment and tissue repair.

## 10 Conclusion

It is interesting to reflect that the link between tissue repair and cancer is evident in the observation that long-lived organisms and tissues with high rates of cycling stem cells—such as the prostate, breast, gut, and blood—also exhibit the highest cancer incidence. In contrast, juvenile cancers often arise in rapidly proliferating neural tissues (e.g., neuroblastoma, medulloblastoma) and parenchymal organs (e.g., nephroblastoma, hepatoblastoma) undergoing postnatal tissue expansion as a continuation of embryonic organogenesis (De Visser & Joyce, 2023). Understanding how tissue regeneration may contribute to cancer necessitates a paradigm shift, viewing regenerative processes that mediate tissue restoration as potential avenues for carcinogenesis. Exploring the intersection between regeneration and tumorigenesis offers promising potential for cancer treatment. While much remains to be discovered about the molecular pathways regulating growth and differentiation, the ongoing exploration holds considerable promise. Such insights may lead to innovative therapeutic strategies rooted in a comprehensive understanding of tumorigenesis, human development, and disease.

**Acknowledgments** I would like to sincerely acknowledge Dr. Ming-Hai Wang M.D., PhD for the invaluable knowledge and guidance he has imparted, which enabled me to write this chapter. I am also deeply grateful for his unwavering support and for granting me the opportunity to serve as the corresponding author for this chapter.

## References

Aguilar, C., & Gardiner, D. M. (2015). DNA methylation dynamics regulate the formation of a regenerative wound epithelium during axolotl limb regeneration. *PLoS One, 10*(8), e0134791. https://doi.org/10.1371/journal.pone.0134791

Aspriţoiu, V. M., Stoica, I., Bleotu, C., & Diaconu, C. C. (2021). Epigenetic regulation of angiogenesis in development and tumors progression: Potential implications for cancer treatment. *Frontiers in Cell and Developmental Biology, 9*, 689962. https://doi.org/10.3389/fcell.2021.689962

Baddour, J. A., Sousounis, K., & Tsonis, P. A. (2012). Organ repair and regeneration: An overview. *Birth Defects Research Part C: Embryo Today: Reviews, 96*(1), 1–29. https://doi.org/10.1002/bdrc.21006

Baramiya, M. G., & Baranov, E. (2020). From cancer to rejuvenation: Incomplete regeneration as the missing link (part I: The same origin, different outcomes). *Future Science OA, 6*(3), FSO450. https://doi.org/10.2144/fsoa-2019-0119

Bareyre, F. M., Garzorz, N., Lang, C., Misgeld, T., Büning, H., & Kerschensteiner, M. (2011). In vivo imaging reveals a phase-specific role of STAT3 during central and peripheral nervous system axon regeneration. *Proceedings of the National Academy of Sciences, 108*(15), 6282–6287. https://doi.org/10.1073/pnas.1015239108

Bastakoty, D., & Young, P. P. (2016). Wnt/β-catenin pathway in tissue injury: Roles in pathology and therapeutic opportunities for regeneration. *The FASEB Journal, 30*(10), 3271–3284. https://doi.org/10.1096/fj.201600502R

Bhat, G. R., Sethi, I., Sadida, H. Q., Rah, B., Mir, R., Algehainy, N., Albalawi, I. A., Masoodi, T., Subbaraj, G. K., Jamal, F., Singh, M., Kumar, R., Macha, M. A., Uddin, S., Akil, A. S. A.-S., Haris, M., & Bhat, A. A. (2024). Cancer cell plasticity: From cellular, molecular, and genetic mechanisms to tumor heterogeneity and drug resistance. *Cancer and Metastasis Reviews, 43*(1), 197–228. https://doi.org/10.1007/s10555-024-10172-z

Boutilier, A. J., & Elsawa, S. F. (2021). Macrophage polarization states in the tumor microenvironment. *International Journal of Molecular Sciences, 22*(13), 6995. https://doi.org/10.3390/ijms22136995

Cervellini, I., Galino, J., Zhu, N., Allen, S., Birchmeier, C., & Bennett, D. L. (2018). Sustained MAPK/ERK activation in adult Schwann cells impairs nerve repair. *The Journal of Neuroscience, 38*(3), 679–690. https://doi.org/10.1523/JNEUROSCI.2255-17.2017

Charni, M., Aloni-Grinstein, R., Molchadsky, A., & Rotter, V. (2017). P53 on the crossroad between regeneration and cancer. *Cell Death & Differentiation, 24*(1), 8–14. https://doi.org/10.1038/cdd.2016.117

Chen, F., Castranova, V., & Shi, X. (2001). New insights into the role of nuclear factor-κB in cell growth regulation. *The American Journal of Pathology, 159*(2), 387–397. https://doi.org/10.1016/S0002-9440(10)61708-7

Coussens, L. M., & Werb, Z. (2002). Inflammation and cancer. *Nature, 420*(6917), 860–867. https://doi.org/10.1038/nature01322

Das, S., Dey, M. K., Devireddy, R., & Gartia, M. R. (2023). Biomarkers in cancer detection, diagnosis, and prognosis. *Sensors, 24*(1), 37. https://doi.org/10.3390/s24010037

Davis, P. K., & Brachmann, R. K. (2003). Chromatin remodeling and cancer. *Cancer Biology & Therapy, 2*(1), 23–30. https://doi.org/10.4161/cbt.189

De Visser, K. E., & Joyce, J. A. (2023). The evolving tumor microenvironment: From cancer initiation to metastatic outgrowth. *Cancer Cell, 41*(3), 374–403. https://doi.org/10.1016/j.ccell.2023.02.016

Eming, S. A., Wynn, T. A., & Martin, P. (2017). Inflammation and metabolism in tissue repair and regeneration. *Science, 356*(6342), 1026–1030. https://doi.org/10.1126/science.aam7928

Erratum. (2022). *MedComm, 3*(2), e135. https://doi.org/10.1002/mco2.135

Galliot, B., Tanaka, E., & Simon, A. (2008). Molecular and cellular basis of regeneration and tissue repair: Regeneration and tissue repair: Themes and variations. *Cellular and Molecular Life Sciences, 65*(1), 1. https://doi.org/10.1007/s00018-007-7424-0

Gilbert, S. F. (2000). *Regeneration. In developmental biology* (6th ed.). Sinauer Associates. https://www.ncbi.nlm.nih.gov/books/NBK9971/

Grivennikov, S. I., Greten, F. R., & Karin, M. (2010). Immunity, inflammation, and cancer. *Cell, 140*(6), 883–899. https://doi.org/10.1016/j.cell.2010.01.025

Hansda, S., & Das, H. (2025). Insights into cancer-associated thrombosis leading towards ischemic stroke. *Biology, 14*(1), 50. https://doi.org/10.3390/biology14010050

Hansda, S., & Ghosh, R. (2022). Bystander effect of ultraviolet a radiation protects A375 melanoma cells by induction of antioxidant defense. *Journal of Environmental Science and Health, Part C, 40*(1), 46–67. https://doi.org/10.1080/26896583.2021.1994820

Hao, D., Lopez, J.-M., Chen, J., Iavorovschi, A. M., Lelivelt, N. M., & Wang, A. (2022). Engineering extracellular microenvironment for tissue regeneration. *Bioengineering, 9*(5), 202. https://doi.org/10.3390/bioengineering9050202

He, Y., Sun, M. M., Zhang, G. G., Yang, J., Chen, K. S., Xu, W. W., & Li, B. (2021). Targeting PI3K/Akt signal transduction for cancer therapy. *Signal Transduction and Targeted Therapy, 6*(1), 425. https://doi.org/10.1038/s41392-021-00828-5

He, J., Zhou, Y., Qian, C., Wang, D., Yang, Z., Huang, Z., Sun, J., Ni, R., Yang, Q., Chen, J., & Luo, L. (2022). DNA methylation maintenance at the p53 locus initiates biliary-mediated liver regeneration. *NPJ Regenerative Medicine, 7*(1), 21. https://doi.org/10.1038/s41536-022-00217-8

Howlader, M. S. I., Prateeksha, P., Hansda, S., Naidu, P., Das, M., Barthels, D., & Das, H. (2024). Secretory products of DPSC mitigate inflammatory effects in microglial cells by targeting

MAPK pathway. *Biomedicine & Pharmacotherapy, 170*, 115971. https://doi.org/10.1016/j.biopha.2023.115971

Jing, J., Wu, Z., Wang, J., Luo, G., Lin, H., Fan, Y., & Zhou, C. (2023). Hedgehog signaling in tissue homeostasis, cancers and targeted therapies. *Signal Transduction and Targeted Therapy, 8*(1), 315. https://doi.org/10.1038/s41392-023-01559-5

Li, J., Lv, X., Zhang, X., Zhao, X., Meng, Y., Liu, S., Fu, S., & Sun, J. (2024). Notch signaling regulates limb regeneration through Hes1 and HeyL in the Chinese mitten crab. *Insect Biochemistry and Molecular Biology, 175*, 104209. https://doi.org/10.1016/j.ibmb.2024.104209

Liu, J., Xiao, Q., Xiao, J., Niu, C., Li, Y., Zhang, X., Zhou, Z., Shu, G., & Yin, G. (2022). Wnt/β--catenin signalling: Function, biological mechanisms, and therapeutic opportunities. *Signal Transduction and Targeted Therapy, 7*(1), 3. https://doi.org/10.1038/s41392-021-00762-6

Martinez-Redondo, P., & Izpisua Belmonte, J. C. (2020). Tailored chromatin modulation to promote tissue regeneration. *Seminars in Cell & Developmental Biology, 97*, 3–15. https://doi.org/10.1016/j.semcdb.2019.04.015

McCabe, M. T., Brandes, J. C., & Vertino, P. M. (2009). Cancer DNA methylation: Molecular mechanisms and clinical implications. *Clinical Cancer Research, 15*(12), 3927–3937. https://doi.org/10.1158/1078-0432.CCR-08-2784

McCusker, C., & Gardiner, D. M. (2011). The axolotl model for regeneration and aging research: A mini-review. *Gerontology, 57*(6), 565–571. https://doi.org/10.1159/000323761

Nakamura, M., Yoshida, H., Moriyama, Y., Kawakita, I., Wlizla, M., Takebayashi-Suzuki, K., Horb, M. E., & Suzuki, A. (2021). TGF-β1 signaling is essential for tissue regeneration in the Xenopus tadpole tail. *Biochemical and Biophysical Research Communications, 565*, 91–96. https://doi.org/10.1016/j.bbrc.2021.05.082

Neely, K. (2002). The complexity of chromatin remodeling and its links to cancer. *Biochimica et Biophysica Acta (BBA) – Reviews on Cancer, 1603*(1), 19–29. https://doi.org/10.1016/S0304-419X(02)00067-7

Nguyen-Lefebvre, A. T., Selzner, N., Wrana, J. L., & Bhat, M. (2021). The hippo pathway: A master regulator of liver metabolism, regeneration, and disease. *The FASEB Journal, 35*(5), e21570. https://doi.org/10.1096/fj.202002284RR

Nieto, M. A. (2013). Epithelial plasticity: A common theme in embryonic and cancer cells. *Science, 342*(6159), 1234850. https://doi.org/10.1126/science.1234850

Okamoto, R., & Watanabe, M. (2004). Molecular and clinical basis for the regeneration of human gastrointestinal epithelia. *Journal of Gastroenterology, 39*(1), 1–6. https://doi.org/10.1007/s00535-003-1259-8

Oviedo, N. J., & Beane, W. S. (2009). Regeneration: The origin of cancer or a possible cure? *Seminars in Cell & Developmental Biology, 20*(5), 557–564. https://doi.org/10.1016/j.semcdb.2009.04.005

Park, M., & Hong, J. (2016). Roles of NF-κB in cancer and inflammatory diseases and their therapeutic approaches. *Cells, 5*(2), 15. https://doi.org/10.3390/cells5020015

Petrova, R., & Joyner, A. L. (2014). Roles for hedgehog signaling in adult organ homeostasis and repair. *Development, 141*(18), 3445–3457. https://doi.org/10.1242/dev.083691

Prateeksha, P., Sariful Islam Howlader, M., Hansda, S., Naidu, P., Das, M., Abo-Aziza, F., Das, H., & Department of Pharmaceutical Sciences, Jerry H. Hodge School of Pharmacy, Texas Tech University Health Sciences Center, Amarillo, Texas 79106, USA. (2023). Secretome of dental pulp-derived stem cells reduces inflammation and proliferation of glioblastoma cells by deactivating Mapk-Akt pathway. *Diseases and Research, 3*(2), 74–86. https://doi.org/10.54457/DR.202302006

Principe, D. R., Doll, J. A., Bauer, J., Jung, B., Munshi, H. G., Bartholin, L., Pasche, B., Lee, C., & Grippo, P. J. (2014). TGF-: Duality of function between tumor prevention and carcinogenesis. *JNCI. Journal of the National Cancer Institute, 106*(2), djt369. https://doi.org/10.1093/jnci/djt369

Qin, S., Jiang, J., Lu, Y., Nice, E. C., Huang, C., Zhang, J., & He, W. (2020). Emerging role of tumor cell plasticity in modifying therapeutic response. *Signal Transduction and Targeted Therapy, 5*(1), 228. https://doi.org/10.1038/s41392-020-00313-5

Reddien, P. W. (2018). The cellular and molecular basis for planarian regeneration. *Cell, 175*(2), 327–345. https://doi.org/10.1016/j.cell.2018.09.021

Riley, S. E., Feng, Y., & Hansen, C. G. (2022). Hippo-yap/Taz signalling in zebrafish regeneration. *npj Regenerative Medicine, 7*(1), 9. https://doi.org/10.1038/s41536-022-00209-8

Riss, J., Khanna, C., Koo, S., Chandramouli, G. V. R., Yang, H. H., Hu, Y., Kleiner, D. E., Rosenwald, A., Schaefer, C. F., Ben-Sasson, S. A., Yang, L., Powell, J., Kane, D. W., Star, R. A., Aprelikova, O., Bauer, K., Vasselli, J. R., Maranchie, J. K., Kohn, K. W., … Barrett, J. C. (2006). Cancers as wounds that do not heal: Differences and similarities between renal regeneration/repair and renal cell carcinoma. *Cancer Research, 66*(14), 7216–7224. https://doi.org/10.1158/0008-5472.CAN-06-0040.

Roberts, C. W. M., & Orkin, S. H. (2004). The SWI/SNF complex—Chromatin and cancer. *Nature Reviews Cancer, 4*(2), 133–142. https://doi.org/10.1038/nrc1273

Sader, F., & Roy, S. (2022). *Tgf-β* superfamily and limb regeneration: *Tgf-β* to start and *bmp* to end. *Developmental Dynamics, 251*(6), 973–987. https://doi.org/10.1002/dvdy.379

Schäfer, M., & Werner, S. (2008). Cancer as an overhealing wound: An old hypothesis revisited. *Nature Reviews Molecular Cell Biology, 9*(8), 628–638. https://doi.org/10.1038/nrm2455

Singh, B. N., Doyle, M. J., Weaver, C. V., Koyano-Nakagawa, N., & Garry, D. J. (2012). Hedgehog and Wnt coordinate signaling in myogenic progenitors and regulate limb regeneration. *Developmental Biology, 371*(1), 23–34. https://doi.org/10.1016/j.ydbio.2012.07.033

Slowik, A. D., & Bermingham-McDonogh, O. (2014). *Notch signaling in mammalian hair cell regeneration.*

Tang, Y., Wang, J., Lian, Y., Fan, C., Zhang, P., Wu, Y., Li, X., Xiong, F., Li, X., Li, G., Xiong, W., & Zeng, Z. (2017). Linking long non-coding RNAs and SWI/SNF complexes to chromatin remodeling in cancer. *Molecular Cancer, 16*(1), 42. https://doi.org/10.1186/s12943-017-0612-0

Tataria, M., Perryman, S. V., & Sylvester, K. G. (2006). Stem cells: Tissue regeneration and cancer. *Seminars in Pediatric Surgery, 15*(4), 284–292. https://doi.org/10.1053/j.sempedsurg.2006.07.008

Volckaert, T., & De Langhe, S. P. (2015). Wnt and FGF mediated epithelial-mesenchymal crosstalk during lung development. *Developmental Dynamics, 244*(3), 342–366. https://doi.org/10.1002/dvdy.24234

Vujic, A., Natarajan, N., & Lee, R. T. (2020). Molecular mechanisms of heart regeneration. *Seminars in Cell & Developmental Biology, 100*, 20–28. https://doi.org/10.1016/j.semcdb.2019.09.003

Wen, X., Jiao, L., & Tan, H. (2022). MAPK/ERK pathway as a central regulator in vertebrate organ regeneration. *International Journal of Molecular Sciences, 23*(3), 1464. https://doi.org/10.3390/ijms23031464

White, R. M., & Zon, L. I. (2008). Melanocytes in development, regeneration, and cancer. *Cell Stem Cell, 3*(3), 242–252. https://doi.org/10.1016/j.stem.2008.08.005

Yu, J. S. L., & Cui, W. (2016). Proliferation, survival and metabolism: The role of PI3K/AKT/mTOR signalling in pluripotency and cell fate determination. *Development, 143*(17), 3050–3060. https://doi.org/10.1242/dev.137075

Yu, J.-Z., Kiss, Z., Ma, W., Liang, R., & Li, T. (2024). Preclinical models for functional precision lung cancer research. *Cancers, 17*(1), 22. https://doi.org/10.3390/cancers17010022

Zhang, S., Xiong, X., & Sun, Y. (2020). Functional characterization of SOX2 as an anticancer target. *Signal Transduction and Targeted Therapy, 5*(1), 135. https://doi.org/10.1038/s41392-020-00242-3

Zhang, C., Sun, C., Zhao, Y., Ye, B., & Yu, G. (2024). Signaling pathways of liver regeneration: Biological mechanisms and implications. *iScience, 27*(1), 108683. https://doi.org/10.1016/j.isci.2023.108683

Zheng, H., Liu, H., Xu, Q., Wang, W., Li, L., Ye, G., Wen, X., Chen, F., & Yu, Y. (2021). PI3K plays an essential role in planarian regeneration and tissue maintenance. *Frontiers in Cell and Developmental Biology, 9*, 649656. https://doi.org/10.3389/fcell.2021.649656

Zhou, B., Lin, W., Long, Y., Yang, Y., Zhang, H., Wu, K., & Chu, Q. (2022). Notch signaling pathway: Architecture, disease, and therapeutics. *Signal Transduction and Targeted Therapy, 7*(1), 95. https://doi.org/10.1038/s41392-022-00934-y

# Biomaterials in Tissue Regeneration

# Musculoskeletal Tissue Regeneration Using Fibrous Biomaterials

**Sampa Halder and George Z. Tan**

**Abstract** The musculoskeletal system, including bones, skeletal muscles, tendons, ligaments, and cartilage, relies heavily on fibrous structures made up of collagen, actin, and other proteins. The fibrous structures provide structural support, flexibility, and strength to ensure the proper functioning of musculoskeletal tissues. Tissue engineering generally requires the use of artificial scaffolds to recreate the in vivo milieu which coordinates a combination of biophysical and biochemical stimuli to regulate cell behaviors. Therefore, biomaterials with fibrous microstructures have drawn extensive attention in engineering musculoskeletal tissues. Multiple techniques have emerged for fabricating synthetic or naturally derived polymers. This chapter summarizes the common materials, fabrication techniques, microstructures, and applications of fibrous biomaterials for regenerating different types of musculoskeletal tissues. We also highlighted recent breakthroughs in creating fibrous patterns tailored to the structural and functional requirements of musculoskeletal tissues. Additionally, we discuss the current challenges and opportunities for future research in designing next-generation fibrous biomaterials. By addressing these aspects, this chapter underscores the transformative potential of fibrous scaffolds in advancing musculoskeletal tissue engineering.

**Keywords** Musculoskeletal · Fibrous structures · Tissue engineering · Scaffolds · Biophysical · Biochemical · Biomaterials

S. Halder (✉) · G. Z. Tan
Department of Industrial Engineering, University of Houston, Houston, TX, USA
e-mail: shalder@CougarNet.UH.EDU



H. Das (ed.), *Tissue Repair and Regeneration*,
https://doi.org/10.1007/978-3-031-93677-7_14

## 1 Introduction

The musculoskeletal tissue serves as the framework and mechanical foundation of the body, analogous to the structural design of a building. It safeguards vital organs, stores essential minerals, and facilitates blood cell production, all of which are critical to overall health and survival. Comprising approximately 45% of the total body mass, musculoskeletal tissues form the integral components of the musculoskeletal system which includes bones, cartilages, ligaments, tendons, skeletal muscle, and joints (Kwee & Mooney, 2017). This system not only provides structural support and stability but also enables coordinated movement and physical activity while protecting the body from injury. In recent era, musculoskeletal injury is a global health concern which is affecting a substantial portion of the population.

Musculoskeletal conditions are among the leading causes of disability worldwide, affecting a significant proportion of the global population. According to the Global Burden of Disease 2019, approximately 1.71 billion individuals globally live with musculoskeletal conditions, including low back pain, neck pain, fractures, and other injuries that can lead to long-term disability (Liu et al., 2022a). These conditions encompass over 150 distinct diseases and disorders that impair the musculoskeletal system, involving muscles, bones, joints, and adjacent connective tissues. Such impairments often result in temporary or lifelong limitations in functionality and participation.

A variety of factors contribute to the manifestation of musculoskeletal injuries, including age-related degeneration, engagement in intense physical activities or sports, poor posture, falls, trauma, and accidents. These conditions profoundly affect quality of life and, if not managed adequately, can lead to progressive disability. In 2020, musculoskeletal disorders were ranked as the second highest cause of nonfatal disability globally, underscoring their prevalence and severity (James et al., 2020). Given the widespread impact of these conditions, there is a growing emphasis on developing advanced tissue engineering strategies. Such approaches aim to reconstruct or regenerate musculoskeletal tissues, thereby reducing the intensity of disability and improving quality of life. As the prevalence of musculoskeletal injuries continues to rise, the necessity for innovative tissue engineering solutions becomes increasingly critical.

The conventional approaches to tissue and organ replacement, including the use of autografts and allografts, bioactive agents, and synthetic materials, are hindered by limitations such as limited availability, immunogenicity, and suboptimal functionality, thus necessitating the development of innovative strategies for tissue engineering and regenerative medicine (Deng et al., 2012).

Tissue engineering aims to create tissues in vitro to be implanted into the human body (Li et al., 2021a). This strategy offers new solutions for repairing bone, cartilage, tendon, and muscle injuries. Engineering musculoskeletal tissues often requires fibrous scaffolds because they closely mimic the natural structure, mechanical properties, and biological functions of the extracellular matrix (ECM) found in these tissues. Musculoskeletal tissues such as tendons, ligaments, and cartilage have a fibrous architecture composed of collagen fibers arranged in specific orientations (Zhou et al., 2021). The alignment of fibers can provide directional strength to withstand mechanical loads, tensile forces, and shear stresses (Tan et al., 2021). Aligned fibers are also important to direct stem cells or progenitor cells to differentiate into specific lineages, such as osteoblasts (bone), chondrocytes (cartilage), or myoblasts (muscle) (Subramony et al., 2013). Overall fibrous scaffolds emulate this architecture, providing a framework that closely resembles the natural ECM, enhancing cell attachment, proliferation, and differentiation. Over the past decades, considerable progress has been made in developing biomimetic scaffolds with desirable mechanical, structural, and biochemical characteristics to repair native musculoskeletal tissues. This chapter summarizes the current advancements in musculoskeletal tissue regeneration utilizing fibrous materials, with an emphasis on the materials, fabrication processes, and their performance in both in vivo and in vitro studies.

# 2 Musculoskeletal Tissues

## *2.1 Bones*

Bone consists of 70% hydroxyapatite (HAP), 20% organic matrix (mainly type I collagen), and 10% water. Collagen supports structure, facilitates mineralization, and ensures elasticity, while non-collagenous proteins regulate osteoblast activity. Bone remodeling involves osteoblasts, osteocytes, osteoclasts, and lining cells (LeGeros et al., 2008). Structurally, bone comprises a hard outer layer known as cortical bone, which bears most of the mechanical stress, and an inner layer called cancellous bone, which is weaker but plays a critical role in load distribution and shock absorption while providing structural support (Safadi et al., 2009). Bone diseases and injuries, commonly referred to as fractures, osteoporosis, osteomalacia (soft bone disease), and tumors, represent significant clinical challenges. According to a study by Clynes et al., osteoporosis-related fractures alone impose an economic burden of approximately $17.9 billion in the United States and £4 billion in the United Kingdom annually (Clynes et al., 2020). The conventional treatment for severe trauma or cancer involving bone is bone grafting, which, while effective, is expensive and associated with limitations. As an alternative, bone tissue engineering has emerged, leveraging natural and synthetic fibrous materials for bone regeneration due to their structural resemblance to the natural extracellular matrix (ECM) (Wingender et al., 2016). Bone tissue scaffolds prepared from fibrous materials are highly desirable due to their structural similarity to the natural extracellular matrix (ECM), which consists of interwoven protein fibers (tens to hundreds of nanometers) that support cell growth (Dorozhkin, 2010). This makes fibrous biomaterials particularly suitable for bone tissue engineering applications.

## 2.2 *Muscle*

Skeletal muscle constitutes approximately 40% of total body weight, contains 50–75% of the body's total proteins, and accounts for 30–50% of whole-body protein turnover. It is primarily composed of organized myofibers (muscle cells) and associated connective tissues, playing a critical role in voluntary movement, posture, and metabolic processes (Frontera & Ochala, 2015). Common pathological conditions affecting skeletal muscles include fibrosis, inflammation, dystrophy, strains, and atrophy (Morgan & Partridge, 2020; Mukund & Subramaniam, 2020). In the United States, the economic burden of muscular dystrophy has been substantial, with the average annual cost reported at $18,930 in 2004 (Ouyang et al., 2008). Specific conditions such as Duchenne muscular dystrophy (DMD) and myotonic dystrophy (DM) were associated with higher costs, averaging $50,952 and $32,236 in 2010, respectively (Larkindale et al., 2014). Tissue engineering approaches to skeletal muscle regeneration have increasingly focused on fibrous biomaterials such as chitosan, collagen, gelatin, polycaprolactone (PCL), and polyesterurethane, which have shown promise as scaffolds for muscle repair and regeneration (Jana et al., 2016).

## 2.3 *Cartilage*

Cartilage, a specialized soft tissue, consists of sparsely distributed chondrocytes embedded within a dense extracellular matrix (ECM) and functions as a cushion to support flexibility and elasticity in joints, facilitating load-bearing between bones (Basu & Ghosh, 2017; Chung & Burdick, 2008). This tissue is highly susceptible to damage caused by injury, disease, and degeneration, which compromises the shock transmission and absorption capabilities of the underlying subchondral and trabecular bone (Bashaireh et al., 2015). Due to its avascular and non-innervated nature, cartilage exhibits a limited capacity for self-repair, making natural recovery inadequate and positioning it as an ideal candidate for tissue engineering approaches (Chung & Burdick, 2008). A 2012 study reported that approximately 27 million Americans, primarily individuals over the age of 40, suffer from knee osteoarthritis, a common form of cartilage injury (Brooks, 2014). Current treatments for cartilage injuries predominantly focus on pain management and are often unsatisfactory in addressing the underlying tissue damage (Armiento et al., 2019). To advance cartilage regeneration, researchers have been exploring fibrous biomaterials, which provide a conducive environment for chondrocyte growth. Examples include PCL, polylactic acid (PLA), collagen, and cellulose. These biomaterials are designed to mimic the structural and functional properties of native cartilage tissue (Wasyłeczko et al., 2020; Eftekhari et al., 2020; Jiang et al., 2022).

### 2.4 *Tendons/Ligaments*

Tendons and ligaments are essential connective tissues that play critical roles in force transmission and joint stabilization. Tendons connect muscle to bone, whereas ligaments connect bone to bone, with ligaments exhibiting lower tensile strength and greater elasticity compared to tendons (Birch et al., 2013). Injuries to tendons and ligaments are generally classified as acute, chronic, or acute on chronic. These injuries may occur in healthy tissue following acute trauma or in degenerated tissue due to normal activity or minor trauma. Tendon and ligament injuries represent significant clinical challenges; for instance, approximately two million Achilles tendon injuries are reported annually worldwide, primarily related to sporting activities (Baldino et al., 2016). The annual economic burden associated with these injuries is estimated to be approximately $30 billion (Breidenbach et al., 2014). Rotator cuff tears (RCTs) are another prevalent tendon injury, affecting 20.7% of 1366 individuals in the general population according to a 2010 study, particularly those aged 50–80 years or older (Yamamoto et al., 2010). Current treatment approaches for tendon and ligament injuries typically focus on pain management but often result in limited functional recovery. These limitations are attributed to the poor regenerative capacity of these tissues, a high risk of infections, and significant retear rates following repair (Shiroud Heidari et al., 2021). To address these challenges, extensive research is being conducted to develop innovative strategies for regenerative tendon and ligament tissue engineering. Various polymers, including poly(α-esters), polycaprolactone (PCL), polylactic acid (PLA), poly(lactic-co-glycolic acid) (PLGA), collagen, methacrylated gelatin (mGLT), and poly(L-lactide) acid (PLLA), are being utilized to create fibrous biomaterials aimed at facilitating the tissue engineering process for tendon and ligament repair (Shiroud Heidari et al., 2021).

## 3 Polymer-Based Fibrous Biomaterials

Natural polymers, such as collagen, alginate (ALG), chitosan, silk, gelatin, and elastin, are widely utilized in bone tissue engineering due to their excellent biocompatibility, although their mechanical strength and processability often fall short of synthetic polymers (Costa et al., 2018; Liu et al., 2017). In skeletal muscle tissue engineering, materials like gelatin methacryloyl (GelMA), collagen, fibrin, and gelatin are preferred for their favorable properties, including cell attachment and proliferation (Kim et al., 2010; García-Lizarribar et al., 2018; Beier et al., 2009; Page et al., 2011). For cartilage tissue engineering, natural biopolymers such as silk fibroin (SF), chitosan, hyaluronic acid (HA), collagen, and gelatin are used to produce fibrous scaffolds due to their biocompatibility and shape retention properties (Mirahmadi et al., 2013; Liu et al., 2016; Zang et al., 2011; Kim et al., 2008; Xue et al., 2013), while collagen and silk are also employed in tendon and ligament regeneration (Xu et al., 2014; Oryan et al., 2014; Sahoo et al., 2010; Naghashzargar et al., 2015).

In contrast, synthetic polymers like polycaprolactone (PCL), polylactic acid (PLA), poly(glycolic acid) (PGA), and poly(lactic-co-glycolic acid) (PLGA) offer superior mechanical properties, tunable degradation rates, and ease of modification, although they generally lack inherent cell affinity (Sun et al., 2006; Athanasiou et al., 1996; Zhu et al., 1990; Gentile et al., 2014). To address these limitations, the blending of natural and synthetic polymers, such as chitosan/PCL or silk fibroin/collagen, has been explored to achieve a balance of biocompatibility, mechanical strength, and ECM-mimetic properties. Common blends for bone tissue engineering include PCL/polyurethane (PU), PCL/gelatin, and silk fibroin/chitosan (Domingues et al., 2016; Tiwari et al., 2017). For skeletal muscle engineering, materials like poly(butylene 1,4-cyclohexanedicarboxylate-co-triethylene cyclohexanedicarboxylate) (P(BCE-TECE)) demonstrate excellent flexibility and biodegradability (Bloise et al., 2018), while poly(ε-caprolactone), PLGA, and polyethylene glycol (PEG) are used for their mechanical properties and promotion of myoblast confluency (Ladd et al., 2011). Additionally, electrically conductive composite fibers, such as polyaniline (PANi) and poly(L-lactide-co-ε-caprolactone) (PLCL), are being explored to support cell adhesion and proliferation (Jun et al., 2009). Shah et al. used phosphate-based glass fibers to fabricate craniofacial muscle tissue scaffold which encouraged cell attachment and proliferation (Shah et al., 2005). For cartilage tissue engineering, PCL is frequently used due to its favorable mechanical properties and biocompatibility. The combination of PCL with natural biopolymers such as silk, collagen, and HA enhances chondrogenic effects, offering new possibilities for cartilage regeneration (Liu et al., 2022b; Garrigues et al., 2014). Likewise, blending PLLA with silk polymers has been shown to improve mechanical properties and promote chondrocyte adhesion, further advancing their application in cartilage engineering (Liu et al., 2016). In tendon and ligament tissue engineering, various combinations of PCL, PLCL, PLGA, and Poly(3-hydroxybutyrate) (P3HB) with natural biopolymers are being investigated to optimize mechanical properties and cytocompatibility. (Xu et al., 2014; Sahoo et al., 2010; Naghashzargar et al., 2015; Vaquette et al., 2010). A summary of commonly used biopolymers is shown in Table 1.

## 4 Fabrication of Fibrous Biomaterials

The biomimicking approach in tissue engineering aims to replicate the intricate structural features of the natural extracellular matrix (ECM), a crucial aspect of developing functional tissue scaffolds. In musculoskeletal tissue engineering, this approach focuses on creating synthetic ECM-like materials that emulate the mechanical properties and structural complexity of natural skeletal tissues. Given that the natural ECM is primarily composed of collagen fibers organized in specific

**Table 1** Commonly used biomaterials for musculoskeletal tissue regeneration

| Focused musculoskeletal system part | Natural biomaterial | Synthetic biomaterial |
|---|---|---|
| Bone | Gelatin (Samadian et al., 2021), keratin (Zhang & Liu, 2013), fibroin (Wu et al., 2019; Melke et al., 2016), fibrin (Perumcherry et al., 2011), chitosan (Wang et al., 2020), cellulose (Patel et al., 2020), GelMA (Qiao et al., 2020), alginate (Jeong et al., 2010) | Poly(L-lactide) acid (PLLA) (Ribeiro et al., 2019), polycaprolactone (PCL) (Ribeiro et al., 2019; Siddiqui et al., 2021), poly(lactic-co-glycolic acid) (PLGA) (Kwak et al., 2016), poly(vinyl alcohol) (PVA) (Abdal-Hay et al., 2013), polyethylene glycol (PEG) (Udomluck et al., 2019), nylon-6 (N6) (Abdal-Hay et al., 2013), hydroxyapatite (HAP), tricalcium phosphate (TCP) |
| Muscle | GelMA (García-Lizarribar et al., 2018; Kim & Kim, 2020), collagen (Beier et al., 2009), gelatin (Kim et al., 2010), fibrin (Page et al., 2011), sodium alginate (Gilbert-Honick et al., 2018) | Poly(butylene 1,4-cyclohexandicarboxylate-co-triethylene cyclohexanedicarboxylate) (P(BCE-TECE)) (Bloise et al., 2018), polyethylene glycol (PEG), poly(ε-caprolactone) (Abarzúa-Illanes et al., 2017), PLLA (Ladd et al., 2011), PLGA (Aviss et al., 2010), polyaniline (PANi) + poly(L-lactide-co-epsilon-caprolactone) (PLCL) (Jun et al., 2009), glass fibers (Shah et al., 2005), polycaprolactone (PCL) |
| Cartilage | Silk fibroin (Mirahmadi et al., 2013; Liu et al., 2016), chitosan (Zang et al., 2011), hyaluronic acid (HA) (Kim et al., 2008), collagen (Kim et al., 2008), gelatin (Xue et al., 2013) | Polycaprolactone (PCL) (Liu et al., 2022b; Garrigues et al., 2014), poly(L-lactide) acid (PLLA) (Liu et al., 2016), polylactic acid (PLA) (Chen et al., 2016) |
| Tendon/ligament | Collagen (Xu et al., 2014; Oryan et al., 2014), silk (Sahoo et al., 2010; Naghashzargar et al., 2015) | Polycaprolactone (PCL) (Naghashzargar et al., 2015; Vaquette et al., 2010), poly(L-lactide-co-ε-caprolactone) (PLCL) (Xu et al., 2014), PLGA (Sahoo et al., 2010), poly(3-hydroxybutyrate) (P3HB) (Naghashzargar et al., 2015), poly(L-lactide) acid (PLLA) (Liu et al., 2013) |

orientations, the fabrication of fibrous scaffolds with controlled fiber alignment has become a critical challenge. Developing advanced fibrous biomaterials that can mimic the orientation, mechanical strength, and biological cues of the natural matrix is essential for effective musculoskeletal tissue regeneration. As such, research continues to explore innovative fabrication techniques to optimize the structure and performance of fibrous scaffolds for musculoskeletal applications. A summary of recent techniques to fabricate fibrous biomaterials and scaffolds are given in Table 2.

**Table 2** Recent techniques to fabricate fibrous biomaterials for musculoskeletal tissue regeneration

| Musculoskeletal system part | Techniques | Biomaterials | | Benefits |
|---|---|---|---|---|
| Bone | Electrospinning | In vivo | Silk fibroin (SF) | Efficient vascularization bridging defect of lamellar bone; improved cell growth (Hofmann et al., 2013) |
| | | | SF, PLGA nanofibers incorporated with calcium phosphate cement (CPC) scaffolds | Growth of new bone in rat calvarial defect area after 4 weeks (Yao et al., 2020) |
| | | | Hybrid nanofiber aerogels from electrospun PLGA-collagen-gelatin (PCG) and Sr-cu codoped bioactive glass (BG) fibers | Greater bone volume, bone formation, and complete aerogel degradation in vivo (Weng et al., 2018) |
| | | In vitro | Polycaprolactone (PCL), hydroxyapatite (HAP), tricalcium phosphate (TCP) | Larger pore size; enhanced cell growth (Patlolla et al., 2010) |
| | | | PCL, collagen | Promotes cell proliferation (Lee et al., 2011) |
| | | | Cellulose, hydroxyapatite | Superior mechanical properties, thermal stability, and biocompatibility (Ao et al., 2017) |
| | | | Poly(D,L-lactide-co-trimethylene carbonate) (PLMC) | Facilitates cell adhesion, proliferation, biomineralization, cytocompatibility, and shape memory (Bao et al., 2014) |
| | | | Gelatin, PCL, bone powder | Enhanced cell growth, adhesion, and proliferation (Rong et al., 2016) |

| Musculoskeletal system part | Techniques | Biomaterials | | Benefits |
|---|---|---|---|---|
| Muscle | 3D bioprinting | In vivo | Alginate, PEG-fibrinogen (PF) | Significant formation of parallel aligned long-range myotube (Costantini et al., 2017) |
| | Electrospinning | | Poly(butylene 1,4-cyclohexandicarboxylate-co-triethylene cyclohexanedicarboxylate) (P(BCE-TECE)) | Facilitates integration with host muscles with high differentiation of cells into myotubes (Bloise et al., 2018) |
| | | | Collagen, gelatin | Significant endothelial interactions, bone mineralization, and tissue reconstruction (Jha et al., 2011) |
| | | | Sodium alginate (SA), fibrin | Exhibits high myofiber and vascular densities (Gilbert-Honick et al., 2018) |
| | | In vitro | Poly(L-lactide) acid (PLLA), collagen | Promotes myotube formation and cytocompatibility (Ladd et al., 2011) |
| | | | Collagen, PCL | Longer differentiated myotubes on aligned fibers (Choi et al., 2008) |
| | | | Gelatin, PCL | Enhances expression of myosin heavy chain, myogenin, and troponin-T genes; supports cell proliferation (Kim et al., 2010) |
| | | | Poly(lactic-co-glycolic acid) (PLGA) | Stimulates myoblasts to adopt elongated morphologies (Aviss et al., 2010) |
| | | | Polyaniline (PANi), poly(L-lactide-co-ε-caprolactone) (PLCL) | Enhances myotube number and length (Jun et al., 2009) |
| | Microthread extrusion | In vivo | Fibrin | Supports wound regeneration with deposition of collagen to grow new muscle tissue (Page et al., 2011; Cornwell & Pins, 2007) |
| | Fiber pulling technique | In vitro | Glass fibers | Enables in vitro myoblast attachment and proliferation (Shah et al., 2005) |

**Table 2** (Continued)

| Musculoskeletal system part | Techniques | Biomaterials | | Benefits |
|---|---|---|---|---|
| Cartilage | Electrospinning | In vivo | Gelatin, PCL | Exhibits good mechanical properties and supports chondrocyte proliferation (Xue et al., 2013) |
| | | | Gelatin, PLA | Displays high water absorption, excellent cytocompatibility, and elasticity (Chen et al., 2016) |
| | | | PCL nanofibers incorporated with collagen, fibrin | Enhances mechanical properties and produces ECM in vivo (Xu et al., 2012) |
| | | In vitro | Hyaluronic acid (HA), collagen | Enhances cellular adhesion and proliferation (Kim et al., 2008) |
| | | | Reinforced electrospun silk fibers in chitosan | Improves mechanical properties with significant chondrocyte proliferation (Mirahmadi et al., 2013) |
| | | | PCL with cartilage-derived matrix | Demonstrates chondrogenic effects (Garrigues et al., 2014) |
| | | | Silk, poly(L-lactide) acid (PLLA) | Exhibits low cytotoxicity and strong cytocompatibility (Liu et al., 2016) |
| | Electrohydrodynamic-jetting | In vitro | PCL | Supports chondrocyte expansion with 200 μm PCL fibers (Liu et al., 2022b) |

| Musculoskeletal system part | Techniques | Biomaterials | | Benefits |
|---|---|---|---|---|
| Tendon/ligament | Electrospinning | In vivo | Poly(L-lactide-co-ε-caprolactone) (PLCL), collagen | Promotes tenogenic differentiation (Xu et al., 2014) |
| | | | Poly(L-lactide) acid (PLLA) | Exhibits anti-adhesion of tissue and tendon healing (Liu et al., 2013) |
| | | | Poly(L-lactide-co-ε-caprolactone) (PLCL) | Delivers mesenchymal stem cells (MSCs) to enhance tendon healing (Liao et al., 2018) |
| | | | Poly(lactic-Co-glycolic acid) (PLGA) | Improves glycosaminoglycan deposition, collagen organization, and biomechanical strength (Zhao et al., 2014) |
| | | In vitro | Poly(lactic-co-glycolic acid) (PLGA), silk | Enhances collagen production and mechanical properties (Sahoo et al., 2010) |
| | | | Poly(3-hydroxybutyrate) (P3HB), PCL, twisted silk fibroin | Exhibits good cytocompatibility with fibroblasts (Naghashzargar et al., 2015) |
| | Microthread extrusion | In vitro | Collagen incorporated with resilin | Improves mechanical properties for tendon-like tissue structure (Sanami et al., 2015) |
| | Electrohydrodynamic-jetting | In vitro | PCL, chitosan | Produces scaffolds with precise fiber orientation and micrometer-sized porosity; supports significant cell proliferation (Wu et al., 2016) |

### *4.1 Electrospinning*

Electrospinning is a scalable and versatile technique for producing nonwoven nanoscale fibers with large surface areas, superior mechanical properties, and customizable functionality. Electrospinning involves applying a high electric field to a polymer solution or melt, generating a charged jet that forms continuous fibers resembling extracellular matrix components shown as Fig. 1. The high electric field deforms the polymer solution at the nozzle into a Taylor cone, propelling jets toward a counter-electrode collector under whipping motion. During this process, solvent evaporation results in the deposition of solid fibers, ranging from nanometer to micrometer scales, on the collector (Agarwal et al., 2008). Polymers used in electrospinning are classified as natural, synthetic, or blends. Blending natural and synthetic polymers enhances electrospun mats' mechanical strength and biocompatibility, combining the strengths of both types (Tiwari et al., 2018). These nanofibrous mats are widely used in musculoskeletal tissue engineering due to their tunable porosity, high surface area-to-volume ratio, and adaptable composition. To achieve both mechanical and cellular function abilities, more than one material goes through electrospinning process as well.

Nanofibers are predominantly fabricated using the electrospinning technique, which produces scaffolds with fiber diameters ranging from nanometers to micrometers, closely mimicking the physicochemical properties of natural extracellular matrix (ECM). Electrospun scaffolds have a high surface-area-to-volume ratio, promoting efficient cell adhesion, proliferation, and differentiation, essential for osteoblast activity.

For bone tissue engineering applications, scaffold pore sizes ranging between 100 and 500 μm are considered optimal for balancing cell accommodation and binding site availability (Wang et al., 2018). Recent advancements in

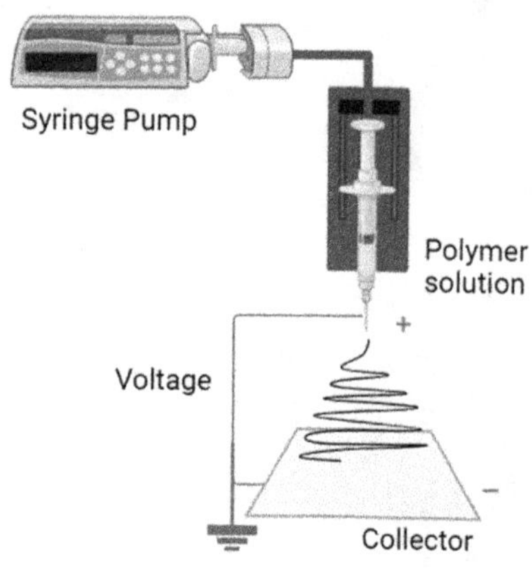


**Fig. 1** Schematic diagram of electrospinning process

electrospinning techniques have enabled the fabrication of fibrous scaffolds to mimic bone tissue using various biomaterial combinations. Patlolla et al. developed electrospun fibrous mats with a diameter of 800 nm, composed of polycaprolactone (PCL), hydroxyapatite (HAP), and tricalcium phosphate (TCP). These scaffolds exhibited good porosity, cell viability, and human mesenchymal stem cell (hMSC) growth over 14 days, presenting a promising strategy for bone tissue engineering (Patlolla et al., 2010). Hofmann et al. fabricated electrospun silk fibroin scaffolds with varied pore sizes, seeded with hMSCs. These scaffolds differentiated into bone-like structures and remodeled into functional bone, demonstrating vascularization and bone marrow ingrowth after implantation in mouse calvarial defects over 8 weeks (Hofmann et al., 2013). Yao et al. investigated calcium phosphate cement (CPC) scaffolds incorporated with electrospun silk fibroin (SF) and poly(lactic-co-glycolic acid) (PLGA) nanofibers to enhance osteoinductivity and biocompatibility. In a rat calvarial defect model, new bone formation was observed at the implant edges after 4 weeks. The scaffolds exhibited enhanced alkaline phosphatase activity (ALP), calcium deposition, osteogenic gene expression, and bone regeneration, demonstrating their potential for bone defect repair (Yao et al., 2020). Similarly, Lee et al. created hybrid scaffolds by combining PCL and collagen through melt-plotting and electrospinning. The scaffolds, optimized for pore sizes of 200 ± 20 μm and 300 ± 27 μm, exhibited enhanced mechanical strength and biological activity, with smaller pores supporting greater cell proliferation due to increased surface area (Lee et al., 2011). Ao et al. produced nanofibrous scaffolds from cotton cellulose and nano-hydroxyapatite (nano-HA) via electrospinning onto a rotating collector (300 m/min), followed by water bath coagulation. The aligned fibers (50–500 nm) demonstrated tensile strength of 70.6 MPa, a Young's modulus of 3.12 GPa, and excellent biocompatibility with human dental follicle cells (HDFCs), highlighting their potential in bone tissue engineering (Ao et al., 2017). Bao et al. fabricated electrospun poly(D,L-lactide-co-trimethylene carbonate) (PLMC) scaffolds with tunable fiber diameters (680–1500 nm), shape memory properties (recovery >94%, ~10 s at 39 °C), and cytocompatibility. These scaffolds supported osteoblast adhesion, proliferation, and biomineralization, making them suitable for bone defect repair and guided bone regeneration (Bao et al., 2014). Additionally, Rong et al. developed gelatin/PCL/bone powder electrospun scaffolds incorporating bioactive factors. These scaffolds demonstrated superior biocompatibility and promoted the adhesion, growth, and proliferation of adipose-derived stem cells, advancing their application in bone tissue engineering (Rong et al., 2016). Weng et al. developed hybrid aerogels combining Sr-Cu codoped bioactive glass (BG) nanofibers and PLGA-collagen-gelatin (PCG) nanofibers, fabricated via electrospinning and subsequent sintering or crosslinking. The aerogels were evaluated in vivo in a mouse cranial defect model, showing enhanced bone volume, bone formation, and complete degradation. These findings highlight the aerogels' potential for cranial bone tissue regeneration (Weng et al., 2018).

To fabricate skeletal muscle tissue fibrous materials, the most efficient technology for now is electrospinning as well. It helps to make aligned fibers, which is a need for this type of tissue. Bloise et al. fabricated P(BCE-co-TECE) electrospun fibers, enhancing cell attachment and biological interactions. Using an in-house electrospinning setup, they produced micrometric and nanometric fibers, treated with ethanol to preserve morphology, and compression-molded films at Tm + 40 °C under 2 ton/$m^2$ pressure. The scaffolds exhibited excellent C2C12 cell proliferation over 14 days in vitro (Bloise et al., 2018). Ladd et al. demonstrated that co-electrospinning natural and synthetic biomaterials enhances biocompatibility while maintaining good mechanical properties (Kim et al., 2008). Polymer solutions (PLLA/collagen and PCL/collagen blends) were co-electrospun onto a rotating mandrel to create distinct PCL/collagen, PLLA/collagen, and overlap regions, with scaffolds crosslinked using glutaraldehyde vapor for improved mechanical integrity. The resulting fibers (452–549 nm diameter) exhibited cytocompatibility, supported cell attachment, and facilitated myotube formation, underscoring their potential for muscle tissue engineering applications (Ladd et al., 2011). Choi et al. also fabricated unidirectional electrospun nanofiber meshes from blend of PCL and collagen, supporting myotube formation for potential implantable skeletal muscle tissues in large defects. The meshes were electrospun at 20 kV, crosslinked with 2.5% glutaraldehyde vapor, sterilized, and pre-incubated in culture medium before cell seeding (Choi et al., 2008). Kim et al. fabricated electrospun nanofibers (250–350 nm) composed of polycaprolactone (PCL), gelatin, and genipin, which demonstrated enhanced mechanical properties due to increased gelatin content. These nanofibers promoted higher expression levels of myogenin, myosin heavy chain, and troponin T genes. C2C12 cells seeded onto the fibers exhibited excellent proliferation over a one-week period, highlighting their potential for muscle tissue engineering applications (Kim et al., 2010). Aviss et al. fabricated aligned electrospun PLGA fibers (100 nm–1.4 μm diameter) to promote elongation and alignment of C2C12 murine myoblasts, enhancing differentiation. The ultraviolet (UV)-sterilized meshes supported myoblast differentiation over 14 days, highlighting their potential for muscle tissue engineering (Aviss et al., 2010). Jun et al. developed electrically conductive composite electrospun fibers from poly(L-lactide-co-3-caprolactone) (PLCL) and polyaniline (PANi), which induced myoblasts to form myotubes without the need for additional electrical stimulation. The fibers, with uniform diameters and interconnected pores, supported significant elongation of C2C12 cells over an 8-day in vitro period (Jun et al., 2009). Jha et al. fabricated electrospun type I collagen (rEC) and gelatin (rEG) scaffolds with 1 μm fibers, optimizing them for tissue engineering through vapor crosslinking and sterilization. Myoblast-laden collagen scaffolds promoted skeletal muscle regeneration in rats, while guinea pigs treated with collagen scaffolds showed superior dermal wound healing compared to gelatin scaffolds. Notably, fibers larger than 750–1000 nm exhibited enhanced wound regeneration within 16 days (Jha et al., 2011). Gilbert-Honick et al. developed electrospun fibrin-SA scaffolds seeded with murine myoblasts, achieving significant muscle regeneration with high myofiber and vascular densities in mouse tibialis anterior volumetric muscle loss (VML) defects within 2–4 weeks. The scaffolds mimic native muscle stiffness, alignment cues, and are biodegradable and robust. Their scalable design supports implantable vasculature formation, promising for larger models (Gilbert-Honick et al., 2018).

Mirahmadi et al. developed electrospun silk fibers (diameter ~ 510 nm) for incorporation into reinforced chitosan hydrogel scaffolds, successfully mimicking the behavior of articular cartilage. These scaffolds exhibited enhanced mechanical properties, significant cytocompatibility, and supported chondrocyte proliferation over a 21-day in vitro study (Mirahmadi et al., 2013). Kim et al. fabricated a 3D macroporous and nanofibrous scaffold combining hyaluronic acid (HA) and collagen using electrospinning and salt leaching techniques, demonstrating improved cellular adhesion and proliferation with increased collagen content, while maintaining the characteristic round morphology of chondrocytes, highlighting its suitability for cartilage tissue engineering (Melke et al., 2016). Xue et al. engineered precise 3D cartilage structures using electrospun gelatin and polycaprolactone (PCL) fibrous membranes in a sandwich model with chondrocytes, producing cartilage-like tissues with 91.41% shape retention and excellent mechanical properties after both in vitro and in vivo cultivation, underscoring its potential for complex tissue fabrication (Xue et al., 2013). Chen et al. developed electrospun gelatin/PLA nanofiber-based 3D scaffolds, with hyaluronic acid crosslinking enhancing their mechanical properties and stability for cartilage tissue regeneration. The scaffolds exhibited high water absorption, cytocompatibility, and superior cartilage repair in an in vivo rabbit model, making them a promising candidate for cartilage tissue engineering (Chen et al., 2016). Liu et al. optimized electrospun silk fibroin (SF)/poly(L-lactide acid) (PLLA) scaffolds for surface morphology, tensile strength, and degradation rate, with fiber diameters ranging from 0.95 μm to 1.77 μm, and demonstrated good chondrocyte adhesion and low cytotoxicity in vitro, suggesting the scaffold's promise for cartilage tissue engineering (Liu et al., 2016). Garrigues et al. demonstrated enhanced chondrogenic effects with electrospun PCL fiber mats incorporating cartilage-derived matrix (CDM), where multilayer scaffolds improved cell infiltration despite a reduction in the elastic modulus, highlighting their potential for cartilage regeneration (Garrigues et al., 2014). Xu et al. developed hybrid 3D inkjet-printed scaffolds of collagen and fibrin with PCL nanofibers for cartilage tissue engineering. After 8 weeks in vivo, the constructs showed over 80% cell viability, with cells proliferating and forming cartilage-like tissues both in vitro and in immunodeficient mice (Xu et al., 2012).

Xu et al. investigated tendon-derived stem cells (TDSCs) seeded on electrospun poly(L-lactide-co-ε-caprolactone) (PLCL) and collagen fiber scaffolds. Under mechanical stimulation in vitro, the scaffolds promoted tenogenic differentiation, as evidenced by the expression of tendon-related extracellular matrix (ECM) genes and proteins, indicating the successful production of functional tissue-engineered tendons. The scaffolds were fabricated by electrospinning PLCL and collagen into nanofibers, twisting them into yarns, and collecting them onto a rotating mandrel (Xu et al., 2014). Liao et al. developed electrospun PLCL mesh scaffolds and evaluated their performance in a 3-week in vivo rabbit model. The scaffolds demonstrated reduced adhesion formation and improved tendon healing outcomes (Liao et al., 2018). Sahoo et al. developed a biohybrid scaffold by electrospinning ultrafine PLGA fibers containing basic fibroblast growth factor (bFGF) onto robust, degummed knitted silk microfibrous scaffolds coated with a 2% aqueous silk solution. The optimized scaffold system promoted mesenchymal stem cell (MSC) attachment, proliferation, and tenogenic differentiation, enhancing ECM protein gene expression, collagen production, and mechanical properties. These biohybrid scaffolds demonstrated strong layer integration and significant potential for ligament/tendon repair applications (Sahoo et al., 2010). Naghas et al. developed a biodegradable nano/micro hybrid scaffold for tendon and ligament tissue engineering by electrospinning P3HB or PCL nanofibers (diameter range 50–500 nm) onto twisted silk fibroin (SF) yarns. The hybrid scaffolds demonstrated improved mechanical properties, tightly wrapped nanofibers, and good cytocompatibility with fibroblasts. This fabrication method shows potential for functional scaffold development in tendon and ligament regeneration, although further in vitro and in vivo testing is needed. (Naghashzargar et al., 2015). Liu et al. fabricated electrospun PLLA fibers (~0.8 μm diameter) encapsulated with growth factors and proteins. In a 21-day rabbit model, the scaffolds promoted cell proliferation and improved tendon healing outcomes (Liu et al., 2013). Zhao et al. developed bFGF-loaded PLGA fibrous membranes using electrospinning for rotator cuff tear (RCT) repair in a rat model. The membranes demonstrated excellent biocompatibility, sustained bFGF release, and significantly enhanced tendon–bone healing by improving glycosaminoglycan deposition, collagen organization, and biomechanical strength (Zhao et al., 2014).

## 4.2 *Microthread Extrusion*

Microthread extrusion is a technique for producing fine, thread-like structures from polymers, proteins, or hydrogels by extruding prepared solutions or gels through fine nozzles under controlled pressure or flow. Stabilization is achieved through crosslinking or solidification mechanisms, followed by collection, tension-drying, and storage for applications in tissue engineering and materials science (Pins et al., 1997). Cornwell et al. and Page et al. coextruded fibrinogen and thrombin solutions at 37 °C through a blending applicator into a buffer bath for stabilization, followed by drying and sterilization; carbon spheres (8–12 μm) were added for in vivo tracking in some cases. Uniform cell seeding was achieved by rotating 2 cm thread segments with cell suspensions, demonstrating significant cell proliferation and muscle tissue regeneration in vivo (both studies) and in vitro (Page et al., 2011; Cornwell & Pins, 2007). Sanami et al. investigated the effects of adding resilin-like protein to mildly crosslinked collagen fibers, finding improvements in mechanical properties for tendon tissue engineering, including increased stress and strain at break, without compromising fibroblast cell metabolic activity. Collagen/resilin solutions were extruded, crosslinked, and processed to form fibers with wet and dry diameters of around 250 and 75 μm, respectively. The study demonstrates enhanced biomechanical properties while maintaining biological functionality, making the fibers suitable for tissue engineering applications (Sanami et al., 2015).

## 4.3 3D Bioprinting

3D bioprinting is an advanced manufacturing process that uses bioinks, composed of cells and biomaterials, to fabricate complex, tissue-like structures layer by layer. This technique involves precise deposition of bioink through a controlled system, often coupled with crosslinking mechanisms such as chemical, thermal, or light-induced stabilization to maintain the structural integrity of the printed construct shown as Fig. 2. By adjusting parameters like printing speed, nozzle size, and layer height, bioprinting enables the creation of constructs with specific geometries and cellular organization (García-Lizarribar et al., 2018). Costantini et al. developed a photocurable PEG-fibrinogen (PF) bioink by cleaving fibrinogen, reducing disulfide bonds, and conjugating PEG-diacrylate, resulting in a semi-synthetic polymer suitable for 3D bioprinting. The bioink, combined with alginate (ALG), was extruded through a coaxial dispensing system to produce cell-laden hydrogel fibers with a diameter of approximately 250 μm, with gelation occurring instantaneously upon contact with a crosslinking solution. The study highlighted the enhanced myogenic differentiation of C2C12 muscle cells, observing significant myotube formation by day 21, demonstrating the potential of 3D bioprinting in the creation of artificial skeletal muscles (Costantini et al., 2017).

## 4.4 Electrohydrodynamic-Jetting (EHDJ)

Electrohydrodynamic jetting (EHDJ) is a precision deposition technique derived from electrospinning, employing electrostatic forces to generate controlled liquid jets for micro- and nanomanufacturing (Fig. 3). Utilizing low voltage (2–3 kV) between the nozzle and substrate, EHDJ achieves aligned fibers as thin as 5 μm, making it suitable for fabricating biomimetic scaffolds with anisotropic mechanical properties (Wang et al., 2024; Vijayavenkataraman et al., 2018). Electrohydrodynamic

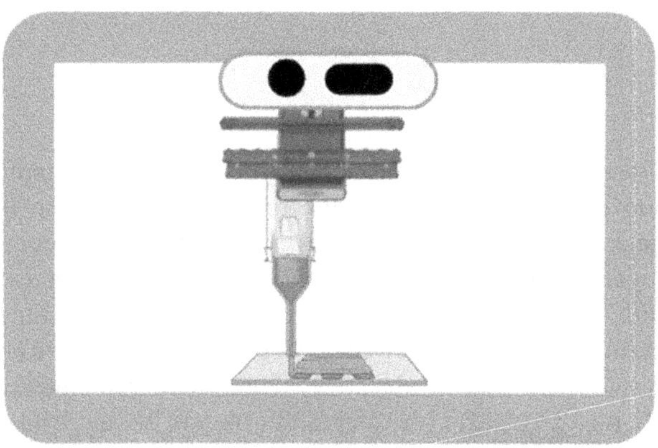

**Fig. 2** Schematic diagram of extrusion-based 3D bioprinting

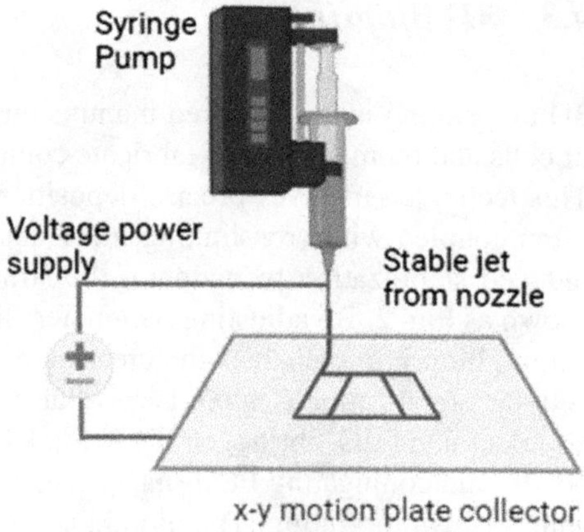


**Fig. 3** Schematic diagram of electrohydrodynamic jetting

jet 3D printing (EHD-3D printing), a subset of EHDJ, constructs complex three-dimensional structures layer by layer, leveraging electric fields for precise fabrication of microscale and nanoscale objects used in tissue engineering and sensor applications (Wu et al., 2016).

Wu et al. fabricated 3D PCL scaffolds using an in-house EHD-jet printing system, creating 20-layer fibrous structures with fibers approximately 21 μm in diameter oriented perpendicularly. In vitro studies with human embryonic stem cells (hESC)-derived fibroblasts demonstrated significant cell proliferation on PCL/chitosan scaffolds over 14 days, suggesting potential for skeletal ligament and connective tissue regeneration (Wu et al., 2016). Similarly, Liu et al. showed that 200 μm PCL fibers produced via EHD-3D printing inhibit chondrocyte dedifferentiation, potentially preventing osteoarthritis and advancing cartilage tissue regeneration. Scaffolds with spacings ranging from 100 to 300 μm were fabricated, with a PCL membrane serving as the control group (Liu et al., 2022b). Wang et al. prepared PCL fibrous scaffolds using EHD-3D printing with a fiber diameter of approximately 13 μm, observing fibroblast cell attachment and proliferation over 7 days (Wang et al., 2019).

## 5 Future Scope

Fibrous biomaterials have emerged as promising candidates in musculoskeletal tissue engineering, offering the potential to replicate the structural complexity of native tissues while providing tunable mechanical properties essential for functional regeneration. Despite these advantages, the challenges in fibrous scaffolds are significant as well, including low durability to withstand physiological loads (Hinz et al., 2022), insufficient biological signals to promote tissue-specific cellular responses (Tonti et al., 2021), lack of microarchitecture to support angiogenesis (Lee et al., 2021), fast-degrading before tissue is fully formed (Uiterwijk et al., 2024), byproducts that trigger inflammation (Zhou et al., 2022), among others. With regard to material selections, balancing the mechanical properties and bioactivities is always a challenge. Synthetic polymers often lack bioactivity, while natural materials may not provide sufficient mechanical strength. These problems lead to poor integration between scaffolds and native tissues, therefore hindering the clinical applicability of engineered tissues.

The future of musculoskeletal tissue scaffolds hinges on the advancement of fibrous biopolymer composite materials that closely replicate the biomechanical properties and biofunctions of natural tissues. A common strategy to balance the mechanical properties and biological properties of fiber materials is to integrate natural and synthetic biopolymers, including the selection of optimal crosslinking agents and the incorporation of mineralized materials. Computational modeling coupled with machine learning and artificial intelligence will be more extensively applied in material sciences to design scaffolds tailored to specific tissues. Meanwhile, advances in fabrication technology are essential to improving the design, production, and functionality of fibrous scaffolds for tissue engineering. Future development needs to enhance the reproducibility and scalability of producing complex hierarchical structures in scaffolds. Technologies like advanced 3D bioprinting and automated electrospinning systems are enabling more precise and reproducible scaffolds, with a focus on cost-effective scalability. Furthermore, the integration of hybrid fiber fabrication techniques, such as needleless electrospinning, electrospinning with external stimuli, and 3D electrospinning (Keirouz et al., 2023; Li et al., 2021b; Klapstova et al., 2024), offers significant potential to enhance both the mechanical performance and scalability of fibrous scaffolds for musculoskeletal tissue regeneration.

## 6 Conclusion

Fibrous biomaterials are central to musculoskeletal tissue engineering, integrating biology, material science, and advanced manufacturing to create scaffolds with hierarchical structures that mimic natural tissues. Recent advancements in fabrication techniques have enabled the development of scalable platforms for producing these biomimetic scaffolds, although few 3D constructs have been successfully translated to clinical applications. Future fibrous scaffolds must possess dynamic functions that respond to mechanical and physiological cues for long-term regeneration and clinical success. Emerging hybrid biofabrication technologies, such as 3D electrospinning and needleless electrospinning, offer the potential for creating complex composite constructs with enhanced performance and efficient bench-to-bedside translation. Advances in fabrication technology are vital to overcoming current limitations in tissue engineering, particularly in creating scaffolds that can precisely replicate the structure and function of native tissues. Through improvements in precision, material innovation, scalability, and bioactivity, fibrous scaffolds will continue to progress toward their full potential in clinical applications.

## References

Abarzúa-Illanes, P. N., et al. (2017). Improving myoblast differentiation on electrospun poly (ε-caprolactone) scaffolds. *Journal of Biomedical Materials Research Part A, 105*(8), 2241–2251.

Abdal-Hay, A., Pant, H. R., & Lim, J. K. (2013). Super-hydrophilic electrospun nylon-6/hydroxyapatite membrane for bone tissue engineering. *European Polymer Journal, 49*(6), 1314–1321.

Agarwal, S., Wendorff, J. H., & Greiner, A. (2008). Use of electrospinning technique for biomedical applications. *Polymer, 49*(26), 5603–5621.

Ao, C., et al. (2017). Fabrication and characterization of electrospun cellulose/nano-hydroxyapatite nanofibers for bone tissue engineering. *International Journal of Biological Macromolecules, 97*, 568–573.

Armiento, A. R., Alini, M., & Stoddart, M. J. (2019). Articular fibrocartilage-why does hyaline cartilage fail to repair? *Advanced Drug Delivery Reviews, 146*, 289–305.

Athanasiou, K. A., Niederauer, G. G., & Agrawal, C. M. (1996). Sterilization, toxicity, biocompatibility and clinical applications of polylactic acid/polyglycolic acid copolymers. *Biomaterials, 17*(2), 93–102.

Aviss, K., Gough, J., & Downes, S. (2010). Aligned electrospun polymer fibres for skeletal muscle regeneration. *European Cells & Materials, 19*(1), 193–204.

Baldino, L., et al. (2016). Regeneration techniques for bone-to-tendon and muscle-to-tendon interfaces reconstruction. *British Medical Bulletin, 117*(1), 25–37.

Bao, M., et al. (2014). Electrospun biomimetic fibrous scaffold from shape memory polymer of PDLLA-co-TMC for bone tissue engineering. *ACS Applied Materials & Interfaces, 6*(4), 2611–2621.

Bashaireh, K., et al. (2015). Efficacy and safety of cross-linked hyaluronic acid single injection on osteoarthritis of the knee: A post-marketing phase IV study. In *Drug design, development and therapy* (p. 2063).

Basu, B., & Ghosh, S. (2017). *Biomaterials for musculoskeletal regeneration*. Springer.

Beier, J. P., et al. (2009). Collagen matrices from sponge to nano: New perspectives for tissue engineering of skeletal muscle. *BMC Biotechnology, 9*, 1–14.

Birch, H. L., et al. (2013). *Tendon and ligament physiology*. In K. W. Hinchcliff, A. J. Kaneps, & R. J. Geor (Eds.), *Equine sports medicine and surgery* (2nd ed., pp. 167–188).

Bloise, N., et al. (2018). Ether-oxygen containing electrospun microfibrous and sub-microfibrous scaffolds based on poly (butylene 1, 4-cyclohexanedicarboxylate) for skeletal muscle tissue engineering. *International Journal of Molecular Sciences, 19*(10), 3212.

Breidenbach, A. P., et al. (2014). Functional tissue engineering of tendon: Establishing biological success criteria for improving tendon repair. *Journal of Biomechanics, 47*(9), 1941–1948.

Brooks, K. S. (2014). Osteoarthritic knee braces on the market: A literature review. *JPO: Journal of Prosthetics and Orthotics, 26*(1), 2–30.

Chen, W., et al. (2016). Superabsorbent 3D scaffold based on electrospun nanofibers for cartilage tissue engineering. *ACS Applied Materials & Interfaces, 8*(37), 24415–24425.

Choi, J. S., et al. (2008). The influence of electrospun aligned poly (epsilon-caprolactone)/collagen nanofiber meshes on the formation of self-aligned skeletal muscle myotubes. *Biomaterials, 29*(19), 2899–2906.

Chung, C., & Burdick, J. A. (2008). Engineering cartilage tissue. *Advanced Drug Delivery Reviews, 60*(2), 243–262.

Clynes, M. A., et al. (2020). The epidemiology of osteoporosis. *British Medical Bulletin, 133*(1), 105–117.

Cornwell, K. G., & Pins, G. D. (2007). Discrete crosslinked fibrin microthread scaffolds for tissue regeneration. *Journal of Biomedical Materials Research Part A, 82*(1), 104–112.

Costa, F., Silva, R., & Boccaccini, A. (2018). Fibrous protein-based biomaterials (silk, keratin, elastin, and resilin proteins) for tissue regeneration and repair. *Peptides and Proteins as Biomaterials for Tissue Regeneration and Repair*, 175–204.

Costantini, M., et al. (2017). Microfluidic-enhanced 3D bioprinting of aligned myoblast-laden hydrogels leads to functionally organized myofibers in vitro and in vivo. *Biomaterials, 131*, 98–110.

Deng, M., et al. (2012). Nanostructured polymeric scaffolds for orthopaedic regenerative engineering. *IEEE Transactions on Nanobioscience, 11*(1), 3–14.

Domingues, R. M., et al. (2016). Fabrication of anisotropically aligned nanofibrous scaffolds based on natural/synthetic polymer blends reinforced with cellulose nanocrystals for tendon tissue engineering. *Frontiers in Bioengineering and Biotechnology, 4*(10.3389).

Dorozhkin, S. V. (2010). Bioceramics of calcium orthophosphates. *Biomaterials, 31*(7), 1465–1485.

Eftekhari, A., et al. (2020). The use of nanomaterials in tissue engineering for cartilage regeneration; current approaches and future perspectives. *International Journal of Molecular Sciences, 21*(2), 536.

Frontera, W. R., & Ochala, J. (2015). Skeletal muscle: A brief review of structure and function. *Calcified Tissue International, 96*(3), 183–195.

García-Lizarribar, A., et al. (2018). Composite biomaterials as long-lasting scaffolds for 3D bioprinting of highly aligned muscle tissue. *Macromolecular Bioscience, 18*(10), 1800167.

Garrigues, N. W., et al. (2014). Electrospun cartilage-derived matrix scaffolds for cartilage tissue engineering. *Journal of Biomedical Materials Research Part A, 102*(11), 3998–4008.

Gentile, P., et al. (2014). An overview of poly (lactic-co-glycolic) acid (PLGA)-based biomaterials for bone tissue engineering. *International Journal of Molecular Sciences, 15*(3), 3640–3659.

Gilbert-Honick, J., et al. (2018). Engineering functional and histological regeneration of vascularized skeletal muscle. *Biomaterials, 164*, 70–79.

Hinz, N., et al. (2022). Biomechanical in vitro analysis of a novel flexible implant for pubic symphysis disruption using an ultra-high molecular weight polyethylene fiber cord. *Clinical biomechanics, 95*, 105652.

Hofmann, S., et al. (2013). Remodeling of tissue-engineered bone structures in vivo. *European Journal of Pharmaceutics and Biopharmaceutics, 85*(1), 119–129.

James, S. L., et al. (2020). The global burden of falls: Global, regional and national estimates of morbidity and mortality from the global burden of disease study 2017. *Injury Prevention, 26*(Suppl 2), i3–i11.

Jana, S., Levengood, S. K. L., & Zhang, M. (2016). Anisotropic materials for skeletal-muscle-tissue engineering. *Advanced Materials, 28*(48), 10588–10612.

Jeong, S. I., et al. (2010). Electrospun alginate nanofibers with controlled cell adhesion for tissue engineering a. *Macromolecular Bioscience, 10*(8), 934–943.

Jha, B. S., et al. (2011). Electrospun collagen: A tissue engineering scaffold with unique functional properties in a wide variety of applications. *Journal of Nanomaterials, 2011*, 1–15.

Jiang, Y. H., et al. (2022). Cross-linking methods of type I collagen-based scaffolds for cartilage tissue engineering. *American Journal of Translational Research, 14*(2), 1146–1159.

Jun, I., Jeong, S., & Shin, H. (2009). The stimulation of myoblast differentiation by electrically conductive sub-micron fibers. *Biomaterials, 30*(11), 2038–2047.

Keirouz, A., et al. (2023). The history of electrospinning: Past, present, and future developments. *Advanced Materials Technologies, 8*(11), 2201723.

Kim, W., & Kim, G. (2020). 3D bioprinting of functional cell-laden bioinks and its application for cell-alignment and maturation. *Applied Materials Today, 19*, 100588.

Kim, T. G., Chung, H. J., & Park, T. G. (2008). Macroporous and nanofibrous hyaluronic acid/collagen hybrid scaffold fabricated by concurrent electrospinning and deposition/leaching of salt particles. *Acta Biomaterialia, 4*(6), 1611–1619.

Kim, M. S., et al. (2010). The development of genipin-crosslinked poly (caprolactone)(PCL)/gelatin nanofibers for tissue engineering applications. *Macromolecular Bioscience, 10*(1), 91–100.
Klapstova, A., et al. (2024). Effective needleless electrospinning for the production of tubular scaffolds. *ACS Applied Engineering Materials, 2*(2), 492–500.
Kwak, S., et al. (2016). Micro/nano multilayered scaffolds of PLGA and collagen by alternately electrospinning for bone tissue engineering. *Nanoscale Research Letters, 11*, 1–16.
Kwee, B. J., & Mooney, D. J. (2017). Biomaterials for skeletal muscle tissue engineering. *Current Opinion in Biotechnology, 47*, 16–22.
Ladd, M. R., et al. (2011). Co-electrospun dual scaffolding system with potential for muscle–tendon junction tissue engineering. *Biomaterials, 32*(6), 1549–1559.
Larkindale, J., et al. (2014). Cost of illness for neuromuscular diseases in the United States. *Muscle & Nerve, 49*(3), 431–438.
Lee, H., et al. (2011). Designed hybrid scaffolds consisting of polycaprolactone microstrands and electrospun collagen-nanofibers for bone tissue regeneration. *Journal of Biomedical Materials Research. Part B, Applied Biomaterials, 97*(2), 263–270.
Lee, J.-H., et al. (2021). Materials roles for promoting angiogenesis in tissue regeneration. *Progress in Materials Science, 117*, 100732.
LeGeros, R., G. Daculsi, and J. LeGeros, Musculoskeletal tissue regeneration: Biological materials and methods. 2008.
Li, Z., et al. (2021a). Tissue engineering for musculoskeletal regeneration and disease modeling. *Organotypic Models in Drug Development*, 235–268.
Li, Y., et al. (2021b). Developments of advanced electrospinning techniques: A critical review. *Advanced Materials Technologies, 6*(11), 2100410.
Liao, J. C., et al. (2018). The effects of bi-functional anti-adhesion scaffolds on flexor tendon healing in a rabbit model. *Journal of Biomedical Materials Research Part B: Applied Biomaterials, 106*(7), 2605–2614.
Liu, S., et al. (2013). Tendon healing and anti-adhesion properties of electrospun fibrous membranes containing bFGF loaded nanoparticles. *Biomaterials, 34*(19), 4690–4701.
Liu, W., et al. (2016). Electrospun fibrous silk fibroin/poly (L-lactic acid) scaffold for cartilage tissue engineering. *Tissue Engineering and Regenerative Medicine, 13*, 516–526.
Liu, M., et al. (2017). Injectable hydrogels for cartilage and bone tissue engineering. *Bone Research, 5*(1), 1–20.
Liu, S., et al. (2022a). Global burden of musculoskeletal disorders and attributable factors in 204 countries and territories: A secondary analysis of the global burden of disease 2019 study. *BMJ Open, 12*(6), e062183.
Liu, X., et al. (2022b). Effect of electrohydrodynamic printing scaffold with different spacing on chondrocyte dedifferentiation. *Annals of Translational Medicine, 10*(13), 743.
Melke, J., et al. (2016). Silk fibroin as biomaterial for bone tissue engineering. *Acta Biomaterialia, 31*, 1–16.
Mirahmadi, F., et al. (2013). Enhanced mechanical properties of thermosensitive chitosan hydrogel by silk fibers for cartilage tissue engineering. *Materials Science and Engineering: C, 33*(8), 4786–4794.
Morgan, J., & T. Partridge (2020). Skeletal muscle in health and disease. *Disease Models & Mechanisms, 13*(2): dmm042192.
Mukund, K., & Subramaniam, S. (2020). Skeletal muscle: A review of molecular structure and function, in health and disease. *WIREs Systems Biology and Medicine, 12*(1), e1462.
Naghashzargar, E., et al. (2015). Nano/micro hybrid scaffold of PCL or P3HB nanofibers combined with silk fibroin for tendon and ligament tissue engineering. *Journal of Applied Biomaterials & Functional Materials, 13*(2), 156–168.
Oryan, A., et al. (2014). Implantation of a novel biologic and hybridized tissue engineered bioimplant in large tendon defect: An in vivo investigation. *Tissue Engineering Part A, 20*(3–4), 447–465.

Ouyang, L., Grosse, S. D., & Kenneson, A. (2008). Health care utilization and expenditures for children and young adults with muscular dystrophy in a privately insured population. *Journal of Child Neurology, 23*(8), 883–888.

Page, R. L., et al. (2011). Restoration of skeletal muscle defects with adult human cells delivered on fibrin microthreads. *Tissue Engineering Part A, 17*(21–22), 2629–2640.

Patel, D. K., et al. (2020). Bioactive electrospun nanocomposite scaffolds of poly (lactic acid)/ cellulose nanocrystals for bone tissue engineering. *International Journal of Biological Macromolecules, 162*, 1429–1441.

Patlolla, A., Collins, G., & Arinzeh, T. L. (2010). Solvent-dependent properties of electrospun fibrous composites for bone tissue regeneration. *Acta Biomaterialia, 6*(1), 90–101.

Perumcherry, S. R., et al. (2011). A novel method for the fabrication of fibrin-based electrospun nanofibrous scaffold for tissue-engineering applications. *Tissue Engineering Part C: Methods, 17*(11), 1121–1130.

Pins, G. D., et al. (1997). Self-assembly of collagen fibers. Influence of fibrillar alignment and decorin on mechanical properties. *Biophysical Journal, 73*(4), 2164–2172.

Qiao, Y., et al. (2020). Gelatin templated polypeptide co-cross-linked hydrogel for bone regeneration. *Advanced Healthcare Materials, 9*(1), 1901239.

Ribeiro, N., et al. (2019). Influence of PLLA/PCL/HA scaffold fiber orientation on mechanical properties and osteoblast behavior. *Materials, 12*(23), 3879.

Rong, D., et al. (2016). Fabrication of gelatin/PCL electrospun fiber mat with bone powder and the study of its biocompatibility. *Journal of Functional Biomaterials, 7*(1), 6.

Safadi, F. F., et al. (2009). Bone structure, development and bone biology. *Bone Pathology*, 1–50.

Sahoo, S., Toh, S. L., & Goh, J. C. (2010). A bFGF-releasing silk/PLGA-based biohybrid scaffold for ligament/tendon tissue engineering using mesenchymal progenitor cells. *Biomaterials, 31*(11), 2990–2998.

Samadian, H., et al. (2021). Bioengineered 3D nanocomposite based on gold nanoparticles and gelatin nanofibers for bone regeneration: In vitro and in vivo study. *Scientific Reports, 11*(1), 13877.

Sanami, M., et al. (2015). Biophysical and biological characterisation of collagen/resilin-like protein composite fibres. *Biomedical Materials, 10*(6), 065005.

Shah, R., et al. (2005). Craniofacial muscle engineering using a 3-dimensional phosphate glass fibre construct. *Biomaterials, 26*(13), 1497–1505.

Shiroud Heidari, B., et al. (2021). Biofabrication and signaling strategies for tendon/ligament interfacial tissue engineering. *ACS Biomaterials Science & Engineering, 7*(2), 383–399.

Siddiqui, N., et al. (2021). Electropsun polycaprolactone fibres in bone tissue engineering: A review. *Molecular Biotechnology, 63*, 363–388.

Subramony, S. D., et al. (2013). The guidance of stem cell differentiation by substrate alignment and mechanical stimulation. *Biomaterials, 34*(8), 1942–1953.

Sun, H., et al. (2006). The in vivo degradation, absorption and excretion of PCL-based implant. *Biomaterials, 27*(9), 1735–1740.

Tan, G., Zhou, Y., & Sooriyaarachchi, D. (2021). *Musculoskeletal tissue engineering using fibrous biomaterials* (pp. 31–40). Methods and Protocols.

Tiwari, A. P., et al. (2017). Heterogeneous electrospun polycaprolactone/polyethylene glycol membranes with improved wettability, biocompatibility, and mineralization. *Colloids and Surfaces A: Physicochemical and Engineering Aspects, 520*, 105–113.

Tiwari, A. P., et al. (2018). Nano-nets covered composite nanofibers with enhanced biocompatibility and mechanical properties for bone tissue engineering. *Journal of Nanoscience and Nanotechnology, 18*(1), 529–537.

Tonti, O. R., et al. (2021). Tissue-specific parameters for the design of ECM-mimetic biomaterials. *Acta Biomaterialia, 132*, 83–102.

Udomluck, N., et al. (2019). Recent developments in nanofiber fabrication and modification for bone tissue engineering. *International Journal of Molecular Sciences, 21*(1), 99.

Uiterwijk, M., et al. (2024). Balancing scaffold degradation and neo-tissue formation in in situ tissue engineered vascular grafts. *Tissue Engineering Part A, 30*, 421.

Vaquette, C., et al. (2010). Aligned poly(L-lactic-co-e-caprolactone) electrospun microfibers and knitted structure: A novel composite scaffold for ligament tissue engineering. *Journal of Biomedical Materials Research. Part A, 94*(4), 1270–1282.

Vijayavenkataraman, S., et al. (2018). Electrohydrodynamic-jetting (EHD-jet) 3D-printed functionally graded scaffolds for tissue engineering applications. *Journal of Materials Research, 33*(14), 1999–2011.

Wang, Y., et al. (2018). Fabrication of nanofibrous microcarriers mimicking extracellular matrix for functional microtissue formation and cartilage regeneration. *Biomaterials, 171*, 118–132.

Wang, D., et al. (2019). Microscale scaffolds with diverse morphology via electrohydrodynamic jetting for in vitro cell culture application. *Biomedical Physics & Engineering Express, 5*(2), 025011.

Wang, F., et al. (2020). Advances in electrospinning of natural biomaterials for wound dressing. *Journal of Nanomaterials, 2020*(1), 8719859.

Wang, H., et al. (2024). Investigation of process parameters of electrohydrodynamic jetting for 3D printed PCL fibrous scaffolds with complex geometries. *International Journal of Bioprinting, 2*(1), 63.

Wasyłeczko, M., Sikorska, W., & Chwojnowski, A. (2020). Review of synthetic and hybrid scaffolds in cartilage tissue engineering. *Membranes, 10*(11), 348.

Weng, L., et al. (2018). Novel 3D hybrid nanofiber aerogels coupled with BMP-2 peptides for cranial bone regeneration. *Advanced Healthcare Materials, 7*(10), 1701415.

Wingender, B., et al. (2016). Biomimetic organization of collagen matrices to template bone-like microstructures. *Matrix Biology, 52-54*, 384–396.

Wu, Y., et al. (2016). Fabrication and evaluation of electrohydrodynamic jet 3D printed polycaprolactone/chitosan cell carriers using human embryonic stem cell-derived fibroblasts. *Journal of Biomaterials Applications, 31*(2), 181–192.

Wu, J., et al. (2019). Functionalization of silk fibroin electrospun scaffolds via BMSC affinity peptide grafting through oxidative self-polymerization of dopamine for bone regeneration. *ACS Applied Materials & Interfaces, 11*(9), 8878–8895.

Xu, T., et al. (2012). Hybrid printing of mechanically and biologically improved constructs for cartilage tissue engineering applications. *Biofabrication, 5*(1), 015001.

Xu, Y., et al. (2014). The effect of mechanical stimulation on the maturation of TDSCs-poly (L-lactide-co-e-caprolactone)/collagen scaffold constructs for tendon tissue engineering. *Biomaterials, 35*(9), 2760–2772.

Xue, J., et al. (2013). Engineering ear-shaped cartilage using electrospun fibrous membranes of gelatin/polycaprolactone. *Biomaterials, 34*(11), 2624–2631.

Yamamoto, A., et al. (2010). Prevalence and risk factors of a rotator cuff tear in the general population. *Journal of Shoulder and Elbow Surgery, 19*(1), 116–120.

Yao, J., et al. (2020). Three-dimensional coating of SF/PLGA coaxial nanofiber membranes on surfaces of calcium phosphate cement for enhanced bone regeneration. *ACS Biomaterials Science & Engineering, 6*(5), 2970–2984.

Zang, M., et al. (2011). Perichondrium directed cartilage formation in silk fibroin and chitosan blend scaffolds for tracheal transplantation. *Acta Biomaterialia, 7*(9), 3422–3431.

Zhang, H., & Liu, J. (2013). Electrospun poly (lactic-co-glycolic acid)/wool keratin fibrous composite scaffolds potential for bone tissue engineering applications. *Journal of Bioactive and Compatible Polymers, 28*(2), 141–153.

Zhao, S., et al. (2014). Biological augmentation of rotator cuff repair using bFGF-loaded electrospun poly (lactide-co-glycolide) fibrous membranes. *International Journal of Nanomedicine, 9*, 2373–2385.

Zhou, Y., et al. (2021). Biomimetic strategies for fabricating musculoskeletal tissue scaffolds: A review. *The International Journal of Advanced Manufacturing Technology, 112*, 1211–1229.

Zhou, X., et al. (2022). Regulated exogenous/endogenous inflammation via "inner-outer" medicated electrospun fibers for promoting tissue reconstruction. *Advanced Healthcare Materials, 11*(10), 2102534.

Zhu, K., Xiangzhou, L., & Shilin, Y. (1990). Preparation, characterization, and properties of polylactide (PLA)–poly (ethylene glycol)(PEG) copolymers: A potential drug carrier. *Journal of Applied Polymer Science, 39*(1), 1–9.

# Pharmacological Compounds in Regeneration

# Role of the Phenolic Compounds and Their Derivatives in the Regeneration Process

**Anoushka Bhat and Nadezhda A. German**

**Abstract** Bone regeneration is a complex physiological process involving multiple signaling pathways, such as oxidative stress, inflammation, and cellular differentiation. Phenolic compounds have emerged as promising bioactive agents due to their antioxidant, anti-inflammatory, and osteogenic properties. This chapter is focused on describing molecular mechanisms by which phenolic compounds regulate bone regeneration, including the modulation of the nuclear factor kappa-B (NF-κB) and Nrf2 pathways. A short overview of chemical characteristics and biological activities of multiple subclasses of phenolic compounds is provided. The discussion extends to the integration of polyphenols into biomaterials-based applications, such as metal-phenolic networks (MPNs) and bioactive scaffolds, highlighting significant potential of these molecules in enhancing bone regeneration. Several challenges related to bioavailability and clinical translation have been identified, along with the recent advancements in the improvement of drug-like properties of phenolic compounds. Future research should focus on optimizing the Absorption, Distribution, Metabolism, and Excretion (ADME) properties of phenolic agents and understanding the molecular bases of their biological activities to develop innovative biomaterial-based strategies for bone tissue engineering and regenerative medicine.

**Keywords** Bone regeneration · Phenolic compounds · Oxidative stress · Bioactive biomaterials · Osteogenesis

A. Bhat (✉) · N. A. German
Department of Pharmaceutical Sciences, Jerry H. Hodge School of Pharmacy, Texas Tech University Health Sciences Center, Amarillo, TX, USA
e-mail: Anoushka.Bhat@ttuhsc.edu; Nadezhda.German@ttuhsc.edu



H. Das (ed.), *Tissue Repair and Regeneration*,
https://doi.org/10.1007/978-3-031-93677-7_15

# 1 Introduction

To address the complexity of bone regeneration, various therapeutics approaches were developed to enhance bone healing, including bone implants, small molecules, natural products, gene therapy, and biomaterials. Each therapeutic approach comes with its own pros and cons, necessitating continuous research in the field aiming the development of alternative solutions. Among these, natural compounds, particularly phenolic compounds, have emerged as promising candidates due to their antioxidant, anti-inflammatory, and osteogenic properties. This chapter explores the role of phenolic agents in the bone regeneration, and reviews their mechanism of action, classification, and potential applications in this therapeutic field.

# 2 Overview of Bone Regeneration

Bone regeneration is a challenging area to tackle with pharmacological approaches, due to the limited efficacy of current drugs in fully restoring bone integrity and functions. Autologous bone grafts harvested from patient's own body are considered the gold standard due to their regeneration-enhancing properties like osteoconductivity, osteogenesis, and osteoinduction (Gillman & Jayasuriya, 2021). Although immunogenicity and infection are minimal in this case, this approach suffers from limitations, including high cost and the need of invasive surgical procedures that can result in donor site morbidity, such as higher fracture risk, nerve damage, infection risk, and others (Schmidt, 2021; Younger & Chapman, 1989). Some of these drawbacks are addressed by allograft bone implants derived from a donor, but this strategy faces its own risks of donor scarcity, immune rejection, possible infections, reduced osteoinductive potential, and high cost of the procedure (Amini et al., 2012). Nevertheless, allograft remains a valuable alternative to autografts, particularly in the cases where patients have insufficient autologous bone available or cannot undergo additional invasive procedures due to the medical conditions. The recent advancements in the field include use of synthetic biomaterials as bioscaffolds (Dimitriou et al., 2011) that mimic natural bone leading to improved differentiation in bone cells and optimized bone regeneration. Bioscaffold approach focuses on enhancing the activity of multiple key players in bone formation and protection, for example, mesenchymal stem cells (MSCs). These biomaterials provide supportive extracellular matrix (ECM)-like environment that promotes MSC adhesion, proliferation, and differentiation into osteoblasts (Lin et al., 2020). By enhancing this interaction with pharmacological agents, one can promote cell migration, angiogenesis, and cell differentiation inducing the repairing of bone matrix and faster bone regeneration (Benavides-Castellanos et al., 2020; Iaquinta et al., 2019). Enhancing MSC activity via natural products like phenolic compounds has become an important research area.

Current therapeutics that promote bone regeneration include antiresorptive and anabolic agents, such as bisphosphonates, glucocorticoids, and monoclonal antibodies. However, they are often criticized for their long-term adverse effects, including thrombotic events, kidney disease, gastrointestinal issues, and cardiovascular complications, limiting their long-term use (Baldo, 2022; Diel et al., 2007; Pofi et al., 2023; Weinstein, 2012). Natural products, especially phytochemical compounds, are often considered as a better pharmacological approach to bone regeneration process due to their biocompatibility and ability to modulate multiple biological pathways (Szwed-Georgiou et al., 2023; Valentino et al., 2021). The evaluation of their safety profiles is subject to ongoing or future clinical trials (Ilyas et al., 2024). Natural products are known to promote bone healing through antioxidant (Wang et al., 2024) and anti-inflammatory activities (Miranda et al., 2019), as well as osteogenic activities, providing strong scientific premises for preclinical and clinical testing of these agents as enhancer of bone healing (Suvarna et al., 2018). Their stimulating effects were noted on MSCs, hematopoietic stem cells (HSC), and immune cells, events that lead to osteoblast proliferation and protection of bone microenvironment. Additionally, these compounds were shown to attenuate nuclear factor kappa-B (NF-κB) pathway, reduce levels of proinflammatory cytokines, and directly affect macrophage polarization by inhibiting the proinflammatory M1 macrophages and stimulating anti-inflammatory M2 macrophages (Iaquinta et al., 2019; Ilyas et al., 2024; Hanga-Farcaş et al., 2023) (Fig. 1). Natural products, like phenolic compounds, are known to alleviate oxidative stress in bone tissue and control cellular differentiation pathways by modulating the level of reactive oxygen species (ROS). Such simultaneous modulation of multiple biological parameters via the polypharmacological activity of phenolic compounds may offer an advantage in enhancing the complex biological process involved in bone regeneration, ultimately resulting in more efficient and sustained bone healing.

# 3 Mechanism of Action of Phenolic Compounds

## *3.1 Antioxidant Activity*

Chemically, phenolic compounds are a large group of molecules that have aromatic hydroxyl groups known to play a crucial role in biological functions of these chemicals. Their classification of phenolic compounds is based on their monophenol or polyphenolic scaffolds and source, either nature-derived or synthetically obtained. The phenol group is a key determinant of the antioxidant properties of phenolic agents (Chen et al., 2024a) responsible for the reduction of the oxidative stress in bone tissue cells (BTCs) and facilitation of osteogenesis (Li et al., 2022a). The antioxidant activity of phenolic agents was shown to be mediated by two primary mechanisms: direct free-radical scavenging and the upregulation of endogenous antioxidant defense mechanisms (Yasuhisa et al., 1993) (Fig. 2). Cyclic

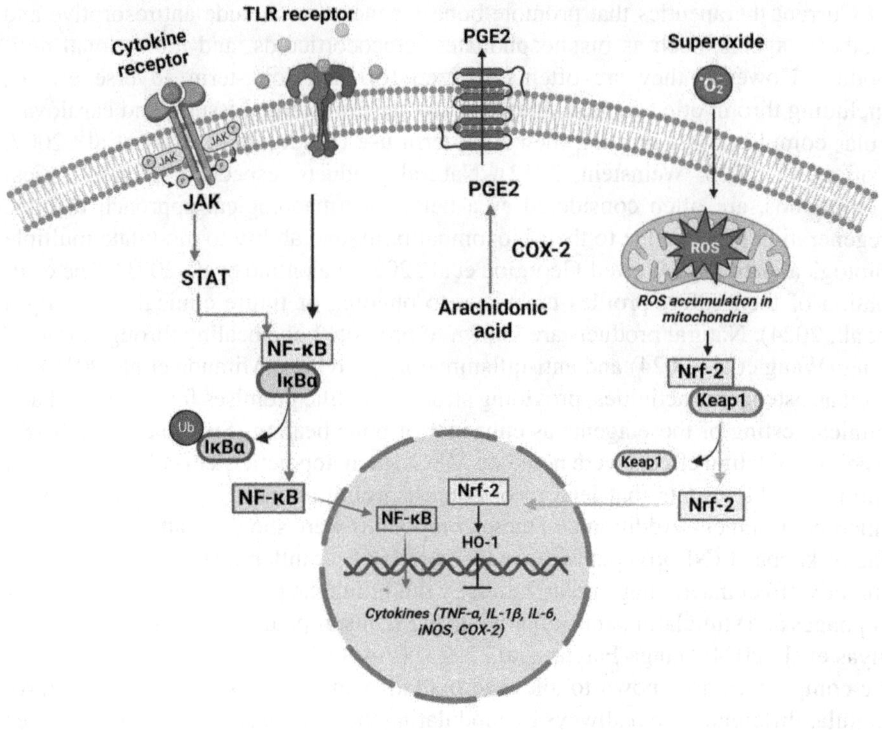


**Fig. 1** Interplay between NF-κB pathway and Nrf2 pathways in inflammation and oxidative stress regulation during bone regeneration

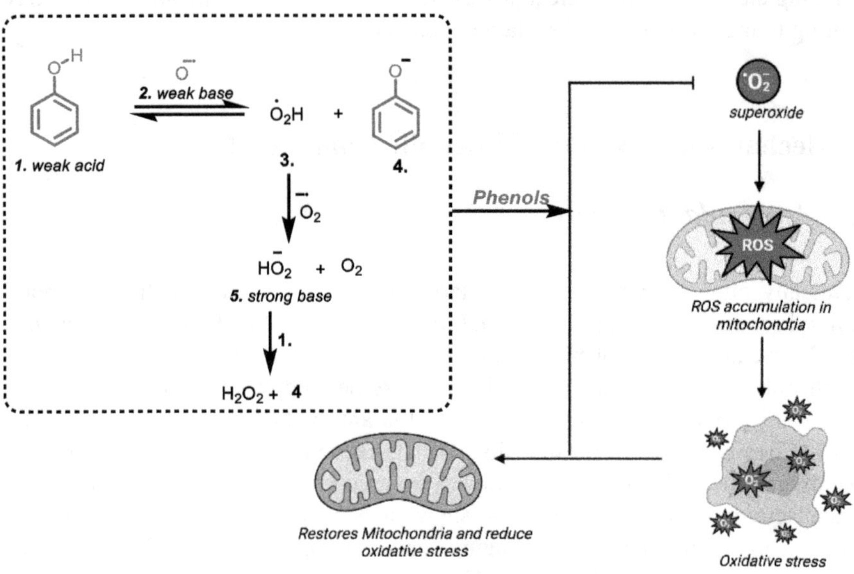


**Fig. 2** Mechanism of Phenol in reducing oxidative stress

voltammetry studies confirmed the intrinsic reducibility of phenolic compounds by demonstrating their ability to undergo redox cycling, efficiency in electron donation, and stabilization of radical species (Andrés et al., 2023). These electrochemical properties are fundamental in contributing to the strong antioxidant activity of phenolic compounds (Chen et al., 2024a; Andrés et al., 2023). This is particularly important in the oxidative microenvironment of BTCs, during the healing process, where high levels of ROS can impair osteoblast function and delay bone regeneration (Wang et al., 2024). The scavenging of toxic superoxide or peroxide radicles occurs via a hydrogen transfer or single electron transfer mechanisms, where the phenol group donates a hydrogen atom or an electron, stabilizing ROS and effectively terminating the oxidative chain reaction (Zeb, 2020). Generated hydroperoxyl radicles or hydrogen superoxide are quenched by phenol to yield non-toxic hydrogen peroxide and a resonance-stabilized phenoxide ion (Andrés et al., 2023; Zeb, 2020). Overall, this process reduces the level of ROS, alleviates the oxidative stress, and minimizes oxidative damage in bone tissue, creating a more favorable microenvironment for bone regeneration. The presence of multiple hydroxyl groups in polyphenolic compounds improves their antioxidant potential by inducing a more rapid and sustained quenching of ROS species (Checinska et al., 2022). This intrinsic property explains the superior antioxidant activity of polyphenolic compounds over monophenols.

The second mechanism of antioxidative activity of phenolic compounds is linked to their ability to enhance the expression levels of antioxidant enzymes, like heme oxygenase-1 (HO-1) and glutathione peroxidase-4 (GPX4) (Chen et al., 2024a). These effects are primarily mediated by Nrf2/KEAP1 system, a well-recognized cellular defense mechanism against oxidative stress (Chen et al., 2024a; Choi et al., 2016). Upregulation of HO and GPX4 levels causes significant reduction in the level of ROS, ultimately leading to the prevention of ROS-induced apoptosis and ferroptosis in BTCs and promotion of osteogenesis.

## 3.2 *Anti-inflammatory and Osteogenic Activity of Phenolic Compounds*

Staying true to their polypharmacological nature, phenolic compounds can modulate inflammatory responses through multiple mechanisms as well (Fig. 1). These compounds exhibit their anti-inflammatory effect by suppressing M1 polarized macrophages, causing reduction in the expression of receptor activator of nuclear factor kappa-B ligand (RANKL) and osteoclast production (Chen et al., 2024a; Marcadet et al., 2022). The key molecular pathways involved in this process are Mitogen-Activated Protein Kinase/Activator Protein 1 (MAPK/AP-1), NF-κB, NOD-, LRR-, and Pyrin Domain-Containing Protein 3 (NLRP-3), and c-Jun N-terminal Kinase/Signal Transducer and Activator of Transcription (JNK/STAT), all of which are critical regulators of inflammation-driven bone resorption (Lee et al., 2021; Tao et al., 2021; Zhang et al., 2022a). For example, the NF-κB pathway regulates osteoclast differentiation and activation (Li et al., 2019), while MAPK and JNK/STAT pathways mediate pro-inflammatory signaling contributing to the loss of the bone (Thummuri et al., 2017). By targeting these pathways, phenolic compounds enhance bone remodeling and preserve skeletal integrity. Autophagy plays a key role in bone regeneration facilitating the recycling of damaged cell components (Xiao & Xiao, 2019), maintaining the cellular homeostasis (Yin et al., 2019), and promoting the formation of bone matrix (Wang et al., 2023). Phenolic compounds have been reported to modulate autophagy through various signaling pathways, like AMP-Activated Protein Kinase/Mechanistic Target of Rapamycin (AMPK/mTOR) and Nrf2, which all are essential for balancing osteoblast and osteoclast activity (Wei et al., 2023; Gong et al., 2022). There are some reports showing the selective ability of phenolic compounds to enhance osteoblast autophagy while inhibiting osteoclast autophagy, ultimately enhancing osteogenesis (Wei et al., 2023; Gong et al., 2022). Oxidative stress is known to non-selectively induce autophagy by upregulating ROS levels and, at the same time, to promote the production of both osteoblast and osteoclasts. All of these processes lead to the loss of the bone remodeling homeostasis. Phenols have been shown to restore this balance, by attenuating osteoclast differentiation and enhancing osteoblast differentiation (Fig. 2.). This phenols-controlled outcome is mediated through the inhibition of multiple pathways, such as Nrf2/NF-κB, and MAPK/JNK/ERK/p38, causing reduction in excessive bone resorption.

### *3.3 Bactericidal Activity of Phenolic Agents*

The bone regeneration microenvironment is crucial in determining the outcomes of bone remodeling process. Postsurgical bacterial infections are common complications in patients undergoing bone implant procedures, potentially slowing the bone healing process (Croes et al., 2019; Seebach & Kubatzky, 2019). These infections can be caused by Gram-positive pathogens, like *Staphylococcus aureus*, including methicillin-resistant *Staphylococcus aureus* (MRSA) and *Staphylococcus epidermidis*, and Gram-negative organisms, like *Pseudomonas aeruginosa* and *Escherichia coli* (Seebach & Kubatzky, 2019). All of these pathogens have biofilm-forming ability and higher resistance to the treatment (Sharma et al., 2023). Bacterial colonization causes cytotoxins and stimulates excessive inflammatory cytokine production, exacerbating local tissue inflammation, disrupting osteogenic signaling, and impairing new bone formation (Croes et al., 2019). Different classes of phenolic compounds exhibit potent bactericidal properties by targeting a wide range of bacterial survival mechanisms (Takó et al., 2020a). They can compromise bacterial cell membrane integrity, disrupt biofilm formation, and inhibit the respiratory chain, effectively blocking energy metabolism and bacterial proliferation (Takó et al., 2020a). Moreover, the bactericidal activity of phenolic agents can be different from the common mechanisms used by antimicrobial agents, suggesting possible use of phenolic compounds against resistant strains of pathogens.

# 4 Classification of Natural Polyphenols

Polyphenols, the predominant type of phenolic compounds synthesized by plants and marine organisms (Matulja et al., 2022; Zagoskina et al., 2023), exhibit significantly stronger antioxidant properties compared to monophenols, due to their multiple hydroxyl groups. This structural feature allows for more efficient electron scavenging ability. Polyphenols have been studied for their bone protective properties, demonstrating their ability to restore bone remodeling balance through the increased osteogenesis and inhibited osteoclasts differentiation (Li et al., 2022a). Furthermore, natural polyphenols exhibit cardioprotective (Iqbal et al., 2023) and hepatoprotective (Machado et al., 2023) properties. These advantages in the functional activity, coupled with the relatively low toxicity profile (Cory et al., 2018), make polyphenols promising drug candidates for long-term therapeutic applications in patients undergoing bone regeneration interventions. Natural polyphenols can be classified into various subclasses based on their structural characteristics and functional bioactivity, including flavonoids, phenolic acids, stilbenes, lignans, and tannins, each of which produce distinct biological effects on bone metabolism and regeneration processes (Fig. 3).

## *4.1 Flavonoids*

Flavonoids are a diverse class of plant-derived natural polyphenols that are commonly found in fruits, vegetables, grains, and other dietary sources (Panche et al., 2016; Roy et al., 2022). Structurally, they consist of a 15- carbon chemical backbone in the form of two benzene rings (A and B) connected by a heterocyclic pyran ring (C). The substitutions on the pyran ring and the degree of oxidation of pyran ring determine the difference in bioactivity of flavonoids (Kumar & Pandey, 2013), deriving additional subclasses like flavones, flavonols, flavanones, isoflavones, flavanols, and anthocyanins. Flavonoids are well-known for their broad spectrum of biological activities, including antioxidant (Panche et al., 2016; Kopustinskiene et al., 2020), antitumor (Kopustinskiene et al., 2020), and anti-inflammatory properties (Ginwala et al., 2019). In bone regeneration process, along with the effective reduction of oxidative stress and inflammation, they also enhance osteoblast differentiation in bone marrow mesenchymal stem cells (BMSCs) (Shanmugavadivu et al., 2022). Several flavonoid compounds, including icariin, quercetin, naringin, silibinin, and genistein, have demonstrated preclinical efficacy in preventing bone loss by increasing bone morphogenetic protein-2 (BMP-2) expression and promoting osteoblast proliferation, while inhibiting osteoclast differentiation

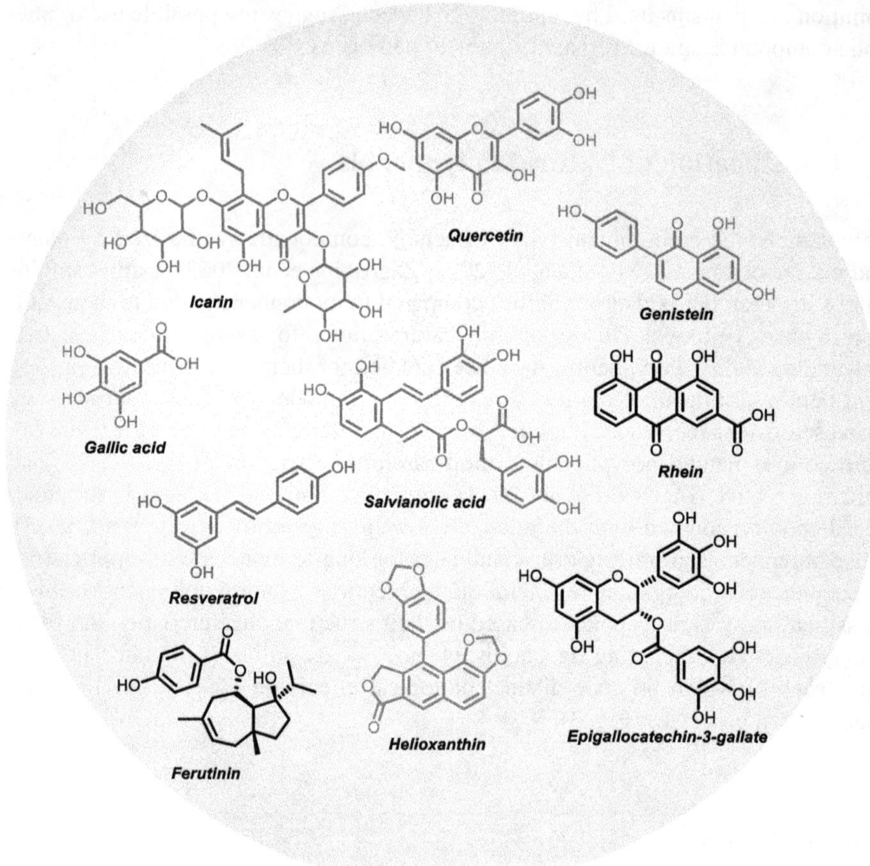


**Fig. 3** Representative structures of phytochemical compounds having bone regenerative properties: flavonoids (blue); phenolic acids (red); stilbenes (green); lignans (magenta); and miscellaneous (black)

(Shanmugavadivu et al., 2022). They were shown to enhance angiogenesis and osteogenic features by driving the expression of key proteins like BMP-2/4/7, runt-related transcription factor 2 (RUNX2), alkaline phosphatase (ALP), collagen type-1 (COL-1), and osteocalcin (OCN) (Shanmugavadivu et al., 2022; Huang et al., 2018; Sharifi et al., 2020; Torre, 2017). In addition, flavonoids can modulate multiple signaling pathways responsible for bone healing process in BMSCs, including BMP-A1/2, Wnt/β-Catenin, Vascular Endothelial Growth Factor Receptor/Receptor 2 (VEGFR/VEGFR-2), Phosphoinositide 3-Kinase/Protein Kinase B/Endothelial Nitric Oxide Synthase (PI3K/AKT/eNOS), Mitogen-Activated Protein Kinase Kinase/Extracellular Signal-Regulated Kinase (MEK/ERK), and NF-κB (Li et al., 2022a; Shanmugavadivu et al., 2022). However, despite demonstrated significant efficacy in both in vitro and in vivo models of bone regeneration, there clinical translation is hindered by limitations like poor bioavailability, rapid metabolism, and low systemic retention (D'Archivio et al., 2007; Manach et al., 2005; Thilakarathna & Rupasinghe, 2013). Currently, innovative delivery systems are being developed to overcome these limitations and enhance flavonoids' pharmacokinetic properties (Thilakarathna & Rupasinghe, 2013; Zverev & Rykunova, 2022).

### 4.2 *Phenolic Acids*

Phenolic acids are a subclass of phenolic compounds characterized by an aromatic phenyl ring substituted with at least one hydroxyl (-OH) and carboxylic acid (-COOH) functional group. The variability in chemical properties and bioactivities of phenolic acids comes from different substitution patterns of these molecules around the phenyl ring (Pereira et al., 2009). Several phenolic acids, including salvianolic acid, gallic acid, and hydroxycinnamic acids, have been investigated for their potential role in bone regeneration (Cao et al., 2025; Liu et al., 2023; Yamaguchi et al., 2013; Yamaguchi & Weitzmann, 2012; Zhang et al., 2022b) and both in vitro and in vivo studies have confirmed their osteoprotective and regenerative properties (Torre, 2017), making these compounds promising therapeutic candidates (Li et al., 2022a). The osteogenic effect of phenolic acids is based on their ability to enhance BMSCs' differentiation and prevent bone resorptions (Li et al., 2022a). These effects are facilitated through modulation of signaling pathways, like p38/MAPK/Wnt/β-catenin, or by regulating osteogenic markers like alkaline phosphatase (ALP), runt-related transcription factor 2 (RUNX2), osteocalcin (OCN), insulin-like growth factor (IGF-1), COL-1, heme oxygenase-1 (HO-1), osterix transcription factor (OSX), and tafazzin (TAZ) (Balagangadharan et al., 2019; Chen et al., 2010; Hou et al., 2025). In addition to their osteogenic potential, phenolic acids exhibit strong antioxidant (Lopez-Corona et al., 2022), antibacterial (Takó et al., 2020b), anti-inflammatory (Lopez-Corona et al., 2022), and peroxynitrite scavenging properties (Kumar & Goel, 2019), demonstrating cytoprotective effect in BMSCs (Nicolin et al., 2019), ultimately enhancing bone regeneration processes.

## 5 Stilbenes

Stilbenes are a class of phenolic compounds that contains two phenyl rings linked via ethylene double bond. They exist as two isomers, cis- or a more stable trans-stilbenes (Likhtenshtein, 2009). Resveratrol is one of the most extensively studied representatives of stilbenes and has a well-documented bone protective properties (Feng et al., 2017; Zou et al., 2024) and potent antioxidant (Faisal et al., 2024) and anti-inflammatory activities (Meng et al., 2021). In addition, resveratrol functions as a selective estrogen receptor modulator (SERM), displaying estrogenic and anti-estrogenic effects in a tissue-dependent manner (Gehm et al., 1997; Bhat et al., 2001). SERM ligands, including resveratrol, have been extensively studied in both in vitro and in vivo models for their effect on bone remodeling. These studies demonstrated their key role in enhancement of bone mineralization, and improved bone mass (Ahmad Hairi et al., 2023; Bo et al., 2018; Murgia et al., 2019). Resveratrol has been shown to induce the pro-autophagic phase, promote osteoblast proliferation, and suppress osteoclast proliferation in human BMSCs using key regulatory pathways like RANKL/RANK/OPG (Ahmad Hairi et al., 2023; Wang et al., 2022a). Additionally, it promotes osteogenic differentiation by upregulating ALP expression and facilitating calcium deposition (Moon et al., 2020). The immunomodulatory properties of resveratrol allow to inhibit inflammatory cytokine release and reduce chronic inflammation associated with bone degradation. As a SERM ligand, it promotes the regeneration of bone precursor cells and enhances osteogenesis. Importantly, resveratrol was shown to improve differentiation of senescent *bone marrow mesenchymal stromal cells* (BMMSCs), suggesting its potential as a therapeutic agent for mitigating age-related osteoporosis (Zhou et al., 2019). Its protective effects against physicochemical bone damage and bone loss are driven by the activation of mitofilin, a mitochondrial inner membrane protein crucial for maintaining mitochondrial homeostasis and promoting bone osteogenesis (Lv et al., 2018). Finally, resveratrol promotes early-stage osteogenic differentiation (Moon et al., 2020) and enhances osteoblasts proliferation by regulating key signaling proteins like Osteoprotegerin (OPG), Sirtuin 1 (SIRT1), B-catenin, and Wnt while inhibiting NF-κB pathway. Such biological activity of resveratrol contributes to bone homeostasis and supports the mineralization (Murgia et al., 2019; Al-Khayri et al., 2023; Inchingolo et al., 2022). Altogether, stilbenes hold an important role in bone regeneration and stilbenes-based therapeutic approaches can have a significant impact in this field (Li et al., 2022a).

## 5.1 Lignans

Lignans are phenolic compounds that contain two linked phenylpropane motifs. Among nature-derived phenolic compounds, lignans remain the least studied in the context of bone regeneration. Nevertheless, flax lignans demonstrated biological activities similar to other phenolic compounds, including notable antioxidant and anti-inflammatory properties (Li et al., 2024). Their most pronounced effect on bone regeneration was observed only in combination with estrogen therapy, suggesting a potential synergistic effect with hormone-modulated osteogenesis (Sacco et al., 2009). Helioxanthin, a structurally distinct lignan, is currently in the early stage of investigation as potential therapy for bone regeneration (Sato et al., 2020). Preliminary in vitro and in vivo studies indicate that it improves osteogenic differentiation in Dental Pulp Stem Cell (DPSCs), promoting bone matrix deposition and accelerating bone healing (Yamakawa et al., 2020). However, more studies are required to validate its initial efficacy and assess true therapeutic potential of this lignan molecule.

## 5.2 *Miscellaneous Phenolic Compounds*

Curcumin, a member of curcuminoids chemical class, consists of aromatic systems linked by a seven-carbon motif and an α,β-unsaturated diketone moiety. Curcumin exhibits significant therapeutic potential because of its well-known antioxidant, anti-inflammatory, and antitumor properties (Wilken et al., 2011). Its biological activities contribute to MSCs' protection from oxidative stress, while simultaneously promoting osteoblasts proliferation and remodeling of extracellular matrix (ECM) (Gorabi et al., 2019; Li et al., 2020a). The ability of curcumin to promote osteogenic differentiation and inhibit adipogenetic differentiation in Human Bone Marrow-Derived Mesenchymal Stem Cell (HBMSCs) is based on modulating activity of multiple key players, such as matrix metalloproteinases (MMP-9), Wnt/β-catenin, and MAPK/Nrf-2/NF-κB/JNK/COX-2/PI3K/Akt signaling pathways (Gorabi et al., 2019). It was shown that curcumin provides osteoprotection by inhibiting osteoclast differentiation, which, in turn, is based on the ability of curcumin to suppress M1 macrophage polarization and inflammatory cytokines release (Chen et al., 2021; Wang, 2024; Yang et al., 2023). Despite the well-established cytoprotective role of curcumin in HBMSCs, its poor aqueous solubility and suboptimal pharmacokinetic profile cause significant challenges for clinical translation (Sohn et al., 2021). Multiple pharmaceutical strategies, such as nanoparticles for the delivery, biomaterials, and other enhancing approaches, have been assessed for their ability to overcome these limitations (John et al., 2013; Karthikeyan et al., 2020; Ghavimi et al., 2020; Dahiya et al., 2024).

Tannic acids belong to a class of phenolic compounds that consists of central glucose molecules esterified with multiple gallic and/or digallic acid residues. They also possess strong antioxidant, anti-inflammatory, and antibacterial properties, mitigating oxidative stress and inflammation, key contributors to bone degenerative disorders (Abolarin & Owoyele, 2024; Winiarska-Mieczan et al., 2023). An important characteristic of tannic acids is its strong concentration-dependent (Tomaszewska et al., 2017), metal-chelating ability (Phiwchai et al., 2018) that plays a crucial role in maintaining the bone regeneration environment (Winiarska-Mieczan et al., 2023). As a chelator, tannic acids have high affinity for toxic metals like cadmium (Cd) and lead (Pb) over calcium (Ca), phosphorus (P), sodium (Na), potassium (K), magnesium (Mg), and iron (Fe) and can be used for detoxifying environment reach with heavy metals (Cd, Pb) (Winiarska-Mieczan et al., 2023; Wu et al., 2023). Tannins, including tannic acid, ellagic acid, and gallic acid, exhibit strong potential for bone regeneration by inhibiting the osteoclast differentiation through the suppression of key markers, including Nuclear Factor of Activated T-Cells, Cytoplasmic 1 (NFATc1), cathepsin K, and Tartrate-Resistant Acid Phosphatase (TRAP) (Laha et al., 2022). They also inhibit autophagy through reduction of intracellular calcium level to affect mitochondria, preventing the formation of osteoclast (Laha et al., 2022). On the other hand, the opposite effects have been reported, where tannic acids can promote proliferation and differentiation of osteoblast (Naidu et al., 2024), suggesting a complicated biological effects induced by this class of compounds.

Multiple other representatives of phenolic compounds have been investigated for their effect on bone regeneration, highlighting the chemical and biological diversity of this class of molecules. One of them are lignins, complex polyphenolic polymers composed of phenylpropanoid units with potential therapeutic application in bone regeneration (Jiang et al., 2024; Wang et al., 2019). These polymers are mechanically strong material with good antioxidant and antibacterial properties (Alzagameem et al., 2019). Because of their ability to improve nutrient transport and durability, lignins are considered to be ideal biomaterials for bone regeneration (Wang et al., 2019; Silva-Barroso et al., 2023). Another representative of phenolic compounds—terpenoids (epigallocatechin-3-gallate)—can control the osteoclastogenesis by downregulating its key differentiation markers and modulating mitophagy (Sarkar et al., 2022). Catechins, such as ferrutinin, have been shown to activate osteoblast differentiation by upregulating the transcription factor, Kruppel-like factor 2 (KLF2), along with autophagy-related molecules (Maity et al., 2022), balancing the bone homeostasis and improving the bone regeneration process (Maity et al., 2022). Anthraquinones, such as rhein and its derivatives, demonstrate strong anti-osteoclastogenesis potential by downregulating key genes involved in osteoclasts differentiation while enhancing osteoblast activity (Jiang et al., 2019). Further research is warranted to explore the full therapeutic potential of these compounds and their derivatives in bone regeneration.

## 6 Phenol Biomaterials

Biomaterials are a modern alternative to the traditional bone grafts, wherein synthetic materials are infused with selected therapeutic agents, like polyphenols, to enhance and accelerate bone regeneration (Gao et al., 2017; Zhao et al., 2021). While natural products exhibit significant bioactivity, their clinical translation is limited by suboptimal pharmacokinetic and limited aqueous solubility, ultimately requiring higher dosages to achieve therapeutic relevance (Hanga-Farcaş et al., 2023). These challenges can be addressed through novel drug delivery approaches, such as the co-administration of synergistic partner, use of nanocarrier-based formulations, and scaffolds-mediated controlled release systems, all aimed at enhancing the bioavailability and ensuring targeted therapeutic delivery (Chen et al., 2024b). Bone regeneration materials include metals, ceramics, polymers, and composites, each using its unique properties to enhance the mimicry of the natural bone architecture and function (Yu et al., 2015). The ideal biomaterial parameters include mechanical integrity, pore architecture, and bioactive surface, aiming to mimic the biochemical, mechanical, and structural parameters of bone tissue (Lutzweiler et al., 2020; Mukasheva et al., 2024). Biodegradable metals, like Fe, Mg, and Zn, have been used in bone regeneration mainly due to their antibacterial, osteogenic, and angiogenic properties and minimal long-term complications (Wang et al., 2022b). The advantage of inorganic ceramics, such as hydroxyapatite, tricalcium phosphate, and biphasic calcium phosphate, is based on their better osteoconductive properties and their ability to mimic the mineralized matrix of native bone, ensuring mechanical stability and biocompatibility (Hou et al., 2022; Baghbani et al., 2012). Biodegradable polymers, either nature-derived (collagen, chitosan, cellulose, alginate, and elastin) or synthetic (polylactic acid, polycaprolactone), provide an adaptable scaffold that mimics the extracellular matrix, promoting cell adhesion, proliferation, and differentiation, all of which are essential for bone regeneration (Camal Ruggieri et al., 2024; Geevarghese et al., 2022; Zhang et al., 2020). However, there are some limitations associated with their use, including increased stiffness, mechanical stress, and potential cytotoxicity (Abdelaziz et al., 2023). Thus, composite biomaterials that combine two or more compounds, or materials from the same or different class, address the limitations of single-component scaffolds and promote bone healing. Scaffold functionalization was also shown to contribute significantly to osteoconduction and osteoinduction, where both of these parameters are needed for effective bone regeneration process (Ilyas et al., 2024; Battafarano et al., 2021).

Bioactive scaffolds incorporating are being extensively studied for potential application in the bone regeneration field (Checinska et al., 2022). Polyphenols show strong antioxidant, anti-inflammatory, and antibacterial properties along with self-healing, hydrophilic, biocompatible, and strong osteoconductive properties, making them ideal for integration into a biomaterial-based therapy (Hobbi et al., 2023). Metal-phenolic network (MPN) has emerged as a promising approach and has been investigated extensively for the bone regeneration properties (Chen et al., 2025; Liu et al., 2024; Wang et al., 2025). As the name suggests, MPNs are constructed through metal-phenol coordination chemistry, yielding biomaterial with enhanced mechanistic properties, biological activity, and immunomodulatory functions (Chen et al., 2025; Wang et al., 2025). MPN can recruit immune cells and facilitate the transition between M1 and M2 macrophages, causing reduced inflammation and promoting optimal environment for bone regeneration (Wang et al., 2025). The hydrophilic nature of MPNs plays a key role in M2 polarization, promoting conformational changes in fibronectin and fibrinogen that lead to anti-inflammatory responses (Atcha et al., 2021). MPNs also induce the osteoblast differentiation and enhance bone tissue vascularization (Liu et al., 2024), overall promoting healing of the bone (Wang et al., 2025; Li et al., 2022b) (Table 1).

In recent years, research has expanded beyond MPNs to explore the potential of other biomaterials in bone regeneration (Yu et al., 2015; Girón et al., 2021; Li et al., 2020b; Lyons et al., 2020; Szwed-Georgiou et al., 2023). Among others, nanomedicines have emerged as a transformative approach, offering precise targeting, controlled release, and improved physicochemical and pharmacokinetic properties with the promise of the enhanced therapeutic efficacy (Lyons et al., 2020). In other approach, polyphenol-functionalized hydrogels have shown promise in tissue adhesion, mitigation of oxidative stress, and improved bone regeneration (Xv et al.,

**Table 1** Classification of major phytochemical compounds useful in bone regeneration

| Compound | Scaffold | Examples | Mechanism of Action (MOA) |
|---|---|---|---|
| Flavonoids<br>(i) Flavones<br>(ii) Flavonols<br>(iii) Flavanones<br>(iv) Isoflavones<br>(v) Flavanols<br>(vi) Anthocyanins | 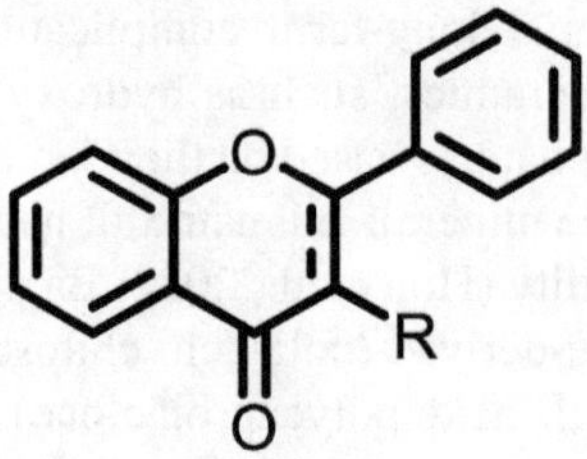 | (i) Luteolin; apigenin<br>(ii) Quercetin; kaempferol; icariin<br>(iii) Hesperetin; naringenin; bavachinin<br>(iv) Diadzein; genistein<br>(v) Catechin; epicatechin<br>(vi) Cyanidin; pelargonidin | Enhance BMP-2 expression, stimulate osteogenesis, and reduce osteoclast activity via Wnt/β-catenin and NF-κB pathways |
| Phenolic acids<br>(i) Hydroxybenzoic acid<br>(ii) Hydrocinnamic acid | 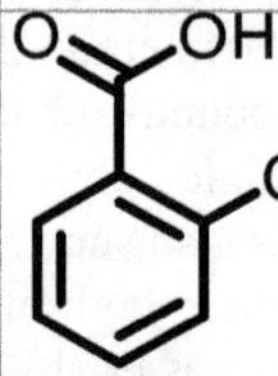 | (i) Gallic acid; vanillic acid<br>(ii) Caffeic acid; ferulic acid; coumaric acid | Promote BMSC differentiation, inhibit bone resorption, and stimulate ALP, RUNX2, and IGF-1 pathways |
| Lignans | 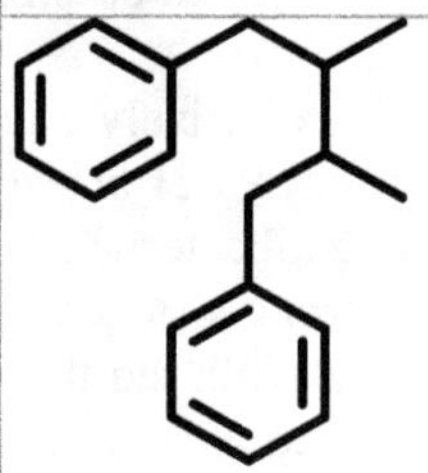 | Secoisolariciresinol; matairesinol; sesamol | Some studies suggest osteogenic differentiation potential in dental pulp stem cells |
| Stilbenes | 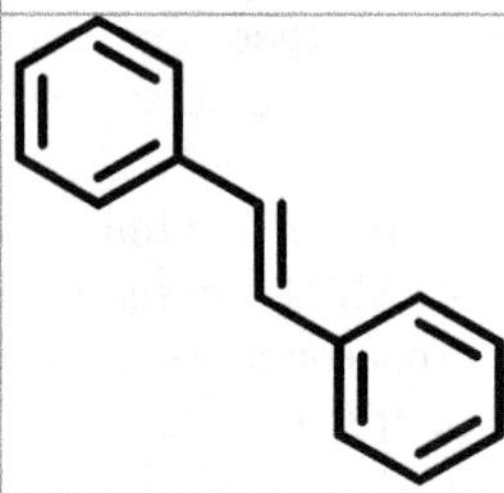 | Viniferin; piceatannol; resveratrol | Act as SERMs, regulate osteoblast autophagy, and enhance bone mineralization |
| Others<br>(i) Tannins<br>(ii) Lignins<br>(iii) Curcumoids<br>(iv) Anthraquinones<br>(v) Catechin<br>(vi)Terpenoids | | Ellagic acid; gallic acid; tannic acid; curcumin ferrutinin; epigallocatechin-3-gallate | Curcuminoids: modulate multiple signaling pathways (Wnt/β-catenin, PI3K/Akt) to support osteoblast proliferation and inhibit osteoclastogenesis<br>Tannins: exhibit strong antioxidant and chelating properties, modulate osteoclast differentiation, and support bone matrix formation |

2025). Another example of advanced biomaterials are photothermal-based strategies that can be used simultaneously for inhibiting bone tumor and enhancing of the bone regenerative properties (Ilyas et al., 2024; Hou et al., 2025; Li et al., 2022b; Xv et al., 2025; Dahiya et al., 2024; Fang et al., 2024; Jin et al., 2023; Liu et al., 2024). Overall, biomaterials in bone regeneration have been in highlight for a few years now, but more research is needed to unlock their full potential in bone regeneration.

# 7 Monophenolic Derivatives

Non-steroidal anti-inflammatory drugs (NSAIDs) in general are known for their use as antipyretic, analgesic, and anti-inflammatory class of drugs. NSAIDs, like salicylic acid, diflunisal, and aspirin, are phenolic derivatives that have been investigated for their role in bone regeneration, both as standalone agents and as biomaterial components, owing to their intrinsic bioactive properties (Hendrix et al., 2016; Liu et al., 2017; Pountos et al., 2011). These compounds demonstrate immunomodulatory effects by promoting osteoblastogenesis while inhibiting osteoclast activity (Fig. 4). The complex modulation of effects is mostly controlled through the inhibition of M1 macrophage polarization via NF-κB and RANKL signaling pathways (Kotake et al., 2010). Furthermore, NSAIDs were shown to stimulate MSCs and

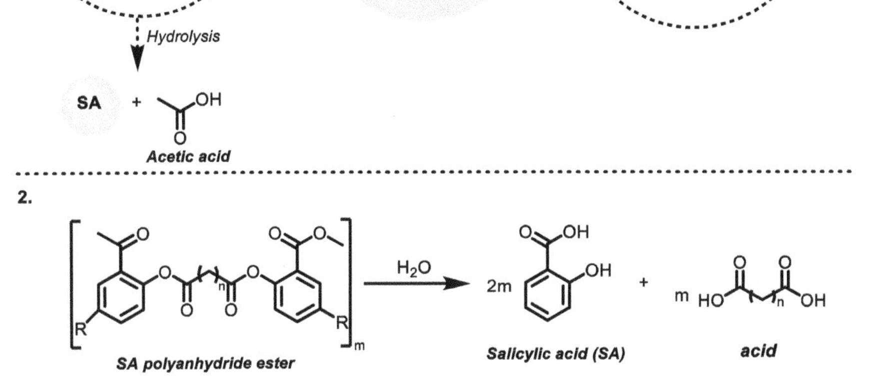


**Fig. 4** NSAIDs used in bone regeneration: (1) NSAIDs (aspirin, salicylic acid, and diflunisal) as phenol and its derivatives, useful in bone regeneration; (2) NSAID biomaterials upon hydrolysis release salicylic acid and carboxylic acid to impart activity

promote osteogenesis through the activation of RUNX2 and Wnt pathways (Kotake et al., 2010). The enhancement of bone healing and the suppression of bone resorption by these molecules have been demonstrated in multiple biomaterial composites, such as aspirin-functionalized mineral particle-coated micro-ribbon scaffolds, MSC-seeded constructs, and calcium phosphate (CaP) hydrogels (Su et al., 2023; Zhu et al., 2022). Several studies have shown that aspirin can regulate bone homeostasis, inhibit osteoclastogenesis in dendritic cells, and promote bone regeneration (Su et al., 2023; Reynolds et al., 2007; Wu et al., 2019). Other phenol-based scaffolds are under investigation for their bone regeneration properties, such as berberine, an isoquinoline alkaloid, and functionalized phenolic compounds, such as flavanones galloyl conjugates (Chen et al., 2023; Lee et al., 2023; Raj Pandey et al., 2021). The integration of these bioactive phenolic derivatives into biomaterial-based therapies hold significant potential for enhancing bone regeneration while preserving their beneficial pharmacological properties, a concept currently under investigation.

## 8 Conclusion

The intrinsic properties of phenolic compounds make them promising candidates for bone regeneration and homeostatic restoration. The concept of their integration into biomaterials has gained significant attention due to their synergistic potential in enhancing bone regeneration through bioactive and structural support mechanisms. Despite their potential, further research is necessary to elucidate their precise mechanisms of action and address any potential limitations related to bioavailability and safety of phenolic compounds, to facilitate clinical translation. Comprehensive investigations into the osteogenic potential of phenolic agents, both individually and in combination, may yield valuable insights, paving the way for novel therapeutic strategies and innovative drug discovery in the field of bone regeneration.

## References

Abdelaziz, A. G., et al. (2023). A review of 3D polymeric scaffolds for bone tissue engineering: Principles, fabrication techniques, immunomodulatory roles, and challenges. *Bioengineering, 10*(2), 204.

Abolarin, P. O., & Owoyele, B. V. (2024). Tannic acid inhibits pain mediators, inflammation and oxidative stress in mice exposed to glyphosate-based herbicide. *Environmental Analysis Health and Toxicology, 39*(2), e2024019.

Ahmad Hairi, H., Jayusman, P. A., & Shuid, A. N. (2023). Revisiting resveratrol as an Osteoprotective agent: Molecular evidence from in vivo and in vitro studies. *Biomedicine, 11*(5), 1453.

Al-Khayri, J. M., et al. (2023). Stilbenes, a versatile class of natural metabolites for inflammation—An overview. *Molecules, 28*(9), 3786.

Alzagameem, A., et al. (2019). Antimicrobial activity of lignin and lignin-derived cellulose and chitosan composites against selected pathogenic and spoilage microorganisms. *Polymers (Basel), 11*(4).

Amini, A. R., Laurencin, C. T., & Nukavarapu, S. P. (2012). Bone tissue engineering: Recent advances and challenges. *Critical Reviews in Biomedical Engineering, 40*(5), 363–408.

Andrés, C. M. C., et al. (2023). Superoxide anion chemistry—Its role at the Core of the innate immunity. *International Journal of Molecular Sciences, 24*(3), 1841.

Atcha, H., et al. (2021). Mechanically activated ion channel Piezo1 modulates macrophage polarization and stiffness sensing. *Nature Communications, 12*(1), 3256.

Baghbani, F., et al. (2012). Biological response of biphasic hydroxyapatite/Tricalcium phosphate scaffolds intended for low load-bearing orthopaedic applications. *Advanced Composites Letters, 21*(1), 096369351202100102.

Balagangadharan, K., et al. (2019). Sinapic acid-loaded chitosan nanoparticles in polycaprolactone electrospun fibers for bone regeneration in vitro and in vivo. *Carbohydrate Polymers, 216*, 1–16.

Baldo, B. A. (2022). Immune- and non-immune-mediated adverse effects of monoclonal antibody therapy: A survey of 110 approved antibodies. *Antibodies, 11*(1), 17.

Battafarano, G., et al. (2021). Strategies for bone regeneration: From graft to tissue engineering. *International Journal of Molecular Sciences, 22*(3).

Benavides-Castellanos, M. P., Garzón-Orjuela, N., & Linero, I. (2020). Effectiveness of mesenchymal stem cell-conditioned medium in bone regeneration in animal and human models: A systematic review and meta-analysis. *Cell Regeneration, 9*(1), 5.

Bhat, K. P., et al. (2001). Estrogenic and antiestrogenic properties of resveratrol in mammary tumor models. *Cancer Research, 61*(20), 7456–7463.

Bo, S., et al. (2018). Effects of resveratrol on bone health in type 2 diabetic patients. A double-blind randomized-controlled trial. *Nutrition and Diabetes, 8*(1).

Camal Ruggieri, I. N., et al. (2024). Biocompatibility and bone regeneration with elastin-like recombinamer-based catalyst-free click gels. *Scientific Reports, 14*(1), 20223.

Cao, B., et al. (2025). Salvianolic acid A promotes bone-fracture healing via balancing osteoblast and osteoclast differentiation. *The FASEB Journal, 39*(2), e70364.

Checinska, K., et al. (2022). Polyphenol-enriched composite bone regeneration materials: A systematic review of in vitro studies. *International Journal of Molecular Sciences, 23*(13).

Chen, J.-R., et al. (2010). Dietary-induced serum phenolic acids promote bone growth via p38 MAPK/β-catenin canonical Wnt signaling. *Journal of Bone and Mineral Research, 25*(11), 2399–2411.

Chen, S., et al. (2021). Curcumin modulates the crosstalk between macrophages and bone mesenchymal stem cells to ameliorate Osteogenesis. *Frontiers in Cell and Developmental Biology,* **9**, 9.

Chen, L., et al. (2023). Berberine-encapsulated poly(lactic-co-glycolic acid)–hydroxyapatite (PLGA/HA) microspheres synergistically promote bone regeneration with DOPA-IGF-1 via the IGF-1R/PI3K/AKT/mTOR pathway. *International Journal of Molecular Sciences, 24*(20), 15403.

Chen, Y., et al. (2024a). Plant molecules reinforce bone repair: Novel insights into phenol-modified bone tissue engineering scaffolds for the treatment of bone defects. *Materials Today Bio, 24*, 100920.

Chen, Q., et al. (2024b). Novel drug delivery systems: An important direction for drug innovation research and development. *Pharmaceutics, 16*(5).

Chen, X., et al. (2025). Regulation of bone remodeling by metal–phenolic networks for the treatment of systemic osteoporosis. *ACS Applied Materials & Interfaces, 17*(4), 5995–6008.

Choi, E. M., et al. (2016). Glabridin alleviates the toxic effects of methylglyoxal on osteoblastic MC3T3-E1 cells by increasing expression of the glyoxalase system and Nrf2/HO-1 signaling and protecting mitochondrial function. *Journal of Agricultural and Food Chemistry, 64*(1), 226–235.

Cory, H., et al. (2018). The role of polyphenols in human health and food systems: A mini-review. *Frontiers in Nutrition, 5*, 5.

Croes, M., van der Wal, B. C. H., & Vogely, H. C. (2019). Impact of bacterial infections on Osteogenesis: Evidence from in vivo studies. *Journal of Orthopaedic Research, 37*(10), 2067–2076.

Dahiya, A., Chaudhari, V. S., & Bose, S. (2024). Bone healing via Carvacrol and curcumin nanoparticle on 3D printed scaffolds. *Small, 20*(50), e2405642.

D'Archivio, M., et al. (2007). Polyphenols, dietary sources and bioavailability. *Annali dell'Istituto Superiore di Sanità, 43*(4), 348–361.

Diel, I. J., Bergner, R., & Grötz, K. A. (2007). Adverse effects of bisphosphonates: Current issues. *The Journal of Supportive Oncology, 5*(10), 475–482.

Dimitriou, R., et al. (2011). Bone regeneration: Current concepts and future directions. *BMC Medicine, 9*(1), 66.

Faisal, Z., et al. (2024). Exploring the multimodal health-promoting properties of resveratrol: A comprehensive review. *Food Science & Nutrition, 12*(4), 2240–2258.

Fang, X., et al. (2024). Polyphenol-mediated redox-active hydrogel with H2S gaseous-bioelectric coupling for periodontal bone healing in diabetes. *Nature Communications, 15*(1), 9071.

Feng, Y.-L., et al. (2017). Resveratrol prevents osteoporosis by upregulating FoxO1 transcriptional activity. *International Journal of Molecular Medicine, 41*, 202.

Gao, C., et al. (2017). Bone biomaterials and interactions with stem cells. *Bone Research, 5*(1), 17059.

Geevarghese, R., et al. (2022). Biodegradable and non-biodegradable biomaterials and their effect on cell differentiation. *International Journal of Molecular Sciences, 23*(24).

Gehm, B. D., et al. (1997). Resveratrol, a polyphenolic compound found in grapes and wine, is an agonist for the estrogen receptor. *Proceedings of the National Academy of Sciences, 94*(25), 14138–14143.

Ghavimi, M. A., et al. (2020). Nanofibrous asymmetric collagen/curcumin membrane containing aspirin-loaded PLGA nanoparticles for guided bone regeneration. *Scientific Reports, 10*(1), 18200.

Gillman, C. E., & Jayasuriya, A. C. (2021). FDA-approved bone grafts and bone graft substitute devices in bone regeneration. *Materials Science and Engineering: C, 130*, 112466.

Ginwala, R., et al. (2019). Potential role of flavonoids in treating chronic inflammatory diseases with a special focus on the anti-inflammatory activity of Apigenin. *Antioxidants (Basel), 8*(2).

Girón, J., et al. (2021). Biomaterials for bone regeneration: An orthopedic and dentistry overview. *Brazilian Journal of Medical and Biological Research, 54*(9), e11055.

Gong, W., et al. (2022). Orcinol glucoside improves senile osteoporosis through attenuating oxidative stress and autophagy of osteoclast via activating Nrf2/Keap1 and mTOR signaling pathway. *Oxidative Medicine and Cellular Longevity, 2022*, 5410377.

Gorabi, A. M., et al. (2019). The effect of curcumin on the differentiation of mesenchymal stem cells into mesodermal lineage. *Molecules, 24*(22), 4029.

Hanga-Farcaş, A., et al. (2023). Phytochemical compounds involved in the bone regeneration process and their innovative administration: A systematic review. *Plants (Basel), 12*(10).

Hendrix, A. S., et al. (2016). Repurposing the nonsteroidal anti-inflammatory drug Diflunisal as an Osteoprotective, Antivirulence therapy for Staphylococcus aureus osteomyelitis. *Antimicrobial Agents and Chemotherapy, 60*(9), 5322–5330.

Hobbi, P., et al. (2023). Fabrication of bioactive polyphenolic biomaterials for bone tissue engineering. *Materials Today Sustainability, 24*, 100541.

Hou, X., et al. (2022). Calcium phosphate-based biomaterials for bone repair. *Journal of Functional Biomaterials, 13*(4).

Hou, X., et al. (2025). Photothermal switch by gallic acid-calcium grafts synthesized by coordination chemistry for sequential treatment of bone tumor and regeneration. *Biomaterials, 312*, 122724.

Huang, Z., et al. (2018). Icariin regulates the osteoblast differentiation and cell proliferation of MC3T3-E1 cells through microRNA-153 by targeting runt-related transcription factor 2. *Experimental and Therapeutic Medicine, 15*, 5159.

Iaquinta, M. R., et al. (2019). Adult stem cells for bone regeneration and repair. *Frontiers in Cell and Developmental Biology, 7*, 7.

Ilyas, S., Lee, J., & Lee, D. (2024). Emerging roles of natural compounds in osteoporosis: Regulation, molecular mechanisms and bone regeneration. *Pharmaceuticals (Basel), 17*(8).

Inchingolo, A. D., et al. (2022). Effects of resveratrol, curcumin and quercetin supplementation on bone metabolism—A systematic review. *Nutrients, 14*(17), 3519.

Iqbal, I., et al. (2023). Plant polyphenols and their potential benefits on cardiovascular health: A review. *Molecules, 28*(17), 6403.

Jiang, M., et al. (2019). A novel Rhein derivative modulates bone formation and resorption and ameliorates estrogen-dependent bone loss. *Journal of Bone and Mineral Research, 34*(2), 361–374.

Jiang, D., et al. (2024). Electrospun lignin-loaded artificial periosteum for bone regeneration and elimination of bacteria. *International Journal of Biological Macromolecules, 282*(Pt 5), 137149.

Jin, S., et al. (2023). Plant-derived polyphenol and LL-37 peptide-modified Nanofibrous scaffolds for promotion of antibacterial activity, anti-inflammation, and type-H vascularized bone regeneration. *ACS Applied Materials & Interfaces, 15*(6), 7804–7820.

John, M. K., et al. (2013). Development and pharmacokinetic evaluation of a curcumin co-solvent formulation. *Anticancer Research, 33*(10), 4285–4291.

Karthikeyan, A., Senthil, N., & Min, T. (2020). Nanocurcumin: A promising candidate for therapeutic applications. *Frontiers in Pharmacology, 11*, 11.

Kopustinskiene, D. M., et al. (2020). Flavonoids as anticancer agents. *Nutrients, 12*, 2.

Kotake, S., et al. (2010). Effects of NSAIDs on differentiation and function of human and murine osteoclasts – Crucial 'human osteoclastology'. *Pharmaceuticals (Basel), 3*(5), 1394–1410.

Kumar, N., & Goel, N. (2019). Phenolic acids: Natural versatile molecules with promising therapeutic applications. *Biotechnology Reports (Amsterdam, Netherlands), 24*, e00370.

Kumar, S., & Pandey, A. K. (2013). Chemistry and biological activities of flavonoids: An overview. *The Scientific World Journal, 2013*(1), 1–16.

Laha, D., et al. (2022). Polyphenolic compounds inhibit osteoclast differentiation while reducing autophagy through limiting ROS and the mitochondrial membrane potential. *Biomolecules, 12*(9).

Lee, E. H., et al. (2021). Sinapic acid controls inflammation by suppressing NLRP3 Inflammasome activation. *Cells, 10*(9).

Lee, S., et al. (2023). Hydroxyapatite microbeads containing BMP-2 and quercetin fabricated via electrostatic spraying to encourage bone regeneration. *Biomedical Engineering Online, 22*(1), 15.

Li, J., et al. (2019). Paeonol attenuates ligation-induced periodontitis in rats by inhibiting osteoclastogenesis via regulating Nrf2/NF-κB/NFATc1 signaling pathway. *Biochimie, 156*, 129–137.

Li, X., et al. (2020a). Curcumin protects osteoblasts from oxidative stress-induced dysfunction via GSK3β-Nrf2 signaling pathway. *Frontiers in Bioengineering and Biotechnology, 8*, 8.

Li, R., et al. (2020b). 3D hybrid nanofiber aerogels combining with nanoparticles made of a biocleavable and targeting Polycation and MiR-26a for bone repair. *Advanced Functional Materials, 30*(49).

Li, Y., et al. (2022a). A mechanistic review of Chinese medicine polyphenols on bone formation and resorption. *Frontiers in Pharmacology, 13*.

Li, Y., et al. (2022b). Recent advances in the development and antimicrobial applications of metal-phenolic networks. *Advanced Science (Weinh), 9*(27), e2202684.

Li, Y., et al. (2024). Research Progress and Prospect analysis of the application of flax Lignans. *Journal of Natural Fibers, 21*(1).

Likhtenshtein, G., Stilbenes preparation and analysis. 2009. 1–41.

Lin, X., et al. (2020). The bone extracellular matrix in bone formation and regeneration. *Frontiers in Pharmacology, 11*.
Liu, Y., et al. (2017). Aspirin inhibits LPS-induced macrophage activation via the NF-κB pathway. *Scientific Reports, 7*(1).
Liu, X., et al. (2023). Salvianolic acid C promotes osteogenic differentiation of bone marrow mesenchymal stem cells in osteoporotic rats through activation of AMPK/SIRT1 pathway. *International Journal of Rheumatic Diseases, 26*(8), 1571–1578.
Liu, Z., et al. (2024). Metal–phenolic networks-reinforced extracellular matrix scaffold for bone regeneration via combining radical-scavenging and photo-responsive regulation of microenvironment. *Advanced Healthcare Materials, 13*(15).
Lopez-Corona, A. V., et al. (2022). Antioxidant, anti-inflammatory and cytotoxic activity of phenolic compound family extracted from raspberries (Rubus idaeus): A general review. *Antioxidants, 11*(6), 1192.
Lutzweiler, G., Ndreu Halili, A., & Engin Vrana, N. (2020). The overview of porous, bioactive scaffolds as instructive biomaterials for tissue regeneration and their clinical translation. *Pharmaceutics, 12*(7).
Lv, Y. J., et al. (2018). Resveratrol counteracts bone loss via mitofilin-mediated osteogenic improvement of mesenchymal stem cells in senescence-accelerated mice. *Theranostics, 8*(9), 2387–2406.
Lyons, J. G., et al. (2020). Nanostructured biomaterials for bone regeneration. *Frontiers in Bioengineering and Biotechnology, **8***, 8.
Machado, I. F., et al. (2023). Targeting oxidative stress with polyphenols to fight liver diseases. *Antioxidants, 12*(6), 1212.
Maity, J., et al. (2022). Ferutinin induces osteoblast differentiation of DPSCs via induction of KLF2 and autophagy/mitophagy. *Cell Death & Disease, 13*(5), 452.
Manach, C., et al. (2005). Bioavailability and bioefficacy of polyphenols in humans. I. Review of 97 bioavailability studies. *The American Journal of Clinical Nutrition, 81*(1 Suppl), 230s–242s.
Marcadet, L., et al. (2022). The roles of RANK/RANKL/OPG in cardiac, skeletal, and smooth muscles in health and disease. *Frontiers in Cell and Development Biology, 10*, 903657.
Matulja, D., et al. (2022). Anticancer activities of marine-derived phenolic compounds and their derivatives. *Molecules, 27*(4), 1449.
Meng, T., et al. (2021). Anti-inflammatory action and mechanisms of resveratrol. *Molecules, 26*(1), 229.
Miranda, L. L., et al. (2019). Plant extracts in the bone repair process: A systematic review. *Mediators of Inflammation, 2019*, 1–22.
Moon, D. K., et al. (2020). Resveratrol can enhance osteogenic differentiation and mitochondrial biogenesis from human periosteum-derived mesenchymal stem cells. *Journal of Orthopaedic Surgery and Research, 15*(1), 203.
Mukasheva, F., et al. (2024). Optimizing scaffold pore size for tissue engineering: Insights across various tissue types. *Frontiers in Bioengineering and Biotechnology, 12*, 12.
Murgia, D., et al. (2019). Advance on resveratrol application in bone regeneration: Progress and perspectives for use in Oral and maxillofacial surgery. *Biomolecules, 9*(3), 94.
Naidu, P., et al. (2024). Ellagic acid (EA) promotes osteoblastic differentiation of dental pulp-derived stem cells (DPSCs) by upregulating KLF2 and by inducing autophagy/mitophagy. *The Journal of Pharmacology and Experimental Therapeutics, 389*, 365–366.
Nicolin, V., et al. (2019). Modulatory effects of plant polyphenols on bone remodeling: A prospective view from the bench to bedside. *Frontiers in Endocrinology, 10*.
Panche, A. N., Diwan, A. D., & Chandra, S. R. (2016). Flavonoids: An overview. *Journal of Nutritional Science, **5***, 5.
Pereira, D. M., et al. (2009). Phenolics: From chemistry to biology. *Molecules, 14*(6), 2202–2211.
Phiwchai, I., et al. (2018). Tannic acid (TA): A molecular tool for chelating and imaging labile iron. *European Journal of Pharmaceutical Sciences, 114*, 64–73.

Pofi, R., et al. (2023). Treating the side effects of exogenous glucocorticoids; can we separate the good from the bad? *Endocrine Reviews, 44*(6), 975–1011.

Pountos, I., et al. (2011). NSAIDS inhibit in vitro MSC chondrogenesis but not osteogenesis: Implications for mechanism of bone formation inhibition in man. *Journal of Cellular and Molecular Medicine, 15*(3), 525–534.

Raj Pandey, A., et al. (2021). Synthesis and evaluation of Galloyl conjugates of flavanones as BMP-2 Upregulators with promising bone anabolic and fracture healing properties. *Journal of Medicinal Chemistry, 64*(17), 12487–12505.

Reynolds, M. A., et al. (2007). Non-steroidal anti-inflammatory drug (NSAID)-derived poly(anhydride-esters) in bone and periodontal regeneration. *Current Drug Delivery, 4*(3), 233–239.

Roy, A., et al. (2022). Flavonoids a bioactive compound from medicinal plants and its therapeutic applications. *BioMed Research International, 2022*(1), 1–9.

Sacco, S. M., et al. (2009). Flaxseed combined with low-dose estrogen therapy preserves bone tissue in ovariectomized rats. *Menopause, 16*(3), 545–554.

Sarkar, J., et al. (2022). Epigallocatechin-3-gallate inhibits osteoclastic differentiation by modulating mitophagy and mitochondrial functions. *Cell Death and Disease, 13*(10).

Sato, M., et al. (2020). Bone regeneration potential of human dental pulp stem cells derived from elderly patients and osteo-induced by a Helioxanthin derivative. *International Journal of Molecular Sciences, 21*(20).

Schmidt, A. H. (2021). Autologous bone graft: Is it still the gold standard? *Injury, 52*, S18–S22.

Seebach, E., & Kubatzky, K. F. (2019). Chronic implant-related bone infections—Can immune modulation be a therapeutic strategy? *Frontiers in Immunology, 10*.

Shanmugavadivu, A., Balagangadharan, K., & Selvamurugan, N. (2022). Angiogenic and osteogenic effects of flavonoids in bone regeneration. *Biotechnology and Bioengineering, 119*(9), 2313–2330.

Sharifi, S., et al. (2020). Phytochemicals impact on osteogenic differentiation of mesenchymal stem cells. *BioFactors, 46*(6), 874–893.

Sharma, S., et al. (2023). Microbial biofilm: A review on formation, infection, antibiotic resistance, control measures, and innovative treatment. *Microorganisms, 11*(6), 1614.

Silva-Barroso, A. S., et al. (2023). Lignin-enriched tricalcium phosphate/sodium alginate 3D scaffolds for application in bone tissue regeneration. *International Journal of Biological Macromolecules, 239*, 124258.

Sohn, S. I., et al. (2021). Biomedical applications and bioavailability of curcumin-an updated overview. *Pharmaceutics, 13*(12).

Su, N., et al. (2023). Aspirin synergizes with mineral particle-coated macroporous scaffolds for bone regeneration through immunomodulation. *Theranostics, 13*(13), 4512–4525.

Suvarna, V., et al. (2018). Bone health and natural products- an insight. *Frontiers in Pharmacology, 9*.

Szwed-Georgiou, A., et al. (2023). Bioactive materials for bone regeneration: Biomolecules and delivery systems. *ACS Biomaterials Science & Engineering, 9*(9), 5222–5254.

Takó, M., et al. (2020a). Plant Phenolics and phenolic-enriched extracts as antimicrobial agents against food-contaminating microorganisms. *Antioxidants, 9*(2), 165.

Takó, M., et al. (2020b). Plant Phenolics and phenolic-enriched extracts as antimicrobial agents against food-contaminating microorganisms. *Antioxidants (Basel), 9*(2).

Tao, H., et al. (2021). Urolithin a suppresses RANKL-induced osteoclastogenesis and postmenopausal osteoporosis by, suppresses inflammation and downstream NF-κB activated pyroptosis pathways. *Pharmacological Research, 174*, 105967.

Thilakarathna, S., & Rupasinghe, H. (2013). Flavonoid bioavailability and attempts for bioavailability enhancement. *Nutrients, 5*(9), 3367–3387.

Thummuri, D., Naidu, V. G. M., & Chaudhari, P. (2017). Carnosic acid attenuates RANKL-induced oxidative stress and osteoclastogenesis via induction of Nrf2 and suppression of NF-κB and MAPK signalling. *Journal of Molecular Medicine (Berlin, Germany), 95*(10), 1065–1076.

Tomaszewska, E., et al. (2017). The effect of tannic acid on bone mechanical and geometric properties, bone density, and trabecular histomorphometry as well as the morphology of articular and growth cartilages in rats co-exposed to cadmium and lead is dose dependent. *Toxicology and Industrial Health, 33*(11), 855–866.

Torre, E. (2017). Molecular signaling mechanisms behind polyphenol-induced bone anabolism. *Phytochemistry Reviews, 16*(6), 1183–1226.

Valentino, A., et al. (2021). Bioactivity and delivery strategies of phytochemical compounds in bone tissue regeneration. *Applied Sciences, 11*(11), 5122.

Wang, K. (2024). The potential therapeutic role of curcumin in osteoporosis treatment: Based on multiple signaling pathways. *Frontiers in Pharmacology, 15*.

Wang, D., et al. (2019). "Tree to bone": Lignin/Polycaprolactone nanofibers for hydroxyapatite biomineralization. *Biomacromolecules, 20*(7), 2684–2693.

Wang, Y., et al. (2022a). Resveratrol synergistically promotes BMP9-induced osteogenic differentiation of mesenchymal stem cells. *Stem Cells International, 2022*, 1–13.

Wang, N., et al. (2022b). Mg-, Zn-, and Fe-based alloys with antibacterial properties as orthopedic implant materials. *Frontiers in Bioengineering and Biotechnology, 10*, 888084.

Wang, J., et al. (2023). The role of autophagy in bone metabolism and clinical significance. *Autophagy, 19*(9), 2409–2427.

Wang, J., et al. (2024). Application of antioxidant compounds in bone defect repair. *Antioxidants, 13*(7), 789.

Wang, Y., et al. (2025). Metal-phenolic network biointerface-mediated cell regulation for bone tissue regeneration. *Materials Today Bio, 30*, 101400.

Wei, L., et al. (2023). Resveratrol protects osteocytes against oxidative stress in ovariectomized rats through AMPK/JNK1-dependent pathway leading to promotion of autophagy and inhibition of apoptosis. *Cell Death Discovery, 9*(1), 16.

Weinstein, R. S. (2012). Glucocorticoid-induced osteoporosis and osteonecrosis. *Endocrinology and Metabolism Clinics of North America, 41*(3), 595–611.

Wilken, R., et al. (2011). Curcumin: A review of anti-cancer properties and therapeutic activity in head and neck squamous cell carcinoma. *Molecular Cancer, 10*(1), 12.

Winiarska-Mieczan, A., et al. (2023). The impact of tannic acid consumption on bone mineralization. *Metabolites, 13*(10), 1072.

Wu, L., et al. (2019). Aspirin inhibits RANKL-induced osteoclast differentiation in dendritic cells by suppressing NF-κB and NFATc1 activation. *Stem Cell Research and Therapy, 10*(1).

Wu, Z., et al. (2023). Promotion of osteoporotic bone healing by a tannic acid modified strontium-doped biomimetic bone lamella with ROS scavenging capacity and pro-osteogenic effect. *Smart Materials in Medicine, 4*, 590–602.

Xiao, L., & Xiao, Y. (2019). The autophagy in Osteoimmonology: Self-eating, maintenance, and beyond. *Frontiers in Endocrinology, 10*.

Xv, D., et al. (2025). Polyphenols and functionalized hydrogels for osteoporotic bone regeneration. *Macromolecular Rapid Communications, 46*(2), e2400653.

Yamaguchi, M., & Weitzmann, M. N. (2012). The bone anabolic carotenoid p-hydroxycinnamic acid promotes osteoblast mineralization and suppresses osteoclast differentiation by antagonizing NF-κB activation. *International Journal of Molecular Medicine, 30*(3), 708–712.

Yamaguchi, M., et al. (2013). Bioactive flavonoid p-hydroxycinnamic acid stimulates osteoblastogenesis and suppresses adipogenesis in bone marrow culture. *Cell and Tissue Research, 354*(3), 743–750.

Yamakawa, D., et al. (2020). Effects of Helioxanthin derivative-treated human dental pulp stem cells on fracture healing. *International Journal of Molecular Sciences, 21*(23).

Yang, S., et al. (2023). Protective effects of curcumin against osteoporosis and its molecular mechanisms: A recent review in preclinical trials. *Frontiers in Pharmacology, 14*.

Yasuhisa, T., Hideki, H., & Muneyoshi, Y. (1993). Superoxide radical scavenging activity of phenolic compounds. *International Journal of Biochemistry, 25*(4), 491–494.

Yin, X., et al. (2019). Autophagy in bone homeostasis and the onset of osteoporosis. *Bone Research, 7*(1), 28.
Younger, E. M., & Chapman, M. W. (1989). Morbidity at bone graft donor sites. *Journal of Orthopaedic Trauma, 3*(3), 192–195.
Yu, X., et al. (2015). Biomaterials for bone regenerative engineering. *Advanced Healthcare Materials, 4*(9), 1268–1285.
Zagoskina, N. V., et al. (2023). Polyphenols in plants: Structure, biosynthesis, abiotic stress regulation, and practical applications (review). *International Journal of Molecular Sciences, 24*(18).
Zeb, A. (2020). Concept, mechanism, and applications of phenolic antioxidants in foods. *Journal of Food Biochemistry, 44*(9), e13394.
Zhang, Y., et al. (2020). Stem cell-friendly scaffold biomaterials: Applications for bone tissue engineering and regenerative medicine. *Frontiers in Bioengineering and Biotechnology*, 8.
Zhang, Y., et al. (2022a). Bisphenol a induces pyroptotic cell death via ROS/NLRP3/Caspase-1 pathway in osteocytes MLO-Y4. *Food and Chemical Toxicology, 159*, 112772.
Zhang, P., et al. (2022b). Gallic acid inhibits osteoclastogenesis and prevents ovariectomy-induced bone loss. *Frontiers in Endocrinology, 13*.
Zhao, R., et al. (2021). Bone grafts and substitutes in dentistry: A review of current trends and developments. *Molecules, 26*(10).
Zhou, T., et al. (2019). Resveratrol improves osteogenic differentiation of senescent bone mesenchymal stem cells through inhibiting endogenous reactive oxygen species production via AMPK activation. *Redox Report, 24*(1), 62–69.
Zhu, W., et al. (2022). Fabrication of naturally derived double-network hydrogels with a sustained aspirin release system for facilitating bone regeneration. Frontiers. *Chemistry, **10***, 10.
Zou, X., Xu, H., & Qian, W. (2024). The role and current research status of resveratrol in the treatment of osteoarthritis and its mechanisms: A narrative review. *Drug Metabolism Reviews, 56*(4), 399–412.
Zverev, Y. F., & Rykunova, A. Y. (2022). Modern Nanocarriers as a factor in increasing the bioavailability and pharmacological activity of flavonoids. *Applied Biochemistry and Microbiology, 58*(9), 1002–1020.

# Small Molecules in Regenerative Medicine: Advancements, Applications, and Challenges

**Parthiban Chokkalingam, Manoranjan Bera, Sasmita Das, Mariko Saito, and Bhaskar C. Das**

**Abstract** In regenerative medicine, small molecules have become extremely effective tools, providing potential approaches to support cell regeneration, tissue repair, and functional recovery in a range of illnesses. Because of their ability to alter important signaling pathways that control cell survival, proliferation, and differentiation, these compounds are excellent prospects for therapeutic uses. The efficacy of stem cell therapies, gene therapies, and tissue engineering may be improved by recent developments in small molecule-based treatments. Small molecules have opened the door for new methods in bone, cartilage, cardiac, and neuroregeneration by combining them with other regenerative technologies. Despite these developments, obstacles like cost, patient variability, and regulatory barriers still stand in the way of implementing small molecule-based treatments in clinical settings. The creation of tailored treatments, improved drug delivery methods, and the use of small molecules to treat more complicated illnesses are some potential future paths in this area. This review emphasizes how small compounds could revolutionize regenerative medicine and how they will influence future treatment approaches for tissue regeneration and repair.

**Keywords** Small molecules · Regenerative medicine · Cell reprogramming · Tissue engineering

P. Chokkalingam · M. Bera · S. Das · B. C. Das (✉)
Division of Drug and Biotherapeutics Discovery, School of Pharmacy and Pharmaceutical Sciences, University at Buffalo, Buffalo, NY, USA
e-mail: pchokkal@buffalo.edu; manoranj@buffalo.edu; sasmitad@buffalo.edu; bhaskard@buffalo.edu

M. Saito (✉)
Nathan S. Kline Institute for Psychiatric Research, Orangeburg, NY, USA
e-mail: Mariko.Saito@nki.rfmh.org

 
H. Das (ed.), *Tissue Repair and Regeneration*,
https://doi.org/10.1007/978-3-031-93677-7_16

# 1 Introduction

Regenerative medicine has gained significant attention in the past few decades for its potential to repair and regenerate damaged tissues. It is a type of treatment that helps damaged cell repair by utilizing the tissue's own natural healing mechanisms. Regenerative medicines are used to trigger the body's own healing processes, like new blood vessel regeneration and the restoration of damaged tissues (Mao & Mooney, 2015; Jaklenec et al., 2012; Bailey et al., 2014). Regenerative medicine is currently being utilized to treat many ailments, but it has the potential to change how we treat illnesses and injuries thoroughly.

Current advancements in regenerative medicine are driven by the following key technologies (Fig. 1) (Altyar et al., 2023). One of the most essential methods in this area is stem cell treatment, which uses adaptable stem cells from bone marrow, fat, or umbilical cord blood to repair damaged tissues and cure ailments like spinal cord injuries and osteoarthritis. The field of regenerative medicine has experienced significant growth in recent years, driven by remarkable advancements in stem cell (SC) biology and regeneration. Damaged cells, tissues, and organs can be repaired and healed by applying stem cells due to their ability to develop into any cell in the

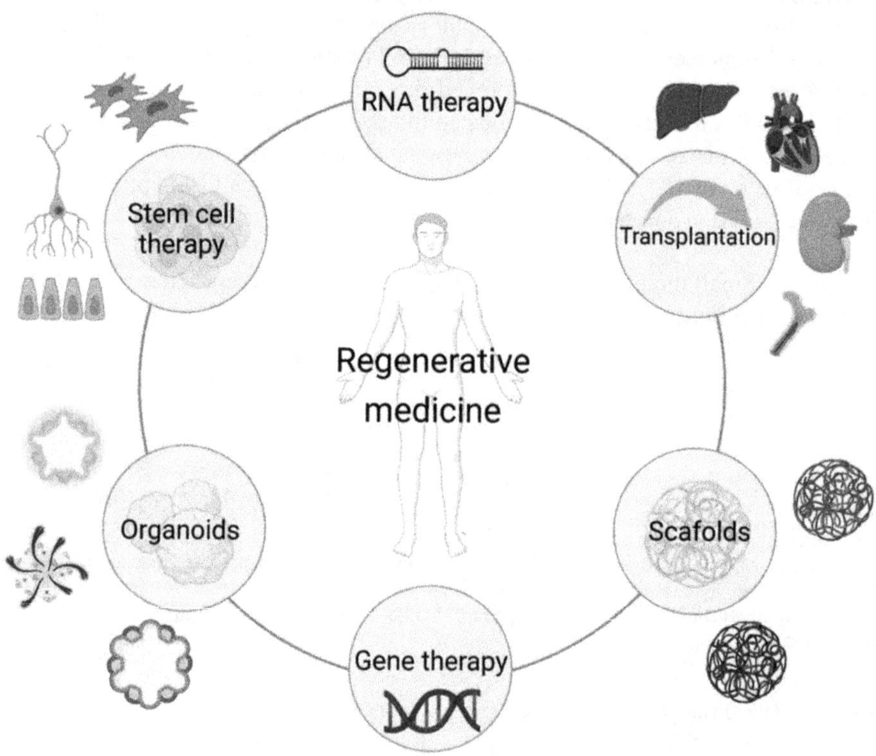


**Fig. 1** Major technologies used in regenerative medicine at present

body (Barry & Murphy, 2013; Cao et al., 2017; Reinders et al., 2010; Rivera-Izquierdo et al., 2019). Another significant area of regenerative medicine is tissue engineering, which is the technique of creating tissues outside of the body. Tissue engineering produces tissues that can regenerate or replace diseased or damaged tissues using growth factors, scaffolds, and cells. By combining cells, engineering techniques, biochemical components, and biomaterials, scientists can create functional tissues such as skin grafts and lab-grown organs (Iatridis et al., 2013; Mason, 2008; Muñoz Ruiz & Regueiro, 2012; Bonifacio et al., 2018; Martinotti et al., 2019; Ng et al., 2020). Injuries to tendons, ligaments, muscles, and joints can be effectively treated with platelet-rich plasma (PRP) therapy, which employs concentrated platelets from the patient's own blood to release growth factors and speed healing (Puskas et al., 2009).

Prolotherapy is a nonsurgical injection technique used to treat arthritis and degenerative disc disease. It involves injecting a dextrose-based irritant solution into painful or degenerated areas like tendons, joints, and ligaments to promote tissue healing. Other ingredients may include polidocanol, manganese, or human growth hormone (Linetsky & Manchikanti, 2005; Goswami, 2012). Techniques like autologous chondrocyte implantation and microfracture surgery aim to restore joint cartilage (Bonifacio et al., 2018; Irie et al., 2008).

Although stem cells are the main focus of cellular therapies, other cell types such as pancreatic islet cells, fibroblasts, chondrocytes, hepatocytes, immune cells (T cells, dendritic cells, and natural killer cells), and fibroblasts are also used in various cellular therapies, depending on the target disease and intended therapeutic outcome, such as hematopoietic stem cell transplantation for blood diseases (El-Kadiry et al., 2021). Gene therapy can heal hereditary illnesses and promote tissue regeneration by altering or adding genes. Utilizing cells to make a suitable protein, inhibiting the overproduction of proteins, and genetically altering and restoring cell activities that may impact disease situations are the objectives of gene therapy (Hosseinkhani et al., 2023). These innovative therapies show the revolutionary potential of regenerative medicine in treating degenerative diseases and injuries; some are currently in clinical use, while others are in research and clinical trials.

Through their ability to alter biological processes and cellular pathways, small molecules are essential to regenerative medicine because they facilitate tissue repair, regeneration, and functional restoration (Green & Lee, 2013). They are crucial for producing the specialized cells needed for treatments because they affect the behavior of stem cells, including their proliferation, differentiation, and reprogramming. Small molecules improve the healing of damaged tissues, including bones, cartilage, and cardiovascular structures, by promoting endogenous repair mechanisms. They also establish favorable microenvironments by modifying the stem cell niche, encouraging angiogenesis, lowering inflammation, or modifying immune responses. These substances also increase the effectiveness of cellular therapies by enhancing the integration, homing, and survival of transplanted cells. Their capacity to be supplied orally or minimally invasively improves therapeutic practicality, and their cost-effectiveness, scalability, and chemical stability offer significant advantages over protein-based medicines. Additionally, small compounds speed up drug development and the comprehension of regeneration mechanisms by enabling high-throughput screening in research. Small molecules are revolutionizing regenerative medicine by providing precise, reversible, and non-immunogenic solutions that overcome the drawbacks of previous therapies and spur creative methods for successfully treating degenerative illnesses and injuries.

Small molecules are essential to regenerative medicine because they alter biological pathways to encourage tissue repair and regeneration. By influencing stem cell behavior, such as proliferation, differentiation, and reprogramming these low-molecular-weight substances can improve the body's natural healing mechanisms (Fig. 2) (Liu et al., 2016). Reprogramming somatic cells to become induced pluripotent stem cells (iPSCs) is a noteworthy use of small molecules. Certain small compounds can decrease the danger of cancer and increase the effectiveness of reprogramming by imitating transcription factors (TFs) and converting adult cells back into a pluripotent state without undergoing genetic change (Federation et al., 2013). For example, it has been demonstrated that a mixture of small compounds may induce pluripotent stem cells from mouse somatic cells with an induction effectiveness that is on par with conventional methods for producing iPSCs (Shi et al., 2008a).

Furthermore, stem cells' development into particular cell types is guided by small molecules, which is crucial for producing targeted tissues for repair. Additionally, they can improve transplanted cells' survival and assimilation into

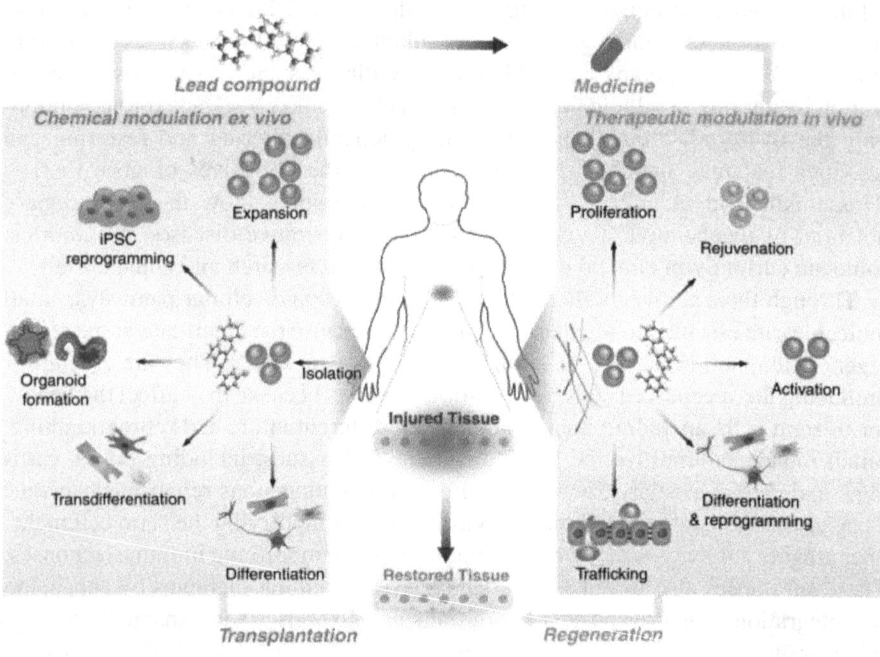


**Fig. 2** Chemical modulation of cell fate in stem cell therapeutics and regenerative medicine. (Liu et al., 2016)

host tissues, boosting the effectiveness of cell-based treatments. By encouraging angiogenesis or modifying immune responses, small compounds can control the microenvironment in tissue engineering to support tissue regeneration (Lukaszewicz et al., 2010). Small molecules' potential in craniofacial regenerative engineering is demonstrated by their application, for instance, in controlling the regeneration of cranial and periodontal bone. Furthermore, compared to protein-based treatments, small molecules have advantages in terms of stability, scalability, and controlled distribution. They are crucial tools in developing regenerative medicine strategies meant to restore function to damaged tissues and organs because they can precisely control biological pathways.

Small molecules' many uses and capacity to alter cellular and molecular processes are revolutionizing regenerative medicine. These substances are critical in directing stem cell differentiation into particular cell types like neurons and cardiomyocytes necessary for mending injured tissues. Additionally, they allow somatic cells to be reprogrammed into induced pluripotent stem cells (iPSCs), providing safer and more effective ways to produce pluripotent cells without undergoing genetic mutation. By focusing on pathways that promote the growth and differentiation of tissue-specific stem cells, small molecules improve tissue repair and regeneration and aid in recovering organs such as the liver, heart, and brain. Furthermore, encouraging angiogenesis, lowering inflammation, and modifying immune responses enhance the regenerative milieu and foster repair (Gao et al., 2024). By mobilizing endogenous stem cells to injury sites and enhancing natural repair mechanisms, these substances improve the survival and integration of transplanted cells in cellular treatments. Beyond treatments, small compounds are essential for drug discovery and disease modeling, especially in stem cell-based platforms, where they help find novel therapeutic targets and promote creativity. Additionally, they are included in scaffolds and biomaterials in tissue engineering to promote cell development and control material degradation, and they are directly delivered in treatments to initiate regenerative pathways, such as bone and heart tissue regeneration. Because of their accuracy, scalability, and affordability, small molecules are essential for developing regenerative medicine and treating complicated degenerative illnesses and injuries.

## 2 Mechanisms of Action

Small molecules are key factors in regenerative medicine because they modulate cellular pathways to improve tissue repair, stem cell differentiation, and organ regeneration. These compounds work through various mechanisms, including activating or inhibiting specific signaling pathways, modifying transcription factors, and interacting with epigenetic regulators. For example, small molecules that target the Wnt, Notch, and Hedgehog pathways can promote stem cell proliferation and differentiation into desired lineages (Takebe et al., 2015). At the same time, epigenetic modifiers such as DNA methyltransferase and histone deacetylase inhibitors can reprogram cells to a pluripotent or progenitor state. Furthermore, small molecules can improve tissue repair by promoting angiogenesis, lowering inflammation, or preventing apoptosis in damaged tissues. They are desirable prospects for creating novel regenerative treatments due to their scalable nature, adjustable specificity, and ease of production.

(a) *Encouragement of Stem Cell Differentiation and Proliferation*

Stem cells, known for their capacity for self-renewal and differentiation into various cell types, are fundamental to regenerative therapy (Fig. 3) (Sultana et al., 2022). It involves transforming stem cells into specialized cell types with specific structures and functions, such as neurons, cardiomyocytes, or hepatocytes. Small molecules act as modulators of key signaling pathways influencing stem cell fate, enhancing their therapeutic potential.

- Wnt/β-Catenin Pathway Activation: The Wnt pathway regulates stem cell self-renewal and differentiation. Small molecules like CHIR99021, a potent glycogen synthase kinase-3β (GSK-3β) inhibitor, promote mesodermal lineage commitment and proliferation (Law & Zheng, 2022). This mechanism is critical for generating cardiomyocytes and other tissue-specific cells for regenerative applications.

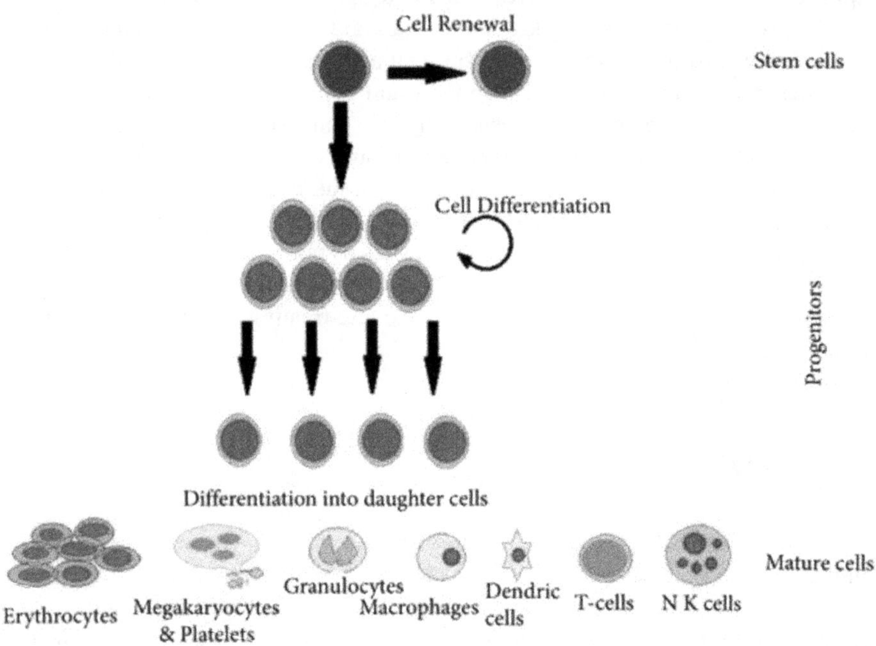


**Fig. 3** Stem cell proliferation shows the ability to self-renew and replicate into mature cells. This is an infinite process, and stem cells are the only ones that can replicate themselves many times, unlike other cells. Unusual proliferation or accelerated cell division can cause leukemia or tumors (Sultana et al., 2022)

- TGF-β and BMP Modulation: Transforming growth factor-β (TGF-β) and bone morphogenetic protein (BMP) signaling is essential for stem cell differentiation. Molecules like SB431542 inhibit TGF-β signaling (Tan et al., 2014), directing mesenchymal stem cells (MSCs) toward osteoblast differentiation. Conversely, dorsomorphin suppresses BMP signaling, favoring neural lineage commitment.
- Epigenetic Modifiers: Histone deacetylase inhibitors (HDACi) (e.g., valproic acid and sodium butyrate) can modify chromatin structure, enhancing the expression of lineage-specific genes (Paino et al., 2014). Such modifications can promote differentiation into cell types like neurons, hepatocytes, or cardiomyocytes, advancing cell-based therapies.

(b) *Induction of Cellular Reprogramming*

Cellular reprogramming is the process of converting one cell type into another, usually a more versatile or desirable cell type, by changing its gene expression and cellular identity. This cutting-edge technique enables the production of patient-specific cells for tissue healing, disease modeling, and medication discovery (Fig. 4) (Mall & Wernig, 2017). Pioneer work by Shinya Yamanaka demonstrated that somatic cells transform into induced pluripotent stem cells (iPSCs) or directly into other functional cell types (Aguirre et al., 2023). These cells have the potential to differentiate into any cell type. This approach circumvents ethical concerns and compatibility issues associated with embryonic stem cells (ESCs).

- Epigenetic Reprogramming: Cellular reprogramming often involves resetting the epigenetic landscape, including DNA methylation and histone modifications, to alter gene expression patterns. Compounds like 5-azacitidine and RG108 inhibit DNA methyltransferases, erasing epigenetic marks and facilitating pluripotency (Pasano Bojang & Ramos, 2013). Thus, these molecules enhance the efficiency and reliability of reprogramming processes.
- Cell Signaling Modulation: Forskolin (diterpene), an adenylate cyclase activator, elevates intracellular cyclic adenosine monophosphate (cAMP) levels, enhancing reprogramming efficiency by activating cyclic AMP-responsive element binding protein (CREB)-mediated transcription (Kapoor et al., 2022). Synergistic combinations of forskolin with other small molecules improve iPSC generation.
- Metabolic Regulation: Cellular metabolism shifts during reprogramming. Small molecules like 2-deoxy-D-glucose favor glycolysis, a hallmark of pluripotency (Yu et al., 2016). These metabolic shifts support the transition from a somatic to a pluripotent state.

(c) *Encouragement of Angiogenesis and Tissue Repair*

Angiogenesis is commonly known for producing new blood vessels from existing ones. It is crucial because developing new blood vessels ensure that oxygen, nutrients, and signaling molecules are delivered to the tissues during regeneration. This important biological process is also involved in wound healing and tissue repair (Fig. 5) (Veith et al., 2019).

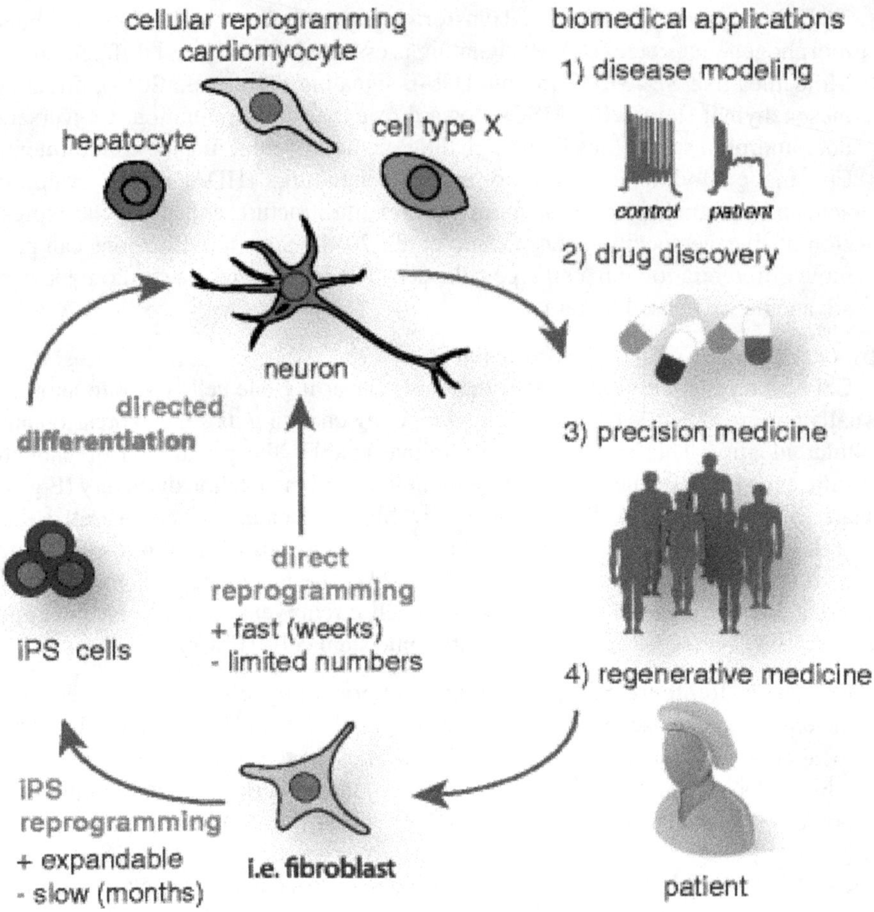


**Fig. 4** Current cellular reprogramming technologies and future biomedical applications (Mall & Wernig, 2017)

- VEGF Pathway Activation: The vascular endothelial growth factor (VEGF) signaling pathway is essential for angiogenesis, which develops new blood vessels from existing ones. Molecules like deferoxamine stabilize hypoxia-inducible factor-1α (HIF-1α), upregulating VEGF expression and fostering vascular growth (Shibuya, 2011).
- Anti-inflammatory Effects: Inflammation is essential to the body's natural healing process; regulating inflammation helps enable successful blood vessel development (angiogenesis) and tissue regeneration. Chronic inflammation impedes tissue repair. Small molecules such as resveratrol and aspirin reduce inflammation, creating a conducive environment for regeneration (Cooke, 2019).
- Stem Cell Homing: One crucial ability of stem cells is to actively navigate to the site of injury and initiate the healing process. Mobilizing stem cells to injury sites is crucial for effective tissue repair. Plerixafor (AMD3100), a CXCR4 antagonist,

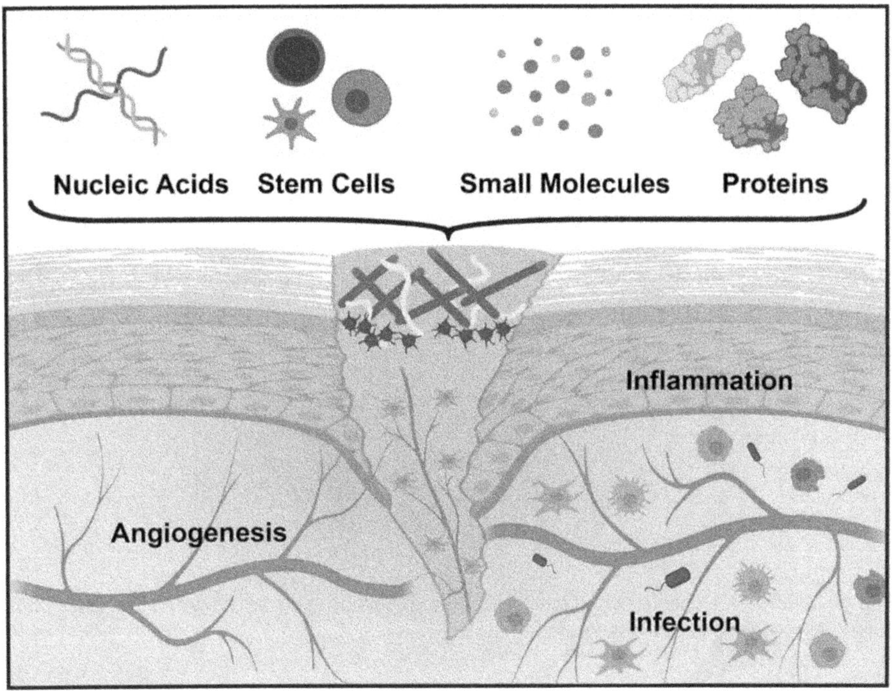


**Fig. 5** Summary diagram of the cellular and molecular components of wound healing and wound angiogenesis (Veith et al., 2019)

disrupts the CXCR4-SDF-1α axis, enhancing stem cell recruitment and tissue regeneration (Dillmann et al., 2009).

(d) *Neural Regeneration and Neuroprotection*

The process by which injured neurons (nerve cells) in the central or peripheral nervous systems (PNS) grow back, fix, or replace lost components, like axons and dendrites, in order to reestablish their functional connections is known as neural regeneration. Regenerating neurons is a significant challenge because the central nervous system (CNS) has a limited regenerative capacity. Neuroprotection involves strategies to prevent or slow neuronal damage or death caused by injury, disease, or degenerative processes (Johnson et al., 2022). Neurodegeneration involves neuronal loss and axonal damage, while immature "N2" neutrophils promote neuroregeneration by secreting growth factors. It is unclear if their effects are direct on neurons or indirect via astrocytes, Müller glia, or microglia. Though observed in the optic nerve and spinal cord, further research is needed to confirm this in other CNS regions and across different life stages (Fig. 6) (Voskuhl, 2020). Small molecules help modulate pathways that support neural repair and protection.

- Neurotrophic Factor Mimicry: Neurotrophic factors are thought to be a component of the treatment plan for neurological conditions such as traumatic brain

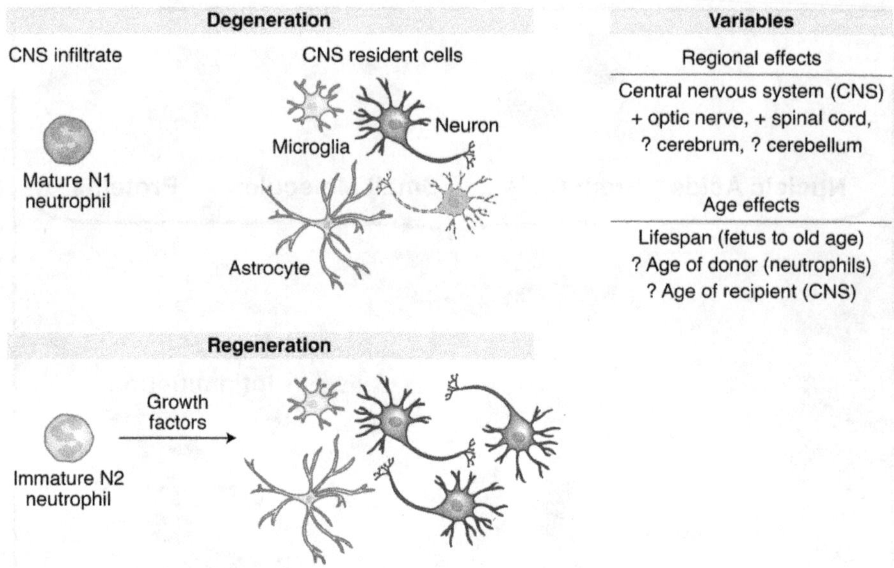


**Fig. 6** Mature "N1" neutrophils confer neurodegeneration, while immature "N2" neutrophils induce neuroregeneration (Voskuhl, 2020)

injury, dementia, and stroke. Cerebrolysin is a neuropeptide that can mimic neurotrophic factors to promote neuronal survival and axonal growth.

- Inhibition of Neural Scarring: In the central nervous system (CNS), glial scars are thought to be a significant barrier to axonal regeneration because they create inhibitory molecules that stop injured axons from growing back and reestablishing contact with their targets; in other words, they essentially act as a physical barrier that impedes the regeneration process. Small molecules like Y-27632 inhibit Rho-associated protein kinase (ROCK), reducing scar formation and supporting neural repair.
- Enhancement of Synaptic Plasticity: Synaptic plasticity is an essential mechanism that governs learning, memory, and recovery from brain damage. Synaptic plasticity enhancement aims to improve the brain's ability to adjust and adapt synaptic strength or efficacy in response to activity, learning, or damage (Lauterborn et al., 2008). Ampakines and other compounds improve synaptic function and plasticity, aiding recovery in damaged brain regions.

(e) *Modulation in the Immune System*

Immunomodulators can alter our body's immunological response. Effective regenerative therapies often require immunomodulation to mitigate inflammation and prevent transplant rejection. Small molecules can serve as crucial agents in this context. The use of biomaterials and the regeneration of all tissues are frequently hampered by the immune response. This immune response can be triggered by cell transplantation, biomaterial scaffold implantation, or inductive factor delivery.

Although the immune response serves several essential functions, including eliminating injury-induced cellular debris and preventing infection, the first inflammatory reaction to the injury may result in further tissue damage (1), which may impede regeneration. It has been demonstrated that preventing macrophage infiltration results in more severe tissue damage and a reduced capacity for regeneration (2). To improve regeneration, techniques that maximize the immune response's positive elements while minimizing its potentially harmful ones may need to be developed (Donnelly & Popovich, 2008).

- Suppression of T cell Activation: A vital part of adaptive immunity is T cell activation, which can cause immunological rejection or persistent inflammation, impairing the integration and performance of designed biomaterials or transplanted tissues. Regenerative medicines can foster a more lenient environment for tissue regeneration and repair by particularly inhibiting T cell activation (Giacomini et al., 2023). For example, cyclosporine and other calcineurin inhibitors suppress T cell activation, minimizing immune-mediated tissue damage and reducing tissue rejection. Cells such as mesenchymal stem cells (MSCs) can actively inhibit T cell activation and proliferation through a variety of methods, including cytokine release and direct cell-to-cell contact (Fig. 7) (Moheb-Alian et al., 2016).
- Macrophage Polarization: Macrophage polarization, the ability of macrophages to transition between pro-inflammatory (M1) and anti-inflammatory (M2) phenotypes, affecting tissue repair and regeneration depending on the surrounding environment, makes it essential in regenerative medicine (Brown et al., 2012). Polarizing macrophages to the M2 phenotype using interleukin-4 (IL-4) and IL-13 analogs fosters a pro-regenerative microenvironment, reducing fibrosis and enhancing tissue repair.

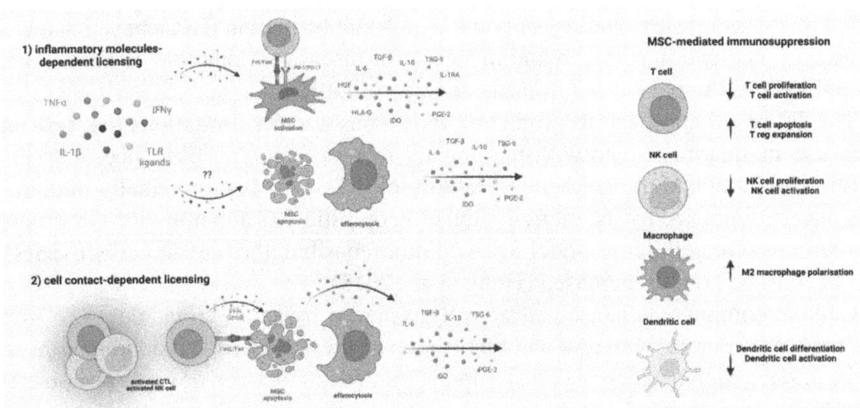


Fig. 7 Mechanisms of MSC-mediated immunosuppression (Giacomini et al., 2023)

## 3 Use of Small Molecules in Regenerative Medicine

Small molecules have emerged as a significant tool in the field of regenerative medicine due to their ability to influence cellular behavior and tissue regeneration. Regenerative medicine focuses on repairing or replacing damaged tissues and organs, and small molecules offer a promising approach to promote cell regeneration, tissue repair, and functional recovery. Here are key aspects of how small molecules are used in regenerative medicine:

### *3.1 Cell Reprogramming and Heart Repair*

Cardiomyocytes, the beating cells of heart tissue, are typically severely reduced in heart attacks. Rebuilding injured cardiac tissue may be possible with cell transplantation, but it needs a lot of cell sources. Because pluripotent stem cells (PSCs) are capable of developing into any kind of cell, including cardiomyocytes, they have enormous potential for use in cardiac repair and rehabilitation. Human embryonic stem cells (hESCs) that are pluripotent are well-established to have a therapeutic promise. Yet, their origin is contentious because it invariably entails the alteration of living embryos (Passier et al., 2008; Burridge et al., 2012; Harvey & Leinwand, 2011).

Takahashi and Yamanaka's ground-breaking reprogramming technique ushered in an entirely novel phase of stem cell biology. They showed that a certain collection of transcription factors (TFs), including Oct4, Sox2, Klf4, and c-Myc (OSKM), may "reprogramme" murine germinal centers into induced pluripotent stem cells (iPSCs). Human somatic cells could be reprogrammed using the same technique, and the resulting human iPSCs could develop into cells in all three layers of germ cells. The development of iPSC technology offers an intriguing prospective allogeneic cell source for cellular-based regenerative therapies while avoiding the moral and political issues surrounding human ESCs. Notably, human iPSCs have begun to be used in medication research and modeling of diseases (Takahashi & Yamanaka, 2006; Engle & Puppala, 2013; Merkle & Eggan, 2013).

The TF-based approach to producing iPSCs has notable limitations that restrict their use in treatments, notwithstanding its revolutionary performance. Cancer-causing potential and the genomic instability of iPSCs are two practically intolerable concerns imposed by the participation of neoplastic TFs and mutations in genes. Furthermore, for cellular remodeling to be more feasible, the rate and effectiveness of it need to be greatly increased (Hong et al., 2013).

Genetic components can be attractively replaced by small molecules. The first two provide accurate oversight and fine-tuning of the sequence and degree of those intracellular actions, which may be temporary and dependent on dosage. The practically limitless potential because there are many chances to enhance the pharmacologic characteristics, discrimination, and strength of small molecules due to structural differences. Highly specific bioavailable small molecules have been shown to be useful chemical probes for studying biological activities. Furthermore, such benefits make small compounds especially well-suited for the translational discovery of pharmaceuticals (Frye, 2010).

The most effective research for small compounds has been to enhance and/or facilitate cell reprogramming into pluripotency. This strategy's advancements have been thoroughly reviewed elsewhere. The attempts to create iPSCs from somatic cells by substituting small molecules for TFs are the primary topic of this review (by others?). We will only emphasize the key takeaways from some of the latest, noteworthy developments concerning human and mouse cell reprogramming. In order to finally realize chemically caused, therapeutically beneficial human PSCs (hPSCs), particular consideration will need to be given to the connections among small molecules' molecular actions and how they are involved in creating pluripotency (Nie et al., 2012; Li et al., 2013).

A summary of the emergence of scientifically specified circumstances to preserve hPSCs will also be provided. The use of small compounds in heart regeneration treatment is an additional topic of emphasis for this overview. We will focus on chemical strategies to increase the production and transplanting of heart cells made from PSCs. We will also talk about possible prospects for small molecule-based approaches to in situ cardiac reconstruction.

### 3.1.1 PSC Activation Using Small Molecules

Despite having nearly similar genomes, PSCs and somatic cells vary mostly in how their genes are expressed. Genetic expression and epigenetic indicators play a major role in identifying PSCs and all other cells. Somatic cells need to undergo major epigenetic alterations (such as DNA methylation and histone modifications) upon reprogramming to acquire ESC-like phenotypes. Conversely, epigenetic alterations impact the activity of genes essential for cell reprogramming by enabling appropriate chromatin structural modifications. That is why small compounds that alter the activity of enzymes associated with epigenetic changes can have a significant impact on cell reprogramming (Spivakov & Fisher, 2007).

Among the most prevalent epigenetic characteristics are posttranslational changes to histones. In general, transcriptional stimulation has been linked to acetylated histones. Drugs that inhibit of histone deacetylase (HDAC) are thought to aid in maintaining a high degree of histone acetylation, which in turn promotes the replication of genes linked to pluripotency that are essential for the reprogramming procedure (Mikkelsen et al., 2008). In the lack of exogenously expressed c-Myc, that was shown to attract various histone acetylase combinations to the genome and possibly transform the chromatin structure of somatic cells into an opened, functioning state typical of PSCs, valproic acid (VPA), an HDAC inhibitor, has been shown to improve the restructuring of mice embryonic fibroblasts (MEFs) (Huangfu et al., 2008a). Reprogramming in the absence of c-Myc was feasible but ineffective.

This slow operation was improved by VPA. VPA has also been shown to encourage human fibroblast reprogramming when Klf4 and c-Myc are not present. In the reprogramming of somatic cells, TFs were additionally substituted by small compounds that modulated histone and/or DNA methylations. The addition of either RG108 (a DNA methyltransferase inhibitor), Bayk-8644 (an L-type calcium channel agonist), or BIX-01294, or a drug that inhibits the H3K9 histone methyltransferase G9a, allowed Oct4/Klf4 (OK)-mediated reprogramming of MEFs. When neural progenitor cells from mice are converted to iPSCs by Sox2/Klf4/Myc (SKM), BIX-01294 may even make up for the lack of ectopic Oct4 (Shi et al., 2008b; Huangfu et al., 2008b; Wernig et al., 2008).

The coordinated communication pathway is essential for establishing pluripotency during TF-mediated reprogramming. It was discovered that small compounds that modify routes of signaling can improve reprogramming effectiveness or operationally substitute TFs in iPSC reprogramming. A process within cells essential for inducing pluripotency, the mesenchymal-to-epithelial transition (MET), is inhibited by stimulation of the converting growth factor-β (TGF-β) regulatory cascade. Murine fibroblasts usually exhibit MET at the very beginning stages of reprogramming, and they are distinguished by the overexpression of epithelial genes similar to E-cadherin, a reduction in mesenchymal genes like Snail, and the establishment of an epithelial-like shape (Li et al., 2010). Human ESCs that mimic epithelial cells additionally show significant levels of e-cadherin expression. Reprogramming toward immortality might be facilitated by using small molecules to block TGF-β signaling. Both PD0325901 (PD), a synthetic inhibitor of MEK, and SB431542 (a TGF-β transmission inhibitor) markedly increased the rate as well as effectiveness of human iPSC production. In MEF reprogramming, E-616542, a small-molecule inhibitor of TGF-β signaling, is capable of replacing Sox2 (Tables 1 and 2) (Yuan et al., 2011). Both Sox2 and c-Myc may be operationally substituted by TGF-β signaling inhibitors. In the meantime, MEFs transformed with Oct4 were utilized to create iPSCs, which were subsequently treated with the small compounds AMI-5, a protein methyltransferase inhibitor, and A83-01, a TGF-β receptor inhibitor (Tables 1 and 2) (Maherali & Hochedlinger, 2009).

Additionally, cell reprogramming has benefited from the effects of various other signaling networks. The signaling pathway of Wnt is necessary for adult stem cell regeneration and ESC pluripotency maintenance. In the lack of Sox2 and cMyc, CHIR99021 (CHIR), a synthetic inhibitor of glycogen synthase kinase 3β (GSK-3β), promotes Wnt signaling and greatly increases the effectiveness of MEF reprogramming. Particularly, keratinocytes from humans were transformed into iPSCs with exogenous expression of Oct4 and Klf4 via the combination of CHIR with Parnate, a synthetic inhibitor of lysine-specific demethylase 1 (Tables 1 and 2). In an additional investigation, Klf4 was substituted by the GSK-3β inhibitor kenpaullone in the reprogramming of MEFs (Tables 1 and 2) (Lyssiotis et al., 2009; Reya & Clevers, 2005).

**Table 1** Typical small molecules for cell reprogramming and hPSC maintenance

| Sl. no. | Name | Structure | Functions |
|---|---|---|---|
| 1 | AMI-5 | 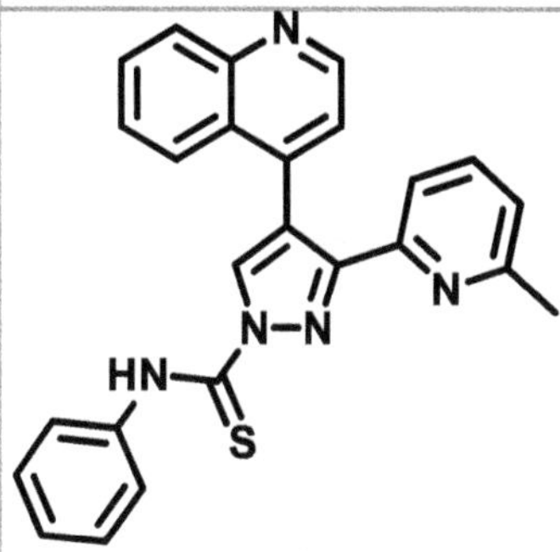 | Protein arginine N-methyltransferase inhibitor |
| 2 | A83-01 | 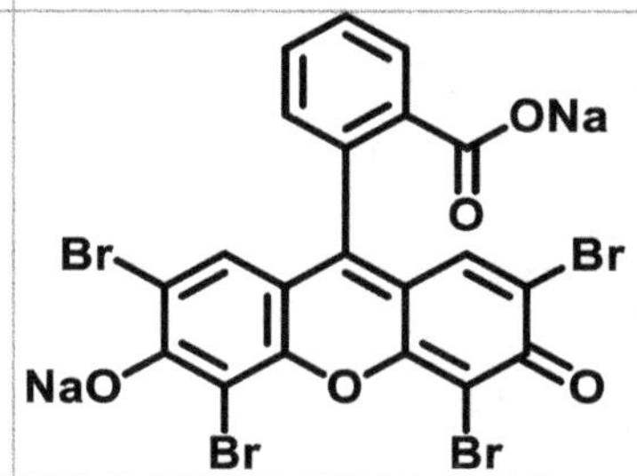 | TGF-β receptor ALK4/5/7 inhibitor |
| 3 | BIX-01294 (BIX) | 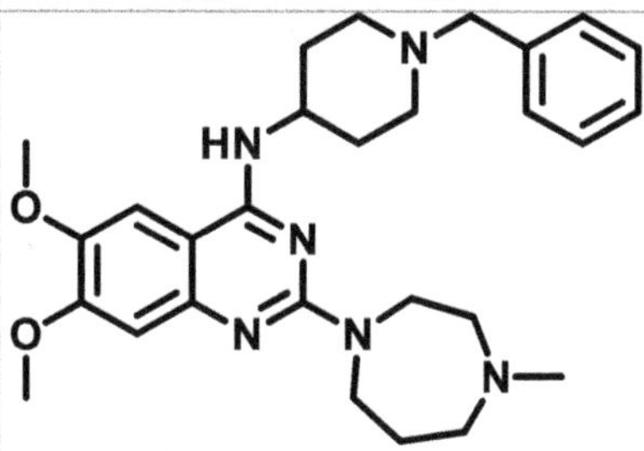 | Histone methyltransferase G9a inhibitor |
| 4 | (±)BayK 8644 (BayK) | 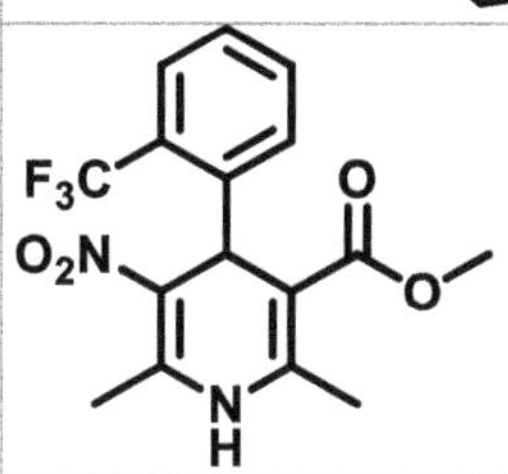 | L-type calcium channel agonist |
| 5 | CHIR99021 (CHIR) | 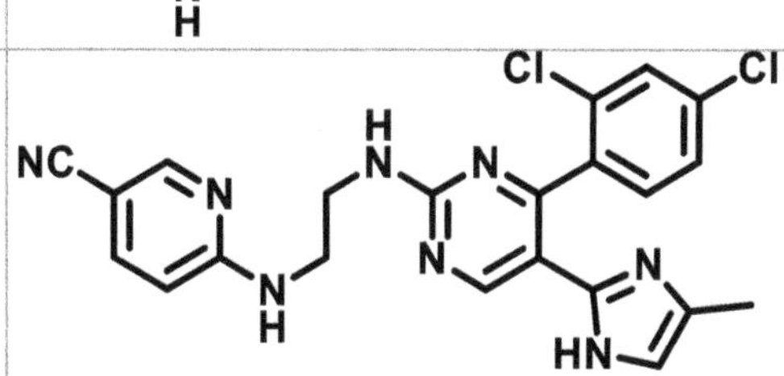 | Glycogen synthase kinase 3β inhibitor |

(continued)

**Table 1** (continued)

| Sl. no. | Name | Structure | Functions |
|---|---|---|---|
| 6 | 3-Deazanepla-nocin A (DZNep) | 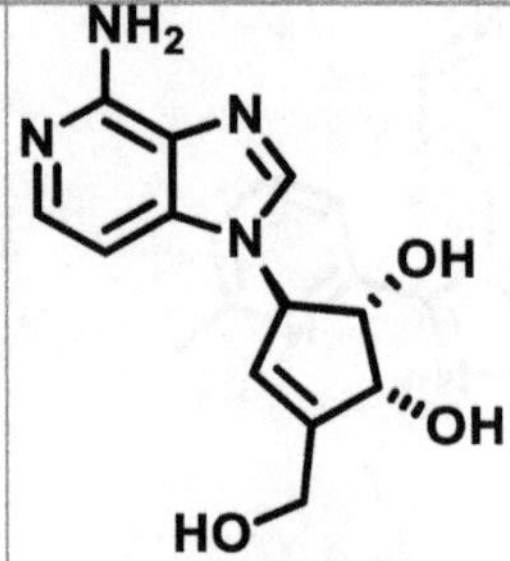 | Histone methylation inhibitor |
| 7 | Forskolin | 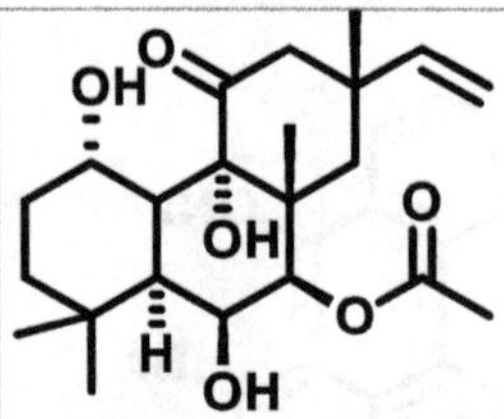 | PKA activator |
| 8 | Kenpaullone | 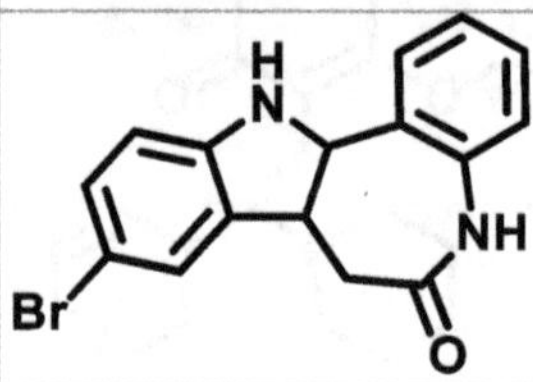 | GSK3 and CDK1/cyclin B inhibitor |
| 9 | Parnate | 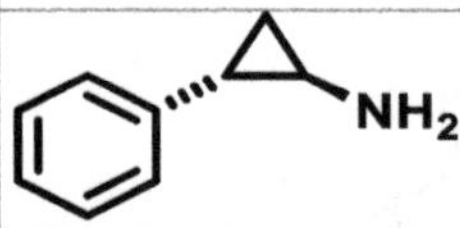 | Lysine-specific demethylase 1 inhibitor |
| 10 | PD0325901 (PD) | 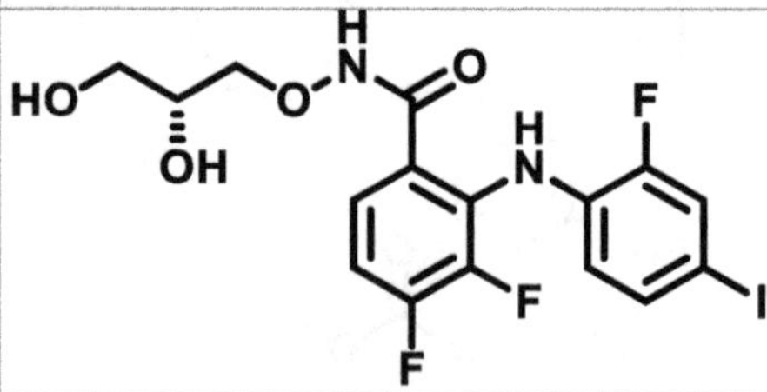 | MEK inhibitor |
| 11 | PS48 | 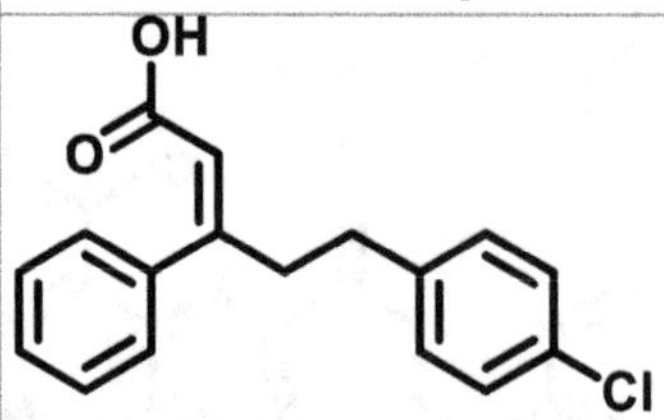 | PDK1 activator |

(continued)

**Table 1** (continued)

| Sl. no. | Name | Structure | Functions |
|---|---|---|---|
| 12 | E-616452 | 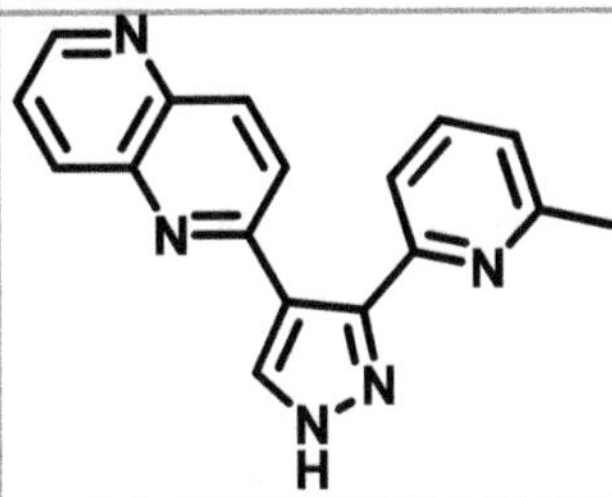 | TGF-β receptor I kinase inhibitor |
| 13 | RG108 | 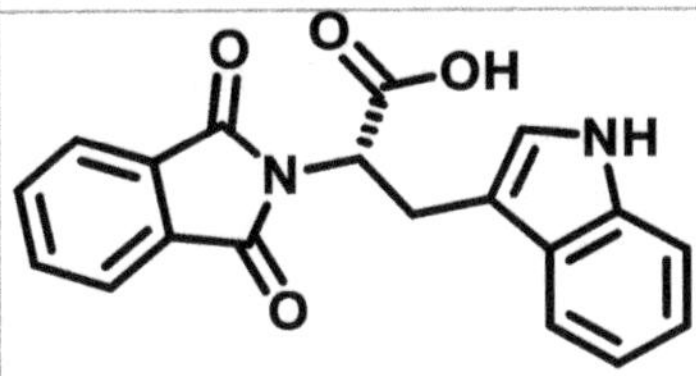 | DNA methyl-transferase inhibitor |
| 14 | SB431542 | 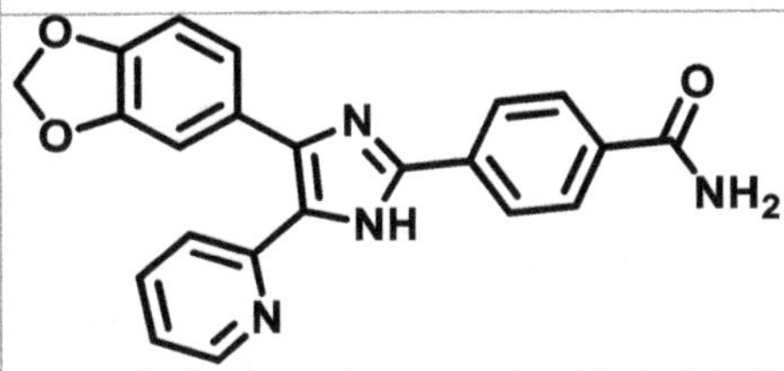 | Activin receptor-like kinase 4/5/7 inhibitor |
| 15 | Sodium butyrate (NaB) | 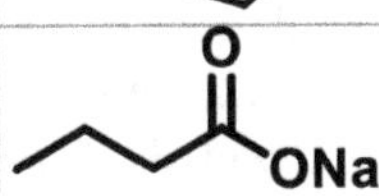 | Histone deacetylase inhibitor |
| 16 | Thiazovivin (Tzv) | 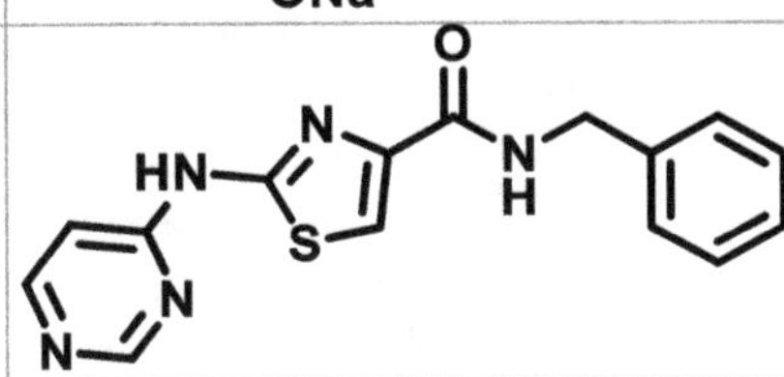 | ROCK inhibitor |
| 17 | Pyrintegrin (Ptn) | 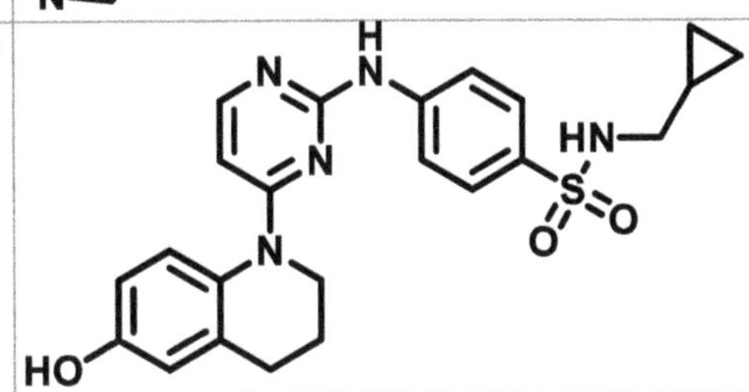 | Unknown |
| 18 | Valproic acid (VPA) | 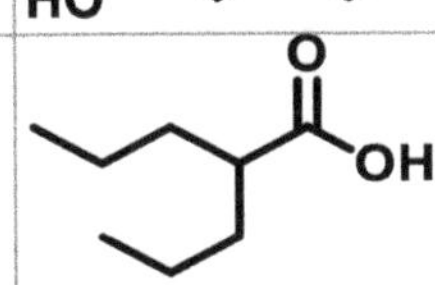 | Histone deacetylase inhibitor |

**Table 2** TFs are replaced by small molecules in somatic cell reprogramming

| Small-molecule combination | Starting cells | TFs required | References |
|---|---|---|---|
| VPA | MEFs | OSK | Huangfu et al. (2008a) |
| BIX, BayK or RG108 | MEFs | OK | Shi et al. (2008b) |
| VPA | Primary human fibroblast | OS | Huangfu et al. (2008b) |
| A83-01, AMI-5 | MEFs | O | Yuan et al. (2011) |
| Kenpaullone | MEFs | OSM | Lyssiotis et al. (2009) |
| NaB, A83-01, PS48, PD | Neonatal human epidermal keratinocytes | O | Zhu et al. (2010a) |
| VPA, CHIR, E-616542, Parnate | MEFs | O | Li et al. (2011) |
| VPA, CHIR, E-616542, Parnate, Forskolin, DZNep | MEFs | – | Hou et al. (2013) |
| E-616542 | MEFs | OKM | Ichida et al. (2009) |
| CHIR, Parnate | Human keratinocytes | OK | Li et al. (2009) |

Cell reprogramming was greatly enhanced by the synergistic use of cocktails of functionally distinct small molecules. In order to reprogram human adult keratinocytes transduced utilizing Oct4 alone, *Zhu* et al. devised an effective mix of small molecules. When Oct4 and Klf4 are transduced into newborn human keratinocytes, PD0325901 and A83-01 cause pluripotency (Zhu et al., 2010a). Subsequent screening revealed two substances that let Klf4 omit: sodium butyrate (NaB) and PS48. It was shown that NaB was a better HDAC inhibitor than VPA. Remarkably, the metabolic transition from mitochondrial oxidation to glycolysis was identified as a key step in reprogramming by mechanistic analysis of PS48. In order to accomplish O-mediated reprogramming of adult human epidermal keratinocytes, Parnate, and CHIR were lastly added to the mixture (Zhu et al., 2010a).

In one recent research, TFs are entirely replaced by small compounds. The scientists aimed to substitute the last TF with chemical substances based on their earlier discoveries that four smaller molecules (VPA, CHIR, E-616542, and Parnate) allowed the reprogramming of MEF under Oct4 alone circumstances. They discovered that the chemicals' combination with Forskolin allowed MEFs to dedifferentiate within the early stages of reprogramming, as evidenced by the upregulation of E-cadherin and the genes linked to pluripotency, Sall4 and Sox2. In order to promote the proliferation of Oct4 and provide completely reprogrammed cells, 3-Deazaneplanocin A, a global histone methylation inhibitor, was introduced at the end. About 0.2% of the initial MEFs in this investigation were transformed into iPSCs (Tables 1 and 2) (Li et al., 2011; Hou et al., 2013).

## 3.2 *Cardiac Differentiation and Grafts*

### 3.2.1 Enhancing Cardiac Differentiation

Human PSCs' genuine capacity for cardiomyogenic distinction has long been acknowledged. Only when directed by suitable external factors do human PSCs develop become cardiomyocytes (Zwi et al., 2009). A lot of work has gone into creating circumstances that can effectively, consistently, and easily promote cardiac differentiation from human PSCs. In the following chapters, small-molecule strategies will be covered. Finding small compounds that encourage the cardiac development of mice or human PSCs was accomplished using phenotypic cell-based examination. One of the first substances found to promote the cardiogenic genesis of ESC was ascorbic acid (Table 3). Additionally, it restored iPSCs with cell line-dependent cardiogenic deficit. Other cardiogenic small-molecule compounds, such as cardiogenols, isoxazolyl-serine-based inhibitors of receptors activated by peroxisome proliferator (PPARs), verapamil, SB203580, sulfonylhydrazones, and cinchona alkaloid derivatives, were also found as a result of high-throughput screening processes (Table 3). Although difficult, attempts to identify the cellular targets of active, cardiogenic small compounds found in phenotypic screening assays will provide additional insight into the mechanisms behind cardiac morphogenesis (Cao et al., 2012; Berkessel et al., 2010; Sachinidis et al., 2006).

Chemical strategies that methodically attack the key signaling associated with all phases of cardiogenesis have shown to be incredibly effective to replicate embryonic heart regeneration. Mesoderm, cardiac progenitors, and cardiomyocytes are formed in an orderly way during the structured procedure of embryonic heart growth. Numerous signaling pathways fine-tune this sequential approach. The effectiveness of chemically driven cardiac differentiation depends on precise signaling regulation employing small molecules at the right time. When delivered at the right moment, a variety of small compounds that specifically inhibit BMP, TGF-β, and Wnt allowed PSCs—particularly those of human origin—to differentiate into heart cells efficiently. Our comprehension of the logic of heart growth is subsequently enhanced by the insight gleaned by these findings (Willems et al., 2012; Kattman et al., 2011; Wang et al., 2011).

### 3.2.2 Chemically Marked Cardiac Development Conditions

A straightforward, dependable, and affordable method for chemically inducing cardiac differentiation from PSCs has been made possible by the discovery of strong cardiogenic compounds that improve differentiation efficacy and substitute complex, and ill-defined aspects (such as serum, growth factors, hormones, and extracellular matrix). Gonzalez et al. developed a step-by-step procedure to produce cardiomyocytes from hESCs with an efficiency of >50% in the absence of growth factors and serum (Gonzalez et al., 2011).. They discovered that IWR-1-endo (a

**Table 3** Typical small molecules improving graft integration, cardiac regeneration, and differentiation

| Sl. no. | Name | Structure | Function | References |
|---|---|---|---|---|
| 1 | AG1478 | 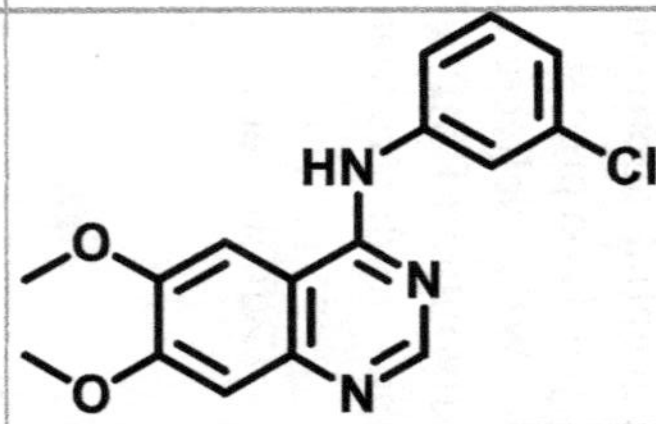 | EGFR tyrosine kinase inhibitor | Zhang et al. (2011) |
| 2 | Ascorbic acid | 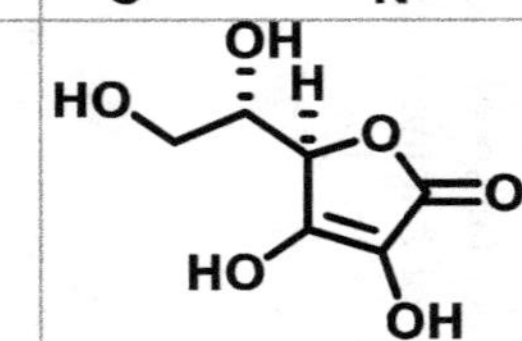 | Multiple | Takahashi et al. (2003) and Cao et al. (2012) |
| 3 | BMS-189453 | 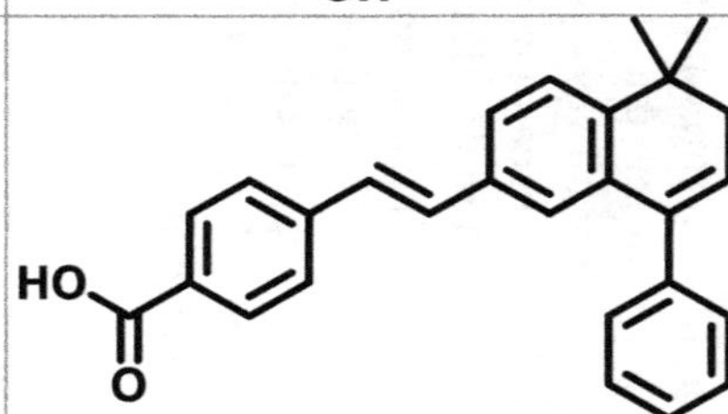 | Pan-retinoic acid receptor antagonist | Zhang et al. (2011) |
| 4 | Cardiogenol C | 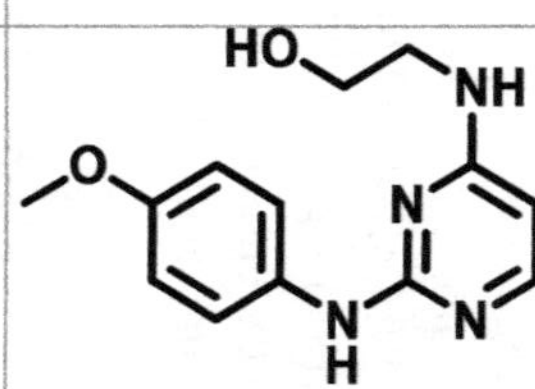 | Unknown | Wu et al. (2004) |

| | | | | |
|---|---|---|---|---|
| 5 | Cinchona alkaloid derivative | 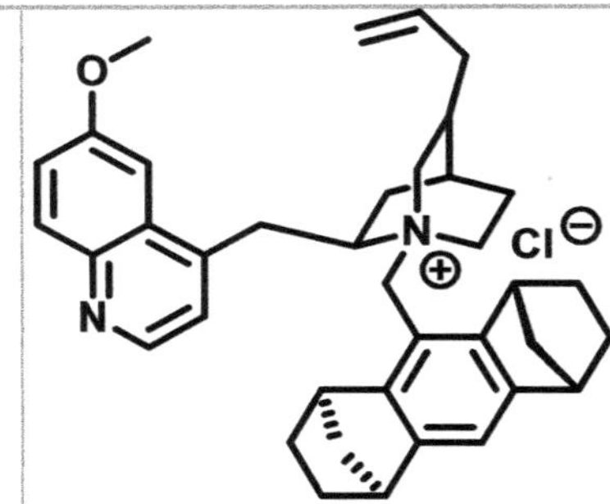 | Unknown | Berkessel et al. (2010) |
| 6 | Diazoxide | 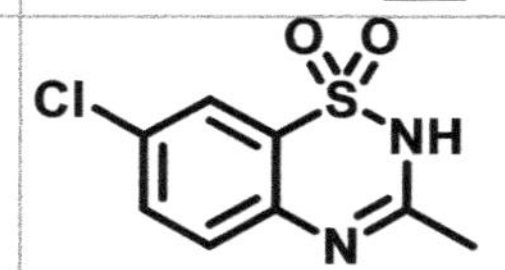 | ATP-sensitive activator $K^+$ channel | Niagara et al. (2007) |
| 7 | 1-EBIO | 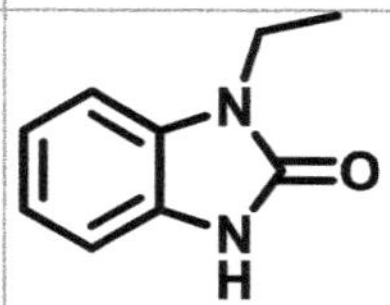 | Agonist of $Ca^{2+}$ -activated $K^+$ channels | Kleger et al. (2010) |

(continued)

**Table 3** (continued)

| Sl. no. | Name | Structure | Function | References |
|---|---|---|---|---|
| 8 | Isoxazolyl-serine derivative | 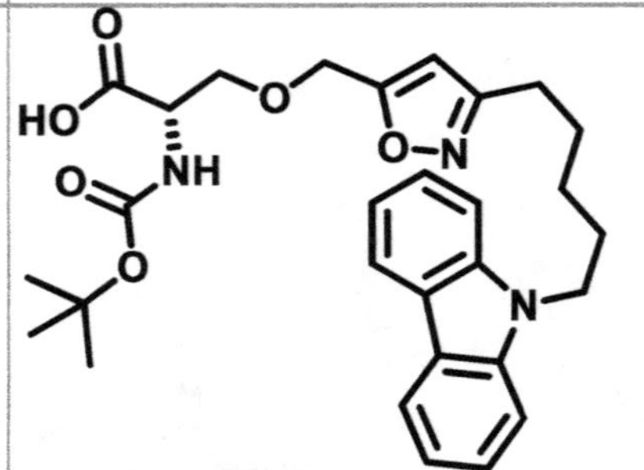 | PPAR agonist | Wei et al. (2004) |
| 9 | Isx1 (isoxazole) | 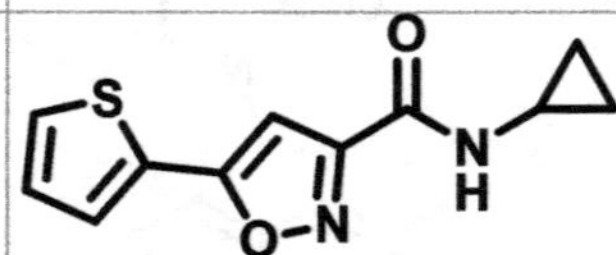 | Cardiac muscle gene activator<br>Neuronal reporter genes activator | Russell et al. (2013) |
| 10 | IWR-1-endo | 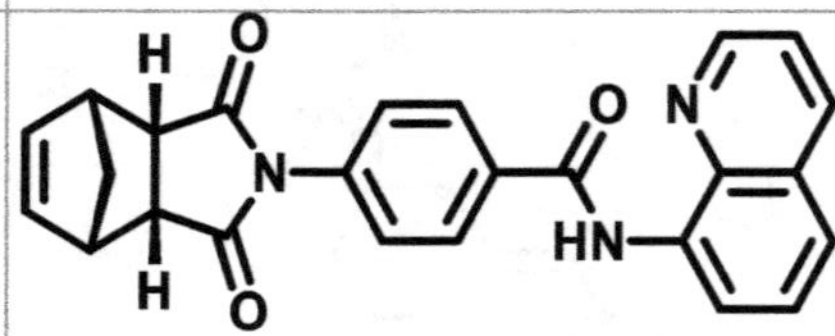 | Wnt inhibitor | Gonzalez et al. (2011) |

| | | | | |
|---|---|---|---|---|
| 11 | IWP-2 | 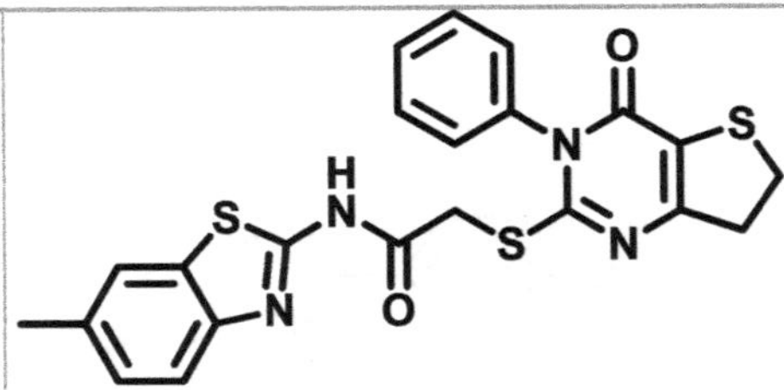 | Wnt inhibitor | Lian et al. (2012) |
| 12 | IWP-4 | 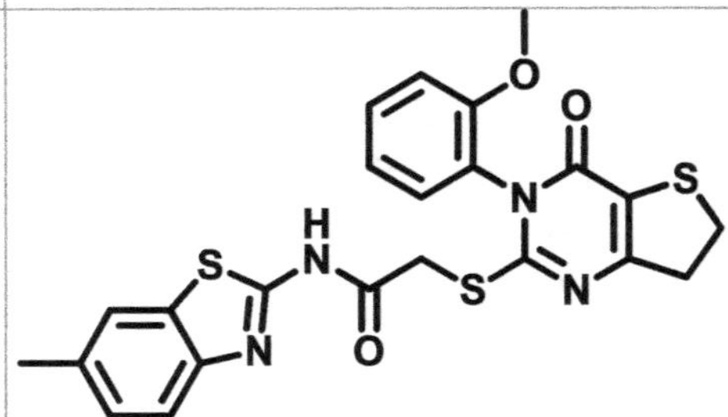 | Wnt inhibitor | |
| 13 | KY02111 | 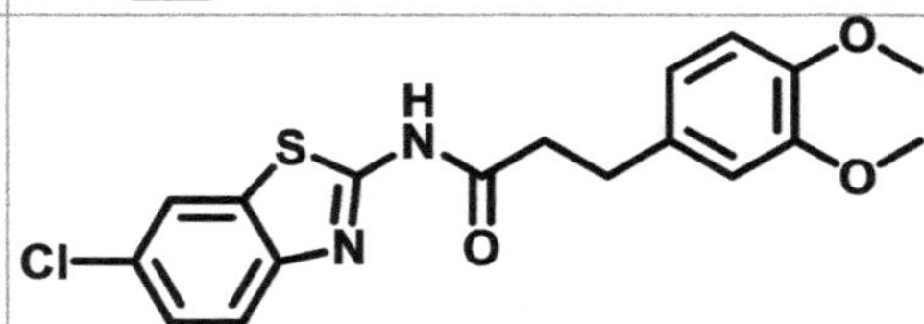 | Wnt inhibitor | Minami et al. (2012) |

(continued)

**Table 3** (continued)

| Sl. no. | Name | Structure | Function | References |
|---|---|---|---|---|
| 14 | Pinacidil | 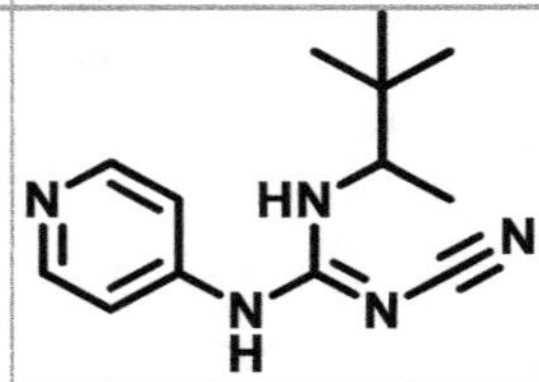 | ATP-dependent $K^+$ channel opener | Laflamme et al. (2007) |
| 15 | Pioglitazone | 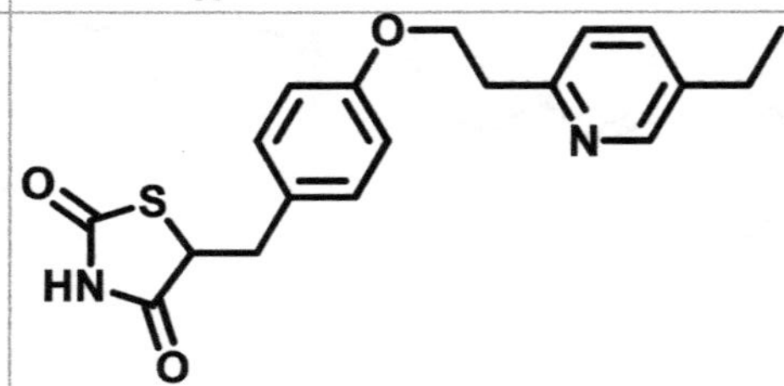 | PPAR-γ activator | Shinmura et al. (2011) |
| 16 | RA | 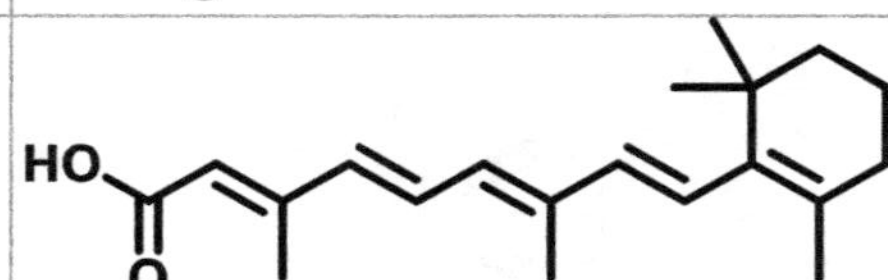 | Natural ligand of RA receptors | Zhang et al. (2011) |
| 17 | SB203580 | 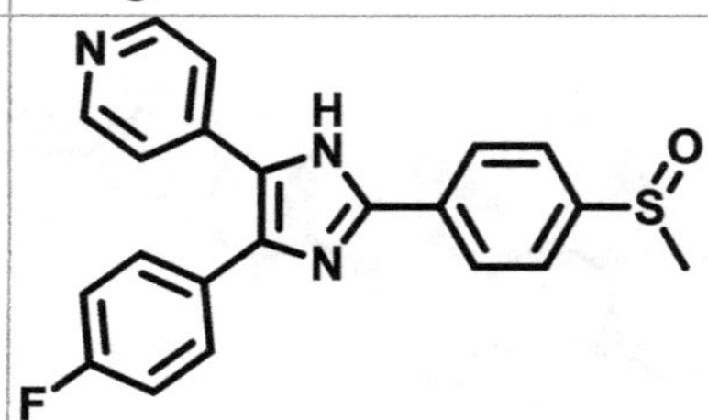 | p38 MAPK inhibitor | Graichen et al. (2008) |

| | | | | |
|---|---|---|---|---|
| 18 | Shz-1 | 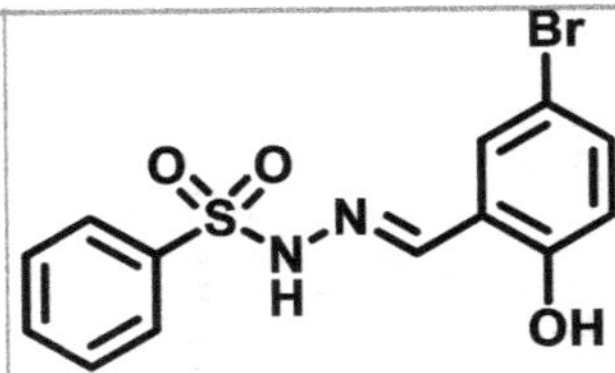 | Activator of gene *Nkx2.5* | Sadek et al. (2008) |
| 19 | Verapamil | 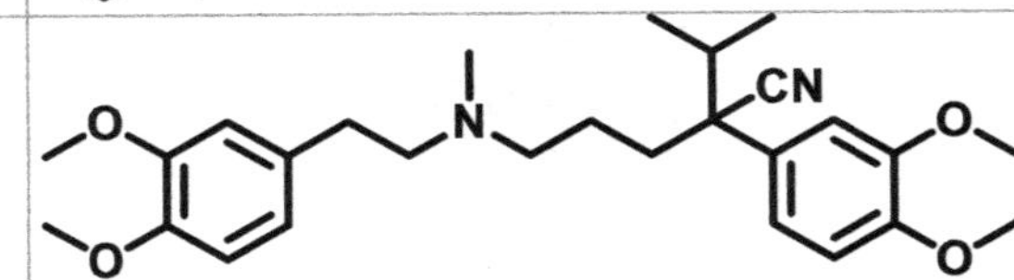 | L-type $Ca^{2+}$ channel blocker | Sachinidis et al. (2006) |
| 20 | ZVAD-fmk | 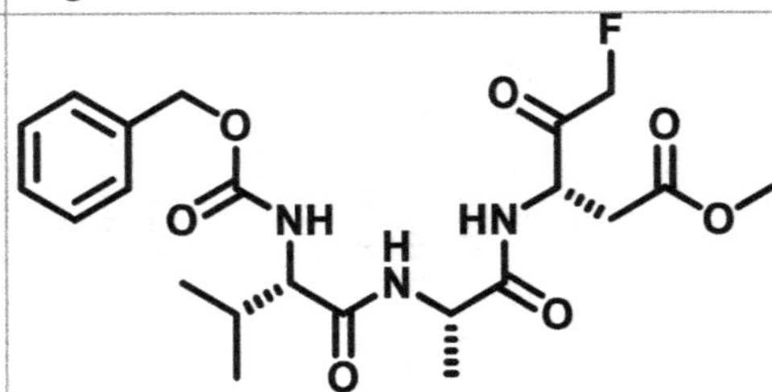 | Caspase inhibitor | Laflamme et al. (2007) |

Wnt antagonist), purmorphamine (a Sonic Hedgehog signaling agonist), and SB431542 are small compounds that promote the differentiation of hESCs into cardiomyocytes after conducting a rigorous analysis of 300 known signal transduction modulators (Table 3). This investigation shows that small compounds may successfully substitute exogenous developmental factors for cardiac differentiation, although the substances they employed—MEF conditioned media and Matrigel—were not entirely chemically defined. A chemically defined cardiac differentiation system has been developed by Lian et al. They demonstrated that effective cardiac production of hPSCs (up to 98% efficiency) under specific growth factor-free environments requires prompt regulation of Wnt signaling, which is stimulated by CHIR during the first 24 hours and then suppressed by inhibitors of Wnt production 4 (IWP-2) or IWP-4 during days 3–5 (Table 3). Likewise, using a xeno-free and serum-free medium, recombinant cytokines, or hormones, Minami et al. discovered a possible Wnt inhibitor KY02111 that, when paired with other Wnt modulators, caused rapid ventricular differentiation of hPSCs (up to 98% efficiency) (Table 3). These important developments cleared the path for reliable, efficient, and economical procedures for de novo cardiomyocyte production from hPSCs on a clinically pertinent scale, even though the endurance and repeatability of these procedures still need to be evaluated on additional cell lines and might require additional adjustments (Minami et al., 2012; Lian et al., 2012).

### 3.2.3 Minimizing Cardiomyocyte Heterogeneity

Strategies that favor a particular subtype of cardiomyocyte, such as atrial-, ventricular-, or nodal-like cells, remain absent despite significant advancements in the development of effective and well-defined procedures for cardiac differentiation of PSCs. The only product of common hPSC differentiation techniques is a combination of the three main cardiomyocyte subtypes 3A. Tragically, its application in cell-based therapeutics and medical research is hampered by its varied cellular composition. Pharmacologic methods have been used to make a number of advances in the area of particular cardiac differentiation situations over the past decade. An agonist of $Ca^{2+}$-activated potassium channels, 1-ethyl-2-benzimidazolinone (1-EBIO), was discovered by Kleger et al. to stimulate cardiogenesis of murine ESCs and to significantly enrich nodal-like cells (from 7.2 to 57.8%) (Table 3) (Kleger et al., 2010). Through the use of subtype-specific biomarker expression, triggering of genetic mark, and immediate possibility of action phenotyping, Zhao and coworkers (Zhu et al., 2010b) showed that NRG-1β/ERBB signaling controls the proportion of nodal to ventricular-type cells in hESC-derived cardiomyocytes. The nodal-like cells were considerably enriched (21–52%) when NRG-1β/ERBB signaling was inhibited by its antagonist AG1478 (Table 3). Zhang et al. also showed within the cardiac differentiation of hESCs, retinoic acid (RA) signaling controls atrial versus ventricular specificity. Eighty-three percent of the cardiomyocytes exhibited ventricular-like characteristics once the RA receptors antagonist BMS-189453 was introduced to cultures, while 94% of the cells revealed atrial-like phenotypes when RA was administered (Table 3). All things considered; these results demonstrate the possibility of small-molecule-based strategies for selective PSC differentiation into particular cardiac subtypes. More clarification of the principles underlying cardiac subtype determination and the discovery of new compounds that enhance these techniques are required in order to effectively and specifically produce cardiomyocytes of superior quality (Zhang et al., 2011).

#### 3.2.4 Enhancing Cardiac Grafts

In addition to producing enough high-quality cardiomyocytes, integrating transplanted cells into tissues is a problem of equal importance. hESC-derived cardiomyocytes have been engrafted and integrated into the hearts of rodents and pigs in a number of promising experiments. However, the majority of cardiomyocytes disappeared soon after transplantation. A short time later, a significant percentage of the cells that were still present in the infarcted myocardium also experienced programmed cell death. A significant barrier to the long-term use of PSC-derived cardiomyocytes in regenerative treatment is the low survival rate of transplanted cells. There have been initiatives to increase cardiomyocyte graft longevity using small-molecule techniques. Diazoxide, a medication that activates mitochondrial ATP-dependent potassium channels (mitoKATP) in a myocardial infarction (MI) model, enhanced survival in cells transplanted prior to treatment (Table 3) (van Laake et al., 2007; Shiba et al., 2012). The lifespan of transplanted mesenchymal stem cell-derived cardiomyocytes in mouse experiments was markedly increased by the small drug pioglitazone, an activator of PPAR-γ signaling (Table 3). Furthermore, cells' longevity both before and after transplantation is increased by small-molecule inhibitors of p38 MAPK and Rho-associated kinase. In addition to small compounds, growth factors like erythropoietin, TGF-β2, and insulin-like growth factor 1 (IGF-1) have been employed to shield transplanted cells. Matrigel, IGF-1, a Bcl XL peptide, pinacidil, peptide cyclosporin A, and the caspase inhibitor ZVAD-fmk comprised the "prosurvival cocktail" that Laflamme et al. created (Table 3) (Laflamme et al., 2007). By combining these several pro-survival elements, a rat MI model showed a sevenfold increase in graft size. In addition to other tactics like preconditioning, immunosuppression, and bioengineering, subsequent efforts to further improve transplant survival will probably entail looking for new combinations that combine small molecules and prosurvival factors.

### *3.3 In Situ Heart Repair and Regeneration*

To overcome the difficulties that cell transplantation therapy is now facing, scientists have worked to create fresh approaches to in situ cardiac regeneration and repair.

Table 4 lists the benefits and limitations of several therapy approaches for heart regenerative medicine. The following sections will concentrate on developments in new, promising approaches in this field. An initial list of possible uses for small-molecule methods in these tactics will be provided.

### 3.3.1 Dedifferentiation and Proliferation of Cardiomyocytes

The cardiovascular system expands primarily by the multiplication of cardiomyocytes at early mammalian developmental stages, but shortly after birth, it transitions to hypertrophic growth. Later rises in cardiac mass are mostly caused by a spike in cardiomyocyte size rather than quantity since cardiomyocytes quit the cell cycle at this stage. Adult mammalian hearts normally have very little ability for renewal and are unable to repair injured myocardium following damage, in contrast to the cardiovascular systems found in certain lower organisms, like zebrafish, that exhibit a robust regenerative reaction around injury primarily via cardiomyocyte dedifferentiation and expansion (Ahuja et al., 2007).

While small amount of differentiated cardiomyocytes are insufficient for fixing an infarcted heart, mammalian cardiomyocytes may gently regenerate themselves and revenue under physiological conditions, most likely through the division of preexisting cardiomyocytes rather than the differentiation of residual progenitor

**Table 4** Comparing in situ cardiac regeneration with PSC-based cardiac cell therapy

| Issues | PSC-based cardiac cell therapy | In situ heart regeneration |
|---|---|---|
| Therapeutic mechanisms | Transplanting cardiac cells produced in vitro into the heart to replace the damaged myocardium | Reprogramming or mimicking endogenous cells to alter the cardiovascular system's natural regenerative capability |
| Cell sources | In theory, there are no limits. Regulated cell quality and type | Generally, cell type, quality, and quantity are limited and context-dependent |
| In vitro bioengineering | Relevant | Not applicable |
| Cellular maturation | Characteristics of fetal or neonatal cardiomyocytes | Frequently mature cardiomyocyte-like characteristics |
| Risk of tumor formation | Potential because of leftover pluripotent cells | Possible as a result of transgenes altering the host genome and unchecked transgenic expression |
| Risk of immune rejection | iPS technology makes it possible yet improvable | Unlikely |
| Risk of arrhythmias | Possible as a result of graft cell autorhythmia, immaturity, and disarray | Possible as a result of incomplete and possibly erratic reconfiguration |
| Graft survival and host-graft integration | Challenging | Not essential |
| Ease of implementation | Low | Reasonably high |
| Cost | Reasonably high | Reasonably low |

cells. Following myocardial infarction, heart function has become better if genetic and pharmacological techniques that increase the intrinsic regeneration capability work better. Because of the high-content testing methodology and the quick advancements in stem cell biology, systematic scans have been carried out to find compounds that promote cardiomyocyte proliferation. To fully realize the potential benefits of this regenerative approach, future studies into new small compounds that reliably cause cardiomyocyte dedifferentiation and proliferation will be extremely valuable (Uosaki et al., 2013; Eulalio et al., 2012; Xin et al., 2013).

### 3.3.2 Rapid Differentiation of Non-myocytes in the Cardiac System

Investigation on the immediate reprogramming of somatic cells between a specific lineage toward another without approaching the pluripotent stage—a phenomenon commonly referred to as transdifferentiation—has been rekindled by the development and advancements in iPSC technology. One intriguing, cutting-edge regeneration strategy is the transdifferentiation of endogenous or explanted fibroblasts. Up to 50% of the cells in a grownup human heart are fibroblasts (Baudino et al., 2006). Fibroblasts overproliferate following heart injury, causing fibrosis and the development of scars in the injured area. The many fibroblasts, when appropriately converted into cardiomyocytes, can be a desirable cell source to restore the cardiac muscle. Ectopic expression of several cardiovascular-enriched transcription components (Gata4, Mef2c, and Tbx5) has demonstrated an effective cardiac reprogramming of fibroblasts in a proof-of-principle manner. However, Efe et al. developed the cell-activation and signaling-directed (CASD) cardiac reprogramming, which combines the modeling of cardiovascular pathways with the temporary manifestation of Yamanaka factors. Transdifferentiation and reprogramming to pluripotency have a same paradigm, as demonstrated by the effectiveness of CASD lineage transformation. By transfecting cardiac fibroblasts with microRNAs (miR-1, miR-133, miR-208, and miR-499), cardiogenic transdifferentiation was also achieved (Efe et al., 2011). Remarkably, by transforming local cardiac fibroblasts into cardiomyocytes, in vivo administration of reprogramming factors into infarcted animal hearts restored the injured myocardial muscle in situ. Enhanced cardiac activity and a grown up, adult-like phenotype was demonstrated by the inspired cardiomyocytes produced in their natural habitat. Additionally, there was testimony to an electrical connection to the host cardiac tissue. Notwithstanding the positive outcomes, a number of safety issues, involving as the viral delivery of transgenes along with partially reprogrammed cells that could potentially throw off the heart rhythm, remain to be addressed. The effective transfer of this alluring approach from bench to bedside will likely be greatly aided by small compounds that can prevent the use of transcription variables and/or promote in situ transdifferentiation (Qian et al., 2012; Song et al., 2012).

### 3.3.3 Endogenous Cardiac Progenitor Cell Activation

Over the past ten years, there has been ample evidence of the presence of resident cardiac progenitor cells (CPCs) in both the adult human and mouse cardiac systems. Despite several attempts to identify, grow, and differentiate these cells in vitro, much has been learned concerning their functions and activities in the receptive heart habitat in both healthy and pathological circumstances (Sturzu & Wu, 2011). Therefore, it is appropriate to envision a CPC-based therapy that permits the appropriate mobilization of resident CPCs to substitute the missing or injured cells in situ, preventing the issues of constrained graft survival, constrained cooperation to the surrounding tissue, and possible immune being dismissed, given their well-characterized cardiogenesis ability within in vitro and in vivo. In the past few decades, the viability of this strategy has been successfully demonstrated. Loffredo et al. showed via genetic destiny tracing that exogenously administered bone marrow-derived cells might activate indigenous CPCs and encourage endogenous cardiomyocyte renewal (Loffredo et al., 2011). Interestingly, Smart et al. reported that thymosin β4, a known pro-angiogenic peptide, activated a population of progenitors with an epicardial origin, caused myocardial tissue regeneration and cardiac differentiation at the same time, and eventually enhanced cardiac activity after myocardial infarction (Smart et al., 2011). In an animal model of MI, Zangi et al. discovered that intramyocardial injection of synthetic modified RNA encoding human VEGF-A led to the growth and guided differentiation of endogenous CPCs and significantly enhanced cardiac function (Zangi et al., 2013). There was a latest report on a strategy to use small substances to target CPCs. According to Russell et al., a 3,5-disubstituted isoxazole called Isx1 could stimulate the production of CPCs and stimulate the transcription of cardiac genes in multipotent, residential Notch-activated epicardium-derived cells in vivo (Table 3). Regretfully, MI caused fibrosis and reversed the cardiogenic benefits of Isx1. However, there is no denying the attractive potential for creating new small-molecule tools that will enable the safe, reliable, and economical stimulation of CPCs for heart regeneration.

Although small molecules have made great strides in controlling the destiny and function of cardiac cells, their potential for in situ heart regeneration has not yet been fully investigated. Alternative, intriguing approaches for regenerative therapy will be made possible by the potential to turn local non-cardiomyocytes into myocardial and to stimulate and/or improve the intrinsic regeneration ability of cardiac cells by pharmacologic techniques. High-throughput drug discovery will remain a potent tactic to find more novel compounds with desirable properties in the pursuit of these end goals. Utilizing small molecule combinations to achieve their combined effect has already shown benefits. Continued attempts will be undertaken to find the best combinations for particular medical applications. Finally, stem cell biology and regenerative therapies will eventually benefit from a deeper comprehension of these cell reprogramming and growth mechanisms.

## 3.4 RPE Cell Reprogramming with Small Molecules in Regenerative Medicine for Degenerative Eye Disease

The third most common cause of blindness worldwide and the main reason for it in developed nations is age-related macular degeneration (AMD). Age is an important risk indicator for AMD growth, and as longevity rises, AMD occurrence is predicted to rise as well. By 2040, 288 million individuals are predicted to acquire an AMD (Surendran et al., 2021). In its most severe phases, AMD causes central vision loss by affecting the macular portion of the retina. There are two types of AMD: wet (exudative) and dry (atrophic). A breakdown of the retinal pigment epithelium (RPE) and the buildup of drusen between Bruch's membrane and the RPE layer are hallmarks of AMD in its early stages. The loss of the retina, which rely on the RPE for trophic aid, and spatial atrophy with RPE degeneration are characteristics of the late stage. When fresh blood vessels proliferate into the retina, causing the RPE and photoreceptors to be destroyed, wet AMD presents as choroidal neovascularization. There is currently no cure for atrophic AMD, despite the encouraging outcomes of vascular endothelial growth factor (VEGF) suppression in the therapy of wet AMD (Surendran et al., 2021).

Retinitis pigmentosa frequently manifests in the early years of life, whereas AMD affects mainly older people. One in three thousand to seven thousand people suffer with retinal pigmentosa, a diverse set of hereditary illnesses. Between the ages of 40 and 50, people who have retinitis pigmentosa suffer from serious vision loss. Retinitis pigmentosa is characterized by rod degeneration followed by cone loss in the optic nerve (Parmeggiani et al., 2011). Vision loss at night and blurry vision are common signs of retinitis pigmentosa, which may ultimately end in total blindness. With current therapeutic approaches, it is practically impossible to cure these conditions along with other degenerative retinal disorders. The cell replacement approach is one of the many novel approaches to visual structure regeneration that are currently being actively sought for. Nowadays, it is easy to differentiate between two approaches: molecular medicine and regenerative medicine?. While conventional molecular medicine has traditionally focused on employing small molecule drugs (small molecules), regenerative medicine seeks to reclaim normal operation by repairing or rebuilding human cells. Nonetheless, there have been numerous attempts in recent years to find small compounds with specific applications in regenerative medicine, taking into account the potential for their pharmacological regulation and accessibility (Lu & Atala, 2014).

### 3.4.1 Chemical Reprogramming of Cells Regulated by Small Molecules

Selections on cell destiny are usually governed by TFs that influence gene expression in conjunction with epigenetic changes. Reprogramming one cell into another is the outcome of changing data via epigenetic alteration and/or gene expression level programming. Direct genetically mediated reprogramming, which aims to produce the ectopic expression of particular TFs, cannot produce full-fledged retinal cells from a grownup human's RPE, according to a thorough review of the global literature (Hsieh & Zhao, 2016). Many scientists have proposed that both mammals and humans have highly evolved epigenetic systems for regulating cell identity in their RPE. Small-molecule chemical reprogramming is an alternate technique for genetically modifying cells. The chemical reprogramming approach is predicated on how small chemicals impact signaling, metabolic, or epigenome pathways, which in turn impact the activity of TFs that dictate cell fate (Table 5).

Under the impact of small chemical cocktails, cells that have terminally differentiated can now produce a variety of cells both in vitro and in vivo, including neurons, cardiomyocytes, neural progenitors, and pluripotent stem cells. For each cell, the low-molecular or chemical reprogramming paradigm applies (Huang et al., 2024). The majority of small molecules fall into one of several categories, including metabolic regulators, regulators that support ESC self-regeneration, genetic modulators, modulators of epithelial-mesenchymal and mesenchymal-epithelial transitions, and molecules performing on multiple pathways of cell signaling (Table 5).

## 4 Small Molecules in Tissue Regeneration

Tissue regeneration, the process by which damaged or lost tissues are repaired or replaced, is a dynamic area of research that holds promise for treating a wide range of injuries and degenerative diseases. Small molecules have emerged as pivotal tools in modulating in vivo aspects of tissue regeneration, including adult stem cell activation, maintenance of stem cell niches, and the regulation of cellular homing, mobilization, engraftment, and proliferation (Fig. 8). Herein, we explore the current progress in leveraging small molecules for these applications.

### *4.1 Adult Stem Cells and Tissue Regeneration*

Adult stem cells, for example, mesenchymal stem cells (MSCs), hematopoietic stem cells (HSCs), and neural stem cells (NSCs), are central to tissue regeneration because of their ability to differentiate into multiple cell types. Small molecules can modulate the behavior of these stem cells by targeting key signaling pathways that regulate self-renewal and differentiation (Roato et al., 2018). One prominent example is the Wnt/β-catenin signaling pathway, essential for stem cell maintenance and tissue repair. Small molecules like CHIR99021, a glycogen synthase kinase-3β (GSK-3β) inhibitor, have been shown to enhance Wnt signaling (Law & Zheng, 2022), thereby promoting the proliferation and differentiation of MSCs. Similarly, modulators of the Hedgehog and Notch pathways, such as SAG (a smoothened agonist) and DAPT (a γ-secretase inhibitor), have demonstrated potential in controlling stem cell fate in vivo (Firestone & Chen, 2010).

**Table 5** The molecular structure and suggested mode of action of small compounds that are used to completely chemically induce reprogramming or transdifferentiation of several cell types into brain cells

| Sl. no. | Name | Structure | Function | References |
|---|---|---|---|---|
| *Epigenetic regulators* | | | | |
| 1 | VPA | 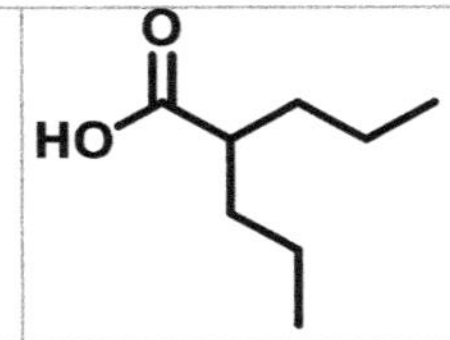 | Inhibitor of HDAC | Hu et al. (2015) |
| 2 | NaB | 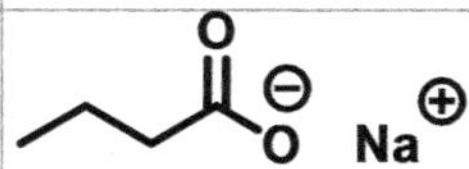 | Inhibitor of HDAC | Zhu et al. (2013) |
| 3 | SAHA | 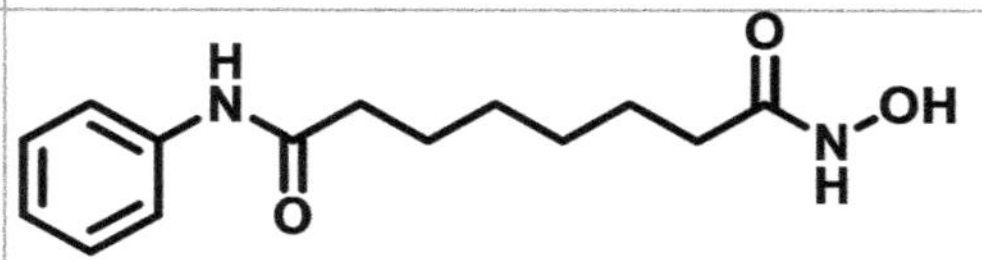 | Inhibitor of I/II HDAC | Dong et al. (2023) |
| 4 | I-BET151 | 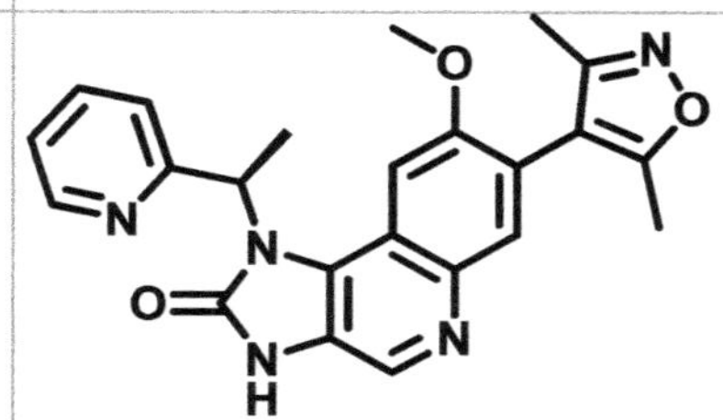 | Inhibitor of BET bromodomain | Li et al. (2015) |

(continued)

**Table 5** (continued)

| Sl. no. | Name | Structure | Function | References |
|---|---|---|---|---|
| 5 | 5-aza-dc | 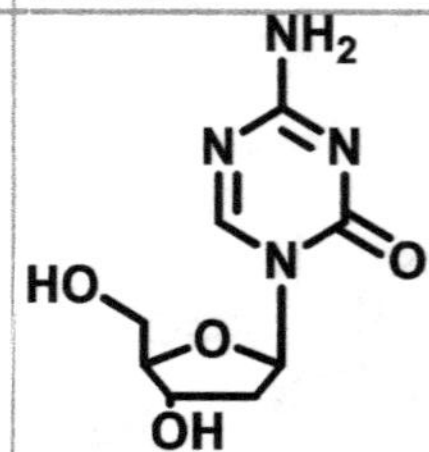 | DNMT inhibitor | Xie et al. (2017) |
| 6 | DZNep | 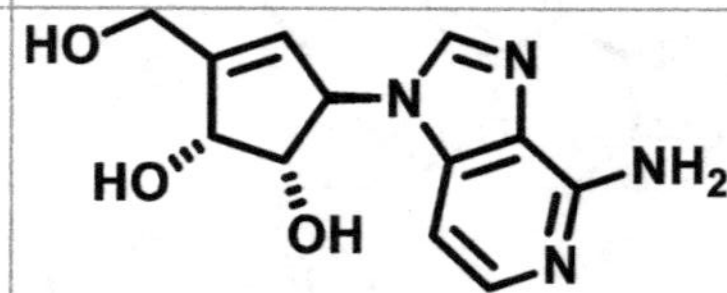 | S-adenosylhomocysteine Synthesis inhibitor & histone methyltransferase EZH2 inhibitor | Xie et al. (2017) |
| 7 | RG108 | 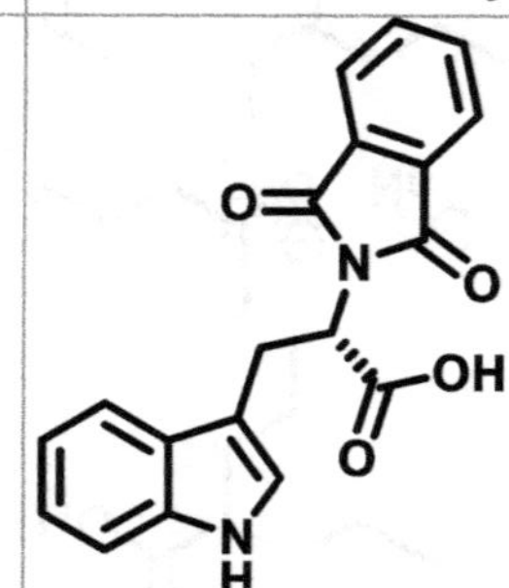 | Inhibitor of DNA methyltransferases | Zhang et al. (2016a) |

| | | | | |
|---|---|---|---|---|
| 8 | SGC0946 | 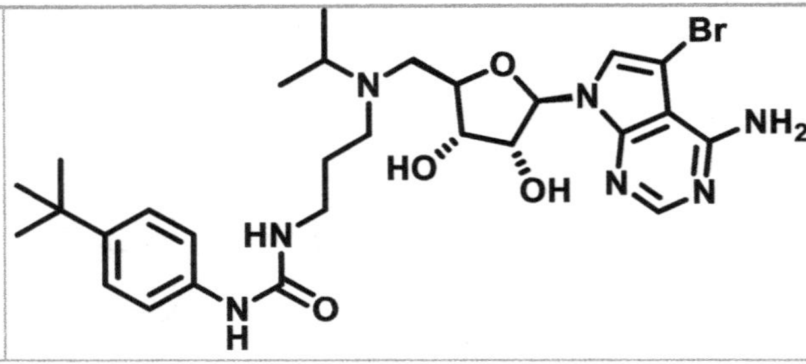 | Inhibitor of DOT1L | Xie et al. (2017) |
| *Acting on signaling pathways* | | | | |
| 9 | Rolipram | 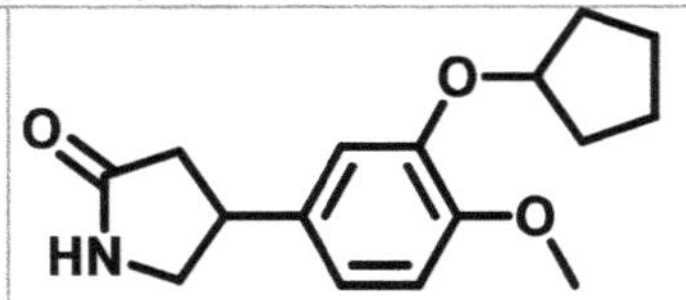 | Phosphodiesterase-4 inhibitor | Zhu et al. (2013) |
| 10 | Ch 55 | 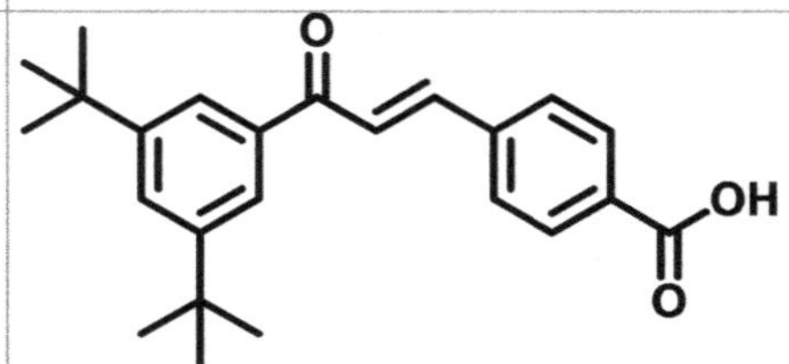 | RAR agonist | Xie et al. (2017) |

(continued)

Table 5 (continued)

| Sl. no. | Name | Structure | Function | References |
|---|---|---|---|---|
| 11 | TTNPB | 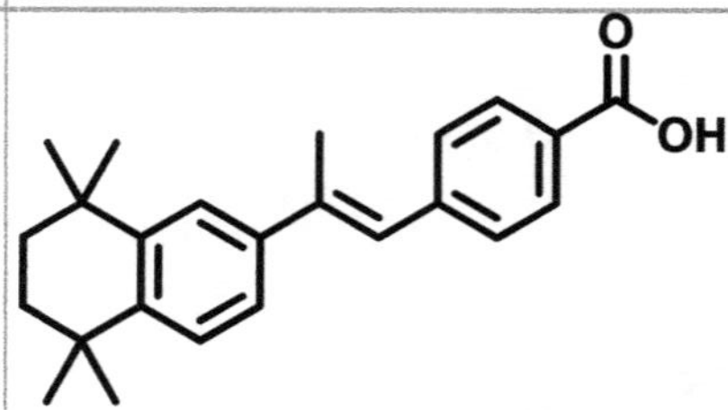 | Potent selective RAR agonist | Zhang et al. (2015) |
| 12 | AM580 | 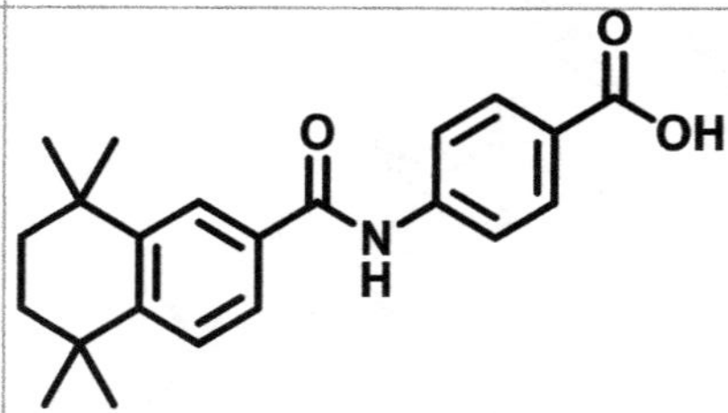 | Selective RARα agonist | Xie et al. (2017) |
| 13 | TWS119 | 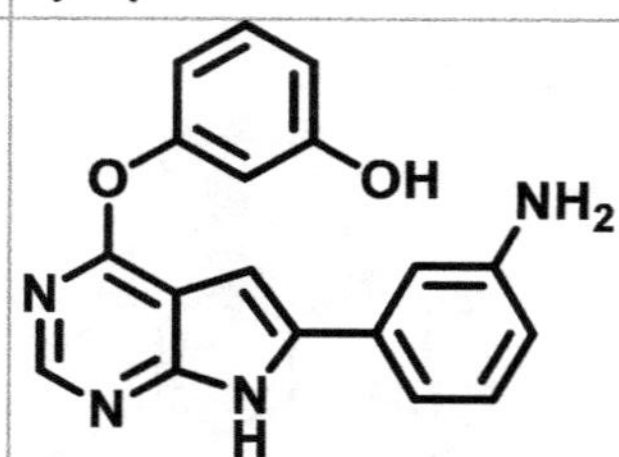 | Inhibitor of GSK-3β | Lu and Atala (2014) |

| | | | | |
|---|---|---|---|---|
| 14 | SP600125 | 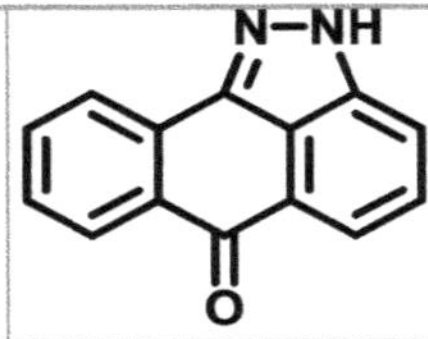 | Pan-JNK inhibitor | Hu et al. (2015) and Zhu et al. (2013) |
| 15 | ISX9 | 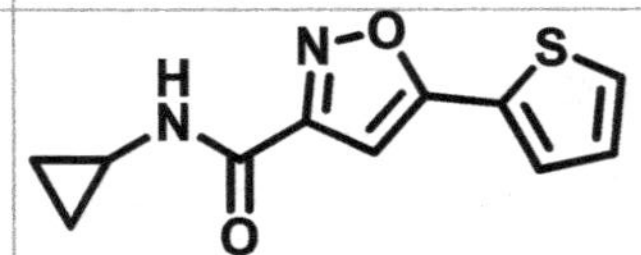 | WNT pathway activator | Li et al. (2015) |
| 16 | SMER28 | 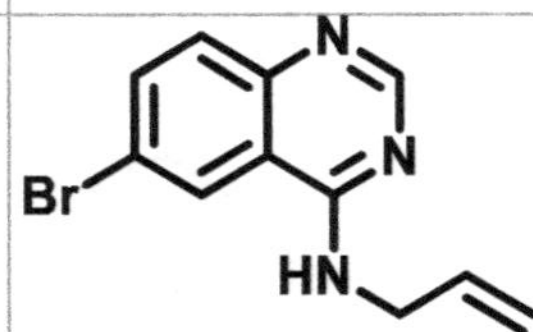 | Enhancer of autophagy | Zhang et al. (2016a) |
| 17 | Tranilast | 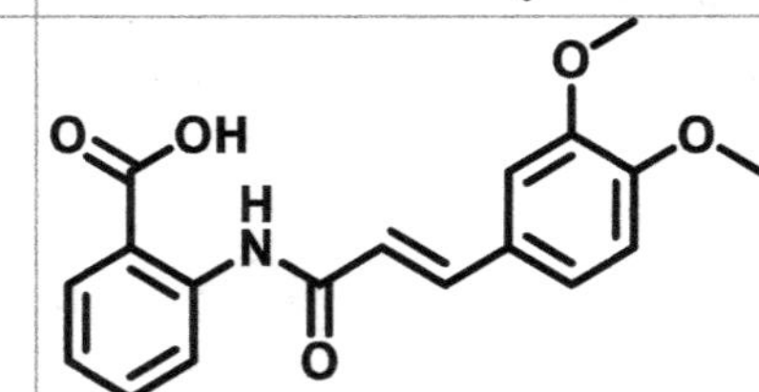 | Inhibitor of TGF-β1 | Cheng et al. (2014) |

(continued)

**Table 5** (continued)

| Sl. no. | Name | Structure | Function | References |
|---|---|---|---|---|
| *Promote survival, proliferation, and function of reprogrammable cells* | | | | |
| 18 | HA-1077 | 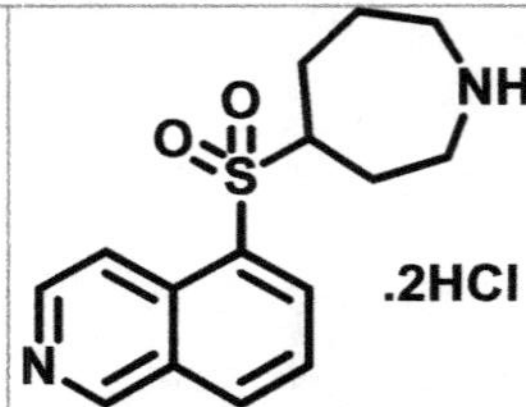 | ROCK inhibitor | Li et al. (2015) |
| 19 | SB203580 | 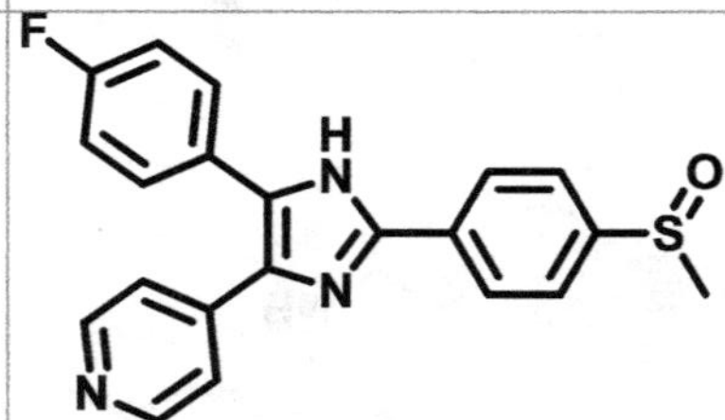 | p38 MAPK inhibitor | Li et al. (2015) |
| 20 | CompA | 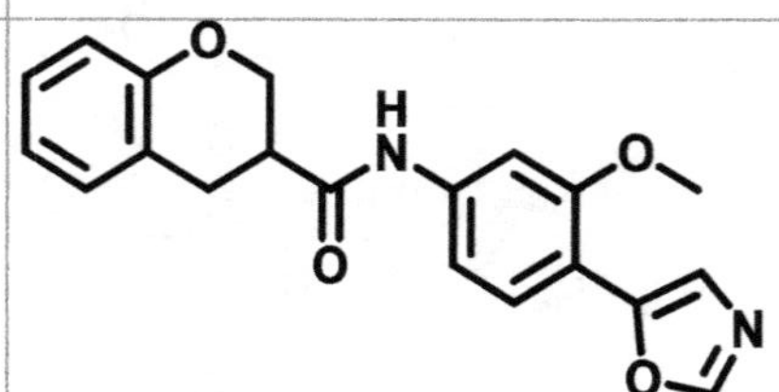 | Inhibitor of p70S6K, ROCK1, & PKG | Mokady et al. (2024) |

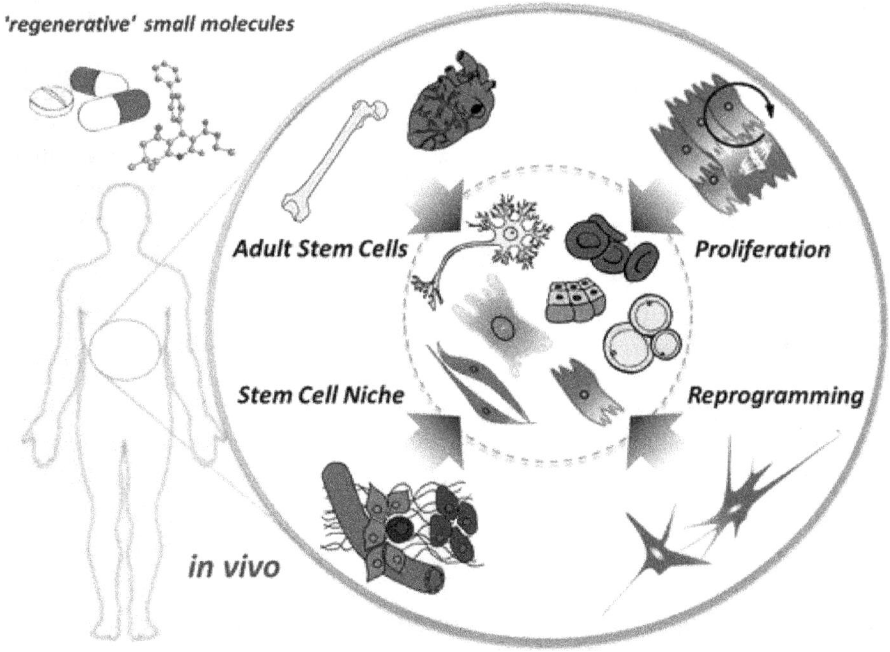


**Fig. 8** Overview of possible approaches, strategies, and targets for the identification and developing small molecules for regenerative medicine (Längle et al., 2014)

## 4.2 *Stem Cell Niches*

The microenvironment or niche stem cells reside in is crucial for regulating their activity. Small molecules can modulate the niche to create a conducive environment for regeneration (Lane et al., 2014). For example, PGE2, a prostaglandin, has been shown to enhance HSC proliferation by altering the bone marrow niche. Similarly, small molecules targeting hypoxia-inducible factor-1α (HIF-1α) can mimic low-oxygen conditions favorable for stem cell maintenance and expansion. Extracellular matrix (ECM) is another critical aspect, which provides structural and biochemical cues to stem cells. Small molecules that modulate ECM components, such as matrix metalloproteinase (MMP) inhibitors, can regulate tissue remodeling and support regeneration. For instance, doxycycline, an MMP inhibitor, has been reported to reduce scar formation and enhance tissue repair by preserving ECM integrity (Yue, 2014).

### 4.3 Mechanisms of Homing, Mobilization, and Engraftment

Homing, mobilization, and engraftment are essential steps for successful tissue regeneration, particularly in the context of stem cell therapies. These processes involve the migration of stem cells to the injury site, their integration into the host tissue, and their subsequent survival and function (Fong et al., 2011). Small molecules targeting chemokine receptors, such as CXCR4, have shown promise in enhancing stem cell homing. Plerixafor (AMD3100), a CXCR4 antagonist, mobilizes HSCs from the bone marrow to the peripheral blood, facilitating their collection for transplantation. Conversely, molecules that activate CXCR4, such as the small peptide analogs of stromal cell-derived factor-1 (SDF-1), can promote stem cell homing to injury sites. Small molecules that inhibit apoptosis and enhance cell survival are being explored to improve engraftment. For instance, ROCK inhibitors like Y-27632 have been reported to increase the survival of transplanted cells by reducing detachment-induced apoptosis (anoikis). Small molecules such as FK506, a calcineurin inhibitor, can suppress immune rejection and promote graft survival (Centonze et al., 2022).

### 4.4 Somatic Cell Proliferation and Tissue Regeneration

Somatic cell proliferation plays a significant role in tissue regeneration, in addition to stem cells. Small molecules that target growth factors (particularly epidermal growth factor (EGF) and fibroblast growth factor (FGF) and their signaling pathways can stimulate somatic cell proliferation and tissue repair (Farooq et al., 2021). For example, small molecules like PD0332991, a selective cyclin-dependent kinase 4/6 (CDK4/6) inhibitor, can stimulate controlled cell cycle re-entry in somatic cells, promoting regeneration without the risk of uncontrolled proliferation. Similarly, the small molecule BMP-7 has been shown to enhance bone and cartilage repair by activating the bone morphogenetic protein (BMP) signaling pathway (Goel et al., 2018).

Though small molecules for tissue regeneration are a promising area for researchers, also it brings challenges too. The specificity of small molecules, potential off-target effects, and the complexity of in vivo environments must be addressed to ensure safety and efficacy. For identification of new small molecules with high selectivity and potency, we need to improve high-throughput screening and computational drug design area. Furthermore, combining small molecule therapies with other regenerative strategies, such as biomaterials and gene editing, could enhance outcomes. For instance, controlled release systems for small molecules can provide localized and sustained delivery, maximizing their regenerative potential while minimizing systemic side effects. Small molecules offer a versatile and powerful approach to modulating in vivo mechanisms of tissue regeneration. By targeting adult stem cells, their niches, and processes like homing, mobilization, and engraftment, these compounds have the potential to revolutionize regenerative medicine.

Continued research and development in this field will pave the way for innovative therapies that restore function and improve the quality of life for patients with injuries and degenerative diseases.

## 5 Small Molecules in Bone Regeneration

An essential biological process called bone regeneration develops new bone tissue to replace damaged bone due to aging, illness, or trauma. Larger and more complicated bone injuries frequently necessitate medical intervention, even if small fractures or abnormalities can be repaired by natural bone healing mechanisms. Allografts, autografts, synthetic bone substitutes, and surgery have been major components of traditional bone regeneration techniques. Nevertheless, these techniques have drawbacks including immunological rejection, donor site morbidity, and inadequate host tissue integration. By altering the molecular pathways involved in bone formation, small molecule-based therapies provide a new strategy to improve bone regeneration by encouraging osteogenesis, boosting tissue repair, and improving bone healing.

### *5.1 Promoting Osteogenesis with Small Molecules*

The process of creating new bone from precursor cells, especially mesenchymal stem cells (MSCs), is known as osteogenesis. By altering important signaling pathways that regulate MSC development into osteoblasts, the cells that make bones, small compounds have been demonstrated to affect osteogenesis. A number of small compounds, including as those that target the Notch signaling system, bone morphogenetic proteins (BMPs), and the Wnt/β-catenin pathway, have surfaced as possible treatments for osteogenic differentiation.

*Wnt/β-Catenin Signaling Modulation*: The Wnt signaling pathway encourages MSC development into osteoblasts, which is essential for controlling bone production. Small compounds can improve bone formation in human cells and preclinical animals by activating Wnt/β-catenin signaling. One GSK-3β inhibitor that stabilizes β-catenin and promotes osteoblast development is BIO (6-bromoindirubin-3′-oxime) (Chen et al., 2009). By preventing β-catenin from degrading, other small molecules like IWP (inhibitors of Wnt production) increase osteogenesis. These compounds activate endogenous regenerating pathways, providing intriguing treatment approaches for osteoporosis and bone abnormalities (Yu et al., 2018).

(a) *Bone Morphogenetic Proteins (BMP) Pathway Modulation*
The growth factor family known as BMPs is essential to osteogenesis. Osteogenic differentiation can be enhanced by small compounds that mimic or improve BMP signaling. For example, it has been demonstrated that lysophos-

phatidic acid (LPA) and L-aspirin (L-asp) increase BMP signaling, which in turn promotes osteoblast mineralization and development (Shah et al., 2017). Furthermore, in animal models of bone defect healing, BMP activity can be upregulated by small molecules like BMS-345541, a specific inhibitor of IκB kinase, which improves bone formation. By enhancing the normal process of bone growth, these chemicals offer a more focused method of bone regeneration (Liu et al., 2019).

(b) *Notch Signaling Pathway Activation*
A key modulator of MSC fate, Notch signaling regulates the ratio of adipogenic to osteogenic differentiation. The potential of small compounds to promote osteogenesis through activation or inhibition of Notch signaling has been studied. For instance, the γ-secretase inhibitor DAPT promotes osteoblast development by preventing Notch receptor cleavage. By favoring osteogenesis over adipogenesis, small compounds that target Notch signaling can stimulate bone regeneration. This is especially significant in the context of osteoporosis and age-related bone loss.

## 5.2 *Applications in Fracture Healing*

A series of processes, including inflammation, bone production, and bone remodeling, are involved in fracture healing. The majority of bone fractures can be successfully repaired by the body's natural healing processes; however, bigger or nonunion fractures could need further medical attention. Potential therapeutic agents that could speed up bone repair, shorten the healing period, and enhance the quality of the regenerated bone include small molecules.

(a) *Statins for Accelerating Bone Healing*
Although statins are best recognized for lowering cholesterol, they have also been shown to increase bone growth and hasten the healing of fractures. The Wnt/β-catenin signaling pathway and BMP-2 activation are responsible for these effects. Simvastatin and atorvastatin have been demonstrated in studies to improve bone mineral density and osteoblast activity. Additionally, vascularization—a critical component of bone healing—can be enhanced by statins. Statins have shown promise in accelerating healing times and improving outcomes in clinical trials examining their usage in patients with delayed fracture healing (Crawford et al., 2014).

(b) *Parathyroid Hormone (PTH) and Teriparatide*
A recombinant version of parathyroid hormone (PTH), teriparatide, has been authorized to treat osteoporosis and promote bone repair in fractures. By activating the PTH receptor, it promotes osteogenesis by increasing bone mineral density and osteoblast activity. In animal models, teriparatide has been demonstrated to enhance bone quality and hasten the healing of nonunion fractures. Teriparatide has been utilized as a treatment for osteoporosis in clinical set-

tings, especially for patients who are at a high risk of fractures (Wein & Kronenberg, 2018).

## 5.3 *Bone Tissue Engineering and Small Molecule Delivery*

When it comes to major deformities or critical-size bone damage, tissue engineering has shown promise as a method of bone regeneration. In order to promote osteogenesis and tissue regeneration, small molecules can be included into scaffolds or biomaterial-based systems.

(a) *Combination of Small Molecules and Biomaterial Scaffolds*
Biomaterials like hydrogels, ceramics, or synthetic polymers are frequently used in bone tissue engineering to create a scaffold for tissue creation and cell proliferation. These scaffolds' capacity for regeneration can be further increased by adding small chemicals that encourage osteogenesis. For instance, to improve bone tissue regeneration, hydrogels can contain naringenin, a flavonoid that encourages osteoblast development. In a similar manner, collagen-based scaffolds and the TGF-β receptor inhibitor SB431542 have been used to encourage osteogenic differentiation of stem cells both in vitro and in vivo. These methods seek to establish conditions that actively encourage bone growth at the molecular level in addition to offering structural support (Jin et al., 2015).

(b) *Nanoparticle-Based Delivery Systems for Small Molecules*
One of the biggest obstacles to bone regeneration is the efficient transport of small molecules. Liposomes, polymeric nanoparticles, and dendrimers are examples of nanoparticle-based delivery systems that have been investigated as small molecule carriers. These technologies offer benefits such targeted locations, controlled release, and enhanced bioavailability. For instance, BMP-2 mimetics have been delivered via silica nanoparticles to encourage osteogenesis in bone defects. By guaranteeing that therapeutic concentrations are reached at the site of the bone defect, these systems offer a novel platform for the targeted administration of small molecules, improving the results of bone regeneration (Zhao et al., 2020).

## 5.4 *Enhancing Vascularization for Bone Regeneration*

The development of a strong vascular network to provide nutrients and oxygen to the newly developing tissue is just as important to successful bone regeneration as the production of new bone tissue. The formation of new blood vessels, or angiogenesis, is a crucial process for the repair of major bone defects and the healing of bone fractures. Bone tissue may become necrotic and healing may be hampered if there is insufficient vascularization. Enhancing angiogenesis, which is essential for encouraging bone regeneration, has been demonstrated to be possible with small compounds.

(a) *Small Molecules for Angiogenesis Promotion*

In the context of bone regeneration, a number of small compounds have been discovered as angiogenic agents. For example, small compounds that trigger VEGF signaling pathways and vascular endothelial growth factor (VEGF) mimetics have been shown to improve vascularization in bone defects. By triggering VEGF signaling, thalidomide has been demonstrated to promote angiogenesis and enhance bone repair in animal models, despite its long-standing controversy regarding its teratogenic effects (Seidkhani-Nahal et al., 2021). Similar to this, it has been discovered that squalamine, a small chemical with broad-spectrum antibacterial activity, promotes angiogenesis in models of bone regeneration by stimulating the proliferation of endothelial cells (Grosso et al., 2017).

Prolyl hydroxylase inhibitors, such as dimethyloxalylglycine (DMOG), which stabilize HIF-1$\alpha$ (hypoxia-inducible factor 1-alpha), are another promising strategy. A transcription factor called HIF-1$\alpha$ controls a number of genes involved in osteogenesis and angiogenesis when oxygen levels are low. In bone tissue engineering applications, small compounds like DMOG that stabilize HIF-1$\alpha$ have been shown to improve bone regeneration by increasing vascularization (Zhang et al., 2016b).

(b) *Incorporating Angiogenic Small Molecules into Scaffolds*

Scaffolds are frequently employed in bone tissue engineering to provide nascent bone tissue with a structural underpinning. Scaffolds' capacity to promote bone healing may be further improved by the addition of small angiogenic molecules. For example, by promoting endothelial cell proliferation and osteoblast development, PLGA (poly(lactic-co-glycolic acid)) scaffolds loaded with small compounds such as AMPK (AMP-activated protein kinase) agonists might improve both angiogenesis and osteogenesis. Critical-size bone defects may be repaired more quickly thanks to the supporting environment these multipurpose scaffolds can offer for the creation of blood vessels and bone (Liu et al., 2014).

## 5.5 *Anti-inflammatory and Immunomodulatory Small Molecules*

Inflammation can considerably slow down the regenerative process in chronic bone illnesses and disorders that affect bone repair, such as osteoarthritis or osteoporosis. Pro-inflammatory cytokines and immune cell activation brought on by inflammation can have a detrimental effect on bone resorption and creation. Another intriguing treatment option for bone regeneration is the use of small compounds that target immunological responses and inflammation.

(a) *Targeting Inflammatory Pathways*

Since the NF-κB pathway is essential for modulating inflammation and immunological responses, NF-κB inhibitors are a promising class of small compounds for bone regeneration. A naturally occurring substance found in turmeric, curcumin is a strong NF-κB inhibitor that has demonstrated promise in increasing bone production and decreasing osteoclast activity. Curcumin's anti-inflammatory properties speed up the process of bone regeneration by fostering an environment that is more conducive to osteoblast formation and function (Bharti et al., 2004).

Another illustration is the polyphenol resveratrol, which is present in red wine and has been demonstrated to improve osteogenic differentiation of stem cells while regulating the inflammatory response. Because it inhibits inflammatory cytokines like tumor necrosis factor-α (TNF-α) and IL-6, resveratrol can help mend bone fractures and may be used to treat inflammatory bone illnesses like rheumatoid arthritis (Yuan et al., 2020).

(b) *Immunomodulation via Small Molecules*

Small compounds that alter the immune system can be employed to improve bone regeneration in addition to lowering inflammation. It has been demonstrated that the mTOR inhibitor rapamycin promotes the activity of anti-inflammatory macrophages while suppressing the development of pro-inflammatory macrophages (Mercalli et al., 2013). In addition to lowering excessive inflammation, this double action fosters a pro-regenerative environment that aids in bone regeneration. By altering immune cell function and lowering oxidative stress, other immunomodulatory small compounds, like tocopherols (vitamin E analogs), have been shown to improve osteoblast development (Ahn et al., 2004).

## 5.6 Combining Small Molecules with Stem Cell Therapy

By enabling the development of new bone tissue from a patient's own stem cells, stem cell treatment has completely transformed regenerative medicine. Small compounds can be added to stem cell therapy to improve its effectiveness and provide a stronger regenerative response when treating bone abnormalities.

(a) *Synergistic Effects of Small Molecules and Stem Cells*

The drawbacks of utilizing stem cells alone may be lessened with the use of small molecules in conjunction with stem cells. Before being transplanted, stem cells can be preconditioned in vitro with small molecules to encourage osteogenic differentiation. For instance, by activating the retinoic acid receptor (RAR) signaling pathway, retinoic acid, a bioactive product of vitamin A, has been employed in conjunction with MSCs to promote osteogenesis. By increasing the effectiveness of stem cell-based bone regeneration, this pre-differentiation technique promotes more efficient bone mending (Green et al., 2017).

Furthermore, stem cell homing and engraftment—two crucial processes for effective bone regeneration—can be influenced by small chemicals. MSCs have been used in conjunction with CXCR4 agonists, which facilitate the migration of stem cells to injury sites and enhance the recruitment of stem cells to bone defects. The amount and caliber of new bone tissue produced in the afflicted location are improved by this cooperative strategy (Shen et al., 2009).

## 5.7 *Optimizing Bone Regeneration in Osteoporosis and Other Bone Diseases*

Clinical bone regeneration is significantly hampered by osteoporosis, a disorder marked by decreased bone mass and an elevated risk of fracture. It is frequently linked to aging and a decrease in osteoblast activity, which causes an imbalance between the production and breakdown of bone. Current medications including bisphosphonates and parathyroid hormone (PTH) analogs can help maintain bone density, but they might not be able to completely restore bone strength or regenerate bone in critical-size lesions. By increasing bone density and the mechanical integrity and functionality of bone tissue, small molecules have great potential to improve the treatment of osteoporosis and related bone disorders.

(a) *Small Molecule Therapeutics in Osteoporosis*

The potential of a number of small compounds to cure osteoporosis and encourage bone regeneration has been studied. Romosozumab and other sclerostin inhibitors are among the most promising treatment approaches for osteoporosis. The Wnt/β-catenin signaling pathway is naturally inhibited by sclerostin, which increases osteoblast activity and inhibits osteoclast differentiation to promote bone formation. In individuals with osteoporosis, romosozumab, an antibody that targets sclerostin, has demonstrated clinical efficacy in boosting bone mass and lowering fracture risk. Research on direct small-molecule inhibitors of sclerostin or small compounds that imitate sclerostin may provide more affordable or locally administered options to the biologic treatments that are now available (Saag et al., 2017).

(b) *Osteoclast Regulation via Small Molecules*

Osteoblast-mediated bone production and osteoclast-mediated bone resorption must be balanced for bone regeneration to occur. In conditions including osteoporosis, rheumatoid arthritis, and Paget's disease, excessive osteoclast activity can cause bone loss and hinder the body's ability to regenerate. By reducing bone resorption, small compounds that target osteoclast development and activity can improve bone regeneration.

For example, denosumab, a monoclonal antibody that targets receptor activator of NF-κB ligand (RANKL), has been created to reduce bone resorption by blocking osteoclast development and activity. Small molecule inhibitors of RANKL or its downstream signaling pathways, including aspirin and other nonsteroidal anti-inflammatory drugs (NSAIDs), have also demonstrated the capacity to inhibit osteoclasts in preclinical studies, despite the fact that denosumab is a biologic drug. Small medicines that target the transcription factor NFATc1 (Nuclear Factor of

Activated T-cells cytoplasmic 1), which is specific to osteoclasts, have recently come to light as intriguing options for lowering osteoclastogenesis in bone disorders and enhancing the results of bone regeneration (Saag et al., 2017).

## *5.8 Bone Regeneration in Critical-Size Defects*

Bone flaws that are too big to mend on their own without assistance are known as critical-size bone defects (CSBD). They are frequently brought on by infection, trauma, or tumor removal. Small molecule-assisted bone tissue engineering and regenerative medicine are being investigated more and more to treat these major abnormalities.

(a) *Small Molecule-Based Scaffolds for Critical-Size Defects*

One of the most important parts of repairing critical-size bone defects is designing scaffolds that can promote bone regeneration. An environment that is favorable for tissue growth can be produced by combining biomaterial scaffolds with osteogenic small molecules. For instance, scaffolds have been modified to speed up bone regeneration by including transforming growth factor-beta 1 (TGF-β1), a crucial molecule involved in the control of osteogenesis and angiogenesis. Small compounds can improve bone regeneration in critical-size defects by activating Smad signaling, which occurs downstream of TGF-β. SB431542, a small drug that selectively inhibits the TGF-β receptor type I, has demonstrated potential for promoting osteogenesis and enhancing scaffold performance in vivo (Wu et al., 2016). Furthermore, even though BMP-2 is a growth factor, its functions can be mimicked via medications that trigger BMP receptor signaling or small molecule mimetics. L68 and other BMP-2 mimic small compounds have been developed to directly activate BMP signaling pathways at the site of the bone defect, providing a more stable and affordable substitute for recombinant BMP proteins (Najafi et al., 2022).

(b) *Local Delivery Systems for Small Molecules*

Delivering small molecules to the damage region in a regulated way is one of the obstacles in repairing critical-size bone defects. To overcome this difficulty, local delivery methods such as hydrogels, microspheres, and nanoparticles have been created. These delivery methods can guarantee the gradual release of small molecules at therapeutic concentrations, promoting improved tissue healing and long-term osteogenesis.

Because polymeric nanoparticles may be designed to release their cargo in a regulated manner, they hold great promise for the delivery of small molecules. For instance, when combined with 3D printed bone scaffolds, poly(lactic-co-glycolic acid) (PLGA) nanoparticles loaded with osteoinductive small compounds, like GSK-3β inhibitors like BIO, have been demonstrated to improve MSC differentiation into osteoblasts. This method maximizes the therapeutic effects of small molecules by allowing for their localized, extended administration to the bone defect (AlMuraikhi et al., 2023).

### 5.9 *Gene Delivery and Small Molecule Synergy*

Combining small compounds with gene therapy is another cutting-edge use of them in bone regeneration. Delivering particular genes to cells in order to stimulate the expression of growth factors, signaling proteins, or transcription factors essential in bone regeneration is the goal of gene therapy. The effects of gene delivery vectors, such as viral or nonviral carriers, can be amplified by combining them with small compounds. For instance, bone-related genes carried via viral vectors can have their expression increased by small compounds like BIO or IWP-2 that activate the Wnt/β-catenin signaling pathway. A more targeted approach to regenerative bone therapy is possible by combining small medicines with gene-editing tools like CRISPR/Cas9 to modify particular genes involved in osteogenesis. By combining these technologies, bone regeneration treatments may be more effective and specific.

## 6 Small Molecules in Neural Regeneration

The process by which injured or destroyed neurons and their connections in the nervous system are repaired and restored is known as neural regeneration. The regenerative capacity of the central nervous system (CNS) and peripheral nervous system (PNS) differ significantly (Nagappan et al., 2020); the PNS can recover to some extent following injury, but the CNS frequently undergoes little to no regeneration (Huebner & Strittmatter, 2009). As a result, neurological conditions and injuries including stroke, traumatic brain injury (TBI), spinal cord injury (SCI), and neurodegenerative disorders like Parkinson's and Alzheimer's pose serious problems for medical research. The use of small compounds to encourage brain repair and encourage the regeneration of neurons and their axonal connections has been the focus of recent developments in regenerative medicine.

Since they can alter the signaling pathways that control cell division, proliferation, axonal development, and survival, small molecules have drawn many studies. The main small compounds that have demonstrated promise in fostering neuronal regeneration are highlighted in this review along with their modes of action and their therapeutic uses (Cooke et al., 2022).

### 6.1 *Mechanisms of Neural Regeneration and the Role of Small Molecules*

The intricate process of neural regeneration includes several steps: (1) endogenous repair mechanisms are activated; (2) axonal growth and neuronal survival; (3) synapse creation; and (4) functional recovery. Small compounds can have an impact on the signaling pathways and molecular components that tightly govern these stages.

Promoting neurogenesis, overcoming inhibitory signals, and improving the survival and development of injured neurons are the main processes involved in brain regeneration.

(a) *Neurogenesis and Neural Stem Cells (NSCs)*: Neural stem cells (NSCs) must frequently be activated in order to develop into neurons, glia, and other neural cell types in order for neurons to regenerate in the brain and spinal cord. Neural regeneration may be aided by small compounds that promote neurogenesis by controlling NSC differentiation and proliferation. In animal models of SCI, for instance, it has been demonstrated that the anticonvulsant valproic acid increases neurogenesis by boosting the number of neural progenitor cells and encouraging the development of NSCs into neurons (Chu et al., 2015).
(b) *Axonal Growth and Guidance*: Another essential component of brain recovery, particularly following injury, is axonal regeneration. Axonal growth in the adult central nervous system is restricted by inhibitory substances in the extracellular matrix (ECM) and myelin. Potential treatments for improving axonal regeneration have been found to utilize small compounds that alter the signaling pathways involved in axonal guidance. For example, by eliminating the inhibiting scar tissue following spinal cord damage, the enzyme chondroitinase ABC, which breaks down chondroitin sulfate proteoglycans (CSPGs) in the extracellular matrix, might encourage axonal development and repair. Furthermore, it has been investigated how to reverse myelin-mediated suppression of axonal development by using small compounds that target receptors such as Rho-kinase or Nogo receptor. By improving cytoskeletal dynamics and neuronal growth, the Rho-kinase inhibitor Y-27632 has been demonstrated to promote axonal outgrowth (Park et al., 2020).
(c) *Neuroprotective Small Molecules*: When it helps maintain neuronal function and stop more damage after injury, neuroprotection is a crucial part of brain regeneration. In order to stop neuronal loss and encourage recovery, small compounds that support neuronal survival and offer defense against excitotoxicity, oxidative stress, and inflammation are crucial. Small compounds that increase the expression of neurotrophic factors, such brain-derived neurotrophic factor (BDNF), have therapeutic potential and are important mediators of neuroprotection. It has been demonstrated that the mood stabilizer lithium increases BDNF expression by inhibiting GSK-3β (glycogen synthase kinase-3 beta), which in turn helps injured neurons survive (Snitow et al., 2021).

Furthermore, because of its anti-inflammatory and antioxidant qualities, curcumin, a naturally occurring polyphenolic molecule, has been investigated for its neuroprotective effects. Curcumin lowers inflammation and oxidative stress, two major factors that lead to brain deterioration in conditions like Alzheimer's disease (Azzini et al., 2024). In addition to enhancing survival, these neuroprotective small compounds also improve the conditions for brain regeneration.

## *6.2 Small Molecules in Spinal Cord Injury and Other Neurodegenerative Diseases*

(a) *Spinal Cord Injury (SCI)*: Because of the CNS's low intrinsic regenerative capacity, SCI is one of the most difficult conditions in the field of brain regeneration. Permanent functional deficits are caused by glial scar formation, axonal damage, and neuronal death following SCI. Small compounds, on the other hand, have demonstrated promise in improving SCI recovery. To prevent the inhibitory effects of Nogo-A, a myelin protein that restricts axonal development, Nogo receptor antagonists (Cao et al., 2008), such NEP1-40, have been created. NEP1-40 has been demonstrated in preclinical research to support axonal regeneration and functional recovery in animal models of SCI. Another repulsive guidance molecule a (RGMa), which is increased following SCI and prevents axonal regrowth, is another intriguing molecule. Anti-RGMa antibodies and other small compounds that block RGMa have been demonstrated to enhance axonal development and healing following SCI (Hata et al., 2006).

(b) *Parkinson's Disease*: Dopaminergic neurons in the substantia nigra gradually disappear in Parkinson's disease (PD), a neurodegenerative condition. Dopamine replacement therapy is one example of a current medication that relieves symptoms but does not prevent or reverse neuronal loss. Treatment for Parkinson's disease may benefit greatly from small compounds that either increase the survival of dopaminergic neurons or encourage the dopaminergic differentiation of stem cells (Stocchi et al., 2015). Though small compounds like rasagiline, a monoamine oxidase-B (MAO-B) inhibitor, have lately been investigated for their neuroprotective and neurogenic qualities, L-dopa (levodopa) is still the gold standard for treating symptoms. In animal models of Parkinson's disease, rasagiline has been demonstrated to enhance dopamine release and support neuroprotection.

(c) *Alzheimer's Disease*: Amyloid plaques and tau tangles build up in Alzheimer's disease (AD), which cause neurons to gradually degenerate. As possible treatments, small compounds that target tau phosphorylation or amyloid precursor protein (APP) processing have been studied. In animal models, bexarotene, a small chemical that stimulates retinoid X receptors (RXRs), has been demonstrated to facilitate the removal of amyloid plaques, which may enhance cognitive performance in AD (Tousi, 2015). Similar to this, small compounds like Tideglusib, which block GSK-3, a crucial regulator of tau phosphorylation, have demonstrated promise in lowering tau tangles and enhancing brain function in AD animals (Domínguez et al., 2012).

### *6.3 Promoting Synaptic Plasticity and Neural Circuit Repair*

In addition to restoring neuronal cell populations, neural regeneration also entails reestablishing appropriate neural circuits and synaptic connections. After a nervous system injury, synapses must be reformed to restore functional neural circuits, and axons must grow to reconnect with target neurons. However, a key factor in brain repair is synaptic plasticity, which is the process by which synapses get stronger or weaker over time. Therefore, by boosting the development of functional synapses and promoting neuronal transmission, small compounds that support synaptic plasticity may hasten brain regeneration.

(a) *Synaptic Plasticity Enhancement Through Small Molecules*

Small compounds have been found to be promising therapeutic agents for brain repair because they alter important signaling pathways involved in synaptic plasticity. Synaptic plasticity depends on AMPA receptors (α-amino-3-hydroxy-5-methyl-4-isoxazolepropionic acid receptors), which mediate rapid synaptic transmission. By promoting the development and fortification of synaptic connections, small compounds that alter AMPA receptor trafficking or increase synaptic receptor expression may aid in brain regeneration. Ampakines are a class of small compounds that promote synaptic plasticity by increasing AMPA receptor activation (Diering & Huganir, 2018). In animal models of neurodegenerative illnesses like Alzheimer's disease, several substances have demonstrated promise in enhancing cognitive function and stimulating neurogenesis.

Another important protein involved in synaptic plasticity and neuronal survival is brain-derived neurotrophic factor (BDNF). It has been demonstrated that a number of small compounds, such as retinoic acid and lithium, raise BDNF expression, improving synaptic plasticity and neural network repair. By blocking GSK-3β, lithium can increase the production of BDNF and improve synaptic function, both of which are critical in neurodegenerative illnesses and following neuronal injury (Jastrzębski et al., 2024). Similarly, by activating nuclear receptors that control the expression of BDNF, retinoic acid, a metabolite of vitamin A, has been found to be a powerful amplifier of synaptic plasticity (Thapliyal et al., 2022).

(b) *Small Molecules for Neural Circuit Repair in Neurodegenerative Diseases*

Neuronal circuits and neurons gradually disappear in neurodegenerative illnesses like Parkinson's and Alzheimer's diseases. Apart from encouraging neuronal survival and regeneration, small chemicals that improve neural circuit repair may be a viable method for regaining cognitive function. Gamma-aminobutyric acid, or GABA, is an inhibitory neurotransmitter that is essential for maintaining appropriate circuit function and regulating brain activity. The potential for improving circuit stability and function has been studied for small compounds that alter GABAergic signaling. The potential of gabapentin, a GABA analogue, to lessen excitotoxicity and promote synaptic function in neurodegenerative disease models has been investigated (Cai et al., 2012).

## 6.4 Modulating the Extracellular Matrix (ECM) to Promote Neural Regeneration

The extracellular matrix (ECM) is crucial for controlling cellular activity and preserving the structural integrity of brain tissue. Changes in the extracellular matrix (ECM) following brain damage can either encourage or impede healing. Axonal development can be physically inhibited by the deposition of inhibitory substances like tenascin-C and chondroitin sulfate proteoglycans (CSPGs), especially in the central nervous system. The obstacles to brain regeneration may be overcome by small compounds that target ECM remodeling or alter inhibitory ECM components.

(a) *Small Molecules for ECM Remodeling*

Using small molecules that can change the extracellular matrix is one of the most promising ways to improve brain regeneration. One of the most researched substances that can break down CSPGs in the extracellular matrix and promote axonal development is the bacterial enzyme chondroitinase ABC. Chondroitinase ABC treatment significantly enhanced axonal regeneration and functional recovery in animal models of neurological injury. By dissolving CSPGs that build up at injury sites and prevent axonal development, this enzyme creates an environment that is more favorable for regeneration (Hu et al., 2021).

To encourage ECM remodeling, several small compounds that target matrix metalloproteinases (MMPs) are being investigated in addition to chondroitinase ABC. The activity of MMPs, which are enzymes that break down different ECM components, is essential for tissue repair. MMP activity control is complicated, though, because both too much and too little MMP activity might impede tissue regeneration. The potential of small compounds that regulate MMP production, like the broad-spectrum MMP inhibitor GM6001, to aid in the healing process following CNS injury has been investigated. These substances may alter the extracellular matrix (ECM) at the site of damage and promote brain regeneration when combined with other regenerative therapies (Lee et al., 2006).

(b) *Inhibition of ECM Inhibitory Factors*

Small compounds that block the action of ECM-associated axonal growth inhibitors are used in another tactic. In the adult central nervous system, it has been demonstrated that the myelin protein Nogo-A inhibits axonal development. In preclinical models of SCI, it has been demonstrated that small compounds that inhibit the interaction between Nogo-A and its receptors improve axonal regeneration (Hirt et al., 2023). It has been shown that small molecule inhibitors such as NEP1-40 and anti-Nogo-A antibodies support axonal development and functional recovery following injury. Similar to this, small molecule inhibitors like anti-RGMa antibodies can target the RGMa pathway, another strong inhibitor of axonal growth, which has been demonstrated to enhance recovery following spinal cord injury by encouraging axonal regrowth.

## 7 Challenges in Clinical Translation

Regenerative medicine uses the natural ability of the body to renew and repair tissue to treat illnesses and injuries in a revolutionary way. Small molecules have great potential to support regeneration, especially in fields like tissue repair, wound healing, and organ regeneration, although stem cells, gene therapies, and tissue engineering have received the majority of attention. However, development of these compounds from preclinical success to clinical use comes with a number of difficult obstacles. Biological complexity, clinical translation, regulatory frameworks, and medication development are all areas where these difficulties arise.

(a) *Off-Target Effects and a Lack of Specificity*
The specificity of small molecules is one of the main issues in regenerative medicine. Even though small compounds are usually made to alter particular biological pathways, it can still be difficult to achieve the intended therapeutic benefits without producing unintended side effects. Small compounds used in regenerative medicine frequently seek to either activate or inhibit particular signaling pathways that are involved in tissue repair, cellular differentiation, or proliferation. But because these routes are often linked together, changing one may have unforeseen effects on other pathways. For instance, small chemicals that encourage tissue regeneration or stem cell differentiation may unintentionally cause oncogenesis or unchecked cell proliferation. When using regenerative techniques to treat diseases like cancer or neurodegeneration, where the delicate balance between regeneration and malignant transformation is involved, this might be especially challenging. Therefore, maintaining safety and effectiveness in clinical settings requires to strike the correct balance between specificity and activity. Furthermore, small molecules may interact with a wide variety of targets in cells, tissues, or organs, resulting in off-target actions that may impair the process of regeneration or cause harmful side effects. Drugs that target signaling molecules, such as growth factors, transcription factors, or epigenetic modifiers, for example, may have detrimental effects on nontarget tissues. These unforeseen outcomes might jeopardize the small molecule's medicinal potential in addition to posing safety risks to patients.

(b) *Pharmacokinetic and Pharmacodynamic Challenges*
Pharmacokinetics (PK) and pharmacodynamics (PD) profiles, which outline the drug's absorption, distribution, metabolism, and excretion (ADME) as well as its effects on the target tissue over time, are crucial to a small molecule's effectiveness in regenerative medicine. Because the disorders being treated frequently include diverse tissue settings, it can be particularly difficult to get the ideal PK/PD profile in regenerative medicine. For instance, the goal of many regenerative therapies is to restore injured tissues with inadequate blood flow or poor vascularization, which can hinder the delivery of small molecules to the intended location. Additionally, the body's capacity to eliminate these molecules may lessen their effectiveness before they can start to have a regenerative effect, especially in the case of chronic illnesses. Certain situations call for the

direct delivery of small molecules to the site of damage or illness, necessitating the creation of innovative delivery methods like scaffolds, hydrogels, or nanoparticles that can precisely target the area that needs to be regenerated. Furthermore, the regeneration capability of small molecules in circulation may be influenced by their half-life. Long-term or high drug concentrations may also raise the risk of toxicity, yet many small compounds need constant exposure to retain their therapeutic benefits. On the other hand, low doses might not be able to initiate the intended regeneration processes. Therefore, it is essential to optimize the delivery methods and dose schedule in order to maximize therapeutic advantages and minimize side effects.

(c) *Biological Complexity and Tissue-Specific Responses*
The ability of small molecules to interact with intricate biological systems is crucial to regenerative medicine. Because the human body is not a straightforward, uniform environment, the behavior of tissues, and cells can differ significantly based on the particular microenvironment. For instance, cell type, tissue context, and the presence of inflammatory signals can all affect how stem cells react to signaling molecules. Because different tissue types have different receptors, communication pathways, and extracellular matrix components, tissues like bone, cartilage, or neuron may react differently to the same small chemical. The regenerative process frequently entails a dynamic interaction between different cell types, growth factors, cytokines, and extracellular matrix elements, all of which the small molecule can control. Due to variations in various microenvironments, a chemical that effectively stimulates tissue regeneration in one organ may not work in another. The tissue's reaction to regenerative small molecules is further complicated by the aging process and the underlying pathology of many degenerative disorders. The same dosage of a small molecule that is useful in young, healthy tissues may be less effective or even dangerous in older, damaged tissues. For example, aged tissues may have limited regeneration capacity due to compromised vascularization or decreased cellular plasticity.

(d) *Regulatory and Manufacturing Barriers*
Small compounds in regenerative medicine present a number of intricate and varied regulatory challenges. To guarantee their safety and effectiveness, small compounds used in regenerative therapies must pass stringent preclinical and clinical testing. However, because the therapeutic benefits of regenerative small molecules frequently depend on their interaction with intricate biological systems that are poorly understood, this approach can be much more difficult than for conventional pharmacological drugs. Preclinical research and early-stage clinical trials must specifically carefully assess the possibility of long-term or unexpected side effects, such as tumorigenesis or fibrosis. In addition to comprehensive information on the manufacturing process, regulatory bodies such as the United States Food and Drug Administration (US FDA) and European Medicines Agency (EMA) need solid data on the molecule's pharmacokinetics, toxicity, and biological effects. It can also be difficult to manufacture small compounds for regenerative applications, particularly when large-scale

production under strict good manufacturing practice (GMP) requirements is required. Additionally, compared to traditional medications, small molecules created as part of regenerative therapies might be governed by distinct regulatory pathways. Because of their innovative use in regenerative environments, they could occasionally call for further regulatory concerns. Regulatory bodies can, for instance, demand long-term research to evaluate how long the regeneration effects last and to keep an eye out for side effects that might appear only after extended exposure.

(e) *Need for Novel Drug Delivery Systems*

The design of efficient drug delivery systems is one of the main obstacles to the practical translation of small compounds in regenerative medicine. Small molecules must be administered in a targeted and regulated way in order to have their healing effects. Regenerative therapies frequently call for localized distribution to certain tissues or organs, in contrast to traditional medications that primarily treat systemic disorders. This makes medication distribution more complicated than simple pharmacokinetics and calls for the development of specific platforms for targeted tissue regeneration. Heart muscle following myocardial infarction or cartilage following osteoarthritis are two examples of tissues used in regenerative medicine that frequently have low vascularity or impaired healing ability. To overcome the difficulties caused by limited blood flow and tissue penetration, small molecules that promote regeneration must be given to these areas in sufficient amounts. For this, conventional oral or intravenous administration techniques are frequently insufficient.

To improve the bioavailability of small compounds at the site of injury, novel delivery systems such as hydrogels, liposomes, nanoparticles, and microspheres have been created. To ensure prolonged exposure to the target area, these systems have the ability to encapsulate the medication and release it gradually over time. In order to treat neurodegenerative illnesses, for instance, nanoparticles may be designed to precisely target cell surface receptors or to pass across the blood-brain barrier. These delivery techniques do have certain drawbacks, though, such maintaining formulation stability, avoiding immune system identification, and avoiding off-target effects that can worsen inflammation or tissue damage. Another interesting approach is the combination of small chemical therapy and gene delivery mechanisms, including viral vectors. When small molecules and gene delivery are combined, the molecules have the potential to improve tissue healing by causing both short-term regenerative effects and long-term cellular alterations. Nevertheless, this hybrid technique makes regulatory approval more complicated and calls for further concerns regarding the stability and safety of the vectors being employed.

(f) *Ethical Considerations and Patient Populations*

The use of small molecules in regenerative medicine brings up a number of moral and societal issues. The main worry is that regenerative technology could be abused, especially when it comes to aging or cosmetic improvements. Treatments that are used outside of their intended therapeutic setting, such as anti-aging medicines or regenerative small molecules to address cosmetic issues in otherwise healthy people, can provide ethical challenges. The issue of equity in access to regenerative medicines is another. By definition, regenerative medicine is an expensive discipline that frequently calls both cutting-edge technology and specific clinical knowledge. Because of this, not all patient populations may be able to afford the expense of treatments that use small molecules for regenerative reasons, especially in settings with limited resources or among poor communities. Because access to experimental medicines may be restricted to specific demographic groups, clinical trials themselves frequently make these inequities worse. As regenerative small molecules progress from laboratory to clinical application, ethical considerations in patient recruitment and consent will be crucial, especially with disadvantaged groups.

The long-term effects of regeneration therapy raise another ethical concern. Although there is potential for small molecules to help repair injured tissue, much remains unknown regarding their long-term impacts. Will their effects wear off with time and necessitate more treatments, or will they promote long-lasting regeneration? What effects may chronic therapies have on a patient's health, expenses, and total healthcare burden? Recurrent exposure to regenerating small molecules may occasionally even cause side effects like fibrosis or aberrant tissue growth. To fully grasp the promise of these treatments, long-term patient monitoring will be required, and clinicians and researchers must be open and honest about the hazards.

(g) *Biomarkers and Personalized Medicine*

Finding and validating biomarkers that can forecast treatment success is a crucial task in the translation of small molecules for regenerative medicine. Biomarkers are crucial for tracking the efficacy of treatment, determining whether the regenerative processes are proceeding as planned, and anticipating possible negative effects. It might be challenging to evaluate the effectiveness of treatment or to tailor therapy to each patient's particular biological profile in the absence of trustworthy biomarkers. When paired with developments in proteomics, metabolomics, and genomes, personalized medicine has immense potential in regenerative therapies. Personalized methods would optimize treatment for increased efficacy and safety by customizing small molecule medicines to each patient's unique genetic and molecular profile. Small compounds that target specific pathways that are dysregulated in their illness, for example, may be beneficial for patients with particular genetic mutations or epigenetic fingerprints. However, the process of finding and clinically validating these biomarkers is time-consuming and labor-intensive. For routine clinical usage, biomarkers need to be accessible and reasonably priced in addition to being trustworthy predictors of treatment effectiveness. Furthermore, the ethical ramifications of genetic testing and possible privacy issues are brought up by biomarker-based patient stratification, particularly when sensitive health data is involved.

(h) *Safety and Long-Term Efficacy*

One of the biggest barriers to successful clinical translation in regenerative medicine is the safety and long-term effectiveness of small molecules. The goal of many small-molecule regenerative therapies is to encourage tissue regeneration, frequently via activating stem cells or triggering cell reprogramming. The safety of these treatments in humans, especially over long periods of time, is yet unknown, despite their impressive preclinical model results. The danger of cancer is one of the main issues. The possibility of mutations or aberrant cell growth, which may result in cancer, can be increased by small chemicals that promote stem cell proliferation or reprogramming. Certain chemicals that promote tissue regeneration have been linked to a higher incidence of cancers in preclinical animals, underscoring the necessity of strict safety monitoring. In a similar vein, nothing is known about small molecules' long-term capacity for regeneration. Will the regenerated tissues continue to regenerate over time, or will they become dysfunctional or experience maladaptive alterations like scarring or fibrosis?

The possibility of immune system rejection or activation is another factor to take into account. The application of small chemicals to encourage regeneration may cause unforeseen changes in the immune system, possibly leading to autoimmune reactions or persistent inflammation. This is especially crucial for treating chronic illnesses like autoimmune disorders or chronic inflammatory diseases, where the immune system is already dysregulated.

(i) *Translational Challenges in Human Trials*

Finally, the transition from animal models to human trials is an intrinsic obstacle for the clinical translation of small compounds in regenerative medicine. Particularly when it comes to intricate tissue regeneration, animal models of illness or injury are frequently not entirely realistic in human biology. For example, some species, such as zebrafish or axolotls, may have a considerably higher ability for regeneration than humans, which could cause the therapeutic potential of the small molecule to be overestimated. Furthermore, because human beings are significantly more complicated than other species, factors like dose escalation, treatment plans, and long-term monitoring must be carefully formulated when translating small compounds into clinical trials.

Furthermore, patient variability frequently places restrictions on human clinical studies in regenerative medicine. Chronic disease patients may not react to treatments the same way as younger, healthy people, especially if they have extensive pathology or genetic predispositions. The interpretation of trial data and the generalizability of therapeutic outcomes can be further complicated by variations in immune responses, comorbidities, and previous treatments.

## 8 Future Directions and Opportunities

Small compounds have a bright future in regenerative medicine, despite the significant obstacles. The more efficient clinical use of small molecules in tissue regeneration and repair is being made possible by ongoing developments in medicinal chemistry, drug delivery technologies, and biomarker identification.

Significant opportunities consist of:

- *Combination Therapies*: Regenerative results may be improved by combining small medicines with additional treatment approaches including gene therapy, stem cell therapy, or tissue-engineered scaffolds. For instance, priming stem cells or tissue scaffolds with a small chemical could increase their ability to regenerate after being inserted into the body.
- *Targeted Small Molecules*: More precise molecular targets within the regenerative pathways are being found thanks to developments in molecular biology. For regenerative therapies to be successful, small compounds that precisely target these pathways while reducing off-target effects are essential.
- *Regenerative Medicine for Aging*: As the population ages, regenerative medicine with small molecules has the potential to treat age-related illnesses like osteoarthritis and neurological diseases. Enhancing quality of life and lowering the burden of age-related diseases are two ways that small compounds that support tissue regeneration and repair could contribute to addressing the global aging challenge.
- *Cellular Reprogramming*: New pathways for producing autologous cells for medical applications may be opened by small compounds that aid in cellular reprogramming, which has the ability to transform adult cells back into pluripotent stem cells. This could help get around a lot of immune rejection problems and ethical issues with using embryonic stem cells.

## 9 Conclusion

Regenerative medicine's use of small molecules is a rapidly growing field with enormous therapeutic potential. These compounds provide a flexible and effective way to alter biological functions, encourage tissue repair, and improve the efficacy of different stem cell and gene treatments. As previously mentioned, by focusing on particular signaling pathways involved in cell differentiation, survival, and tissue remodeling, small molecules have shown promise in a variety of fields, including bone, cartilage, cardiac, and neuroregeneration. The transition of small molecule-based treatments into clinical practice is still fraught with difficulties, despite notable preclinical achievements. To enable wider clinical application, challenges such guaranteeing long-term safety, overcoming patient variability, and resolving cost and accessibility issues need to be carefully managed. However, the effectiveness and accessibility of regeneration therapies could be improved by combining small molecules with biomaterials, sophisticated drug delivery systems, and customized medicine techniques. Regenerative therapy based on small molecules appears to have a bright future. More efficient and customized treatment approaches are probably in store as a result of advancements in precision medication delivery, patient-specific models, and genomic profiling. It is anticipated that as research progresses, small molecules will become more and more important in creating novel therapies for a variety of illnesses, ultimately advancing regenerative medicine and providing hope to patients with conditions for which there are few available treatments.

## References

Aguirre, M., Escobar, M., Amézquita, S. F., Cubillos, D., Rincón, C., Vanegas, P., Tarazona, M. P., Escobar, S. A., Blanco, J. C., & Celis, L. G. (2023). Application of the Yamanaka transcription factors Oct4, Sox2, Klf4, and c-Myc from the laboratory to the clinic. *Genes, 14*(9), 1697.

Ahn, H. J., Kim, Y. S., Kim, J. U., Han, S. M., Shin, J. W., & Yang, H. O. (2004). Mechanism of taxol-induced apoptosis in human SKOV3 ovarian carcinoma cells. *Journal of Cellular Biochemistry, 91*(5), 1043–1052.

Ahuja, P., Sdek, P., & MacLellan, W. R. (2007). Cardiac myocyte cell cycle control in development, disease, and regeneration. *Physiological Reviews, 87*, 521–544.

AlMuraikhi, N., Binhamdan, S., Alaskar, H., Alotaibi, A., Tareen, S., Muthurangan, M., & Alfayez, M. (2023). Inhibition of GSK-3β enhances osteoblast differentiation of human mesenchymal stem cells through Wnt Signalling overexpressing Runx2. *International Journal of Molecular Sciences, 24*(8), 7164.

Altyar, A. E., El-Sayed, A., Abdeen, A., Piscopo, M., Mousa, S. A., Najda, A., & Abdel-Daim, M. M. (2023). Future regenerative medicine developments and their therapeutic applications. *Biomedicine & Pharmacotherapy, 158*, 114131.

Azzini, E., Peña-Corona, S. I., Hernández-Parra, H., Chandran, D., Saleena, L. A. K., Sawikr, Y., & Calina, D. (2024). Neuroprotective and anti-inflammatory effects of curcumin in Alzheimer's disease: Targeting neuroinflammation strategies. *Phytotherapy Research, 38*, 3169.

(a) Bailey, A. M., Mendicino, M., & Au, P. (2014). An FDA perspective on clinical development of cell-based regenerative medicine products. *Nature Biotechnology, 32*(8), 721–723; (b) Mendelson, A., & Frenette, P. S. (2014). Hematopoietic stem cell niche maintenance during homeostasis and regeneration. *Nature Medicine*, *20*(8), 833–846.

Barry, F., & Murphy, M. (2013). Mesenchymal stem cells in joint disease and repair. *Nature Reviews Rheumatology, 9*, 584–594.

Baudino, T. A., Carver, W., Giles, W., & Borg, T. K. (2006). Cardiac fibroblasts: Friend or foe? *American Journal of Physiology. Heart and Circulatory Physiology, 291*, H1015–H1026.

Berkessel, A., Seelig, B., Schwengberg, S., Hescheler, J., & Sachinidis, A. (2010). Chemically induced cardiomyogenesis of mouse embryonic stem cells. *Chembiochem, 11*, 208–217.

Bharti, A. C., Takada, Y., & Aggarwal, B. B. (2004). Curcumin (diferuloylmethane) inhibits receptor activator of NF-κB ligand-induced NF-κB activation in osteoclast precursors and suppresses osteoclastogenesis. *The Journal of Immunology, 172*(10), 5940–5947.

Bonifacio, M. A., Cometa, S., Cochis, A., Gentile, P., Ferreira, A. M., Azzimonti, B., Procino, G., Ceci, E., De Rimondini, L., & Giglio, E. (2018). Antibacterial effectiveness meets improved mechanical properties: Manuka honey/gellan gum composite hydrogels for cartilage repair. *Carbohydrate Polymers, 198*, 462–472.

Brown, B. N., Ratner, B. D., Goodman, S. B., Amar, S., & Badylak, S. F. (2012). Macrophage polarization: An opportunity for improved outcomes in biomaterials and regenerative medicine. *Biomaterials, 33*(15), 3792.

Burridge, P. W., Keller, G., Gold, J. D., & Wu, J. C. (2012). Production of de novo cardiomyocytes: Human pluripotent stem cell differentiation and direct reprogramming. *Cell Stem Cell, 10*, 16–28.

Cai, K., Nanga, R. P., Lamprou, L., Schinstine, C., Elliott, M., Hariharan, H., & Epperson, C. N. (2012). The impact of gabapentin administration on brain GABA and glutamate concentrations: A 7T 1H-MRS study. *Neuropsychopharmacology, 37*(13), 2764–2771.

Cao, Y., Shumsky, J. S., Sabol, M. A., Kushner, R. A., Strittmatter, S., Hamers, F. P. T., & Murray, M. (2008). Nogo-66 receptor antagonist peptide (NEP1-40) administration promotes functional recovery and axonal growth after lateral funiculus injury in the adult rat. *Neurorehabilitation and Neural Repair, 22*(3), 262–278.

Cao, N., Liu, Z., Chen, Z., Wang, J., Chen, T., Zhao, X., Ma, Y., Qin, L., Kang, J., Wei, B., Wang, L., Jin, Y., & Yang, H.-T. (2012). Ascorbic acid enhances the cardiac differentiation of induced pluripotent stem cells through promoting the proliferation of cardiac progenitor cells. *Cell Research, 22*, 219–236.

Cao, Y., Gang, X., Sun, C., & Wang, G. (2017). Mesenchymal stem cells improve healing of diabetic foot ulcer. *Journal Diabetes Research, 1*, 9328347.

Centonze, G., Centonze, S., Ponzone, L., & Calautti, E. (2022). ROCK 'n TOR: An outlook on keratinocyte stem cell expansion in regenerative medicine via protein kinase inhibition. *Cells, 11*(7), 1130.

Chen, B., Dodge, M. E., Tang, W., Lu, J., Ma, Z., Fan, C. W., & Lum, L. (2009). Small molecule–mediated disruption of Wnt-dependent signaling in tissue regeneration and cancer. *Nature Chemical Biology, 5*(2), 100–107.

Cheng, L., Hu, W., Qiu, B., Zhao, J., Yu, Y., Guan, W., Wang, M., Yang, W., & Pei, G. (2014). Generation of neural progenitor cells by chemical cocktails and hypoxia. *Cell Research, 24*, 665–679.

Chu, T., Zhou, H., Lu, L., Kong, X., Wang, T., Pan, B., & Feng, S. (2015). Valproic acid-mediated neuroprotection and neurogenesis after spinal cord injury: From mechanism to clinical potential. *Regenerative Medicine, 10*(2), 193–209.

Cooke, J. P. (2019). Inflammation and its role in regeneration and repair: A caution for novel anti-inflammatory therapies. *Circulation Research., 124*(8), 1166.

Cooke, P., Janowitz, H., & Dougherty, S. E. (2022). Neuronal redevelopment and the regeneration of neuromodulatory axons in the adult mammalian central nervous system. *Frontiers in Cellular Neuroscience, 16*, 872501.

Crawford, B., Kim, D. G., Moon, E. S., Johnson, E., Fields, H. W., Palomo, J. M., & Johnston, W. M. (2014). Cervical vertebral bone mineral density changes in adolescents during orthodontic treatment. *American Journal of Orthodontics and Dentofacial Orthopedics, 146*(2), 183–189.

Diering, G. H., & Huganir, R. L. (2018). The AMPA receptor code of synaptic plasticity. *Neuron, 100*(2), 314–329.

Dillmann, F., Veldwijk, M. R., Laufs, S., Sperandio, M., Calandra, G., Wenz, F., & Fruehauf, S. (2009). Plerixafor inhibits chemotaxis toward SDF-1 and CXCR4-mediated stroma contact in a dose-dependent manner resulting in increased susceptibility of BCR-ABL+ cell to Imatinib and Nilotinib. *Leukemia & Lymphoma., 50*(10), 1676–1686.

Domínguez, J. M., Fuertes, A., Orozco, L., del Monte-Millán, M., Delgado, E., & Medina, M. (2012). Evidence for irreversible inhibition of glycogen synthase kinase-3β by tideglusib. *Journal of Biological Chemistry, 287*(2), 893–904.

Dong, Y., Yan, J., Xu, W., Paquet-Durand, F., Hu, Z., & Jiao, K. (2023). HDAC inhibition delays photoreceptor loss in Pde6b mutant mice of retinitis Pigmentosa: Insights from ScRNA-Seq and CUT & tag. *PeerJ, 11*, e15659.

(a) Donnelly, D. J., & Popovich, P. G. (2008). Inflammation and its role in neuroprotection, axonal regeneration and functional recovery after spinal cord injury. *Experimental Neurology, 209*, 378–388; (b) Tidball, J. G., Wehling-Henricks, M. (2007). Macrophages promote muscle membrane repair and muscle fiber growth and regeneration during modified muscle loading in mice in vivo. *The Journal of Physiology, 578*, 327–336.

Efe, J. A., Hilcove, S., Kim, J., Zhou, H., Ouyang, K., Wang, G., Chen, J., & Ding, S. (2011). Conversion of mouse fibroblasts into cardiomyocytes using a direct reprogramming strategy. *Nature Cell Biology, 13*, 215–222.

(a) El-Kadiry, A.E-H., Rafei, M., & Shammaa, R. (2021). Cell therapy: Types, regulation, and clinical benefits. *Frontiers in Medicine, 8*, 756029; (b). Mount, N.M. Ward, S. J. Kefalas, P., & Hyllner, J. (2015). Cell-based therapy technology classifications and translational challenges. *Philosophical Transactions of the Royal Society B – Journals, 370*, 20150017; (c). Golchin, A., & Farahany, T. Z. (2019). Biological products: Cellular therapy and FDA-approved products. *Stem Cell Reviews and Reports 15*, 166–175; (d) De Pieri, A. Rochev, Y., & Zeugolis, D. I. (2021). Scaffold-free cell-based tissue engineering therapies: Advances, shortfalls, and forecast. *NPJ Regenerative Medicine, 6*, 1–15.

Engle, S. J., & Puppala, D. (2013). Integrating human pluripotent stem cells into drug development. *Cell Stem Cell, 12*, 669–677.

Eulalio, A., Mano, M., Dal Ferro, M., Zentilin, L., Sinagra, G., Zacchigna, S., & Giacca, M. (2012). Functional screening identifies miRNAs inducing cardiac regeneration. *Nature, 492*, 376–381.

Farooq, M., Khan, A. W., Kim, M. S., & Choi, S. (2021). The role of fibroblast growth factor (FGF) signaling in tissue repair and regeneration. *Cells, 10*(11), 3242.

Federation, A. J., Bradner, J. E., & Meissner, A. (2013). The use of small molecules in somatic-cell reprogramming. *Trends in Cell Biology, 24*(3), 179.

Firestone, A. J., & Chen, J. K. (2010). Controlling destiny through chemistry: Small-molecule regulators of cell fate. *ACS Chemical Biology, 5*(1), 15–34.

Fong, E. L., Chan, C. K., & Goodman, S. B. (2011). Stem cell homing in musculoskeletal injury. *Biomaterials, 32*(2), 395–409.

Frye, S. V. (2010). The art of the chemical probe. *Nature Chemical Biology, 6*, 159–161.

Gao, M., Guo, H., Dong, X., Wang, Z., Yang, Z., Shang, Q., & Wang, Q. (2024). Regulation of inflammation during wound healing: The function of mesenchymal stem cells and strategies for therapeutic enhancement. *Frontiers in Pharmacology., 15*, 1345779.

Giacomini, C., Granéli, C., Hicks, R., & Dazzi, F. (2023). The critical role of apoptosis in mesenchymal stromal cell therapeutics and implications in homeostasis and normal tissue repair. *Cellular & Molecular Immunology, 20*(6), 570–582.

Goel, S., DeCristo, M. J., McAllister, S. S., & Zhao, J. J. (2018). CDK4/6 inhibition in cancer: Beyond cell cycle arrest. *Trends in Cell Biology, 28*(11), 911–925.

Gonzalez, R., Lee, J. W., & Schultz, P. G. (2011). Stepwise chemically induced cardiomyocyte specification of human embryonic stem cells. *Angewandte Chemie International Edition, 50*, 11181–11185.

Goswami, A. (2012). Prolotherapy. *Journal of Pain & Palliative Care Pharmacotherapy, 26*, 376–378.

Green, E. M., & Lee, R. T. (2013). Proteins and small molecules for cellular regenerative medicine. *Physiological Reviews, 93*(1), 311–325.

Green, A. C., Kocovski, P., Jovic, T., Walia, M. K., Chandraratna, R. A. S., Martin, T. J., & Purton, L. E. (2017). Retinoic acid receptor signalling directly regulates osteoblast and adipocyte differentiation from mesenchymal progenitor cells. *Experimental Cell Research, 350*(1), 284–297.

Grosso, A., Burger, M. G., Lunger, A., Schaefer, D. J., Banfi, A., & Di Maggio, N. (2017). It takes two to tango: Coupling of angiogenesis and osteogenesis for bone regeneration. *Frontiers in Bioengineering and Biotechnology, 5*, 68.

Harvey, P. A., & Leinwand, L. A. (2011). The cell biology of disease: Cellular mechanisms of cardiomyopathy. *The Journal of Cell Biology, 194*, 355–365.

Hata, K., Fujitani, M., Yasuda, Y., Doya, H., Saito, T., Yamagishi, S., & Yamashita, T. (2006). RGMa inhibition promotes axonal growth and recovery after spinal cord injury. *The Journal of Cell Biology, 173*(1), 47–58.

Hirt, J., Khanteymoori, A., Hohenhaus, M., Kopp, M. A., Howells, D. W., Schwab, J. M., & Watzlawick, R. (2023). Inhibition of the Nogo-pathway in experimental spinal cord injury: A meta-analysis of 76 experimental treatments. *Scientific Reports, 13*(1), 22898.

Hong, S. G., Dunbar, C. E., & Winkler, T. (2013). Assessing the risks of genotoxicity in the therapeutic development of induced pluripotent stem cells. *Molecular Therapy, 2*, 272–281.

Hosseinkhani, H., Domb, A. J., Sharifzadeh, G., & Nahum, V. (2023). Gene therapy for regenerative medicine. *Pharmaceutics., 6,15(3)*, 856.

Hou, P., Li, Y., Zhang, X., Liu, C., Guan, J., Li, H., Zhao, T., Ye, J., Yang, W., Liu, K., Ge, J., Xu, J., Zhang, Q., Zhao, Y., & Deng, H. (2013). Pluripotent stem cells induced from mouse somatic cells by small-molecule compounds. *Science, 341*, 651–654.

Hsieh, J., & Zhao, X. (2016). Genetics and epigenetics in adult neurogenesis. *Cold Spring Harbor Perspectives in Biology, 8*, a018911.

Hu, J., Rodemer, W., Zhang, G., Jin, L. Q., Li, S., & Selzer, M. E. (2021). Chondroitinase ABC promotes axon regeneration and reduces retrograde apoptosis signaling in lamprey. *Frontiers in Cell and Developmental Biology, 9*, 653638.

Hu, W., Qiu, B., Guan, W., Wang, Q., Wang, M., Li, W., Gao, L., Shen, L., Huang, Y., & Xie, G. (2015). Direct conversion of normal and Alzheimer's disease human fibroblasts into neuronal cells by small molecules. *Cell Stem Cell, 17*, 204–212.

Huang, L., Lai, X., Liang, X., Chen, J., Yang, Y., Xu, W., Qin, Q., Qin, R., Huang, X., & Xie, M. (2024). A promise for neuronal repair: Reprogramming astrocytes into neurons in vivo. *Bioscience Reports, 44*, BSR20231717.

Huangfu, D., Maehr, R., Guo, W., Eijkelenboom, A., Snitow, M., Chen, A. E., & Melton, D. A. (2008a). Induction of pluripotent stem cells by defined factors is greatly improved by small-molecule compounds. *Nature Biotechnology, 26*, 795–797.

Huangfu, D., Osafune, K., Maehr, R., Guo, W., Eijkelenboom, A., Chen, S., Muhlestein, W., & Melton, D. A. (2008b). Induction of pluripotent stem cells from primary human fibroblasts with only Oct4 and Sox2. *Nature Biotechnology, 26*, 1269–1275.

Huebner, E. A., & Strittmatter, S. M. (2009). Axon regeneration in the peripheral and central nervous systems. *Cell Biology of the Axon*, 305–360.

Iatridis, J. C., Nicoll, S. B., Michalek, A. J., Walter, B. A., & Gupta, M. S. (2013). Role of biomechanics in intervertebral disc degeneration and regenerative therapies, what needs repairing in the disc and what are promising biomaterials for its repair? *The Spine Journal, 13*(3), 243–262.

Ichida, J. K., Blanchard, J., Lam, K., Son, E. Y., Chung, J. E., Egli, D., Loh, K. M., Carter, A. C., Di Giorgio, F. P., Koszka, K., Huangfu, D., Akutsu, H., Liu, D. R., Rubin, L. L., & Eggan, K. (2009). A small molecule inhibitor of TGF-β signaling replaces Sox2 in reprogramming by inducing nanog. *Cell Stem Cell, 5*, 491–503.

Irie, Y., Mizumoto, H., Fujino, S., & Kajiwara, T. (2008). Development of articular cartilage grafts using organoid formation techniques. *Transplantation Proceedings, 40*, 631–633.

Jaklenec, A., Stamp, A., Deweerd, E., Sherwin, A., & Langer, R. (2012). Progress in the tissue engineering and stem cell industry "are we there yet?". *Tissue Engineering. Part B, Reviews, 18*(3), 155–166.

Jastrzębski, M. K., Wójcik, P., Stępnicki, P., & Kaczor, A. A. (2024). Effects of small molecules on neurogenesis: Neuronal proliferation and differentiation. *Acta Pharmaceutica Sinica B, 14*(1), 20–37.

Jin, Y., Tian, T., Ma, Y., Xu, H., & Du, Y. (2015). Simultaneous determination of ginsenoside Rb1, naringin, ginsenoside Rb2 and oridonin in rat plasma by LC–MS/MS and its application to a pharmacokinetic study after oral administration of Weifuchun tablet. *Journal of Chromatography B, 1000*, 112–119.

Johnson, T. V., Polo, A. D., Sahel, A., & Schuman, J. S. (2022). Neuroprotection, Neuroenhancement, and Neuroregeneration of the retina and optic nerve. *Ophthalmology Science, 2*(3), 100216.

Kapoor, T., Mehan, S., Suri, M., Sharma, N., Kumar, N., Narula, A. S., Alshammari, A., Alasmari, A. F., Alharbi, M., Assiri, M. A., & Kalfin, R. (2022). Forskolin, an Adenylcyclase/cAMP/CREB signaling activator restoring myelin-associated oligodendrocyte destruction in experimental ethidium bromide model of multiple sclerosis. *Cells, 11*(18), 2771.

Kattman, S. J., Witty, A. D., Gagliardi, M., Dubois, N. C., Niapour, M., Hotta, A., Ellis, J., & Keller, G. (2011). Stage-specific optimization of activin/nodal and BMP signaling promotes cardiac differentiation of mouse and human pluripotent stem cell lines. *Cell Stem Cell, 8*, 228–240.

Kleger, A., Seufferlein, T., Malan, D., Tischendorf, M., Storch, A., Wolheim, A., Latz, S., Protze, S., Porzner, M., Proepper, C., Brunner, C., Katz, S. F., Pusapati, G. V., Bullinger, L., Franz, W. M., Koehntop, R., Giehl, K., Spyrantis, A., Wittekindt, O., Lin, Q., Zenke, M., Fleischmann, B. K., Wartenberg, M., Wobus, A. M., Boeckers, T. M., & Liebau, S. (2010). Modulation of calcium-activated potassium channels induces cardiogenesis of pluripotent stem cells and enrichment of pacemaker-like cells. *Circulation, 122*, 1823–1836.

Laflamme, M. A., Chen, K. Y., Naumova, A. V., Muskheli, V., Fugate, J. A., Dupras, S. K., Reinecke, H., Xu, C., Hassanipour, M., Police, S., O'Sullivan, C., Collins, L., Chen, Y., Minami, E., Gill, E. A., Ueno, S., Yuan, C., Gold, J., & Murry, C. E. (2007). Cardiomyocytes derived from human embryonic stem cells in pro-survival factors enhance function of infarcted rat hearts. *Nature Biotechnology, 25*, 1015–1024.

Lane, S. W., Williams, D. A., & Watt, F. M. (2014). Modulating the stem cell niche for tissue regeneration. *Nature Biotechnology, 32*(8), 795–803.
Längle, D., Halver, J., Rathmer, B., Willems, E., & Schade, D. (2014). Small molecules targeting in vivo tissue regeneration. *ACS Chemical Biology, 9*(1), 57–71.
Lauterborn, J. C., Pineda, E., Chen, L. Y., Ramirez, E. A., Lynch, G., & Gall, C. M. (2008). Ampakines cause sustained increases in BDNF signaling at excitatory synapses without changes in AMPA receptor subunit expression. *Neuroscience, 159*(1), 283.
Law, S. M., & Zheng, J. J. (2022). Premise and peril of Wnt signaling activation through GSK-3β inhibition. *iScience, 25*(4), 104159.
Lee, S. R., Kim, H. Y., Rogowska, J., Zhao, B. Q., Bhide, P., Parent, J. M., & Lo, E. H. (2006). Involvement of matrix metalloproteinase in neuroblast cell migration from the subventricular zone after stroke. *Journal of Neuroscience, 26*(13), 3491–3495.
Li, W., Zhou, H., Abujarour, R., Zhu, S., Young, J. J., Lin, T., Hao, E., Schöler, H. R., Hayek, A., & Ding, S. (2009). Generation of human-induced pluripotent stem cells in the absence of exogenous Sox2. *Stem Cells, 27*, 2992–3000.
Li, R., Liang, J., Ni, S., Zhou, T., Qing, X., Li, H., He, W., Chen, J., Li, F., Zhuang, Q., Qin, B., Xu, J., Li, W., Yang, J., Gan, Y., Qin, D., Feng, S., Song, H., Yang, D., Zhang, B., Zeng, L., Lai, L., Esteban, M. A., & Pei, D. (2010). A mesenchymal-to-epithelial transition initiates and is required for the nuclear reprogramming of mouse fibroblasts. *Cell Stem Cell, 7*, 51–63.
Li, Y., Zhang, Q., Yin, X., Yang, W., Du, Y., Hou, P., Ge, J., Liu, C., Zhang, W., Zhang, X., Wu, Y., Li, H., Liu, K., Wu, C., Song, Z., Zhao, Y., Shi, Y., & Deng, H. (2011). Generation of iPSCs from mouse fibroblasts with a single gene, Oct4, and small molecules. *Cell Research, 21*, 196–204.
Li, W., Li, K., Wei, W., & Ding, S. (2013). Chemical approaches to stem cell biology and therapeutics. *Cell Stem Cell, 13*, 270–283.
Li, X., Zuo, X., Jing, J., Ma, Y., Wang, J., Liu, D., Zhu, J., Du, X., Xiong, L., & Du, Y. (2015). Small-molecule-driven direct reprogramming of mouse fibroblasts into functional neurons. *Cell Stem Cell, 17*, 195–203.
Lian, X., Hsiao, S., Wilson, G., Zhu, K., Hazeltine, L. B., Azarin, S. M., Raval, K., Zhang, J., Kamp, T. J., & Palecek, S. P. (2012). Robust cardiomyocyte differentiation from human pluripotent stem cells via temporal modulation of canonical Wnt signaling. *Proceedings. National Academy of Sciences. United States of America, 109*, E1848–E1857.
Linetsky, F. S., & Manchikanti, L. (2005). Regenerative injection therapy for axial pain. *Tech Reg Anesth Pain Manage., 9*, 40–49.
Liu, X., Tu, Y., Zhang, L., Qi, J., Ma, T., & Deng, L. (2014). Prolyl hydroxylase inhibitors protect from the bone loss in ovariectomy rats by increasing bone vascularity. *Cell Biochemistry and Biophysics, 69*, 141–149.
(a) Liu, K., Yu, C., Xie, M., Li, K., & Ding, S. (2016). Chemical modulation of cell fate in stem cell therapeutics and regenerative medicine. *Cell Chemical Biology*, *23*(8), 893–916; (b). Hou, P., Li, Y., Zhang, X., Liu, C., Guan, J., Li, H., Zhao, T., Ye, J., Yang, W., Liu, K., Ge, J., Xu, J., Zhang, Q., Zhao, Y., & Deng, H. (2013). Pluripotent stem cells induced from mouse somatic cells by small-molecule compounds. *Science.*
Liu, L., Lauro, B. M., Ding, L., Rovere, M., Wolfe, M. S., & Selkoe, D. J. (2019). Multiple BACE1 inhibitors abnormally increase the BACE1 protein level in neurons by prolonging its half-life. *Alzheimer's & Dementia, 15*(9), 1183–1194.
Loffredo, F. S., Steinhauser, M. L., Gannon, J., & Lee, R. T. (2011). Bone marrow-derived cell therapy stimulates endogenous cardiomyocyte progenitors and promotes cardiac repair. *Cell Stem Cell, 8*, 389–398.
Lu, B., & Atala, A. (2014). Small molecules and small molecule drugs in regenerative medicine. *Drug Discovery Today, 19*, 801–808.
Lukaszewicz, A. I., Mcmillan, M. K., & Kahn, M. S. (2010). Perspective: Small molecules and stem cells, potency, and lineage commitment: The new quest for the fountain of youth. *Journal of Medicinal Chemistry, 53*(9), 3439.

Lyssiotis, C. A., Foreman, R. K., Staerk, J., Garcia, M., Mathur, D., Markoulaki, S., Hanna, J., Lairson, L. L., Charette, B. D., Bouchez, L. C., Bollong, M., Kunick, C., Brinker, A., Cho, C. Y., Schultz, P. G., & Jaenisch, R. (2009). Reprogramming of murine fibroblasts to induced pluripotent stem cells with chemical complementation of Klf4. *Proceedings. National Academy of Sciences. United States of America, 106*, 8912–8917.

Maherali, N., & Hochedlinger, K. (2009). Tgfβ signal inhibition cooperates in the induction of iPSCs and replaces Sox2 and cMyc. *Current Biology, 19*, 1718–1723.

Mall, M., & Wernig, M. (2017). The novel tool of cell reprogramming for applications in molecular medicine. *Journal of Molecular Medicine, 95*, 695–703.

Mao, A. S., & Mooney, D. J. (2015). Regenerative medicine: Current therapies and future directions. *National Academy of Sciences of the United States of America, 112*(47), 14452–14459.

Martinotti, S., Bucekova, M., Majtan, J., & Ranzato, E. (2019). Honey: An effective regenerative medicine product in wound management. *Current Medicinal Chemistry, 26*, 5230–5240.

Mason, C. D. (2008). A brief definition of regenerative medicine. *Regenerative Medicine, 3*, 1–5.

Mercalli, A., Calavita, I., Dugnani, E., Citro, A., Cantarelli, E., Nano, R., & Piemonti, L. (2013). Rapamycin unbalances the polarization of human macrophages to M 1. *Immunology, 140*(2), 179–190.

Merkle, F. T., & Eggan, K. (2013). Modeling human disease with pluripotent stem cells: From genome association to function. *Cell Stem Cell, 12*, 656–668.

Mikkelsen, T. S., Hanna, J., Zhang, X., Ku, M., Wernig, M., Schorderet, P., Bernstein, B. E., Jaenisch, R., Lander, E. S., & Meissner, A. (2008). Dissecting direct reprogramming through integrative genomic analysis. *Nature, 454*, 49–55.

Minami, I., Yamada, K., Otsuji, T. G., Yamamoto, T., Shen, Y., Otsuka, S., Kadota, S., Morone, N., Barve, M., Asai, Y., Tenkova- Heuser, T., Heuser, J. E., Uesugi, M., Aiba, K., & Nakatsuji, N. (2012). A small molecule that promotes cardiac differentiation of human pluripotent stem cells under defined, cytokine- and xeno-free conditions. *Cell Reports, 2*, 1448–1160.

Moheb-Alian, A., Forouzesh, F., Rostami-Nejad, M., & Rostami, K. (2016). Mesenchymal stem cells as potential therapeutic approaches in celiac disease. *Gastroenterology and Hepatology from Bed to Bench, 9*(Suppl1), S1.

Mokady, D., Charish, J., Barretto-Burns, P., Grisé, K. N., Coles, B. L. K., Raab, S., Ortin-Martinez, A., Müller, A., Fasching, B., & Jain, P. (2024). Small-molecule-directed endogenous regeneration of visual function in a mammalian retinal degeneration model. *International Journal of Molecular Sciences, 25*, 1521.

Muñoz Ruiz, M., & Regueiro, J. R. (2012). New tools in regenerative medicine: Gene therapy. In C. López-Larrea, A. López-Vázquez, & B. Suárez-Álvarez (Eds.), *Stem cell transplantation. Advances in experimental medicine and biology* (Vol. 741, pp. 254–275). Springer.

Nagappan, P. G., Chen, H., & Wang, D. Y. (2020). Neuroregeneration and plasticity: A review of the physiological mechanisms for achieving functional recovery postinjury. *Military Medical Research, 7*, 1–16.

Najafi, S., Barasa, L., Huang, S. Y., Yoganathan, S., & Perron, J. C. (2022). Discovery of a novel class of benzimidazoles as highly effective agonists of bone morphogenetic protein (BMP) receptor signaling. *Scientific Reports, 12*(1), 12146.

Ng, J. Y., Obuobi, S., Chua, M. L., Zhang, C., Hong, S., Kumar, Y., Gokhale, R., & Ee, P. L. R. (2020). Biomimicry of microbial polysaccharide hydrogels for tissue engineering and regenerative medicine – a review. *Carbohydrate Polymers, 241*, 116345.

Nie, B., Wang, H., Laurent, T., & Ding, S. (2012). Cellular reprogramming: A small molecule perspective. *Current Opinion in Cell Biology, 24*, 784–792.

Paino, F., Tirino, V., Naddeo, P., Desiderio, V., Pirozzi, G., Rosa, A. D., Laino, L., Altucci, L., & Papaccio, G. (2014). Histone deacetylase inhibition with Valproic acid downregulates Osteocalcin gene expression in human dental pulp stem cells and osteoblasts: Evidence for HDAC2 involvement. *Stem Cells (Dayton, Ohio), 32*(1), 279.

Park, S. Y., An, J. M., Seo, J. T., & Seo, S. R. (2020). Y-27632 induces neurite outgrowth by activating the NOX1-mediated AKT and PAK1 phosphorylation cascades in PC12 cells. *International Journal of Molecular Sciences, 21*(20), 7679.

Parmeggiani, F. S., Sorrentino, F., Ponzin, D., Barbaro, V., Ferrari, S., & Di Iorio, E. (2011). Retinitis pigmentosa: Genes and disease mechanisms. *Current Genomics, 12*, 238.

Pasano Bojang, J., & Ramos, K. S. (2013). The promise and failures of epigenetic therapies for cancer treatment. *Cancer Treatment Reviews, 40*(1), 153. https://doi.org/10.1016/j.ctrv.2013.05.009

Passier, R., van Laake, L. W., & Mummery, C. L. (2008). Stem cell-based therapy and lessons from the heart. *Nature, 453*, 322–329.

Puskas, F. T. E., Mandelbaum, B. L., Gerhardt, B. R., & M.B. and Rodeo, S.A. (2009). Platelet-rich plasma: From basic science to clinical applications. *The American Journal of Sports Medicine, 37*, 2259–2272.

Qian, L., Huang, Y., Spencer, C. I., Foley, A., Vedantham, V., Liu, L., Conway, S. J., Fu, J.-D., & Srivastava, D. (2012). In vivo reprogramming of murine cardiac fibroblasts into induced cardiomyocytes. *Nature, 485*, 593–598.

Reinders, M. E. J., Fibbe, W. E., & Rabelink, T. J. (2010). Multipotent mesenchymal stromal cell therapy in renal disease and kidney transplantation. *Nephrology, Dialysis, Transplantation, 25*, 17–24.

Reya, T., & Clevers, H. (2005). Wnt signalling in stem cells and cancer. *Nature, 434*, 843–850.

Rivera-Izquierdo, M., Cabeza, L., Ramos-Bossini, A., Quesada, L., Perazzoli, R., Alvarez, G., Prados, P., & Melguizo, J. (2019). An updated review of adipose derived-mesenchymal stem cells and their applications in musculoskeletal disorders. *Expert Opinion on Biological Therapy, 19*, 233–248.

Roato, I., Belisario, D. C., Compagno, M., Verderio, L., Sighinolfi, A., Mussano, F., & Ferracini, R. (2018). Adipose-derived stromal vascular fraction/xenohybrid bone scaffold: An alternative source for bone regeneration. *Stem Cells International, 1*, 4126379.

Saag, K. G., Petersen, J., Brandi, M. L., Karaplis, A. C., Lorentzon, M., Thomas, T., & Grauer, A. (2017). Romosozumab or alendronate for fracture prevention in women with osteoporosis. *New England Journal of Medicine, 377*(15), 1417–1427.

Sachinidis, A., Schwengberg, S., Hippler-Altenburg, R., Mariappan, D., Kamisetti, N., Seelig, B., Berkessel, A., & Hescheler, J. (2006). Identification of small signalling molecules promoting cardiac-specific differentiation of mouse embryonic stem cells. *Cellular Physiology and Biochemistry, 18*, 303–314.

Seidkhani-Nahal, A., Khosravi, A., Mirzaei, A., Basati, G., Abbasi, M., & Noori-Zadeh, A. (2021). Serum vascular endothelial growth factor (VEGF) levels in ischemic stroke patients: A systematic review and meta-analysis of case–control studies. *Neurological Sciences, 42*, 1811–1820.

Shah, T. A., Zhu, Y., Shaikh, N. N., Harris, M. A., Harris, S. E., & Rogers, M. B. (2017). Characterization of new bone morphogenetic protein (bmp)-2 regulatory alleles. *Genesis, 55*(7), e23035.

Shen, X., Wan, C., Ramaswamy, G., Mavalli, M., Wang, Y., Duvall, C. L., & Gilbert, S. R. (2009). Prolyl hydroxylase inhibitors increase neoangiogenesis and callus formation following femur fracture in mice. *Journal of Orthopaedic Research, 27*(10), 1298–1305.

Shi, Y., Desponts, C., Do, J. T., Hahm, H. S., Schöler, H. R., & Ding, S. (2008a). Induction of pluripotent stem cells from mouse embryonic fibroblasts by Oct4 and Klf4 with small-molecule compounds. *Cell Stem Cell, 3*(5), 568–574.

Shi, Y., Do, J. T., Desponts, C., Hahm, H. S., Schöler, H. R., & Ding, S. (2008b). A combined chemical and genetic approach for the generation of induced pluripotent stem cells. *Cell Stem Cell, 2*, 525–528.

Shiba, Y., Fernandes, S., Zhu, W.-Z., Filice, D., Muskheli, V., Kim, J., Palpant, N. J., Gantz, J., Moyes, K. W., Reinecke, H., Van Biber, B., Dardas, T., Mignone, J. L., Izawa, A., Hanna, R., Viswanathan, M., Gold, J. D., Kotlikoff, M. I., Sarvazyan, N., Kay, M. W., Murry, C. E., & Laflamme, M. A. (2012). Human ES-cell-derived cardiomyocytes electrically couple and suppress arrhythmias in injured hearts. *Nature, 489*, 322–325.

Shibuya, M. (2011). Vascular endothelial growth factor (VEGF) and its receptor (VEGFR) signaling in angiogenesis: A crucial target for anti- and pro-Angiogenic therapies. *Genes & Cancer., 2*(12), 1097.

Smart, N., Bollini, S., Dubé, K. N., Vieira, J. M., Zhou, B., Davidson, S., Yellon, D., Riegler, J., Price, A. N., Lythgoe, M. F., Pu, W. T., & Riley, P. R. (2011). De novo cardiomyocytes from within the activated adult heart after injury. *Nature, 474*, 640–644.

Snitow, M. E., Bhansali, R. S., & Klein, P. S. (2021). Lithium and therapeutic targeting of GSK-3. *Cells, 10*(2), 255.

Song, K., Nam, Y. J., Luo, X., Qi, X., Tan, W., Huang, G. N., Acharya, A., Smith, C. L., Tallquist, M. D., Neilson, E. G., Hill, J. A., Bassel-Duby, R., & Olson, E. N. (2012). Heart repair by reprogramming non-myocytes with cardiac transcription factors. *Nature, 485*, 599–604.

Spivakov, M., & Fisher, A. G. (2007). Epigenetic signatures of stem-cell identity. *Nature Reviews. Genetics, 8*, 263–271.

Stocchi, F., Fossati, C., & Torti, M. (2015). Rasagiline for the treatment of Parkinson's disease: An update. *Expert Opinion on Pharmacotherapy, 16*(14), 2231–2241.

Sturzu, A. C., & Wu, S. M. (2011). Developmental and regenerative biology of multipotent cardiovascular progenitor cells. *Circulation Research, 108*, 353–364.

(a) Sultana, M., Khan, F. S., Khalid, M., Ali, A. H., & Bazighifan, O. (2022). Comparison of predator-prey model and hawk-dove game for modelling leukemia. *Computational Intelligence and Neuroscience, 2022*(1), 9957514; (b). Poliwoda, S., Noor, N., Downs, E., Schaaf, A., Cantwell, A., Ganti, L., Kaye, A. D., Mosel, L. I., Carroll, C. B., Viswanath, O., & Urits, I. (2022). Stem cells: A comprehensive review of origins and emerging clinical roles in medical practice. *Orthopedic Reviews, 14*(3), 37498.

Surendran, H., Nandakumar, S., Reddy, K. V. B., Stoddard, J., Mohan, K. V., Upadhyay, P. K., McGill, T. J., & Pal, R. (2021). Transplantation of retinal pigment epithelium and photoreceptors generated concomitantly via small molecule-mediated differentiation rescues visual function in rodent models of retinal degeneration. *Stem Cell Research & Therapy, 12*, 70.

Takahashi, K., & Yamanaka, S. (2006). Induction of pluripotent stem cells from mouse embryonic and adult fibroblast cultures by defined factors. *Cell, 126*, 663–676.

Takebe, N., Miele, L., Harris, P. J., Jeong, W., Bando, H., Kahn, M., Yang, S. X., & Ivy, S. P. (2015). Targeting notch, hedgehog, and Wnt pathways in cancer stem cells: Clinical update. *Nature Reviews Clinical Oncology, 12*(8), 445.

Tan, F., Qian, C., Tang, K., Abd-Allah, S. M., & Jing, N. (2014). Inhibition of transforming growth factor β (TGF-β) signaling can substitute for Oct4 protein in reprogramming and maintain pluripotency. *The Journal of Biological Chemistry, 290*(7), 4500.

Thapliyal, S., Arendt, K. L., Lau, A. G., & Chen, L. (2022). Retinoic acid-gated BDNF synthesis in neuronal dendrites drives presynaptic homeostatic plasticity. *eLife, 11*, e79863.

Tousi, B. (2015). The emerging role of bexarotene in the treatment of Alzheimer's disease: Current evidence. *Neuropsychiatric Disease and Treatment, 11*, 311–315.

Uosaki, H., Magadum, A., Seo, K., Fukushima, H., Takeuchi, A., Nakagawa, Y., Moyes, K. W., Narazaki, G., Kuwahara, K., Laflamme, M., Matsuoka, S., Nakatsuji, N., Nakao, K., Kwon, C., Kass, D. A., Engel, F. B., & Yamashita, J. K. (2013). Identification of chemicals inducing cardiomyocyte proliferation in developmental stage-specific manner with pluripotent stem cells. *Circulation. Cardiovascular Genetics, 6*, 624–633.

van Laake, L. W., Passier, R., Monshouwer-Kloots, J., Verkleij, A. J., Lips, D. J., Freund, C., den Ouden, K., Ward-van Oostwaard, D., Korving, J., Tertoolen, L. G., van Echteld, C. J., Doevendans, P. A., & Mummery, C. L. (2007). Human embryonic stem cell-derived cardiomyocytes survive and mature in the mouse heart and transiently improve function after myocardial infarction. *Stem Cell Research, 1*, 9–24.

Veith, A. P., Henderson, K., Spencer, A., Sligar, A. D., & Baker, A. B. (2019). Therapeutic strategies for enhancing angiogenesis in wound healing. *Advanced Drug Delivery Reviews, 146*, 97–125.

Voskuhl, R. (2020). A new cell subtype that confers neuroprotection. *Nature Immunology, 21*, 1477–1478.

Wang, H., Hao, J., & Hong, C. C. (2011). Cardiac induction of embryonic stem cells by a small molecule inhibitor of Wnt/betacatenin signaling. *ACS Chemical Biology, 6*, 192–197.

Wein, M. N., & Kronenberg, H. M. (2018). Regulation of bone remodeling by parathyroid hormone. *Cold Spring Harbor Perspectives in Medicine, 8*, a031237.

Wernig, M., Meissner, A., Cassady, J. P., & Jaenisch, R. (2008). C-Myc is dispensable for direct reprogramming of mouse fibroblasts. *Cell Stem Cell, 2*, 10–12.

Willems, E., Cabral-Teixeira, J., Schade, D., Cai, W., Reeves, P., Bushway, P. J., Lanier, M., Walsh, C., Kirchhausen, T., Izpisua Belmonte, J. C., Cashman, J., & Mercola, M. (2012). Small molecule mediated TGF-β type II receptor degradation promotes cardiomyogenesis in embryonic stem cells. *Cell Stem Cell, 11*, 242–252.

Wu, M., Chen, G., & Li, Y. P. (2016). TGF-β and BMP signaling in osteoblast, skeletal development, and bone formation, homeostasis and disease. *Bone Research, 4*(1), 1–21.

Xie, X., Fu, Y., & Liu, J. (2017). Chemical reprogramming and Transdifferentiation. *Current Opinion in Genetics & Development, 46*, 104–113.

Xin, M., Olson, E. N., & Bassel-Duby, R. (2013). Mending broken hearts: Cardiac development as a basis for adult heart regeneration and repair. *Nature Reviews. Molecular Cell Biology, 14*, 529–541.

Yu, L., Chen, X., Wang, L., & Chen, S. (2016). The sweet trap in tumors: Aerobic glycolysis and potential targets for therapy. *Oncotarget, 7*(25), 38908.

Yu, Y., Shen, X., Luo, Z., Hu, Y., Li, M., Ma, P., & Cai, K. (2018). Osteogenesis potential of different titania nanotubes in oxidative stress microenvironment. *Biomaterials, 167*, 44–57.

Yuan, X., Wan, H., Zhao, X., Zhu, S., Zhou, Q., & Ding, S. (2011). Brief report: Combined chemical treatment enables oct4- induced reprogramming from mouse embryonic fibroblasts. *Stem Cells, 29*, 549–553.

Yuan, J., Wang, X., Ma, D., Gao, H., Zheng, D., & Zhang, J. (2020). Resveratrol rescues TNF-α-induced inhibition of osteogenesis in human periodontal ligament stem cells via the ERK1/2 pathway. *Molecular Medicine Reports, 21*(5), 2085–2094.

Yue, B. (2014). Biology of the extracellular matrix: An overview. *Journal of Glaucoma, 23*, S20–S23.

Zangi, L., Lui, K. O., von Gise, A., Ma, Q., Ebina, W., Ptaszek, L. M., Später, D., Xu, H., Tabebordbar, M., Gorbatov, R., Sena, B., Nahrendorf, M., Briscoe, D. M., Li, R. A., Wagers, A. J., Rossi, D. J., Pu, W. T., & Chien, K. R. (2013). Modified mRNA directs the fate of heart progenitor cells and induces vascular regeneration after myocardial infarction. *Nature Biotechnology, 31*, 898–907.

Zhang, Q., Jiang, J., Han, P., Yuan, Q., Zhang, J., Zhang, X., Xu, Y., Cao, H., Meng, Q., Chen, L., Tian, T., Wang, X., Li, P., Hescheler, J., Ji, G., & Ma, Y. (2011). Direct differentiation of atrial and ventricular myocytes from human embryonic stem cells by alternating retinoid signals. *Cell Research, 21*, 579–587.

Zhang, L., Yin, J. C., Yeh, H., Ma, N. X., Lee, G., Chen, X. A., Wang, Y., Lin, L., Chen, L., & Jin, P. (2015). Small molecules efficiently reprogram human Astroglial cells into functional neurons. *Cell Stem Cell, 17*, 735–747.

Zhang, M., Lin, Y. H., Sun, Y. J., Zhu, S., Zheng, J., Liu, K., Cao, N., Li, K., Huang, Y., & Ding, S. (2016a). Pharmacological reprogramming of fibroblasts into neural stem cells by signaling-directed transcriptional activation. *Cell Stem Cell, 18*, 653–667.

Zhang, J., Guan, J., Qi, X., Ding, H., Yuan, H., Xie, Z., & Huang, Y. (2016b). Dimethyloxaloylglycine promotes the angiogenic activity of mesenchymal stem cells derived from iPSCs via activation of the PI3K/Akt pathway for bone regeneration. *International Journal of Biological Sciences, 12*(6), 639.

Zhao, J., Santino, F., Giacomini, D., & Gentilucci, L. (2020). Integrin-targeting peptides for the design of functional cell-responsive biomaterials. *Biomedicine, 8*(9), 307.

Zhu, S., Li, W., Zhou, H., Wei, W., Ambasudhan, R., Lin, T., Kim, J., Zhang, K., & Ding, S. (2010a). Reprogramming of human primary somatic cells by OCT4 and chemical compounds. *Cell Stem Cell, 7*, 651–655.

Zhu, W. Z., Xie, Y., Moyes, K. W., Gold, J. D., Askari, B., & Laflamme, M. A. (2010b). Neuregulin/ErbB signaling regulates cardiac subtype specification in differentiating human embryonic stem cells. *Circulation Research, 107*, 776–786.

Zhu, S., Ambasudhan, R., Sun,W., Kim, H.J., Talantova, M., Wang, X., Zhang, M., Zhang, Y., Laurent, T., and Parker, J. Small molecules enable OCT4-mediated direct reprogramming into expandable human neural stem cells. Cell Research 2013, 24, 126–129.

Zwi, L., Caspi, O., Arbel, G., Huber, I., Gepstein, A., Park, I. H., & Gepstein, L. (2009). Cardiomyocyte differentiation of human induced pluripotent stem cells. *Circulation, 120*, 1513–1523.

# Compound Delivery for Effective Regeneration

# Nanoparticle-Mediated Delivery of Materials for Effective Regeneration

**Archana Shrestha, Md Anamul Haque, and George Mattheolabakis**

**Abstract** The field of regenerative medicine is vastly growing, with its applications ranging from many acute to chronic disease conditions. While the body's natural regenerative and healing properties play a vital function as a part of the defense mechanism, external influence to boost the body's regenerative traits has gained significant traction in recent years. The incorporation of nanoparticles in regenerative drug delivery aims to enhance bioavailability, improve biological barrier crossing, create sustained drug release, and deliver compounds to desired sites. Numerous polymeric, organic, and inorganic nanomaterials are explored due to their wide applicability and significant success in delivering biomolecules. The range of nanomaterials to choose from also allows the delivery of a diverse list of molecules, such as small biomolecules, peptides, and genes for tissue engineering and regeneration. Here, we discuss a repertoire of nanocarriers designated for tissue regeneration and their extensive therapeutic applications in this field. We also explore further the current trends in using various nanocarriers based on their physicochemical characteristics and compatibility with the therapeutic molecules for tissue engineering and regeneration.

**Keywords** Tissue regeneration · Polymeric nanoparticles · Organic nanoparticles · Inorganic nanoparticles · Biocompatibility · Drug delivery · Targeted delivery

---

A. Shrestha · M. A. Haque · G. Mattheolabakis (✉)
School of Basic Pharmaceutical and Toxicological Sciences, College of Pharmacy, University of Louisiana at Monroe, Monroe, LA, USA
e-mail: shresthaar@warhawks.ulm.edu; haquema@warhawks.ulm.edu; matthaiolampakis@ulm.edu



H. Das (ed.), *Tissue Repair and Regeneration*,
https://doi.org/10.1007/978-3-031-93677-7_17

# 1 Introduction

Modern medicine and technological advancements have allowed for significant improvements in our quality of life, reducing mortality rates at younger ages, and eventually leading to an overall significant extension of the average life expectancy of humans. Simultaneously, as the human population has, on average, increased its age, chronic diseases have become prominent in the new extended human lifespan. The primary approach for chronic diseases has been to "handle the symptoms." That includes providing medical attention to limit the manifestation or progression of diseases that frequently may not be curable, such as diabetes, heart disease, or cancer. Regenerative medicine aims to heal or replace damaged tissue and organs to counter trauma, age, or diseases (Mao & Mooney, 2015).

As the human body has a remarkable capacity for self-healing, regenerative medicine focuses on harnessing the human body's capability to regenerate. Regenerative medicine may incorporate a variety of methodologies or targeted tissues, including tissue engineering, scaffolding, or drug administration, for targeting tissues and organs, including teeth, nerves, bones, or skin regeneration, among others (Mao & Mooney, 2015; Takeo et al., 2015; van Rijt & Habibovic, 2017; Zhang & Yelick, 2021).

While tissue regeneration can incorporate stem-cell treatments, tissue transplantation, or drug-based therapies, here we will focus on the latter and evaluate different drug delivery approaches that have been developed, with a particular interest in the nanosized-based delivery systems. In fact, significant advancements in drug delivery through the development of a plethora of nanosized carriers has allowed for the administration of molecules with widely varying physicochemical properties and the possibility of targeting specific tissues (Mattheolabakis et al., 2012). Biodegradable and biocompatible materials based on lipids and polymers or inorganic substances have emerged to improve drug administration, which would otherwise be difficult to do. With an increasing number of nanosized-based formulations being US Food and Drug Administration (FDA)-approved or evaluated in clinical trials (Abdellatif & Alsowinea, 2021; Patra et al., 2018), more complex drug structures, including labile-sensitive molecules, such as nucleic acids, are being assessed.

Nanoparticles and nanomedicines have traditionally been associated with cancer treatments. As tumors potentially develop their microenvironment and blood vessels, nanoparticles, sub-micron particles commonly sized below 200 nm, can utilize the enhanced permeability and retention effect to potentially improve drug uptake in the tumor area (Lahooti et al., 2023). In contrast, there have been significant advancements in active targeting as well. Nanoparticles present a large surface area that allows chemical modifications and the attachment of molecules. Hence, ligands, peptides, antibodies, sugars, and other molecules that can target specific receptors on specific cell types have been conjugated on nanoparticles to induce active targeting to these cells. This can enhance drug accumulation specificity to target cells/tissues, potentially diminishing dosing requirements or side effects to non-targeted tissues (Clemons et al., 2018; Lahooti et al., 2023; Mattheolabakis et al., 2012; Poudel et al., 2021; Shrestha et al., 2022). Cumulatively, nanoparticles have contributed to the protection and delivery of labile, challenging-to-dissolve compounds, improving pharmacokinetics and biodistribution, and potentially enhancing drug efficacy.

Similarly, tissue regeneration and regenerative medicine have been the focus of the utilization of nanocarriers. The properties of nanoparticles to regulate drug release rates and deliver complicated compounds to desired tissues can be significantly advantageous in regenerative medicine, as these treatments can require extensive therapeutic periods and selectivity of treatment to desired tissues involving complex molecules. In this chapter, we will present representative therapeutic applications of this vast field, including nanocarriers for tissue regeneration. We will present examples of inorganic and organic nanocarriers, their application, and the future potential of the convergence of the fields of nanotechnology and regenerative medicine.

## 2 Drug Delivery Using Nanoparticles in Regenerative Medicine

Regenerative medicine and tissue engineering is a broad therapeutic field targeting a variety of tissues in the body. As with any other drug-based therapy, drug-based regenerative medicine will only be limited by the successful localization of the drugs to the desired tissues. Simply stated, the complexity of drug delivery for regenerative medicine can be appreciated by the variety of molecules used and the tissues targeted currently under research. Nonetheless, even in the case of the utilization of (stem-) cell-based therapies (Chen et al., 2012), tissue regeneration can be as successful as the drug therapy that may be associated with it. Thus, the drug delivery of agents, such as hormones, proteins, genes, growth factors, and others, takes a central role in the success of tissue regeneration. The objective for the drug delivery would be to control the release of the drug at the site of interest over an extended period, while the carrier itself would be degraded or absorbed by the body once drug release is completed (Tabata, 2008).

Nanotechnology is utilized to overcome several challenges related to drug delivery. As with every other therapeutic objective, the identification of the proper nanocarrier for each application remains a question, as different carriers will present different capabilities, or more importantly, different drugs will require different release or delivery mechanisms. Hence, "the right carrier for the right drug for the right application" would be a proper description. For example, protein molecules, commonly used in regenerative medicine, have been reported to lose their functionality due to denaturation or deactivation because of the formulation preparation process in certain nanocarriers (Wee & Gombotz, 1998). Similarly, nucleic acid-based products, such as mRNAs, DNAs or short nucleic acids, must be delivered and released inside the cells, i.e., the cell's cytoplasm, to exert their action. Otherwise, their release outside the cells would most likely result to their degradation (unless appropriately modified) (Shrestha et al., 2023). Thus, the choice of a drug carrier depends on the site of application, administration route, and type of drug.

As such, a variety of materials has been utilized for drug delivery in regenerative medicine. The materials for the drug carriers have been adopted from other therapeutic applications as well, such as the treatment of cancer or inflammation, as there are apparent similarities in treatments with regenerative medicine.

Representatively, bone regeneration is an important study area within regenerative medicine. The different types of drugs used for bone regeneration include nucleic acids (Sidharthan et al., 2022), nonsteroidal anti-inflammatory drugs, (Fattahi et al., 2022), small molecules (Gurtner & Chapman, 2016), growth factors (Yun et al., 2012), proteins (El Bialy et al., 2017; Kuroda et al., 2019), angiogenesis factors (Zhang et al., 2023), and hormones (Wojda & Donahue, 2018), among others. Accordingly, polymeric, lipid-based, or inorganic nanoparticles or respective combinations of these materials have been utilized to deliver these compounds (Burdusel et al., 2022; Filipczak et al., 2021; Fu et al., 2021; Lei et al., 2019; Ogay et al., 2020). Hence, the applications multiply by recognizing that numerous types of compounds can be used in other fields of regenerative medicine, which can include skin regeneration and wound healing (Gurtner & Chapman, 2016; Shimizu et al., 2022), bone regeneration (Battafarano et al., 2021), spinal cord injury (Ashammakhi et al., 2019), Alzheimer's disease (Qin et al., 2022), neuroregeneration in general, diabetes (Brovkina & Dashinimaev, 2020), cardiovascular diseases (Lee & Walsh, 2016), cancer (Mansouri et al., 2021), and antiangiogenic therapies (Jahani et al., 2020), among other.

Not surprisingly, the broad spectrum of diseases and molecular targets addressed in regenerative medicine highlights how nanocarriers developed for other therapeutic areas can also be applied in this field. Furthermore, the diversity of materials and applications would allow us to provide broader descriptions for the desired properties of the carriers, which would correlate with applications outside regenerative medicine. Therefore, typically desirable properties would include:

(i) Biocompatible and biodegradable
(ii) Minimal toxicity and immunogenicity
(iii) Ability to protect and deliver the active compound
(iv) Maximize the ratio of drug to carrier material
(v) Simple chemistry and formulation for upscaling (Din et al., 2017; Filipczak et al., 2021)

Below, we describe applications associated with different types of nanocarriers, focusing on regenerative medicine.

# 3 Drug Delivery Carriers and Their Applications

In the following sections, we will describe some of the most commonly used nanocarrier systems that have been applied for regenerative medicine and provide representative examples of their utilization. As the field is extensive, the carrier systems described here may not represent an exhaustive list. Our presentation groups the different systems under organic and inorganic carriers, though combinations of the two types of carriers are also present in the literature.

## *3.1 Organic Carriers*

### 3.1.1 Polymeric Nanoparticles

Polymeric nanoparticles (PNPs) are particles of <1000 nm in diameter composed of natural or synthetic polymers for drug entrapment or encapsulation (Zielińska et al., 2020). PNPs can be categorized in nanospheres, polymeric micelles, nanocapsules, dendrimers, and nanogel, according to their morphological characteristics (Fathi-Achachelouei et al., 2019). Natural and synthetic polymers, such as poly (lactide) (PLA), poly (lactide-co-glycolide) (PLGA), poly ε-caprolactone (PCL), polyethylene glycol (PEG), chitosan, collagen and hyaluronic acid, have gained attention over polymers, such as poly (methyl methacrylate) (PMMA), polystyrene or polyacrylates, due to their biodegradability, biocompatibility, and limited or absent chronic toxicity or induction of inflammatory responses (Banik et al., 2016; Tang et al., 2014). PNPs' characteristics include small size, chemically modifiable surface for endowing additional functionalities to the carriers, ability to protect biomolecules or drugs against harsh physiological environments, biocompatibility, and biodegradability, control release property and capacity to carry a large variety of active compounds, spanning from small molecules to nucleic acids. Thus, PNPs have been exploited in regenerative medicine. Here, we discuss some of the PNPs' applications in different areas of regenerative medicine.

Polymeric nanoparticles have a significant presence in bone and cartilage tissue regeneration. Bone regeneration entails osteogenic cells' adhesion, proliferation, and attachment, a bio-degradable extracellular matrix (ECM) to provide structural support, a highly porous interconnected network for nutrient and metabolic waste transport, and mechanical strength to facilitate regeneration (BaoLin & Ma, 2014). Scaffolds in bone regeneration have been extensively utilized to meet the described goals. Furthermore, nanoparticles have been incorporated into the scaffolds to ensure proper drug release. For example, Wang et al. (2016) reported on bovine serum albumin (BSA) nanoparticles for the sustained release of bone morphogenetic protein-2 (BMP-2). The nanoparticles were coated with chitosan and incorporated via a layer-by-layer self-assembling biomimetic ECM. Moreover, a combination of BMP-2 and glycine-arginine-glycine-aspartate-serine (GRGDS), a cell binding protein, in chitosan-alginate-BSA matrix favored bone marrow stem cell adhesion, proliferation and differentiation, which can enhance bone formation. In another study, BMP-2 was loaded in chitosan/chondroitin sulfate nanoparticles to promote bone formation via their incorporation on a biphasic calcium phosphate scaffold. The researchers indicated the system promoted ectopic bone formation in vivo (Wang et al., 2015). Similarly, a fiber mesh scaffold prepared with chitosan and polyethylene oxide was used to incorporate two types of nanocapsules, PLGA nanocapsules with entrapped BMP-2 for early release of the molecule and poly (3-hydroxybytyrate-co-3-hydroxyvalerate) nanocapsules with entrapped BMP-7 for prolonged release of the molecule (Yilgor et al., 2009).

Gene delivery has been explored for bone and cartilage regeneration, among others. Similarly to other gene delivery efforts using polymeric nanoparticles, a variety of polymers have been utilized, including chitosan, PLGA, polyethyleneimine (PEI), and poly (amidoamine), among others (Labatut & Mattheolabakis, 2018; Shakeel et al., 2022; Yang et al., 2020a). In another study utilizing the biocompatible and biodegradable PLGA polymer, the authors coated PEI/plasmid complexes on Runt-related transcription factor 2 (RUNX2) protein-loaded PLGA nanoparticles. The plasmid would express the BMP-2 gene for the co-delivery of BMP-2 and RUNX2 to human mesenchymal stem cells for osteogenesis induction (Park et al., 2016). In a similar approach, Park et al. (2017) reported on the formulation of PLGA nanoparticles coated a PEI/plasmid complex. The plasmid was designed to express SOX5, SOX6, and SOX9 genes. The authors mentioned that the transfection of human mesenchymal stem cells with the nanoparticles carrying multicistronic genes exhibited higher chondrogenic differentiation than each single gene transfection.

Another area in which polymeric nanoparticles have significantly contributed is in nerve regeneration. The nervous system is divided into the central nervous system (CNS) and the peripheral nervous system (PNS). CNS is composed of the brain and spinal cord, whereas nerves branching off the spinal cord and are present in the different body parts are components of the PNS. The two systems and the respective nerves present different responses to injury, with PNS nerves being able to regenerate, while axons of CNS being limited by the environment and molecules identified that limit their regeneration. Also, macrophages contribute more to the regeneration process of the former than the latter. Similarly in bone regeneration, various types of natural polymers, such as hyaluronic acid, fibrin, chitosan, alginate, silk, and synthetic polymers, such as poly (lactide) (PLA), poly (glycolic acid) (PGA), PLGA, PEG, and poly [N-(2-hydroxypropyl) methacryamide] (PHPMA), among others, or combinations of the mentioned polymers, have been evaluated for treating nervous system injury (Amani et al., 2019; Boni et al., 2018; Huebner & Strittmatter, 2009; Schmidt & Leach, 2003). Indicatively, control release of the nerve growth factor (NGF) from chitosan nanoparticles indicated neuronal differentiation of canine bone marrow-derived mesenchymal stem cells in vitro (Mili et al., 2017). Neural stem cells (NSC) can potentially treat neurodegenerative diseases, with their capability to produce neurons, replace injured cells and provide growth factors. The presence of NSCs in an injured area is beneficial for the healing process, and their recruitment can be triggered by the stromal cell-derived factor 1 (SDF-1) protein. Post-traumatic brain injury injection of the nanoparticles with encapsulated SDF-1 in mice revealed migration of NSC to the injured area. Furthermore, increased neuroblasts were detected around the lesion when nanoparticles with SDF-1 were used compared to other groups, indicating that the system could contribute towards treating brain injury lesions (Zamproni et al., 2017). In a study by Teixeira et al. (2016), the authors injected anti-inflammatory drug diclofenac loaded chitosan/poly-(γ-glutamic acid) nanoparticles in a bovine intervertebral disc (IVD) organ culture under pro-inflammatory conditions. The nanoparticles were uptaken by the IVD cells, resulting in suppressed pro-inflammatory cytokines (i.e., interleukin-6 (IL-6), IL-8, matrix metalloproteinase 1 (MMP-1), and MMP-3) and prostaglandin E2 release. Furthermore, the authors identified increased ECM proteins, including collagen type II and aggrecan.

Finally, skin tissue regeneration has been extensively studied using PNPs. Skin is considered the largest organ in the body that acts as a first-line barrier against the external environment. It maintains body homeostasis, protects the internal organs from infection and exposure to toxic substances and radiation, prevents water loss, and regulates body temperature. The skin is frequently exposed to intrinsic and extrinsic factors that may lead to its injury and degeneration, while aging and diseases may further impact the skin's function (Walker, 2022). The wound healing process is complicated and involves interaction between pathological events, such as inflammatory responses, angiogenesis, and oxygenation (Mir et al., 2018). PNPs can efficiently deliver growth factors, nucleic acids, and drugs to treat infections or inflammation at the wound site (Shalaby et al., 2022).

PLGA nanoparticles entrap the recombinant human epidermal growth factor (rhEGF). The rhEGF was released by the PLGA nanoparticles over a 24 hours period, and the nanoparticles promoted fibroblast proliferation and the fastest healing rate, compared to the other treatment groups, in a diabetic rat model with full-thickness wounds (Chu et al., 2010). In a study by Jung et al. (2015), the authors presented that ceramide entrapped in chitosan/PGLA nanoparticles presented a controlled release of ceramide, which enhanced the rejuvenation of the stratum corneum in a rat atopic dermatitis model. Curcumin (CUR) is known for its anti-inflammatory effects. However, it presents low bioavailability and stability. Karri et al. [63] encapsulated CUR in chitosan nanoparticles, which were impregnated in collagen/alginate scaffolds to overcome these limitations. The final product was administered at the lesion sites of diabetic rats. The formulation exhibited anti-inflammatory effects, fast wound contraction, and complete epithelialization, and tissue formation.

In another study, LL37, an endogenous human host defense peptide that modulates angiogenesis and wound healing, was entrapped in PLGA nanoparticles. PLGA degradation releases lactate/lactic acid, which can also promote neovascularization and wound healing. The nanoparticles promoted wound healing compared to LL37 or PLGA alone, represented by advanced granulation tissue formation and collagen deposition. Furthermore, the nanoparticles accelerated angiogenesis and stimulated vascular endothelial growth factor a (VEGFa) and interleukin 6 (IL-6) expression (Chereddy et al., 2014).

### 3.1.2 Lipid-Based Nanoparticles

Lipid-based nanoparticles, such as liposomes, lipid nanoparticles (LNPs), nanoemulsions, and micelles have been extensively studied due to their potential to reliably deliver hydrophilic and hydrophobic drugs. Furthermore, the lack of apparent toxicity for the majority of the studied lipids, along with their biocompatibility, makes lipid-based nanoparticles excellent candidates for utilization in regenerative medicine (Filipczak et al., 2021).

Liposomes are nano- to micro-meter spherical vesicles having one or more lipid bilayers, depending on their preparation procedure. They can entrap hydrophilic solutions or drugs in their inner aqueous compartment/shell and incorporate lipophilic compounds within the phospholipid bilayers (Filipczak et al., 2021; Mattheolabakis et al., 2012; Monteiro et al., 2014). Their importance in drug delivery can be readily appreciated through the numerous FDA-approved applications used in patient treatment (Bulbake et al., 2017). Likewise liposomes are evaluated in regenerative medicine.

Indicatively, liposomes were prepared with entrapped epidermal growth factor (EGF) and were incorporated into a chitosan gel. In a rat model of second-degree burn wounds, the formulation was administered daily to the burn wounds and biopsies collected at predetermined time points indicated that the formulation significantly increased cell proliferation and epithelization rate (Değim et al., 2011). Madecassoside (MA), a triterpenoid known for anti-inflammatory and antioxidant action, was entrapped in liposomes to enhance transdermal and wound healing effects. The authors concluded that PEG-modified liposomes prepared using a double-emulsion method presented enhanced MA delivery for improving wound healing (Li et al., 2009, 2016).

In a study by Choi et al. (2017), the authors developed cDNA products of low molecular weight protamine (LMWP) with EGF, insulin-like growth factor-I (IGF-I), and platelet-derived growth factor-A (PGDF-A), by adding the codons for LMWP into the respective gene sequences, and the PCR products were sub-cloned into a TOPO vector, and cloned into a pET-41b(+) expression vector to transform *Escherichia coli* cells. Following protein expression, the LMWP-EGF, LMWP-IGF-I, or LMWP-PDGF-A were collected and purified via affinity chromatography and size exclusion chromatography. The molecules were complexed with hyaluronic acid (HA), and a mixture of the LMWP-growth factors (GFs)-HA was introduced to cationic elastic liposomes. The LMWP-modified GFs presented higher skin permeation compared to the unmodified GFs, and the LMWP-GFs-HA complexes entrapped in the liposomes accelerated wound closure using a diabetic mouse model. Todorovic et al. (2018) developed a liposomal formulation by entrapping coenzyme Q10 to evaluate wound healing after tooth extraction. The authors indicated that the liposomal formulation significantly promoted wound healing and reduced inflammation after tooth extraction in rats.

Focusing on the nervous system, surface serotonin modulator (SM) and apolipoprotein E (ApoE) were grafted on liposomes for β-amyloid targeting and the delivery of nerve growth factor (NGF) across the blood-brain barrier (BBB). Using

SK-N-MC-cells in vitro and Aβ-insulted Wistar rats in vivo as Alzheimer's disease models, the liposomal formulation was evaluated. The results depicted that SM and ApoE facilitated as permeation enhancers for NGF to penetrate across the BBB and induced phosphorylated neurotrophic tyrosine kinase receptor type 1 expression, decreased acetylcholinesterase and malondialdehyde secretions, and protected hippocampal neurons from apoptosis (Kuo & Lee, 2016). In another study, epigallocatechin-3-gallate (EGCG) was loaded in liposomes, and upon intra-nigral injection in a Parkinsonian syndrome rat model, the formulation restored motor impairment and decreased LPS-induced pro-inflammatory cytokines (Cheng et al., 2021a). In an interesting study, liposomes were prepared with a tetrapeptide (CAQK) modification to encapsulate docetaxel and a brain-derived neurotrophic factor. The prepared liposomes were placed into a thermosensitive, heparin-modified, affinity-bound acidic fibroblast growth factor poloxamer hydrogel. The final formulation was locally administered to the spinal cord injury site in a rat model. The formulation promoted nerve regeneration, neuronal survival, and plasticity (Wang et al., 2018).

Liposomal delivery of thrombolytic drugs could be a promising approach to manage thrombosis associated complications, including stroke, myocardial infarction, and embolism (Bruch et al., 2019; Levchenko et al., 2012; Shen et al., 2021b). Streptokinase-loaded liposomes were prepared for targeting thrombus areas. In vivo evaluation of the formulation reduced the clot dissolution time compared to the streptokinase solution only in thrombus mouse model (Vaidya et al., 2016). Another thrombolytic drug, urokinase, was encapsulated in cyclic RGD (cRGD) functionalized liposomes. In vivo, using a mesenteric thrombosis mouse model, the formulation presented strong thrombolytic activity (Zhang et al., 2018). Takahama et al. (2013) reported that amiodarone, one of the most effective anti-arrhythmic drug, can be delivered specifically to ischemic myocardium using liposomal delivery. Such an approach improved the therapeutic effects and minimized the side effects in an ischemia/reperfusion rat model.

LNPs have been extensively used in gene delivery and have found applicability in regenerative medicine. Representatively, in a recent study by Evers et al. (2022), the authors evaluated the LNPs' ability to deliver mRNA to ischemic myocardium. Using an ischemia/reperfusion injury-induced mouse model, mRNA LNP accumulation was observed in the heart, based on fluorescent labeling. Firefly luciferase encoding mRNA in LNPs was administered intravenous (i.v.) to mice and luciferase expression was detected in the heart upon myocardial reperfusion, and subsequent analysis using Cre-modRNA and Cre-reporter mice indicated fibroblast, cardiomyocyte, and macrophage targeting at the infarct region and the epicardial layer upon LNP infusion.

## *3.2 Inorganic Nanoparticles in Tissue Regeneration*

Inorganic nanoparticles have been extensively used in drug delivery applications, including small drugs, biologicals, and nucleic acids. Understandably, their utilization as biomaterials in regenerative medicine and tissue engineering applications has attracted significant attention, especially for their unique properties. For example, they can deliver bioactive molecules with ease, tunability, biocompatibility, and favorable physicochemical properties (Brokesh & Gaharwar, 2020). They can also be modified with biomolecules, such as peptides, nucleic acids, and carbohydrates, to enhance their biological activity or function (Bianchi et al., 2022). Similarly to other nanosized materials, inorganic nanoparticles possess a high surface area for modifications, using peptides, polysaccharides, proteins, polymers, and others, endowing them with functions such as active targeting, long-circulation, protection against the immune system, when necessary. Additionally, they can carry and release a wide variety of molecules under various environments in the body, depending upon the inorganic element being used, or modulate cell-specific functions including cell proliferation, differentiation, adhesion, migration, and viability, among others. Naturally, these materials have found application in tissue healing and regeneration (Urie et al., 2018). The field of inorganic nanocarriers or inorganic materials used in regenerative medicine is extensive and multifaceted. Here, we will discuss representative examples of the most commonly used inorganic nanocarriers and materials in this extensive field.

### 3.2.1 Gold Nanoparticles

Gold nanoparticles (AuNPs) are excellent inorganic materials due to their inert nature, size-dependent optical properties, physicochemical properties, and accessible functionalization properties with various molecules (Sakthi Devi et al., 2022). AuNPs have been favored for tissue engineering and regenerative medicine due to their low toxicity, colloidal stability, and superior physicochemical properties (Corsi et al., 2020; Sandhu et al., 2002; Thomas & Klibanov, 2003). Owing to these advantages, AuNPs have found extensive applications as drug delivery systems. Their surface favors conjugation for tuning the particle's characteristics, allowing them to deliver therapeutic agents and biomolecules, including genes, peptides, and others (Han et al., 2006; Kim et al., 2009).

AuNPs also present optical properties attributed to quantum effects, suitable for imaging applications. Furthermore, AuNPs can modulate biological mechanisms, such as migration, proliferation, and cell differentiation. For example, AuNPs can interact with cardiomyocytes and promote electrophysiological responses, mimicking the native myocardium (Esmaeili et al., 2022). AuNPs have also been involved in applications for myocardial regeneration using stem cells. Various studies have explored the combination of mesenchymal stem cells (MSCs) with functionalized AuNPs incorporated into different materials, such as chitosan (Wu et al., 2023b), or various extracellular matrix (ECM) scaffolds (Dong et al., 2020). For example, Sridhar et al. (2015) reported on nanofibrous scaffolds of gold nanoparticles and PCL, vitamin B12, aloe vera, and silk fibroin for the differentiation of mesenchymal stem cells (MSCs) into cardiac lineage. The formulation with the gold nanoparticles enhanced MSCs proliferation and differentiation. Similarly, AuNPs incorporated in a hyaluronic acid hydrogel matrix was used to encapsulate human-induced pluripotent stem cells-derived cardiomyocytes. The transplanted formulation formed robust gap junctions in a mouse model of infarcted heart, induced potent angiogenesis, and resynchronized electrical conduction of the ventricle (Li et al., 2021a).

Similarly, AuNPs have been utilized in bone (Li et al., 2020a), skin (Toczek et al., 2022), and nerve regeneration (Paviolo & Stoddart, 2015). For example, Ko et al. (2015) reported that gold nanoparticles of various sizes affected the differentiation of human adipose-derived stem cells into osteoblasts, while AuNPs with sizes of 30 nm and 50 nm induced the strong osteogenic differentiation of the cells. Similarly, Arafa et al. (2018) reported on the formulation of a thermosensitive gel that contained AuNPs and its use in vivo against burn healing in a mouse model of burn-induced wounds. The formulation presented strong healing properties. On the other hand, although AuNPs have been used for nerve regeneration (Papastefanaki et al., 2015) and regeneration medicine in general, there have been concerns regarding their safety, with reports stressing that AuNPs may potentially be detrimental to the nervous system (Siddiqi et al., 2012).

In Table 1, we summarize representative literature on using AuNPs in regenerative medicine.

### 3.2.2 Silver Nanoparticles (AgNPs)

AgNPs are among the most commonly used metallic nanoparticles, with intrinsic antimicrobial and anti-inflammatory activities (Fathi-Achachelouei et al., 2019). AgNPs possess versatile functionality, can be easily synthesized with a variety of inexpensive and environmentally friendly methods, and can be further modified for the needs of the respective applications. They are a vital class of inorganic NPs extensively used in promoting skin regeneration in acute or chronic wounds inflicted by burns, diabetes, or external factors (Burduşel et al., 2018; Choudhury et al., 2020). They can be favorably interspersed and conjugated with various polymers, metallic or nonmetallic nanoparticles, and bioactive compounds to enhance or synergize their characteristics in tissue regeneration. Therefore, these combination formulations with AgNPs can enhance bone (Choi et al., 2023; Han et al., 2021; He et al., 2020), nerve (Ding et al., 2015a; Huang & Wang, 2023), skin (Abdel-Mohsen et al., 2017; Antosiak-Iwańska et al., 2020), cardiovascular (Adibkia et al., 2021), liver (Ali et al., 2019), and skeletal muscle (Raimondo & Mooney, 2018) regeneration. In the following Table 2, we summarize a few representative examples of AgNPs in regenerative medicine.

### 3.2.3 Magnetic NPs (MNPs) or Iron Oxide Nanoparticles (FeNPs)

MNPs are inert, safe, and highly biocompatible materials, which are typically iron oxide that are classified as single metal MNPs, metal oxide MNPs (iron oxide, and ferrites), and alloy MNPs (FeCo, FePt) (Bianchi et al., 2022; Farinha et al., 2021).

**Table 1** Representative studies on AuNPs and their utilization in regenerative medicine

| | | |
|---|---|---|
| **Cardiac muscle regeneration** | | |
| Cardiac patches composed of three-dimensional nanocomposites of gold nanowires (AuNWs) within alginate scaffolds | Repaired infarcted heart muscle | Dvir et al. (2011) |
| Homogeneously dispersed AuNPs in thiol-2-hydroxyethyl methacrylate (thiol-HEMA)/HEMA scaffolds | Increased connexin 43 expression | You et al. 2011) |
| AuNPs conjugated with reverse thermal gel (RTG)-based cardiac cell scaffold | Targeted survival of neonatal rat ventricular myocytes | Peña et al. (2019) |
| AuNPs with lysozyme nanofibrils (LNFs) incorporated in gelatin hyaluronic acid hydrogels | Myocardial regeneration | Carvalho et al. (2023) |
| Cardiac patches composed of functionalized AuNPs with extracellular matrix/silk protein scaffolds | Cardiomyocyte cell proliferation | Dong et al. (2020) |
| AuNPs-hyaluronic acid hydrogel matrix encapsulating hiPS-cardiomyocytes with cell-penetrating peptide | Promotion of robust gap junctions in infarcted mice hearts and resynchronized electrical conduction in ventricular myocardial infarction | Li et al. (2021a) |
| Electroactive Au and laponite (lap) NPs loaded with ECM as injectable hydrogels | Improved cell compatibility and phenotype maturation of cardiac-specific proteins | Zhang et al. (2019b) |
| AuNPs are incorporated with polycaprolactone (PCL), vitamin B12, aloe vera, and silk fibroin scaffolds by electrospinning | Enhanced MSCs proliferation and differentiation | Sridhar et al. (2015) |
| AuNPs embedded in polyvinyl alcohol (PVA)/bovine serum albumin (BSA) scaffolds | Enhanced cardiomyogenic differentiation of MSCs | Ravichandran et al. (2014) |
| **Bone regeneration** | | |
| Silk fibroin/nanohydroxyapatite (HAp) hydrogel modified with AgNPs/AuNPs | Improved antimicrobial effects and repair of bone defects | Ribeiro et al. (2017) |
| AuNP-loaded mesoporous silica NPs | Enhanced preosteoblastic cell differentiation and accelerated new bone formation in critical-sized cranial defect site in rats | Liang et al. (2019) |
| PLA-PCL matrix polymer with AuNPs and gelatin nanofibers incorporated using thermal-induced phase separation | Increased neo-bone formation, osteocyte in lacuna woven bone formation, and angiogenesis | Samadian et al. (2021) |
| Pretreated human periodontal ligament stem cells (hPDLSCs) with Au-nanocomplex embedded in type-1 collagen hydrogel | Prevention of alveolar bone loss, enhanced bone remodeling and regeneration | Zong et al. (2023) |

(continued)

**Table 1** (continued)

| | | |
|---|---|---|
| **Skin regeneration** | | |
| AuNPs combined with epigallocatechin gallate and alpha-lipoic acid for topical application | Enhanced antioxidative and anti-inflammatory effects in cutaneous wound healing | Leu et al. (2012) |
| PEG-Au nanorods (AuNRs), polyallyl amine AuNRs loaded in poloxamer 407 hydrogel for skin wound healing | Prolonged Au release with excellent colloidal stability accelerated wound healing and anti-inflammatory effect | Mahmoud et al. (2019) |
| AuNP gel in conjunction with microcurrent therapy | Improved mitochondrial functioning and oxidative stress facilitating burn wound healing | Silveira et al. (2014) |
| Phytochemically stabilized AuNPs coated on hydrocolloid membrane | Accelerated wound repair with enhanced angiogenesis, connective tissue formation, and antioxidant effects | Kim et al. (2015) |
| Polydopamine (PDA)-coated HAp/AuNPs | | Xu et al. (2018) |
| *Nerve regeneration* | | |
| Poly (ε-caprolactone) (PCL) conduits with Au and AuCoO NPs in poly-PEG membrane | Potentiated sciatic nerve regeneration in rat model | Hazer Rosberg et al. (2021) |
| AuNPs and brain-derived neurotrophic factor (BDNF)-encapsulated chitosan/laminin-coated PLGA conduit | | Jahromi et al. (2021) |
| Au nanocrystals delivered orally or cocultured with central nervous system cells | Improved myelination as a treatment for multiple sclerosis | Robinson et al. (2020) |
| AuNPs with chitosan/hyaluronic acid nanocomposite film | Enhanced MSCs differentiation | Shen et al. (2021a) |

These NPs have been commonly used for various diagnostic purposes, such as magnetic resonance imaging (MRI), drug delivery, and tissue targeting methods. In recent years, MNPs have gained much attention in tissue engineering and regenerative medicine, given their physicochemical properties and diverse characteristics depending upon their synthesis method. MNPs present magnetic properties, responding to external magnetic fields, called superparamagnetism. This class of MNPs is also called superparamagnetic iron oxide NPs (SPIONs). This unique feature allows MNPs to maintain their colloidal stability and avoid clustering (Farinha et al., 2021).

Additionally, they show superior thermal conductivity and are easy to functionalize. SPIONs-incorporating scaffolds can potentiate cell differentiation and proliferation and show overall improvement of biological processes for tissue regeneration. Furthermore, as iron is one of the most common trace elements in the brain, important for neurotransmission, DNA synthesis, and normal cerebral function, attributed to its role as a co-enzyme factor (Guo et al., 2022), it is an excellent material to lend itself to neural regenerative medicine. Guo et al. (2022) investigated the role of iron in nerve regeneration by employing liposomal-encapsulated deferoxamine (DFO) NPs and found that the DFO NPs could reverse the plasticity changes in cerebral cortex function followed by stroke and that the increase in iron aided in endogenous

**Table 2** Representative studies on AgNPs and their utilization in regenerative medicine

| | | |
|---|---|---|
| **Skin regeneration** | | |
| AgNPs incorporated with HAp and PEI were evaluated in WEHI-164 cells | Improved bacteriostatic activity in wound healing | Antosiak-Iwańska et al. (2020) |
| Evaluation of Hyaluron/AgNPs nanocomposite fabrics as a protective for would and ulcer | Enhanced wound and chronic ulcer treatment with increased antimicrobial effects | Abdel-Mohsen et al. (2017) |
| Chitosan-dextran hybrid hydrogel encapsulating AgNPs for sustained Ag + release | Increased recruitment of anti-inflammatory cytokines and immune cells in wound regions | Shi et al. (2019) |
| Polyacrylonitrile (PAN) and urchin-like Ag-Au NPs or AgNPs nonwoven mats | Increased neovascularization and compact collagen fibers promoting skin tissue regeneration | Bai et al. (2021) |
| **Dental regeneration** | | |
| Chitosan-AgNPs-polyurethane nanocomposite fibers using electrospinning technique as a dental barrier membrane | Improved biocompatibility, antimicrobial effects and effective tissue regeneration | Lee et al. (2018) |
| 2-methylacryloyloxyethyl phosphorylcholine (MPC)-AgNPs-dimethylaminohexadecyl methacrylate (DMAHDM) incorporated resin-modified glass ionomer cement (RMGI) | Reduced enamel demineralization, plaque buildup, and enhanced antibacterial activity against biofilms | Zhang et al. (2016) |
| AgNPs incorporated with heat-cured acrylic resin | Increased viscoelastic and physicomechanical properties of denture base material | Mahross and Baroudi (2015) |
| **Bone regeneration** | | |
| AgNPs loaded with gelatin hydrogel | Promoted osteoblast cell survival and spreading | Han et al. (2021) |
| AgNP-loaded tetra calcium phosphate-dicalcium phosphate dihydrate (TTCP-DCPD) cement for treatment of MRSA (methicillin-resistant *Staphylococcus aureus*) induced-osteomyelitis | Improved antimicrobial effects and osteoconduction and osteointegration | Choi et al. (2023) |
| AgNPs decorated β-tricalcium phosphate porous scaffolds coated with poly(3-hydroxybutyrate) | Increased neovascularization and compact collagen fibers | Bai et al. (2021) |
| AgNP-loaded PLGA/PCL scaffold using electrospinning method | Enhanced alveolar bone regeneration in mouse periodontal disease model with increased osteogenic and antimicrobial properties | Qian et al. (2019) |
| Polydopamine-functionalized mesoporous silicon NP loaded with AgNP and osteogenic dexamethasone in poly-lactic acid (PLLA) scaffolds | Enhanced osteogenic differentiation of mouse-derived bone mesenchymal stem cells (mBMSCs) and improved drug release for osteomyelitis therapy | Wu et al. (2023a) |

(continued)

**Table 2** (continued)

| | | |
|---|---|---|
| PLA/AgNP nanocomposite fiber polymerized with dopamine and chitosan/AgNP coating | Potentiated long-term antibacterial effects with sustained release of Ag + with improved osteoinductive and angiogenic properties | Liu et al. (2021) |
| **Nerve regeneration** | | |
| Crosslinked nanosilver-collagen scaffolds | Promoted reparation of sciatic nerve injury in rat model | Ding et al. (2015a) |
| Nanosilver-collagen scaffolds coated with laminin and fibronectin | | Ding et al. (2011) |
| Polyvinylpyrrolidone (PVP)-AgNPs-graphene embedded poly (glycerol sebacate) acrylate matrix | Enhanced cell growth, proliferation of Schwann cells | Huang and Wang 2023) |

repair. Similarly, FeNPs covalently conjugated with neurotrophic factors prevented the proteins from proteolytic degradation and increased their stability, enhancing their activity in forming axonal networks and early neurite outgrowth (Ziv-Polat et al., 2014).

Superparamagnetic FeNPs have also been employed as inorganic labeling particles. For example, they can be used alongside chitosan-driven self-assembly of mesenchymal stem cells (MSC) spheroids and brain-derived neurotrophic factor (BDNF) in nerve regeneration for visualization under MRI in vivo (Tseng & Hsu, 2014). Moreover, the magnetic targeting capabilities of FeNPs have been extensively explored in the delivery of drugs, genes, and various biomolecules. Similarly, dextran-coated SPIONs combined with curcumin can enhance the effect of the compound for neurogenesis. Interestingly, they identified that the SPIONS alone had a similar ability to curcumin in increasing neuronal branching, and their combination enhanced the pharmacological functions (Zarei et al., 2022).

Furthermore, flavonoids such as quercetin with nerve growth factor combined with SPIONs exhibit favorable neuritogenesis by promoting and producing complex neurite branching attributable to FeNPs (Katebi et al., 2019). In a study by another group (Soto et al., 2021), the authors used adipose-derived MSC loaded with SPIONs. The cells were transplanted and magnetically recruited to the injured sciatic nerve, promoting cell recruitment and post-traumatic sciatic nerve regeneration by a noninvasive method. Yuan et al. (2019) demonstrated the applicability of SPIONS-AuNPs coated nerve growth factor with LED irradiation for noninvasive neuron regeneration.

FeNPs' role in skin wound healing has also been reported in several studies (Arakha et al., 2015; Ismail et al., 2015; Lozano Chamizo et al., 2021; Yadav et al., 2021; Zangeneh et al., 2020). $Fe_3O_4$ combined with chitosan using an ionotropic gelation technique where cationic chitosan and anionic tripolyphosphate (TPP) underwent quick ionic cross-linking and thus can provide synergistic effects of tissue regeneration and antibacterial activity (Lozano Chamizo et al., 2021). In the following Table 3, we summarize a few representative examples of FeNPs in regenerative medicine.

**Table 3** Representative studies on FeNPs and their utilization in regenerative medicine

| | | |
|---|---|---|
| **Nerve regeneration** | | |
| SPION-Au core-shell NPs with porous coordination cages (PCCs) loaded with retinoic acid (RA) | Enhanced intracellular uptake of NPs by PC-12 neuron-like cells and controlled RA release | Yuan et al. (2021) |
| SPION ($Fe_3O_4$-Rhodamine-PDA) particles cultured with Schwann cells | Promoted sciatic nerve regeneration and functional recovery of injured peripheral nerves | Liu et al. (2022) |
| MNPs embedded in aligned fibrin hydrogel (AFG) | Enhanced neural regeneration in spinal cord injury | Yang et al. (2023) |
| SOS-1 cat-loaded MNPs with magnetic guidance from the cytoplasm to the neurite tip | Potentiated neuroprotective and neurite-inducing biological pathway | Raudzus et al. (2020) |
| **Bone regeneration** | | |
| Mesoporous silica-coated $Fe_3O_4$ NPs | Enhanced osteogenic differentiation of MSCs | Jia et al. (2019) |
| Magnetite/HAp NPs porous scaffolds | Enhanced bone tissue formation in critical femoral defect in rabbits | Russo et al. (2018) |
| Pristine-SPIONs incorporated in chitosan/PVA/hap | Increased cell adhesion, proliferation, and alkaline phosphatase expression | Tavares et al. (2023) |
| $Fe_3O_4$ NP with hap-NP functionalized with miR-21/124 | Promoted cell growth and survival of preosteoblast and preosteoclast cells with improved ECM differentiation. Improved bone healing in cranial defect rat model | Marycz et al. (2021) |
| $Mg_2SiO_4$-$CoFe_2O_4$ nanocomposite | Enhanced antimicrobial activity, increased cell adherence and bioactivity | Bigham et al. (2019) |
| **Skin regeneration** | | |
| Chitosan-TPP hydrogel incorporating $Fe_3O_4$ NPs using ionotropic gelation | Enhanced wound healing in fibroblast model and increased antimicrobial effect | Lozano Chamizo et al. (2021) |
| FeNPs ointment synthesized using *Allium saralicum* extracts in green synthesis of MNP | Increased antioxidant, antimicrobial, and wound healing properties | Zangeneh et al. (2020) |
| Curcumin-loaded SPIONs with photobiomodulation technique | Antimicrobial effects against MRSA-wound infection in diabetic wound healing | Hekmat et al. (2023) |
| $CoFe_2O_4$ NPs grafted onto cellulose fibers using layer-by-layer method as a smart wound dressing material | Efficient wound healing monitoring technology | Williams et al. (2019) |
| MNP-embedded alginate/poly-L-ornithine/gelatin grooved hydrogel sheet | Enhanced neovascularization and skin wound reparation | Noh et al. (2019) |
| **Muscle regeneration** | | |
| Iron oxide NPs embedded in gelatin/polyurethane scaffold with applied magnetic fields as a magnetic actuated strategy | Enhanced recovery of skeletal muscle injury with increased myoblast differentiation and angiogenic activity | Hu et al. (2022) |
| MNPs entrapping basic fibroblast growth factor (bEGF) | Increased skeletal muscle contractility and stress relaxation in rat's skeletal muscles | Jin et al. (2022) |

(continued)

Table 3 (continued)

| | | |
|---|---|---|
| MNP-incorporated liposomes (mag-AL) and atecollagen and magnetization with MSCs | Reduced tissue inflammation, reduced toxicity and enhanced cellular uptake | Kono, 2022) |
| Fibrin scaffolds loaded with $Fe_3O_4$ NPs with magnetically induced stretch conditions | Enhanced cardiovascular tissue conditioning | Hogan et al. (2016) |

### 3.2.4 Ceramic Nanoparticles

Ceramic NPs are most commonly used to bolster the mechanical properties and control biological responses in scaffolds made of biodegradable synthetic polymers. They are widely characterized by inorganic materials, including silica, titania, and alumina (Singh et al., 2016). Their mechanical strength, excellent biocompatibility, high porosity and surface area, and long-term effect make them suitable for tissue engineering. Based on their structural differences, nanoceramics can be categorized as ceramic nanoparticles, nanoscaffolds, and nanoclays (Singh et al., 2016). Several studies have explored the use of ceramic NPs in bone tissue (Bagchi et al., 2014) and skin regeneration (Niu et al., 2021; Sharifi et al., 2022; Sharifi et al., 2023). Noninvasive techniques for bone reconstruction are among the most preferred options for patients suffering from bone diseases, such as cancers, abnormalities in bone tissue differentiation, and impaired bone regeneration. For this purpose, bioceramics are increasingly selected to form customizable 3D-printed bone grafts (Belluomo et al., 2023). They resemble the bone's architecture and mechanical properties, while they have similar bioactivity and physical properties compared to bones, making them suitable for this purpose.

Porous silica nanofibers have been widely used, and they can be doped with other inorganic materials and polymeric substances for tissue regeneration. Das et al. (2016) developed highly porous Si nanofibers on inert glass to potentiate the carbonate hydroxyapatite (HAp) layer and enhance cell proliferation. The mesoporous or high porosity silica has demonstrated applicability in regenerative medicine due to its enhanced tunability and ease of functionalization. Silica can be also used for controlled drug release and is a promising candidate for biomedical applications, while it is generally regarded as biodegradable under certain conditions (Croissant & Brinker, 2018; Hu et al., 2021; Zhang et al., 2019a). In muscle regeneration, mesoporous silica NPs can potentiate myocardial cell viability (Ornelas-Soto et al., 2017), and silica coating of MNPs can further accentuate their bioactivity to alleviate myocardial injury (Shen et al., 2019). Furthermore, mesoporous silica NPs are also being used to treat peripheral nerve injuries by either combining them with various bioactive molecules, such as nerve growth factor (Sun et al., 2016) and genetic materials (Lei et al., 2019), or entrapping drugs to alleviate pain and improve nerve regeneration (Sant'Anna et al., 2019; Wang et al., 2021a).

Zirconia NPs are another important class of bioceramics widely used in bone regeneration due to their high tensile strength and fracture toughness. They are also combined in varied concentrations with different materials, such as HAp, to promote bone healing and enhance osteoconductivity (Abedi et al., 2014; Balagangadharan et al., 2018). Zirconia NPs also showed synergistic action when combined with calcium phosphate in bone restoration in an animal model (Taha et al., 2023). They can also be used for dental restorative purposes, enabling favorable dentin coating and antimicrobial effects (Fathima et al., 2017).

HAp falls under the class of inorganic materials composed of calcium phosphate crystals. Due to its chemical resemblance to mammalian hard tissue, such as like bones and teeth, it exhibits osteoconductive effects attributed to its role in bone growth and osseointegration (Ignjatović et al., 2013). It is widely used in tissue engineering, drug delivery, and diagnostic probes. Furthermore, rare earth doped-HApNPS display high brightness, contrast, photostability, and narrow emission bands compared to other fluorescent probes (Gu et al., 2022). It can be combined with numerous elements to strengthen their physicochemical properties (Dittler et al., 2021; Frasnelli et al., 2017). Due to the biocompatible nature of HAp, the HAp-NPs have an extensive presence in regenerative medicine (Bonany et al., 2022; Klačić et al., 2022), as further described in Table 4.

### 3.2.5 Glass Nanoparticles

Glass nanoparticles or bioactive glass (BG) nanoparticles (BGNPs) belong to another class of inorganic materials, which can be doped or combined with additional inorganic materials, including ceramic, zinc (Kim et al., 2021; Naruphontjirakul et al., 2023; Paramita et al., 2021), strontium (Gentleman et al., 2010; Hoppe et al., 2014; Kargozar et al., 2019b), or copper (Zhou et al., 2023), among others, for applications in tissue regeneration. BGs are silicate-, borate-, boro-silicate- or phosphate-based glasses (Kargozar et al., 2017). Their mechanism in bone regeneration is due to the release of silicate and calcium ions that aid in osteogenic differentiation, bone metabolism, and hemostasis (Pajares-Chamorro & Chatzistavrou, 2020). Introducing bioactive glass to the body forms a carbonated hydroxyapatite (HCA) layer with chemical similarity to the inorganic bone phase, enhancing biocompatibility (Hoppe et al., 2011). It also presents bonding features to soft [49–51] and hard tissues [52], when exposed to the physiological environment (Baino et al., 2016). Furthermore, BGNPs' role in muscle regeneration has been elucidated in different studies (Kargozar et al., 2017; Kargozar et al., 2019a; Qi et al., 2020; Shi et al., 2021; Winston et al., 2023). Similarly, cerium-containing BGNPs can increase endothelial cellular migration, and promote angiogenesis and collagen deposition to promote wound healing (Chen et al., 2021). Similarly, BGNPs with copper and sodium alginate have also promoted wound healing and skin regeneration by regulating early angiogenesis (Li et al., 2020b). Intriguingly, the BGNPs can also be combined with different active ingredients for sustained release to achieve rapid wound healing and pain relief (El-Okaily et al., 2023; Tomar et al., 2023).

In the following Table 5, we present a few representative examples of bioactive glass in regenerative medicine.

**Table 4** Representative studies on HAp in regenerative medicine

| | | |
|---|---|---|
| **Dental regeneration** | | |
| Proanthocyanidin-functionalized HApNPs | Increased dentin mineral content and collagen cross-links with improved dentin mechanical properties | Enrich-Essvein et al. (2021) |
| Topical HApNPs and fluoride | Decreased dentin erosion on simultaneous and sequential application of HApNPs and fluoride | Rodemer et al. (2022) |
| Polydopamine (PDA)-coated silver nanoparticle-laden HAp nanowires hydrogel | Enhanced flexural strength and modulus of dental composites and increased antimicrobial effect due to sustained Ag release | Ai et al. (2017) |
| **Bone regeneration** | | |
| Si/Sr-doped HAp-collagen NPs incorporated into poly anhydride matrix | Enhanced proliferation of MSCs and osteoblast differentiation | Dasgupta et al. (2021) |
| HApNPs prepared via the sol-gel method | Facilitation of bone formation in sagittal sutures and increased osteoblast differentiation | Liang et al. (2021) |
| Mesoporous HApNPs loaded with bone morphogenetic protein-2 (BMP-2) were incorporated in silk fibroin/chitosan scaffolds | Enhanced osteoinductivity and osteogenic differentiation of bone-derived MSCs (BMSCs) | Qiu et al. (2020) |
| Injectable hydrogel made of HApNPs and tricalcium phosphate NPs with gellan gum derivatives | Promoted bone healing and viable preosteoblastic cells | Pitarresi et al. (2022) |
| Nanocomposite scaffolds made of PLLA- HApNPs modified with dopamine and hexamethylenediamine | Enhanced osteogenic differentiation and osteoconductivity | Yang et al. (2018) |
| Self-assembling peptide amphiphilic nanofibers integrated with HApNPs | Improved bone healing mediated by RGDS ligand signaling and osteogenic differentiation | Anderson et al. (2011) |

### 3.2.6 Titanium Oxide

Titanium is one of the most suitable materials for applications in bone and dental tissue regeneration, due to its resistance to corrosion, inertness, higher tensile strength, and ease of integration with bones (Brammer et al., 2012; Chakraborty et al., 2011; Nair & Elizabeth, 2015). Titanium exhibits notable bone-bonding characteristics and imparts promising bioactivity with bone compared to other materials (Azzawi et al., 2018; Brammer et al., 2009). Several studies have demonstrated the incorporation of $TiO_2$ nanoparticles with different polymeric scaffolds and biomaterials to improve stem cell differentiation. Ashraf et al. (2020b) described a sol-gel technique for $TiO_2$ NPs synthesis, which were used with an electrospinning method to incorporate them in chitosan and collagen nanofibers scaffolds. Subsequently, the authors cultured the samples with mouse-derived MSCs to investigate the MSC differentiation into osteogenic lineage. In another study, bio-nanocomposite chitosan/PVA/nano-HAp films were doped with $TiO_2$, presenting excellent elasticity and

**Table 5** Representative studies on bioactive glass and their utilization in regenerative medicine

| | | |
|---|---|---|
| **Bone regeneration** | | |
| Poly(citrate-siloxane) elastomer hybridized with BGNPs | Enhanced proliferation and osteoblastic differentiation. Decreased in vivo inflammatory response | Li et al. (2018) |
| BG (45S5)-based 3D scaffolds coated with metal-doped HAp nanoparticles | Promoted initial adhesion through upregulation of integrins and spreading of rat MSCs | Dittler et al. (2019) |
| BGNPs interacting with bone marrow mesenchymal stem cells | Reduced bone loss in osteoporotic mouse model by inhibiting osteoclastic differentiation | Yang et al. (2022) |
| BGNPs in hydrogel scaffolds | Enhanced human endometrial stromal cell EnSC differentiation into endothelial lineage | Barabadi et al. (2016) |
| Boron-doped mesoporous BGNPs | Increased C2C12 myogenic cell differentiation and viability | Ege et al. (2022) |
| **Skin regeneration** | | |
| Polyethylene glycol diacrylate (PEGDA) forming scaffolds with BGNPs containing Cu and sodium alginate | Promoted the viability, proliferation, and angiogenic ability of endothelial progenitor cells. Restored blood vessels through enhancing HIF-1α/VEGF expression | Li et al. (2020b) |
| Ce-BGNPs formed by sol-gel method in gelatin methacryloyl hydrogel (GelMA) | Improved wound healing in diabetic rats. Accelerated angiogenesis, collagen deposition, and the formation of granulation tissue | Chen et al. (2021) |
| Gelatin methacrylate and Cu-loaded BGNPs form a hydrogel bio-ink | A portable handheld SkinPen for clinical wound management | Zhou et al. (2023) |
| BG-AuNPs composites ointment | Potentiated skin reparation in rat model and strong vascularization | Mârza et al. (2019) |
| ZnO-doped BG in chitosan/alginate composite hydrogel for wound dressing | Enhanced antimicrobial, angiogenic effects and increased collagen deposition | Zhu et al. (2019) |
| Electrospun 70S30C BG/polyvinyl alcohol (PVA)/AgNPs mats with curcumin | Imparts antibacterial activity with increased biodegradability in wound dressing application | Mahmoudi et al. (2023) |
| PCL fibers containing BGNPs and PCL with $Ag_2O$ and CoO-doped BGNPs | Enhanced mechanical properties with antibacterial and angiogenic properties | Moura et al. (2017) |
| Electrospun PCL nanofibers with BGNPs loaded atorvastatin with ethyl cellulose coating | Potentiated wound healing in a human skin fibroblast cell line | El-Okaily et al. (2023) |
| Gelatin hydrogel with chitosan, polydopamine-coated BG loaded with diclofenac sodium, and curcumin-capped silver nanoparticles | Potentiated anti-inflammatory, angiogenic, and pain-relieving properties to alleviate skin wound healing | Tomar et al. (2023) |

(continued)

**Table 5** (continued)

| | | |
|---|---|---|
| Thiolated hyaluronic acid/silk fibroin with BGNPs | Increased neo-vascularization and stimulated migration of fibroblasts and human umbilical vein endothelial cells (HUVECS) | Yu et al. (2022b) |
| BG mesoporous nanoparticles incorporated with silver (Ag) | Enhanced antimicrobial effects with intact apatite-forming ability in skin wound healing | Zheng et al. (2019) |

antibacterial properties (Khan et al., 2020). Similarly, $TiO_2$ NPs have been combined with AgNPs for improved antibacterial and bioactivity properties (Ashraf et al., 2020a; Cheng et al., 2014; Sun et al., 2023).

Titanium can also be modified with other inorganic materials, including calcium and strontium, to enhance bone regeneration. More innovative approaches have also been introduced to increase the success rate of bone regeneration experiments. Electrochemical $TiO_2$ anodization creates a nanotubular structure that contributes to rapid cell growth and MSCs' osteo-differentiation, which assists bone formation and accelerates bone regeneration (Brammer et al., 2012). In the following Table 6, we accumulated representative examples of titanium oxide nanoparticles in regenerative medicine.

### 3.2.7 Cerium Oxide Nanoparticles (CeNPs)

CeNPs show significant antioxidant properties with two oxidation states $Ce^{3+}$ and $Ce^{4+}$. In skin regeneration, cerium oxide acts as a superoxide scavenger, aiding in rapid wound healing and enhancing antibacterial, anti-inflammatory, and anti-oxidative properties, and can promote skin regeneration and scar reduction (Allu et al., 2023). Moreover, CeNPs applied locally for wound healing can enhance cell proliferation and migration of skin tissues and cells. Several studies have used cerium oxide in skin regeneration and wound healing experiments (Augustine et al., 2021; Bhattacharya et al., 2019; Blinova et al., 2021; Cheng et al., 2021b; Corsi et al., 2023; Cui et al., 2021; Gong et al., 2022; Legon'kova et al., 2017; Peloi et al., 2020; Silina et al., 2021; Singh et al., 2022; Wu et al., 2018; Ye et al., 2023). Chigurupati et al. (2013) formulated a cerium oxide nanoparticle, Nanoceria, and treated primary skin cells at different concentrations, demonstrating increased keratinocyte and fibroblast migration at micromolar concentrations. They further validated their analysis in mice by exhibiting enhanced wound healing. CeNPs can also be conjugated with microRNAs, as demonstrated by various study groups (Niemiec et al., 2020; Zgheib et al., 2019) against diabetic wound healing. In a study by Singh et al. (2021), CeNPs were incorporated in a biomatrix obtained from goat tissue for effective tissue engineering purposes. Finally, CeNPs' use in nerve regeneration has also been highlighted (Rahimi et al., 2023), where scaffolds containing $CeO_2$ NPs were formulated to investigate nerve cell regeneration in a spinal cord injury rat model.

**Table 6** Representative studies on $TiO_2$ in regenerative medicine

| | | |
|---|---|---|
| **Bone tissue regeneration** | | |
| $TiO_2$ NPs and silk fibroin (SF) porous nanocomposite scaffolds | Promoted osteoblast/ fibroblast cell attachment and proliferation on these scaffolds | Johari et al. (2017, 2018), Kim et al. (2016) and Kim et al. (2014) |
| PCL/gelatin nanofiber encapsulated with $TiO_2$ NPs and metformin-loaded mesoporous silica NPs | Increased in vitro expansion of human adipose-derived cells | Pourpirali et al. (2021) |
| Poly(3-hydroxybutyrate-co-3-hydroxyvalerate) core-shell fibers encapsulated $TiO_2$ NPs as an organic-inorganic composite scaffold model | Antibacterial and increased bone regeneration | Karbowniczek et al. (2023) |
| Poly-sulfone-silica microfiber fabricated through the inversion phase method with N-methyl pyrrolidone grafted with $TiO_2$ NPs as microspheres | Antimicrobial activity and bone regeneration | Nechifor et al. (2019) |
| $TiO_2$ NPs combined with bovine-derived HAp | Increased mineralized matrix deposition and osteoblast cell viability | Fernandes et al. (2022) |
| **Skin regeneration** | | |
| Polyhydroxybutyrate (PHB)/PCL mats modified with hybrid melanin-encapsulated $TiO_2$ NPs | Mimicked ECM and, demonstrated materials-cell interaction, and antibacterial properties | Avossa et al. (2021) |
| Gelatin-loaded $TiO_2$ NPs and $Ag_2O$ NPs | Burn wound healing and enhanced neovascularization in a second-degree burn | Eldebany et al. (2021) |
| Anatase $TiO_2$ NPs colloidal dispersion formed by sol-gel method | Reduction in burn wound area and absence of scar tissue | Seisenbaeva et al. (2017) |

### 3.2.8 Zinc Oxide NPs (ZnONPs)

Zinc is an essential trace element found in soft and hard tissues. ZnONPs are low-cost, low-toxic inorganic materials demonstrating antimicrobial effects, which, even in low concentrations, can potentiate tissue integration by cellular processes, including angiogenesis and cell adherence (Carofiglio et al., 2020; Laurenti & Cauda, 2017). Moreover, releasing zinc ions contributes to antioxidant and antimicrobial effects significant for wound healing and skin regeneration. ZnO nanostructures like ZnO nanorods, nanowires, and nanoflowers can present physicochemical properties, such as biodegradation, biocompatibility, and biosafety at cellular levels (Li et al., 2008; Zhou et al., 2006). A considerable body of work has been conducted in the field of bone tissue engineering (Carofiglio et al., 2020; Feng et al., 2014; Park et al., 2010) and wound healing (Díez-Pascual & Díez-Vicente, 2015) using ZnONPs. In the following Table 7, we presented a few representative examples of zinc oxide in regenerative medicine.

**Table 7** Representative studies on zinc oxide in regenerative medicine

| **Skin regeneration** | | |
|---|---|---|
| Chitosan-modified ZnONPs | Increased antimicrobial activities against gram-negative bacteria and enhanced wound healing | Díez-Pascual and Díez-Vicente (2015) |
| Azithromycin-loaded ZnONPs incorporated into hydroxypropyl methyl cellulose (HPMC) gel | Enhanced bacterial clearance, epidermal regeneration, and skin tissue formation | Saddik et al. (2022) |
| ZnO-Ag bimetallic NP-loaded nanofibers | Potent antibacterial activity, enhanced tissue regeneration, and wound healing, with antioxidant and anti-inflammatory effects | Borges Rosa de Moura et al. (2022), Hu et al. (2018, and Li et al. (2021b) |
| Carboxymethylcellulose (CMC), hyaluronic acid, and ZnO NPs | Cell proliferation and antibacterial properties | Paramadini et al. (2023) |
| Rhamnolipid-coated ZnO NPs | Promising wound healing efficacy with antibacterial activity in a rat model | Malakar et al. (2023) |
| Cefazolin loaded ZnO NPs with gelatin nanofiber mats | Antibacterial activity, accelerated wound healing, enhanced cell adhesion, epithelial migration, and efficient collagen synthesis | Rath et al. (2016) |
| Dopamine-ZnO NPs doped poly(N-hydroxyethyl acrylamide) (p(HEAA))/ agar dual network hydrogels | Wound healing and antibacterial activity | Sun et al. (2021) |
| Polyethylene oxide (PEO) and ZnO NPs | Adhesion and proliferation of fibroblasts on the scaffold | Nosrati et al. (2020) |
| Potassium-doped ZnO NPs in cross-linked chitosan | Stimulated angiogenesis | Shahzadi et al. (2018) |
| Heparinized ZnO NPs in chitosan hydrogels | Increased antibacterial synergistic effect, accelerated wound closure, re-epithelialization, and decreased collagen deposition | Khorasani et al. (2021) |
| ZnO NPs are covalently attached with acacia and alginate for topical application | Topical application accelerated full-thickness excision wound healing in rabbits | Manuja et al. (2020) |
| Sericin-fabricated microneedle tips integrated ZnO NPs as antibacterial modules | Antibacterial activity, improved skin repair via hair follicle regeneration and angiogenesis | Deng et al. (2022) |
| Nonanimal carboxymethyl chitosan as natural polymer stabilizing in ZnO nanocomposite ultrasonic preparation | Exhibited synergetic antibacterial properties, hemostatic properties with good hemocompatibility | Rao et al. (2020) |
| Electro spun hyaluronic acid/polyvinyl alcohol/polyethylene oxide nanofibers embedding a ZnO NP/cinnamon essential oil (CEO) antimicrobial combination | Wound healing and antibacterial properties | El-Aassar et al. (2021) |

(continued)

**Table 7** (continued)

| **Bone regeneration** | | |
|---|---|---|
| PCL-glycerin-tea tree essential oil-glycerin-PLA with ZnONP nanocomposites for subdermal implantation | Biocompatibility and biodegradability for 60 days of implantation, resorption of implanted material | Grande-Tovar et al. (2022) |
| ZnONP, in conjunction with poly-octane diol citrate (POC) polymer scaffolds, cocultured in hypoxic conditions with bovine chondrocyte cells | Preservation of cartilage homeostasis with chondrocyte proliferation upregulation | Mirza et al. (2015) |
| ZnONP-doped scaffolds of β-tricalcium phosphate | Robust mechanical properties of the scaffold and formation of bone-like apatite layer indicating osteoinduction and osteoconduction | Feng et al. (2014) |

### 3.2.9 Quantum Dots

Quantum dots (QD) are semiconductor materials with excellent optical properties and nanosize dimensions, which are composed of groups II–VI or III–V elements in the periodic table and are extensively used in biolabeling due to their optical attributes, high extinction coefficient, high fluorescent quantum yield, and high fluorescence emittance compared to typical organic fluorescence probes (Kargozar et al., 2020; Yukawa et al., 2023). They are one of the advanced fluorescent NPs with sizes of ≤20 nm, which is why the electrons are bound in this small area, generating the quantum effects (Yukawa et al., 2023). Moreover, QDs are exploited as luminescent probes due to their narrow emission bandwidths and resistance to photobleaching and as drug delivery carriers in theranostics and regenerative medicine (Yukawa et al., 2013). Unlike other inorganic biomaterials, QDs are often introduced intracellularly instead of being incorporated with the stem cells or differentiated cells to allow them to generate stable imaging. The advancement of in vivo imaging for visualization of delivered biomolecules in various tissue regeneration studies has been significantly facilitated by QDs. Similarly, in vivo tracking of transplanted stem cells, QDs are deemed to show excellent outcomes as they can detect transplanted stem cells in a cellular level (Yukawa et al., 2010).

## *3.3 Carbon-Based Particles*

### 3.3.1 Carbon Nanotubes

Bones carry the body's weight, maintain movement, store calcium and phosphorus, and are composed of osteoblasts, bone lining cells, osteocytes, and osteoclasts (Florencio-Silva et al., 2015). Although bones have the capability of self-healing, larger fractures are more likely to remain in a nonunion state. Healing of bone fracture is an intricate biological process that entails specific regenerative patterns and involves the release of several pro-inflammatory molecules and mesenchymal stem cells to remodel and restore the normal bone structure (Marsell & Einhorn, 2011). Bone grafting, which can include autografts, allografts, and xenografts, can be used for bone defects, but their limitations include pain at the area of harvest, while immune responses might restrict their use (Oryan et al., 2014). Hydroxyapatite is widely used as an artificial scaffold for bones, but its brittle nature limits its uses. These challenges pave the way for a growing number of studies on carbon nanotube (CNT)-based composites for bone regeneration. Carbon nanotubes (CNTs) are cylindrical fullerenes, with nano-range diameter. They present multiple useful characteristics, which include high yield strength, high surface area, and electric conductivity. The two main types of CNTs are single-walled CNTs (SWCNTs) and multiwalled CNTs (MWCNTs) (Maheswaran & Shanmugavel, 2022). Supronowicz et al. (2002) reported that CNT/PLA nanocomposites increased cell proliferation and extracellular calcium levels, mRNA expression upregulation for collagen type-I and osteoblast formation, following electric stimulation with 10 microA at 10 Hz, for various periods of time over 6 h/day. In 2017, Das et al. (2017) observed that COOH-functionalized CNTs promoted both osteogenic and chondrogenic differentiation in canine bone-derived mesenchymal stem cells (MSC).

Recognizing the extensive literature on CNTs and bone regeneration, we present representative examples on the topic in Table 8.

Electrical stimulation can facilitate the repair of damages by various nervous system diseases. The high surface area, electrical conductivity, and porous structure of CNTs can be exploited in neural tissue regeneration alone or in scaffolds (Xiang et al., 2020). Mattson et al. (2000) reported that neurons can attach and grow on the surface of MWCNTs, and increased branching of embryonic rat brain-derived neurons was detected when the neurons were grown on 4-hydroxynonenal coated CNTs. The high electrical conductivity of CNTs can augment the efficacy of neural signal transmission by facilitating dendrite elongation and cell adhesion (Lovat et al., 2005). It has also been reported that CNTs functionalized with PEI or coated with a nerve growth factor can increase the regeneration properties, such as neurite outgrowth, branching, and attachment (Hu et al., 2005; Matsumoto et al., 2007). Ahn et al. (2015) implanted CNTs-PGF (phosphate glass fiber) scaffold in between the proximal and distal stumps of completely transected sciatic nerves in 12-week-old rats and found restoration of motor function electrophysiologically along with re-innervation of scaffolds crossing axons. Biocompatible chitosan-CNTs composite hydrogel also was found to increase neuronal cell adhesion, proliferation, and neurite outgrowth (Wu et al., 2017). One recent research revealed that transparent CNTs can improve the recovery of lesioned entorhinal-hippocampal complex organotypic cultures and significantly increase signal synchronization and fiber sprouting between cortex and hippocampus (Pampaloni et al., 2020). CNTs yarns were used as a scaffold to repair 15 mm rat sciatic nerve defect and peripheral nerve regeneration occurred 8 weeks after surgery in a in vivo study conducted by Kunisaki et al. (2021), which depicted the clinical application of CNTs in nerve regeneration approaches.

**Table 8** Representative studies of CNT utilization for the management of bone tissue regeneration, repair, and fracture healing

| Materials used with CNTs | Study details | Outcome | References |
|---|---|---|---|
| –COOH, polyethylene glycol (PEG), poly amino benzene sulfonic acid | Functionalized SWCNTs and MWCNTs were sprayed on preheated glass coverslips, air dried, irradiated overnight, and evaluated with osteosarcoma ROS 17/2.8 cells | CNTs assisted in the growth of osteoblasts and bone formation that may help the regeneration of fractured bones | Zanello et al. (2006) |
| Collagen | Recombinant human bone morphogenic protein-2 (rhBMP-2) solution was mixed with MWCNTs and type I collagen | Restoration of periosteal tissue with little inflammation, formation of new ectopic bone 3 weeks after implantation | Usui et al. (2008) |
| PEG | MWCNTs modified with PEG sprayed onto preheated coverslips. Finally, air dried to get the composite | Accelerated differentiation and proliferation of hMSCs without any inducers or growth factors | Nayak et al. (2010) |
| Chitosan | Chitosan-fMWCNT scaffold was prepared by dispersing varying amounts of f-MWCNTs in chitosan solutions and lyophilization | Cell proliferation was higher than chitosan alone | Venkatesan et al. (2012) |
| –COOH, polyvinyl alcohol (PVA), biomimetic apatite | Modification of MWCNTs was performed using nitric acid to produce MWCNT-COOH. PVA-modified MWCNTs were prepared by mixing PVA solution with carboxylated MWCNTs. Dipping of PVA-MWCNTs in simulated body fluid (SBF) for 14 days formed MWCNTs-apatite | No morphologic changes in MG-63 human osteosarcoma cells were observed with the formulation<br>Higher cell viability was found with biomimetic MWCNTs-apatite | Zhang et al. (2012) |
| Chitosan | Different concentration of MWCNTs was mixed with chitosan to form chitosan-MWCNTs films | The nanocomposite presented improved tensile modulus and strength compared to chitosan, while no toxic effect was detected on osteoblasts | Aryaei et al. (2014) |
| Chitosan, PLA | MWCNTs were mixed with chitosan and PLA and underwent freeze | The composite promoted cell adhesion, proliferation, and differentiation on osteoblast cells | Orozco et al. (2016) |

(continued)

**Table 8** (continued)

| Materials used with CNTs | Study details | Outcome | References |
|---|---|---|---|
| Polyacrylonitrile (PAN), polymethyl-methacrylate (PMMA) | PAN was mixed with MWCNTs and PMMA microspheres were added. Two-step heating at 300 and 1200 °C took place to obtain porous CNT (CNTp) | Improved osteoconduction, cell proliferation and attachment and new bone formation when implanted with rh BMP-2 | Tanaka et al. (2017) |
| Native bacterial cellulose (BC) | Homogenous dispersion of BC and functionalized MWCNTs were obtained by high intensity mixing. Aqueous $CaCl_2$ was added to facilitate cross linking and final formulation was lyophilized | Promoted osteoblast adhesion, viability, and proliferation, with reduced toxicity | Gutiérrez-Hernández et al. (2017) |
| Polyetheretherketone (PEEK), bioactive glass (BG) | A dispersion of MWCNTs added dropwise to a PEEK powder and BG particles dispersion. PEEK-BG-MWCNTs bio composite was obtained following drying | Formulation demonstrated improved mechanical properties, increased cell adhesion, differentiation, proliferation, and metabolic vitality using MC3T3-E1 cells | Cao et al. (2019) |
| Chitosan (CS), Gelatin (GEL), nanohydroxyapatite (nHAp) | A CS/GEL/(nHAp)/ MWCNTs mix was prepared by adding GEL and nHAp in CS solution, followed by crosslinking and freeze drying | Improved adhesion, proliferation, and osteogenic differentiation of osteoblast MC3T3-E1 cells in vitro | Xu et al. (2019) |
| – | SWCNTs and MWCNTs were dispersed in 4-O-methyl-α-d glucuronoxylan (GX) and pour drop wise onto 12 mm glass disk and dried | Rat bone marrow mesenchymal stem cells showed high activities of alkaline phosphatase | Mori et al. (2020) |
| –COOH Poly ε-caprolactone (PCL), Keratin (Kr) | Kr and PCL separately dissolved were mixed under stirring and MWCNTs-COOH was added. The suspension was sonicated and electro-spun to form a Kr-PCL-MWCNT scaffold | The addition of MWCNTs-COOH with K-PCL enhanced the mechanical properties of the scaffold. Cell adhesion, proliferation, and osteogenic differentiation significantly improved in mesenchymal stem cells | Mahmoodi et al. (2021) |

(continued)

**Table 8** (continued)

| Materials used with CNTs | Study details | Outcome | References |
|---|---|---|---|
| 1-Ethyl-3-(3-dimethyl aminopropyl) carbodiimide (EDC) | SWCNT suspension was mixed with an aqueous solution of EDC yielding EDC-functionalized SWCNTs | Showed biocompatibility, enhanced proliferation and differentiation of bone mesenchymal stem cells (BMSCs). Promoted angiogenesis and bone regeneration | Luo et al. (2022) |
| –COOH | MWCNTs-COOH having length of 20–30 μm and diameter 20 nm were studied with bone mesenchymal stem cells (BMSCs), osteoblast (MC3T3-E1), macrophages (RAW264.7), human umbilical vein endothelial cells (HUVECs), and bone marrow mononuclear cells (BMMNCs) | Separate treatment of BMSCs and MC3T3-E1s with MWCNTs-COOH hindered their osteogenic differentiation. However, coculture of BMSCs with macrophages increased the osteogenesis of BMSCs through M2 polarization of macrophages. Facilitated angiogenesis of HUVECs and demonstrated antiinflammatory effect on BMSCs to enhance bone formation | Lin et al. (2023) |

CNTs have found extensive applications in a large spectrum of applications. One important aspect to mention is dentistry, with the objective of preparing restorative material and bone defect regeneration. Conventional dental composite resin used in decayed teeth repair is made of bisphenol A-glycidyl methacrylate (BIS-GMA) and silica (Bhattacharya & Seong, 2019). Zhang et al. (2008) developed a process for improving dental resin-based composites, using functionalized (oxidized) SWCNTs coated with nano-$SiO_2$ followed by surface modification with allyl terminated silane layer and blended with resin monomer urethane dimethacrylate to form the SWCNTs-$SiO_2$-ATES-resin composite. The final product presented enhanced flexural strength. In their study, Akasaka et al. [335] observed that CNTs selectively adhere to dentin and cementum surfaces, possibly due to adhesion to exposed collagen fibers. Furthermore, the CNT coating did not alter the dentin adhesive's tensile bond strength. In another study, a suspension of PLA, MWCNTs and HAp was electrospun for developing a guided tissue regeneration membrane. The membrane promoted adhesion and proliferation of periodontal ligament cells, while it inhibited the adhesion and proliferation of gingival epithelial cells (Mei et al., 2007).

CNTs have great potential in regenerative medicine due to their unique chemical, mechanical, and electrical properties. However, there is a strong urge for proper evaluation of CNTs' toxicity, teratogenicity, and carcinogenicity. MWCNTs are physiochemically stable and less susceptible to degradation (Elgrabli et al., 2015), and studies have supported no negative impacts of CNTs in cells (Chetyrkina et al., 2022). Nonetheless, there have also been reports that present potential CNT cytotoxicity in various cell types (Hirano et al., 2010; Mohanta et al., 2019). Thus, more research is necessary to successfully translate the CNTs' numerous applications to clinical practice.

### 3.3.2 Graphene NPs

Graphene is one of the most robust materials characterized by $sp^2$ hybridized carbon monolayers (Ding et al., 2015b; Tiwari et al., 2020). It possesses remarkable physicochemical properties with high elasticity, large surface area, high thermal and electric conductivity, and notable biocompatibility. Moreover, according to their planar structure, they can entrap several molecules and substances, including biomolecules, fluorescent probes, drugs, and cells. Due to these features, graphene has become one of the desirable materials for biomedical applications, cancer research, targeted drug delivery, regenerative medicine, and tissue engineering. Graphene nanomaterials can be divided into different categories such as few-layer graphenes (FLG), graphene nanosheets, graphene oxide (GO), and reduced GO (rGO) (Ding et al., 2015b; Lee et al., 2011; Nayak et al., 2011; Tiwari et al., 2020). In recent years, extensive literature associated with graphene NPs can be found in neuronal (Akhavan et al., 2014; Kang et al., 2018; Qian et al., 2021; Şelaru et al., 2022; Yang et al., 2014; Yao et al., 2023), bone (Abedin Dargoush et al., 2022; Chen et al., 2019; Duan et al., 2015; Kang et al., 2015; La et al., 2013; Nair et al., 2015; Nishida et al., 2014; Qi et al., 2014; Tatavarty et al., 2014; Yu et al., 2022a; Zhao et al., 2018), liver (Kim et al., 2023; Melgar-Lesmes et al., 2018), skin (Chen et al., 2022; Yang et al., 2020b), cardiac (Han et al., 2018; Park et al., 2014; Soltani et al., 2022; Wang et al., 2021b), skeletal muscles (Chaudhuri et al., 2014; Ciriza et al., 2015; Dai et al., 2016; Jo et al., 2017; Jo et al., 2020; Krueger et al., 2016; Lee et al., 2016; Park et al., 2019; Patel et al., 2018; Shin et al., 2018; Wierzbicki et al., 2020), and overall tissue regeneration (Choe et al., 2019; Ciriza et al., 2018; Ciriza et al., 2015; Jagiełło et al., 2019; Karimi Hajishoreh et al., 2022; Mokhtari et al., 2019; Yu et al., 2022a), with the main outcomes being enhanced cell proliferation, viability, differentiation, and overall biological activity essential for regeneration. Representative examples of the extensive applications of graphene in regenerative medicine are presented in Table 9.

## 4 Conclusion and Future Directions

Regenerative medicine and tissue engineering have predominately relied on developing stem cell therapies to address the needs of the various diseases it aims to treat. Nonetheless, cell-based or tissue-based therapies can frequently be effective if the environment they are placed in promotes their proliferation and eventual differentiation for the desired outcome. Active pharmaceutical compounds can enhance, promote, or sustain such environments. Thus, cell-based therapies may be as good as

**Table 9** Representative studies on graphene in regenerative medicine

| | | |
|---|---|---|
| **Bone regeneration** | | |
| Graphenes containing poly(L-Lactide) (PLLA) nanofibrous scaffolds | Enhanced of bone MSC differentiation | Duan et al. (2015) |
| Pluronic F127 coated GO-calcium phosphate (GO-CaP) | Osteogenesis of hMSCs and enhancement of Ca deposition by osteoblast | Tatavarty et al. (2014) |
| Methacrylamidated hyaluronic acid with methacrylamidated gelatin and GO/ alginate encapsulation system for rat bone marrow-derived macrophages and rat bone mesenchymal stem cells | Promoted osteogenic differentiation of BMSCs, early M2-type polarization of macrophages, avoiding excessive inflammation, facilitating bone repair | Yu et al. (2022a) |
| Baicalin/GO/demineralized bone matrix scaffold | Enhanced M2-type polarization, angiogenesis and osteogenesis | Guo et al. (2021) |
| 3D-printed scaffolds of 2D-hetero nanostructures containing GO/ black phosphorous (BP) nanosheets loaded with interleukin (IL)-4 cytokines | Enhanced pro-healing environment for bone regeneration | . (Liu et al. (2023) |
| Immortalized mouse-adipose-derived MSCs in citrate-based hydrogel with GO | Potentiated BMP9-induced expression shown by enhanced bone mineralization and vascularization of trabecular bone | Zhao et al. (2018) |
| Strontium(Sr)-GO nanocomposites modified collagen scaffolds | Prolonged and sustained release of $Sr^{2+}$ ions, increased osteogenesis and angiogenesis via MAPK signaling pathway in large bone defect | Chen et al. (2019) |
| **Dental tissue regeneration** | | |
| Composite films of GO, silk fibroin (SF), and a combination of both | Increased proliferation of periodontal ligament stem cells (PDLSCs) | Rodríguez-Lozano et al. (2014) |
| | Promoted osteo/cementoblast differentiation of PDLSCs | Vera-Sánchez et al. (2016) |
| Liquid-phase exfoliated graphene (LPEG) films | Stimulated neuro-differentiation of stem cells from apical papilla (SCAP) | Simonovic et al. (2022) |
| GO nano powder mixed with calcium phosphate bone cement (CPC)-chitosan paste | Enhanced antimicrobial effect against biofilms and increased human dental pulp stem cells (hDPSCs) viability, attachment and growth | Wu et al. (2021) |
| **Cardiac muscle regeneration** | | |
| Graphene films fabricated by chemical vapor deposition on Cu foils | MSCs cocultured graphene sheets modulate genes regulating cardiomyogenic differentiation | Park et al. (2014) |
| MSCs encapsulated GO/alginate and rGO/ alginate nanocomposite microgels | Decreased infarction area and improvement of cardiac function in infarcted heart | Choe et al. (2019) |

(continued)

**Table 9** (continued)

| | | |
|---|---|---|
| Near-infrared (NIR) sensitive 4D cardiac constructs made of PEGDA hydrogels doped with graphene nanoplates | Uniformly distributed cells with enhanced myocardial maturation | Wang et al. (2021b) |
| Injectable shear-thinning hydrogel made of Adamantane-GO and alginate-cyclodextrin | Enhanced differentiation of rat MSCs into cardiomyocyte-like cells | Soltani et al. (2022) |
| **Skin wound healing and regeneration** | | |
| GO-based hydrogel matrix made of poly-D, L-lactic acid-polyethylene glycol-poly D-lactic acid (PDLLA-PEG-PDLLA) | Enhanced secretion of small extracellular vesicles for diabetic wound healing associated with miR-21 upregulation and suppression of PVT1/PTEN/IL-17 axis | Chen et al. (2022) |
| GO dispersion cocultured with adipose-derived stem cells | Promoted diabetic wound healing by increasing cell apoptosis by regulating miR-7977 and linc00324 expression | Ji et al. (2022) |
| GO/Cu nanocomposites with chitosan/HA as bioactive dressing scaffold | Improved angiogenesis and reduced inflammation in bacteria-infected wound | Yang et al. (2020b) |
| **Liver regeneration** | | |
| Nanosized GO crosslinked with porcine decellularized extracellular matrix (dECM) scaffold | Inhibitory effects of MMP catalysis and enhanced M2-polarization | Kim et al. (2023) |
| Graphene nanostars with PAMAM-G5 dendrimer | Reduced liver fibrosis, and improved hepatic restoration in cirrhotic mice | Melgar-Lesmes et al. (2018) |
| **Graphene in neuronal regeneration** | | |
| Electrospun nano-gelatin meshes containing GO and magnetic nanoparticles | Enhanced differentiation of neuron-like cells derived from hASCs | Şelaru et al. (2022) |
| Reduced-GO/PCL scaffolds anisotropic microfibers with nano grooves loaded with adipose-derived stem cells | Improved nerve conduction recovery | Yao et al. (2023) |

the associated pharmacological treatments. From that perspective, drug delivery can provide necessary solutions for compounds to reach targeted tissues, sustain their release over prolonged periods, and protect them from degradation. Nanotechnological approaches in drug delivery that are extensively evaluated predominately for cancer treatment present the potential to meet the drug delivery needs for regenerative medicine. In fact, most of the advantages that nanoparticles present and have been utilized against cancer, inflammation or other diseases are the driving forces for using nanoparticles in regenerative medicine. However, similarly to any other disease, nanocarriers need to be evaluated using in vitro and in vivo models to identify their potential for adaptation, especially if we recognize the vast variety of tissue types and drug molecules that regenerative medicine is targeting or utilizing. The existing interdisciplinary nature of regenerative medicine only reinforces how this need must be met, as multiple aspects must converge properly for successful treatments using drug carriers alone. This includes identifying the type of nanocarrier, its material's compatibility, stability, biodegradation, along with its ability to meet the pharmacological therapeutic expectations. While the different carriers above offer present similar or distinct advantages, significant work remains to translate the described approaches from the bench to large-scale clinical applications. Although the limitations described here represent only a partial view of the broader challenges, the clinical relevance and translational potential for tissue regeneration is undeniable. Thus, regenerative medicine holds the potential to gradually reshape our understanding and perspective on the future health care and medical treatments.

## References

Abdel-Mohsen, A. M., Jancar, J., Abdel-Rahman, R. M., Vojtek, L., Hyršl, P., Dušková, M., & Nejezchlebová, H. (2017). A novel in situ silver/hyaluronan bio-nanocomposite fabrics for wound and chronic ulcer dressing: In vitro and in vivo evaluations. *International Journal of Pharmaceutics, 520*(1), 241–253. https://doi.org/10.1016/j.ijpharm.2017.02.003

Abdellatif, A. A. H., & Alsowinea, A. F. (2021). Approved and marketed nanoparticles for disease targeting and applications in COVID-19. *Nanotechnology Reviews, 10*(1), 1941–1977. https://doi.org/10.1515/ntrev-2021-0115

Abedi, G., Jahanshahi, A., Fathi, M. H., Haghdost, I. S., & Veshkini, A. (2014). Study of nano-hydroxyapatite/zirconia stabilized with yttria in bone healing: Histopathological study in rabbit model. *Polish Journal of Pathology, 65*(1), 40–47. https://doi.org/10.5114/pjp.2014.42668

Abedin Dargoush, S., Hanaee-Ahvaz, H., Irani, S., Soleimani, M., Khatami, S. M., & Sohi, A. N. (2022). A composite bilayer scaffold functionalized for osteochondral tissue regeneration in rat animal model. *Journal of Tissue Engineering and Regenerative Medicine, 16*(6), 559–574. https://doi.org/10.1002/term.3297

Adibkia, K., Ehsani, A., Jodaei, A., Fathi, E., Farahzadi, R., & Barzegar-Jalali, M. (2021). Silver nanoparticles induce the cardiomyogenic differentiation of bone marrow derived mesenchymal stem cells via telomere length extension. *Beilstein Journal of Nanotechnology, 12*, 786–797. https://doi.org/10.3762/bjnano.12.62

Ahn, H. S., Hwang, J. Y., Kim, M. S., Lee, J. Y., Kim, J. W., Kim, H. S., Shin, U. S., Knowles, J. C., Kim, H. W., & Hyun, J. K. (2015). Carbon-nanotube-interfaced glass fiber scaffold for regeneration of transected sciatic nerve. *Acta Biomaterialia, 13*, 324–334. https://doi.org/10.1016/j.actbio.2014.11.026

Ai, M., Du, Z., Zhu, S., Geng, H., Zhang, X., Cai, Q., & Yang, X. (2017). Composite resin reinforced with silver nanoparticles-laden hydroxyapatite nanowires for dental application. *Dental Materials, 33*(1), 12–22. https://doi.org/10.1016/j.dental.2016.09.038

Akhavan, O., Ghaderi, E., Abouei, E., Hatamie, S., & Ghasemi, E. (2014). Accelerated differentiation of neural stem cells into neurons on ginseng-reduced graphene oxide sheets. *Carbon, 66*, 395–406. https://doi.org/10.1016/j.carbon.2013.09.015

Ali, M. S., Anuradha, V., Yogananth, N., & Krishnakumar, S. (2019). Heart and liver regeneration in zebrafish using silver nanoparticle synthesized from Turbinaria conoides—In vivo. *Biocatalysis and Agricultural Biotechnology, 17*, 104–109. https://doi.org/10.1016/j.bcab.2018.10.013

Allu, I., Kumar Sahi, A., Kumari, P., Sakhile, K., Sionkowska, A., & Gundu, S. (2023). A brief review on cerium oxide (CeO(2)NPs)-based scaffolds: Recent advances in wound healing applications. *Micromachines (Basel), 14*(4). https://doi.org/10.3390/mi14040865

Amani, H., Kazerooni, H., Hassanpoor, H., Akbarzadeh, A., & Pazoki-Toroudi, H. (2019). Tailoring synthetic polymeric biomaterials towards nerve tissue engineering: A review. *Artificial Cells, Nanomedicine, and Biotechnology, 47*(1), 3524–3539. https://doi.org/10.1080/21691401.2019.1639723

Anderson, J. M., Patterson, J. L., Vines, J. B., Javed, A., Gilbert, S. R., & Jun, H. W. (2011). Biphasic peptide amphiphile nanomatrix embedded with hydroxyapatite nanoparticles for stimulated osteoinductive response. *ACS Nano, 5*(12), 9463–9479. https://doi.org/10.1021/nn203247m

Antosiak-Iwańska, M., Bącal, P., Kazimierczak, B., Kwiatkowska, A., Godlewska, E., Grzeczkowicz, A., Stachowiak, R., Bielecki, J., & Granicka, L. (2020). Polyelectrolyte membrane with hydroxyapatite and silver nanoparticles as a material for modern wound dressings. *Journal of Biomedical Nanotechnology, 16*(5), 702–714. https://doi.org/10.1166/jbn.2020.2907

Arafa, M. G., El-Kased, R. F., & Elmazar, M. M. (2018). Thermoresponsive gels containing gold nanoparticles as smart antibacterial and wound healing agents. *Scientific Reports, 8*(1), 13674. https://doi.org/10.1038/s41598-018-31895-4

Arakha, M., Pal, S., Samantarrai, D., Panigrahi, T. K., Mallick, B. C., Pramanik, K., Mallick, B., & Jha, S. (2015). Antimicrobial activity of iron oxide nanoparticle upon modulation of nanoparticle-bacteria interface. *Scientific Reports, 5*, 14813. https://doi.org/10.1038/srep14813

Aryaei, A., Jayatissa, A. H., & Jayasuriya, A. C. (2014). Mechanical and biological properties of chitosan/carbon nanotube nanocomposite films. *Journal of Biomedical Materials Research. Part A, 102*(8), 2704–2712. https://doi.org/10.1002/jbm.a.34942

Ashammakhi, N., Kim, H. J., Ehsanipour, A., Bierman, R. D., Kaarela, O., Xue, C., Khademhosseini, A., & Seidlits, S. K. (2019). Regenerative therapies for spinal cord injury. *Tissue Engineering. Part B, Reviews, 25*(6), 471–491. https://doi.org/10.1089/ten.TEB.2019.0182

Ashraf, R., Sofi, H. S., Akram, T., Rather, H. A., Abdal-Hay, A., Shabir, N., Vasita, R., Alrokayan, S. H., Khan, H. A., & Sheikh, F. A. (2020a). Fabrication of multifunctional cellulose/TiO(2) / Ag composite nanofibers scaffold with antibacterial and bioactivity properties for future tissue engineering applications. *Journal of Biomedical Materials Research. Part A, 108*(4), 947–962. https://doi.org/10.1002/jbm.a.36872

Ashraf, R., Sofi, H. S., & Sheikh, F. A. (2020b). Methodology involved in the osteogenic differentiation of mesenchymal stem cells on chitosan-collagen nanofibers incorporated with titanium dioxide nanoparticles. *Methods in Molecular Biology, 2125*, 85–94. https://doi.org/10.1007/7651_2019_257

Augustine, R., Zahid, A. A., Hasan, A., Dalvi, Y. B., & Jacob, J. (2021). Cerium oxide nanoparticle-loaded gelatin Methacryloyl hydrogel wound-healing patch with free radical scavenging activity. *ACS Biomaterials Science & Engineering, 7*(1), 279–290. https://doi.org/10.1021/acsbiomaterials.0c01138

Avossa, J., Pota, G., Vitiello, G., Macagnano, A., Zanfardino, A., Di Napoli, M., Pezzella, A., D'Errico, G., Varcamonti, M., & Luciani, G. (2021). Multifunctional mats by antimicrobial nanoparticles decoration for bioinspired smart wound dressing solutions. *Materials Science & Engineering. C, Materials for Biological Applications, 123*, 111954. https://doi.org/10.1016/j.msec.2021.111954

Azzawi, Z. G. M., Hamad, T. I., Kadhim, S. A., & Naji, G. A. (2018). Osseointegration evaluation of laser-deposited titanium dioxide nanoparticles on commercially pure titanium dental implants. *Journal of Materials Science. Materials in Medicine, 29*(7), 96. https://doi.org/10.1007/s10856-018-6097-6

Bagchi, A., Meka, S. R., Rao, B. N., & Chatterjee, K. (2014). Perovskite ceramic nanoparticles in polymer composites for augmenting bone tissue regeneration. *Nanotechnology, 25*(48), 485101. https://doi.org/10.1088/0957-4484/25/48/485101

Bai, M. Y., Ku, F. Y., Shyu, J. F., Hayashi, T., & Wu, C. C. (2021). Evaluation of Polyacrylonitrile nonwoven Mats and Silver-gold bimetallic nanoparticle-decorated nonwoven Mats for potential promotion of wound healing in vitro and in vivo and bone growth in vitro. *Polymers (Basel), 13*(4). https://doi.org/10.3390/polym13040516

Baino, F., Fiorilli, S., & Vitale-Brovarone, C. (2016). Bioactive glass-based materials with hierarchical porosity for medical applications: Review of recent advances. *Acta Biomaterialia, 42*, 18–32. https://doi.org/10.1016/j.actbio.2016.06.033

Balagangadharan, K., Viji Chandran, S., Arumugam, B., Saravanan, S., Devanand Venkatasubbu, G., & Selvamurugan, N. (2018). Chitosan/nano-hydroxyapatite/nano-zirconium dioxide scaffolds with miR-590-5p for bone regeneration. *International Journal of Biological Macromolecules, 111*, 953–958. https://doi.org/10.1016/j.ijbiomac.2018.01.122

Banik, B. L., Fattahi, P., & Brown, J. L. (2016). Polymeric nanoparticles: The future of nanomedicine. *WIREs Nanomedicine and Nanobiotechnology, 8*(2), 271–299. https://doi.org/10.1002/wnan.1364

BaoLin, G., & Ma, P. X. (2014). Synthetic biodegradable functional polymers for tissue engineering: A brief review. *Science China. Chemistry, 57*(4), 490–500. https://doi.org/10.1007/s11426-014-5086-y

Barabadi, Z., Azami, M., Sharifi, E., Karimi, R., Lotfibakhshaiesh, N., Roozafzoon, R., Joghataei, M. T., & Ai, J. (2016). Fabrication of hydrogel based nanocomposite scaffold containing bioactive glass nanoparticles for myocardial tissue engineering. *Materials Science and Engineering: C, 69*, 1137–1146. https://doi.org/10.1016/j.msec.2016.08.012

Battafarano, G., Rossi, M., De Martino, V., Marampon, F., Borro, L., Secinaro, A., & Del Fattore, A. (2021). Strategies for bone regeneration: From graft to tissue engineering. *International Journal of Molecular Sciences, 22*(3). https://doi.org/10.3390/ijms22031128

Belluomo, R., Khodaei, A., & Amin Yavari, S. (2023). Additively manufactured Bi-functionalized bioceramics for reconstruction of bone tumor defects. *Acta Biomaterialia, 156*, 234–249. https://doi.org/10.1016/j.actbio.2022.08.042

Bhattacharya, D., Tiwari, R., Bhatia, T., Purohit, M. P., Pal, A., Jagdale, P., Mudiam, M. K. R., Chaudhari, B. P., Shukla, Y., Ansari, K. M., Kumar, A., Kumar, P., Srivastava, V., & Gupta, K. C. (2019). Accelerated and scarless wound repair by a multicomponent hydrogel through simultaneous activation of multiple pathways. *Drug Delivery and Translational Research, 9*(6), 1143–1158. https://doi.org/10.1007/s13346-019-00660-z

Bhattacharya, M., & Seong, W.-J. (2019). Chapter 3 – Carbon nanotube-based materials—Preparation, biocompatibility, and applications in dentistry. In K. Subramani & W. Ahmed (Eds.), *Nanobiomaterials in clinical dentistry (second edition) (pp. 41–76)*. Elsevier. https://doi.org/10.1016/B978-0-12-815886-9.00003-6

Bianchi, E., Vigani, B., Viseras, C., Ferrari, F., Rossi, S., & Sandri, G. (2022). Inorganic nanomaterials in tissue engineering. *Pharmaceutics, 14*(6). https://doi.org/10.3390/pharmaceutics14061127

Bigham, A., Aghajanian, A. H., Behzadzadeh, S., Sokhani, Z., Shojaei, S., Kaviani, Y., & Hassanzadeh-Tabrizi, S. A. (2019). Nanostructured magnetic mg(2)SiO(4)-CoFe(2)O(4) composite scaffold with multiple capabilities for bone tissue regeneration. *Materials Science & Engineering. C, Materials for Biological Applications, 99*, 83–95. https://doi.org/10.1016/j.msec.2019.01.096

Blinova, E., Pakhomov, D., Shimanovsky, D., Kilmyashkina, M., Mazov, Y., Demura, T., Drozdov, V., Blinov, D., Deryabina, O., Samishina, E., Butenko, A., Skachilova, S., Sokolov, A., Vasilkina, O., Alkhatatneh, B. A., Vavilova, O., Sukhov, A., Shmatok, D., Sorokvasha, I., Tumutolova, O., & Lobanova, E. (2021). Cerium-containing N-Acetyl-6-Aminohexanoic acid formulation accelerates wound reparation in diabetic animals. *Biomolecules, 11*(6). https://doi.org/10.3390/biom11060834

Bonany, M., Perez-Berna, A. J., Ducic, T., Pereiro, E., Martin-Gomez, H., Mas-Moruno, C., van Rijt, S., Zhao, Z., Espanol, M., & Ginebra, M. P. (2022). Hydroxyapatite nanoparticles-cell interaction: New approaches to disclose the fate of membrane-bound and internalised nanoparticles. *Biomaterials Advances, 142*, 213148. https://doi.org/10.1016/j.bioadv.2022.213148

Boni, R., Ali, A., Shavandi, A., & Clarkson, A. N. (2018). Current and novel polymeric biomaterials for neural tissue engineering. *Journal of Biomedical Science, 25*(1), 90. https://doi.org/10.1186/s12929-018-0491-8

Borges Rosa de Moura, F., Antonio Ferreira, B., Helena Muniz, E., Benatti Justino, A., Gabriela Silva, A., de Azambuja Ribeiro, R. I. M., Oliveira Dantas, N., Lisboa Ribeiro, D., de Assis Araújo, F., Salmen Espindola, F., Silva, C. A., & A., & Carla Tomiosso, T. (2022). Antioxidant, anti-inflammatory, and wound healing effects of topical silver-doped zinc oxide and silver oxide nanocomposites. *International Journal of Pharmaceutics, 617*, 121620. https://doi.org/10.1016/j.ijpharm.2022.121620

Brammer, K. S., Frandsen, C. J., & Jin, S. (2012). TiO2 nanotubes for bone regeneration. *Trends in Biotechnology, 30*(6), 315–322. https://doi.org/10.1016/j.tibtech.2012.02.005

Brammer, K. S., Oh, S., Cobb, C. J., Bjursten, L. M., Heyde, H., & v. d., & Jin, S. (2009). Improved bone-forming functionality on diameter-controlled TiO2 nanotube surface. *Acta Biomaterialia, 5*(8), 3215–3223. https://doi.org/10.1016/j.actbio.2009.05.008

Brokesh, A. M., & Gaharwar, A. K. (2020). Inorganic biomaterials for regenerative medicine. *ACS Applied Materials & Interfaces, 12*(5), 5319–5344. https://doi.org/10.1021/acsami.9b17801

Brovkina, O., & Dashinimaev, E. (2020). Advances and complications of regenerative medicine in diabetes therapy. *PeerJ, 8*, e9746. https://doi.org/10.7717/peerj.9746

Bruch, G. E., Fernandes, L. F., Bassi, B. L. T., Alves, M. T. R., Pereira, I. O., Frezard, F., & Massensini, A. R. (2019). Liposomes for drug delivery in stroke. *Brain Research Bulletin, 152*, 246–256. https://doi.org/10.1016/j.brainresbull.2019.07.015

Bulbake, U., Doppalapudi, S., Kommineni, N., & Khan, W. (2017). Liposomal formulations in clinical use: An Updated Review. *Pharmaceutics, 9*(2). https://doi.org/10.3390/pharmaceutics9020012

Burdusel, A. C., Gherasim, O., Andronescu, E., Grumezescu, A. M., & Ficai, A. (2022). Inorganic nanoparticles in bone healing applications. *Pharmaceutics, 14*(4). https://doi.org/10.3390/pharmaceutics14040770

Burdușel, A. C., Gherasim, O., Grumezescu, A. M., Mogoantă, L., Ficai, A., & Andronescu, E. (2018). Biomedical applications of silver nanoparticles: An up-to-date overview. *Nanomaterials (Basel), 8*(9). https://doi.org/10.3390/nano8090681

Cao, J., Lu, Y., Chen, H., Zhang, L., & Xiong, C. (2019). Preparation, properties and in vitro cellular response of multi-walled carbon nanotubes/bioactive glass/poly(etheretherketone) biocomposite for bone tissue engineering. *International Journal of Polymeric Materials and Polymeric Biomaterials, 68*(8), 433–441. https://doi.org/10.1080/00914037.2018.1455679

Carofiglio, M., Barui, S., Cauda, V., & Laurenti, M. (2020). Doped zinc oxide nanoparticles: Synthesis, characterization and potential use in Nanomedicine. *Applied Sciences (Basel), 10*(15), 5194. https://doi.org/10.3390/app10155194

Carvalho, T., Bártolo, R., Pedro, S. N., Valente, B. F. A., Pinto, R. J. B., Vilela, C., Shahbazi, M. A., Santos, H. A., & Freire, C. S. R. (2023). Injectable nanocomposite hydrogels of gelatin-hyaluronic acid reinforced with hybrid lysozyme Nanofibrils-gold nanoparticles for the regeneration of damaged myocardium. *ACS Applied Materials & Interfaces, 15*(21), 25860–25872. https://doi.org/10.1021/acsami.3c03874

Chakraborty, A., Kundu, B., Basu, D., Pal, T., & Nandi, S. (2011). <i>in vivo</i> bone response and interfacial properties of titanium-alloy implant with different designs in rabbit model with time [original research]. *Indian Journal of Dental Research, 22*(2), 277–284. https://doi.org/10.4103/0970-9290.84304

Chaudhuri, B., Bhadra, D., Mondal, B., & Pramanik, K. (2014). Biocompatibility of electrospun graphene oxide–poly(ε-caprolactone) fibrous scaffolds with human cord blood mesenchymal stem cells derived skeletal myoblast. *Materials Letters, 126*, 109–112. https://doi.org/10.1016/j.matlet.2014.04.008

Chen, F. M., Zhao, Y. M., Jin, Y., & Shi, S. (2012). Prospects for translational regenerative medicine. *Biotechnology Advances, 30*(3), 658–672. https://doi.org/10.1016/j.biotechadv.2011.11.005

Chen, X., Peng, Y., Xue, H., Liu, G., Wang, N., & Shao, Z. (2022). MiR-21 regulating PVT1/PTEN/IL-17 axis towards the treatment of infectious diabetic wound healing by modified GO-derived biomaterial in mouse models. *Journal of Nanobiotechnology, 20*(1), 309. https://doi.org/10.1186/s12951-022-01516-4

Chen, Y., Zheng, Z., Zhou, R., Zhang, H., Chen, C., Xiong, Z., Liu, K., & Wang, X. (2019). Developing a strontium-releasing graphene oxide-/collagen-based organic-inorganic Nanobiocomposite for large bone defect regeneration via MAPK signaling pathway. *ACS Applied Materials & Interfaces, 11*(17), 15986–15997. https://doi.org/10.1021/acsami.8b22606

Chen, Y. H., Rao, Z. F., Liu, Y. J., Liu, X. S., Liu, Y. F., Xu, L. J., Wang, Z. Q., Guo, J. Y., Zhang, L., Dong, Y. S., Qi, C. X., Yang, C., & Wang, S. F. (2021). Multifunctional injectable hydrogel loaded with cerium-containing bioactive glass nanoparticles for diabetic wound healing. *Biomolecules, 11*(5). https://doi.org/10.3390/biom11050702

Cheng, C. Y., Barro, L., Tsai, S. T., Feng, T. W., Wu, X. Y., Chao, C. W., Yu, R. S., Chin, T. Y., & Hsieh, M. F. (2021a). Epigallocatechin-3-Gallate-loaded liposomes favor anti-inflammation of microglia cells and promote neuroprotection. *International Journal of Molecular Sciences, 22*(6). https://doi.org/10.3390/ijms22063037

Cheng, H., Li, Y., Huo, K., Gao, B., & Xiong, W. (2014). Long-lasting in vivo and in vitro antibacterial ability of nanostructured titania coating incorporated with silver nanoparticles. *Journal of Biomedical Materials Research. Part A, 102*(10), 3488–3499. https://doi.org/10.1002/jbm.a.35019

Cheng, H., Shi, Z., Yue, K., Huang, X., Xu, Y., Gao, C., Yao, Z., Zhang, Y. S., & Wang, J. (2021b). Sprayable hydrogel dressing accelerates wound healing with combined reactive oxygen species-scavenging and antibacterial abilities. *Acta Biomaterialia, 124*, 219–232. https://doi.org/10.1016/j.actbio.2021.02.002

Chereddy, K. K., Her, C.-H., Comune, M., Moia, C., Lopes, A., Porporato, P. E., Vanacker, J., Lam, M. C., Steinstraesser, L., Sonveaux, P., Zhu, H., Ferreira, L. S., Vandermeulen, G., & Préat, V. (2014). PLGA nanoparticles loaded with host defense peptide LL37 promote wound healing. *Journal of Controlled Release, 194*, 138–147. https://doi.org/10.1016/j.jconrel.2014.08.016

Chetyrkina, M. R., Fedorov, F. S., & Nasibulin, A. G. (2022). In vitro toxicity of carbon nanotubes: A systematic review. *RSC Advances, 12*(25), 16235–16256. https://doi.org/10.1039/d2ra02519a

Chigurupati, S., Mughal, M. R., Okun, E., Das, S., Kumar, A., McCaffery, M., Seal, S., & Mattson, M. P. (2013). Effects of cerium oxide nanoparticles on the growth of keratinocytes, fibroblasts and vascular endothelial cells in cutaneous wound healing. *Biomaterials, 34*(9), 2194–2201. https://doi.org/10.1016/j.biomaterials.2012.11.061

Choe, G., Kim, S. W., Park, J., Park, J., Kim, S., Kim, Y. S., Ahn, Y., Jung, D. W., Williams, D. R., & Lee, J. Y. (2019). Anti-oxidant activity reinforced reduced graphene oxide/alginate microgels: Mesenchymal stem cell encapsulation and regeneration of infarcted hearts. *Biomaterials, 225*, 119513. https://doi.org/10.1016/j.biomaterials.2019.119513

Choi, J. U., Lee, S. W., Pangeni, R., Byun, Y., Yoon, I. S., & Park, J. W. (2017). Preparation and in vivo evaluation of cationic elastic liposomes comprising highly skin-permeable growth factors combined with hyaluronic acid for enhanced diabetic wound-healing therapy. *Acta Biomaterialia, 57*, 197–215. https://doi.org/10.1016/j.actbio.2017.04.034

Choi, Y. S., Kim, Y. H., An, H. M., Bae, S. K., & Lee, Y. K. (2023). Efficacy of silver nanoparticles-loaded bone cement against an MRSA induced-osteomyelitis in a rat model. *Medicina (Kaunas, Lithuania), 59*(4). https://doi.org/10.3390/medicina59040811

Choudhury, H., Pandey, M., Lim, Y. Q., Low, C. Y., Lee, C. T., Marilyn, T. C. L., Loh, H. S., Lim, Y. P., Lee, C. F., Bhattamishra, S. K., Kesharwani, P., & Gorain, B. (2020). Silver nanoparticles: Advanced and promising technology in diabetic wound therapy. *Materials Science & Engineering. C, Materials for Biological Applications, 112*, 110925. https://doi.org/10.1016/j.msec.2020.110925

Chu, Y., Yu, D., Wang, P., Xu, J., Li, D., & Ding, M. (2010). Nanotechnology promotes the full-thickness diabetic wound healing effect of recombinant human epidermal growth factor in diabetic rats. *Wound Repair and Regeneration, 18*(5), 499–505. https://doi.org/10.1111/j.1524-475X.2010.00612.x

Ciriza, J., Saenz Del Burgo, L., Gurruchaga, H., Borras, F. E., Franquesa, M., Orive, G., Hernández, R. M., & Pedraz, J. L. (2018). Graphene oxide enhances alginate encapsulated cells viabil-

ity and functionality while not affecting the foreign body response. *Drug Delivery, 25*(1), 1147–1160. https://doi.org/10.1080/10717544.2018.1474966
Ciriza, J., Saenz del Burgo, L., Virumbrales-Muñoz, M., Ochoa, I., Fernandez, L. J., Orive, G., Hernandez, R. M., & Pedraz, J. L. (2015). Graphene oxide increases the viability of C2C12 myoblasts microencapsulated in alginate. *International Journal of Pharmaceutics, 493*(1–2), 260–270. https://doi.org/10.1016/j.ijpharm.2015.07.062
Clemons, T. D., Singh, R., Sorolla, A., Chaudhari, N., Hubbard, A., & Iyer, K. S. (2018). Distinction between active and passive targeting of nanoparticles dictate their overall therapeutic efficacy. *Langmuir, 34*(50), 15343–15349. https://doi.org/10.1021/acs.langmuir.8b02946
Corsi, F., Carotenuto, F., Di Nardo, P., & Teodori, L. (2020). Harnessing inorganic nanoparticles to direct macrophage polarization for skeletal muscle regeneration. *Nanomaterials, 10*(10), 1963. https://www.mdpi.com/2079-4991/10/10/1963
Corsi, F., Di Meo, E., Lulli, D., Deidda Tarquini, G., Capradossi, F., Bruni, E., Pelliccia, A., Traversa, E., Dellambra, E., Failla, C. M., & Ghibelli, L. (2023). Safe-shields: Basal and anti-UV protection of human keratinocytes by redox-active cerium oxide nanoparticles prevents UVB-induced mutagenesis. *Antioxidants (Basel), 12*(3). https://doi.org/10.3390/antiox12030757
Croissant, J. G., & Brinker, C. J. (2018). Biodegradable silica-based nanoparticles: Dissolution kinetics and selective bond cleavage. *Enzyme, 43*, 181–214. https://doi.org/10.1016/bs.enz.2018.07.008
Cui, F. M., Wu, Z. J., Zhao, R., Chen, Q., Liu, Z. Y., Zhao, Y., Yan, H. B., Shen, G. L., Tu, Y., Zhou, D. H., Diwu, J., Hou, J., Hu, L., & Wang, G. J. (2021). Development and characterization of a novel hydrogel for the decontaminating of radionuclide-contaminated skin wounds. *Macromolecular Bioscience, 21*(5), e2000399. https://doi.org/10.1002/mabi.202000399
Dai, Z., Wang, Y., Liu, L., Liu, X., Tan, P., Xu, Z., Kuang, J., Liu, Q., Lou, J., & Zhang, Z. (2016). Hierarchical graphene-based films with dynamic self-stiffening for biomimetic artificial muscle. *Advanced Functional Materials, 26*(38), 7003–7010. https://doi.org/10.1002/adfm.201503917
Das, I., De, G., Hupa, L., & Vallittu, P. K. (2016). Porous SiO2 nanofiber grafted novel bioactive glass-ceramic coating: A structural scaffold for uniform apatite precipitation and oriented cell proliferation on inert implant. *Materials Science & Engineering. C, Materials for Biological Applications, 62*, 206–214. https://doi.org/10.1016/j.msec.2016.01.053
Das, K., Madhusoodan, A. P., Mili, B., Kumar, A., Saxena, A. C., Kumar, K., Sarkar, M., Singh, P., Srivastava, S., & Bag, S. (2017). Functionalized carbon nanotubes as suitable scaffold materials for proliferation and differentiation of canine mesenchymal stem cells. *International Journal of Nanomedicine, 12*, 3235–3252. https://doi.org/10.2147/ijn.S122945
Dasgupta, S., Mondal, S., Ray, S., Singh, Y. P., & Maji, K. (2021). Hydroxyapatite-collagen nanoparticles reinforced polyanhydride based injectable paste for bone substitution: Effect of dopant addition in vitro. *Journal of Biomaterials Science. Polymer Edition, 32*(10), 1312–1336. https://doi.org/10.1080/09205063.2021.1916867
Değim, Z., Çelebi, N., Alemdaroğlu, C., Deveci, M., Öztürk, S., & Özoğul, C. (2011). Evaluation of chitosan gel containing liposome-loaded epidermal growth factor on burn wound healing. *International Wound Journal, 8*(4), 343–354. https://doi.org/10.1111/j.1742-481X.2011.00795.x
Deng, Y., Yang, C., Zhu, Y., Liu, W., Li, H., Wang, L., Chen, W., Wang, Z., & Wang, L. (2022). Lamprey-teeth-inspired oriented antibacterial Sericin microneedles for infected wound healing improvement. *Nano Letters, 22*(7), 2702–2711. https://doi.org/10.1021/acs.nanolett.1c04573
Díez-Pascual, A. M., & Díez-Vicente, A. L. (2015). Wound healing bionanocomposites based on Castor oil polymeric films reinforced with chitosan-modified ZnO nanoparticles. *Biomacromolecules, 16*(9), 2631–2644. https://doi.org/10.1021/acs.biomac.5b00447
Din, F. U., Aman, W., Ullah, I., Qureshi, O. S., Mustapha, O., Shafique, S., & Zeb, A. (2017). Effective use of nanocarriers as drug delivery systems for the treatment of selected tumors. *International Journal of Nanomedicine, 12*, 7291–7309. https://doi.org/10.2147/IJN.S146315
Ding, T., Lu, W. W., Zheng, Y., Li, Z., Pan, H., & Luo, Z. (2011). Rapid repair of rat sciatic nerve injury using a nanosilver-embedded collagen scaffold coated with laminin and fibronectin. *Regenerative Medicine, 6*(4), 437–447. https://doi.org/10.2217/rme.11.39

Ding, T., Yin, J. B., Hao, H. P., Zhu, C., Zhang, T., Lu, Y. C., Wang, L. Y., Wang, Z., & Li, Y. Q. (2015a). Tissue engineering of nanosilver-embedded peripheral nerve scaffold to repair nerve defects under contamination conditions. *The International Journal of Artificial Organs, 38*(9), 508–516. https://doi.org/10.5301/ijao.5000439

Ding, X., Liu, H., & Fan, Y. (2015b). Graphene-based materials in regenerative medicine. *Advanced Healthcare Materials, 4*(10), 1451–1468. https://doi.org/10.1002/adhm.201500203

Dittler, M. L., Unalan, I., Grünewald, A., Beltrán, A. M., Grillo, C. A., Destch, R., Gonzalez, M. C., & Boccaccini, A. R. (2019). Bioactive glass (45S5)-based 3D scaffolds coated with magnesium and zinc-loaded hydroxyapatite nanoparticles for tissue engineering applications. *Colloids and Surfaces. B, Biointerfaces, 182*, 110346. https://doi.org/10.1016/j.colsurfb.2019.110346

Dittler, M. L., Zelís, P. M., Beltrán, A. M., Destch, R., Grillo, C. A., Gonzalez, M. C., & Boccaccini, A. R. (2021). Magnetic 3D scaffolds for tissue engineering applications: Bioactive glass (45S5) coated with iron-loaded hydroxyapatite nanoparticles. *Biomedical Materials, 16*(5), 055006. https://doi.org/10.1088/1748-605X/ac14cc

Dong, Y., Hong, M., Dai, R., Wu, H., & Zhu, P. (2020). Engineered bioactive nanoparticles incorporated biofunctionalized ECM/silk proteins based cardiac patches combined with MSCs for the repair of myocardial infarction: In vitro and in vivo evaluations. *Science of the Total Environment, 707*, 135976. https://doi.org/10.1016/j.scitotenv.2019.135976

Duan, S., Yang, X., Mei, F., Tang, Y., Li, X., Shi, Y., Mao, J., Zhang, H., & Cai, Q. (2015). Enhanced osteogenic differentiation of mesenchymal stem cells on poly(l-lactide) nanofibrous scaffolds containing carbon nanomaterials. *Journal of Biomedical Materials Research Part A, 103*(4), 1424–1435. https://doi.org/10.1002/jbm.a.35283

Dvir, T., Timko, B. P., Brigham, M. D., Naik, S. R., Karajanagi, S. S., Levy, O., Jin, H., Parker, K. K., Langer, R., & Kohane, D. S. (2011). Nanowired three-dimensional cardiac patches. *Nature Nanotechnology, 6*(11), 720–725. https://doi.org/10.1038/nnano.2011.160

Ege, D., Nawaz, Q., Beltrán, A. M., & Boccaccini, A. R. (2022). Effect of boron-doped mesoporous bioactive glass nanoparticles on C2C12 cell viability and differentiation: Potential for muscle tissue application. *ACS Biomaterials Science & Engineering, 8*(12), 5273–5283. https://doi.org/10.1021/acsbiomaterials.2c00876

El Bialy, I., Jiskoot, W., & Reza Nejadnik, M. (2017). Formulation, delivery and stability of bone morphogenetic proteins for effective bone regeneration. *Pharmaceutical Research, 34*(6), 1152–1170. https://doi.org/10.1007/s11095-017-2147-x

El-Aassar, M. R., El-Beheri, N. G., Agwa, M. M., Eltaher, H. M., Alseqely, M., Sadik, W. S., & El-Khordagui, L. (2021). Antibiotic-free combinational hyaluronic acid blend nanofibers for wound healing enhancement. *International Journal of Biological Macromolecules, 167*, 1552–1563. https://doi.org/10.1016/j.ijbiomac.2020.11.109

El-Okaily, M. S., Mostafa, A. A., Dulnik, J., Denis, P., Sajkiewicz, P., Mahmoud, A. A., Dawood, R., & Maged, A. (2023). Nanofibrous Polycaprolactone membrane with bioactive glass and atorvastatin for wound healing: Preparation and characterization. *Pharmaceutics, 15*(7). https://doi.org/10.3390/pharmaceutics15071990

Eldebany, N., Abd Elkodous, M., Tohamy, H., Abdelwahed, R., El-Kammar, M., Abou-Ahmed, H., & Elkhenany, H. (2021). Gelatin loaded titanium dioxide and silver oxide nanoparticles: Implication for skin tissue regeneration. *Biological Trace Element Research, 199*(10), 3688–3699. https://doi.org/10.1007/s12011-020-02489-x

Elgrabli, D., Dachraoui, W., Ménard-Moyon, C., Liu, X. J., Bégin, D., Bégin-Colin, S., Bianco, A., Gazeau, F., & Alloyeau, D. (2015). Carbon nanotube degradation in macrophages: Live nanoscale monitoring and understanding of biological pathway. *ACS Nano, 9*(10), 10113–10124. https://doi.org/10.1021/acsnano.5b03708

Enrich-Essvein, T., Rodríguez-Navarro, A. B., Álvarez-Lloret, P., Cifuentes-Jiménez, C., Bolaños-Carmona, M. V., & González-López, S. (2021). Proanthocyanidin-functionalized hydroxyapatite nanoparticles as dentin biomodifier. *Dental Materials, 37*(9), 1437–1445. https://doi.org/10.1016/j.dental.2021.07.002

Esmaeili, H., Patino-Guerrero, A., Hasany, M., Ansari, M. O., Memic, A., Dolatshahi-Pirouz, A., & Nikkhah, M. (2022). Electroconductive biomaterials for cardiac tissue engineering. *Acta Biomaterialia, 139*, 118–140. https://doi.org/10.1016/j.actbio.2021.08.031

Evers, M. J. W., Du, W., Yang, Q., Kooijmans, S. A. A., Vink, A., van Steenbergen, M., Vader, P., de Jager, S. C. A., Fuchs, S. A., Mastrobattista, E., Sluijter, J. P. G., Lei, Z., & Schiffelers, R. (2022). Delivery of modified mRNA to damaged myocardium by systemic administration of lipid nanoparticles. *Journal of Controlled Release, 343*, 207–216. https://doi.org/10.1016/j.jconrel.2022.01.027

Farinha, P., Coelho, J. M. P., Reis, C. P., & Gaspar, M. M. (2021). A comprehensive updated review on magnetic nanoparticles in diagnostics. *Nanomaterials (Basel), 11*(12). https://doi.org/10.3390/nano11123432

Fathi-Achachelouei, M., Knopf-Marques, H., Ribeiro da Silva, C. E., Barthès, J., Bat, E., Tezcaner, A., & Vrana, N. E. (2019). Use of nanoparticles in tissue engineering and regenerative medicine. *Frontiers in Bioengineering and Biotechnology, 7*, 113. https://doi.org/10.3389/fbioe.2019.00113

Fathima, J. B., Pugazhendhi, A., & Venis, R. (2017). Synthesis and characterization of ZrO(2) nanoparticles-antimicrobial activity and their prospective role in dental care. *Microbial Pathogenesis, 110*, 245–251. https://doi.org/10.1016/j.micpath.2017.06.039

Fattahi, R., Mohebichamkhorami, F., Khani, M. M., Soleimani, M., & Hosseinzadeh, S. (2022). Aspirin effect on bone remodeling and skeletal regeneration: Review article. *Tissue & Cell, 76*, 101753. https://doi.org/10.1016/j.tice.2022.101753

Feng, P., Wei, P., Shuai, C., & Peng, S. (2014). Characterization of mechanical and biological properties of 3-D scaffolds reinforced with zinc oxide for bone tissue engineering. *PLoS One, 9*(1), e87755. https://doi.org/10.1371/journal.pone.0087755

Fernandes, P. H. M., Bordini, E. A. F., Cassiano, F. B., de Azevedo-Silva, L. J., Ferrairo, B. M., Lisboa-Filho, P. N., Fortulan, C. A., Soares Dos Passos, D. G., & Borges, A. F. S. (2022). TiO(2) nanoparticles added to dense bovine hydroxyapatite bioceramics increase human osteoblast mineralization activity. *Dental Materials, 38*(11), e275–e283. https://doi.org/10.1016/j.dental.2022.08.007

Filipczak, N., Yalamarty, S. S. K., Li, X., Khan, M. M., Parveen, F., & Torchilin, V. (2021). Lipid-based drug delivery Systems in Regenerative Medicine. *Materials (Basel), 14*(18). https://doi.org/10.3390/ma14185371

Florencio-Silva, R., Sasso, G. R. D. S., Sasso-Cerri, E., Simões, M. J., & Cerri, P. S. (2015). Biology of bone tissue: Structure, function, and factors that influence bone cells. *BioMed Research International, 2015*, 421746. https://doi.org/10.1155/2015/421746

Frasnelli, M., Cristofaro, F., Sglavo, V. M., Dirè, S., Callone, E., Ceccato, R., Bruni, G., Cornaglia, A. I., & Visai, L. (2017). Synthesis and characterization of strontium-substituted hydroxyapatite nanoparticles for bone regeneration. *Materials Science & Engineering. C, Materials for Biological Applications, 71*, 653–662. https://doi.org/10.1016/j.msec.2016.10.047

Fu, Y., Cui, S., Luo, D., & Liu, Y. (2021). Novel inorganic nanomaterial-based therapy for bone tissue regeneration. *Nanomaterials (Basel), 11*(3). https://doi.org/10.3390/nano11030789

Gentleman, E., Fredholm, Y. C., Jell, G., Lotfibakhshaiesh, N., O'Donnell, M. D., Hill, R. G., & Stevens, M. M. (2010). The effects of strontium-substituted bioactive glasses on osteoblasts and osteoclasts in vitro. *Biomaterials, 31*(14), 3949–3956. https://doi.org/10.1016/j.biomaterials.2010.01.121

Gong, X., Luo, M., Wang, M., Niu, W., Wang, Y., & Lei, B. (2022). Injectable self-healing ceria-based nanocomposite hydrogel with ROS-scavenging activity for skin wound repair. *Regenerative Biomaterials, 9*(1), rbab074. https://doi.org/10.1093/rb/rbab074

Grande-Tovar, C. D., Castro, J. I., Valencia Llano, C. H., Tenorio, D. L., Saavedra, M., Zapata, P. A., & Chaur, M. N. (2022). Polycaprolactone (PCL)-Polylactic acid (PLA)-glycerol (Gly) composites incorporated with zinc oxide nanoparticles (ZnO-NPs) and tea tree essential oil (TTEO) for tissue engineering applications. *Pharmaceutics, 15*(1). https://doi.org/10.3390/pharmaceutics15010043

Gu, M., Li, W., Jiang, L., & Li, X. (2022). Recent progress of rare earth doped hydroxyapatite nanoparticles: Luminescence properties, synthesis and biomedical applications. *Acta Biomaterialia, 148*, 22–43. https://doi.org/10.1016/j.actbio.2022.06.006

Guo, B., Feng, X., Wang, Y., Wang, X., & He, Y. (2021). Biomimetic and immunomodulatory baicalin-loaded graphene oxide-demineralized bone matrix scaffold for in vivo bone regeneration. *Journal of Materials Chemistry B, 9*(47), 9720–9733. https://doi.org/10.1039/d1tb00618e

Guo, X., Jin, X., Han, K., Kang, S., Tian, S., Lv, X., Feng, M., Zheng, H., Zuo, Y., Xu, G., Hu, M., Xu, J., Lv, P., & Chang, Y.-Z. (2022). Iron promotes neurological function recovery in mice with ischemic stroke through endogenous repair mechanisms. *Free Radical Biology and Medicine, 182*, 59–72. https://doi.org/10.1016/j.freeradbiomed.2022.02.017

Gurtner, G. C., & Chapman, M. A. (2016). Regenerative medicine: Charting a new course in wound healing. *Advances in Wound Care (New Rochelle), 5*(7), 314–328. https://doi.org/10.1089/wound.2015.0663

Gutiérrez-Hernández, J. M., Escobar-García, D. M., Escalante, A., Flores, H., González, F. J., Gatenholm, P., & Toriz, G. (2017). In vitro evaluation of osteoblastic cells on bacterial cellulose modified with multi-walled carbon nanotubes as scaffold for bone regeneration. *Materials Science & Engineering. C, Materials for Biological Applications, 75*, 445–453. https://doi.org/10.1016/j.msec.2017.02.074

Han, G., You, C. C., Kim, B. J., Turingan, R. S., Forbes, N. S., Martin, C. T., & Rotello, V. M. (2006). Light-regulated release of DNA and its delivery to nuclei by means of photolabile gold nanoparticles. *Angewandte Chemie (International Ed. in English), 45*(19), 3165–3169. https://doi.org/10.1002/anie.200600214

Han, J., Kim, Y. S., Lim, M. Y., Kim, H. Y., Kong, S., Kang, M., Choo, Y. W., Jun, J. H., Ryu, S., Jeong, H. Y., Park, J., Jeong, G. J., Lee, J. C., Eom, G. H., Ahn, Y., & Kim, B. S. (2018). Dual roles of graphene oxide to attenuate inflammation and elicit timely polarization of macrophage phenotypes for cardiac repair. *ACS Nano, 12*(2), 1959–1977. https://doi.org/10.1021/acsnano.7b09107

Han, X., He, J., Wang, Z., Bai, Z., Qu, P., Song, Z., & Wang, W. (2021). Fabrication of silver nanoparticles/gelatin hydrogel system for bone regeneration and fracture treatment. *Drug Delivery, 28*(1), 319–324. https://doi.org/10.1080/10717544.2020.1869865

Hazer Rosberg, D. B., Hazer, B., Stenberg, L., & Dahlin, L. B. (2021). Gold and cobalt oxide nanoparticles modified poly-propylene poly-ethylene glycol membranes in poly (ε-Caprolactone) conduits enhance nerve regeneration in the sciatic nerve of healthy rats. *International Journal of Molecular Sciences, 22*(13). https://doi.org/10.3390/ijms22137146

He, W., Zheng, Y., Feng, Q., Elkhooly, T. A., Liu, X., Yang, X., Wang, Y., & Xie, Y. (2020). Silver nanoparticles stimulate osteogenesis of human mesenchymal stem cells through activation of autophagy. *Nanomedicine, 15*(4), 337–353. https://doi.org/10.2217/nnm-2019-0026

Hekmat, M., Ahmadi, H., Baniasadi, F., Ashtari, B., Naserzadeh, P., Mirzaei, M., Omidi, H., Mostafavinia, A., Amini, A., Hamblin, M. R., Chien, S., & Bayat, M. (2023). Combined use of Photobiomodulation and curcumin-loaded iron oxide nanoparticles significantly improved wound healing in diabetic rats compared to either treatment alone. *Journal of Lasers in Medical Sciences, 14*, e18. https://doi.org/10.34172/jlms.2023.18

Hirano, S., Fujitani, Y., Furuyama, A., & Kanno, S. (2010). Uptake and cytotoxic effects of multi-walled carbon nanotubes in human bronchial epithelial cells. *Toxicology and Applied Pharmacology, 249*(1), 8–15. https://doi.org/10.1016/j.taap.2010.08.019

Hogan, M., Chen, Y. T., Kolhatkar, A. G., Candelari, C. J., Madala, S., Lee, T. R., & Birla, R. (2016). Conditioning of cardiovascular tissue using a noncontact magnetic stretch bioreactor with embedded magnetic nanoparticles. *ACS Biomaterials Science & Engineering, 2*(9), 1619–1629. https://doi.org/10.1021/acsbiomaterials.6b00375

Hoppe, A., Güldal, N. S., & Boccaccini, A. R. (2011). A review of the biological response to ionic dissolution products from bioactive glasses and glass-ceramics. *Biomaterials, 32*(11), 2757–2774. https://doi.org/10.1016/j.biomaterials.2011.01.004

Hoppe, A., Sarker, B., Detsch, R., Hild, N., Mohn, D., Stark, W. J., & Boccaccini, A. R. (2014). In vitro reactivity of Sr-containing bioactive glass (type 1393) nanoparticles. *Journal of Non-Crystalline Solids, 387*, 41–46. https://doi.org/10.1016/j.jnoncrysol.2013.12.010

Hu, H., Ni, Y., Mandal, S. K., Montana, V., Zhao, B., Haddon, R. C., & Parpura, V. (2005). Polyethyleneimine functionalized single-walled carbon nanotubes as a substrate for neuronal growth. *The Journal of Physical Chemistry. B, 109*(10), 4285–4289. https://doi.org/10.1021/jp0441137

Hu, M., Li, C., Li, X., Zhou, M., Sun, J., Sheng, F., Shi, S., & Lu, L. (2018). Zinc oxide/silver bimetallic nanoencapsulated in PVP/PCL nanofibres for improved antibacterial activity. *Artificial Cells, Nanomedicine, and Biotechnology, 46*(6), 1248–1257. https://doi.org/10.1080/21691401.2017.1366339

Hu, X., Liu, W., Sun, L., Xu, S., Wang, T., Meng, J., Wen, T., Liu, Q., Liu, J., & Xu, H. (2022). Magnetic Nanofibrous scaffolds accelerate the regeneration of muscle tissue in combination with extra magnetic fields. *International Journal of Molecular Sciences, 23*(8). https://doi.org/10.3390/ijms23084440

Hu, Y., Bai, S., Wu, X., Tan, S., & He, Y. (2021). Biodegradability of mesoporous silica nanoparticles. *Ceramics International, 47*(22), 31031–31041. https://doi.org/10.1016/j.ceramint.2021.08.129

Huang, W. J., & Wang, J. (2023). Development of 3D-printed, biodegradable, conductive PGSA composites for nerve tissue regeneration. *Macromolecular Bioscience, 23*(3), e2200470. https://doi.org/10.1002/mabi.202200470

Huebner, E. A., & Strittmatter, S. M. (2009). Axon regeneration in the peripheral and central nervous systems. *Results and Problems in Cell Differentiation, 48*, 339–351. https://doi.org/10.1007/400_2009_19

Ignjatović, N., Ajduković, Z., Savić, V., Najman, S., Mihailović, D., Vasiljević, P., Stojanović, Z., Uskoković, V., & Uskoković, D. (2013). Nanoparticles of cobalt-substituted hydroxyapatite in regeneration of mandibular osteoporotic bones. *Journal of Materials Science. Materials in Medicine, 24*(2), 343–354. https://doi.org/10.1007/s10856-012-4793-1

Ismail, R. A., Sulaiman, G. M., Abdulrahman, S. A., & Marzoog, T. R. (2015). Antibacterial activity of magnetic iron oxide nanoparticles synthesized by laser ablation in liquid. *Materials Science & Engineering. C, Materials for Biological Applications, 53*, 286–297. https://doi.org/10.1016/j.msec.2015.04.047

Jagiełło, J., Sekuła-Stryjewska, M., Noga, S., Adamczyk, E., Dźwigońska, M., Kurcz, M., Kurp, K., Winkowska-Struzik, M., Karnas, E., Boruczkowski, D., Madeja, Z., Lipińska, L., & Zuba-Surma, E. K. (2019). Impact of graphene-based surfaces on the basic biological properties of human umbilical cord mesenchymal stem cells: Implications for ex vivo cell expansion aimed at tissue repair. *International Journal of Molecular Sciences, 20*(18). https://doi.org/10.3390/ijms20184561

Jahani, M., Rezazadeh, D., Mohammadi, P., Abdolmaleki, A., Norooznezhad, A., & Mansouri, K. (2020). Regenerative medicine and angiogenesis, challenges and opportunities. *Advanced Pharmaceutical Bulletin, 10*(4), 490–501. https://doi.org/10.34172/apb.2020.061

Jahromi, M., Razavi, S., Seyedebrahimi, R., Reisi, P., & Kazemi, M. (2021). Regeneration of rat sciatic nerve using PLGA conduit containing rat ADSCs with controlled release of BDNF and gold nanoparticles. *Journal of Molecular Neuroscience, 71*(4), 746–760. https://doi.org/10.1007/s12031-020-01694-6

Ji, Z., Wang, J., Yang, S., Tao, S., Shen, C., Wei, H., Li, Q., & Jin, P. (2022). Graphene oxide accelerates diabetic wound repair by inhibiting apoptosis of ad-MSCs via Linc00324/miR-7977/STK4 pathway. *The FASEB Journal, 36*(11), e22623. https://doi.org/10.1096/fj.202201079RR

Jia, Y., Zhang, P., Sun, Y., Kang, Q., Xu, J., Zhang, C., & Chai, Y. (2019). Regeneration of large bone defects using mesoporous silica coated magnetic nanoparticles during distraction osteogenesis. *Nanomedicine, 21*, 102040. https://doi.org/10.1016/j.nano.2019.102040

Jin, B., Zhang, L., Wang, X., & Jin, D. (2022). Research on orientation of basic fibroblast growth factor with magnetic nanoparticles (MNPs) on regeneration and recovery of rats' dampened skeletal muscle and expressed level of matrix metalloproteinase. *Journal of Biomedical Nanotechnology, 18*(2), 557–564. https://doi.org/10.1166/jbn.2022.3260

Jo, H., Sim, M., Kim, S., Yang, S., Yoo, Y., Park, J. H., Yoon, T. H., Kim, M. G., & Lee, J. Y. (2017). Electrically conductive graphene/polyacrylamide hydrogels produced by mild chemical reduction for enhanced myoblast growth and differentiation. *Acta Biomaterialia, 48*, 100–109. https://doi.org/10.1016/j.actbio.2016.10.035

Jo, S. B., Erdenebileg, U., Dashnyam, K., Jin, G. Z., Cha, J. R., El-Fiqi, A., Knowles, J. C., Patel, K. D., Lee, H. H., Lee, J. H., & Kim, H. W. (2020). Nano-graphene oxide/polyurethane nanofibers: Mechanically flexible and myogenic stimulating matrix for skeletal tissue engineering. *Journal of Tissue Engineering, 11*, 2041731419900424. https://doi.org/10.1177/2041731419900424

Johari, N., Madaah Hosseini, H. R., & Samadikuchaksaraei, A. (2017). Optimized composition of nanocomposite scaffolds formed from silk fibroin and nano-TiO(2) for bone tissue engineering. *Materials Science & Engineering. C, Materials for Biological Applications, 79*, 783–792. https://doi.org/10.1016/j.msec.2017.05.105

Johari, N., Madaah Hosseini, H. R., & Samadikuchaksaraei, A. (2018). Novel fluoridated silk fibroin/ TiO(2) nanocomposite scaffolds for bone tissue engineering. *Materials Science & Engineering. C, Materials for Biological Applications, 82*, 265–276. https://doi.org/10.1016/j.msec.2017.09.001

Jung, S. M., Yoon, G. H., Lee, H. C., Jung, M. H., Yu, S. I., Yeon, S. J., Min, S. K., Kwon, Y. S., Hwang, J. H., & Shin, H. S. (2015). Thermodynamic insights and conceptual Design of Skin-Sensitive Chitosan Coated Ceramide/PLGA Nanodrug for regeneration of stratum Corneum on atopic dermatitis. *Scientific Reports, 5*, 18089. https://doi.org/10.1038/srep18089

Kang, E. S., Kim, D. S., Suhito, I. R., Lee, W., Song, I., & Kim, T. H. (2018). Two-dimensional material-based bionano platforms to control mesenchymal stem cell differentiation. *Biomaterials Research, 22*, 10. https://doi.org/10.1186/s40824-018-0120-3

Kang, S., Park, J. B., Lee, T.-J., Ryu, S., Bhang, S. H., La, W.-G., Noh, M.-K., Hong, B. H., & Kim, B.-S. (2015). Covalent conjugation of mechanically stiff graphene oxide flakes to three-dimensional collagen scaffolds for osteogenic differentiation of human mesenchymal stem cells. *Carbon, 83*, 162–172. https://doi.org/10.1016/j.carbon.2014.11.029

Karbowniczek, J. E., Berniak, K., Knapczyk-Korczak, J., Williams, G., Bryant, J. A., Nikoi, N. D., Banzhaf, M., de Cogan, F., & Stachewicz, U. (2023). Strategies of nanoparticles integration in polymer fibers to achieve antibacterial effect and enhance cell proliferation with collagen production in tissue engineering scaffolds. *Journal of Colloid and Interface Science, 650*(Pt B), 1371–1381. https://doi.org/10.1016/j.jcis.2023.07.066

Kargozar, S., Hamzehlou, S., & Baino, F. (2017). Potential of bioactive glasses for cardiac and pulmonary tissue engineering. *Materials (Basel), 10*(12). https://doi.org/10.3390/ma10121429

Kargozar, S., Hamzehlou, S., & Baino, F. (2019a). Can bioactive glasses be useful to accelerate the healing of epithelial tissues? *Materials Science & Engineering. C, Materials for Biological Applications, 97*, 1009–1020. https://doi.org/10.1016/j.msec.2019.01.028

Kargozar, S., Hoseini, S. J., Milan, P. B., Hooshmand, S., Kim, H. W., & Mozafari, M. (2020). Quantum dots: A review from concept to clinic. *Biotechnology Journal, 15*(12), e2000117. https://doi.org/10.1002/biot.202000117

Kargozar, S., Lotfibakhshaeish, N., Ebrahimi-Barough, S., Nazari, B., & Hill, R. G. (2019b). Stimulation of osteogenic differentiation of induced pluripotent stem cells (iPSCs) using bioactive glasses: An in vitro study. *Frontiers in Bioengineering and Biotechnology, 7*, 355. https://doi.org/10.3389/fbioe.2019.00355

Karimi Hajishoreh, N., Baheiraei, N., Naderi, N., Salehnia, M., & Razavi, M. (2022). Left ventricular geometry and angiogenesis improvement in rat chronic ischemic cardiomyopathy following injection of encapsulated mesenchymal stem cells. *Cell Journal, 24*(12), 741–747. https://doi.org/10.22074/cellj.2022.557257.1040

Katebi, S., Esmaeili, A., Ghaedi, K., & Zarrabi, A. (2019). Superparamagnetic iron oxide nanoparticles combined with NGF and quercetin promote neuronal branching morphogenesis of

PC12 cells. *International Journal of Nanomedicine, 14*, 2157–2169. https://doi.org/10.2147/ijn.S191878

Khan, S., Garg, M., Chockalingam, S., Gopinath, P., & Kundu, P. P. (2020). TiO(2) doped chitosan/poly (vinyl alcohol) nanocomposite film with enhanced mechanical properties for application in bone tissue regeneration. *International Journal of Biological Macromolecules, 143*, 285–296. https://doi.org/10.1016/j.ijbiomac.2019.11.246

Khorasani, M. T., Joorabloo, A., Adeli, H., Milan, P. B., & Amoupour, M. (2021). Enhanced antimicrobial and full-thickness wound healing efficiency of hydrogels loaded with heparinized ZnO nanoparticles: In vitro and in vivo evaluation. *International Journal of Biological Macromolecules, 166*, 200–212. https://doi.org/10.1016/j.ijbiomac.2020.10.142

Kim, C. K., Ghosh, P., Pagliuca, C., Zhu, Z. J., Menichetti, S., & Rotello, V. M. (2009). Entrapment of hydrophobic drugs in nanoparticle monolayers with efficient release into cancer cells. *Journal of the American Chemical Society, 131*(4), 1360–1361. https://doi.org/10.1021/ja808137c

Kim, D., Shim, Y. S., An, S. Y., & Lee, M. J. (2021). Role of zinc-doped bioactive glass encapsulated with microspherical gelatin in localized supplementation for tissue regeneration: A contemporary review. *Molecules, 26*(7). https://doi.org/10.3390/molecules26071823

Kim, D.-H., Kim, M.-J., Kwak, S.-Y., Jeong, J., Choi, D., Choi, S. W., Ryu, J., & Kang, K.-S. (2023). Bioengineered liver crosslinked with nano-graphene oxide enables efficient liver regeneration via MMP suppression and immunomodulation. *Nature Communications, 14*(1), 801. https://doi.org/10.1038/s41467-023-35941-2

Kim, J. E., Lee, J., Jang, M., Kwak, M. H., Go, J., Kho, E. K., Song, S. H., Sung, J. E., Lee, J., & Hwang, D. Y. (2015). Accelerated healing of cutaneous wounds using phytochemically stabilized gold nanoparticle deposited hydrocolloid membranes. *Biomaterials Science, 3*(3), 509–519. https://doi.org/10.1039/c4bm00390j

Kim, J. H., Kim, D. K., Lee, O. J., Ju, H. W., Lee, J. M., Moon, B. M., Park, H. J., Kim, D. W., Lee, J. H., & Park, C. H. (2016). Osteoinductive silk fibroin/titanium dioxide/hydroxyapatite hybrid scaffold for bone tissue engineering. *International Journal of Biological Macromolecules, 82*, 160–167. https://doi.org/10.1016/j.ijbiomac.2015.08.001

Kim, J. H., Sheikh, F. A., Ju, H. W., Park, H. J., Moon, B. M., Lee, O. J., & Park, C. H. (2014). 3D silk fibroin scaffold incorporating titanium dioxide (TiO2) nanoparticle (NPs) for tissue engineering. *International Journal of Biological Macromolecules, 68*, 158–168. https://doi.org/10.1016/j.ijbiomac.2014.04.045

Klačić, T., Peranić, N., Radatović, B., & Kovačević, D. (2022). Biocompatible hydroxyapatite nanoparticles as templates for the preparation of thin film polyelectrolyte multilayer nanocapsules. *Colloids and Surfaces A: Physicochemical and Engineering Aspects, 648*, 129385. https://doi.org/10.1016/j.colsurfa.2022.129385

Ko, W. K., Heo, D. N., Moon, H. J., Lee, S. J., Bae, M. S., Lee, J. B., Sun, I. C., Jeon, H. B., Park, H. K., & Kwon, I. K. (2015). The effect of gold nanoparticle size on osteogenic differentiation of adipose-derived stem cells. *Journal of Colloid and Interface Science, 438*, 68–76. https://doi.org/10.1016/j.jcis.2014.08.058

Kono, Y. (2022). Preparation of magnetized mesenchymal stem cells using magnetic liposomes to enhance their retention in targeted tissue -evaluation of retention and anti-inflammatory efficiency in skeletal muscle. *Yakugaku Zasshi, 142*(11), 1145–1151. https://doi.org/10.1248/yakushi.22-00132

Krueger, E., Chang, A. N., Brown, D., Eixenberger, J., Brown, R., Rastegar, S., Yocham, K. M., Cantley, K. D., & Estrada, D. (2016). Graphene foam as a three-dimensional platform for Myotube growth. *ACS Biomaterials Science & Engineering, 2*(8), 1234–1241. https://doi.org/10.1021/acsbiomaterials.6b00139

Kunisaki, A., Kodama, A., Ishikawa, M., Ueda, T., Lima, M. D., Kondo, T., & Adachi, N. (2021). Carbon-nanotube yarns induce axonal regeneration in peripheral nerve defect. *Scientific Reports, 11*(1), 19562. https://doi.org/10.1038/s41598-021-98603-7

Kuo, Y. C., & Lee, Y. J. (2016). Rescuing cholinergic neurons from apoptotic degeneration by targeting of serotonin modulator-and apolipoprotein E-conjugated liposomes to the hippocampus. *International Journal of Nanomedicine, 11*, 6809–6824. https://doi.org/10.2147/ijn.S123442

Kuroda, Y., Kawai, T., Goto, K., & Matsuda, S. (2019). Clinical application of injectable growth factor for bone regeneration: A systematic review. *Inflammation and Regeneration, 39*, 20. https://doi.org/10.1186/s41232-019-0109-x

La, W. G., Park, S., Yoon, H. H., Jeong, G. J., Lee, T. J., Bhang, S. H., Han, J. Y., Char, K., & Kim, B. S. (2013). Delivery of a therapeutic protein for bone regeneration from a substrate coated with graphene oxide. *Small, 9*(23), 4051–4060. https://doi.org/10.1002/smll.201300571

Labatut, A. E., & Mattheolabakis, G. (2018). Non-viral based miR delivery and recent developments. *European Journal of Pharmaceutics and Biopharmaceutics, 128*, 82–90. https://doi.org/10.1016/j.ejpb.2018.04.018

Lahooti, B., Akwii, R. G., Zahra, F. T., Sajib, M. S., Lamprou, M., Alobaida, A., Lionakis, M. S., Mattheolabakis, G., & Mikelis, C. M. (2023). Targeting endothelial permeability in the EPR effect. *Journal of Controlled Release, 361*, 212–235. https://doi.org/10.1016/j.jconrel.2023.07.039

Laurenti, M., & Cauda, V. (2017). ZnO nanostructures for tissue engineering applications. *Nanomaterials (Basel), 7*(11). https://doi.org/10.3390/nano7110374

Lee, D., Lee, S. J., Moon, J.-H., Kim, J. H., Heo, D. N., Bang, J. B., Lim, H.-N., & Kwon, I. K. (2018). Preparation of antibacterial chitosan membranes containing silver nanoparticles for dental barrier membrane applications. *Journal of Industrial and Engineering Chemistry, 66*, 196–202. https://doi.org/10.1016/j.jiec.2018.05.030

Lee, J. H., Lee, Y., Shin, Y. C., Kim, M. J., Park, J. H., Hong, S. W., Kim, B., Oh, J.-W., Park, K. D., & Han, D.-W. (2016). In situ forming gelatin/graphene oxide hydrogels for facilitated C2C12 myoblast differentiation. *Applied Spectroscopy Reviews, 51*(7–9), 527–539. https://doi.org/10.1080/05704928.2016.1165686

Lee, R. T., & Walsh, K. (2016). The future of cardiovascular regenerative medicine. *Circulation, 133*(25), 2618–2625. https://doi.org/10.1161/CIRCULATIONAHA.115.019214

Lee, W. C., Lim, C. H., Shi, H., Tang, L. A., Wang, Y., Lim, C. T., & Loh, K. P. (2011). Origin of enhanced stem cell growth and differentiation on graphene and graphene oxide. *ACS Nano, 5*(9), 7334–7341. https://doi.org/10.1021/nn202190c

Legon'kova, O. A., Ushakova, T. A., Savchenkova, I. P., Perova, N. V., Belova, M. S., Torkova, A. A., Baranchikov, A. E., Ivanova, O. S., Korotaeva, A. I., & Ivanov, V. K. (2017). Experimental study of the effects of Nanodispersed ceria on wound repair. *Bulletin of Experimental Biology and Medicine, 162*(3), 395–399. https://doi.org/10.1007/s10517-017-3624-2

Lei, L., Liu, Z., Yuan, P., Jin, R., Wang, X., Jiang, T., & Chen, X. (2019). Injectable colloidal hydrogel with mesoporous silica nanoparticles for sustained co-release of microRNA-222 and aspirin to achieve innervated bone regeneration in rat mandibular defects. *Journal of Materials Chemistry B, 7*(16), 2722–2735. https://doi.org/10.1039/c9tb00025a

Leu, J.-G., Chen, S.-A., Chen, H.-M., Wu, W.-M., Hung, C.-F., Yao, Y.-D., Tu, C.-S., & Liang, Y.-J. (2012). The effects of gold nanoparticles in wound healing with antioxidant epigallocatechin gallate and α-lipoic acid. *Nanomedicine: Nanotechnology, Biology and Medicine, 8*(5), 767–775. https://doi.org/10.1016/j.nano.2011.08.013

Levchenko, T. S., Hartner, W. C., & Torchilin, V. P. (2012). Liposomes in diagnosis and treatment of cardiovascular disorders. *Methodist DeBakey Cardiovascular Journal, 8*(1), 36–41. https://doi.org/10.14797/mdcj-8-1-36

Li, H., Gong, X., Zhang, L., Zhang, Z., Luo, F., Zhou, Q., Chen, J., & Wan, J. (2009). Madecassoside attenuates inflammatory response on collagen-induced arthritis in DBA/1 mice. *Phytomedicine, 16*(6–7), 538–546. https://doi.org/10.1016/j.phymed.2008.11.002

Li, H., Pan, S., Xia, P., Chang, Y., Fu, C., Kong, W., Yu, Z., Wang, K., Yang, X., & Qi, Z. (2020a). Advances in the application of gold nanoparticles in bone tissue engineering. *Journal of Biological Engineering, 14*, 14. https://doi.org/10.1186/s13036-020-00236-3

Li, H., Yu, B., Yang, P., Zhan, J., Fan, X., Chen, P., Liao, X., Ou, C., Cai, Y., & Chen, M. (2021a). Injectable AuNP-HA matrix with localized stiffness enhances the formation of gap junction in engrafted human induced pluripotent stem cell-derived cardiomyocytes and promotes cardiac repair. *Biomaterials, 279*, 121231. https://doi.org/10.1016/j.biomaterials.2021.121231

Li, M., Hu, M., Zeng, H., Yang, B., Zhang, Y., Li, Z., Lu, L., & Ming, Y. (2021b). Multifunctional zinc oxide/silver bimetallic nanomaterial-loaded nanofibers for enhanced tissue regeneration and wound healing. *Journal of Biomedical Nanotechnology, 17*(9), 1840–1849. https://doi.org/10.1166/jbn.2021.3152

Li, Y., Guo, Y., Niu, W., Chen, M., Xue, Y., Ge, J., Ma, P. X., & Lei, B. (2018). Biodegradable multifunctional bioactive glass-based nanocomposite elastomers with controlled biomineralization activity, real-time bioimaging tracking, and decreased inflammatory response. *ACS Applied Materials & Interfaces, 10*(21), 17722–17731. https://doi.org/10.1021/acsami.8b04856

Li, Y., Xu, T., Tu, Z., Dai, W., Xue, Y., Tang, C., Gao, W., Mao, C., Lei, B., & Lin, C. (2020b). Bioactive antibacterial silica-based nanocomposites hydrogel scaffolds with high angiogenesis for promoting diabetic wound healing and skin repair. *Theranostics, 10*(11), 4929–4943. https://doi.org/10.7150/thno.41839

Li, Z., Liu, M., Wang, H., & Du, S. (2016). Increased cutaneous wound healing effect of biodegradable liposomes containing madecassoside: Preparation optimization, in vitro dermal permeation, and in vivo bioevaluation. *International Journal of Nanomedicine, 11*, 2995–3007. https://doi.org/10.2147/ijn.S105035

Li, Z., Yang, R., Yu, M., Bai, F., Li, C., & Wang, Z. L. (2008). Cellular level biocompatibility and biosafety of ZnO nanowires. *The Journal of Physical Chemistry C, 112*(51), 20114–20117. https://doi.org/10.1021/jp808878p

Liang, H., Jin, C., Ma, L., Feng, X., Deng, X., Wu, S., Liu, X., & Yang, C. (2019). Accelerated bone regeneration by gold-nanoparticle-loaded mesoporous silica through stimulating immunomodulation. *ACS Applied Materials & Interfaces, 11*(44), 41758–41769. https://doi.org/10.1021/acsami.9b16848

Liang, W., Ding, P., Li, G., Lu, E., & Zhao, Z. (2021). Hydroxyapatite nanoparticles facilitate osteoblast differentiation and bone formation within sagittal suture during expansion in rats. *Drug Design, Development and Therapy, 15*, 905–917. https://doi.org/10.2147/dddt.S299641

Lin, R., Ge, K., Fan, D., Li, J., Zhou, G., Zhang, K., Huang, Y., Ma, L., & Zhang, J. (2023). Multi-walled carbon nanotubes reversing the bone formation of bone marrow stromal cells by activating M2 macrophage polarization. *Regenerative Biomaterials, 10*, rbad042. https://doi.org/10.1093/rb/rbad042

Liu, F., Cheng, X., Xiao, L., Wang, Q., Yan, K., Su, Z., Wang, L., Ma, C., & Wang, Y. (2021). Inside-outside Ag nanoparticles-loaded polylactic acid electrospun fiber for long-term antibacterial and bone regeneration. *International Journal of Biological Macromolecules, 167*, 1338–1348. https://doi.org/10.1016/j.ijbiomac.2020.11.088

Liu, T., Wang, Y., Lu, L., & Liu, Y. (2022). SPIONs mediated magnetic actuation promotes nerve regeneration by inducing and maintaining repair-supportive phenotypes in Schwann cells. *Journal of Nanobiotechnology, 20*(1), 159. https://doi.org/10.1186/s12951-022-01337-5

Liu, X., Gaihre, B., Park, S., Li, L., Dashtdar, B., Astudillo Potes, M. D., Terzic, A., Elder, B. D., & Lu, L. (2023). 3D-printed scaffolds with 2D hetero-nanostructures and immunomodulatory cytokines provide pro-healing microenvironment for enhanced bone regeneration. *Bioactive Materials, 27*, 216–230. https://doi.org/10.1016/j.bioactmat.2023.03.021

Lovat, V., Pantarotto, D., Lagostena, L., Cacciari, B., Grandolfo, M., Righi, M., Spalluto, G., Prato, M., & Ballerini, L. (2005). Carbon nanotube substrates boost neuronal electrical signaling. *Nano Letters, 5*(6), 1107–1110. https://doi.org/10.1021/nl050637m

Lozano Chamizo, L., Luengo Morato, Y., Ovejero Paredes, K., Contreras Caceres, R., Filice, M., & Marciello, M. (2021). Ionotropic gelation-based synthesis of chitosan-metal hybrid nanoparticles showing combined antimicrobial and tissue regenerative activities. *Polymers (Basel), 13*(22). https://doi.org/10.3390/polym13223910

Luo, F., Li, R., Zheng, H., Xu, Y., Yang, L., Qu, C., Hong, G., & Wan, Q. (2022). Differentiation of bone mesenchymal stem cells into vascular endothelial cell-like cells using functionalized single-walled carbon nanotubes. *Frontiers in Bioengineering and Biotechnology, 10*, 913080. https://doi.org/10.3389/fbioe.2022.913080

Maheswaran, R., & Shanmugavel, B. P. (2022). A critical review of the role of carbon nanotubes in the Progress of next-generation electronic applications. *Journal of Electronic Materials, 51*(6), 2786–2800. https://doi.org/10.1007/s11664-022-09516-8

Mahmoodi, M., Haghighi, V., Mirhaj, M., Tavafoghi, M., Shams, F., & Darabi, A. (2021). Highly osteogenic and mechanically strong nanofibrous scaffolds based on functionalized multi-walled carbon nanotubes-reinforced electrospun keratin/poly(ε-caprolactone). *Materials Today Communications, 27*, 102401. https://doi.org/10.1016/j.mtcomm.2021.102401

Mahmoud, N. N., Hikmat, S., Abu Ghith, D., Hajeer, M., Hamadneh, L., Qattan, D., & Khalil, E. A. (2019). Gold nanoparticles loaded into polymeric hydrogel for wound healing in rats: Effect of nanoparticles' shape and surface modification. *International Journal of Pharmaceutics, 565*, 174–186. https://doi.org/10.1016/j.ijpharm.2019.04.079

Mahmoudi, M., Alizadeh, P., & Soltani, M. (2023). Wound healing performance of electrospun PVA/70S30C bioactive glass/Ag nanoparticles mats decorated with curcumin: In vitro and in vivo investigations. *Biomaterials Advances, 153*, 213530. https://doi.org/10.1016/j.bioadv.2023.213530

Mahross, H. Z., & Baroudi, K. (2015). Effect of silver nanoparticles incorporation on viscoelastic properties of acrylic resin denture base material. *European Journal of Dentistry, 9*(2), 207–212. https://doi.org/10.4103/1305-7456.156821

Malakar, C., Kashyap, B., Kalita, M. C., & Deka, S. (2023). Wound healing efficacy of rhamnolipid-coated zinc oxide nanoparticle along with its in vivo antibacterial efficacy against Staphylococcus aureus. *Experimental Dermatology, 32*(2), 154–164. https://doi.org/10.1111/exd.14692

Mansouri, V., Beheshtizadeh, N., Gharibshahian, M., Sabouri, L., Varzandeh, M., & Rezaei, N. (2021). Recent advances in regenerative medicine strategies for cancer treatment. *Biomedicine & Pharmacotherapy, 141*, 111875. https://doi.org/10.1016/j.biopha.2021.111875

Manuja, A., Raguvaran, R., Kumar, B., Kalia, A., & Tripathi, B. N. (2020). Accelerated healing of full thickness excised skin wound in rabbits using single application of alginate/acacia based nanocomposites of ZnO nanoparticles. *International Journal of Biological Macromolecules, 155*, 823–833. https://doi.org/10.1016/j.ijbiomac.2020.03.221

Mao, A. S., & Mooney, D. J. (2015). Regenerative medicine: Current therapies and future directions. *Proceedings of the National Academy of Sciences of the United States of America, 112*(47), 14452–14459. https://doi.org/10.1073/pnas.1508520112

Marsell, R., & Einhorn, T. A. (2011). The biology of fracture healing. *Injury, 42*(6), 551–555. https://doi.org/10.1016/j.injury.2011.03.031

Marycz, K., Śmieszek, A., Kornicka-Garbowska, K., Pielok, A., Janeczek, M., Lipińska, A., Nikodem, A., Filipiak, J., Sobierajska, P., Nedelec, J. M., & Wiglusz, R. J. (2021). Novel Nanohydroxyapatite (nHAp)-based scaffold doped with iron oxide nanoparticles (IO), functionalized with small non-coding RNA (miR-21/124) modulates expression of runt-related transcriptional factor 2 and Osteopontin, promoting regeneration of osteoporotic bone in bilateral cranial defects in a senescence-accelerated mouse model (SAM/P6). PART 2. *International Journal of Nanomedicine, 16*, 6049–6065. https://doi.org/10.2147/ijn.S316240

Mârza, S. M., Magyari, K., Bogdan, S., Moldovan, M., Peştean, C., Nagy, A., Tăbăran, F., Licarete, E., Suarasan, S., Dreanca, A., Baia, L., & Papuc, I. (2019). Skin wound regeneration with bioactive glass-gold nanoparticles ointment. *Biomedical Materials, 14*(2), 025011. https://doi.org/10.1088/1748-605X/aafd7d

Matsumoto, K., Sato, C., Naka, Y., Kitazawa, A., Whitby, R. L., & Shimizu, N. (2007). Neurite outgrowths of neurons with neurotrophin-coated carbon nanotubes. *Journal of Bioscience and Bioengineering, 103*(3), 216–220. https://doi.org/10.1263/jbb.103.216

Mattheolabakis, G., Rigas, B., & Constantinides, P. P. (2012). Nanodelivery strategies in cancer chemotherapy: Biological rationale and pharmaceutical perspectives. *Nanomedicine (London, England), 7*(10), 1577–1590. https://doi.org/10.2217/nnm.12.128

Mattson, M. P., Haddon, R. C., & Rao, A. M. (2000). Molecular functionalization of carbon nanotubes and use as substrates for neuronal growth. *Journal of Molecular Neuroscience, 14*(3), 175–182. https://doi.org/10.1385/jmn:14:3:175

Mei, F., Zhong, J., Yang, X., Ouyang, X., Zhang, S., Hu, X., Ma, Q., Lu, J., Ryu, S., & Deng, X. (2007). Improved biological characteristics of poly(L-lactic acid) electrospun membrane by incorporation of multiwalled carbon nanotubes/hydroxyapatite nanoparticles. *Biomacromolecules, 8*(12), 3729–3735. https://doi.org/10.1021/bm7006295

Melgar-Lesmes, P., Luquero, A., Parra-Robert, M., Mora, A., Ribera, J., Edelman, E. R., & Jiménez, W. (2018). Graphene-dendrimer Nanostars for targeted macrophage overexpression of metalloproteinase 9 and hepatic fibrosis precision therapy. *Nano Letters, 18*(9), 5839–5845. https://doi.org/10.1021/acs.nanolett.8b02498

Mili, B., Das, K., Kumar, A., Saxena, A. C., Singh, P., Ghosh, S., & Bag, S. (2017). Preparation of NGF encapsulated chitosan nanoparticles and its evaluation on neuronal differentiation potentiality of canine mesenchymal stem cells. *Journal of Materials Science Materials in Medicine, 29*(1), 4. https://doi.org/10.1007/s10856-017-6008-2

Mir, M., Ali, M. N., Barakullah, A., Gulzar, A., Arshad, M., Fatima, S., & Asad, M. (2018). Synthetic polymeric biomaterials for wound healing: A review. *Progress in Biomaterials, 7*(1), 1–21. https://doi.org/10.1007/s40204-018-0083-4

Mirza, E. H., Pan-Pan, C., Wan Ibrahim, W. M., Djordjevic, I., & Pingguan-Murphy, B. (2015). Chondroprotective effect of zinc oxide nanoparticles in conjunction with hypoxia on bovine cartilage-matrix synthesis. *Journal of Biomedical Materials Research Part A, 103*(11), 3554–3563. https://doi.org/10.1002/jbm.a.35495

Mohanta, D., Patnaik, S., Sood, S., & Das, N. (2019). Carbon nanotubes: Evaluation of toxicity at biointerfaces. *Journal of Pharmaceutical Analysis, 9*(5), 293–300. https://doi.org/10.1016/j.jpha.2019.04.003

Mokhtari, H., Kharaziha, M., Karimzadeh, F., & Tavakoli, S. (2019). An injectable mechanically robust hydrogel of kappa-carrageenan-dopamine functionalized graphene oxide for promoting cell growth. *Carbohydrate Polymers, 214*, 234–249. https://doi.org/10.1016/j.carbpol.2019.03.030

Monteiro, N., Martins, A., Reis, R. L., & Neves, N. M. (2014). Liposomes in tissue engineering and regenerative medicine. *Journal of the Royal Society Interface, 11*(101), 20140459. https://www.ncbi.nlm.nih.gov/pubmed/25401172

Mori, H., Ogura, Y., Enomoto, K., Hara, M., Maurstad, G., Stokke, B. T., & Kitamura, S. (2020). Dense carbon-nanotube coating scaffolds stimulate osteogenic differentiation of mesenchymal stem cells. *PLoS One, 15*(1), e0225589. https://doi.org/10.1371/journal.pone.0225589

Moura, D., Souza, M. T., Liverani, L., Rella, G., Luz, G. M., Mano, J. F., & Boccaccini, A. R. (2017). Development of a bioactive glass-polymer composite for wound healing applications. *Materials Science & Engineering. C, Materials for Biological Applications, 76*, 224–232. https://doi.org/10.1016/j.msec.2017.03.037

Nair, M., & Elizabeth, E. (2015). Applications of Titania nanotubes in bone biology. *Journal of Nanoscience and Nanotechnology, 15*(2), 939–955. https://doi.org/10.1166/jnn.2015.9771

Nair, M., Nancy, D., Krishnan, A. G., Anjusree, G. S., Vadukumpully, S., & Nair, S. V. (2015). Graphene oxide nanoflakes incorporated gelatin-hydroxyapatite scaffolds enhance osteogenic differentiation of human mesenchymal stem cells. *Nanotechnology, 26*(16), 161001. https://doi.org/10.1088/0957-4484/26/16/161001

Naruphontjirakul, P., Kanchanadumkerng, P., & Ruenraroengsak, P. (2023). Multifunctional Zn and Ag co-doped bioactive glass nanoparticles for bone therapeutic and regeneration. *Scientific Reports, 13*(1), 6775. https://doi.org/10.1038/s41598-023-34042-w

Nayak, T. R., Andersen, H., Makam, V. S., Khaw, C., Bae, S., Xu, X., Ee, P. L., Ahn, J. H., Hong, B. H., Pastorin, G., & Özyilmaz, B. (2011). Graphene for controlled and accelerated osteogenic differentiation of human mesenchymal stem cells. *ACS Nano, 5*(6), 4670–4678. https://doi.org/10.1021/nn200500h

Nayak, T. R., Jian, L., Phua, L. C., Ho, H. K., Ren, Y., & Pastorin, G. (2010). Thin films of functionalized multiwalled carbon nanotubes as suitable scaffold materials for stem cells proliferation and bone formation. *ACS Nano, 4*(12), 7717–7725. https://doi.org/10.1021/nn102738c

Nechifor, G., Eftimie Totu, E., Nechifor, A. C., Isildak, I., Oprea, O., & Cristache, C. M. (2019). Non-Resorbable nanocomposite membranes for guided bone regeneration based on Polysulfone-quartz fiber grafted with Nano-TiO(2). *Nanomaterials (Basel), 9*(7). https://doi.org/10.3390/nano9070985

Niemiec, S. M., Louiselle, A. E., Hilton, S. A., Dewberry, L. C., Zhang, L., Azeltine, M., Xu, J., Singh, S., Sakthivel, T. S., Seal, S., Liechty, K. W., & Zgheib, C. (2020). Nanosilk increases the strength of diabetic skin and delivers CNP-miR146a to improve wound healing. *Frontiers in Immunology, 11*, 590285. https://doi.org/10.3389/fimmu.2020.590285

Nishida, E., Miyaji, H., Takita, H., Kanayama, I., Tsuji, M., Akasaka, T., Sugaya, T., Sakagami, R., & Kawanami, M. (2014). Graphene oxide coating facilitates the bioactivity of scaffold material for tissue engineering. *Japanese Journal of Applied Physics, 53*(6S), 06JD04. https://doi.org/10.7567/JJAP.53.06JD04

Niu, W., Chen, M., Guo, Y., Wang, M., Luo, M., Cheng, W., Wang, Y., & Lei, B. (2021). A multifunctional bioactive glass-ceramic Nanodrug for post-surgical infection/cancer therapy-tissue regeneration. *ACS Nano, 15*(9), 14323–14337. https://doi.org/10.1021/acsnano.1c03214

Noh, M., Choi, Y. H., An, Y. H., Tahk, D., Cho, S., Yoon, J. W., Jeon, N. L., Park, T. H., Kim, J., & Hwang, N. S. (2019). Magnetic nanoparticle-embedded hydrogel sheet with a groove pattern for wound healing application. *ACS Biomaterials Science & Engineering, 5*(8), 3909–3921. https://doi.org/10.1021/acsbiomaterials.8b01307

Nosrati, H., Khodaei, M., Banitalebi-Dehkordi, M., Alizadeh, M., Asadpour, S., Sharifi, E., Ai, J., & Soleimannejad, M. (2020). Preparation and characterization of poly(ethylene oxide)/zinc oxide nanofibrous scaffold for chronic wound healing applications. *Polimery w Medycynie, 50*(1), 41–51. https://doi.org/10.17219/pim/128378

Ogay, V., Mun, E. A., Kudaibergen, G., Baidarbekov, M., Kassymbek, K., Zharkinbekov, Z., & Saparov, A. (2020). Progress and prospects of polymer-based drug delivery Systems for Bone Tissue Regeneration. *Polymers (Basel), 12*(12). https://doi.org/10.3390/polym12122881

Ornelas-Soto, N., Rubio-Govea, R., Guerrero-Beltrán, C. E., Vázquez-Garza, E., Bernal-Ramírez, J., García-García, A., Oropeza-Almazán, Y., García-Rivas, G., & Contreras-Torres, F. F. (2017). Enhancing internalization of silica particles in myocardial cells through surface modification. *Materials Science & Engineering. C, Materials for Biological Applications, 79*, 831–840. https://doi.org/10.1016/j.msec.2017.05.092

Orozco, A., Vega, F. J., Martel-Estrada, S.-A., Aguilar, A., Mendoza-Duarte, M. E., Chavarría-Gaytán, M.-C., Rodríguez-González, C. A., & Olivas-Armendáriz, I. (2016). Development of chitosan/poly(L-Lactide)/multiwalled carbon nanotubes scaffolds for bone tissue engineering.

Oryan, A., Alidadi, S., Moshiri, A., & Maffulli, N. (2014). Bone regenerative medicine: Classic options, novel strategies, and future directions. *Journal of Orthopaedic Surgery and Research, 9*(1), 18. https://doi.org/10.1186/1749-799X-9-18

Pajares-Chamorro, N., & Chatzistavrou, X. (2020). Bioactive glass nanoparticles for tissue regeneration. *ACS Omega, 5*(22), 12716–12726. https://doi.org/10.1021/acsomega.0c00180

Pampaloni, N. P., Rago, I., Calaresu, I., Cozzarini, L., Casalis, L., Goldoni, A., Ballerini, L., & Scaini, D. (2020). Transparent carbon nanotubes promote the outgrowth of enthorino-dentate projections in lesioned organ slice cultures. *Developmental Neurobiology, 80*(9–10), 316–331. https://doi.org/10.1002/dneu.22711

Papastefanaki, F., Jakovcevski, I., Poulia, N., Djogo, N., Schulz, F., Martinovic, T., Ciric, D., Loers, G., Vossmeyer, T., Weller, H., Schachner, M., & Matsas, R. (2015). Intraspinal delivery of polyethylene glycol-coated gold nanoparticles promotes functional recovery after spinal cord injury. *Molecular Therapy, 23*(6), 993–1002. https://doi.org/10.1038/mt.2015.50

Paramadini, A. W., Chinavinijkul, P., Meemai, A., Thongkam, P., Apasuthirat, A., & Nasongkla, N. (2023). Fabrication and in vitro characterization of zinc oxide nanoparticles and hyaluronic acid-containing carboxymethylcellulose gel for wound healing application. *Pharmaceutical Development and Technology, 28*(1), 95–108. https://doi.org/10.1080/10837450.2022.2164304

Paramita, P., Ramachandran, M., Narashiman, S., Nagarajan, S., Sukumar, D. K., Chung, T.-W., & Ambigapathi, M. (2021). Sol–gel based synthesis and biological properties of zinc integrated nano bioglass ceramics for bone tissue regeneration. *Journal of Materials Science: Materials in Medicine, 32*(1), 5. https://doi.org/10.1007/s10856-020-06478-3

Park, J., Choi, J. H., Kim, S., Jang, I., Jeong, S., & Lee, J. Y. (2019). Micropatterned conductive hydrogels as multifunctional muscle-mimicking biomaterials: Graphene-incorporated hydrogels directly patterned with femtosecond laser ablation. *Acta Biomaterialia, 97*, 141–153. https://doi.org/10.1016/j.actbio.2019.07.044

Park, J., Park, S., Ryu, S., Bhang, S. H., Kim, J., Yoon, J. K., Park, Y. H., Cho, S. P., Lee, S., Hong, B. H., & Kim, B. S. (2014). Graphene-regulated cardiomyogenic differentiation process of mesenchymal stem cells by enhancing the expression of extracellular matrix proteins and cell signaling molecules. *Advanced Healthcare Materials, 3*(2), 176–181. https://doi.org/10.1002/adhm.201300177

Park, J. K., Kim, Y. J., Yeom, J., Jeon, J. H., Yi, G. C., Je, J. H., & Hahn, S. K. (2010). The topographic effect of zinc oxide nanoflowers on osteoblast growth and osseointegration. *Advanced Materials, 22*(43), 4857–4861. https://doi.org/10.1002/adma.201002255

Park, J. S., Yi, S. W., Kim, H. J., Kim, S. M., Kim, J. H., & Park, K. H. (2017). Construction of PLGA nanoparticles coated with Polycistronic SOX5, SOX6, and SOX9 genes for Chondrogenesis of human mesenchymal stem cells. *ACS Applied Materials & Interfaces, 9*(2), 1361–1372. https://doi.org/10.1021/acsami.6b15354

Park, J. S., Yi, S. W., Kim, H. J., Kim, S. M., & Park, K. H. (2016). Regulation of cell signaling factors using PLGA nanoparticles coated/loaded with genes and proteins for Osteogenesis of human mesenchymal stem cells. *ACS Applied Materials & Interfaces, 8*(44), 30387–30397. https://doi.org/10.1021/acsami.6b08343

Patel, A., Xue, Y., Hartley, R., Sant, V., Eles, J. R., Cui, X. T., Stolz, D. B., & Sant, S. (2018). Hierarchically aligned fibrous hydrogel films through microfluidic self-assembly of graphene and polysaccharides. *Biotechnology and Bioengineering, 115*(10), 2654–2667. https://doi.org/10.1002/bit.26801

Patra, J. K., Das, G., Fraceto, L. F., Campos, E. V. R., Rodriguez-Torres, M. D. P., Acosta-Torres, L. S., Diaz-Torres, L. A., Grillo, R., Swamy, M. K., Sharma, S., Habtemariam, S., & Shin, H. S. (2018). Nano based drug delivery systems: Recent developments and future prospects. *Journal of Nanobiotechnology, 16*(1), 71. https://doi.org/10.1186/s12951-018-0392-8

Paviolo, C., & Stoddart, P. R. (2015). Metallic nanoparticles for peripheral nerve regeneration: Is it a feasible approach? *Neural Regeneration Research, 10*(7), 1065–1066. https://doi.org/10.4103/1673-5374.160083

Peloi, K. E., Contreras Lancheros, C. A., Nakamura, C. V., Singh, S., Neal, C., Sakthivel, T. S., Seal, S., & Lautenschlager, S. O. S. (2020). Antioxidative photochemoprotector effects of cerium oxide nanoparticles on UVB irradiated fibroblast cells. *Colloids and Surfaces. B, Biointerfaces, 191*, 111013. https://doi.org/10.1016/j.colsurfb.2020.111013

Peña, B., Maldonado, M., Bonham, A. J., Aguado, B. A., Dominguez-Alfaro, A., Laughter, M., Rowland, T. J., Bardill, J., Farnsworth, N. L., Alegret Ramon, N., Taylor, M. R. G., Anseth, K. S., Prato, M., Shandas, R., McKinsey, T. A., Park, D., & Mestroni, L. (2019). Gold nanoparticle-functionalized reverse thermal gel for tissue engineering applications. *ACS Applied Materials & Interfaces, 11*(20), 18671–18680. https://doi.org/10.1021/acsami.9b00666

Pitarresi, G., Palumbo, F. S., Fiorica, C., Bongiovì, F., Martorana, A., Federico, S., Chinnici, C. M., & Giammona, G. (2022). Composite hydrogels of alkyl functionalized Gellan gum derivative and hydroxyapatite/Tricalcium phosphate nanoparticles as injectable scaffolds for bone regeneration. *Macromolecular Bioscience, 22*(2), e2100290. https://doi.org/10.1002/mabi.202100290

Poudel, S., Napit, P. R., Briski, K. P., & Mattheolabakis, G. (2021). Oral delivery of nucleic acids with passive and active targeting to the intestinal tissue using polymer-based Nanocarriers. *Pharmaceutics, 13*(7), 1075. https://www.mdpi.com/1999-4923/13/7/1075

Pourpirali, R., Mahmoudnezhad, A., Oroojalian, F., Zarghami, N., & Pilehvar, Y. (2021). Prolonged proliferation and delayed senescence of the adipose-derived stem cells grown on the electrospun composite nanofiber co-encapsulated with TiO(2) nanoparticles and metformin-loaded mesoporous silica nanoparticles. *International Journal of Pharmaceutics, 604*, 120733. https://doi.org/10.1016/j.ijpharm.2021.120733

Qi, Q., Zhu, Y., Liu, G., Yuan, Z., Li, H., & Zhao, Q. (2020). Local intramyocardial delivery of bioglass with alginate hydrogels for post-infarct myocardial regeneration. *Biomedicine & Pharmacotherapy, 129*, 110382. https://doi.org/10.1016/j.biopha.2020.110382

Qi, W., Yuan, W., Yan, J., & Wang, H. (2014). Growth and accelerated differentiation of mesenchymal stem cells on graphene oxide/poly-l-lysine composite films [10.1039/C4TB00856A]. *Journal of Materials Chemistry B, 2*(33), 5461–5467. https://doi.org/10.1039/C4TB00856A

Qian, Y., Wang, X., Song, J., Chen, W., Chen, S., Jin, Y., Ouyang, Y., Yuan, W. E., & Fan, C. (2021). Preclinical assessment on neuronal regeneration in the injury-related microenvironment of graphene-based scaffolds. *NPJ Regenerative Medicine, 6*(1), 31. https://doi.org/10.1038/s41536-021-00142-2

Qian, Y., Zhou, X., Zhang, F., Diekwisch, T. G. H., Luan, X., & Yang, J. (2019). Triple PLGA/PCL scaffold modification including silver impregnation, collagen coating, and electrospinning significantly improve biocompatibility, antimicrobial, and osteogenic properties for orofacial tissue regeneration. *ACS Applied Materials & Interfaces, 11*(41), 37381–37396. https://doi.org/10.1021/acsami.9b07053

Qin, C., Wang, K., Zhang, L., & Bai, L. (2022). Stem cell therapy for Alzheimer's disease: An overview of experimental models and reality. *Animal Models and Experimental Medicine, 5*(1), 15–26. https://doi.org/10.1002/ame2.12207

Qiu, Y., Xu, X., Guo, W., Zhao, Y., Su, J., & Chen, J. (2020). Mesoporous hydroxyapatite nanoparticles mediate the release and bioactivity of BMP-2 for enhanced bone regeneration. *ACS Biomaterials Science & Engineering, 6*(4), 2323–2335. https://doi.org/10.1021/acsbiomaterials.9b01954

Rahimi, B., Behroozi, Z., Motamednezhad, A., Jafarpour, M., Hamblin, M. R., Moshiri, A., Janzadeh, A., & Ramezani, F. (2023). Study of nerve cell regeneration on nanofibers containing cerium oxide nanoparticles in a spinal cord injury model in rats. *Journal of Materials Science. Materials in Medicine, 34*(2), 9. https://doi.org/10.1007/s10856-023-06711-9

Raimondo, T. M., & Mooney, D. J. (2018). Functional muscle recovery with nanoparticle-directed M2 macrophage polarization in mice. *Proceedings of the National Academy of Sciences of the United States of America, 115*(42), 10648–10653. https://doi.org/10.1073/pnas.1806908115

Rao, K. M., Suneetha, M., Park, G. T., Babu, A. G., & Han, S. S. (2020). Hemostatic, biocompatible, and antibacterial non-animal fungal mushroom-based carboxymethyl chitosan-ZnO nanocomposite for wound-healing applications. *International Journal of Biological Macromolecules, 155*, 71–80. https://doi.org/10.1016/j.ijbiomac.2020.03.170

Rath, G., Hussain, T., Chauhan, G., Garg, T., & Goyal, A. K. (2016). Development and characterization of cefazolin loaded zinc oxide nanoparticles composite gelatin nanofiber mats for postoperative surgical wounds. *Materials Science & Engineering. C, Materials for Biological Applications, 58*, 242–253. https://doi.org/10.1016/j.msec.2015.08.050

Raudzus, F., Schöneborn, H., Neumann, S., Secret, E., Michel, A., Fresnais, J., Brylski, O., Ménager, C., Siaugue, J. M., & Heumann, R. (2020). Magnetic spatiotemporal control of SOS1 coupled nanoparticles for guided neurite growth in dopaminergic single cells. *Scientific Reports, 10*(1), 22452. https://doi.org/10.1038/s41598-020-80253-w

Ravichandran, R., Sridhar, R., Venugopal, J. R., Sundarrajan, S., Mukherjee, S., & Ramakrishna, S. (2014). Gold nanoparticle loaded hybrid nanofibers for cardiogenic differentiation of stem cells for infarcted myocardium regeneration. *Macromolecular Bioscience, 14*(4), 515–525. https://doi.org/10.1002/mabi.201300407

Ribeiro, M., Ferraz, M. P., Monteiro, F. J., Fernandes, M. H., Beppu, M. M., Mantione, D., & Sardon, H. (2017). Antibacterial silk fibroin/nanohydroxyapatite hydrogels with silver and gold nanoparticles for bone regeneration. *Nanomedicine, 13*(1), 231–239. https://doi.org/10.1016/j.nano.2016.08.026

Robinson, A. P., Zhang, J. Z., Titus, H. E., Karl, M., Merzliakov, M., Dorfman, A. R., Karlik, S., Stewart, M. G., Watt, R. K., Facer, B. D., Facer, J. D., Christian, N. D., Ho, K. S., Hotchkin, M. T., Mortenson, M. G., Miller, R. H., & Miller, S. D. (2020). Nanocatalytic activity of clean-surfaced, faceted nanocrystalline gold enhances remyelination in animal models of multiple sclerosis. *Scientific Reports, 10*(1), 1936. https://doi.org/10.1038/s41598-020-58709-w

Rodemer, T., Pütz, N., & Hannig, M. (2022). Influence of hydroxyapatite nanoparticles on the formation of calcium fluoride surface layer on enamel and dentine in vitro. *Scientific Reports, 12*(1), 17612. https://doi.org/10.1038/s41598-022-21545-1

Rodríguez-Lozano, F. J., García-Bernal, D., Aznar-Cervantes, S., Ros-Roca, M. A., Algueró, M. C., Atucha, N. M., Lozano-García, A. A., Moraleda, J. M., & Cenis, J. L. (2014). Effects of composite films of silk fibroin and graphene oxide on the proliferation, cell viability and mesenchymal phenotype of periodontal ligament stem cells. *Journal of Materials Science. Materials in Medicine, 25*(12), 2731–2741. https://doi.org/10.1007/s10856-014-5293-2

Russo, A., Bianchi, M., Sartori, M., Boi, M., Giavaresi, G., Salter, D. M., Jelic, M., Maltarello, M. C., Ortolani, A., Sprio, S., Fini, M., Tampieri, A., & Marcacci, M. (2018). Bone regeneration in a rabbit critical femoral defect by means of magnetic hydroxyapatite macroporous scaffolds. *Journal of Biomedical Materials Research. Part B, Applied Biomaterials, 106*(2), 546–554. https://doi.org/10.1002/jbm.b.33836

Saddik, M. S., Elsayed, M. M. A., El-Mokhtar, M. A., Sedky, H., Abdel-Aleem, J. A., Abu-Dief, A. M., Al-Hakkani, M. F., Hussein, H. L., Al-Shelkamy, S. A., Meligy, F. Y., Khames, A., & Abou-Taleb, H. A. (2022). Tailoring of novel azithromycin-loaded zinc oxide nanoparticles for wound healing. *Pharmaceutics, 14*(1). https://doi.org/10.3390/pharmaceutics14010111

Sakthi Devi, R., Girigoswami, A., Siddharth, M., & Girigoswami, K. (2022). Applications of gold and silver nanoparticles in Theranostics. *Applied Biochemistry and Biotechnology, 194*(9), 4187–4219. https://doi.org/10.1007/s12010-022-03963-z

Samadian, H., Khastar, H., Ehterami, A., & Salehi, M. (2021). Bioengineered 3D nanocomposite based on gold nanoparticles and gelatin nanofibers for bone regeneration: In vitro and in vivo study. *Scientific Reports, 11*(1), 13877. https://doi.org/10.1038/s41598-021-93367-6

Sandhu, K. K., McIntosh, C. M., Simard, J. M., Smith, S. W., & Rotello, V. M. (2002). Gold nanoparticle-mediated transfection of mammalian cells. *Bioconjugate Chemistry, 13*(1), 3–6. https://doi.org/10.1021/bc015545c

Sant'Anna, M. B., Lopes, F. S. R., Kimura, L. F., Giardini, A. C., Sant'Anna, O. A., & Picolo, G. (2019). Crotoxin conjugated to SBA-15 nanostructured mesoporous silica induces Long-last analgesic effect in the neuropathic pain model in mice. *Toxins (Basel), 11*(12). https://doi.org/10.3390/toxins11120679

Schmidt, C. E., & Leach, J. B. (2003). Neural tissue engineering: Strategies for repair and regeneration. *Annual Review of Biomedical Engineering, 5*, 293–347. https://doi.org/10.1146/annurev.bioeng.5.011303.120731

Seisenbaeva, G. A., Fromell, K., Vinogradov, V. V., Terekhov, A. N., Pakhomov, A. V., Nilsson, B., Ekdahl, K. N., Vinogradov, V. V., & Kessler, V. G. (2017). Dispersion of TiO2 nanoparticles improves burn wound healing and tissue regeneration through specific interaction with blood serum proteins. *Scientific Reports, 7*(1), 15448. https://doi.org/10.1038/s41598-017-15792-w

Şelaru, A., Mocanu-Dobranici, A. E., Olăreţ, E., Ginghină, R. E., Stancu, I. C., Costache, M., & Dinescu, S. (2022). Gelatin meshes enriched with graphene oxide and magnetic nanoparticles support and enhance the proliferation and neuronal differentiation of human adipose-derived stem cells. *International Journal of Molecular Sciences, 24*(1). https://doi.org/10.3390/ijms24010555

Shahzadi, L., Chaudhry, A. A., Aleem, A. R., Malik, M. H., Ijaz, K., Akhtar, H., Alvi, F., Khan, A. F., Rehman, I. U., & Yar, M. (2018). Development of K-doped ZnO nanoparticles encapsulated crosslinked chitosan based new membranes to stimulate angiogenesis in tissue engineered skin grafts. *International Journal of Biological Macromolecules, 120*(Pt A), 721–728. https://doi.org/10.1016/j.ijbiomac.2018.08.103

Shakeel, A., Dash, S., Kumar, V. K., & Mohanty, S. (2022). Role of polymeric nanomaterial in regenerative medicine and stem cell biology. In M. Dash (Ed.), *Biomimetic biomaterials for tissue regeneration and drug delivery* (pp. 75–98). Springer Singapore. https://doi.org/10.1007/978-981-16-4566-2_4

Shalaby, M. A., Anwar, M. M., & Saeed, H. (2022). Nanomaterials for application in wound healing: Current state-of-the-art and future perspectives. *Journal of Polymer Research, 29*(3), 91. https://doi.org/10.1007/s10965-021-02870-x

Sharifi, E., Sadati, S. A., Yousefiasl, S., Sartorius, R., Zafari, M., Rezakhani, L., Alizadeh, M., Nazarzadeh Zare, E., Omidghaemi, S., Ghanavatinejad, F., Jami, M. S., Salahinejad, E., Samadian, H., Paiva-Santos, A. C., De Berardinis, P., Shafiee, A., Tay, F. R., Pourmotabed, S., & Makvandi, P. (2022). Cell loaded hydrogel containing Ag-doped bioactive glass-ceramic nanoparticles as skin substitute: Antibacterial properties, immune response, and scarless cutaneous wound regeneration. *Bioengineering & Translational Medicine, 7*(3), e10386. https://doi.org/10.1002/btm2.10386

Sharifi, E., Yousefiasl, S., Laderian, N., Rabiee, N., Makvandi, P., Pourmotabed, S., Ashrafizadeh, M., Familsattarian, F., & Fang, W. (2023). Cell-loaded genipin cross-linked collagen/gelatin skin substitute adorned with zinc-doped bioactive glass-ceramic for cutaneous wound regeneration. *International Journal of Biological Macromolecules, 251*, 125898. https://doi.org/10.1016/j.ijbiomac.2023.125898

Shen, C. C., Yang, M. Y., Chang, K. B., Tseng, C. H., Yang, Y. P., Yang, Y. C., Kung, M. L., Lai, W. Y., Lin, T. W., Hsieh, H. H., & Hung, H. S. (2021a). Fabrication of hyaluronic acid-gold nanoparticles with chitosan to modulate neural differentiation of mesenchymal stem cells. *Journal of the Chinese Medical Association, 84*(11), 1007–1018. https://doi.org/10.1097/jcma.0000000000000589

Shen, M., Wang, Y., Hu, F., Lv, L., Chen, K., & Xing, G. (2021b). Thrombolytic agents: Nanocarriers in targeted release. *Molecules, 26*(22). https://doi.org/10.3390/molecules26226776

Shen, Y., Gong, S., Li, J., Wang, Y., Zhang, X., Zheng, H., Zhang, Q., You, J., Huang, Z., & Chen, Y. (2019). Co-loading antioxidant N-acetylcysteine attenuates cytotoxicity of iron oxide nanoparticles in hypoxia/reoxygenation cardiomyocytes. *International Journal of Nanomedicine, 14*, 6103–6115. https://doi.org/10.2147/ijn.S209820

Shi, G., Chen, W., Zhang, Y., Dai, X., Zhang, X., & Wu, Z. (2019). An antifouling hydrogel containing silver nanoparticles for modulating the therapeutic immune response in chronic wound healing. *Langmuir, 35*(5), 1837–1845. https://doi.org/10.1021/acs.langmuir.8b01834

Shi, M., Cao, X., Zhuang, J., & Chen, X. (2021). The cardioprotective effect and mechanism of bioactive glass on myocardial reperfusion injury. *Biomedical Materials, 16*(4), 045044. https://doi.org/10.1088/1748-605X/ac067e

Shimizu, Y., Ntege, E. H., & Sunami, H. (2022). Current regenerative medicine-based approaches for skin regeneration: A review of literature and a report on clinical applications in Japan. *Regenerative Therapy, 21*, 73–80. https://doi.org/10.1016/j.reth.2022.05.008

Shin, Y. C., Kim, C., Song, S. J., Jun, S., Kim, C. S., Hong, S. W., Hyon, S. H., Han, D. W., & Oh, J. W. (2018). Ternary aligned nanofibers of RGD peptide-displaying M13 bacteriophage/PLGA/graphene oxide for facilitated Myogenesis. *Nano, 2*(2), 144–156. https://doi.org/10.7150/ntno.22433

Shrestha, A., Haque, M. A., & Mattheolabakis, G. (2023). Pulmonary delivery for miRs: Present and future potential. *PRO, 11*(6), 1788. https://www.mdpi.com/2227-9717/11/6/1788

Shrestha, A., Lahooti, B., Mikelis, C. M., & Mattheolabakis, G. (2022). Chlorotoxin and lung cancer: A targeting perspective for drug delivery. *Pharmaceutics, 14*(12). https://doi.org/10.3390/pharmaceutics14122613

Siddiqi, N. J., Abdelhalim, M. A., El-Ansary, A. K., Alhomida, A. S., & Ong, W. Y. (2012). Identification of potential biomarkers of gold nanoparticle toxicity in rat brains. *Journal of Neuroinflammation, 9*, 123. https://doi.org/10.1186/1742-2094-9-123

Sidharthan, D. S., Abhinandan, R., Balagangadharan, K., & Selvamurugan, N. (2022). Advancements in nucleic acids-based techniques for bone regeneration. *Biotechnology Journal, 17*(2), e2100570. https://doi.org/10.1002/biot.202100570

Silina, E. V., Stupin, V. A., Suzdaltseva, Y. G., Aliev, S. R., Abramov, I. S., & Khokhlov, N. V. (2021). Application of polymer drugs with cerium dioxide Nanomolecules and mesenchymal stem cells for the treatment of skin wounds in aged rats. *Polymers (Basel), 13*(9). https://doi.org/10.3390/polym13091467

Silveira, P. C. L., Venâncio, M., Souza, P. S., Victor, E. G., de Souza Notoya, F., Paganini, C. S., Streck, E. L., da Silva, L., Pinho, R. A., & Paula, M. M. S. (2014). Iontophoresis with gold nanoparticles improves mitochondrial activity and oxidative stress markers of burn wounds. *Materials Science and Engineering: C, 44*, 380–385. https://doi.org/10.1016/j.msec.2014.08.045

Simonovic, J., Toljic, B., Lazarevic, M., Markovic, M. M., Peric, M., Vujin, J., Panajotovic, R., & Milasin, J. (2022). The effect of liquid-phase exfoliated graphene film on Neurodifferentiation of stem cells from apical papilla. *Nanomaterials (Basel), 12*(18). https://doi.org/10.3390/nano12183116

Singh, D., Singh, S., Sahu, J., Srivastava, S., & Singh, M. R. (2016). Ceramic nanoparticles: Recompense, cellular uptake and toxicity concerns. *Artificial Cells, Nanomedicine, and Biotechnology, 44*(1), 401–409. https://doi.org/10.3109/21691401.2014.955106

Singh, H., Bashir, S. M., Purohit, S. D., Bhaskar, R., Rather, M. A., Ali, S. I., Yadav, I., Makhdoomi, D. M., Din Dar, M. U., Gani, M. A., Gupta, M. K., & Mishra, N. C. (2022). Nanoceria laden decellularized extracellular matrix-based curcumin releasing nanoemulgel system for full-thickness wound healing. *Biomaterials Advances, 137*, 212806. https://doi.org/10.1016/j.bioadv.2022.212806

Singh, H., Purohit, S. D., Bhaskar, R., Yadav, I., Bhushan, S., Gupta, M. K., Gautam, S., Showkeen, M., & Mishra, N. C. (2021). Biomatrix from goat-waste in sponge/gel/powder form for tissue engineering and synergistic effect of nanoceria. *Biomedical Materials, 16*(2), 025008. https://doi.org/10.1088/1748-605X/abdb74

Soltani, S., Emadi, R., Haghjooy Javanmard, S., Kharaziha, M., Rahmati, A., Thakur, V. K., & Lotfian, S. (2022). Development of an injectable shear-thinning nanocomposite hydrogel for cardiac tissue engineering. *Gels, 8*(2). https://doi.org/10.3390/gels8020121

Soto, P. A., Vence, M., Piñero, G. M., Coral, D. F., Usach, V., Muraca, D., Cueto, A., Roig, A., van Raap, M. B. F., & Setton-Avruj, C. P. (2021). Sciatic nerve regeneration after traumatic injury using magnetic targeted adipose-derived mesenchymal stem cells. *Acta Biomaterialia, 130*, 234–247. https://doi.org/10.1016/j.actbio.2021.05.050

Sridhar, S., Venugopal, J. R., Sridhar, R., & Ramakrishna, S. (2015). Cardiogenic differentiation of mesenchymal stem cells with gold nanoparticle loaded functionalized nanofibers. *Colloids and Surfaces. B, Biointerfaces, 134*, 346–354. https://doi.org/10.1016/j.colsurfb.2015.07.019

Sun, B., Taing, A., Liu, H., Nie, G., Wang, J., Fang, Y., Liu, L., Xue, Y., Shi, J., Liao, Y. P., Ku, J., Xia, T., & Liu, Y. (2016). Nerve growth factor-conjugated mesoporous silica nanoparticles promote neuron-like PC12 cell proliferation and neurite growth. *Journal of Nanoscience and Nanotechnology, 16*(3), 2390–2393. https://doi.org/10.1166/jnn.2016.10958

Sun, F., Li, R., Jin, F., Zhang, H., Zhang, J., Wang, T., & Feng, Z. Q. (2021). Dopamine/zinc oxide doped poly(N-hydroxyethyl acrylamide)/agar dual network hydrogel with super self-healing, antibacterial and tissue adhesion functions designed for transdermal patch. *Journal of Materials Chemistry B, 9*(27), 5492–5502. https://doi.org/10.1039/d1tb00822f

Sun, L., Chen, X., Chen, R., Ji, Z., Mu, H., Liu, C., Yu, J., Wang, J., Xia, R., Zhang, S., Xu, Y., Ma, K., & Xia, L. (2023). Balancing the antibacterial and osteogenic effects of double-layer TiO(2) nanotubes loaded with silver nanoparticles for the osseointegration of implants. *Nanoscale, 15*(6), 2911–2923. https://doi.org/10.1039/d2nr06154f

Supronowicz, P. R., Ajayan, P. M., Ullmann, K. R., Arulanandam, B. P., Metzger, D. W., & Bizios, R. (2002). Novel current-conducting composite substrates for exposing osteoblasts to alternating current stimulation. *Journal of Biomedical Materials Research, 59*(3), 499–506. https://doi.org/10.1002/jbm.10015

Tabata, Y. (2008). Current status of regenerative medical therapy based on drug delivery technology. *Reproductive Biomedicine Online, 16*(1), 70–80. https://doi.org/10.1016/s1472-6483(10)60558-5

Taha, S. K., Hassan, E. A., Mousa, S., El-Bassyouni, G. T., Shalash, H. N., & Abdel Hamid, M. A. (2023). Biphasic calcium phosphate doped with zirconia nanoparticles for reconstruction of induced mandibular defects in dogs: Cone-beam computed tomographic and histopathologic evaluation. *Journal of Materials Science. Materials in Medicine, 34*(6), 27. https://doi.org/10.1007/s10856-023-06731-5

Takahama, H., Shigematsu, H., Asai, T., Matsuzaki, T., Sanada, S., Fu, H. Y., Okuda, K., Yamato, M., Asanuma, H., Asano, Y., Asakura, M., Oku, N., Komuro, I., Kitakaze, M., & Minamino, T. (2013). Liposomal amiodarone augments anti-arrhythmic effects and reduces hemodynamic adverse effects in an ischemia/reperfusion rat model. *Cardiovascular Drugs and Therapy, 27*(2), 125–132. https://doi.org/10.1007/s10557-012-6437-6

Takeo, M., Lee, W., & Ito, M. (2015). Wound healing and skin regeneration. *Cold Spring Harbor Perspectives in Medicine, 5*(1), a023267. https://doi.org/10.1101/cshperspect.a023267

Tanaka, M., Sato, Y., Zhang, M., Haniu, H., Okamoto, M., Aoki, K., Takizawa, T., Yoshida, K., Sobajima, A., Kamanaka, T., Kato, H., & Saito, N. (2017). In Vitro and in vivo evaluation of a three-dimensional porous multi-walled carbon nanotube scaffold for bone regeneration. *Nanomaterials (Basel), 7*(2). https://doi.org/10.3390/nano7020046

Tang, X., Thankappan, S. K., Lee, P., Fard, S. E., Harmon, M. D., Tran, K., & Yu, X. (2014). Chapter 21 - polymeric biomaterials in tissue engineering and regenerative medicine. In S. G. Kumbar, C. T. Laurencin, & M. Deng (Eds.), *Natural and synthetic biomedical polymers* (pp. 351–371). Elsevier. https://doi.org/10.1016/B978-0-12-396983-5.00022-3

Tatavarty, R., Ding, H., Lu, G., Taylor, R. J., & Bi, X. (2014). Synergistic acceleration in the osteogenesis of human mesenchymal stem cells by graphene oxide–calcium phosphate nanocomposites [10.1039/C4CC02442G]. *Chemical Communications, 50*(62), 8484–8487. https://doi.org/10.1039/C4CC02442G

Tavares, F., Soares, P. I. P., Silva, J. C., & Borges, J. P. (2023). Preparation and in vitro characterization of magnetic CS/PVA/HA/pSPIONs scaffolds for magnetic hyperthermia and bone regeneration. *International Journal of Molecular Sciences, 24*(2). https://doi.org/10.3390/ijms24021128

Teixeira, G. Q., Leite Pereira, C., Castro, F., Ferreira, J. R., Gomez-Lazaro, M., Aguiar, P., Barbosa, M. A., Neidlinger-Wilke, C., & Goncalves, R. M. (2016). Anti-inflammatory chitosan/poly-γ-glutamic acid nanoparticles control inflammation while remodeling extracellular matrix in degenerated intervertebral disc. *Acta Biomaterialia, 42*, 168–179. https://doi.org/10.1016/j.actbio.2016.06.013

Thomas, M., & Klibanov, A. M. (2003). Conjugation to gold nanoparticles enhances polyethylenimine's transfer of plasmid DNA into mammalian cells. *Proceedings of the National Academy of Sciences of the United States of America, 100*(16), 9138–9143. https://doi.org/10.1073/pnas.1233634100

Tiwari, S., Patil, R., Dubey, S. K., & Bahadur, P. (2020). Graphene nanosheets as reinforcement and cell-instructive material in soft tissue scaffolds. *Advances in Colloid and Interface Science, 281*, 102167. https://doi.org/10.1016/j.cis.2020.102167

Toczek, J., Sadlocha, M., Major, K., & Stojko, R. (2022). Benefit of silver and gold nanoparticles in wound healing process after endometrial cancer protocol. *Biomedicine, 10*(3). https://doi.org/10.3390/biomedicines10030679

Todorovic, K., Jovanovic, G., Todorovic, A., Mitic, A., Stojiljkovic, N., Ilic, S., Stojanovic, N., & Stojnev, S. (2018). Effects of coenzyme Q10 encapsulated in nanoliposomes on wound healing processes after tooth extraction. *Journal of Dental Sciences, 13*(2), 103–108. https://doi.org/10.1016/j.jds.2017.10.004

Tomar, S., Pandey, R., Surya, P., Verma, R., Mathur, R., Gangenahalli, G., & Singh, S. (2023). Multifunctional, adhesive, and PDA-coated bioactive glass reinforced composite hydrogel for regenerative wound healing. *ACS Biomaterials Science & Engineering, 9*(3), 1520–1540. https://doi.org/10.1021/acsbiomaterials.2c01223

Tseng, T. C., & Hsu, S. H. (2014). Substrate-mediated nanoparticle/gene delivery to MSC spheroids and their applications in peripheral nerve regeneration. *Biomaterials, 35*(9), 2630–2641. https://doi.org/10.1016/j.biomaterials.2013.12.021

Urie, R., Ghosh, D., Ridha, I., & Rege, K. (2018). Inorganic nanomaterials for soft tissue repair and regeneration. *Annual Review of Biomedical Engineering, 20*, 353–374. https://doi.org/10.1146/annurev-bioeng-071516-044457

Usui, Y., Aoki, K., Narita, N., Murakami, N., Nakamura, I., Nakamura, K., Ishigaki, N., Yamazaki, H., Horiuchi, H., Kato, H., Taruta, S., Kim, Y. A., Endo, M., & Saito, N. (2008). Carbon nanotubes with high bone-tissue compatibility and bone-formation acceleration effects. *Small, 4*(2), 240–246. https://doi.org/10.1002/smll.200700670

Vaidya, B., Nayak, M. K., Dash, D., Agrawal, G. P., & Vyas, S. P. (2016). Development and characterization of highly selective target-sensitive liposomes for the delivery of streptokinase: In vitro/in vivo studies. *Drug Delivery, 23*(3), 801–807. https://doi.org/10.3109/10717544.2014.916770

van Rijt, S., & Habibovic, P. (2017). Enhancing regenerative approaches with nanoparticles. *Journal of the Royal Society Interface, 14*(129), 20170093. https://doi.org/10.1098/rsif.2017.0093

Venkatesan, J., Ryu, B., Sudha, P. N., & Kim, S.-K. (2012). Preparation and characterization of chitosan–carbon nanotube scaffolds for bone tissue engineering. *International Journal of Biological Macromolecules, 50*(2), 393–402. https://doi.org/10.1016/j.ijbiomac.2011.12.032

Vera-Sánchez, M., Aznar-Cervantes, S., Jover, E., García-Bernal, D., Oñate-Sánchez, R. E., Hernández-Romero, D., Moraleda, J. M., Collado-González, M., Rodríguez-Lozano, F. J., & Cenis, J. L. (2016). Silk-fibroin and graphene oxide composites promote human periodontal ligament stem cell spontaneous differentiation into Osteo/Cementoblast-like cells. *Stem Cells and Development, 25*(22), 1742–1754. https://doi.org/10.1089/scd.2016.0028

Walker, M. (2022). Human skin through the ages. *International Journal of Pharmaceutics, 622*, 121850. https://doi.org/10.1016/j.ijpharm.2022.121850

Wang, H., Zhang, Y., Xu, X., & Wang, A. (2021a). An injectable mesoporous silica-based analgesic delivery system prolongs the duration of sciatic nerve block in mice with minimal toxicity. *Acta Biomaterialia, 135*, 638–649. https://doi.org/10.1016/j.actbio.2021.09.008

Wang, Q., Zhang, H., Xu, H., Zhao, Y., Li, Z., Li, J., Wang, H., Zhuge, D., Guo, X., Xu, H., Jones, S., Li, X., Jia, X., & Xiao, J. (2018). Novel multi-drug delivery hydrogel using scar-homing liposomes improves spinal cord injury repair. *Theranostics, 8*(16), 4429–4446. https://doi.org/10.7150/thno.26717

Wang, Y., Cui, H., Wang, Y., Xu, C., Esworthy, T. J., Hann, S. Y., Boehm, M., Shen, Y. L., Mei, D., & Zhang, L. G. (2021b). 4D printed cardiac construct with aligned Myofibers and adjustable curvature for myocardial regeneration. *ACS Applied Materials & Interfaces, 13*(11), 12746–12758. https://doi.org/10.1021/acsami.0c17610

Wang, Z., Dong, L., Han, L., Wang, K., Lu, X., Fang, L., Qu, S., & Chan, C. W. (2016). Self-assembled biodegradable nanoparticles and polysaccharides as biomimetic ECM nanostructures for the synergistic effect of RGD and BMP-2 on bone formation. *Scientific Reports, 6*, 25090. https://doi.org/10.1038/srep25090

Wang, Z., Wang, K., Lu, X., Li, M., Liu, H., Xie, C., Meng, F., Jiang, O., Li, C., & Zhi, W. (2015). BMP-2 encapsulated polysaccharide nanoparticle modified biphasic calcium phosphate scaffolds for bone tissue regeneration. *Journal of Biomedical Materials Research. Part A, 103*(4), 1520–1532. https://doi.org/10.1002/jbm.a.35282

Wee, S., & Gombotz, W. R. (1998). Protein release from alginate matrices. *Advanced Drug Delivery Reviews, 31*(3), 267–285. https://doi.org/10.1016/s0169-409x(97)00124-5

Wierzbicki, M., Hotowy, A., Kutwin, M., Jaworski, S., Bałaban, J., Sosnowska, M., Wójcik, B., Wędzińska, A., Chwalibog, A., & Sawosz, E. (2020). Graphene oxide scaffold stimulates differentiation and proangiogenic activities of myogenic progenitor cells. *International Journal of Molecular Sciences, 21*(11). https://doi.org/10.3390/ijms21114173

Williams, S., Okolie, C. L., Deshmukh, J., Hawco, L., McNeil, J., Nganou Assonkeng, A. C., Bennett, C., & Mkandawire, M. (2019). Magnetizing cellulose fibers with CoFe(2)O(4) nanoparticles for smart wound dressing for healing monitoring capability. *ACS Applied Bio Materials, 2*(12), 5653–5662. https://doi.org/10.1021/acsabm.9b00731

Winston, D. D., Li, T., & Lei, B. (2023). Bioactive nanoglass regulating the myogenic differentiation and skeletal muscle regeneration. *Regenerative Biomaterials, 10*. https://doi.org/10.1093/rb/rbad059

Wojda, S. J., & Donahue, S. W. (2018). Parathyroid hormone for bone regeneration. *Journal of Orthopaedic Research, 36*(10), 2586–2594. https://doi.org/10.1002/jor.24075

Wu, H., Li, F., Wang, S., Lu, J., Li, J., Du, Y., Sun, X., Chen, X., Gao, J., & Ling, D. (2018). Ceria nanocrystals decorated mesoporous silica nanoparticle based ROS-scavenging tissue adhesive for highly efficient regenerative wound healing. *Biomaterials, 151*, 66–77. https://doi.org/10.1016/j.biomaterials.2017.10.018

Wu, S., Duan, B., Lu, A., Wang, Y., Ye, Q., & Zhang, L. (2017). Biocompatible chitin/carbon nanotubes composite hydrogels as neuronal growth substrates. *Carbohydrate Polymers, 174*, 830–840. https://doi.org/10.1016/j.carbpol.2017.06.101

Wu, S., Shuai, Y., Qian, G., Peng, S., Liu, Z., Shuai, C., & Yang, S. (2023a). A spatiotemporal drug release scaffold with antibiosis and bone regeneration for osteomyelitis. *Journal of Advanced Research, 54*, 239. https://doi.org/10.1016/j.jare.2023.01.019

Wu, S., Weir, M. D., Lei, L., Liu, J., & Xu, H. H. K. (2021). Novel nanographene oxide-calcium phosphate cement inhibits enterococcus faecalis biofilm and supports dental pulp stem cells. *Journal of Orthopaedic Surgery and Research, 16*(1), 580. https://doi.org/10.1186/s13018-021-02736-4

Wu, Z., Li, W., Cheng, S., Liu, J., & Wang, S. (2023b). Novel fabrication of bioengineered injectable chitosan hydrogel loaded with conductive nanoparticles to improve therapeutic potential of mesenchymal stem cells in functional recovery after ischemic myocardial infarction. *Nanomedicine, 47*, 102616. https://doi.org/10.1016/j.nano.2022.102616

Xiang, C., Zhang, Y., Guo, W., & Liang, X. J. (2020). Biomimetic carbon nanotubes for neurological disease therapeutics as inherent medication. *Acta Pharmaceutica Sinica B, 10*(2), 239–248. https://doi.org/10.1016/j.apsb.2019.11.003

Xu, J., Hu, X., Jiang, S., Wang, Y., Parungao, R., Zheng, S., Nie, Y., Liu, T., & Song, K. (2019). The application of multi-walled carbon nanotubes in bone tissue repair hybrid scaffolds and the effect on cell growth in vitro. *Polymers (Basel), 11*(2). https://doi.org/10.3390/polym11020230

Xu, X., Liu, X., Tan, L., Cui, Z., Yang, X., Zhu, S., Li, Z., Yuan, X., Zheng, Y., Yeung, K. W. K., Chu, P. K., & Wu, S. (2018). Controlled-temperature photothermal and oxidative bacteria killing and acceleration of wound healing by polydopamine-assisted Au-hydroxyapatite nanorods. *Acta Biomaterialia, 77*, 352–364. https://doi.org/10.1016/j.actbio.2018.07.030

Yadav, E., Yadav, P., & Verma, A. (2021). In silico study of Trianthema portulacastrum embedded iron oxide nanoparticles on glycogen synthase kinase-3β: A possible contributor to its enhanced in vivo wound healing potential. *Frontiers in Pharmacology, 12*, 664075. https://doi.org/10.3389/fphar.2021.664075

Yang, C. Y., Meng, Z., Yang, K., He, Z., Hou, Z., Yang, J., Lu, J., Cao, Z., Yang, S., Chai, Y., Zhao, H., Zhao, L., Sun, X., Wang, G., & Wang, X. (2023). External magnetic field non-invasively stimulates spinal cord regeneration in rat via a magnetic-responsive aligned fibrin hydrogel. *Biofabrication, 15*(3), 035022. https://doi.org/10.1088/1758-5090/acdbec

Yang, D., Li, T., Xu, M., Gao, F., Yang, J., Yang, Z., & Le, W. (2014). Graphene oxide promotes the differentiation of mouse embryonic stem cells to dopamine neurons. *Nanomedicine (London, England), 9*(16), 2445–2455. https://doi.org/10.2217/nnm.13.197

Yang, R., Chen, F., Guo, J., Zhou, D., & Luan, S. (2020a). Recent advances in polymeric biomaterials-based gene delivery for cartilage repair. *Bioactive Materials, 5*(4), 990–1003. https://doi.org/10.1016/j.bioactmat.2020.06.004

Yang, W. F., Long, L., Wang, R., Chen, D., Duan, S., & Xu, F. J. (2018). Surface-modified hydroxyapatite nanoparticle-reinforced Polylactides for three-dimensional printed bone tissue engineering scaffolds. *Journal of Biomedical Nanotechnology, 14*(2), 294–303. https://doi.org/10.1166/jbn.2018.2495

Yang, Y., Dong, Z., Li, M., Liu, L., Luo, H., Wang, P., Zhang, D., Yang, X., Zhou, K., & Lei, S. (2020b). Graphene oxide/copper Nanoderivatives-modified chitosan/hyaluronic acid dressings for facilitating wound healing in infected full-thickness skin defects. *International Journal of Nanomedicine, 15*, 8231–8247. https://doi.org/10.2147/ijn.S278631

Yang, Z., Liu, X., Zhao, F., Yao, M., Lin, Z., Yang, Z., Liu, C., Liu, Y., Chen, X., & Du, C. (2022). Bioactive glass nanoparticles inhibit osteoclast differentiation and osteoporotic bone loss by activating lncRNA NRON expression in the extracellular vesicles derived from bone marrow mesenchymal stem cells. *Biomaterials, 283*, 121438. https://doi.org/10.1016/j.biomaterials.2022.121438

Yao, X., Zhan, L., Yan, Z., Li, J., Kong, L., Wang, X., Xiao, H., Jiang, H., Huang, C., Ouyang, Y., Qian, Y., & Fan, C. (2023). Non-electric bioelectrical analog strategy by a biophysical-driven nano-micro spatial anisotropic scaffold for regulating stem cell niche and tissue regeneration in a neuronal therapy. *Bioactive Materials, 20*, 319–338. https://doi.org/10.1016/j.bioactmat.2022.05.034

Ye, J., Li, Q., Zhang, Y., Su, Q., Feng, Z., Huang, P., Zhang, C., Zhai, Y., & Wang, W. (2023). ROS scavenging and immunoregulative EGCG@cerium complex loaded in antibacterial polyethylene glycol-chitosan hydrogel dressing for skin wound healing. *Acta Biomaterialia, 166*, 155–166. https://doi.org/10.1016/j.actbio.2023.05.027

Yilgor, P., Tuzlakoglu, K., Reis, R. L., Hasirci, N., & Hasirci, V. (2009). Incorporation of a sequential BMP-2/BMP-7 delivery system into chitosan-based scaffolds for bone tissue engineering. *Biomaterials, 30*(21), 3551–3559. https://doi.org/10.1016/j.biomaterials.2009.03.024

You, J. O., Rafat, M., Ye, G. J., & Auguste, D. T. (2011). Nanoengineering the heart: Conductive scaffolds enhance connexin 43 expression. *Nano Letters, 11*(9), 3643–3648. https://doi.org/10.1021/nl201514a

Yu, K., Huangfu, H., Qin, Q., Zhang, Y., Gu, X., Liu, X., Zhang, Y., & Zhou, Y. (2022a). Application of bone marrow-derived macrophages combined with bone mesenchymal stem cells in Dual-Channel three-dimensional bioprinting scaffolds for early immune regulation and osteogenic induction in rat Calvarial defects. *ACS Applied Materials & Interfaces, 14*(41), 47052–47065. https://doi.org/10.1021/acsami.2c13557

Yu, Y., Yang, B., Tian, D., Liu, J., Yu, A., & Wan, Y. (2022b). Thiolated hyaluronic acid/silk fibroin dual-network hydrogel incorporated with bioglass nanoparticles for wound healing. *Carbohydrate Polymers, 288*, 119334. https://doi.org/10.1016/j.carbpol.2022.119334

Yuan, M., Wang, Y., & Qin, Y. X. (2019). Engineered nanomedicine for neuroregeneration: Light emitting diode-mediated superparamagnetic iron oxide-gold core-shell nanoparticles functionalized by nerve growth factor. *Nanomedicine, 21*, 102052. https://doi.org/10.1016/j.nano.2019.102052

Yuan, M., Yan, T. H., Li, J., Xiao, Z., Fang, Y., Wang, Y., Zhou, H. C., & Pellois, J. P. (2021). Superparamagnetic iron oxide-gold nanoparticles conjugated with porous coordination cages: Towards controlled drug release for non-invasive neuroregeneration. *Nanomedicine, 35*, 102392. https://doi.org/10.1016/j.nano.2021.102392

Yukawa, H., Kagami, Y., Watanabe, M., Oishi, K., Miyamoto, Y., Okamoto, Y., Tokeshi, M., Kaji, N., Noguchi, H., Ono, K., Sawada, M., Baba, Y., Hamajima, N., & Hayashi, S. (2010). Quantum dots labeling using octa-arginine peptides for imaging of adipose tissue-derived stem cells. *Biomaterials, 31*(14), 4094–4103. https://doi.org/10.1016/j.biomaterials.2010.01.134

Yukawa, H., Sato, K., & Baba, Y. (2023). Theranostics applications of quantum dots in regenerative medicine, cancer medicine, and infectious diseases. *Advanced Drug Delivery Reviews, 200*, 114863. https://doi.org/10.1016/j.addr.2023.114863

Yukawa, H., Suzuki, K., Kano, Y., Yamada, T., Kaji, N., Ishikawa, T., & Baba, Y. (2013). Induced pluripotent stem cell labeling using quantum dots. *Cell Medicine, 6*(1–2), 83–90. https://doi.org/10.3727/215517913X674270

Yun, Y. R., Jang, J. H., Jeon, E., Kang, W., Lee, S., Won, J. E., Kim, H. W., & Wall, I. (2012). Administration of growth factors for bone regeneration. *Regenerative Medicine, 7*(3), 369–385. https://doi.org/10.2217/rme.12.1

Zamproni, L. N., Mundim, M. V., Porcionatto, M. A., & des Rieux, A. (2017). Injection of SDF-1 loaded nanoparticles following traumatic brain injury stimulates neural stem cell recruitment. *International Journal of Pharmaceutics, 519*(1–2), 323–331. https://doi.org/10.1016/j.ijpharm.2017.01.036

Zanello, L. P., Zhao, B., Hu, H., & Haddon, R. C. (2006). Bone cell proliferation on carbon nanotubes. *Nano Letters, 6*(3), 562–567. https://doi.org/10.1021/nl051861e

Zangeneh, A., Zangeneh, M. M., & Moradi, R. (2020). Ethnomedicinal plant-extract-assisted green synthesis of iron nanoparticles using *Allium saralicum* extract, and their antioxidant, cytotoxicity, antibacterial, antifungal and cutaneous wound-healing activities. *Applied Organometallic Chemistry, 34*(1), e5247. https://doi.org/10.1002/aoc.5247

Zarei, M., Esmaeili, A., Zarrabi, A., & Zarepour, A. (2022). Superparamagnetic iron oxide nanoparticles and curcumin equally promote neuronal branching morphogenesis in the absence of nerve growth factor in PC12 cells. *Pharmaceutics, 14*(12). https://doi.org/10.3390/pharmaceutics14122692

Zgheib, C., Hilton, S. A., Dewberry, L. C., Hodges, M. M., Ghatak, S., Xu, J., Singh, S., Roy, S., Sen, C. K., Seal, S., & Liechty, K. W. (2019). Use of cerium oxide nanoparticles conjugated with MicroRNA-146a to correct the diabetic wound healing impairment. *Journal of the American College of Surgeons, 228*(1), 107–115. https://doi.org/10.1016/j.jamcollsurg.2018.09.017

Zhang, B., Liu, Q., Liu, M., Shi, P., Zhu, L., Zhang, L., & Li, R. (2019a). Biodegradable hybrid mesoporous silica nanoparticles for gene/chemo-synergetic therapy of breast cancer. *Journal of Biomaterials Applications, 33*(10), 1382–1393. https://doi.org/10.1177/0885328219835490

Zhang, F., Weidmann, A., Nebe, J. B., & Burkel, E. (2012). Osteoblast cell response to surface-modified carbon nanotubes. *Materials Science and Engineering: C, 32*(5), 1057–1061. https://doi.org/10.1016/j.msec.2010.07.007

Zhang, F., Xia, Y., Xu, L., & Gu, N. (2008). Surface modification and microstructure of single-walled carbon nanotubes for dental resin-based composites. *Journal of Biomedical Materials Research. Part B, Applied Biomaterials, 86*(1), 90–97. https://doi.org/10.1002/jbm.b.30991

Zhang, M., Fukushima, Y., Nozaki, K., Nakanishi, H., Deng, J., Wakabayashi, N., & Itaka, K. (2023). Enhancement of bone regeneration by coadministration of angiogenic and osteogenic factors using messenger RNA. *Inflamm Regen, 43*(1), 32. https://doi.org/10.1186/s41232-023-00285-3

Zhang, N., Li, C., Zhou, D., Ding, C., Jin, Y., Tian, Q., Meng, X., Pu, K., & Zhu, Y. (2018). Cyclic RGD functionalized liposomes encapsulating urokinase for thrombolysis. *Acta Biomaterialia, 70*, 227–236. https://doi.org/10.1016/j.actbio.2018.01.038

Zhang, N., Melo, M. A. S., Antonucci, J. M., Lin, N. J., Lin-Gibson, S., Bai, Y., & Xu, H. H. K. (2016). Novel dental cement to combat biofilms and reduce acids for orthodontic applications to avoid enamel demineralization. *Materials (Basel), 9*(6). https://doi.org/10.3390/ma9060413

Zhang, W., & Yelick, P. C. (2021). Tooth repair and regeneration: Potential of dental stem cells. *Trends in Molecular Medicine, 27*(5), 501–511. https://doi.org/10.1016/j.molmed.2021.02.005

Zhang, Y., Fan, W., Wang, K., Wei, H., Zhang, R., & Wu, Y. (2019b). Novel preparation of Au nanoparticles loaded Laponite nanoparticles/ECM injectable hydrogel on cardiac differentiation of resident cardiac stem cells to cardiomyocytes. *Journal of Photochemistry and Photobiology. B, 192*, 49–54. https://doi.org/10.1016/j.jphotobiol.2018.12.022

Zhao, C., Zeng, Z., Qazvini, N. T., Yu, X., Zhang, R., Yan, S., Shu, Y., Zhu, Y., Duan, C., Bishop, E., Lei, J., Zhang, W., Yang, C., Wu, K., Wu, Y., An, L., Huang, S., Ji, X., Gong, C., Yuan, C., Zhang, L., Liu, W., Huang, B., Feng, Y., Zhang, B., Dai, Z., Shen, Y., Wang, X., Luo, W., Oliveira, L., Athiviraham, A., Lee, M. J., Wolf, J. M., Ameer, G. A., Reid, R. R., He, T. C., & Huang, W. (2018). Thermoresponsive citrate-based graphene oxide scaffold enhances bone regeneration from BMP9-stimulated adipose-derived mesenchymal stem cells. *ACS Biomaterials Science & Engineering, 4*(8), 2943–2955. https://doi.org/10.1021/acsbiomaterials.8b00179

Zheng, K., Balasubramanian, P., Paterson, T. E., Stein, R., MacNeil, S., Fiorilli, S., Vitale-Brovarone, C., Shepherd, J., & Boccaccini, A. R. (2019). Ag modified mesoporous bioactive glass nanoparticles for enhanced antibacterial activity in 3D infected skin model. *Materials Science & Engineering. C, Materials for Biological Applications, 103*, 109764. https://doi.org/10.1016/j.msec.2019.109764

Zhou, F., Xin, L., Wang, S., Chen, K., Li, D., Wang, S., Huang, Y., Xu, C., Zhou, M., Zhong, W., Wang, H., Chen, T., & Song, J. (2023). Portable handheld "SkinPen" loaded with biomaterial ink for in situ wound healing. *ACS Applied Materials & Interfaces, 15*(23), 27568–27585. https://doi.org/10.1021/acsami.3c02825

Zhou, J., Xu, N. S., & Wang, Z. L. (2006). Dissolving behavior and stability of ZnO Wires in biofluids: A study on biodegradability and biocompatibility of ZnO nanostructures. *Advanced Materials, 18*(18), 2432–2435. https://doi.org/10.1002/adma.200600200

Zhu, J., Jiang, G., Song, G., Liu, T., Cao, C., Yang, Y., Zhang, Y., & Hong, W. (2019). Incorporation of ZnO/bioactive glass nanoparticles into alginate/chitosan composite hydrogels for wound closure. *ACS Applied Bio Materials, 2*(11), 5042–5052. https://doi.org/10.1021/acsabm.9b00727

Zielińska, A., Carreiró, F., Oliveira, A. M., Neves, A., Pires, B., Venkatesh, D. N., Durazzo, A., Lucarini, M., Eder, P., Silva, A. M., Santini, A., & Souto, E. B. (2020). Polymeric nanoparticles: Production, characterization, toxicology and ecotoxicology. *Molecules, 25*(16). https://doi.org/10.3390/molecules25163731

Ziv-Polat, O., Shahar, A., Levy, I., Skaat, H., Neuman, S., Fregnan, F., Geuna, S., Grothe, C., Haastert-Talini, K., & Margel, S. (2014). The role of neurotrophic factors conjugated to iron oxide nanoparticles in peripheral nerve regeneration: In vitro studies. *BioMed Research International, 2014*, 267808. https://doi.org/10.1155/2014/267808

Zong, C., Van Holm, W., Bronckaers, A., Zhao, Z., Čokić, S., Aktan, M. K., Castro, A. B., Van Meerbeek, B., Braem, A., Willems, G., & Cadenas de Llano-Pérula, M. (2023). Biomimetic periodontal ligament transplantation activated by gold nanoparticles protects alveolar bone. *Advanced Healthcare Materials, 12*(15), e2300328. https://doi.org/10.1002/adhm.202300328

# Mechanisms of Tissue Regeneration

# Tissue Regeneration: Cellular and Molecular Mechanisms

**Surajit Hansda and Hiranmoy Das**

**Abstract** The ability to repair and regenerate damaged tissues and/or organs after illness, injury, or aging is vital for the survival of any living organism. The molecular and cellular mechanisms of tissue regeneration have been studied using a variety of animal models, ranging from invertebrates to vertebrates. The findings are serving as the foundation for developing new clinical interventions that enhance people's ability to heal and regenerate. Recent findings have given us a basic understanding of the molecular mechanisms underlying effective and flawless tissue regeneration and repair, which is summarized in this chapter. Briefly, we describe the molecular mechanisms involved in this regeneration process, including inflammatory responses, regeneration in various organisms, and signaling pathways that may extend our current knowledge to enable the development of novel potential therapies.

**Keywords** Tissue regeneration · Wound healing · Molecular mechanism · Signaling pathways

## 1 Introduction

The process of an organism's reaction to damage and eventually repair the morphological and functional integrity of tissues and organs to ensure their life is an important factor in the biomedical field of study. Most mammals, including humans, are incapable of regenerating complex tissues or healing without scars when they reach adulthood. Conversely, even as adults, several nonmammalian vertebrates can still

S. Hansda · H. Das (✉)
Department of Pharmaceutical Sciences, Jerry H. Hodge School of Pharmacy, Texas Tech University Health Sciences Center, Amarillo, TX, USA
e-mail: hiranmoy.das@ttuhsc.edu

 
H. Das (ed.), *Tissue Repair and Regeneration*,
https://doi.org/10.1007/978-3-031-93677-7_18

regenerate different organs and appendages and recover without leaving scars. For instance, urodele amphibians can flawlessly repair wounds and regenerate a variety of sophisticated tissues and organs, including limbs, tails, lenses, and retinas, while fish may renew their fins and hearts. Replacing missing bodily components or tissues is a remarkable occurrence. The earliest known account of this phenomenon dates back to Trembley's demonstration of an animal's capacity to regrow missing parts following amputation (Gómez & Echeverri, 2021). He gave this animal the name Hydra, after a serpent from Greek mythology that, when severed, regenerates several heads. The process of regeneration occurs differently in each animal phylum. The capacity to regenerate the entire organism has been seen in certain animals, whereas it is limited to particular organs or tissue types in many others. The ability to regenerate the entire body is present in the earliest diverse multicellular creatures, including sponges (Alibardi, 2023). Throughout the phylum, chordates exhibit varying degrees of regeneration capacity. Among these, urochordates exhibit significant regeneration capacities, particularly ascidians like *Ciona intestinalis* (partial body) and *Botrylloides violaceus* (full body). Certain members of the Pisces have been known to be able to regenerate, and zebrafish research has shown that they can regenerate various organs and fins. The ability to regenerate damaged limbs in amphibians, including newts and axolotls, is a rather complicated feature that is exclusive to amphibians and has not been observed in any other animal. Mammals and Aves have very limited capacity for regeneration; in their mature lives, they can regenerate relatively simple tissue types. Appendages and other more complicated structures cannot be regenerated by them. Numerous deer species possess the intriguing ability to renew their antlers (Reddy et al., 2019).

Tissue regeneration is the process of the tissue regrowing and repairing when it has been injured or lost and hinders the tissue from performing its intended function at its most efficient state. Tissue regeneration occurs when the tissue is damaged or is partially missing. Tissue regeneration is most effective during embryonic development compared to when the body is fully developed. Embryos can regenerate any tissue damage flawlessly, while mature developed bodies cannot regenerate tissues to the same effect and leave scarring (Khan et al., 2020). Tissues, including epithelial, fibrous, cartilage, bone, blood vessels, muscle, and nerve tissues, are all tissues that go through repair when damaged (Tucker et al., 2023). This repair process is a heavily researched area of study as, currently, humans lack the same full capabilities of tissue regeneration as found in other species. Organisms such as axolotls and nematodes exhibit full regenerative capabilities of organs and limbs and are a major interest for research identifying how such animals can regenerate the exact structures (McCusker et al., 2015). Studies have been conducted to determine the sources for these new cells, the molecular mechanisms that trigger tissue regeneration, and why humans and, more broadly, mammals lack this evolutionary ability. From this study of tissue regeneration, a broad field of research in regenerative medicine has emerged, including fields such as tissue engineering. Further research into the mechanisms of tissue regeneration in both humans and other organisms at a molecular level is required. Understanding these molecular mechanisms behind the variety of tissue regeneration will provide greater insight into how humans can achieve tissue regeneration capabilities as found in other organisms.

## 2 Importance of Tissue Regeneration

There are two essential stages of tissue regeneration: regeneration and repair. Regeneration is when the damaged or lost tissue is replaced by the proliferation of undamaged cells and tissues. Repair is the regeneration of cells or tissues that restore the normal tissue structure or by healing, which can also form a scar mark. The epidermis tissue, gastrointestinal tract epithelium, and the hematopoietic system (blood-forming tissues) can be almost or completely restored due to its high proliferative capacity. When the tissue cannot be fully regenerated, the body uses a repair process that uses collagen deposition and scar formation. This can occur when the extracellular matrix (ECM) or the tissue has a limited capacity to proliferate. The result is a scar, which may restore some strength and integrity to the tissue but does not replicate the original structure and functions. Regenerative medicine can hold the promise of regenerating damaged tissues and organs in the body by replacing damaged tissue or by stimulating the body's repair mechanisms to heal tissues or organs. It also allows scientists to grow tissues and organs in the laboratory as an alternative for patients who are unable to heal themselves. Nearly one in three Americans could potentially benefit from regenerative medicine. In addition, tissue regeneration and reprogramming bring back homeostasis in the organism in the event of a physical injury. The body's natural tissue regenerative ability starts to decrease with age (Mao & Mooney, 2015).

## 3 Stage of Tissue Regeneration

Both animals and plants can regenerate their natural maintenance of tissues and organs. Tissue regeneration is critical for the survival of a multicellular organism. Multicellular organisms have evolved to have specialized tissues that have a specific purpose and function. Some animals have extensive regenerative properties, such as the axolotl or Mexican salamander (Gómez & Echeverri, 2021). Other complex animals have limited regenerative abilities, such as regrowing hair and skin, building new tissues to knit bone pieces together, and forming thick scars to promote healing wounds and burns. Tissue regeneration occurs in several distinct stages, typically following tissue injury. These stages are part of the body's natural repair process and involve various cellular and molecular mechanisms. The stages are the hemostasis (immediate response) phase, inflammatory phase, proliferative phase, and remodeling (maturation) phase (Paul & Sharma, 2021).

## 4 Regeneration Processes

Tissue regeneration involves a complex interplay of cytokines, inflammatory cells, and signaling molecules, with initially limited inflammation playing a key role in the healing process. The tissues involved require certain mechanical and structural properties for proper functioning. The term has also been applied to efforts to perform specific biochemical functions using cells within an artificially created support system. Tissue repair is a coordinated process that can either restore normal function after an injury or result in the formation of a fibrous scar, leading to loss of normal tissue function. This process involves interactions between cells, interactions between cells and the ECM, and the secretion of growth factors and other mediators by endothelial cells, damaged parenchymal cells, and inflammatory cells (King, 2006). The extent to which normal function can be restored depends on several factors: the nature and extent of the injury, the type of tissue involved, the patient's genetic makeup, and the presence of other medical conditions. Tissue repair primarily aims to restore mechanical stability to tissues and prevent the loss of blood or other fluids. The outcome of tissue repair can vary widely: it may restore normal function, result in diminished or abnormal function, or lead to a total loss of function. In some cases, excessive fibrosis can damage or destroy the normal function of minimally damaged tissue. To mitigate this, the use of corticosteroids may help limit fibrosis and preserve normal tissue function (Ye et al., 2023).

Cell division is essential for growth, tissue repair, regeneration, and reproduction, as new cells are produced by cell division. In eukaryotic cells, cells go through a process called mitosis, where new cells are generated (Ong & Torres, 2019). In the cell cycle, there are two main phases where DNA and new cells are formed. The first phase is called interphase. In interphase, the first step is G1 or Gap 1 phase, where the cell grows and is metabolically active. The second stage is the S phase, also called the synthesis phase. The cell's DNA is replicated in this stage, and the centriole is also duplicated. The final phase in interphase is G2 or Gap 2, where the synthesis of proteins and the cell continues to grow. The second phase of mitosis is called the M phase, where the cell will split into two identical daughter cells.

Inflammation is a defense mechanism in response to infections and tissue injury (Hannoodee & Nasuruddin, 2020). Inflammation is an active factor involved in the repair of damaged tissues and restoration of homeostatic conditions after an injury to the tissue. There are also different types of intensities of the inflammatory response and types of cells involved that affect the quality of tissue repair. Immune cell-derived factors, including cytokines and growth factors, play a crucial role in modulating the proliferation of local stem cells and progenitor cells, thereby promoting tissue regeneration (Fig. 1).

Wound healing is a dynamic process consisting of four continuous, overlapping, and precisely programmed phases. In humans, effective wound healing involves several key steps: 1) rapid achievement of hemostasis, 2) controlled inflammation response, 3) differentiation, proliferation, and migration of mesenchymal cells to the wound site, 4) effective angiogenesis to restore blood supply, 5) timely

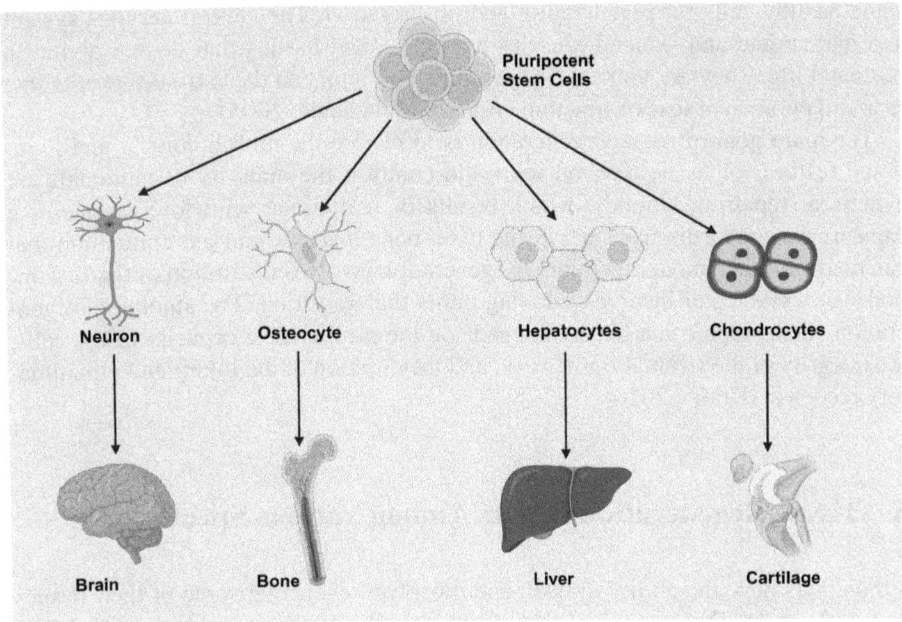


**Fig. 1** Differentiation of stem cells in response to specific growth factors or ligands

reepithelialization, where new epithelial tissue covers the wound, and 6) proper collagen synthesis, crosslinking, and alignment to ensure the strength and integrity of the repaired tissue (Guo & DiPietro, 2010).

A biomedical approach developed as a regenerative medicine for tissue regeneration using various stem cells. Several types of applications include cell therapies (injection of stem cells or progenitor cells), immunomodulation therapy (regeneration by biologically active molecules administered alone or as secretions by infused cells), and tissue engineering (transplantation of laboratory-grown organs and tissues).

## 5 Cells or Tissues Are Involved in the Regeneration Process

There are three main types of cells in the body depending upon their regenerative capacity: labile, stable, and permanent (Apte & Mehendale, 2005). The labile cells are under continuous active division and replace the cells that are lost from the body. An example of labile cells is the epidermis. Any injury to labile cells is rapidly repaired due to an aggressive tissue repair response. The stable cells have a long life span and divide at a slow rate. For example, the parenchymal cells of the solid glandular organs such as the liver and kidney. They can undergo rapid division upon injury to the organ to replace the dead cells. Regeneration after an injury requires some healthy cells for regeneration and proliferation. The central nervous system and the cardiac and skeletal muscles are permanent tissues that do not divide in postnatal life. They are unable to regenerate, and injury to these tissues results in a scar and permanent loss of function (Apte & Mehendale, 2005).

The main goal of tissue regeneration is to restore the morphological and functional features of tissue after an injury. In contrast, the majority of mammals are experts at repairing wounds, which results in scar tissue with low regenerative capacity except for tissues such as the liver, bone marrow, and gut epithelium that can recover after a moderate injury. Regeneration involves restitution of tissue components, while repair involves patching rather than restoring. The amount of regeneration versus repair that occurs depends on the proliferative capacity of the cells, the integrity of the stromal framework, and the duration of the injury and inflammatory response (Krafts, 2010).

## 6 Tissue Regeneration Differs Among Various Species

All animals have the ability to heal, and most can regenerate some of their tissues and body parts. The great diversity of regenerative capabilities among animals is striking, from the feeble ability of humans to the remarkable capacity of some worms to re-form an entire body from a small clump of cells. However, the evolutionary relationships or the complexity of the body do not necessarily predict regenerative ability. Some vertebrates, such as amphibians, are quite good at regeneration, while *Caenorhabditis elegans*, a simple roundworm, has no regenerative ability. Salamanders are recognized for their ability to regenerate a broad range of tissues. They have also been used for hundreds of years for classical developmental biology studies because of their large, accessible embryos. The range of tissues these animals can regenerate is fascinating, from full limbs to parts of the brain or heart, a potential that is missing in humans. Among vertebrates, salamanders are one of the organisms that have the outstanding ability to regenerate different tissues and organs such as the limb, heart, spinal cord, and lens (Yun, 2021). To date, researchers have made important discoveries regarding some cellular and molecular mechanisms that promote regenerative responses in salamanders. There is a nerve dependency on regeneration; remodeling of the ECM plays a crucial, possibly instructive role in directing cells, and the timing of the immune cells arriving at the injury site is important.

## 7 Tissue Regeneration in Response to Injuries

The skin is the largest organ of the body, which meets the environment directly. Thus, the skin is vulnerable to various damages, particularly burn injuries. Skin wound healing requires interaction between cell types, cytokines, mediators, neurovascular system, and matrix remodeling. Tissue regeneration technology remarkably enhances skin repair via reepidermalization, epidermal-stromal cell interactions, angiogenesis, and inhabitation of hypertrophic scars and keloids. The success rates of skin healing for burn injuries have significantly increased with the use of various skin substitutes. Skin damage causes the death of skin cells, leading to an enormous loss of body fluids that is followed by dehydration, electrolyte imbalance, and renal and circulatory failure. The survival rates of patients with burns have significantly improved due to the application of various skin grafts over the last decades. Despite wide use, autologous skin grafts are deficient in the treatment of severe burns for patients (Shpichka et al., 2019).

## 8 Development and Regeneration

The regeneration of organs is often proposed to mimic development with the reactivation of master regulators and developmental signaling pathways during repair. To develop therapeutic strategies for organ regeneration, studies have focused on understanding how organs develop normally during embryogenesis. However, a common question that emerges is the extent to which regeneration recapitulates embryonic development (Chatzeli et al., 2021). The future of regenerative medicine for these disorders lies in understanding how the brain develops and maintains itself throughout life. Regeneration following injury requires cells from surrounding tissue to proliferate and replenish tissues with the correct cell identities in the right pattern. Regeneration seems to make use of many transcriptional regulators and signaling pathways that also specify tissues during development. However, it is unclear how these are deployed to regenerate missing structures, particularly given that the starting point for regeneration is unpredictable. Though during leg development and regeneration, the temporal patterns of gene expression and the underlying mechanisms are different in both processes, there is a large overlap in gene usage (Sinigaglia et al., 2022).

In recent years, there has been a significant shift in the understanding of the role of metabolic transitions in cellular processes beyond being solely a response to energy demands. Instead, it is now recognized that cellular metabolism plays a central role in regulating cell fate during various developmental stages, neonatal growth, and tissue repair and regeneration. The existence of common mechanisms between development and regeneration suggests that metabolic transitions may also play a conserved role in regulating these processes (Mahmoud, 2023). Regeneration can be mediated via activation and proliferation of quiescent adult stem cell populations, or via cellular proliferation of preexisting cells to replenish the damaged tissue. Interestingly, cellular metabolism plays an important role in maintaining stem cell quiescence, as well as subsequent activation and proliferation during regeneration. For example, long-term hematopoietic stem cells (HSC) depend on anaerobic glycolysis during quiescence as they reside in a hypoxic niche, whereas loss of quiescence and differentiation of HSC is accompanied by a metabolic switch to mitochondrial metabolism. This is further evident in the dependence of HSC function on their metabolic state (Wang et al., 2014). Furthermore, recent evidence shows that inhibition of fatty acid oxidation by malonyl-CoA can promote the proliferation and differentiation of both mouse and human hematopoietic stem/progenitor cells (HSPC). Similarly, increased fatty acid oxidation, as well as metabolic reprogramming to glycolysis, regulates intestinal stem cell function in homeostasis and aging (Mahmoud, 2023). In some contexts, the metabolic regulation of stem cell quiescence and activation is complex. For example, during adult hippocampal neurogenesis, which is crucial for the learning and memory functions of the mammalian brain, the transition of neural stem cells (NSC) from quiescence to proliferation is modulated by changes in fatty acid oxidation levels. Specifically, elevated levels of malonyl-CoA promote de novo lipogenesis and NSC activation and proliferation. Additionally, the activation of the NSC is induced by lower mitochondrial pyruvate metabolism. Subsequent differentiation of the activated NSC to an intermediate progenitor cell is driven by an increase in oxidative phosphorylation (Mahmoud, 2023). In reconsidering regeneration as a set of fundamental processes starting with wound healing, it becomes apparent that to say regeneration merely recapitulates development is an oversimplification. Since early animal regeneration experiments and evidentiary support of cell theory swept aside preformationist notions of development, a major point of inquiry has been the degree to which development and regeneration represent similar events deployed during different life stages. This more recent mode of inquiry has operated under the assumption that the evolution of the embryo predates the evolution of regeneration, and thus, researchers contextualize their studies by asking what developmental pathways are redeployed during regeneration. While the evolution of molecular mechanisms underlying regeneration and embryonic development may echo the conundrum of the chicken or the egg, considering alternative hypotheses about their relationship has value for understanding how regenerative ability is missing or curtailed in some animals (Seifert et al., 2023).

## 9 Dedifferentiation in Regeneration

Despite being one of the most common terms used to define regeneration, dedifferentiation remains ambiguously defined conceptually and experimentally. Dedifferentiation is a fascinating and crucial concept in the field of regenerative biology. This process is essential for many organisms to repair or regenerate damaged tissues or organs. It refers to the process where specialized, differentiated cells revert to a more primitive or less specialized state (Lord & Ficz, 2022). During dedifferentiation, these specialized cells lose their unique characteristics and revert to a more stem-like state. This process can enable them to divide and give rise to a variety of cell types that are necessary for regeneration. When used in the context of a cell participating in regeneration, dedifferentiation originally referred to a loss of the terminally differentiated state in favor of a reacquired capacity to proliferate.

Experimental evidence for the extended definition of dedifferentiation, as defined here, is supported in part by studies leveraging single-cell RNA sequencing comparisons of tissue undergoing development versus those undergoing regeneration, beginning to resemble embryonic cells found in the developmental anlagen (Lin et al., 2021). Dedifferentiation often involves changes in gene expression. Genes associated with differentiation are downregulated, while those associated with a more stem-like state are upregulated. Several signaling pathways are involved in dedifferentiation, including those related to growth factors, cytokines, and ECM components. These signals help reprogram the differentiated cells back to a more versatile state.

Understanding dedifferentiation has significant implications for regenerative medicine and tissue engineering. By harnessing or inducing dedifferentiation, scientists hope to develop new therapies for repairing damaged tissues and organs. Abnormal dedifferentiation can also contribute to cancer development, as cancer cells may revert to a more primitive state that allows them to proliferate uncontrollably. Studying this process can provide insights into tumor biology and potential treatments. In summary, dedifferentiation is a key mechanism in the ability of certain organisms to regenerate tissues and organs.

## 10 Inflammation in Regeneration and Repair

Inflammation is a complex biological response to tissue injury, infection, or harmful stimuli. While often viewed negatively due to its association with chronic diseases, initially, limited inflammation plays a crucial role in regeneration and tissue repair.

- *Initial response to injury:* Blood vessels dilate, allowing immune cells and nutrients to reach the injury site. This helps in clearing debris and pathogens. Neutrophils are usually the first responders, followed by monocytes that differentiate into macrophages. These cells help clear damaged tissue and pathogens (Landén et al., 2016).
- *Macrophages:* Macrophages can polarize into M1 (pro-inflammatory) and M2 (anti-inflammatory) phenotypes. M1 macrophages combat pathogens, while M2 macrophages promote healing and tissue regeneration. M2 macrophages secrete factors like interleukin (IL)-10, transforming growth factor-beta (TGF-β), and vascular endothelial growth factor (VEGF) that promote cell proliferation, angiogenesis (formation of new blood vessels), and tissue remodeling (Strizova et al., 2023).
- *Cytokines and growth factors:* Such as IL-1, IL-6, and tumor necrosis factor-alpha (TNF-α), help mediate the inflammatory response and can influence tissue repair. Platelet-derived growth factor (PDGF), fibroblast growth factor (FGF), and transforming growth factor-beta (TGF-β) are crucial for cell migration, proliferation, and differentiation (Zgheib et al., 2014) (Hansda & Das, 2025).

- *Inflammation and stem cells:* Inflammatory signals can attract stem cells to the site of injury. These stem cells can differentiate into various cell types needed for tissue repair. The inflammatory milieu can influence stem cell fate, promoting their proliferation and differentiation into specific lineages necessary for regeneration (Maxson et al., 2012).
- *Tissue remodeling:* These enzymes are involved in the breakdown of extracellular matrix (ECM) components, allowing tissue remodeling as they heal. Following inflammation, fibroblasts synthesize collagen and other ECM components, restoring tissue integrity and function (Hansda et al., 2022) (Diller & Tabor, 2022) (Fig. 2).
- *Tissue repair and pathology:* For regeneration to occur effectively, inflammation must resolve appropriately without unnecessary delay. Failure to do so can lead to chronic inflammation and fibrosis, impairing tissue function (Caballero-Sánchez et al., 2024).
- *Tissue repair mechanisms:* After initial inflammatory signaling, fibroblasts, endothelial cells, and epithelial cells proliferate to replace lost cells. Cells migrate to the injury site, guided by gradients of cytokines and growth factors. Fibroblasts produce collagen and other matrix proteins, providing structural support. Matrix metalloproteinases (MMPs) and their inhibitors regulate the degradation and remodeling of ECM, allowing for the dynamic nature of healing. The formation of new blood vessels is crucial for delivering oxygen and nutrients to

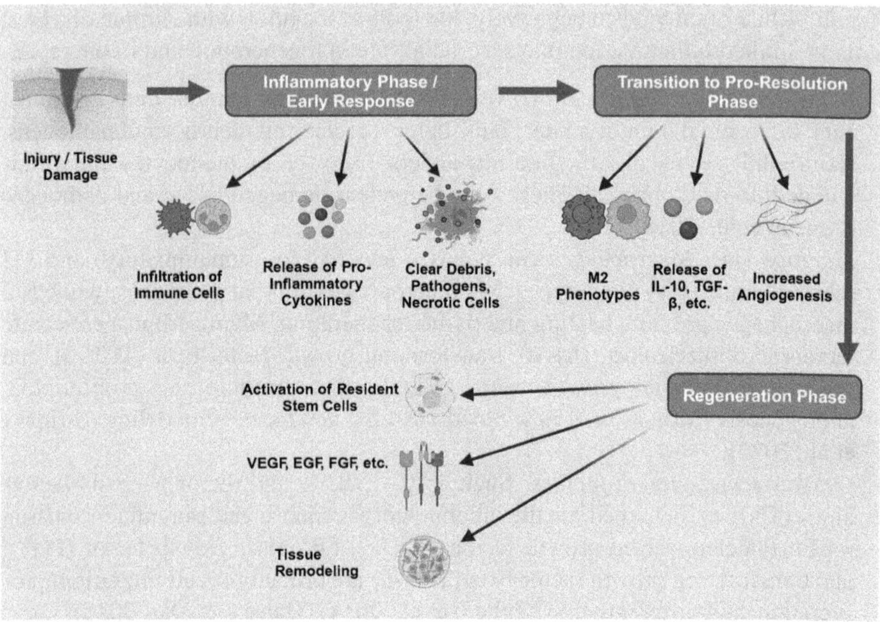


**Fig. 2** The role of inflammation in the regeneration process

the healing tissue. This process is driven by factors like VEGF and is tightly regulated to ensure proper repair (Schultz et al., 2011).

- *Resolution of inflammation:* Following the resolution phase, many immune cells undergo apoptosis to prevent excessive tissue damage. Macrophages play a key role in phagocytosing apoptotic cells and extracellular debris, facilitating a return to homeostasis. The final stage of healing is where the repaired tissue is restructured to restore its normal function, often involving a shift from a pro-inflammatory to a reparative environment (Maxson et al., 2012).
- *Clinical implications:* Persistent inflammation can lead to pathological conditions such as fibrosis, autoimmune diseases, and cancer. Targeting specific inflammatory pathways can enhance tissue repair in conditions like wounds, ischemic injuries, and degenerative diseases. Therapies that modulate inflammation, such as corticosteroids, nonsteroidal anti-inflammatory drugs (NSAIDs), and biologics, are commonly used to manage inflammatory responses (Hansda et al., 2020) (Ghosh & Hansda, 2021) (Benjamin et al., 2018).

Inflammation is a double-edged sword in the context of tissue repair. While acute inflammation is essential for initiating the healing process, chronic inflammation can be detrimental. A balanced inflammatory response is critical for successful regeneration, making it a key focus in both basic research and clinical applications. Understanding the role of inflammation in tissue repair has implications for regenerative therapies, such as stem cell therapy and the use of biomaterials to modulate inflammation and enhance healing for a variety of injuries and diseases.

## 11 Krüppel-Like Factors (KLFs) and Regeneration

Transcription factors (TFs) play a pivotal role in the regeneration of tissues by regulating gene expression in response to various signals during the regenerative process (Guan et al., 2021). Transcription factors are essential orchestrators of the regeneration process. They regulate stem cell dynamics, coordinate responses to injury, modulate inflammation, and guide tissue-specific repair mechanisms. Understanding their roles provides valuable insights into regenerative biology and may inform therapeutic strategies for enhancing regeneration in various tissues. Continued research in this area holds the potential for advancements in regenerative medicine, stem cell therapies, and the treatment of degenerative diseases. Krüppel-like factors (KLFs) are a family of transcription factors that play significant roles in various biological processes, including development, differentiation, and regeneration in animal cells. They are characterized by their zinc finger DNA-binding domains and can regulate gene expression in response to cellular signals (Pollak et al., 2018).

- *Stem cell function and KLFs:* KLFs, particularly KLF4 and KLF5, are involved in maintaining the pluripotency of stem cells and promoting self-renewal. They help regulate the balance between stem cell proliferation and differentiation (Park et al., 2019) (Maity et al., 2020).

- *Tissue repair and regeneration and KLFs:* KLF4 is essential for skin keratinocyte proliferation and migration during wound healing (Cai et al., 2023). KLF15 plays a role in muscle regeneration by regulating myogenic differentiation and satellite cell activation. KLF6 is involved in liver regeneration, promoting hepatocyte proliferation in response to liver injury (Sydor et al., 2017).
- *Bone remodeling and KLFs:* Several KLF members, notably KLF2, KLF4, and KLF5, have been implicated in the regulation of bone metabolism. KLF2 suppresses the differentiation and activity of osteoclasts, the cells responsible for bone resorption. It negatively regulates the receptor activator of nuclear factor-kappa B ligand (RANKL) signaling, which is critical for osteoclast differentiation or osteoclastogenesis process. KLF2 promotes osteoblast differentiation by enhancing the expression of genes essential for bone matrix production (Hansda et al., 2024) (Maity et al., 2022) (Laha et al., 2019) (Rolph & Das, 2020) (Prateeksha et al., 2023). KLF4 is highly expressed in osteoblasts and promotes their differentiation and mineralization activities. KLF5 supports osteoblast proliferation and differentiation by regulating genes involved in cell cycle progression and matrix production (Aksoy et al., 2014).
- *Environmental stimuli and KLFs:* KLFs can respond to external signals such as hypoxia, oxidative stress, and inflammation. For instance, KLF2 and KLF4 regulate genes involved in the response to oxidative stress, promoting cell survival and regeneration (Niu et al., 2019).
- *Angiogenesis and KLFs:* KLFs, such as KLF5, are involved in the regulation of angiogenic factors, promoting new blood vessel formation, which is vital for supplying nutrients and oxygen to regenerating tissues (Xie et al., 2021).
- *Inflammatory responses and KLFs:* Certain KLFs modulate inflammatory responses, balancing the pro-inflammatory and anti-inflammatory signals that influence tissue repair. For example, KLF4 can promote macrophage polarization toward an M2 phenotype, supporting tissue healing (Liao et al., 2011) (Howlader et al., 2024) (Hansda et al., 2024).
- *Transcriptional regulation and KLFs:* KLFs bind to specific DNA sequences in target gene promoters, activating or repressing their expression. This regulation is essential for controlling the genes involved in cell cycle progression, differentiation, and apoptosis (Zhang et al., 2023).
- *Proteins and KLFs:* KLFs often interact with other transcription factors and co-regulators, enhancing or inhibiting their activity. This collaborative action is critical for precise gene expression control during regeneration (Santoyo-Suarez et al., 2023).
- *Epigenetic modifications and KLFs:* KLFs can influence the epigenetic landscape of cells, altering chromatin structure and accessibility to transcription machinery, thereby affecting gene expression patterns crucial for regeneration (Moonen et al., 2022).
- *Clinical implications:* Understanding the role of KLFs in regeneration opens potential therapeutic avenues. KLFs could be targeted or manipulated to enhance regenerative capacities in tissues following injury or disease. Modulating KLF

expression may improve the efficiency of stem cell differentiation protocols for tissue engineering. Strategies to boost KLF activity could enhance healing processes in chronic wounds or injuries (Raghuram et al., 2020).

KLFs are vital regulators in the complex processes of cell and tissue regeneration. Their multifaceted roles in stem cell maintenance, tissue repair, response to injury, and modulation of inflammation underscore their importance in regenerative biology. Continued research on KLFs could lead to innovative approaches in regenerative medicine and the development of therapies for various diseases.

## 12 Mechanisms of Regeneration

Tissue regeneration in vertebrates involves several processes: hemostasis (stopping bleeding), inflammation, reepithelialization, activation of local progenitor cells, tissue morphogenesis, and replacement of injured tissue. In contrast, most mammals heal injuries through fibrotic repair, characterized by limited cell proliferation and extensive collagen deposition, resulting in scar tissue. When trauma or infection disrupts tissue architecture, both regeneration and fibrotic repair occur simultaneously with a multiphasic immune response. This response promotes hemostasis, initiates inflammation, protects against microbial infection, stimulates reepithelialization, and triggers a local fibrotic response. In regenerative processes (like in limbs, fins, or digit tips), inflammation is resolved as resident progenitor cells accumulate, reenter the cell cycle, and self-organize for morphogenesis. This transition from inflammation to morphogenesis leads to the formation of a regenerative blastema. Therefore, injured tissue must carefully coordinate interactions among cells and factors (such as cytokines, chemokines, and growth factors) in the injury microenvironment to resolve inflammation and promote blastema formation (Gawriluk et al., 2020). Tissue regeneration is a complex process involving a variety of molecular mechanisms. These mechanisms enable damaged tissues to repair and restore their function, and they are crucial in both acute and chronic healing processes. Here is an overview of the key molecular mechanisms involved:

I. *Stem and progenitor cells*

- *Stem cells*: These undifferentiated cells can differentiate into various mature cell types. In regeneration, they can be recruited or activated to replace damaged or lost cells. Examples include mesenchymal stem cells (MSC) and hematopoietic stem cells (HSC).
- *Progenitor cells*: These are more specialized than stem cells but can still give rise to a limited range of cell types. They are often involved in repairing tissues by differentiating into the required cell types (Picerno et al., 2021).

II. *ECM*

- *Components of ECM*: Proteins like collagen, elastin, and glycosaminoglycans provide structural support and biochemical cues that regulate cell behavior. During regeneration, the ECM is remodeled to create a suitable environment for cell growth and tissue formation.
- *MMPs*: These enzymes degrade ECM components, allowing for tissue remodeling. Their activity must be carefully regulated to balance tissue repair and prevent excessive scarring (Khoswanto, 2023).

III. *Cell migration and adhesion.*

- *Cell migration*: Cells move to the injury site via chemotactic attraction following chemical gradients of growth factors and interaction with various signaling molecules. Integrins and other adhesion molecules play a role in guiding cells to the damaged area and ensuring proper tissue formation (Parisi et al., 2020).
- *Cell adhesion*: Adhesion molecules such as cadherins and selectins facilitate cell-cell interactions and help organize cells into tissues. Proper adhesion is crucial for maintaining tissue structure and function during regeneration.

IV. *Gene expression and epigenetics*

- *Transcription factors*: Proteins like Sox2, Oct4, and Nanog are critical for regulating the gene expression programs that control cell differentiation and reprogramming (Kashyap et al., 2009). They play a role in activating regenerative processes by modulating the expression of genes involved in cell proliferation and tissue repair.
- *Epigenetic modifications*: DNA methylation and histone modification influence gene expression without altering the DNA sequence. These modifications can affect cell behavior and contribute to tissue regeneration by regulating the accessibility of genes involved in repair and regeneration (Radak & Fallahi, 2023).

V. *Inflammation and immune response*

- *Inflammatory mediators*: Cytokines and chemokines are involved in the initial inflammatory response to injury. They recruit immune cells to the site of damage, which help clear debris and release growth factors that promote healing.
- *Resolution of inflammation*: Proper resolution of inflammation is crucial for successful regeneration. Chronic inflammation can impede healing and lead to pathological conditions like fibrosis (Mescher et al., 2017).

VI. *Angiogenesis*

- *Blood vessel formation*: New blood vessels are formed from the existing vessels. This is crucial for supplying nutrients and oxygen to regenerating tissues. Key factors involved include vascular endothelial growth factor (VEGF) and angiopoietins (Liu et al., 2023).

VII. *Regenerative models and mechanisms*

- *Model organisms*: Research in model organisms like zebrafish, salamanders, and mice provides insights into regeneration mechanisms. For example, salamanders can regenerate entire limbs due to their ability to reprogram cells at the injury site.
- *Synthetic biology*: Advances in synthetic biology aim to engineer cells and tissues to enhance regenerative capabilities, such as creating bioengineered tissues or using gene editing technologies to promote repair.

VIII. *Signaling pathways*

- *Wnt pathway*: The Wnt signaling pathway plays a critical role in tissue regeneration by influencing cell fate decisions and promoting the growth of progenitor cells. It is involved in regulating stem cell maintenance and proliferation, too. The Wnt pathway can be divided into canonical Wnt pathway (β-catenin-dependent) and noncanonical Wnt pathway (β-catenin-independent). The canonical is involved in the regeneration of various tissues, including the liver, intestine, and limb. For instance, during liver regeneration, Wnt signaling promotes hepatocyte proliferation. It plays a key role in maintaining the pluripotency of stem cells and in the differentiation of stem cells into various cell types during tissue regeneration. The noncanonical is important for processes that require coordinated cell movement, such as during limb regeneration in amphibians. It is involved in the orientation of cells and tissues during regeneration, particularly in processes like neural tube closure and limb development. The Wnt signaling pathway is fundamental to the regeneration of various tissues in animals (Houschyar et al., 2018). Its ability to regulate cell proliferation, differentiation, and survival makes it a key player in maintaining tissue homeostasis and facilitating repair processes. Understanding these mechanisms is essential for developing regenerative therapies.
- *Hedgehog pathway*: This pathway influences cell growth, differentiation, and tissue architecture and is involved in processes like limb regeneration in some animals. Hedgehog signaling plays a vital role in activating and maintaining stem cells in various tissues, including hair follicles, neural tissue, and skeletal muscle (Jing et al., 2023). In amphibians, the Hedgehog pathway is crucial for limb regeneration. It helps in the formation of the blastema (a mass of cells capable of growth and regeneration) and directs the differentiation of these cells into the appropriate tissue types. In the central nervous system, Hedgehog signaling is involved in the repair processes after injury, promoting the proliferation of neural progenitor cells and facilitating axon regeneration (Chen et al., 2024). The noncanonical Hedgehog signaling promotes the migration of progenitor and stem cells to sites of injury, aiding in tissue repair. It helps regulate the spatial orga-

nization of cells during regeneration, ensuring that new tissues form correctly. This pathway is involved in establishing body patterns during embryonic development and is reactivated during tissue regeneration in adults. This is evident in the limb regeneration of salamanders and the healing of wounds. The Hedgehog pathway interacts with other signaling pathways, such as Wnt, Notch, and FGF, to coordinate complex regenerative processes (Singh et al., 2015). This crosstalk is essential for integrating multiple signals that regulate tissue repair. Dysregulation of Hedgehog signaling is implicated in various cancers. Understanding its role in regeneration could lead to therapeutic strategies that harness its regenerative properties while avoiding tumorigenesis.

- *Notch signaling pathway*: This pathway regulates cell fate decisions and is important for maintaining the balance between stem cell proliferation and differentiation during tissue repair (Wang et al., 2009). Notch signaling plays a crucial role in maintaining stem cell populations in various tissues. It helps balance self-renewal and differentiation, ensuring a reservoir of progenitor cells is available for tissue repair. In the hair follicle, Notch signaling regulates the fate of stem cells, promoting hair follicle cycling and regeneration. It is activated in satellite cells (muscle stem cells) during muscle injury. It regulates their activation, proliferation, and differentiation into myoblasts, facilitating muscle repair. In the central nervous system, Notch signaling is involved in maintaining neural stem cells and influencing their differentiation after injury. It can help modulate the repair process and influence neuronal survival. Notch signaling also regulates angiogenesis (the formation of new blood vessels) by controlling the differentiation of endothelial cells. This process is critical during tissue regeneration, as adequate blood supply is necessary for healing. Notch signaling influences the fate of various progenitor cells in developing and regenerating tissues. For example, in the intestine, the Notch regulates the balance between secretory and absorptive cell lineages, crucial for maintaining intestinal homeostasis and repair (Akil et al., 2021) (Demitrack & Samuelson, 2016).

The Notch receptor is a protein that is involved in cell communication. It helps regulate processes like cell growth, differentiation, and tissue development. It is initially processed in the endoplasmic reticulum and transported to the Golgi complex so it can be activated. In the Golgi complex, Notch receptors change to ensure it is functioning properly. With the help of Furin, an enzyme, Notch receptors will undergo S1 cleavage and be put into two parts: Notch extracellular domain (NECD) and transmembrane/intracellular domain (van Tetering & Vooijs, 2011). Both the NECD and transmembrane domain are bonded together and create a fully functioning Notch receptor. Notch receptors can be modified by glycosylation so they can interact with neighboring ligands such as Delta, Serrate, or Lag-2 on neighboring cells. When a ligand on a neighboring cell binds to the receptor, a second cleavage occurs using a disintegrin and metalloproteinase (ADAM); another enzyme will cut the Notch receptor in half for S3. When the receptor on one cell binds to the ligand on another cell, this is called trans-interactions, and this leads to an activated Notch signaling pathway; however, if the receptor of one cell is also bonded to a ligand on the same cell, this is called cis-interactions and this leads to an inactive Notch signaling pathway (Chillakuri et al., 2012) (Kuintzle et al., 2024). This is important for proper cell function. γ-Secretase (which includes presenilin) cuts Notch receptors for the last time, which will release the Notch intracellular domain (NICD) to travel to the nucleus and regulate gene expression while the external part of Notch receptors goes to a neighboring cell with the ligand that was bonded. The Notch signaling pathway is fundamental in regulating tissue regeneration, influencing stem cell maintenance, cell fate decisions, and tissue repair mechanisms. Its role in coordinating cellular responses makes it a key target for therapeutic strategies aimed at enhancing regeneration in various tissues.

- *TGF-β*: This superfamily of proteins regulates various cellular processes, including cell growth, differentiation, and ECM production, which are essential for tissue repair. The canonical TGF-β signaling is crucial for the differentiation of various cell types, including myofibroblasts, which are essential for wound healing and tissue repair (Tie et al., 2022). TGF-β stimulates the production of ECM components (e.g., collagen), which provide structural support during tissue regeneration. While TGF-β is necessary for normal healing, excessive activation can lead to fibrosis, where excessive ECM deposition occurs, hindering proper regeneration. The noncanonical TGF-β pathways modulate inflammatory responses, which are critical during the initial phases of tissue repair. These pathways promote the migration of cells to injury sites and support the proliferation of progenitor cells essential for regeneration. TGF-β interacts with pathways like Wnt, Notch, and Hedgehog, integrating signals that regulate cell fate and tissue homeostasis (Deng et al., 2024). The balance between TGF-β and other signaling pathways is crucial to prevent excessive fibrosis and promote effective regeneration. The TGF-β signaling pathway plays a pivotal role in tissue regeneration by not only regulating differentiation and ECM production but also regulating inflammatory responses (Li et al., 2017). While essential for healing, its dysregulation can lead to fibrosis and impaired regeneration. Understanding TGF-β pathways can inform therapeutic strategies for enhancing tissue repair and managing fibrotic diseases.
- *FGF signaling pathway*: This signaling pathway is involved in promoting cell proliferation and differentiation, playing a significant role in limb regeneration and wound healing (Farooq et al., 2021). FGFs promote the regeneration of various tissues, including skin, muscle, and nerve. They enhance the proliferation and migration of progenitor cells and fibroblasts during the healing process. In organisms capable of limb regeneration (like salamanders), FGFs are crucial for initiating and maintaining the

regenerative processes, particularly in forming the blastema. FGF signaling is involved in promoting the survival and proliferation of neural progenitor cells, aiding in the repair of nervous tissue after injury. FGFs are important for the development of cartilage and bone (Farooq et al., 2021). They regulate the differentiation of MSCs into chondrocytes and osteoblasts, which are essential for skeletal regeneration. FGF signaling interacts with pathways like TGF-β, Notch, and Wnt, creating a network of signals that coordinate complex regenerative processes. FGF signaling plays a role in balancing cell proliferation and differentiation, ensuring proper tissue architecture and function during regeneration. FGF signaling pathways are vital for regulating tissue regeneration and repair also (Xie et al., 2020). Their roles in cell proliferation, differentiation, and angiogenesis make them critical players in the healing process.

- *Inflammatory response*: The initial inflammatory response to injury can promote regeneration by releasing growth factors and cytokines that attract stem cells and promote healing. Increased blood flow and vascular permeability allow immune cells and nutrients to reach the site of injury (Zhao et al., 2021). Neutrophils are typically the first responders, followed by macrophages, which play a critical role in phagocytosis and orchestrating further repair processes. Macrophages can polarize into pro-inflammatory (M1) and anti-inflammatory (M2) phenotypes. M2 macrophages promote tissue repair and resolution of inflammation. Programmed cell death of neutrophils helps resolve inflammation and transition to the repair phase. Fibroblasts are recruited and activated, leading to ECM production and collagen deposition. The formation of new blood vessels is promoted by growth factors released during the inflammatory response, ensuring adequate nutrient supply for regeneration (Schultz et al., 2011). The inflammatory response interacts with growth factor pathways (like FGF and TGF-β) to coordinate tissue repair and regeneration. Proper regulation of inflammatory signals is critical to ensure that inflammation resolves and does not lead to chronic inflammation or fibrosis, which can impede healing. The inflammatory response is vital for initiating tissue regeneration and orchestrating the repair process (Thomas & Grisanti, 2020).
- *Neuroregeneration*: In the nervous system, pathways like the mechanistic target of rapamycin (mTOR) pathway and the presence of specific growth factors (e.g., nerve growth factor [NGF]) are essential for neuronal regeneration (Li et al., 2020). The mTOR pathway regulates cell growth and metabolism. It plays a crucial role in neuronal survival, axon regeneration, and synaptic plasticity. Activation of mTOR promotes protein synthesis and cell growth, which are essential for repairing damaged neurons. Nerve growth factor (NGF), brain-derived neurotrophic factor (BDNF), and neurotrophin-3 (NT-3) are vital for neuronal survival and growth (Takei et al., 2014). These factors promote neuron survival, encourage growth cone formation, and support synaptic connections. Genes like c-jun and ATF3 are upregulated after injury. They promote neuronal survival, axon regenera-

tion, and the formation of growth cones, aiding in the repair process. After injury, glial cells become activated and release cytokines and growth factors. While they can promote repair, excessive inflammation can lead to scar formation, which may hinder regeneration (Atala et al., 2010). Semaphorins, netrins, and ephrins direct the growth of axons during regeneration. They help in navigating regenerating axons to their target areas, facilitating functional recovery (De Winter et al., 2002).

Apart from those, the Hippo pathways have been established in *Drosophila melanogaster* as an important regulator of organ size, and this pathway is highly conserved in mammals. The core Hippo pathway consists of the kinase cascade mammalian STE20-like protein kinase 1/2 (MST1/2), and mitogen-activated protein kinases (MAP4Ks) phosphorylate large tumor suppressor kinase 1/2 (LATS1/2), leading to LATS1/2. Activated LATS1/2 then phosphorylate yes associated protein/transcriptional coactivator with Postsynaptic density 95, PSD-85; Discs large, Dlg; Zonula occludens-1, ZO-1 (PDZ)-binding motif (YAP/TAZ), which results in YAP/TAZ inactivation. As the major downstream effectors of the Hippo pathway, unphosphorylated YAP/TAZ translocate to the nucleus and induce target gene transcription by interacting with the transcription factors TEA domain transcription factor (TEAD1–4). In addition, Salvador 1 and mob kinase activator 1 are scaffold proteins for MST1/2 and LATS1/2, respectively, and upstream regulators such as Neurofibromin 2 can also induce LATS1/2 activity. Furthermore, the interaction between YAP/TAZ and TEADs is antagonized by vestigial-like family member 4 (Wang et al., 2017). These pathways are interlinked and can vary significantly across different species and types of tissues.

## 13 Tissue Regeneration in Sea Urchins

The ability to reconstruct appendages that have been damaged or amputated in any way is the major part of the regeneration (Medina-Feliciano & García-Arrarás, 2021), and sea urchins like echinoderms have exceptional regenerative capabilities. Genome analysis of the *Strongylocentrotus purpuratus* and the *Lytechinus variegatus* indicates which genes and proteins are activated during different stages of regeneration in sea urchins. As sea urchins are genetically related to humans, it makes it easier to study how they regrow their parts and help to highlight human regeneration therapies. Sea urchin spines are made from a large crystal of magnesium calcite. Special cells, called skeletogenic cells, are responsible for the biomineralization of the sea urchin spine (Parey et al., 2024). Sea urchin spines start to regenerate by the epidermis reconstructing around the broken spine, and specialized cells, sclerocytes, will calcify in the syncytium. Details about the regenerative process are still unknown, but the genes and signaling pathways are known. Echinoderms can also regrow body parts using two different processes: 1) morphallactic regeneration, which uses existing cells that have little growth and 2) epimorphic regeneration, where cells near the injury site change into simpler form and differentiate into the cells that are needed to regrow the lost tissue (Gilbert, 2017). Despite limited genetic resources for echinoderms, cell pathways, bone morphogenic protein (BMP), Hox, and Wnt signaling pathways are the same for regeneration. Notch signaling is an important biological process that is necessary for some regeneration in animals like frogs, zebrafish, and sea urchins. Notch receptors are active when they interact with neighboring delta or serrate/jagged membrane-bound ligands. When they interact, γ-secretase cleaves the Notch receptor, leaving the Notch intracellular domain (NICD). Finally, the NICD will move into the nucleus and interact with transcriptional regulators to change the expression of genes. NICD specifically influences genes that code for transcription factors in the Hes and Hey families. These transcription factors help inhibit or initiate the transcription of other genes. Notch signaling has been required for the regeneration of the African clawed frog, *Xenopus* larval tails, zebrafish fins, and mammalian skeletal muscles. Recent studies have shown that the Notch signaling pathway is involved in both tube feet and spine regeneration in various animals (Gilbert, 2017; Medina-Feliciano & García-Arrarás, 2021). Because Notch plays an important role in cell function, scientists are trying to understand the expression of stem cell markers called Piwi and Vasa, as they show an important role in multipotent stem cell types in many animals (Mashanov et al., 2020).

## 14 Tissue Regeneration and Repair in *Xenopus*

*Xenopus laevis* are the perfect model system to better investigate and understand the mechanism of their regenerative abilities. Single cell wound healing offers insight into how individual cells repair damage to their plasma membranes. When a cell is damaged, it needs to seal its plasma membrane by exocytosis to fix the damaged area. In studies with *Xenopus* oocytes, researchers found two important components of the cell's structure, F-actin and myosin-2, crucial for wound healing. After a cell is injured, a ring of F-actin and myosin-2 forms around the wound and helps pull the edges of the wound together. Protein RhoA and Cdc42 are at the wound's edge and are activated. Abr, a protein, and flow of calcium ions are both necessary for activating and coordinating the activities for RhoA and Cdc42. In addition, specific lipid kinases called Itpkb and inositol 1,3,4,5-tetrakisphosphate (InsP4) have been shown to speed up the activation of Rho and Cdc42, which can lead to faster wound healing. However, Sph1 and Sph2 are two components that researchers know that slow down the Rho activation leading to slower wound healing. More information regarding the study of wound healing in *Xenopus* oocytes will help researchers and scientists find better treatments for wound healing in humans (Reinardy et al., 2015) (Tseng & Levin, 2008) (Li et al., 2016).

- *The multicellular wound healing in Xenopus embryos:* Unlike single cell wounds, multicellular wound healing requires both intracellular and intercellular signal transduction pathways and behaviors among many cells. Multicellular wounds share similar mechanisms as single cell wounds. They both use calcium flow, activation of Rho and GTPases, and the actomyosin array. Calcium influx is a more complicated mechanism compared to single cell wound healing. When a cell is injured, an increase in calcium not only goes to the injured cell, but it also goes to neighboring cells that then help in the healing process. In addition, Itpkb and its product InsP4 help with Rho activation and the spread of calcium signals to the injury site. Signaling pathways such as Erk activate Rho and myosin-2, which aid in closing open wounds, and another signaling pathway P13k activates proteins Rac and Cdc42 that help to seal the edges. Studying wound healing at the different stages of *Xenopus* embryo will give further insight into the inflammation influences in the healing process (Li & Amaya, 2019) (Li et al., 2016).
- *The tail regeneration in Xenopus: Xenopus* tadpoles can regenerate their tails within 14 days. The regeneration period occurs in three steps: the early stage, intermediate stage, and late stage. In the early stage, the tail is focused on scar-free healing and closing the wound and inflammatory response to protect the site from infection. In the intermediate stage, there is rapid cell growth and cell proliferation called regenerative bud, where the cells will differentiate into muscles, tissues, and skin. In the final stage, the regenerative bud starts to differentiate and *Xenopus* tadpoles have a fully functioning tail again. There is an interesting refractory period where the tadpoles lose their ability to regenerate their tail at stage 46–47 but at stage 48, they regain their regenerative abilities. Researchers think this is due to two changes: metabolic shift, and immune system maturation. Tadpoles rely on yolk for their nutrients, but as they grow, they start eating and digesting food. This change in nutrients occurs during the refractory period. Regeneration is an energy intensive process, so the decrease in energy could be the cause. The other cause that researchers think would lead to the temporary loss of regenerative ability is the immune system maturation, when in stage 46–47 the tadpole's immune system grows more complex, which may interfere with the regenerative process (Tseng & Levin, 2008) (Lara et al., 2024). Further understanding how older tadpoles regain their regenerative abilities can provide further insight into why mammals are unable to regenerate limbs or organs. To gain further insight into the molecular and genetic mechanisms of the tadpole's regeneration, transcriptomic analysis was carried out for the three phases of tail regeneration and shows the gene expression and inflammation. There has been an increase in the expression levels of different genes that code for enzymes of reactive oxygen species (ROS), molecules that are important for cell signaling. Some enzymes generated by reactive oxygen species (ROS) are more active than others such as N-acetylserotonin (NAS) kinase and glucose-6-phosphate dehydrogenase. Nicotinamide adenine dinucleotide (NAD) kinase makes the energy for cells and glucose-6-phosphate dehydrogenase converts nicotinamide adenine dinucleotide phosphate ($NADP^+$) to nicotinamide adenine dinucleotide phosphate hydrogen (NADPH). The enzyme NADPH oxidase uses NADPH to pro-

duce superoxide, a type of ROS. ROS production increases as a tadpole's tail is amputated. ROS is crucial for regeneration because the reduction of ROS affects the regenerative ability. In addition, ROS contributes to growth factor signaling, which help the cells of the tail regrow. In recent studies there have been comparisons between the spinal cords of tadpoles and froglets and their response to injury. Tadpoles can regenerate tails while froglets cannot. Researchers nowadays are looking into the different genes that are activated and have seen that tadpoles can change the activities of certain genes that are linked to cell growth, to fight inflammation and to grow nerve cells. However, froglets show different patterns of gene activity. The Hippo pathway controls growth of cells and researchers have discovered that this pathway is important in the regrowth of a tadpole's tail. The planar cell polarity pathway manages how and when the tail grows to ensure the cells are all properly aligned. Other pathways like Wnt, Notch, BMP, FGF, Shh, and TGF-β aid the cell by sending signals to tell them to stop or start growing. Scientists have been trying to determine what types of cells are responsible for tissue development during regeneration. It has been shown that blastema cells can be formed by either pluripotent or by lineage-restricted stem cells. New technology has allowed scientists to understand *Xenopus*' tail regenerative tissue. They found out that the cells come from lineage-restricted stem cells. This has shown us that *Xenopus* is more similar to mammalian tissue regeneration (Richardson & O'Malley, 2025) (Hayashi et al., 2014).

## 15 The Molecular Aspects of Regeneration Mechanisms in Holothurians

Sea cucumbers, also called holothurians, have incredible regenerative abilities similar to the *Xenopus* in the fact that they can heal wounds and regrow body parts. In addition, they can regenerate internal organs. They also have this ability called evisceration where they can eject their internal organs when stressed. Evisceration can occur through the rear or the front depending on the species. The difference between sea cucumbers and mammals is that they use dedifferentiation, where the mature cells revert to a flexible state to be able to regrow tissues. This process involves a special tissue system called coelomic epithelium, which responds quickly to damage (Dolmatov, 2021).

- *Regeneration after fission:* Echinoderms and holothurians can asexually reproduce through fission, where the organism splits into two parts. Holothurians' outer body is made up of connective tissues that play a role in the fission process. The connective tissue has the ability to change its mechanical properties. This is called mutable collagenous tissue (MCT) or catch connective tissue where the tissues are dependent on factors to determine the stiffness or softness. The stiffness of MCT is regulated by the reactions of three different types of proteins: MMP, tissue inhibitors of metalloproteinases (TIMPs), and crosslink complexes.

The functions of these are: enzymes that break down collagen, proteins that inhibit MMPs, and structures that connect collagen fibers together, respectively. When TIMPs are active they block MMPs leading to strong crosslinks between collagen fibers. When MMPs are more active they break down these crosslinks, allowing collagen fibers to become more flexible. The changes in the properties of these connective tissues allow sea cucumbers to divide. The first signs of regeneration in holothurians becomes evident and the thickening of the connective tissue starts to grow along the mesentery edge. During this process, enterocytes begin to differentiate and undergo mitotic division. The main process involved in holothurian regeneration is the ECM, cell proliferation, and cell dedifferentiation. The connective tissue is the foundation where specific cells migrate to (Dolmatov, 2021) (Laronha & Caldeira, 2020).

- *Structural components of holothurians' connective tissue:* Holothurians have many ECM components that are very similar to multicellular organisms. This includes collagen, proteoglycans, and glycoproteins. However, holothurians lack elastin, but they rely on fibrillin. These fibrillin microfibrils also facilitate signal modulations and ECM remodeling. They interact with growth factors like TGF-β and BMP to help regulate the cell activity of holothurians. In addition, holothurians do not have certain proteins that lack structural integrity that vertebrates have; these include tenascin and fibronectins. However, they express tenascin-like proteins. Other ECM components found in holothurians are fibulins, agrin, laminins, and thrombostondins that stabilize connective tissue. The structural components of connective tissue in holothurians have a variety of genes that code for different proteins involved in modifying the ECM. There are numerous proteins such as serine, cysteine, aspartyl, and metal peptidases as well as their inhibitors. Degeneration of holothurians has focused on matrix metalloproteinases that have a key role in breaking down collagen, a major part of connective tissue. Holothurians have TIMPs proteins that control the activity of MMPs that stop them from breaking down too much collagen. This is important for connective tissue during regeneration. Tensilin is also another protein that sea cucumbers have that is similar to TIMPs and they make connective tissue strong and flexible. In addition, when sea cucumbers divide into two parts, two important signaling pathways, Wnt and Sonic Hedgehog (Shh), become active when internal organs start preparing for regeneration. Because these pathways are not active in undivided sea cucumbers, they suggest that they help regulate the splitting process and prepare the body. There are 26 different transcription factors that are activated when sea cucumbers divide and these transcription factors are involved in forming the tissues and organs. These genes help regulate the formation of the digestive system and the body's internal layers (Dolmatov, 2021) (Kornthong et al., 2021) (Tamori et al., 2006).
- *Early events after evisceration:* When sea cucumbers are injured, this process is called evisceration. They first lose coelomic fluid, which has many coelomocytes, and losing this fluid can induce a lot of stress on their body, which activates the immune system. Many biological processes slow down as a response to an injury and some genes that are related to signaling, metabolism, and the

immune system become less active. The number one priority for sea cucumbers is to replenish their coelomic fluid and start to heal the wounds. Researchers have found that the earliest response is the production of HSPs, which are heat shock proteins that repair tissue. Another example is that the amount of HSP90 increases after injury. Both of these proteins help the immune response of holothurians. In addition, sea cucumbers replenish their coelomocytes and produce new cells that differentiate into different types of immune cells. Another key part of their defense is lectin proteins that inactivate bacteria. Lectins bind to bacteria and help coelomocytes destroy them. After an injury certain cells in the body called coelomic epithelial cells become activated. Specific genes like *TAF1A*, *MYBBP1A*, *PWP1*, and *EEF1A1* help make transport proteins, which control other genes. Their activity increases day one after injury and decreases by day three. These processes were detected in the esophagus remnant. An in-depth look into the phosphorylation and acetylation shows changes in the esophagus. Transcription, translation, and processing of proteins are influenced in the activation of the mesenterial coelomic epithelium. Three days following an injury, the esophagus in sea cucumbers starts gathering nutrients and proteins start to become more active (Huo et al., 2022) (Montgomery et al., 2024) (Cruz-González et al., 2022). This process is called deacetylation and uses a lot of energy to remove certain chemical groups from proteins. The cells in the coelomic epithelium store nutrients as lipoprotein granules. These granules move to other cells and help with regeneration. Furthermore, genes that control the transformation of structures like the cytoskeleton become more active as well as genes that help build connective tissues work to thicken the mesentery and start to help rebuild the damaged esophagus. There are also many genes that control the body's structures such as the HOX family genes. HOX1 and HOX3 are activated after an injury and determine the front of the new gut, and genes like HOX9/10 and HOX11/13 shape the back end of the gut. Additionally, the Wnt signaling pathway aids in regeneration and body orientation. Wnt9 determines the back part of the new digestive system (Poliacikova et al., 2021).

- *Role of retinoic acid signaling:* Retinoic acid controls many processes in animals including regeneration. They attach to special receptors in the nucleus, which turn genes on and off. These receptors are called RXR (retinol receptor) and RAR (retinoic acid receptor). These receptors are found in sea cucumbers. There are two enzymes that manage retinoic acid levels in the body, dehydrogenase reductase 7 and aldehyde dehydrogenase 8A1. They convert retinoic acid into its active form and blocking the activity of the retinoic acid pathway can lead to a decrease in the growth of coelomic epithelial cells. Retinoic acid is imperative for regulating sea cucumber cells. In addition, the Wnt pathway also aids in regeneration, similar to retinoic acid. Both of these pathways control coelomic epithelial cells and thickening of connective tissue (Dolmatov, 2021).
- *The role of Wnt signaling pathway:* The Wnt pathway in sea cucumbers helps regenerate the front part of the body and the gut after evisceration. The Wnt signaling pathway is a communication system to arrange regeneration development and tissue repair. There are several Wnt ligands and their receptors that are

involved in this signaling pathway. These ligands include WntA, Wnt4, Wnt6, Wnt16, and Wnt5 and they trigger the Wnt pathway when they bind to their specific receptors on cells. These receptors are called frizzled receptors. When these Wnt ligands bind to frizzled receptors, there are changes in the gene expression of the cell. Three days after evisceration the first Wnt ligand that starts to become active for the regenerative process is Wnt16. Wnt16 is related to the development of the connective tissue and the ECM. MMP also starts to become active. Five to seven days after evisceration, ambulacral band (AB) and the gut start to form with the expression of WntA, Wnt4, and Wnt6. Also, 10–14 days after evisceration, genes frizzled5 and frizzled8 continue in the late stages of regeneration. The Wnt pathway makes its final adjustments on the tissue to ensure that they are formed correctly and are functioning properly (Alicea-Delgado & García-Arrarás, 2021).

## 16 Regeneration in Echinoderms

Echinoderms are a group of animals that have very extraordinary regenerative capabilities. The purpose of studying these echinoderms is to better understand the genes and pathways that control regeneration for medical treatments. Researchers have looked into regeneration processes of echinoderm larvae. Sea urchin embryos are some of the most well-documented systems for understanding the genes that are involved in regeneration. There are three different categories in molecular studies: single gene studies, gene analysis, and functional gene studies. For single gene studies, researchers are looking into the expression of genes during regeneration. They do this by identifying the genes' mRNA and they measure the level of gene expression in the regenerating tissue versus the healthy tissue. For gene analysis they analyze the expression of genes using RNA sequencing microarrays, which measure the gene expression levels. For functional gene studies, try to understand how these genes function in regeneration by applying it to experiments. There are two types of experiments: pharmaceutical experiments and gene knockdowns. In pharmaceutical experiments, different drugs are used to stop or start the expression of certain genes in the signaling pathways. In gene knockdowns, RNA interference (RNAi) technology is used to decrease the expression of a certain gene. Three main molecular techniques have been used to identify certain regeneration associative genes. These include in situ hybridization, immunohistochemistry, and quantitative reverse transcriptase polymerase chain reaction (qRT-PCR). To locate the expression of mRNA or proteins in a gene, in situ hybridization and immunohistochemistry are used. To quantify the level of gene expression, qRT-PCR is used. For example, crinoids have genes from the TGF and BMP that aid in arm regeneration. In sea urchins, researchers focus on single genes involved in spine and pedicellaria regeneration. These genes include Notch, Piwi, Vasa, and tert. In ophiuroids (brittle stars), Afuni, BMP2/4, and Hox are genes that are involved in regeneration. Some transcription factors that are also involved are alx1, ets1/2, foxN2/3, gataC, nk7, soxE, and twist, which are essential in tissue differentiation (García-Arrarás & Dolmatov, 2010) (Adonin et al., 2021).

## 17 Regeneration in Brittle Star

Molecular studies were performed on arm regeneration in brittle stars, *Amphiura filiformis*. When the arm of a brittle star is amputated, the wound starts to heal and reepithelialize. This injury leads to the formation of blastema. The formation of the blastema signals the regenerative process. New tissues and structures start to regenerate, and key expression patterns are being studied. Molecular studies used cDNA microarrays to help identify gene expressions during different stages of arm regeneration. In the first stage, blastema formation shows high transcriptional activity. Genes associated with transcription, translation, and energy production increased its response. While genes regulated to cell proliferation, division, and apoptosis are also expressed, indicating a higher cell growth is necessary for regeneration. In the next stage, gene expression continues, but the expression starts to decrease over time. In the final stage of gene expression, the new arm begins to take shape, and the BMP1 homolog increases as a response (Parey et al., 2024).

## 18 Regeneration in *Drosophila*

Regeneration in fruit fly, *Drosophila*, is a powerful model for understanding the molecular mechanisms of tissue repair and regrowth (Belacortu & Paricio, 2011). *Drosophila* can regenerate tissues such as imaginal disks, intestines, and other epithelial structures, depending on the developmental stage and tissue type. This regeneration involves multiple molecular mechanisms and signaling pathways coordinated to restore tissue integrity and function. Similar to macrophages in vertebrates, hemocytes (immune cells are recruited to the wound site and release signaling molecules like ROS) and cytokines trigger regeneration. In epithelial tissues, cells at the wound edge undergo cytoskeletal rearrangements to crawl and close the wound through actin-myosin contraction. Signaling pathways like c-Jun N-terminal kinase (JNK), Wnt/Wg, Hippo, Hedgehog, and epidermal growth factor receptor (EGFR)/Ras/MAPK are activated during *Drosophila* tissue regeneration. JNK signaling pathway promotes transcription of regeneration-associated genes, including Puckered (Puc) and matrix metalloproteinases (MMPs), which remodel the extracellular matrix. The Wnt/Wg pathway is essential for activating the proliferation and differentiation of regenerating tissues. The Hippo pathway modulates organ size and promotes regeneration by regulating Yorkie (Yki), a transcriptional coactivator, and Yki activation enhances cell survival and proliferation in response to injury. *Drosophila* regeneration involves a tightly regulated interplay of conserved molecular pathways, cellular behaviors, and epigenetic changes (Mehta & Singh, 2019).

## 19 Recent Advances in Wound Healing and Regeneration

The skin is one of our largest organs and makes up a total of at least 16% of our body weight. It has an important role in keeping balance in the body and acts as a barrier against external factors. There are three different layers of the skin: the epidermis, dermis, and subcutaneous layer. The epidermis is the outer layer that is composed of cells called keratinocytes that produce proteins that strengthen the skin; melanocytes, which produce pigments; Merkel cells for sensing touch; and Langerhans cells, which are involved in the immune response. The dermis is located beneath the epidermis and contains blood vessels, nerves, hair follicles, sweat glands, and oil glands. The subcutaneous layer is made up of fat and connective tissue that helps insulate the body and stores energy. When the skin is damaged, it has to be treated either through traditional methods or through regenerative approaches. In traditional methods, the focus is to prevent infections in the wound, keeping the wound moist, and removing dead tissue. However, new approaches are regenerative treatments that use new technology like bioactive materials, advanced dressings, stem cell therapy, and large-grown skin grafts to heal wounds faster with minimal scarring.

When it comes to healing a wound, it is a complex process that incorporates many different cells and molecular pathways to ensure that a wound is healed completely. It is imperative to reinstate the structural integrity of the skin as it serves as a barrier against outside infections, bacteria, and more, and it prevents the body from dehydration. The process of wound healing involves many steps including proliferation, inflammation, hemostasis, and remodeling. When a skin is cut or damaged in any way, blood vessels in the body are damaged as well. This leads to bleeding. Hemostasis is the body's defense against this bleeding by forming blood clots in the damaged areas. This occurs in three steps: vasoconstriction, primary hemostasis, and secondary hemostasis. In vasoconstriction, the blood vessels will construct to reduce the amount of blood lost. In the hemostasis stage, platelets will start to move to the damaged area and plug the wound. At this time platelets will also interact with fibrinogen that will aid in clotting. In the secondary stage of hemostasis, a coagulation cascade starts that converts fibrinogen into fibrin. After the bleeding has stopped, the inflammatory phase begins. Immune cells like macrophages, neutrophils, and lymphocytes are sent to the wound site and fight various infections (Peña & Martin, 2024) (Eming et al., 2014).

Currently, there are no defined molecular processes that cause enterocyte dedifferentiation despite a significant number of transcriptomic, proteomic, metabolomic, and other molecular genetic investigations. What genes initiate and regulate this process is still unclear. The variety of morphological and molecular information that is currently available indicates that the foundation of regeneration is epithelial morphogenesis and the dedifferentiation of specialized cells of the organ remnant. However, depending on the type of damage, the mechanisms of regeneration in holothurians, sea cucumbers, can vary greatly in terms of how the regeneration process is organized spatially, which cell types are involved, and how deeply their genomes are reprogrammed (dedifferentiation or trans-differentiation) (Dolmatov, 2021).

Lastly, a question about the shared molecular and cellular processes of reparative regeneration across many tissues still importantly stands. A common regeneration notion may be clarified by the peripheral nerve research discussed in this study. Similarly, fresh questions about specific reparative regenerations may be brought up by a contemporary common notion. In particular, the occurrence of a brief inflammatory reaction that induces cell differentiation, reprogramming, trans-differentiation, capillary permeabilization, invasion by blood cells, the formation of a plastic and heterogeneous ECM, cell proliferation, the involvement of stem cells and progenitors, phenomena of polyploidy, cell migrations upregulated by the matrix, with retro-controls of secreted matrix products by migrating cells, and cell differentiation and repair, can all be seen as a tool that offers an open perspective that connects several biological mechanisms involved in regeneration.

Both science and medicine must comprehend the molecular processes underlying tissue regeneration. Nevertheless, little is currently defined about the molecular and cellular processes by which they work. The development of therapeutic approaches targeted at tissue regeneration and repair requires an understanding of all these processes. Although this chapter touched on a lot of those topics, a more thorough experimental analysis is still required. We anticipate that this study will encourage more investigation into this difficult field of regeneration and provide insights into general principles of tissue regeneration and potential therapeutic targets for the development of an effective regenerative therapy.

## References

Adonin, L., Drozdov, A., & Barlev, N. A. (2021). Sea urchin as a universal model for studies of gene networks. *Frontiers in Genetics, 11*, 627259.

Akil, A., Gutiérrez-García, A. K., Guenter, R., Rose, J. B., Beck, A. W., Chen, H., & Ren, B. (2021). Notch signaling in vascular endothelial cells, angiogenesis, and tumor progression: An update and prospective. *Frontiers in Cell and Development Biology, 9*, 642352. https://doi.org/10.3389/fcell.2021.642352

Aksoy, I., Giudice, V., Delahaye, E., Wianny, F., Aubry, M., Mure, M., Chen, J., Jauch, R., Bogu, G. K., & Nolden, T. (2014). Klf4 and Klf5 differentially inhibit mesoderm and endoderm differentiation in embryonic stem cells. *Nature Communications, 5*(1), 3719.

Alibardi, L. (2023). Regeneration or scarring derive from specific evolutionary environmental adaptations of the life cycles in different animals. *Biology, 12*(5), 733.

Alicea-Delgado, M., & García-Arrarás, J. E. (2021). Wnt/β-catenin signaling pathway regulates cell proliferation but not muscle dedifferentiation nor apoptosis during sea cucumber intestinal regeneration. *Developmental Biology, 480*, 105–113.

Apte, U. M., & Mehendale, H. M. (2005). *Tissue repair.*

Atala, A., Irvine, D. J., Moses, M., & Shaunak, S. (2010). Wound healing versus regeneration: Role of the tissue environment in regenerative medicine. *MRS Bulletin, 35*(8). https://doi.org/10.1557/mrs2010.528

Belacortu, Y., & Paricio, N. (2011). Drosophila as a model of wound healing and tissue regeneration in vertebrates. *Developmental Dynamics, 240*(11), 2379–2404.

Benjamin, O., Goyal, A., & Lappin, S. L. (2018). *Disease modifying anti-rheumatic drugs (DMARD).*

Caballero-Sánchez, N., Alonso-Alonso, S., & Nagy, L. (2024). Regenerative inflammation: When immune cells help to re-build tissues. *The FEBS Journal, 291*(8), 1597–1614.

Cai, S., Wang, X., Xu, R., Liang, Z., Zhu, Q., Chen, M., Lin, Z., Li, C., Duo, T., & Tong, X. (2023). KLF4 regulates skeletal muscle development and regeneration by directly targeting P57 and Myomixer. *Cell Death & Disease, 14*(9), 612.

Chatzeli, L., Teshima, T. H., Hajihosseini, M. K., Gaete, M., Proctor, G. B., & Tucker, A. S. (2021). Comparing development and regeneration in the submandibular gland highlights distinct mechanisms. *Journal of Anatomy, 238*(6), 1371–1385.

Chen, Y. L., Feng, X. L., Tam, K. W., Fan, C. Y., Cheung, M. P., Yang, Y. T., Wong, S., Shum, D. K., Chan, Y. S., Cheung, C. W., Cheung, M., & Liu, J. A. (2024). Intrinsic and extrinsic actions of human neural progenitors with SUFU inhibition promote tissue repair and functional recovery from severe spinal cord injury. *NPJ Regen Med, 9*(1), 13. https://doi.org/10.1038/s41536-024-00352-4

Chillakuri, C. R., Sheppard, D., Lea, S. M., & Handford, P. A. (2012). Notch receptor-ligand binding and activation: Insights from molecular studies. *Seminars in Cell & Developmental Biology, 23*(4), 421–428. https://doi.org/10.1016/j.semcdb.2012.01.009

Cruz-González, S., Quesada-Díaz, E., Miranda-Negrón, Y., García-Rosario, R., Ortiz-Zuazaga, H., & García-Arrarás, J. E. (2022). The stress response of the holothurian central nervous system: A transcriptomic analysis. *International Journal of Molecular Sciences, 23*(21), 13393.

De Winter, F., Holtmaat, A. J., & Verhaagen, J. (2002). Neuropilin and class 3 semaphorins in nervous system regeneration. *Advances in Experimental Medicine and Biology, 515*, 115–139. https://doi.org/10.1007/978-1-4615-0119-0_10

Demitrack, E. S., & Samuelson, L. C. (2016). Notch regulation of gastrointestinal stem cells. *The Journal of Physiology, 594*(17), 4791–4803. https://doi.org/10.1113/jp271667

Deng, Z., Fan, T., Xiao, C., Tian, H., Zheng, Y., Li, C., & He, J. (2024). TGF-β signaling in health, disease, and therapeutics. *Signal Transduction and Targeted Therapy, 9*(1), 61. https://doi.org/10.1038/s41392-024-01764-w

Diller, R. B., & Tabor, A. J. (2022). The role of the extracellular matrix (ECM) in wound healing: a review. *Biomimetics, 7*(3), 87.

Dolmatov, I. Y. (2021). Molecular aspects of regeneration mechanisms in holothurians. *Genes (Basel), 12*(2). https://doi.org/10.3390/genes12020250

Eming, S. A., Martin, P., & Tomic-Canic, M. (2014). Wound repair and regeneration: Mechanisms, signaling, and translation. *Science Translational Medicine, 6*(265), 265sr266.

Farooq, M., Khan, A. W., Kim, M. S., & Choi, S. (2021). The role of fibroblast growth factor (FGF) signaling in tissue repair and regeneration. *Cells, 10*(11). https://doi.org/10.3390/cells10113242

García-Arrarás, J. E., & Dolmatov, I. Y. (2010). Echinoderms: Potential model systems for studies on muscle regeneration. *Current Pharmaceutical Design, 16*(8), 942–955.

Gawriluk, T. R., Simkin, J., Hacker, C. K., Kimani, J. M., Kiama, S. G., Ezenwa, V. O., & Seifert, A. W. (2020). Complex tissue regeneration in mammals is associated with reduced inflammatory cytokines and an influx of T cells. *Frontiers in Immunology, 11*, 1695.

Ghosh, R., & Hansda, S. (2021). Targeted and non-targeted effects of radiation in mammalian cells: An overview. *Archives of Biotechnology and Biomedicine, 5*(1), 013–019.

Gilbert, S. F. (2017). Developmental biology, the stem cell of biological disciplines. *PLoS Biology, 15*(12), e2003691. https://doi.org/10.1371/journal.pbio.2003691

Gómez, C. M. A., & Echeverri, K. (2021). Salamanders: The molecular basis of tissue regeneration and its relevance to human disease. *Current Topics in Developmental Biology, 145*, 235–275.

Guan, Y., Yang, Y. J., Nagarajan, P., & Ge, Y. (2021). Transcriptional and signalling regulation of skin epithelial stem cells in homeostasis, wounds and cancer. *Experimental Dermatology, 30*(4), 529–545.

Guo, S. A., & DiPietro, L. A. (2010). Factors affecting wound healing. *Journal of Dental Research, 89*(3), 219–229.

Hannoodee, S., & Nasuruddin, D. N. (2020). Acute inflammatory response.

Hansda, S., & Das, H. (2025). Insights into cancer-associated thrombosis leading towards ischemic stroke. *Biology, 14*(1), 50.

Hansda, S., Ghosh, G., & Ghosh, R. (2020). 9-phenyl acridine photosensitizes A375 cells to UVA radiation. *Heliyon, 6*(9).

Hansda, S., Ghosh, G., & Ghosh, R. (2022). The role of bystander effect in ultraviolet A induced Photoaging. *Cell Biochemistry and Biophysics, 80*(4), 657–664.

Hansda, S., Prateeksha, P., & Das, H. (2024). Krüppel-like factor 2 (KLF2), a potential target for neuroregeneration. *Neural Regeneration Research, 19*(11), 2327–2328.

Hayashi, S., Ochi, H., Ogino, H., Kawasumi, A., Kamei, Y., Tamura, K., & Yokoyama, H. (2014). Transcriptional regulators in the hippo signaling pathway control organ growth in Xenopus tadpole tail regeneration. *Developmental Biology, 396*(1), 31–41. https://doi.org/10.1016/j.ydbio.2014.09.018

Houschyar, K. S., Tapking, C., Borrelli, M. R., Popp, D., Duscher, D., Maan, Z. N., Chelliah, M. P., Li, J., Harati, K., Wallner, C., Rein, S., Pförringer, D., Reumuth, G., Grieb, G., Mouraret, S., Dadras, M., Wagner, J. M., Cha, J. Y., Siemers, F., Lehnhardt, M., & Behr, B. (2018). Wnt pathway in bone repair and regeneration – What do we know so far. *Frontiers in Cell and Development Biology, 6*, 170. https://doi.org/10.3389/fcell.2018.00170

Howlader, M. S. I., Prateeksha, P., Hansda, S., Naidu, P., Das, M., Barthels, D., & Das, H. (2024). Corrigendum to" secretory products of DPSC mitigate inflammatory effects in microglial cells by targeting MAPK pathway" biomed. Pharmacother. 170 (2024) 115971. *Biomedicine & Pharmacotherapy= Biomedecine & Pharmacotherapie, 177*, 117061-117061.

Huo, D., Su, F., Cui, W., Liu, S., Zhang, L., Yang, H., & Sun, L. (2022). Heat stress and evisceration caused lipid metabolism and neural transduction changes in sea cucumber: Evidence from metabolomics. *Marine Pollution Bulletin, 182*, 113993. https://doi.org/10.1016/j.marpolbul.2022.113993

Jing, J., Wu, Z., Wang, J., Luo, G., Lin, H., Fan, Y., & Zhou, C. (2023). Hedgehog signaling in tissue homeostasis, cancers, and targeted therapies. *Signal Transduction and Targeted Therapy, 8*(1), 315. https://doi.org/10.1038/s41392-023-01559-5

Kashyap, V., Rezende, N. C., Scotland, K. B., Shaffer, S. M., Persson, J. L., Gudas, L. J., & Mongan, N. P. (2009). Regulation of stem cell pluripotency and differentiation involves a mutual regulatory circuit of the NANOG, OCT4, and SOX2 pluripotency transcription factors with polycomb repressive complexes and stem cell microRNAs. *Stem Cells and Development, 18*(7), 1093–1108. https://doi.org/10.1089/scd.2009.0113

Khan, I., Neumann, C., & Sinha, M. (2020). Tissue regeneration and reprogramming. In *Wound healing, tissue repair, and regeneration in diabetes* (pp. 515–534). Elsevier.

Khoswanto, C. (2023). Role of matrix metalloproteinases in bone regeneration: Narrative review. *Journal of Oral Biology and Craniofacial Research, 13*(5), 539–543. https://doi.org/10.1016/j.jobcr.2023.06.002

King, T. (2006). *Elsevier's integrated pathology E-book*. Elsevier Health Sciences.

Kornthong, N., Phanaksri, T., Saetan, J., Duangprom, S., Lekskul, B., Vivattanasarn, T., Songkoomkrong, S., Jattujan, P., Cummins, S. F., Sobhon, P., & Suwansa-Ard, S. (2021). Identification and localization of growth factor genes in the sea cucumber, Holothuria scabra. *Heliyon, 7*(11), e08370. https://doi.org/10.1016/j.heliyon.2021.e08370

Krafts, K. P. (2010). Tissue repair: The hidden drama. *Organogenesis, 6*(4), 225–233.

Kuintzle, R., Santat, L. A., & Elowitz, M. B. (2024). Diversity in notch ligand-receptor signaling interactions. *bioRxiv*. https://doi.org/10.1101/2023.08.24.554677

Laha, D., Deb, M., & Das, H. (2019). KLF2 (kruppel-like factor 2 [lung]) regulates osteoclastogenesis by modulating autophagy. *Autophagy, 15*(12), 2063–2075.

Landén, N. X., Li, D., & Ståhle, M. (2016). Transition from inflammation to proliferation: a critical step during wound healing. *Cellular and Molecular Life Sciences, 73*, 3861–3885.

Lara, J., Mastela, C., Abd, M., Pitstick, L., & Ventrella, R. (2024). Tail Tales: What we have learned about regeneration from Xenopus Laevis tadpoles. *International Journal of Molecular Sciences, 25*(21). https://doi.org/10.3390/ijms252111597

Laronha, H., & Caldeira, J. (2020). Structure and function of human matrix metalloproteinases. *Cells, 9*(5). https://doi.org/10.3390/cells9051076

Li, J., & Amaya, E. (2019). Investigating the cellular and molecular mechanisms of wound healing in Xenopus oocytes and embryos. *Cold Spring Harbor Protocols, 2019*(4). https://doi.org/10.1101/pdb.prot100982

Li, J., Zhang, S., & Amaya, E. (2016). The cellular and molecular mechanisms of tissue repair and regeneration as revealed by studies in Xenopus. *Regeneration (Oxf), 3*(4), 198–208. https://doi.org/10.1002/reg2.69

Li, R., Li, D., Wu, C., Ye, L., Wu, Y., Yuan, Y., Yang, S., Xie, L., Mao, Y., Jiang, T., Li, Y., Wang, J., Zhang, H., Li, X., & Xiao, J. (2020). Nerve growth factor activates autophagy in Schwann cells to enhance myelin debris clearance and to expedite nerve regeneration. *Theranostics, 10*(4), 1649–1677. https://doi.org/10.7150/thno.40919

Li, S., Gu, X., & Yi, S. (2017). The regulatory effects of transforming growth factor-β on nerve regeneration. *Cell Transplantation, 26*(3), 381–394. https://doi.org/10.3727/096368916x693824

Liao, X., Sharma, N., Kapadia, F., Zhou, G., Lu, Y., Hong, H., Paruchuri, K., Mahabeleshwar, G. H., Dalmas, E., & Venteclef, N. (2011). Krüppel-like factor 4 regulates macrophage polarization. *The Journal of Clinical Investigation, 121*(7), 2736–2749.

Lin, T.-Y., Gerber, T., Taniguchi-Sugiura, Y., Murawala, P., Hermann, S., Grosser, L., Shibata, E., Treutlein, B., & Tanaka, E. M. (2021). Fibroblast dedifferentiation as a determinant of successful regeneration. *Developmental Cell, 56*(10), 1541–1551. e1546.

Liu, Z. L., Chen, H. H., Zheng, L. L., Sun, L. P., & Shi, L. (2023). Angiogenic signaling pathways and anti-angiogenic therapy for cancer. *Signal Transduction and Targeted Therapy, 8*(1), 198. https://doi.org/10.1038/s41392-023-01460-1

Lord, A., & Ficz, G. (2022). Corrupted devolution: How normal cells are reborn as cancer precursors. *The International Journal of Biochemistry & Cell Biology, 149*, 106263.

Mahmoud, A. I. (2023). Metabolic switches during development and regeneration. *Development, 150*(20), dev202008.

Maity, J., Barthels, D., Sarkar, J., Prateeksha, P., Deb, M., Rolph, D., & Das, H. (2022). Ferutinin induces osteoblast differentiation of DPSCs via induction of KLF2 and autophagy/mitophagy. *Cell Death & Disease, 13*(5), 452.

Maity, J., Deb, M., Greene, C., & Das, H. (2020). KLF2 regulates dental pulp-derived stem cell differentiation through the induction of mitophagy and altering mitochondrial metabolism. *Redox Biology, 36*, 101622.

Mao, A. S., & Mooney, D. J. (2015). Regenerative medicine: Current therapies and future directions. *Proceedings of the National Academy of Sciences, 112*(47), 14452–14459.

Mashanov, V., Akiona, J., Khoury, M., Ferrier, J., Reid, R., Machado, D. J., Zueva, O., & Janies, D. (2020). Active notch signaling is required for arm regeneration in a brittle star. *PLoS One, 15*(5), e0232981. https://doi.org/10.1371/journal.pone.0232981

Maxson, S., Lopez, E. A., Yoo, D., Danilkovitch-Miagkova, A., & LeRoux, M. A. (2012). Concise review: Role of mesenchymal stem cells in wound repair. *Stem Cells Translational Medicine, 1*(2), 142–149.

McCusker, C., Bryant, S. V., & Gardiner, D. M. (2015). The axolotl limb blastema: Cellular and molecular mechanisms driving blastema formation and limb regeneration in tetrapods. *Regeneration, 2*(2), 54–71.

Medina-Feliciano, J. G., & García-Arrarás, J. E. (2021). Regeneration in echinoderms: Molecular advancements. *Frontiers in Cell and Development Biology, 9*, 768641. https://doi.org/10.3389/fcell.2021.768641

Mehta, A. S., & Singh, A. (2019). Insights into regeneration tool box: An animal model approach. *Developmental Biology, 453*(2), 111–129.

Mescher, A. L., Neff, A. W., & King, M. W. (2017). Inflammation and immunity in organ regeneration. *Developmental and Comparative Immunology, 66*, 98–110. https://doi.org/10.1016/j.dci.2016.02.015

Montgomery, E. M., Suhrbier, A. D., & Pearce, C. M. (2024). Biology, ecology, aquaculture, and commercial products of Apostichopus californicus. In *The world of sea cucumbers* (pp. 653–675). Elsevier.

Moonen, J.-R., Chappell, J., Shi, M., Shinohara, T., Li, D., Mumbach, M. R., Zhang, F., Nair, R. V., Nasser, J., & Mai, D. H. (2022). KLF4 recruits SWI/SNF to increase chromatin accessibility and reprogram the endothelial enhancer landscape under laminar shear stress. *Nature Communications, 13*(1), 4941.

Niu, N., Xu, S., Xu, Y., Little, P. J., & Jin, Z.-G. (2019). Targeting mechanosensitive transcription factors in atherosclerosis. *Trends in Pharmacological Sciences, 40*(4), 253–266.

Ong, J. Y., & Torres, J. Z. (2019). Dissecting the mechanisms of cell division. *Journal of Biological Chemistry, 294*(30), 11382–11390.

Parey, E., Ortega-Martinez, O., Delroisse, J., Piovani, L., Czarkwiani, A., Dylus, D., Arya, S., Dupont, S., Thorndyke, M., Larsson, T., Johannesson, K., Buckley, K. M., Martinez, P., Oliveri, P., & Marlétaz, F. (2024). The brittle star genome illuminates the genetic basis of animal appendage regeneration. *Nature Ecology & Evolution, 8*(8), 1505–1521. https://doi.org/10.1038/s41559-024-02456-y

Parisi, L., Toffoli, A., Ghezzi, B., Mozzoni, B., Lumetti, S., & Macaluso, G. M. (2020). A glance on the role of fibronectin in controlling cell response at biomaterial interface. *Japanese Dental Science Review, 56*(1), 50–55. https://doi.org/10.1016/j.jdsr.2019.11.002

Park, C. S., Lewis, A., Chen, T., & Lacorazza, D. (2019). Concise review: Regulation of self-renewal in normal and malignant hematopoietic stem cells by Krüppel-like factor 4. *Stem Cells Translational Medicine, 8*(6), 568–574.

Paul, W., & Sharma, C. P. (2021). Tissue and organ regeneration: An introduction. In *Regenerated organs* (pp. 3–9). Elsevier.

Peña, O. A., & Martin, P. (2024). Cellular and molecular mechanisms of skin wound healing. *Nature Reviews Molecular Cell Biology*, 1–18.

Picerno, A., Stasi, A., Franzin, R., Curci, C., di Bari, I., Gesualdo, L., & Sallustio, F. (2021). Why stem/progenitor cells lose their regenerative potential. *World Journal of Stem Cells, 13*(11), 1714–1732. https://doi.org/10.4252/wjsc.v13.i11.1714

Poliacikova, G., Maurel-Zaffran, C., Graba, Y., & Saurin, A. J. (2021). Hox proteins in the regulation of muscle development. *Frontiers in Cell and Developmental Biology, 9*, 731996.

Pollak, N. M., Hoffman, M., Goldberg, I. J., & Drosatos, K. (2018). Krüppel-like factors: Crippling and uncrippling metabolic pathways. *JACC: Basic to Translational Science, 3*(1), 132–156.

Prateeksha, P., Howlader, M. S. I., Hansda, S., Naidu, P., Das, M., Abo-Aziza, F., & Das, H. (2023). Secretome of dental pulp-derived stem cells reduces inflammation and proliferation of glioblastoma cells by deactivating Mapk-Akt pathway. *Diseases & Research, 3*(2), 74.

Radak, M., & Fallahi, H. (2023). The epigenetic regulation of quiescent in stem cells. *Global Medical Genetics, 10*(4), 339–344. https://doi.org/10.1055/s-0043-1777072

Raghuram, A. C., Roy, P. Y., Lo, A. Y., Sung, C. J., Bircan, M., Thompson, H. J., & Wong, A. K. (2020). Role of stem cell therapies in treating chronic wounds: A systematic review. *World Journal of Stem Cells, 12*(7), 659.

Reddy, P. C., Gungi, A., & Unni, M. (2019). Cellular and molecular mechanisms of hydra regeneration. In *Evo-devo: Non-model species in cell and developmental biology*, pp. 259–290.

Reinardy, H. C., Emerson, C. E., Manley, J. M., & Bodnar, A. G. (2015). Tissue regeneration and biomineralization in sea urchins: Role of notch signaling and presence of stem cell markers. *PLoS One, 10*(8), e0133860. https://doi.org/10.1371/journal.pone.0133860

Richardson, S. R., & O'Malley, G. F. (2025). Glucose-6-phosphate dehydrogenase deficiency. In *StatPearls*. StatPearls Publishing. Copyright © 2025, StatPearls Publishing LLC.

Rolph, D., & Das, H. (2020). Transcriptional regulation of osteoclastogenesis: The emerging role of KLF2. *Frontiers in Immunology, 11*, 937.

Santoyo-Suarez, M. G., Mares-Montemayor, J. D., Padilla-Rivas, G. R., Delgado-Gallegos, J. L., Quiroz-Reyes, A. G., Roacho-Perez, J. A., Benitez-Chao, D. F., Garza-Ocañas, L., Arevalo-

Martinez, G., & Garza-Treviño, E. N. (2023). The involvement of Krüppel-like factors in cardiovascular diseases. *Life, 13*(2), 420.
Schultz, G. S., Chin, G. A., Moldawer, L., & Diegelmann, R. F. (2011). Principles of wound healing. In R. Fitridge & M. Thompson (Eds.), *Mechanisms of vascular disease: A reference book for vascular specialists*. University of Adelaide Press. © The Contributors 2011.
Seifert, A. W., Duncan, E. M., & Zayas, R. M. (2023). Enduring questions in regenerative biology and the search for answers. *Communications Biology, 6*(1), 1139.
Shpichka, A., Butnaru, D., Bezrukov, E. A., Sukhanov, R. B., Atala, A., Burdukovskii, V., Zhang, Y., & Timashev, P. (2019). Skin tissue regeneration for burn injury. *Stem Cell Research & Therapy, 10*, 1–16.
Singh, B. N., Koyano-Nakagawa, N., Donaldson, A., Weaver, C. V., Garry, M. G., & Garry, D. J. (2015). Hedgehog signaling during appendage development and regeneration. *Genes (Basel), 6*(2), 417–435. https://doi.org/10.3390/genes6020417
Sinigaglia, C., Almazán, A., Lebel, M., Sémon, M., Gillet, B., Hughes, S., Edsinger, E., Averof, M., & Paris, M. (2022). Distinct gene expression dynamics in developing and regenerating crustacean limbs. *Proceedings of the National Academy of Sciences, 119*(27), e2119297119.
Strizova, Z., Benesova, I., Bartolini, R., Novysedlak, R., Cecrdlova, E., Foley, L. K., & Striz, I. (2023). M1/M2 macrophages and their overlaps–myth or reality? *Clinical Science, 137*(15), 1067–1093.
Sydor, S., Manka, P., Best, J., Jafoui, S., Sowa, J.-P., Zoubek, M. E., Hernandez-Gea, V., Cubero, F. J., Kälsch, J., & Vetter, D. (2017). Krüppel-like factor 6 is a transcriptional activator of autophagy in acute liver injury. *Scientific Reports, 7*(1), 8119.
Takei, N., Furukawa, K., Hanyu, O., Sone, H., & Nawa, H. (2014). A possible link between BDNF and mTOR in control of food intake. *Frontiers in Psychology, 5*, 1093. https://doi.org/10.3389/fpsyg.2014.01093
Tamori, M., Yamada, A., Nishida, N., Motobayashi, Y., Oiwa, K., & Motokawa, T. (2006). Tensilin-like stiffening protein from Holothuria leucospilota does not induce the stiffest state of catch connective tissue. *The Journal of Experimental Biology, 209*(Pt 9), 1594–1602. https://doi.org/10.1242/jeb.02178
Thomas, T. P., & Grisanti, L. A. (2020). The dynamic interplay between cardiac inflammation and fibrosis. *Frontiers in Physiology, 11*, 529075. https://doi.org/10.3389/fphys.2020.529075
Tie, Y., Tang, F., Peng, D., Zhang, Y., & Shi, H. (2022). TGF-beta signal transduction: Biology, function and therapy for diseases. *Molecular Biomedicine, 3*(1), 45. https://doi.org/10.1186/s43556-022-00109-9
Tseng, A. S., & Levin, M. (2008). Tail regeneration in Xenopus laevis as a model for understanding tissue repair. *Journal of Dental Research, 87*(9), 806–816. https://doi.org/10.1177/154405910808700909
Tucker, E. J., Rojas-Canales, D., Ali, S. R., & Gleadle, J. M. (2023). MicroRNAs in renal development and regeneration. *MicroRNA in Regenerative Medicine*, 293–313.
van Tetering, G., & Vooijs, M. (2011). Proteolytic cleavage of notch: "HIT and RUN". *Current Molecular Medicine, 11*(4), 255–269. https://doi.org/10.2174/156652411795677972
Wang, Y., Yu, A., & Yu, F. X. (2017). The hippo pathway in tissue homeostasis and regeneration. *Protein & Cell, 8*(5), 349–359. https://doi.org/10.1007/s13238-017-0371-0
Wang, Y.-H., Israelsen, W. J., Lee, D., Vionnie, W., Jeanson, N. T., Clish, C. B., Cantley, L. C., Vander Heiden, M. G., & Scadden, D. T. (2014). Cell-state-specific metabolic dependency in hematopoiesis and leukemogenesis. *Cell, 158*(6), 1309–1323.
Wang, Z., Li, Y., Banerjee, S., & Sarkar, F. H. (2009). Emerging role of notch in stem cells and cancer. *Cancer Letters, 279*(1), 8–12. https://doi.org/10.1016/j.canlet.2008.09.030
Xie, Y., Su, N., Yang, J., Tan, Q., Huang, S., Jin, M., Ni, Z., Zhang, B., Zhang, D., Luo, F., Chen, H., Sun, X., Feng, J. Q., Qi, H., & Chen, L. (2020). FGF/FGFR signaling in health and disease. *Signal Transduction and Targeted Therapy, 5*(1), 181. https://doi.org/10.1038/s41392-020-00222-7

Xie, Z., Chen, J., Wang, C., Zhang, J., Wu, Y., & Yan, X. (2021). Current knowledge of Krüppel-like factor 5 and vascular remodeling: Providing insights for therapeutic strategies. *Journal of Molecular Cell Biology, 13*(2), 79–90.

Ye, W., Lv, X., Gao, S., Li, Y., Luan, J., & Wang, S. (2023). Emerging role of m6A modification in fibrotic diseases and its potential therapeutic effect. *Biochemical Pharmacology, 115873.*

Yun, M. H. (2021). Salamander insights into ageing and rejuvenation. *Frontiers in Cell and Developmental Biology, 9*, 689062.

Zgheib, C., Xu, J., & Liechty, K. W. (2014). Targeting inflammatory cytokines and extracellular matrix composition to promote wound regeneration. *Advances in Wound Care, 3*(4), 344–355.

Zhang, Y., Yao, C., Ju, Z., Jiao, D., Hu, D., Qi, L., Liu, S., Wu, X., & Zhao, C. (2023). Krüppel-like factors in tumors: Key regulators and therapeutic avenues. *Frontiers in Oncology, 13*, 1080720.

Zhao, H., Wu, L., Yan, G., Chen, Y., Zhou, M., Wu, Y., & Li, Y. (2021). Inflammation and tumor progression: Signaling pathways and targeted intervention. *Signal Transduction and Targeted Therapy, 6*(1), 263. https://doi.org/10.1038/s41392-021-00658-5

# Index





H. Das (ed.), *Tissue Repair and Regeneration*,
https://doi.org/10.1007/978-3-031-93677-7

GPSR Compliance
The European Union's (EU) General Product Safety Regulation (GPSR) is a set of rules that requires consumer products to be safe and our obligations to ensure this.

If you have any concerns about our products, you can contact us on

ProductSafety@springernature.com

In case Publisher is established outside the EU, the EU authorized representative is:

Springer Nature Customer Service Center GmbH
Europaplatz 3
69115 Heidelberg, Germany

www.ingramcontent.com/pod-product-compliance
Lightning Source LLC
Chambersburg PA
CBHW071955170426
43348CB00003B/39
*9783031936760*